T0203877

Progressive Technologies of Coal, Coalbed Methane, and Ores Mining

Editors

Volodymyr Bondarenko
Department of Underground Mining, National Mining University, Ukraine

Iryna Kovalevs'ka
Department of Underground Mining, National Mining University, Ukraine

Kostiantyn Ganushevych
Department of Underground Mining, National Mining University, Ukraine

CRC Press
Taylor & Francis Group
Boca Raton London New York

CRC Press is an imprint of the
Taylor & Francis Group, an **informa** business

A BALKEMA BOOK

Published by:
CRC Press/Balkema
P.O. Box 447, 2300 AK Leiden, The Netherlands
e-mail: Pub.NL@taylorandfrancis.com
www.crcpress.com – www.taylorandfrancis.com

First issued in paperback 2020

ISBN 13: 978-0-367-57609-7 (pbk)
ISBN 13: 978-1-138-02699-5 (hbk)

Visit the Taylor & Francis Web site at
http://www.taylorandfrancis.com

and the CRC Press Web site at
http://www.crcpress.com

Typeset by Olga Malova, Kostiantyn Ganushevych & Denys Astafiev, Department of Underground Mining, National Mining University, Dnipropetrovs'k, Ukraine

Progressive Technologies of Coal, Coalbed Methane, and Ores Mining – Bondarenko, Kovalevs'ka & Ganushevych (eds)
© 2014 Taylor & Francis Group, London, ISBN: 978-1-138-02699-5

Table of contents

Progressive Technologies of Coal, Coalbed Methane, and Ores Mining – Bondarenko, Kovalevs'ka & Ganushevych (eds)
© 2014 Taylor & Francis Group, London, ISBN: 978-1-138-02699-5

Preface

The present collection of scientific papers is addressed to mining engineers, scientific and research personnel, students, postgraduates and all professionals connected with the coal and ore industry.

The authors of this book have contributed their articles that cover economic aspects of mining companies' development strategies, peculiarities of various mineral deposits development techniques, imitational modeling of mine workings with rock massif, methane extraction technologies during coal mining, geomechanical processes during plow mining, mining transport importance for mineral extraction, massif strain-stress state management using non-explosive destructing materials, surface mining negative influence on the environment.

Taking into account worsening of the mining-geological conditions for conventional extraction of coal, there is much of attention dedicated to borehole underground coal gasification technology at Ukraine's coal deposits. Very intriguing topic is connected with one of the most perspective and abundant sources of energy on the planet – gas hydrates. The question of their prospecting, properties and ways of extraction is also covered in this book.

Examination is given to financial conditions of work and financial strategy of mining industry in Ukraine, Germany, Kazakhstan, Poland, Russia, Guinea and other countries.

Volodymyr Bondarenko
Iryna Kovalevs'ka
Kostiantyn Ganushevych

Progressive Technologies of Coal, Coalbed Methane, and Ores Mining – Bondarenko, Kovalevs'ka & Ganushevych (eds)
© 2014 Taylor & Francis Group, London, ISBN: 978-1-138-02699-5

Main directions in roof bolting technology development at DTEK mines

O. Smirnov & V. Pilyugin
"Donbass Fuel-energy company" Donets'k, Ukraine

ABSTRACT: The article shows the achievements of DTEK in extending technology of anchor and frame-anchor support at coal mines. Main directions of improving normative base were disclosed. Technical and financial results of the anchor program were evaluated.

1 INTRODUCTION

DTEK's coal mines are characterized by the diversity and complexity of geological conditions and significant length of the operating development workings. Nevertheless, until recently they universally applied the technologically outdated 3-link yieldable arch support made of special SVP sections with M-24 locks. The roadway support parameters were determined in line with the instruction (Standart of Ukrainian Companies... 2007).

Practical experience proves that chock supports often fail to ensure proper operational condition of the underground roadways, and sometimes may cause emergency situations. Intensive support deformations and contraction losses are observed in the mine workings, resulting in the need for repair.

Finding a solution to the issue of roadway stability is particularly important in the current situation where many enterprises face the need to increase their mining output significantly due to economic reasons, improving the labour productivity and speeding up mining operations while reducing the prime cost of the mined coal. To achieve this, high throughput capacity and, respectively, good operational condition of the development workings are crucial.

As global experience shows, the most promising and efficient method to improve stability of development workings in the complicated conditions of deep coal mines is the broad and flexible application of combined frame-and-bolt supports. These supports, in comparison with regular (passive) arch supports, have a number of substantial technological advantages:

– they start operating actively immediately after they are installed in the mine working;

– they contribute to maintaining the natural strength and integrity of the surrounding bed;

– they ensure the personnel's safety even after destruction of the marginal rocks and substantial contraction loss.

The combined frame-and-bolt supports are the most universal type of supports. They are used in a wide range of mining geological conditions and have virtually no limitations as for the types and purpose of the mine workings, service life period, shape of cross-section and types of location inside the rock massif. Taking this into consideration, DTEK made a huge technological leap in 2013 introducing the frame-and-bolt supports at its coal mines.

2 PROGRAM FOR DEVELOPMENT OF FRAME-AND-BOLT SUPPORTS AND ROOF BOLTING AT DTEK'S MINES

Before 2013, the roof bolting and frame-and-bolt supports at DTEK's mines were mostly used in Western Donbass due to a number of favourable factors:

– high level of technological development of DTEK Pavlogradugol enterprises as regards the roadway construction;

– the mined coal seams are deposited in the soft surrounding rocks, which enables fast drilling of the boreholes for roof bolting;

– universal use of the roadheader-based development of the roadways and high share of their repeated use.

Against this background, starting in 2010, the frame-and-bolt support technology was successfully mastered and became the new standard. By the end of 2012, the scope of its implementation at 10 mines of DTEK Pavlogradugol made up 67.5 km, which was 58% of the total length of the constructed development roadways.

In preparing the program for further development of roof bolt supports for 2013, the main focus was on the new companies and more technologically problematic mines of Dobropillya and Roven'ky coal districts.

The main objective of the program is the process preparation for a full transition to the roof bolting and frame-and-bolt support technology at all DTEK's mines based on the personnel training, designing of technological passports for the roadway construction, roadheading and control over the quality of the supports installation and operational condition of the roadways.

Teaching the personnel the basic principles of roof bolting support was performed in several stages. First, the technical managers of the mines and mine groups, including the section foremen and computer-based training coaches, were trained by the experts of the Anchor Support Centre of M.S. Polyakov Institute of Geotechnical Mechanics of the National Academy of Sciences of Ukraine. The trainees passed the test and received special certificates entitling them to perform mining operations, to exercise operational control over the quality of support installation and commission the constructed roadway, as well as teach the roof bolting support basics to the workers. The second stage was dedicated to the training of tunnellers and miners engaged in the technological processes.

Development of technological passports for the roof bolted roadway construction in line with standard (Standart of Ukrainian Companies...2007) by the specialists of M.S. Polyakov Institute of Geotechnical Mechanics of the National Academy of Sciences of Ukraine.

DTEK implemented a large-scale program for *roadheading at the pilot sections of the mine workings with roof bolting and frame-and-bolt supports.* The main program parameters are as follows:
– total number of the pilot development faces: 16, including 2 with roof bolting supports only;
– total number of pilot mines for the technology implementation: 12;
– total length of experimental sections: 13.5 km, including 800 m with roof bolting supports only;
– investments in equipment: approximately UAH 4.5 million.

Control over the quality of roof bolting support installation was exercised by the experts of M.S. Polyakov Institute of Geotechnical Mechanics of the National Academy of Sciences of Ukraine. The control was performed by anchor testing for extraction from the rock mass with a special tool – hydraulic bolt puller.

Requirements to the testing of the installed roof bolts:

– frequency: at least 3 times per month at the section advanced as of the testing time;
– number of tested bolts: at least 3 bolts;
– type of bolts: roof bolts;
– maximum load per a bolt during the testing: 70% of yield limit or 100% of the bolt's elastic limit;
– location of measurement sections: in areas of maximum deformation of the working's contour.

Where cases of violation in technology of roof bolting installation were found at the tested sections, work defect report was compiled and reasons for violation were determined. The technical service of the mine then developed measures to bring the roadway to proper operating condition.

Monitoring of the roadways' condition in the course of their subsequent operation was exercised in line with the general requirements to the development workings of the coal mines.

3 APPLICATION OF TWO-LEVEL ROOF BOLTING SYSTEMS INCLUDING ROPE BOLT ANCHORS FOR DEEP SETTING

In recent years, the technology of roof bolting received an additional impulse in its development due to the spread in the mines of rope bolt anchors for deep setting (Standart of Ukrainian Companies... 2008 & Ilyashov 2011). Rope bolt anchors represent pieces of metal rope with the length of 4 to 8 m that are set into the boreholes. The bolt anchor is fixed in the rock massif by means of the polymeric capsule and is tightened from the side of the roadway. Thus, a second level of rock consolidation is made around the bolt-supported area.

The main function of the rope bolt support in the *main* (permanent) roadways supported outside the coal-face work impact area is preservation of integrity for rock above the pre-contour massif bolted with conventional steel and polymer bolt anchors. Two-level roof bolting is especially efficient when coal sheds are found in the roof in close vicinity (up to 4–5 m) to the roadways.

Main functions of rope bolting in *coal extraction workings:*
– reduced lamination of rock massif in the roof above the bolt supported area;
– no need to install prop supports and hydraulic face-end supports in the area of longwall junction with the roadway (face-end area);
– improved performance efficiency of combined frame-and-bolt support of the first level;
– ensuring safe removal and installation of arch support props when moving the longwall conveyor.

The pilot project for two-level roof bolting trial was made in hauling roadway of the 4th longwall,

h_9 seam, Tsentrosoyuz Mine of Sverdlovskoye Mine Group. The roadway construction conditions were as follows: coal grade – A, seam thickness – 1.28 m, depth of work – 420 m, roadway cross-section – 13.8 m^2, the surrounding rock is represented by layers of stable clay and sandy shale, the average weighted rock strength R_e = 83.8 MPa, that of the roof – R_r = 90.8 MPa, strength of bed – R_s = 71.1 MPa, mining method – long-pillar, reverse movement. Extracts from mine workings' layout plan and diagram of two-level frame-and-bolt support installation (in line with recommendations from M.S. Polyakov Institute of Geotechnical Mechanics of the National Academy of Sciences of Ukraine) are given in Figure 1.

(a)

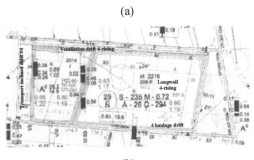

(b)

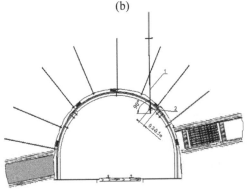

Figure 1. Location of workings of the 4th longwall, h_9 seam, Tsentrsoyuz mine: (a) mining scheme; (b) location of combined two-level frame-and-bolt support elements; 1 – rope bolt anchors; 2 – straight-line section of special SVP profile.

In developing the support pattern, the question of selecting a well-positioned place for installation of rope bolt anchors in respect to the development face arose. Recommendations of the specialized research institute on the matter, according to which bolt anchors should be installed directly inside the face, gave rise to certain doubts both from the point of view of this operation's feasibility and from the point of view of their effective work.

According to the quality certificate for 'die-rolled screw-type steel profile for roof bolting of mine workings', d = 22 mm, produced by Yenakievo Iron and Steel Works of METINVEST in line with the technical specification TU U 27.3-23365425-661-2011, which is used at DTEK mines, the load bearing capacity of the bolt anchors is 200–220 kN, and their ultimate rapture strength reaches 11%. According to the quality certificate of rope bolt anchors AK1 from Russian manufacturer, RANK 2 LLC, which are used for second level bolting, such specifications are 210 kN and 1.3% accordingly. Thus, having almost one and the same ultimate rapture strength, steel and polymer anchors with the length of 2.4 m could extend by 200–250 mm, and the rope bolts with the length of 6.0 m – only by 8 cm. This gives a simple technological conclusion: the second level bolt anchors (rope bolts) will rapture quicker than steel and polymer anchors in case of massif deformations. Therefore, it is advisable to install them after completion of active rock convergence phase behind the development face, i.e. in the area of relative stabilization of the working's contour. This will also make it possible to reduce the timing of support installation and increase the running time of the tunneling machinery.

The pilot implementation of the combined two-level frame-and-bolt support technology in the hauling roadway of the 4th longwall, h_9 seam, resulted in maintaining the cross-section and arch support of the roadway during its advancement and support behind the breakage face (protection with wood chocks). Convergence after the influence of coal-face work amounted to only 20%, the remaining cross-section of the roadway – 10 m^2 minimum, which is significantly higher than the minimal value of 7.5 m^2 stipulated by the safety rules of Ukraine (Razumov et al. 2012). The actual condition of the hauling roadway is shown in Figure 2. The experiment resulted in the economic effect of about UAH 11 million.

Figure 2. The condition of the hauling roadway of the 4th longwall, h_9 seam, Tsentrsoyuz mine, in the area of the roadway junction with the breakage face.

4 IMPROVEMENT OF THE REGULATORY FRAMEWORK

After the first steps in implementation of the program for development of frame-and-bolt support technology at the mines of DTEK, engineering and technical staff of the coal mining enterprises, on the one hand, received the first experience of designing support patterns and carrying out the mining operations, and, on the other hand, identified some omissions in the existing standard of Ukrainian companies 10.1.05411357.010:2008 'System ensuring reliable and safe operation of roof bolted mine roadways. General specifications' (Standart of Ukrainian Companies... 2007). They consist in the following:

1. The guideline does not envisage direct participation of the enterprises' technological functions in the development of support patterns and roadway mining with frame-and-bolt support, and moreover, with roof bolting technology.

2. The standard does not contain the engineering methodology for calculation and selection of frame-and-bolt support's characteristics similar to the existing methodology for design of frame support (Guideline for calculation... 2012).

3. The regulatory framework does not make it possible to forecast the contour line deformations of the roadways with frame-and-bolt support and assess the technical effect from its application at the stage of mid-term and strategic planning.

4. The concept of practical 'blocking of rock deformations with roof bolting support' irrespective of the complexity of mining conditions contained in the standards gives rise to reasonable doubts.

The above mentioned drawbacks have led to the need to revise the standard (Standart of Ukrainian Companies... 2007).

To solve this task DTEK initiated creation of a special task force on development of new standard that brought together some leading experts of the Ministry of Energy and Coal Industry, scientists of specialized, industry specific and academic institutes, as well as representatives of technical departments of DTEK.

Recently, the specialists of the Technical Development Department of DTEK developed the draft 'Guideline on design of the combined frame-and-bolt support at the coal mines of Ukraine'. It was provided to all the members of the task force for review. Further on, it is planned to discuss the draft guideline, to amend it and test it at coal mining enterprises of Ukraine.

5 CONCLUSIONS

1. Improvement of competitive power of DTEK's coal mining enterprises on the energy market requires significant reduction of cost on roadway support and preservation of development workings' stability.

2. The most promising solution for the issue is wide application and switch to the use of roof bolting and frame-and-bolt support technologies.

3. In the course of 2013, trials of roof bolting and frame-and-bolt support technologies were successfully implemented in the roadways of 14 coal mines of DTEK. The total length of the trial sections amounted to 13.5 kilometers. 800 meters of that total length were supported using only roof bolting technology. The annual saving from the technology introduction amounted to UAH 33.4 million.

The first stage of modification of the existing Ukrainian regulatory framework in the area of roof bolting has been accomplished. It resulted in the development of the draft Guideline for design of the combined frame-and-bolt roadway support at the coal mines of Ukraine.

REFERENCES

Standard of Ukrainian Companies 10.1.0018590.011:2007. Development workings in flat-lying seams. Selection of support, means and methods of protection. Kyiv: 113.

*Standard of Ukrainian Companies 10.1.05411357.010:*2008. System ensuring reliable and safe operation of roof bolted mine roadways. General specifications': Kyiv: Ministry of Energy and Coal Industry of Ukraine: 83.

Ilyashov, M.A. 2011. *Efficient reserve for increase of competitive power of the mining fund – secondary use of section workings.* Coal of Ukraine, 1: 15–17.

Razumov, E.A., Grechishkin, P.V., Samok, A.V. & Pozolotin, A.S. 2012. *Experience of rope bolt anchors' application for preservation and secondary use of coal mine workings.* Coal, 6: 10–12.

Guideline for calculation and application of roof bolting technology at coal mines of the Russian Federation. 2012. Russia. Rostekhnadzor: 203.

Influence of the structure and properties of coal-bearing massif on bottom heaving

V. Bondarenko & G. Symanovych
National Mining University, Dnipropetrovs'k, Ukraine

V. Chervatyuk
"Donbass Fuel-energy company" Donets'k, Ukraine

V. Snigur
PrJSC "DTEK Pavlogradugol", Pavlograd, Ukraine

ABSTRACT: Schematic diagrams of bottom heaving development that was created on the basis of distribution field features of each stress components depending on structures and properties of foliated coal-bearing rock massif are shown. Such methodology allowed for establishing three common factors that generates bottom heaving manifestations and presents constantly. However, dominate effect of one of them put in principle of heaving process division on three potential variants of development: influence of tensile vertical stresses σ_y that forming foliation zone of immediate and upper part of main roof; sinking of poor rocks that is under weakening and fracturing forms similarity of abutment pressure zone in sides of mine working – as called "stample effect"; formation in sandstones and more hard siltstones relatively small capacity of quasiplastic joints (under the influence of tensile vertical and increased compressive horizontal stresses σ_x) that increases mobility of given layer and can intensify bottom heaving process of in-seam working.

1 INTRODUCTION

In geomechanics perceptions about origin mechanism and development of bottom heaving confirms that the main factor is increasing rock pressure in sides of mine working. This pressure creates as stamp on bottom sinking. This process is shown on Figure 1 where with help of vectors illustrates direction changing of full displacements of near-the-contour rocks from vertical in roof, practically horizontal in the district of frame support props to inclined and vertical in bottom closer to mine working axis. Similar vector perception was received in real work with help of physical model of rock that represents full deformation diagram. Stages of weakening and fracturing are determinative in the process of plastic yielding development in bottom rocks of in-seam working.

2 DEFINITION OF SCHEME OF NEARBY BOTTOM ROCKS DEFORMATION

Influence uncertainty of bottom structure and properties on its stress-strain state predetermines uncertainty of heaving manifestations in bottom of in-seam working. In regularities explanation complex approach is very necessary. Analysis of bottom dis-

placements diagrams is supported by reasons taken from analysis of distribution field regularities of each stress components. Such analysis methodology allowed for educing and explaining three schematic diagrams of bottom heaving development in mine workings of PrJSC "DTEK Pavlogradugol". It should be noted that established regularities of heaving development that connected with structure and properties of nearby bottom of coal seam do not contradict, but adds and concretizes perceptions about mechanism (Chernyak 1973) of given process behavior. Analysis of stress-strain state of system "massif – in-seam working" allowed to educe three common factors that generates heaving manifestations in mine workings:

– influence of tensile vertical stresses σ_y that forms foliation (according to form of inverse dome of natural equilibrium) zone of immediate and upper part of main bottom rocks;

– sinking of poor rocks that's under the stage of weakening and fracturing in cavity of mine working under the influence of more hard coal seam. This seam forms similarity of abutment pressure zone in sides of mine working – as it called "stample effect";

– formation in more hard siltstones and sandstones relatively small capacity of quasiplastic joints (under the influence of tensile and increased compressive horizontal stresses) that increases mobility of given rock layer and can intensify bottom heaving process of in-seam working.

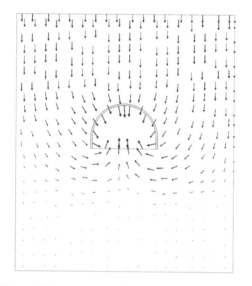

Figure 1. Vector representation diagrams of full displacements of coal-bearing rock layers in nearby of in-seam working.

All enumerated factors are presented constantly in a varying agree. However, dominate effect of one of them put in the principle of bottom heaving division process of mine working on four potential variants of its development that represents real mine-geological conditions of in-seam working supporting.

First group of mine-geological conditions is characterized by enough stable condition of bottom rocks when its nearby layers represented essentially unlike on capacity and mechanical properties lithological differences. But all of them have increased (in mine-geological conditions of Western Donbass) designed drag to compressive $\sigma^r_{comp_i} \geq 20$ MPa.

Here the basic reason of bottom rocks manifestation consists in action of compressive vertical stresses σ_y that forms area of partially weakening rocks that determinates by penetration depth of compressive σ_y, but width is relevant to width of mine working in drivage. Area shape is similar to the dome of natural equilibrium that inverse relatively to vertical axis on 180° (see Figure 2). Area sizes is enough stable regardless of bottom rocks structure.

Here is take place foliation of rock in vertical direction (action of compressive σ_y) that gain by other factors. Actually, side areas of bottom rocks (beyond mine working width) are in stable condition: in the lower part thickness of each layer of compressive stresses σ_x don't appears, in the upper part of each layer compressive and reduced stresses are far of stress value to pressure $\sigma^r_{comp_i}$. Therefore, in given areas bottom rocks actively resist to abutment pressure from the side of coal seam and "stample effect" doesn't observe. From the other side rock layers that bedding below foliation area (as a rule maximum depth in the place of mine working vertical axis do not exceed 2–2.5 m) under mine working also stable on forgoing reasons: in upper part of each layer tensile stresses σ_x don't appears, but in lower part compressive stresses σ_x and reduced stresses σ don't exceed value of $\sigma^r_{comp_i}$.

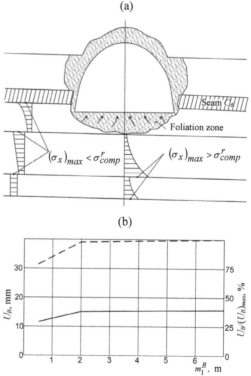

Figure 2. Scheme (a) of nearby bottom rocks deformation of mine working with increased resistance to compression $\sigma^r_{comp\,i}$ and regularities (b) of displacement development U_B with thickness growth m_1^B of first rock layer: —— U_B ; -- $U_B/(U_B)_{max}$.

Consequently, quasiplastic joints in rock layers are absent. Therefore, heaving of mine working bottom generates exceptionally by limited action area of tensile vertical components σ_y that unable for creation essential bottom displacements in mine working cavity and in different variants of bottom structure does not lead to excess $U_B = 20$–30 mm.

Features of bottom rocks structure influence on its displacements consists in following. Distribution area of tensile σ_y is enough constant: at first thin layer tensile σ_y penetrates in second more thickness bottom layer; at more thickness first layer tensile σ_y localizing only here; lithological differences of Western Donbass rocks are resist similarity to tension, so foliation process and rock displacements are changing not essentially. Mudstone and siltstone as more poor rocks (in comparison with sandstone of second and third layers) with increased deformation properties are the main source of bottom rocks deformations. However it stable condition doesn't allow for displacement develop beyond the area of tensile σ_y that is cleanly shown on diagrams (Figure 2b). With siltstone thickness growth from 0.5 to 2.0 m take place increasing of bottom displacements U_B only on few millimeters. This means that dominating contribution has foliation zone from tensile σ_y. Because absolute values U_B are small, so it's relatively increasing (towards maximum $(U_B)_{max}$ at $m_1^B = 7$ m) grows up more essentially on 25–35%. At further increasing of siltstone thickness decreases intensity of bottom displacements growth and practically faded at $m_1^B \geq 4$ m.

Generally we can make conclusion that during calculation (inclusive of weakening factors) resistance to compression of nearby bottom rock layers not less then 20 MPa in it doesn't observes essential heaving manifestations on the districts of in-seam working that situated outside of influence zone of stoping. That's why first group of mine-geological conditions has little practical interest foreshorten of planning measures on heaving prevention. By this factor exploitation condition of mine working anticipates as satisfactory and in further researches we will not observes analyzed group of conditions.

Second group of conditions is linear opposition of first, because reflects mine-geological situation when nearby rock layers are presented by mudstones and siltstones of less rock strength. Potentially it is the most dangerous (from viewpoint of bottom rocks steadiness) group of mine-geological conditions. Relevant mechanism of bottom rocks

deformation is shown on Figure 3 and for systemization of further researches forming perceptions are called "scheme I of heaving development". Roof rocks forms increased load in sides of mine working that perceives coal seam as more solid element and transmit this load on poor bottom rocks. It leads to appearance of extensive areas of limited and unlimited condition. Rock layers feels contortion in cavity of mine working. During this process destruction happens firstly in areas of deflection moment action (Figure 3(a) – in central crosscuts of each layer and in sides – abutment pressure zone) but then on all width of span slabs. Increased tensile horizontal stresses σ_x in bottom under abutment pressure zone promotes to its unstable condition. The main condition of increased compressive stresses appearance is rock transition to limited and unlimited state that substantiates by its low resistant to compression.

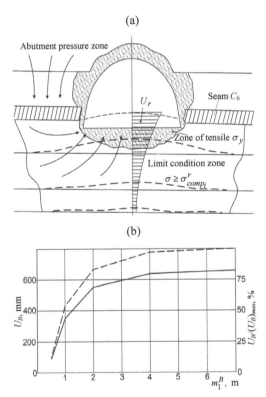

Figure 3. Scheme (a) of deformation of nearby bottom rocks with decreased resistance to compression $\sigma_{comp\,i}^r$ and regularities (b) of displacements development U_B with the growth of thickness m_1^B of first rock layer: —— U_B; -- -- $U_B/(U_B)_{max}$.

Therefore, low rock strength (also by the reason of weakening factors action: water content, cracks content, etc) first of all, creates conditions of rock layers intensive deflection, and then grow up contortion by means of increased horizontal stresses. As a result all volume of rock from one increased rock pressure zone (in left side of mine working) to another (in right side of mine working) transits in limited and unlimited state that characterized not only by weakening, but also rock fracturing which moves in direction of less resistance, i.e. in mine working cavity. This conclusion about formation extension zones of limited (unlimited) condition of bottom rocks confirms by assembly of stress-strain calculations for different variants of structure when value of heaving U_b consistently increase of 300–400 mm (for thickness of first layer $m_1^B = 1.5...2$ m and more) at absence in lower layers of sandstone.

At the same time, if more poor first rock layer has low thickness ($m_1^B = 0.5...0.8$ m) but after that underlie harder mudstone or siltstone, heaving will be enough less (to 150–200 mm). However, with the growth of m_1^B happens sharp heaving intensification (Figure 3b) which substantiated by increasing volume of fracturing rocks. Nevertheless, gradient of function $U_B\left(m_1^B\right)$ growth has tendency for fading at $m_1^B \geq 2$ yet (80–85% of displacement from maximum value $\left(U_B\right)_{max}$ at $m_1^B = 7$ m), but at $m_1^B \geq$ 4 m heaving practically steady. To our opinion, this event substantiates by two reasons: firstly, with the growth of distance from mine working value of perturbations of massif stress-strain state faded; secondly, increasing value of fracturing rocks partially compensates by compression of remote rocks that are less weakened, but has more increased deformation. In such conditions (when calculated resistance to compression is $\sigma_{comp_i}^r \leq 10$ MPa) take place plastic yielding of bottom rocks. This event happens in nearby and more remote rock layers of mine working bottom.

Essentially differents has mechanism of heaving development during enough expanded mine-geological situation of bedding in nearby bottom of in-seam working of mudstone and siltstone and lower more hard sandstone. Main parameters that have influence are: thickness m_1^B of first rock layer (water content siltstone), thickness of second layer from sandstone, resistance to compression $\sigma_{comp_3}^r$ of third layer that represented by siltstone and mudstone. Also

is possible variant of sandstone bedding in third layer after mudstone or siltstone of second layer.

On Figure 4 basic diagram of more characteristic structure of bottom rocks in first layer underlie poor mudstone or siltstone. Given group of mine-geological condition is shown on "scheme II of heaving development". Increased load on abutment pressure zone transmits on immediate roof from poor (for example, water content) siltstone or mudstone. Low calculated resistance to compression of first layer ($\sigma_{comp_1}^r < 10$ MPa) contributes to development in it expansion areas of limited and unlimited conditions that generates essential heaving of bottom rocks. By these reasons resistance to rock pressure in first rock layer decreases and most part of load transmits on sandstone.

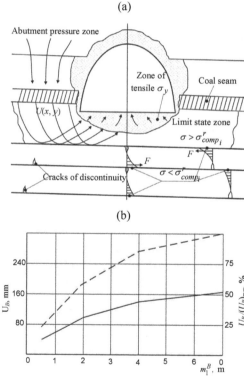

Figure 4. Scheme (a) of essentially nonuniform nearby rock layers deformation and regularities (b) of displacement development U_B with thickness growth m_1^B of first rock layer during sandstone steady preservation: —— U_B ; – – $U_B/\left(U_B\right)_{max}$.

On rock slab from sandstone under the influence of outward load forms concentrations of horizontal stresses σ_x in the district (which characterized contortion) of mine working vertical axis and in its sides: in the middle of span of rock slab tensile σ_x situates in upper part of layer, but increased compressive σ_x – in lower; in side areas of slab (under abutment pressure zones).

Everybody knows that rocks of Western Donbass have low resistance to compression in areas of tensile σ_x action forms cracks discontinuity as its shown on Figure 4(a). Nevertheless, by means of compressive stresses forms couple of resulting force F, that creates stabilizing moment. Slab possess in stable condition, whereby reduced stresses σ (in areas compressive σ_x action) do not exceed resistance to compression of σ_{comp2}^R sandstone. During enough strength of underlying rock layers (third and fourth) happens analogical processes of cracks appearance only in areas of tensile σ_x action.

In this conditions bottom heaving development mechanism of in-seam working essentially converts by reason of situation of poor weakening layer of immediate roof between two hard layers: above – coal seam, underneath – hard sandstone. Weakening siltstone or mudstone have great mobility (similarity of plastic yielding) and under the influence of "hard stample" (coal seam) deforms in vertical and horizontal directions. This event prevents (besides steady rocks in remote areas of mine working sides) integral and hard sandstone in main roof. Displacements of siltstone don't expand in bottom depth but its vector changes direction in sideway of mine working cavity. There take place "reflection" of displacements direction of weakening and fracturing siltstone from hard barrier as sandstone with high thickness. In other words, hard sandstone increases heaving manifestation due to creation of directive stream of broken rock in mine working cavity.

Such representation about mechanism of heaving process streaming is proved by series of calculations for different structure of nearby bottom (Figure 4b, lines 1). Here thickness of first layer m_1^B acts more essential role in comparison with previous schemes, because from parameter m_1^B depends excess volume of weakening and fracturing rock that directed by sandstone in sideway of mine working. Thus, alternatively to previous scheme (Figure 3(b), when at $m_1^B = 2$ m realized 80–85% of displacements U_B from maximum value, but at $m_1^B = 4$ m growth is practically stops), sandstone provokes further growth of U_B: at $m_1^B = 2$ m realized only 55–60%, but at

$m_1^B = 4$ m – near 85% from maximum value. Nevertheless, function $U_B\left(m_1^B\right)$ has tendency to flattering at $m_1^B > 4$ m. Generally, we can make conclusion that main influence on heaving process is substantiated by two factors: activation of poor mudstone and siltstone displacements by means of rigid foundation that represented by sandstone; limited of lower bottom rock layers displacements by hard sandstone. These two factors determines double situation of third heaving development scheme: in comparison with first scheme dominates heaving activation factor; in comparison with second scheme – limited factor of rock displacements that bedding lower than sandstone.

In represented regularities also acts essential role removal value of sandstone from contour of mine working in its situation plan. For example in third layer between broken layer and sandstone bedding layer of harder (unwatered) siltstone or mudstone. Here take place common tendency of "rigid foundation effect" decreasing with the growth of sandstone bedding depth that predetermines decrease of intensify sinking of first layer. For this process adds influence of more steady second layer from mudstone that partially take up deformation of first layer. In total this process leads to decreasing heaving and increasing "realization" of heaving at thicknesses of first layer to 2 m. It means that mechanism of mine working bottom approximates to second scheme.

Another variant of bottom rock heaving development essentially differs from observed and characterized by following structure: in immediate bottom bedding rocks of poor strength, second layer represented by sandstone with thickness to 1.5–2 m , but in third layer bedded poor siltstone or mudstone. Behavior of such structure with alternation of layers of lower and extra strength is characterized by relatively independent (Bondarenko 2007, 2008) transition of each rock layer in limited and unlimited condition. At poor rocks of first and third layers sandstone is represented by rigid slab that loaded from bottom and roof. Distribution of load is enough profitable in relation to sandstone steady. In central part of rock span acts increased vertical load (in direction to mine working) from weakened third rock layer that bending slab and open cracks in the area of tensile σ_x action. In side part of sandstone acts vertical load from plastic yielding of first rock layer (Figure 5a) that forms quasiplastic joints in slab supports. Therefore, appears prominent deflection moments in sandstone that could break it in crosscuts. Here deflection moment reaches to maximum, especially in slab prop or in central part of span. It happens at relatively small thickness of sandstone $m_2^B \leq 1.5...2$ m and from its blocks creates similarity of anchor system. This system has some reaction to displacement resistance of rocks in sideway of mine working bottom. Appearance of anchor system in sandstone is depends upon not only its thickness m_2^B, but also from thicknesses of first m_1^B and third m_3^B poor rock layers. On diagrams (Figure 5b, lines 2) represented influence of first layer thickness m_1^B on value of heaving U_B and this regularity differs by slow growth of function $U_B\left(m_1^B\right)$. Explanation of this regularity is opened by two factors: increased mobility of anchor system in sand stone that intensify displacements of first layer sideway to mine working; intensify movements of anchor system in sandstone by means of plastic yielding in third rock layer.

Therefore, when both factors act in same direction value of heaving is connected with excess value of rocks from fracturing that, first of all, depends upon first rock layer thickness.

In summary results of analysis of strain-stress state system "massif – in-seam working" finds its logical explanation in proposed mechanism of heaving development in mine working that reflected by three principal diagrams inasmuch as single behavior of multivarious and nonuniform bottom structure is impossible here in the nature of geomechanical processes. That's why next necessary stage of researches will be calculation base widening of real different structures and properties of nearby bottom rocks for more full specification and elaboration both mechanism of heaving development and regularities of its streaming.

3 CONCLUSIONS

1. Analysis of stress-strain state calculation results educed uncertainty of influence on heaving process, structure and properties of nearby bottom rocks: on one hand well-known tendency about bottom displacements decreasing U_B during increasing resistance to compression is proved out; from the other hand on process overlay impact of properties nonuniformity (on definite districts of coal fields) of adjoining rock layers (for example, watered siltstone of immediate roof and harder sandstone of main roof); this impact increased at great fluctuations range (on the length of mine working) thickness of adjoining layers and possibilities of breaking low-thickness sandstone. Different tendencies of heaving development with the growth of weakening bottom rocks volumes are stated.

2. Explanation of indoubt tendencies in bottom heaving development realized on the basis of complex approach. Analysis of bottom displacement diagrams is fortified by reasons that were educed during regularities of distribution field of each stress component examination. Such methodology allows for setting three common factors that generates bottom rock manifestation and presents constantly, but dominate effect put into the principle of heaving process division on three possible development variants:

– influence of tensile vertical stresses σ_y that forms rock foliation zone of immediate and upper part of main roof;

– sinking of poor rock during weakening and fracturing in cavity of mine working under the influence of harder sandstone that forms similarity of abutment pressure zone what is "stample effect";

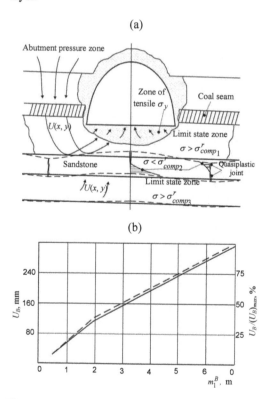

Figure 5. Scheme (a) of essentially nonuniform nearby rock layers deformation and regularities (b) of displacement development U_B with thickness growth m_1^B of first rock layer during quasiplastic joints formation in sandstone: —— U_B; – – $U_B/(U_B)_{max}$.

– formation in sandstones and harder siltstones relatively small power of quasiplastic joints (under the influence of compressive horizontal stresses σ_x) that increased mobility of given layer and intensify bottom rock heaving process.

Indicated factors found its logical explanation in proposed mechanism of bottom heaving development that reflected by three principal schemes which opened nature of geomechanical processes in nonuniform and multivarious structure of bottom rocks.

REFERENCES

Chernyak, I. 1973. *Prevention of bottom heaving in mine workings*. Moscow: Nedra: 273.

Bondarenko, V., Kovalevs'ka, I., Symanovych, G. & Fomychov, V. 2007. *Computer modeling of stress-strain state of fine-grained rock massif around in-seam working*. Book 2. *Limited and unlimited state of system: "rock – support"*. Dnipropetrovs'k: System technologies: 198.

Bondarenko, V., Illyashov, M., Kovalevs'ka, I., Symanovych, G. & Fomychov, V. 2008. *Prediction of in-seam working contour displacement in foliated massif of poor rocks*. Dnipropetrovs'k: System technologies: 193.

Progressive Technologies of Coal, Coalbed Methane, and Ores Mining – Bondarenko, Kovalevs'ka & Ganushevych (eds)
© 2014 Taylor & Francis Group, London, ISBN: 978-1-138-02699-5

Loading mechanism of extraction mine working in front of a stoping face

I. Kovalevs'ka & D. Astafiev
National Mining University, Dnipropetrovs'k, Ukraine

O. Vivcharenko
Department of Coal Industry Ministry of Energy and Coal Industry of Ukraine

O. Malykhin
Shevchenkivs'ka district state administration, Kyiv, Ukraine

ABSTRACT: Specialties of movement mechanism of above-the-coal strata of foliated massif poor rocks, its deformation in near-the-contour part of extraction mine working and supporting system loading of increased abutment rock pressure in front of longwall are reviewed.

1 INTRODUCTION

Common regularities of anomalous zone formation of stress-strain state of above-the-coal strata nearby longwall are developed over several decades of this event researches, and were created basic for geomechanical researches regardless of mine-geological and mine-technical conditions of coal seams development (Kovalevs'ka 2004 & Zborshik 1991). In front of longwall abutment pressure zone forms and moves in same time with stoping face advance; length of this zone (coordinates X) on Western Donbass mines estimates mainly within the range to 20–60 m; in cross section of extraction mine working and nearby rocks (coordinate Z) diagram of vertical rock pressure (component Z) changes in quality and quantity plan as far as longwall approximation. It is for the reason that more objective development of displacement mechanism of above-the-coal strata and supporting system loading of extraction mine working primarily was reviewed spatial field diagram σ_y in abutment pressure zone in front of longwall, its quantity view is given on Figure 1.

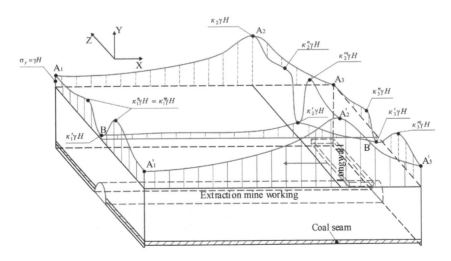

Figure 1. Distribution scheme of anomalous areas of vertical stresses σ_y nearby longwall in abutment rock pressure zone.

On some distance (A_1B) from mine working along extraction production unit A_1A_3 on coordinate X perturbation of vertical rock pressure diagram is characterized by following regularities. Experimentally established (Bondarenko 2012) that on definite distance from stoping face its influence is unimportant and vertical stresses are close to initial value $\sigma_y = \gamma H$ that marked by point A_1. While longwall approximation concentration σ_y increases when reaches to maximum $\sigma_y = \kappa_2\gamma H$ in point A_2 which as a rule lag behind the wall face on 2–15 m (Kovalevs'ka 2004). This distance mainly depends upon the correlation of coal seam rigidities and rock layers of roof. Here the main speciality of Western Donbass mine-geological conditions is combination of increased coal hardness towards hardness of roof rocks and seam bottom which are mainly represented by poor mudstones and siltstones with enough intensive cracks content from 2–5 to 10–15 cracks per meter. Changing of massif initial condition during longwall access way provokes weakening of poor coal-bearing rocks, especially of immediate roof and bottom, as a result sharp increasing of its deformation (decline of deformation modulus). Resistance to compression of coal seam usually in 1.5–4 times more than rocks of immediate roof and bottom. That is why degree of coal seam weakening is far less and manifests basically in local areas. These processes determine significant rigidity exceedence of coal seam under the rigidity of roof rock layers: at together deformation under the influence of rock pressure; movements of coal seam will far less than roof rock layers and on some distance from wall face bulk of vertical movements U_y realized by means of downward movement of easy-deformable roof layers. Such situation of increased coal seam deformations leads to approximation of maximum $\kappa_2\gamma H$ to wall face. It happens because traditional for Donbass mines as called "non-rigid fixing" of rock consoles under the longwall in Western Donbass takes on the properties which approach to term "rigid fixing" (using in building mechanics) with concentration of deflection moment in rock console near wall rock.

Described specialty provokes more intensive growth of grade of rock contours movements of mine working on enough limited producing unit in front of longwall with length to 20–40 m where speed of roof rocks, bottom and sides of mine working sharply increases from shares of mm/day to units and even few dozens of mm / day according to the data of different researches. That is why appropriate measures on strengthening of mine working support realize, as a rule in front of abutment pressure zone.

During analyze of abutment pressure zone at displacement of long section (on coordinate X) to mine working (on coordinate Z) was marked change of vertical stresses field distribution σ_y that substantiates both influence of mine working and face end. At a remove from longwall that characterizes beginning of stoping influence (line A_1A_1'), anomalous of diagrams σ_y determine only by impact of mine working: in its roof (near point B) forms unload area where according to rock layer situation on its height (on coordinate Y) compressive σ_y not only decrease ($\sigma_y < \gamma H$), but near mine working contour transit into tensile with formation of dome of natural equilibrium on the length of mine working (line BB'). In sides of mine working on the distance (on coordinate Z) to one and half – two its widths forms concentration of compressive σ_y around $(1.5...3.0)\gamma H$ that on its own can provokes some local areas of weakening, but during approximation of stoping face increases abutment pressure and happens overlaying of two tensile σ_y concentrations – from extraction mine working to longwall end. This event promotes more intensive development of roof rocks weakening process and as a result, activation of rock pressure manifestations in sides of mine working.

Near and under the longwall, abutment pressure headily decreases (by reason of roof rocks active foliation) down to zero (point A_3) on some removal from conjugation along the longwall length. Value of vertical stresses σ_y decreases (on coordinate Z), but have perturbations (line A_3A_3') which determined by mine working influence and abutment pressure nearby side boundaries of producing unit (line $A_1'A_3'$).

Analysis of vertical stresses σ_y field in abutment pressure zone in front and on sides of longwall (Figure 1) that was built on current quantity representations about movement processes of above-the-coal strata allow to make set of conclusions and states which are necessary during expansion and geomechanical substantiation of supporting system loading mechanism specialties of extraction mine workings in conditions of Western Donbass mines.

Firstly, along the horizontal ZX stresses field σ_y constantly changes by force of two factors influence: perturbations of stress-strain state from mine working cavity and face end cavity. Degree of influence on different producing units variable with impact domination from one to another factors. Thus, in beginning of abutment pressure zone in front of stoping face (line A_1A_1') superceding influence have perturbations σ_y which connected with extraction mine working: if in point A_1 vertical stress approximates to value of untouched massif $\sigma_y = \gamma H$, in sides of mine working appeari concentration to value $\sigma_y = \kappa_1\gamma H$; in roof of mine working take place unload to value $\kappa_1'\gamma H$. On drawing near the longwall its influence become dominant: maximums of abutment pressure (line A_2A_2') increase to value $\sigma_y = \kappa_2''\gamma H$ (point 2, where influence of mine working is unimportant) to value $\kappa_2''\gamma H$ in side of mine working by means of overlaying influence. In roof of mine working concentration $\sigma_y = \kappa_2'\gamma H$ decrease by reason of partial compensation (unload zone from influence of mine working) of concentration σ_y from longwall influence. Further in side of mine working from untouched massif abutment pressure increase again to value $\sigma_y = \kappa_2'''\gamma H$ by means of concentrations σ_y summation from mine working to boundaries of producing unit. On goaf boundary of longwall (A_3A_3') abutment pressure zone converts in unload zone and such longwall influence significantly decrease diagram $\sigma_y(Z)$ with appearing of low concentrations $\kappa_3''\gamma H$ in neighbor (to longwall) side of mine working. Above it in roof unload zone from

mine working overlay on unload zone from longwall and it is possible extensive area formation of tensile stresses σ_y that weaken poor rocks unambiguously. In opposite side of mine working increased concentration $\kappa_3'''\gamma H$ forms as combined impact of mine working and abutment pressure in boundary side of producing unit.

Secondly, variable stresses field generates different intensity of rock pressure manifestations in extraction mine working and also processes of stress arching in sides and bottom.

Thirdly, it is necessary to take account such specialty of rock massif movement in Western Donbass as close proximity to initial cut (wall face, side of mine working) of abutment pressure maximums in front of longwall and producing unit boundaries.

2 MAIN PART

According to changed representations and main specialties conclusions of above-the-coal strata movement mechanism and loading on supporting system of extraction mine working on abutment pressure area in front of longwall were examined on separate pieces (from scheme on Figure 1). First of all, was reviewed deformation mechanism of roof layer rocks of coal seam in extraction mine working crosscut (scheme on Figure 2). Traditional technology of mine workings supporting in abutment pressure zone in front of longwall connected with reinforcement support installation, more often in the form of two-link yielding props from special concave profile (SCP) by two variants: whether installation of one prop in central part of mine working dome, or two props in frame crown.

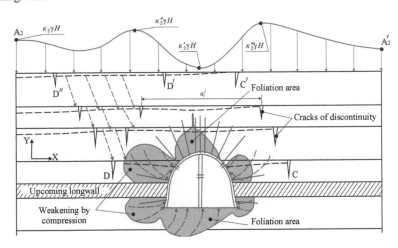

Figure 2. On the intensification mechanism of increased side loads on support in abutment pressure zone.

Reinforcement support has a lot of attention because it props resist to vertical rock pressure (in the aggregate with frame and abutment bolting) but anyhow not increase frame resistance to side loads.

On Figure 2 shown crosscut of mine working in front of longwall in action of abutment pressure maximum (line $A_2 A_2'$ on Figure 1) where geomechanical processes of roof rocks movement manifests more intensively. Around extraction mine working forms series of anomalous areas which connected with rock pressure redistribution and rock weakening processes. It was established by numerous researches, including modeling of near-the-contour rocks condition and support in zone of stoping influence during usage of abutment bolting. Specialties of coal-bearing rocks conditions are as follows.

On one side in near-the-contour roof rocks forms foliation area from tensile vertical stresses σ_y action and also tensile horizontal stresses σ_z and σ_x, which activate weakening process. It is common knowledge that poor rocks of Western Donbass have extremely low resistance to tension (usually to 1.5–3.5 MPa), furthermore owing to impact of rock weakening factors such resistance practically absent. Nevertheless because of dense reinforcement grid of roof by anchors foliation area in form of dome not so essential, but weight of rocks inside dome id far less of frame support resistance to vertical loads. Besides, length of anchors usually exceeds sizes of dome (by means of crown runner and metal lath of interframe protection) can fully retain volume of foliated rocks.

Other specialty of Western Donbass coal-bearing strata of consists in extremely poor binding along the contact surfaces of rock layers; on many units such binding practically absent. Also it is necessary to take in account action of rock pressure geostatic anomalous near mine workings that break poor contacts, so rock layers near extraction mine working deforms in part independently from each other. This movement of rock layers (is shown by stipple lines on Figure 2) can separate on two regularities of two mine workings influence – stoping and extraction. Under the influence of mine working in its roof happens layers contortion with cracks discontinuity formation (in upper part on thickness of each layer) along the lines CC' and DD'. Cracks forms under the influence of tensile stresses σ_z and increase rock layers deformation in direction of mine working cavity. Here take place unload with decreasing of stresses intensity and as a rule, increasing of roof rocks steadiness. Also for this event promotes abutment bolting installation. Longwall influence (even on some distance from wall face) manifests in

development of roof rock layers deformation in abutment pressure zone since before longwall approximation. Sinking of roof layers characterizes by maximal contortion in central (along the longwall length) part of producing unit and minimal contortion near extraction mine working; in view of such contortion direction in side of mine working (from longwall) forms another system of cracks discontinuity which shown by line DD'' (Figure 2).

Now we consider rocks condition in sides of mine working and also how formation of rock blocks with length a_i^j in its roof change rock pressure manifestations in sides and bottom. With help of analytic, computational and experimental methods were established that in side of mine working (generally in rocks of immediate roof and bottom) increased rock pressure weaken some volume of poor rocks that increase it mobility. From the other side, sinking incumbent rock layers load overlaying, also and partially weakening of immediate roof rocks layers volume; rigid coal seam limit vertical movements of side rocks and its movement vector changes direction in mine working cavity with load on props. This process in abutment pressure zone develops more intensively because of increased vertical component of rock pressure; intensification of "stamp effect" happens in rocks of immediate bottom. From longwall side develops similar process of stamping of side rocks and mine working bottom, but it intensify because of rock blocks deformation between lines DD' and DD''. That is why asymmetry of load appears on all perimeter of extraction mine working which is necessary to take into account during geomechanical models generation and calculation of combined supporting system parameters.

Reviewed schematic representation of above-the-coal strata movement mechanism near extraction mine working is supplemented by consideration in its lengthwise direction (flatness YX). Scheme is given on Figure 3. Here chosen crosscut YX on the longwall end where rock pressure diagram changes more dynamically is shown by line $A_1 A_3$ on Figure 1 and 3. Now let's bring to mind regularity of roof rock layers contortion in front of stoping face and its layer by layer collapse in worked-out area behind it with variable length rock consoles formation significant for Western Donbass contacts fracturing between layers of coal-bearing strata and extremely low resistance to tensile forces during cracks content (usually two or three cracks systems) determines some specific of above-the-coal strata deformation process in front of stoping face. Increase of abutment pressure on drawing near the longwall (unit $A_1 A_2$ of diagram σ_y) intensify definitely independent deformation of above-the-coal strata non-uniform layers, during its contortion appear tensile stresses σ_x (in upper part of each layer thickness) on approximation to maximum $\sigma_y = \kappa_2 \gamma H$ (point A_2 on Figure 3) of abutment pressure. Keep in mind that in poor fractured rocks is enough small tensile σ_x for generation and growth of technogenic origin cracks (lines $ББ'$, $ГГ'$, EE' and $ЖЖ'$). In crosscut of layer where forms crack, sharply decrease resistance moment and probably further cracks development (on drawing near the longwall) that finally interlock with broken rocks area (from concentration of tensile σ_x) in lower part of each layer happen its separation on rock blocks with length $b_i^{\,j}$.

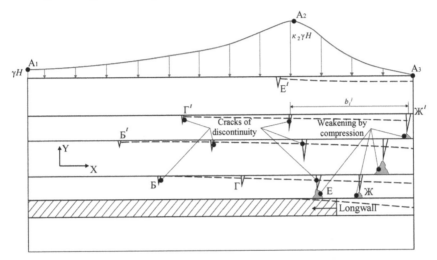

Figure 3. On the mechanism of cracks generation and development in above-the-coal strata rock layers.

From viewpoint of mine working steadiness cracks formation in nearby rock layers in front of longwall increase its deformation and decrease resistance to abutment pressure that unavoidably leads to intensification of rock weakening process in sides and bottom of mine working (Figure 2). That is why for increasing reliability of rock pressure manifestation prediction in developed geomechanical models it is necessary to take into account discrete fracturing of rock layers that increase on drawing near the stoping face. One of the main questions is how we can take into account this discrete fracturing? For this nonce on Figure 4 was described fragment of partially fracturing rock slab deformation with diagrams of horizontal stresses σ_x and σ_z distribution. Because of these stresses action appears reaction of rock pressure resistance. In such representation between partially fracturing rock slabs in abutment pressure zone appears similarity of simplex system that have increased deformation and specific bearing capacity by means of only compressive stresses σ_x and σ_z action on layer thickness. At tensile stresses σ_x and σ_z absence because of cracks in crosscut of rock slab its bearing

capacity realizes by means of recovering moment from arm of equivalent forces on opposite rolling planes YX and YZ of slab as integral values of compressive σ_x and σ_z, on distribution areas of hardway. Such deformation mechanism and interworking of nearby rock layers with discrete spatial fracturing is more authentic than traditional representations about massif continuity (in abutment pressure zone in front of longwall) and appearance of some broken rocks areas.

3 CONCLUSIONS

Eventually, some conclusions and recommendations on creation principles of geomechanical massif movement model and loading of extraction mine working support in zone of stoping influence are represented.

Firstly, partial weakening of above-the-coal rock layers by way of its discrete fracturing in space in tensile stresses areas (generally horizontal σ_x and σ_z) and high concentrations of compressive stresses lead to increasing of rocks that amenable to active deformations. Its division on system of inter-working slabs intensified fracturing process of partially weaken near-the-contour massif in sides and bottom of mine working.

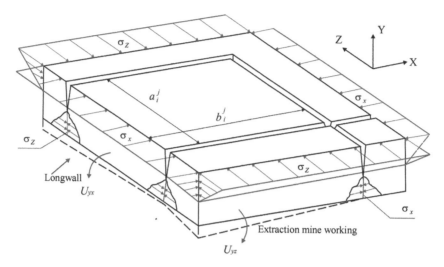

Figure 4. Scheme of rock blocks interworking near extraction mine working.

Secondly, creation and perfection (by way of increasing authenticity and objectivity) of geomechanical model have to get through a lot of consistent stages that reflect real process of generation and development of discrete fracturing in foliated above-the-coal strata. Stage by stage methodology of modeling is more effectually to use by following way:

– on the first stage reflected integral layers of coal-bearing massif nearby extraction mine working and face end and crosscuts took shape, where developed maximal tensile stresses $\left(\sigma_{y,z,x}\right)_{max}$. These stresses generate rock foliation and cracks of discontinuity creation, also maximal values of stresses σ_{max} intensity that according to chosen strengthening theory provokes weakening of rock from compressive forces;

– on the second stage in ascertained crosscuts and areas simulates cracks or weakening local volumes in flexible-and-plastic setting, realizes calculation of stress-strain state of researched spatial geomechanical system; new areas of broken rocks from tensile and compressive loads are established;

– on the third stage procedure repeats with ascertained earlier and again fixed dislocations of geomechanical system wholeness;

– perfection of model deceases during absence of wholeness fracturing of coal-bearing massif and finally its stress-strain state calculates in elastic-visco-plastic setting, in other words with take into account of rheological factor.

Final demand is very important, because experimentally determined connection of end sections of load on mechanized complex with technological parameters of stoping operations that change in time: stoping face advance rate and duration of its break. Influence of roof lowering in face end on loading process of extraction mine workings supporting system was proved, so given technological parameters will also act and these regularities are more objectively for studying in rheological setting. Given situation is proved by complex of experimental researches of stoping face advance rate influence on geomechanical processes in extraction mine workings of Western Donbass mines.

REFERENCES

Kovalevs'ka, I.A. 2004. *Geomechanics of management of spatial system "massif – reinforcement rocks – support of mine workings"*. Dissertation... Doctor of Engineering Science. Dnipropetrovs'k: National mining university: 349.

Zborshik, M.P. & Nazimko, V.V. 1991. *Protection of mine workings on deep mines in unload zone*. Monograph. Kyiv: Tekhnics: 248.

Bondarenko, V.I., Kovalevs'ka, I.A., Symanovych, G.A. and oth. 2012. *Experimental researches of reuse extraction mine workings steadiness on Donbass flat seams*. Monograph. Dnipropetrovs'k: LizunovPress: 426.

Parameters optimization of heat pump units in mining enterprises

G. Pivnyak, V. Samusia, Y. Oksen & M. Radiuk
National Mining University, Dnipropetrovs'k, Ukraine

ABSTRACT: The analysis of the main diagrams of the heat pump waste heat recovery from mine water has been accomplished. The method of calculating the most efficient thermodynamic cycle for the heat pump has been developed. It has been shown that in the superheating cycle at the compressor inlet the increase of the heat transfer agent flow rate beyond a certain limit does not lead to the increase of the energy efficiency of the heat pump.

1 INTRODUCTION

Due to decrease of the fossil fuel resources, environmental and heat pollution the problems of rational fossil fuel and renewable energy resources usage as well as waste heat recovery are becoming urgent (Samusia, Oksen & Komissarov 2012). The mine water pumped from the mines, outlet mine air, the outlet water of the compressor cooling systems are the main sources of the waste heat at the coal and ore mines (Samusia, Oksen, Komissarov & Radiuk 2012). Essential temperature value (15–30 °C), cheapness and availability of those sources make them feasible for heat pump technology utilization for the mine's heat supply (Krasnyck 2005). In 2010 the scientists from the National Mining University developed and implemented the first in Ukraine and CIS the heat pump unit that uses mine water for waste heat recovery at "Blagodatna" mine PJSC "DTEK Pavlogradvugillia". The heat pump heat rate is 800 kW. It allows to heat 120 m^3 of mine's waster for the hot water supply system (HWSS) at 7-hours operation cycle (Samusia 2013). The positive experience of the unit operation justifies the feasibility of the heat pump waste heat recovery of the mine water for HWSS and allows recommending the technology to be used in other mines. Due to this the improvements of this technology taking into account the specifics of its implementation in other mines are increasing.

The purpose of the article is to justify the chosen schemes for the heat pump technology implementation for mine water waste heat recovery for HWSS and estimate the influence of the heat pumps' refrigerants properties and the mine water flow rate on the heat pumps' energy efficiency.

2 PICULIARITIES OF THE HEAT PUMP TECHNOLOGY

The air characteristics of the heat pump technology considered are as follows:

• small seasonal variations in mine water temperature (1–2 °C)

• relatively large flow rate of mine water, which provides the possibility of extracting it from the thermal power that exceeds the mine's need in hot water significantly;

• presence in the mine water suspended dirt particles, and its significant mineralization, which can lead to contamination of heat transfer surfaces of heat exchangers heat pump system;

• presence of hard salts in pure water, heated for HWSS;

• low initial temperature of water heated for HWSS in winter time equals 5–9 °C and in summer time 15–19 °C;

• constancy of required final temperature of water for HWSS in the range of 42–45 °C.

3 THE SELECTION OF THE HEAT PUMP SCHEME

Principally there are two possible main heat pump schemes – with mine water flowing directly via the evaporator and the heating water throw the heat pump's condenser (Figure 1a), and with the intermediate circuits of the heat transfer (Figure 1b). It is also possible that one of the circuits is absent.

The scheme presented on Figure 1(a) is simpler and cheaper. Moreover it is more thermodynamically efficient compared to the scheme on Figure 1(b) because of the absence of the energy losses on the friction when the fluid flows in the intermediate circuits and as a result of the irreversibility of the heat transfer in additional circuits of the heat transfer. However when the unit is working the evaporator is influenced by polluted and aggressive mine water and at the condenser it is possible to see sediments on the heat transfer surfaces contained in the heating water. Thereby it is reasonable to use that scheme in cases when mine and clean water are clear enough and the heat pump's evaporator and condenser are shell-and-tube heat exchanges with the water flowing inside the tubes. The construction of the tubes allows their disassembling and mechanical cleaning of the tubes' internal surfaces.

(a)

(b)

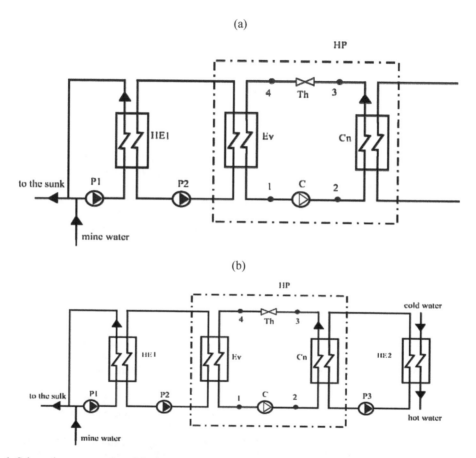

Figure 1. Schematic representation of the heat pump system without intermediate circuits (a); the heat pump system with intermediate circuits (b): HP – heat pump; C – compressor; Cn – condenser; Ev – evaporator; Th – throttle; HE1, HE2 – mine and clean water heat exchangers; P1, P2, P3 – water pumps.

In modern heat pumps the nonseparable brazed plate heat exchangers are used as the evaporator and condenser. Therefore, to protect heat exchangers' surfaces from dirt, salt deposits and corrosion the scheme with intermediate circuits where chemically purified water or low-freezing aqueous solutions, for example, ethylene glycol or propylene glycol circulates (Figure 1b) should be preferable.

4 MAIN PARAMETERS OF THE HEAT PUMP SYSTEM

From the peculiarities of the technology described above the task of heat pump's main parameters choosing can be considered as follows.

The initial t_{h1} and final clean water temperature t_{h2} heating for HWSS, water mass flow rate m_h and initial mine water temperature t_{x1} are chosen as initial parameters. It is required to determine the values of mine water mass flow rate m_x and thermodynamic cycle parameters of the heat pump to provide the maximum energy efficiency of the heat pump unit.

As energy efficiency parameters the coefficient of performance has been chosen

$$k = \frac{Q_h}{N_{el}}, \qquad (1)$$

where Q_h and N_{el} – condenser heat power and the electrical power consumed by the unit's motors.

Let us consider the scheme with intermediate circuits (Figure 1b). To simplify let us assume that as a heat transfer agent in the intermediate circuits water is used, its mass flow rate m_{wx} in the evaporator circuit is equal to the flow rate of the mine water. The mass flow rate of the condenser circuit m_{wh} is equal to the mass flow rate of the clean heating water.

$$m_{wx} = m_x, \qquad (2)$$

$$m_{wh} = m_h. \qquad (3)$$

It has been assumed that the power consumed by electric water pumps of intermediate contours, compared to the power consumed by the heat pump compressor motor is negligible. Electrical and mechanical losses in the drive compressor assign to its internal losses based on the fact that in modern design of the heat pump' compressor the cooling of the rotor is provided with the part of the stream of the cold refrigerant vapor (Kyrychenko 2012). Because of that

$$N_{el} = N_c, \qquad (4)$$

where N_c – internal power of the heat pump's compressor.

The heat load rate q_h is calculated as

$$q_h = q_x + l_c, \qquad (5)$$

where q_x – cold load rate, l_c – specific compressor work.

If to neglect the heat exchange between the refrigerant and the environment through the walls of pipelines and then

$$Q_h = m_a q_h; \qquad (6)$$

$$Q_x = m_a q_x; \qquad (7)$$

$$N_c = m_a l_c; \qquad (8)$$

$$Q_h = Q_x + N_c, \qquad (9)$$

where m_a – mass flow rate of the refrigerant in the heat pump's circuit; Q_x – heat pump evaporator heat power (the heat power that is transferred from mine water).

The required condenser heat load is determined by the given mass flow rate and water temperature mode that is heated by HWSS

$$Q_h = c_w m_h (t_{h2} - t_{h1}), \qquad (10)$$

where c_w – heat capacity of water.

The required heat power that is transferred from mine water is determined by its flow rate and cooling degree in the heat exchanger

$$Q_x = c_w m_x (t_{x1} - t_{x2}), \qquad (11)$$

where t_{x2} – final temperature of mine water.

Mine water mass flow rate influences mine water final temperature and the temperature of the refrigerant in the evaporator that determines the energy efficiency of the heat pump. It is know that to increase the energy efficiency of the heat pumps it is required to decrease exergy losses caused by irreversibility (Goman 2012). To do it is necessary to decrease the temperature difference in the heat exchangers of the heat pumps.

The diagrams of the water and refrigerant changes in the evaporator and condenser for the cases when condensing and evaporating processes occur at constant (R134a) and variable (R407C) temperature of the working fluid are shown on Figure 2.

Since the temperature of the refrigerant is changed stepwise, it is necessary to consider the possibility of approaching the refrigerants temperature profiles not only in the boundary sections of heat exchangers, but also in the interior, corresponding to the beginning and the end of the phase transitions. Consequently, the system of the temperature constraints for the condenser and evaporator of the heat pump are presented as follows

$$t_{wx2} - t_1 \geq \Delta t_{1\,min}; \qquad (12)$$

$$t_{wx1} - t_4 \geq \Delta t_{4\,min}; \qquad (13)$$

$$t_{wx5} - t_1 \geq \Delta t_{5\,min}; \qquad (14)$$

$$t_2 - t_{wh1} \geq \Delta t_{2\,min}; \qquad (15)$$

$$t_3 - t_{wh2} \geq \Delta t_{3\,min}; \qquad (16)$$

$$t_6 - t_{wh6} \geq \Delta t_{6\,min}; \qquad (17)$$

$$t_7 - t_{wh7} \geq \Delta t_{7\,min}, \qquad (18)$$

where t_1, t_2, ..., t_7 – refrigerant temperature at points 1–7 of the cycle; Δt_{1min}, Δt_{2min}, ..., Δt_{7min} – given minimum temperature differences at the heat exchangers sections where the state of the refrigerant is determined by points 1–7 of the cycle. t_{wx5} – water temperature at the section of the evaporator where the refrigerant state is dry-saturated vapor. t_{wx6} and t_{wx7} – water temperature at the sections of the evaporator where process of vapor condensing starts and ends.

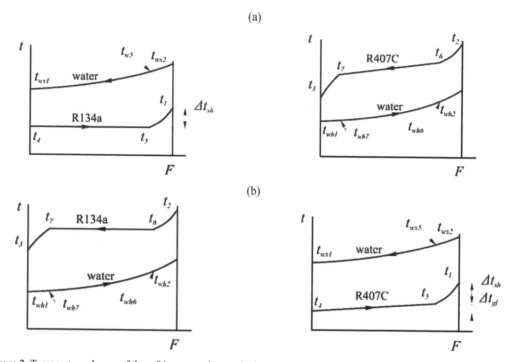

Figure 2. Temperature change of the refrigerant and water in the evaporator (a) and condenser (b) of the heat pump.

From the equation of the heat balance of the individual sections of the heat exchangers (evaporator and condenser)

$$t_{wx5} = t_{wx2} - (t_{wx2} - t_{wx1})\frac{i_1 - i_5}{i_1 - i_4};\qquad (19)$$

$$t_{wh6} = t_{wh1} - (t_{wh1} - t_{wh2})\frac{i_2 - i_6}{i_2 - i_3};\qquad (20)$$

$$t_{wh7} = t_{wh1} - (t_{wh1} - t_{wh2})\frac{i_2 - i_7}{i_2 - i_3},\qquad (21)$$

where $i_1, i_2, ..., i_7$ – refrigerant enthalpy ay points 1–7 of the cycle.

The system of the temperature constraints for mine and clean water heat exchanges:

$$t_{x1} - t_{wx2} \geq \Delta t_{x min};\qquad (22)$$

$$t_{x2} - t_{wx1} \geq \Delta t_{x min};\qquad (23)$$

$$t_{wh1} - t_{h2} \geq \Delta t_{h min};\qquad (24)$$

$$t_{wh2} - t_{h1} \geq \Delta t_{h min},\qquad (25)$$

where Δt_{xmin} and Δt_{hmin} – given minimal temperature differences in the mine and clean water heat exchanges.

The difference of the vapor state at the compressor inlet from dry-saturated vapor is considered by the given value of superheated vapor Δt_{sh} and real process of compression in a compressor from a theoretical one by isoentropic coefficient of the compressor η_s, and because of that

$$t_1 = t_5 + \Delta t_{sh};\qquad (26)$$

$$\Delta i_{1-2} = \frac{\Delta i_s}{\eta_s},\qquad (27)$$

where Δi_{1-2} and Δi_s – the changes of the refrigerant enthalpy in the real process of compression 1–2 and isoentropic process with the same initial and final pressure.

At a given flow rate of mine water the most efficient thermodynamic cycle of the heat pump will be characterized by the fact that at least one of the inequalities (12)–(14) and two of the inequalities (15)–(18) take the lower boundary values and converted to equality.

The algorithm for the calculation of such a cycle, and the program has been written in Matlab R2009b.

With this program the influence of flow rate of the mine cooling water on the heat pump system efficiency has been investigated. While modeling it has been given $Q_x = 200$ kW, $t_{x1} = 16\,°C$, $t_{h1} = 9\,°C$, $t_{h2} = 45\,°C$, $\Delta t_{x min} = \Delta t_{h min} = 4\,°C$, $\Delta t_{1 min} = \Delta t_{4 min} = \Delta t_{5 min} = 4\,°C$, $\Delta t_{2 min} = 10\,°C$, $\Delta t_{3 min} = \Delta t_{6 min} = \Delta t_{7 min} = 5\,°C$, $\Delta t_{sh} = 6\,°C$, $\eta_s = 0.7$. As refrigerants R410A, R134a and R407C have been chosen. Based on the calculation results the influence of mine water mass flow rate on coefficient of the performance of the heat pump with R410A, R134a and R407C (lines 1–3) are shown on the diagrams (Figure 3). The changes of mine water final temperature has also been show (line 4).

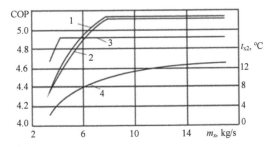

Figure 3. Variation of the coefficient of the performance of heat transformation of the most efficient cycles and mine water final temperatures with various mass flow rates.

Figure 3 shows that in the area of the small mine water mass flow rate for all of the refrigerants considered as mass flow rate increases the coefficient of performance increases. It reaches its maximum k_{max} at the certain value of the flow rate $m_{x lim}$. After that value the flow rate remains constant (line 1–3)

when $m_x \leq m_{x ep}$ $k \leq k_{max}$; (28)

when $m_x \geq m_{x ep}$ $k = k_{max}$. (29)

Such a behavior $k = f(m_x)$ can be explained by the influence of the temperature constraints (12)–(21) on the temperature mode of the evaporator of the heat pump. When $m_x > m_{x lim}$ the constraint (12) is de-

terminative that in this case is converted to the limited equality $\Delta t_1 = \Delta t_{1 min}$ and the constraints (13) and (14) have the view $\Delta t_4 > \Delta t_{4 min}$ and $\Delta t_5 > \Delta t_{5 min}$. On the contrary, when $m_x < m_{x lim}$ because of the more significant decrease of the mine water final temperature (see Figures 3 and 4) the constraint (13) converted to the quality namely $\Delta t_4 = \Delta t_{4 min}$ and the constraint (12) has the view of the inequality $\Delta t_1 > \Delta t_{1 min}$. When $m_x = m_{x lim}$ both of the constraints (12) and (13) converted to the constraint-inequality. In all of the cases the temperature mode of the condenser determined by the simultaneous conversion to the equality of the conditions (16) and (17). Consequently, to get the maximum efficiency of the heat pump unit it is not necessary to tend to the processing of large quantity of mine water and its flow rate can be taken as $m_x = m_{x lim}$ or a bit higher.

From the refrigerants considered freon R134a is a simple substance (tetrafluoroethane), R410A is isoentropic mixture and R407C is zeotropic mixture. Figure 4 shows that the cycles of the heat pumps with refrigerants R134a and R410C boiling at constant temperature are characterized by almost equal efficiency (line 1 and 2). The cycle with R407C has lower efficiency than with R134a and R410C at the large mine water mass flow rate and higher efficiency at the small mass flow rate. The offset of the limiting flow rate value $m_{x lim}$ to the area of lower values in case of R407C use can be explained by the presence of the temperature glide for this refrigerant Δt_{gl} (graphically presented in Figure 3a) since it boils at variable temperature.

5 CONCLUSIONS

The recommendations on the selection of rational circuits' design of the heat pump technology use for mine water waste heat recovery have been developed. The method of the calculation of the most efficient thermodynamic cycle for heat pumps under the conditions of the limitations in mass flow rate and temperatures of the transfer agents of low- and high potential heat.

It has been stated that as mine water mass flow rate increases the coefficient of the performance of the heat pump increases in the area of low flow rate only. When the values of mass flow rate exceed some specific limiting values than their values do not influence the heat pump energy efficiency that in this case will be equal to maximum reachable value for the given temperature mode of the low- and high potential heat transfer agents.

The temperature glide of the zeotropic refrigerants lead to the increase of the heat pump efficiency in the low mine water mass flow rate.

REFERENCES

Samusia. V., Oksen, Y. & Komissarov, Y. 2012. *Evaluation of the energy efficiency of the heat pump system, using complex heat outgoing air flow and waste water.* Mining Electrical Mechanics and Automatics: Proceedings of scientific works. Issue 88: 131–136.

Samusia, V., Oksen, Y., Komissarov, Y. & Radiuk, M. 2012. *Heat pumps for mine water waste heat recovery. Hot water supply unit.* State higher educational Institution National Mining University. Dnipropetrovs'k: Scientific Bulletin of National Mining University, 4: 143–144.

Krasnyck, V., Ostapenko, V. & Ulanov, N. 2005. *Technological possibilities and prospects of water and heat potential use of mine waters of Ukraine.* Coal of Ukraine, Issue 12: 35–37.

Samusia, V., Oksen, Y. & Radiuk, M. 2013. *Heat pumps for mine water waste heat recovery. Proceedings of the international scientific and technical conference "Mining of mineral deposits".* Taylor & Francis group, London, UK: 153–157.

Kyrychenko, Y., Samusia, V. & Kyrychenko, V. 2012. *Software development for the automatic control system of deep water hydrohoist.* Geomechanical processes during underground mining, Taylor & Francis group, – London, ISBN: 978-0-415-66174-4: 81–86.

Goman, Kyrychenko, Y, Samusia, V. & Kyrychenko, V. 2012. *Experimental investigation of aeroelastic and hydroelastic instability parameters of a marine pipeline.* Geomechanical processes during underground mining, Taylor & Francis group, London, UK, ISBN: 978-0-415-66174-4: 163–167.

Progressive Technologies of Coal, Coalbed Methane, and Ores Mining – Bondarenko, Kovalevs'ka & Ganushevych (eds)
© 2014 Taylor & Francis Group, London, ISBN: 978-1-138-02699-5

Physical modeling of waste inclusions stability during mining of complex structured deposits

M. Stupnik, V. Kolosov, V. Kalinichenko & S. Pismennyi
Kryvyi Rih National University, Kryvyi Rih, Ukraine

ABSTRACT: Boundary conditions for the application of development system with an open production areas when powerful complex structured deposits mining in terms of Kryvyi Rih iron ore basin are proposed. Laboratory modeling was carried out the results of which confirm previously defined boundary conditions of application system development with open face.

1 INTRODUCTION

Kryvyi Rih basin deposits are heterogeneous. Within the same area deposits of ore bodies varies from 500 to 3000 m^2, at a power of 10 to 150 m and more. Mineral content in the rock ore ranges from 36 to 64%. Often mine fields have two parallel deposits containing to 70% of total mine field. The remaining 30% of stocks are 3 or more ore bodies (Stupnik & Pismennyi 2012, Hivrenko & Schelkanov 2001, Development of technological... 2012).

In heterogeneous deposits inclusions of barren areas or ore with lower nutrient content are often found. Power of barren areas varies from 2–3 m to 6 m in some places – up to 6–10 m. The specific area of barren areas within the level (sublevel) is 10...15–18%.

Currently deposit mining is carried out by open face systems and systems with the caving of overlying rocks and ores. Deposits in which there is barren area usually mined completely that reduces the quality of the ore mined in 3–5 weight % and as a consequence to increase ore loss of 5–10% from the standard.

2 ANALYSIS OF RESEARCH AND PUBLICATIONS

Shchelkanov V.A and Hivrenko O.J. offered to mine complex ore deposits by systems of ore and overlying rocks caving (Shchelkanov et al. 2001, Shchelkanov & Hivrenko 2001, Byzov 2001). The idea of the proposed option was in ore deposits mining alternately from hanging to the bottom layers, leaving barren inclusion in the stope. Complex deposits rock ore shooting-off is carried out in stages: I – stocks breaking followed by the release of ore located between hanging wall and barren inclusion; II – barren inclusion holing and moving it to the hanging wall

by force of explosion; III – breaking and release of the remaining ore in the bottom wall.

Proposed version of the development system has a number of drawbacks which may include the following: constructive elements parameters of the development system are not based; barren inclusion parameters to ensure its sustainability when developing unit (first stage) are not defined; increasing in the weight of explosives on the rock ore breaking, as well as the cost of deep holes drilling; increasing of lateral ore clogging on 2–5% (third stage).

When stope mining in stable rocks the authors have proposed a variant of the development system with open face. The essence of this version of the development system is the following: unit development is carried from hanging to the bottom wall of the ore deposit, leaving barren inclusion directly in the unit (Stupnik et al. 2013). The minimum allowable power of barren inclusion when developing complex structured ore deposits by development system with open face and subsequent abatement of treatment chambers with the boundary conditions is defined

– for stable strong ores:

$$\begin{cases} m_{b.i} \geq (1.9...2.1) \cdot W \cdot K_t, \\ m_{b.i} \leq (0.1...0.3) \cdot h \cdot K_t \leq m_o, \\ m_{b.i} = \dfrac{H \cdot \gamma \cdot a_c \cdot K_{st} \cdot K_{as} \cdot K_t \cdot cos\,\alpha}{1000 \cdot n_p \cdot f \cdot K_{str.w}} \end{cases},$$

– for stable weak ores:

$$\begin{cases} m_{b.i} \geq (1.5...1.7) \cdot W \cdot K_t, \\ m_{b.i} \leq (0.1...0.3) \cdot h \cdot K_t \leq m_o, \\ m_{b.i} = \dfrac{H \cdot \gamma \cdot a_c \cdot K_{st} \cdot K_{as} \cdot K_t \cdot cos\,\alpha}{1000 \cdot n_p \cdot f \cdot K_{str.w}} \end{cases},$$

where m_{di} – minimally allowable power of barren inclusion, m; W – line of least resistance when longhole stoping, m; K_t – coefficient of barren inclusion stability of its existence period; h – level height, m; m_o – minimum permissible thickness of the ore body, located near barren inclusion, m; H – deep of mining, m; γ – bulk density of rocks forming the sloping pillar, kg/m^3; a_c – chamber width across the strike, m; K_{st} – coefficient depending on the tensile stress and deformation of rocks; K_{as} = 1.5–2.5 – safety factor pillar; α – the angle of incidence of the ore deposit, degree.; n_p – number of longitudinal pillars, corresponding to one camera; f – barren rock strength inclusion; $K_{str.w}$ = 0.5–0.8 – structural weakening coefficient of rock ore.

According to the results of pilot tests determined the dependence of the stability factor of the time of its existence, which is described by the empirical equation with a correlation coefficient of 0.9977.

$$K_t = 0.0014 \cdot t^2 - 0.0306 \cdot t + 1.028 ,$$

where t – existence of pillars in the beginning of carrying out and completion of treatment works, months.

In order to establish the reliability of theoretical research in development of complex ore deposits with the use of the open face and the abatement of the interaction of multimodulus rock ores and rocks, as well as changes to the source of the stress field of a phase-homogeneous rock ore must perform laboratory simulations.

3 PRESENTATION OF THE MATERIAL AND RESULTS

According to a method of modeling on equivalent materials (unit (Stupnik et al. 2013, Kirpichev & Brick 1953, Kulikov 1980) a laboratory model with design parameters 1000×500×850 mm was made. The front wall of the model is transparent, made of glass, the side walls are deaf. The bottom of the model is made of metal and the special outlet for the ore flowing.

Geometric modeling scale 1:200 was adopted, and the time scale of 1:15. Equivalent material, given the scale simulation was selected according to geological and physical- mechanical properties of rocks of the Kryvyi Rih iron ore basin. Load of the caved overlying rocks on the simulated unit area was replaced by an equivalent load, taking into account the scale of the geometric modeling and simulated depth of mining.

A total of 48 series of laboratory experiments, each of which was repeated 3–5 times depending on

the reliability of the results obtained during the simulation were carried out. The values obtained in the simulation differed by no more than 10–15%, which confirms the validity of the results.

Laboratory model consisted of two types of materials: equivalent and full-scale. Equivalent material imitated barren inclusion, and full-scale – caved rock ore. The initial data for modeling were as follows: height of caved ore bed was 80 cm, the angle of incidence of the ore deposit and barren turn was 70 deg., power of barren inclusion was 6 cm, the number of inclusions was 1 pc., the total weight of the material in the model was equivalent to 34 kg, 231 kg of full-scale, depth of 1260 m was assumed to develop.

As a full-scale material was used magnetite quartzite at mine "Pervomaiskaya" with high magnetic properties, which allows to separate the life material of equivalent magnetic separator. Quantitative and qualitative composition of the equivalent and full-scale material is adopted in accordance with the methodology of modeling (Kirpichev & Brick 1953, Kulikov 1980).

Output of ore was produced of the model from hanging to bottom walls without further filling of formed treatment chambers with caved rocks. In the process of manufacture of ore and the full release after with two treatment panels, with stresses pressure sensors appearing in the equivalent material in the modeling process were continuously measured. Testing laboratory model and its behavior observing was carried out in stages.

The first stage. In the model with layers of 5 cm, life material was filled by areas and equivalent one was rolled, forming the geological structure of the rock mass, Figure 1(a).

Every 10 cm in the equivalent material sensors capturing emerging voltage were installed and full-scale material through 5 cm in backfilled chalk glass strips to monitor the release of the full-scale material.

The second stage. Through the outlets located in the lower part of the model for full-scale production of material caving ore from unit by doses was produced. Total weight of the ore released from model at one time does not exceed the 1.2–1.6 kg.

Caving ore produced from the extreme of the equivalent number of material outlets located in the hanging wall up until ellipsoid loosening reaches the upper wall model simulating hanging wall of the ore deposit, Figure 1(c).

Then, loosen the ellipsoid parameters measured and compared with the calculated values by the method of Acad. G.M. Malahova. Volume was made with an extreme number of outlets, as long ellipsoid loosening until it reaches the upper bound of the model. The simulation results band width issue no more than 5–10 cm, which is confirmed by theoretical studies (Kirpichev & Brick 1953, Kulikov 1980).

On the basis of the second stage of modeling it was determined that the qualitative composition of full-scale material has good convergence with theoretical researches (Malakhov et al. 1968) it confirms the selected modeling techniques (Kirpichev & Brick 1953, Kulikov 1980).

Than ore was produced using steady consistent manufacture modes from all draw holes. After the withdrawal of each dose ore was passed through a magnetic separator. Withdrawal was carried until all full-scale material was extracted. Figure 1(d).

As a result of modeling it was confirmed that dirt inclusion keeps its stability during withdrawal.

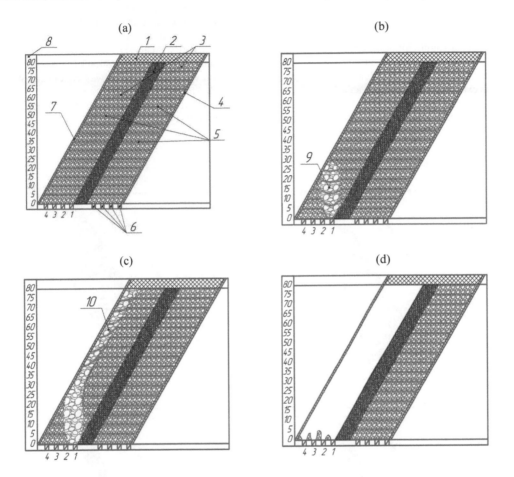

Figure 1. Modeling of output of caved ore when complex structured deposits mining with one rock inclusion and its release in the hanging wall: 1 – external load; 2 – equivalent material (barren inclusion); 3 – life material (caved ore); 4, 7 – wooden wall (respectively lying and hanging wall of deposits); 5 – chalk; 6 – outlet openings; 8 – model; 9, 10 –, respectively ellipsoid of loosening zone.

The third stage. After the formation of the camera at the hanging wall sensor variations were fixed every hour. In the case where the difference in sensor variations varied by more than 3–5% of a laboratory model there was time allowed for stabilization of the stresses in the equivalent material needed for a full-scale material simulating dirt inclusion reached a standstill.

The fourth stage. When the model reached a standstill, withdrawal of simulated broken ore at the lying-wall started, Figure 2. At the withdrawal process loosening ellipsoid parameters were measured, substitute (rock layer).character was under supervision.

The fifth stage. Monitoring the condition of the block for 12 days, it corresponds to 6 months in real conditions.

During the modeling it was found that at the withdrawal of broken ore to the loosening ellipsoid of lying-wall its volume corresponds to the research of Acad. G.M. Malahov (Malakhov et al. 1968), and the error doesn't exceed 10%, it confirms the accuracy of the selected modeling techniques, material and size distribution of its composition.

With the further ore withdrawal (Figure 1) at an exposure of rock stratum, it retains its stability over the lifetime of 8–10 months. During the modeling there was a slight chipping substitute (barren inclusion). Averaged measurement results stress of changes in the rear sight after each stage according to the experiments shown in Table 1.

It is clear from Table 1 that the stress concentration in the barren inclusion (substitute) remains constant. However, in the middle there was an increase in the values of the pressure sensors which corresponds according to the theory studies the emergence of zones of tensile stresses. At this destruction and chipping of substitute is not observed.

With the withdrawal of broken ore from the lying-wall, after reaching ellipsoid loosening 10 of the upper limit of the cleaning chamber (Figure 2), there is chipping barren inclusion in the bottom 1/3 of the height of the simulated block from the waste chamber of the first stage. Also, a change of sensor 2 and 3, Table 2. After stabilization of the sensor values withdrawal of broken ore was continued. With the withdrawal 3/4 of the total reserve there is a slight chipping of substitute in the middle of maintaining its stability.

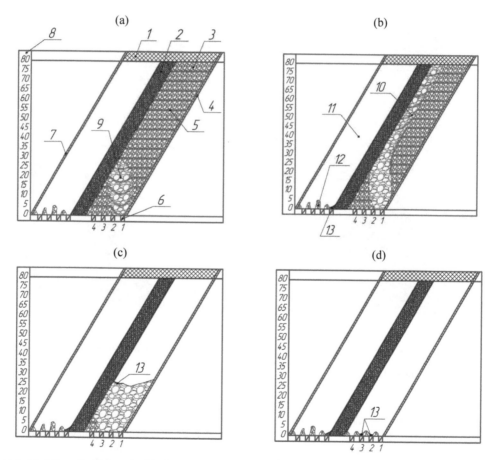

Figure 2. Modeling of withdrawal of broken ore at complex structured deposit mining with dirt inclusion and its withdrawal at the lying-wall: 1–10 graphical symbols given in Figure 1; 11 – graphical symbols given in figure at the lying-wall; 12 – ore loss at kettle back; 13 – rotted substitute.

Table 1. The results of measuring stresses in the modeling.

Name of sensor measurement	# sensor fixed in substitute					Amount of issued substitute	% extract
	1	2	3	4	5		
1	159	159	159	159	159		
2	159	159	159	159	159		
3	159	159	148	159	159	2.352	7.00
4	159	159	155	159	159		
5	159	159	152	159	159		

Note: 1 – pressure sensor values before withdrawal; 2 – values of pressure sensors after the withdrawal of broken ore from block 1; 3 – values of pressure sensors with the withdrawal of block 2 at the formation of loosening zone at hanging wall; 4 – values of pressure sensors after the withdrawal of block 2 3/4 of the total ore reserve of modeled block; 5 – Pressure sensor values after withdrawal.

Thus, at the modeling, it was found that the height of the broken layer 80 cm, there was a slight delamination of rock layer which was 7% of the total weight of the seamed equivalent material. Insignificantly detached pieces of equivalent material simulating breed layers did not change its strength relative to the initial state. Modeling results confirm that at developing complex structured deposits in stable rocks with barren inclusion ensured by extraction with fullness maintain the stability of rock stratum. Value changes in the internal resistance of rock pressure sensors that characterize the stability of a layer of rock from the time of its existence are given in Table 2.

Table 2. Average modeling results of tense changes in barren inclusions from the time of its existence.

Values indicating resistance layer of rock during the day												
0	1	2	3	4	5	6	7	8	9	10	11	12
159.0	158.9	158.5	158.1	157.8	157.8	157.6	157.2	157.1	157.0	156.8	156.3	156.2

Table 2 shows that with increasing time the existence of barren inclusion under pressure sensors strength rock layer varies slightly with time, thereby maintaining its stability. Depending on the pressure sensor values lifetime inclusion on dirt heights broken layer shown in Figure 3. Figure 3 shows that at the height of the cleaning chamber 35, 40 cm tense arising in the block sight remain unchanged, while increasing the height of the camera 2 times the values of sensor values are reduced to 5%, which confirms the possibility of maintain the strength barren inclusion.

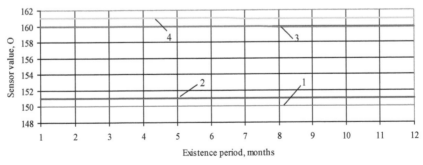

Figure 3. The dependence of the pressure sensors resistance on existence time of block height and camera cleaning: 1 – chamber height 35 cm; 2 – chamber height 40 cm; 3 – chamber height 70 cm; 4 – chamber height 80 cm.

4 CONCLUSIONS

Implementation of the study found that the use of the system of mining with open cleaning space and leaving barren inclusion in the chamber to enhance the quality of extracted ore. Laboratory tests confirmed that the stable rocks barren inclusion retains its stability for 6–8 months. This time is sufficient for the subsequent completion of the treatment chambers formed stowing material.

REFERENCES

Stupnik, N.I. & Pismennyi, S.V. 2012. *Advanced technological options for further iron deposits mining by systems with massive collapse ore*. Kryvyi Rih National University. Reporter: 30: 3–7.

Hivrenko, V.O. & Schelkanov V.A. 2001. *Technological classification complex structured deposits. Ore Mining*. Kryvyi Rih Technical University: 76: 26–29.

Development of technological release, preparation and cleaning schemes for removing hard deposits and structural with further mining stocks over large depth: Report on students research work (final). 2012. #0109U002336, Kryvyi Rih National University: 306.

Shchelkanov, V.A., Hivrenko, O.J. & Hivrenko, V.O. 2001. *Kryvbass complex structured deposits Analysis*. Kryvyi Rih Technical University: Scientific and technical journal "Development of ore deposits": 75: 30–35.

Shchelkanov, V.A. & Hivrenko, O.Y. 2001. *Tehnological classification of complexstructured deposits*. Kryvyi Rih Technical University: Scientific and technical journal "Development of ore deposits": 76: 26–29.

Byzov, V.F. 2001. License number 37982A E 21 C41/16 UA "*Method development high angled ore bodies containing dirt inclusion*" / Byzov, V.F. Storchak, S.O., Syrichko, V.A., Cherednichenko, O.E., Harkusha, A.F. Vitryak, V.A., Plotnikov, V.F., Repin, O.H. Hivrenko, O.Y., Schelkanov, V.A., Andreev, B.M., Hivrenko, V.O. Publish. 15.05.2001. Certificate, 4.

Stupnik, N., Kalinichenko, V. & Pismennyi, S. 2013. *Pillars sizing at magnetite quartzite's room-work*. Mining of Mineral Deposite. A Balkema Book: 11–15.

Kirpichev, M.V. & Brick, M.V. 1953. *Similarity Theory*. Moscow: USSR Academy of Sciences: 95.

Kulikov, V.V. 1980. *Ore Withdrawal*. Moscow: Nedra: 303.

Malakhov, G.M., Bezuh, V.R. & Petrenko, P.D. 1968. *Theory and practice of broken ore*. Moscow: Nedra: 311.

Progressive Technologies of Coal, Coalbed Methane, and Ores Mining – Bondarenko, Kovalevs'ka & Ganushevych (eds)
© 2014 Taylor & Francis Group, London, ISBN: 978-1-138-02699-5

Utilization of the waste products for stowing

V. Buzylo, T. Savelieva, V. Saveliev & T. Morozova
National Mining University, Dnipropetrovs'k, Ukraine

ABSTRACT: Study was carried out with stowing mixtures prepared on the basis of dressing mill tailings to reduce stowing production cost. Methods and results of study of stowing material strength limit under compression are given.

1 INTRODUCTION

Mining of Zavalievs'k graphite deposit is carried out by open-pit method. It is recommended to develop deposit first by combined and then by underground methods using room-and-pillar system with goaf stowing by refinement tailings to preserve Earth surface for national economy and for ecological balance of industrial region.

Conditions of mining ore deposit by underground method get worse with depth increase. The risk of mining increases due to rising intensity of rock pressure manifestation. Expenditures connected with working timbering and retimbering are increased. Chamber reserves and areas being in the state of simultaneous mining are reduced. All these factors are the cause of decreasing technical-and-economic indexes of mining deposit. Mining ore deposit by underground method using systems with overlying rock caving is followed by high losses and ore dilution. One of the efficient ways which enables to improve ore extraction by underground method is mining with stowing.

Designed stowing complex is an independent unit of the plant and it consists of two sections. The first section is on the surface. It comprises complex for receiving, unloading, storing refinement tailings, phosphogypsum and technological complex for preparing consolidating stowing mixture. The second section is underground. The task of this section is to erect bulkheads, assemble stowing pipelines and carry out stowing operations (Kravchenko 1974).

Application of refinement tailings for preparing stowing mixture and goaf stowing requires to carry out the study of mixture strength properties. It is essential part of mining technology under mining method with consolidating stowing.

2 METHODS OF DETERMINING THE STRENGTH OF STOWING MATERIAL

Determining the strength of concrete block is particularly important before working out the nearest chamber as it enables to find out allowed value of outcropping area and choose seismically safety parameters of blasting operations (Dobrovolievsky 1970). Furthermore, this study finally permits to choose optimal composition of stowing mixture which can withstand static loads, to learn the changes of material strength in various cross-sections of stowing block with the aim of further technology improvement.

Determining value of strength limit under compression of stowing material is carried out by crushing blocks of $5.0 \times 5.0 \times 5.0$ cm. If fluidity of stowing mixture was 4 cm, metal forms with metal containment basin were used.

Equipment and materials applied during the study are the following: average sample of test stowing mixture, set of forms, complete set of devices to determine degree of stowing mixture fluidity, bar for poking solution, standard pallet for mixture consolidation, paper, towel, machine grease, knife, comb, press.

Samples from stowing mixtures with fluidity of 4 cm are prepared in metal forms with metal containment basin. Forms are greased inside and filled with test stowing mixture twice over. First, stowing mixture of 3 cm high is placed into forms. Mixture is consolidated by prismatic end of standard pallet (12 pressings). After consolidation of the first layer, the second layer of the mixture is placed into form. It is also consolidated. 24 pallet pressings are made to consolidate both layers in each form. Then excess mixture is cut by the wet knife level with form edge smoothing mixture.

Samples prepared on the basis of hydraulic binding substances are matured within forms up to form removal under $t = 20\pm3$ °C and relative humidity of 90%.

Analyzing test results concerning determination of stowing material strength carried out at current mines the following conclusion was made – three months are required to take the main part of stowing strength .Strength growth of stowing mixtures were determined every 7 days for one month and every 20 days.

Before test samples are carefully cleaned by the comb against adherent to the surface sand particles and solution nubbles. Each sample is measured. Its volume is determined with accuracy up to 10^{-6} m^3, then the sample is weighed and bulk weight is calculated with the accuracy up to 10 kg/m^3 of the mixture.

Sample test is carried out on the press UP-5. For testing sample is placed on foundation plate of the press by edges which turned into form walls. The rate of load rise per second must not exceed 3% from assumed destructive power.

3 CHOICE OF RATIONAL COMPOSITION OF THE STOWING MIXTURE

To choose rational composition of stowing mixture, composition which meets the requirements of strength condition of mining deposit was selected. It can be transported through pipeline and at the same time it is least cost.

Such components as cement "300" and "400" (it is relatively expensive material) and calciphytes – rock with hardness according to the Protodiakonov scale is up to 14 are used for underground mining of Zavalievs'k deposit of graphite ores. It is required to use additional calciphyte capital expenditure on equipment of picker lines of given power.

Therefore, to reduce production cost of 1 m^3 of stowing study was carried out with stowing mixture prepared on the basis of production wastes.

"Tailings" of dressing mill were used during the study as filler. Waste sand of Zavalievs'k graphite plant is a polymineral feldspar – quartz rock with impurity of granite and mica. It is characterized by mineral composition (in %): quartz – 56.0; feldspar – 25.3; mica – 4; granite – 6; graphite – 0.7; ore minerals – 8.0; accessory minerals – 5.

Table 1 shows granular-metriccomposition of waste sand at dressing mill. Table 2 shows physical and mechanical properties of waste sand at dressing mill. It demonstrates that test sand is extra fine and thin according to granular metric composition and size modulus.

Table 1. Granular-metric composition of waste sand at dressing mill.

Sample number	Granular-metric composition, %				
	Size of screen hole, mm				
	1.2	0.6	0.315	0.15	0.15
I	0.00	4.31	38.62	38.88	18.19
	0.00	4.31	42.98	81.81	
II	0.00	4.78	36.37	38.21	20.64
	0.00	4.78	41.15	7936	

Content of dusty and clay fractions is in the limit of 1.42–3.59%. This sand is characterized by significant addition of volume (2.9–15.5%) and high-surface area (0.0182–0.0254 m^2/g). Phosphogypsum is used as binding substance. Wastes of chemical plants which produce phosphate fertilizer are its basis.

Table 2. Physical and mechanical properties of waste sand at dressing mill.

Indices	Sample number	
	I	II
Modulus of elasticity, 10^5 Pa	1.29	1.25
Number of dusty and clay particles, %	3.95	1.42
Incrementation under swelling, %	5.78	2.9
Surface area, m^2 / kg	25.389	18.264
Bulk weight, kg / m^3	1570.0	1580.0

Absence of stowing operation criterion in practice characterizing quality of consolidating mixtures makes difficult objective assessment of their properties. Therefore, strength value of stowing within three months is the basic criterion while choosing rational composition of stowing.

Technology analysis of stowing operations at running mines shows that strength value of stowing is ranged in considerable limits: 2.0–7.0 MPa (Kryvyi Rig basin). However vast majority of mines carries out stowing operations under three months of stowing strength 4.0–6.0 MPa.

Four compositions of mixture were tested to carry out study concerning selection of rational stowing composition (Table 3).

Table 3. Qualitative and quantitative composition of test mixture composition.

Components	Material consumption per I m^3 of stowing			
	I	II	III	IV
	Composition number			
Flotation tailings, kg	1931.0	1867.0	1697.0	1600.0
Phosphogypsum, kg	33.0	64.0	117.0	165.0
Water, l	276.0	267.0	273.0	257.0
Bulk weight of stowing mixture, kg / m^3	2.24	2.20	208.0	2.02

Table 4. Results of laboratory tests of stowing mixtures.

Composition number	Sample number in series	Mass bulk, kg	Sample size, 10^{-6} m^3	Cross-section area, 10^{-4} cm^2	Strength of stowing mixture, 10^5 Pa Time of sample withstanding, day			
					7	14	28	90
I	1	0.207	126	25	1.2	1.8	3.1	7.2
	2	0.206			1.0	1.6	2.9	7.4
	3	0.206			1.0	1.4	3.3	6.6
	4	0.207			1.4	1.6	3.0	6.9
	5	0.205			1.3	1.7	2.8	7.0
	Average	0.206			1.2	1.6	3.0	7.0
II	1	0.206	125	25	2.3	4.4	8.4	18
	2	0.207			2.8	4.2	8.3	18.7
	3	0.206			2.0	4.4	8.3	18.9
	4	0.205			2.4	4.7	8.0	18.5
	5	0.204			2.2	4.8	8.6	18.4
	Average	0.205			2.4	4.5	8.3	18.5
III	1	0.204	125	25	4.2	6.7	12.3	28.3
	2	0.201			3.9	6.6	11.2	28.3
	3	0.205			4.0	6.9	11.4	27.7
	4	0.202			4.0	7.0	14.1	28.1
	5	0.203			3.8	7.3	11.0	27.8
	Average	0.203			4.0	6.9	12.0	28.0
IV	1	0.201	25	25	5.4	10.0	16.4	41.5
	2	0.198			5.5	9.3	15.9	40.8
	3	0.200			5.8	9.5	17.0	40.9
	4	0.199			5.6	9.2	17.4	40.6
	5	0.202			5.6	9.5	16.8	41.2
	Average	0.200			5.6	9.5	16.7	41.0

Tests concerning determination of stowing material strength were carried out according to developed technique. Results of tests are given in the Table 4.

One of the most important characteristics of consolidating stowing is the growth of strength in time. Receiving such dependence enables to determine the character of the process of stowing consolidation and estimate its strength in any time. Furthermore, regularity of stowing strength growth in time is very important while determining minimal period of mining start of neighboring chambers with stowing and influences on selection of mining parameters and blasting operations.

On the basis of results received while testing samples of stowing materials graphic chart of stowing strength (R) depending on consolidation time (t) (Figure 1) was built. After 90 days (maximum time of consolidation) four compositions of mixtures took the strength correspondingly 0.7; 1.85; 2.8; 4.1 MPa. Therefore, the fourth composition of mixture is recommended as stowing material.

Technology of stowing operations was developed on the basis of analysis of stowing operation carried out at running mines taking into account mining-geological and mining conditions of deposit development by combined method. It is rational to use gravity stowing of worked-out area. Pneumatic transport is required to apply while working-out chambers located beyond gravity site.

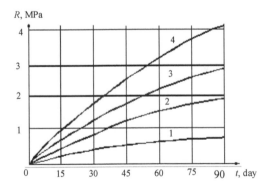

Figure 1. Graphic chart of stowing strength depending on consolidation time.

Filling worked-out chambers by consolidating stowing is recommended to carry out in three stages:
1. Filling chamber bottom by mixture with increased phosphogypsum solution to receive strengthened stowing.
2. Filling chamber all the way high with the stop on the level of bulkheads to prevent emergency leakage of stowing into working.

3. Chamber stowing with increased phosphogypsum content.

Systematic monitoring of stowing level and condition of all bulkheads are carried out while filling chambers with stowing. After chamber filling additional chamber stowing by pipeline with quick-disconnect joints applied in the place of maximal chamber height is performed.

4 CONCLUSIONS

According to given technique study of strength of four compositions of stowing mixture is carried out. They consist of flotation tailings, wastes of chemical plants which produce phosphate fertilizers (phosphogypsum, water). It is recommended to apply the fourth mixture composition given in the Table 3 strength of which was 4.1 MPa after 90 days while working-out Zavalievs'k graphite deposit by combined method using pillar system with goaf stowing. It is recommended to apply gravity stowing of working-out area using additional pneumatic transport within chambers outside gravity stowing.

REFERENCE

Kravchenko, V.P. 1974. *Application of consolidating stowing while mining ore deposits.* Moscow: Nedra: 200.
Dobrovolievsky, V.V. 1970. *Technique of determining physical and mechanical characteristics of stowing material.* Moscow: Institution of Mining named after A.A. Skochinsky: 38.

Progressive Technologies of Coal, Coalbed Methane, and Ores Mining – Bondarenko, Kovalevs'ka & Ganushevych (eds)
© 2014 Taylor & Francis Group, London, ISBN: 978-1-138-02699-5

Ways to achieve the optimal schedule of the mining mode of double subbench mining

S. Moldabayev, Zh. Sultanbekova & Ye. Aben
Kazakh National Technical University after K.I. Satbayev, Almaty, Kazakhstan

B. Rysbaiuly
International Information Technologies University, Almaty, Kazakhstan

ABSTRACT: Solution of the optimization parameters of two subbench technologies with varying levels of working area by nonlinear programming method allows building the design stages of steep deposits mining of complex configuration with providing the best mining operationswith uniform distribution of accessed mineral reserves between two highwalls. Developed algorithm for implementing this method, all formulas are assigned and attached to selected coordinate system and will help solve the problem, which needs to determine some parameters. The specified function does not exceed the fixed values and dedicated objective function reaches a global minimum. Test results show that the width of panels should be determined for mining stages and by hanging and lying sides of deposits when the double subbench technology with varying levels of working area is designed. Its provide obtaining the optimal mining operationswith uniform distribution of accessed reserves on both pit walls.

1 INTRODUCTION

The width of initial cut on each level and its cross section area will be different during double subbench technology usage and if structure of mineral deposit is difficult (Raksihev et al. 2013). The transport berms are provided in each part of mine until pit gets its final depth (Figure 1). The length of initial cut on each level should provide sufficient volume of accessed ore reserves with a minimum amount of stripping operations. Axis position of the initial cut base on the first level depends on ensuring a uniform distribution of accessed reserves on both highwalls until the final depth is reached and stripping volumes minimization until the final contours is reached.

Initial cut position on the next level should create favorable conditions for implementation of double subbench mining technology with mining the benches by cross panels. The main premise is aspiration for a uniform distribution of mining operations and providing sufficient accessed reserves on both highwalls with minimizing current stripping ratio in the initial period of mining (Moldabaev et al. 2013).

Mining stages are isolated on each level including trench cutting and mining the panels on each highwalls. Transport berms with wide that enough to two-traffic of heavy trucks are provides between working and nonworking benches in three-dimensional pit model. Therefore, the position of initial cut in each subsequent level depends on adopted highwall construction. It ensures the equal volumes of mining in each panel and leaving safety berms between subbenches.

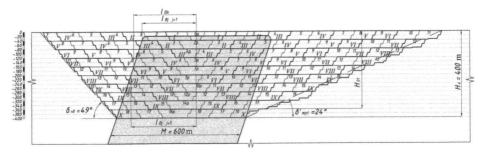

Figure 1. Definition of the axis of initial cut on the first level with equal horizontal width at all mining stages.

Panel width (B_b) is assumed to be 100 m until final contours of pit is reached. Panels, which are adjacent to final contours, have different width. It may be more or less than accepted width. Maintaining of adopted construction of highwalls, which provides independent mining operations on each level determines a joint mining of panels only in one highwall with a lag in the space. For example, panels 2, 4 and 7 will be mined after cutting of trenches 2, 4 and 7 on the second, on the third and on the fifth level on the right highwall. Then panels will be mined on both highwalls. Select of initial cut location on these levels by hanging side of mineral deposits is dictated offset to its lying side during the sinking of mining works. Refusal of joint mining the panels together with a lag in the space with trenches cutting will increase the angle of highwall by the lying side of mineral deposits and worsen the mining mode.

The aspiration to minimization of the gravity center of mining operations to lying side of mineral deposit leads to the fact that width (b_m) of the trenches bottom will be different on each level, but it should be not less than the minimum of its value ($b_{m.min}$)

$$b_m \geq b_{m.min} . \qquad (1)$$

The number of mined panels for trench cutting on each level corresponding to its mining stage depends on the conditions of ensuring the trenches cutting in the underlying level. B_m – transport berm width; h_n – subbench height; B_n – safety berm width, α – subbench slope angle, degree.

The following cases are provided depending on the pit bottom (d) width:

If $d - 2B_m - b_{m.min} - 4h_nctg\alpha - 2B_n < B_b$, then $b_m = d - 2B_m - 4h_nctg\alpha - 2B_n$ and both side of trench will be spaced (numbering of initial cut and panels varies on this level);

If $d - 2B_m - b_{m.min} - 4h_nctg\alpha - 2B_n > B_b$, then $b_m = d - 2B_m - 4h_nctg\alpha - 2B_n - B_b$ and only one of sides of trench will be spaced (numbering of initial cut and panels is the same); if d is not enough for trench cutting the additional panels will be mined in both higwallls on this level.

Placement (which pit wall) of the initial cut in new horizon depends on amount of accessed reserves in each subsequent stage and providing simultaneous mining of block-panels on both pit walls. The best schedule of mining operations is obtained by trenches (initial cuts) forming along the strike of the mineral deposits at each horizon. If trenches are formed across the strike of the mineral deposits accessed reserves reduction takes place un-

til progress in depth of mining works. It occurs due to the displacement of trenches and mining the block-panels to side lying mineral deposits.

Also there are other methods of maintaining sufficient volume of accessed mineral reserves. After the transition of mining operations, block-panels of same horizon will be mined in both pit walls (Figure 4, stage VI) or pit bottom maximizes in certain horizon (stage VII) when pit gets to final contours. Such techniques allow a wide (stage VIII, 13) and narrow (stage IX, 14) initial cut, which will pass closer to the mineral deposits. It doesn't significantly affect to the volume of overburden volume during this period.

Thus, it is advisable to build stages of double subbench mining technology by cross panels with varying levels of work area with reference of mining stages design to mineral deposits and final pit contours from its hanging and lying sides.

2 MINING OPERATIONSOPTIMIZATION BY MINING STAGES

Optimization problem is solved on the basis of cross-sectional pit (Figure 1). Earlier we noted that the problem of nonlinear programming must be solved to determine the optimal stripping ratio by mining. The following steps: a coordinate system is introduced; calculates the coordinates of the points of intersection of horizontal and inclined pit lines; formulas of area of trapezoids, parallelograms and triangles formed after crossing horizontal and diagonal lines are compiled; stripping ratio formulas are compiled. Then each item will be considered separately:

Coordinate system. A cross-sectional pit with depth $H = 400$ m and level height 40 m is shown on Figure 1. The height of sublevel is $m = 20$. Therefore, the total amount of levels is 10, sublevels – 20, and horizontal lines – 21. The horizontal lines, levels, sublevels and mining stages should be fixed in the coordinate system. Moreover, we must distinguish between the coordinates of points hanging and lying sides of deposit. Let describe the point of intersection of horizontal and inclined sublevel lines.

Numbering of sublevels starts from the top $j = 1, 2, ..., 20$. Each sublevel pair form one level. For example, the sublevels $j = 1, 2$ form level numbered $i = 1$, and sublevels $j = 3, 4$ form the level $i = 2$. Knowing the number of the level, you can calculate the sublevel number by the formula $j = 2i - 1$ and $j = 2i$. Each sublevel is between the two horizontal lines – the upper and lower. xk, j, t and yk, j, t based on the above notation were introduced. Where x, y – the coordinates of the hanging and lying pit wall; k – sublevel number of horizontal lines; if

$k = 0$, it is the top line, and if $k = 1$, it is the bottom line ; t – serial number of the stage.

$P_2P_0P_{-1}$ broken line is taken as the origin of the coordinates of the points of horizontal lines. *FB* is parallel to P_2P_0. P_0P_{-1} – perpendicular to horizon *AD* (Figure 2). Part of pit, which lies to the left of the line $P_2P_0P_{-1}$ is assign to the hanging side of deposit, and the right of the $P_2P_0P_{-1}$ – to the lying side of the deposit. Position of the points P_2, P_0, P_{-1} relative to direct FB depends on the panel width B_b and transport berm width B_m. Panel width B_b is determined during solving a nonlinear programming problem. This means that the positions of the points P_2, P_0, P_{-1} becomes a variable in the computation. During solving of nonlinear programming *FB* and *EC* line positions remain constant, thus the coordinates of the hanging side of deposit should be controlled relatively line *BF* and position of lying side of deposits relative to *CE*.

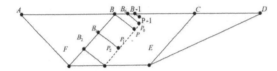

Figure 2. Pits coordinate system.

The work of curvilinear coordinates. The right end of the line *BP* has coordinate equal to the horizontal distance from the point *P* to the line *AF*. End of the line BP will be moved to the left during exploitation. Moving distance is definite. New coordinate of point P will be equal to the previous coordinate minus B_b. The second important aspect of the problem is that we must be able to recognize in which part of the pit there are the points of intersection of horizontal and diagonal lines during the development. Introduce the characteristic functions Lxk, k, j, t and Ryk, j, t. If point is inside the *BCEF*, then $Lxk, j, t = 1$ and $Ryk, j, t = 1$. If the point is outside the scope *BCEF*, then $Lxk, j, t = 0$ and $Ryk, j, t = 0$. Further, we emphasize that all the trapezes separating pit on hanging and lying side of deposits abut on the line P_iP_{i-1} (Figure 2).

Coordinates of the points of intersection of the horizontal and incline lines. Distance from incline *AF* to broken line $P_2P_0P_{-1}$ is denoted as *APj*. Distance from *AF* to *BF* is denoted as *ABj*. Also introduce the following notation: $B_b^l(t)$ – panels width from the hanging side of deposit of stage t; $B_b^r(t)$ – panels width from the lying side of deposit of stage t; $Tr(k, j)$ – initial cut base's width (central trapeze) on sublevel j.

Therefore, for even values of sublevel following formulas

$$x(1, j, t) = AP(j),$$

$$y(1, j, t) = M - x(1, j, t) - Tr(k, j) + CD(j), \tag{2}$$

where $CD(j)$ – horizontal distance from *EC* to *ED*. The coordinates of the odd level are calculated with taking into account the transport berm width and the angle B_r of hanging side *a*. Coordinates of remaining points of the pit from the hanging side of deposits are calculated with using of recursive formula:

$$x0, 2j-1, t = x0, 2j-1, t-1 - B_b^l(t), \tag{3}$$

$$x1, 2j-1, t = x1, 2j-1, t-1 - B_b^l(t), \tag{4}$$

$$x0, 2j, t = x0, 2j, t-1 - B_b^l(t), \tag{5}$$

$$x1, 2j, t = x1, 2j, t-1 - B_b^l(t). \tag{6}$$

Similarly, the coordinates of all points of the pit wall of lying side of deposits are calculated by the formula

$$y0, 2j-1, t = y0, 2j-1, t-1 - B_b^l(t), \tag{7}$$

$$y1, 2j-1, t = y1, 2j-1, t-1 - B_b^l(t), \tag{8}$$

$$y0, 2j, t = y0, 2j, t-1 - B_b^l(t), \tag{9}$$

$$y1, 2j, t = y1, 2j, t-1 - B_b^l(t). \tag{10}$$

Calculation formulas of areas of geometric shapes and the stripping ratio. Denote the area by hanging and lying side of deposits through the figures $S^l(j, t)$, $S^r(j, t)$ formed by the intersection of horizontal and inclined lines on the cross-sectional pit. The characteristic functions of these figures are denoted $LS(j, t)$, $RS(j, t)$. With this notation equalities are compiled

$$S^l(t; capping) = \sum_j S^l y, t LS(j, t), \tag{11}$$

$$S^r(t; capping) = \sum_j S^r y, t RS(j, t), \tag{12}$$

$$S^l(t; ore) = \sum_j S^l y, t(1 - LS(j, t)), \tag{13}$$

$$S^r(t; ore) = \sum_j S^r y, t(1 - RS(j, t)), \tag{14}$$

where $t = 0, 1... -$ number of mining stage; $S^l(t, capping)$, $S^r(t, capping)$ – are of figures located outside $BCEF$ at the stage t; $S^l(t, ore)$, $S^r(t, ore)$ – are of figures located inside $BCEF$ at the stage t. Let $k(t)$ – stripping ratio at the stage t, which is definedby formula

$$k(t) = \frac{S^l(t; capping) + S^r(t; capping)}{S^l(t; ore) + S^r(t; ore)} =$$

$$= \frac{S(t; capping)}{S(t; ore)}. \qquad (15)$$

Stripping ratio optimization. Following condition is required at each stage according to the mining technology of deposits of complex configuration

$$S^l(t, ore) = S^r(t, ore). \qquad (16)$$

Geometry of the domain and feature of mining technology shows that the maximum value of the function $k(t)$ is reached at stage $t = 7$ (the time of mining approach to limiting contours of pit). The practical realization of the mining technology causes inequality (Figure 1)

$$k(t) \le k(t+1),\ t = 2,...,6 ; \qquad (17)$$

$$k(t+1) \le k(t),\ t = 7, 8,...,13 . \qquad (18)$$

Functions are compiled to ensure a smooth change (2) of stripping ratio $k(t)$ during pit's sinking before approaching the limit pit contours

$$V(t) = S(t; capping)S(t+1; ore) -$$

$$- S(t+1; capping)S(t; ore), \qquad (19)$$

when $t = 1, ..., 6$

$$V(t) = S(t+1; capping)S(t; ore) -$$

$$S(t; capping)S(t+1; ore), \qquad (20)$$

when $t = 7, 8,..., 13$, and functional is minimized

$$J(B_b, Tr) = k(7) +$$

$$+ \sum_t V^2(t) + N \sum_t (S^l(t; ore) - S^r(t; ore))^2 . \qquad (21)$$

Minimum of this functional is sought under the following restrictions

$$B_{b,min} \le B_b \le B_{b,max} ,\ V(t) \ge 0. \qquad (22)$$

We want to get a minimum of functions $k(7)$ and try to obtain a smooth change of $k(t)$. It is achieved

by minimizing the value of the sum $\sum_t V^2(t)$. The third term on the right side of the equal sign ensures that the condition (1) for large values of N (penalty). Therefore, the last value is $t = 17$. The main objective of the test is to achieve a uniform distribution of the stripping ratio $k(t)$ depending on the stage of mining by varying elements parameters of the mining system. The implementation of the algorithm is as follows. Initially panel width is fixed from both sides of deposits. Values of parameters B_b^l and B_b^r are listed in Table 1.

Table 1. Parameters of panel width of hanging and lying sides of deposit.

#	1	2	3
B_b^l (m)	60	90	120
B_b^r (m)	60	90	120

Graph of $k(t)$ for different values of the panel width is shown on Figure 3.

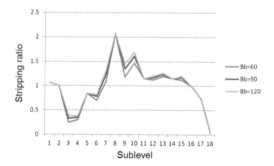

Figure 3. Graph of the current stripping ratio by stages at various values of panel widths.

Implementation of the algorithm shows that the desired result (smooth increasing of the stripping ratio by stages mining) isn't achieved by same panels width of hanging and lying sides of deposits. Therefore, the problem was solved with variable panel widths from both sides of the deposit. So, more uniform distribution of the stripping ratio is provided under these conditions. The results are shown on Figure 4.

The results show that abrupt changes of values of the stripping ratio are excluded with using a variable panel width from both sides of deposit. Analysis of the chart in Figure 4 shows that the maximum value of $K_{max} = 2.1$ with equal widths of panels can be reduced to $K_{max} = 1.4$ with variable widths of panels from both sides of the deposit at each stage of mining.

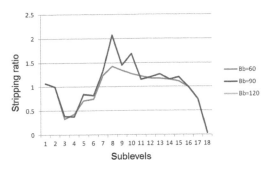

Figure 4. Comparative graph of the stripping ratio at constant and variable panel widths.

Therefore, panel width shoul be defined when by mining stages of hanging and lying sides of deposit when double subbench mining technoligy is used. It provides the optimal mining operations with uniform distribution of accessed reserves on both pit walls. Therefore, the change interval of rational width of panel must be interconnected with technological complexes depending on the necessary desired productivity of the pit.

3 CONCLUSIONS

1. Solution of the parameters optimization of the double subbench technology with varying levels of working are by nonlinear programming method allows to build the mining stages of steep deposits of complex configuration with providing the best mining operationswith the uniform distribution of accessed reserves between the two highwalls.

All the formulas of developed algorithm for implementing this method are assigned and attached to the selected coordinate system. They will help to solve the problem, which is required to determine the value of certain parameters of the specified function and the fixed values are not exceed, and some dedicated objective function reaches a global minimum.

Test results show that the width of the panels should be determined for mining stages and by the hanging and lying sides of deposits when the double subbench technology with varying levels of working area is designed. Its provide obtaining the optimal mining operations with uniform distribution of accessed reserves on both pit walls.

Choice of the order and consistency to solve the problem of parameters optimization of the higwall construction during development of mining in the implementation of the proposed method consists in constructing their parameters nexus between panels at each mining stage and from both pit walls. The quality of rock crushing, trip duration time will affect to their change in the range of change of panel width produced in the implementation of an algorithm that ensures optimal schedule of mining mode.

REFERENCES

Rakishev, B.R., Moldabyaev, S.K., Aben, Ye. & Rysbaiuly, B. 2013. *Establishing optimal contours of stages of double subbench steep mining deposits by cross panels*. Development strategy of mining and metallurgic complexes of Kazakhstan: 55–58.

Moldabayev, S.R., Rysbaiuly, B. & Sultanbekova, Zh.Zh. 2013. *Substantiation of Countours Belonging to the Stages of Mining Steeply Dipping Deposits Using Solution of the Problem of Nonlinear Programming*. Proceeding of the 22nd MPES Conference. Dresden, Germany. Springer, Volume 1: 125–132.

Progressive Technologies of Coal, Coalbed Methane, and Ores Mining – Bondarenko, Kovalevs'ka & Ganushevych (eds)
© 2014 Taylor & Francis Group, London, ISBN: 978-1-138-02699-5

Patterns of landslide processes development in conditions of man-made water exchange of mining complex

Ye. Sherstuk & Yu. Demchenko
National Mining University, Dnipropetrovs'k, Ukraine

Yu. Cherednichenko
"Donbass Fuel-energy company" Donets'k, Ukraine

ABSTRACT: General regularities of deformation of landslide massifs within the slope areas of territories which situated in influence zone of man-made water exchange of mining objects in Kryvyi Rih iron ore basin are investigated in this paper. The effect of changes in deformation properties of carbonate rocks, which occurs as a result of leaching and suffusion in the interval of their flooding, on stability of landslide slopes within the territory of the village Novoselivka Shirokovs'kyi district of Dnipropetrovs'k region a investigated by numerical simulation. Significant reduction in deformation characteristics of crushing zones in carbonate rocks at the base of the landslide slopes leads to violation of stability of rock massifs are scrutinized. In this case the surface violation occurs with prevalence of vertical deformations and forming ground collapse and subsidence.

1 INTRODUCTION

There are areas that affected extraordinarily high technogenic impacts in mining regions of central and eastern Ukraine. The hazard of technogenic influence consists in variety of negative factors, which often results in irreversible changes of hydrogeomechanical and hydrochemical modes.

The results of their action are the development of suffusion and karst phenomena, landslides intensification, flooding and subsidence of ground surface; they are the most developed in the zone of mining objects influence (Lushchyk et al. 1997, Evgrashkina 2003, Tymoshchuk et al. 2008).

2 SITE DESCRIPTION

One of the examples of such an effect is the activation of landslides within the slope of the left bank of the river Ingulets, resulting in a violation of the geomechanical stability of territory of village Novoselivka Shirokovs'kyi district of Dnipropetrovs'k region. Over the past two decades, there is noted an activation of engineering-geological processes that led to a catastrophic deterioration of built-up area and living conditions of local population. Sites of subsidence and ground failures rate of 0.1–0.8 m, horizontal displacement of soil mass with a gap of 0.1–0.3 m, which is typical for landslides were registered within an investigated area by observations

and surveys. Activation of these processes accounts for 1989–1990, 1996, 1998–2004.

Numerical modeling (Tymoshchuk & Sherstuk 2012) of filtration processes taking place in the investigated territory proves the existence of flow of man-made highly mineralized water from sludge depositories into Neogenic aquifer. This in turn leads to elevated levels of groundwater of Neogenic horizon in the territory of the village by about 5 m. It should be noted that technogenic increase the level of groundwater with sulfate-chloride aggressiveness in the Neogenic aquifer occurs in redeposited weathered Neogenic limestone layer. Moreover, water, fresh in the past, now has mineralization up to 7.8 g/l within the village territory and up to 20 g/l in the neighborhood. This fact gives grounds to assume that the karst-suffusion processes have a significant effect on the stability of the slope.

Karst-suffusion processes associated with the dissolution of carbonate rocks, silt removal from the sand layer and fractured limestone by groundwater.

They are often stimulated by violation of geodynamic regime and appear on the surface in the form of depressions, cracks, sinkholes, etc.; lead to the most dangerous and difficult to predict phenomena - the formation of land failures, holes and irregular subsidence of separate sites of ground surface (Sizov 2009).

Investigation (Perkova & Rudakov 2014) of the process of Neogenic limestone dissolution with mineralized water showed a significant rate of their leaching with active porosity increasing at a speed of the order of 10^{-4} 1/day.

It is known that an increase in porosity of soils leads to a significant decrease in their deformation properties. Thus, for various soils (Belikov 2012), increasing in porosity of 30% reduces the modulus of deformation in 2–3 times or more.

Thus, continuous effect of technogenic water on carbonate sediments within territory of Novoselivka village can cause the dissolution of significant part of limestone strata and emergence cavities and weakened zones.

In this regard, one of the objectives of improving the geotechnical stability of this area is to assess the impact of the weakening zones because of the leaching of carbonate deposits that are present in the geological section at the bottom of the landslide slope on the geomechanical stability of the slope.

3 MODELLING

The numerical model of the investigated area in the direction of where landslides are most likely was created to assess the stability of the soil slope.

Modeling technique included a series of numerical solutions, which were carried out on the basis of possible changes in deformation characteristics and strength parameters of rocks under the influence of flooding. Developing of geomechanical model and carrying out numerical experiments performed using a software application for Windows Phase2 version 7.019 (RocSciense Inc, Canada), intended for finite element analysis of geomechanical processes. Simulation of the stress-strain state is performed in elastoplastic formulation, suggesting that the soil mass may experience plastic deformation.

The model structure and the geometry of its elements are set based on data concerning engineering and geological conditions, received by state enterprise "DNEPROGIINTIZ" by the results of research carried out within the landslide slope (Figure 1).

Coulomb-Mohr strength criterion was used as a failure criterion in this model. Input parameters for the model: E and v – elastic parameters of the soil, c and φ – strength parameters of soil, γ – unit weight of soil. Three-noded triangles were used for meshing the geometry into finite elements.

Model dimensions for the plane solution: width of 130 m, height 47 m. Boundary conditions of model contain restricted horizontal displacement at the side borders and fixed displacement along the bottom border. Force interaction between the model elements within simulated area is determined by gravitational stress distribution throughout the slope (Figure 2).

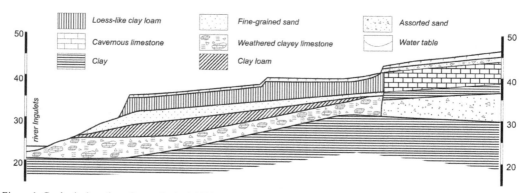

Figure 1. Geological section of area. Scale 1:1000.

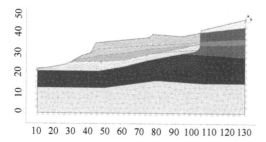

Figure 2. Schematization of the simulated area.

Physical and mechanical parameters of rocks in the area of numerical model characterized by the following value ranges: the values of Poisson's ratio lies within 0.25–0.3.

Young's modulus was 10–30 MPa for loam, 30–35 MPa for sand, 30 MPa for the weathering crust of granite, 30–32 MPa for the clay, and up to 30 MPa for different limestones.

Unit weight of clayey soil was 15.86–19.17 kN/m^3, for sandy soils 15.35–20.01 kN/m^3, for carbonate soils 16.73–19.17 kN/m^3, for granite weathering crust – 27 kN/m^3.

Cohesion for clay soils was 16–126 kPa, for sandy – 0.1–7 kPa, for carbonate – 37–43 kPa, 40 kPa for weathering crust.

Internal friction angle was 11–28 degrees for clay soils, 24–33 degrees for sand, 12–71 for carbonate and 25 degrees for weathering crust.

There was a series of calculations of the soil slope stress-strain state under a steady reduction of the deformation modulus of redeposited limestone (30000, 20000, 15000, 10000, 7000, 5000, 3000 and 1000 kPa, respectively).

Each calculation thus contained 11 stages: on each phase of calculation strength characteristics of redeposited limestone were changed (cohesion c and internal friction angle φ values proportionally decreased by 10% from the initial set).

This solution set allowed analyzing the complex effect of reducing a strength and deformation characteristics of weathered limestone within specified interval of values caused by man-made influence on the stress-strain state of the soil mass as a whole.

The numerical model is complemented by hydraulic boundary conditions, those are determined by ponded water level in the river Ingulets – is 22.5 m, the groundwater elevation upstream – is 31 m, where it increases up to 38 meters on the third stage. Hydraulic properties of rocks composing landslide massif are presented with values: for permeable rocks – 0.4–4 m/day, for aquitards – $1 \cdot 10^{-5}$–$1 \cdot 10^{-2}$ m/day.

The key parameters which were used and analyzed during the interpretation of the simulation results were vertical and horizontal displacements of rock massif while reducing the strength and deformation properties of weathered limestone.

4 RESULTS AND DISCUSSION

The model validity is ensured by inverse modeling of slope in undisturbed conditions and after the occurrence of landslides. The accuracy of simulation is confirmed by convergence of simulation results with the results of field surveys in the village Novoselivka.

It is found by results of modeling that the critical strain (the nonlinearity of displacement dependence on changes in strength and deformation properties) occur with a decrease in strength properties of Sarmatian limestone to a level 40–30% of the initial and deformation properties of 5000 to 3000 kPa (Figures 3–6).

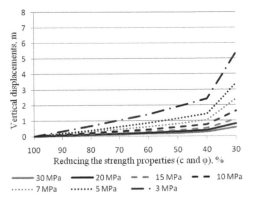

Figure 3. Changing the vertical displacements with a decrease in strength properties of limestone for different values of deformation modulus, m.

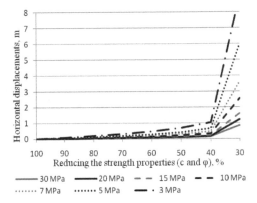

Figure 4. Changing the horizontal displacements with a decrease in strength properties of limestone for different values of deformation modulus, m.

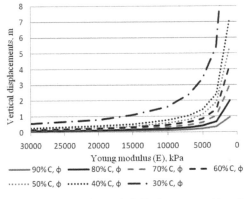

Figure 5. Changing the vertical displacements with a decrease in deformation modulus of limestone for various degrees of strength reduction, m.

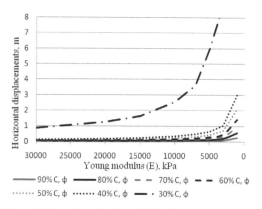

Figure 6. Changing of horizontal displacements with a decrease in deformation modulus of limestone for various degrees of strength reduction, m.

Thus, the vertical displacements of the order of 0.5–0.8 m occur at the test site already with a decrease in deformation modulus of limestone up to 10 MPa in the leaching process. It should be noted that the limestone loses about half of its strength, and the slope keeps its stability as a whole. Soil mass moves in the horizontal direction with an amplitude of 0.1–0.3 m, this occurs when the residual strength is about 70–40% at the $E = 10$MPa, and residual strength 80–60% – when $E = 5$ MPa.

General regularities of rock deformation at the site of landslide slope with a decrease in deformation properties of limestone in the zone of its flooding were found. These include the prevalence of vertical deformations above horizontal, which is accompanied by "failures" formation on the day surface of the massif (Figure 7).

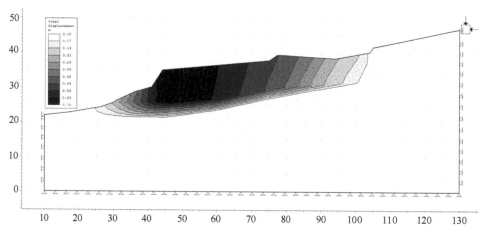

Figure 7. Landslide deformation with a decrease in deformation modulus of limestone to 3.000 kPa and 80% of the remaining strength.

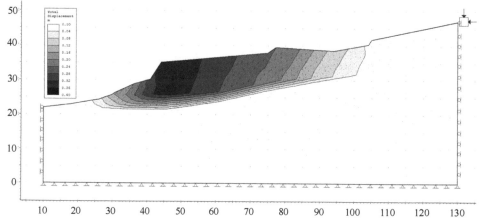

Figure 8. Landslide deformation with a decrease in deformation modulus of limestone to 25.000 kPa and 40% of the remaining strength.

Moreover, a substantial decrease in strength properties of c and φ (reducing to 30%) leads to an intensification of shear strains at the bottom of the landslide slope, and the loss of its stability (Figure 8).

5 CONCLUSIONS

General regularities of deformation of landslide massifs within the slope areas of territories which situated in influence zone of man-made water exchange of mining objects in Kryvyi Rih iron ore basin were investigated.

The effect of changes in deformation properties of carbonate rocks, which occurs as a result of leaching and suffusion in the interval of their flooding, on stability of landslide slopes within the territory of the village Novoselivka Shirokovs'kyi district of Dnipropetrovs'k region was investigated by numerical simulation. Found that a significant reduction in deformation characteristics of crushing zones in carbonate rocks at the base of the landslide slopes leads to violation of stability of rock massifs. In this case the surface violation occurs with prevalence of vertical deformations and forming ground collapse and subsidence.

REFERENCES

Lushchyk, A., Davidenko, I., Shvyrlo, M. & Yakovlev, Y. 1997. *Engineering-geological conditions in the area of Krivoy Rog iron ore basin.* Informational bulletin about situation in environment of Ukraine 1994–1995. Issue 14: 36–41.

Evgrashkina, G. 2003. *Impact of mining on the hydrogeological and soil-reclamation conditions of territories.* Dnipropetrovs'k: Monolit: 200.

Tymoshchuk, V., Demchenko, Yu., Zagritsenko, A. & Sherstiuk, Y. 2008. *The impact of mining objects on hydrogeomechanical state of surrounding areas in Krivoy Rog Basin conditions.* School of underground mining, Dnipropetrovs'k. 234–237.

Tymoshchuk, V. & Sherstiuk, Y. 2012. *Regularities of geofiltration in the area of gravitationally loaded sites of tailings and waste rock.* Dnipropetrovs'k: Scientific bulletin of National mining university, 4: 30–36.

Sizov, A. 2009. *Monitoring and protection of urban lands,* Moscow: 264.

Perkova, T. & Rudakov, D. 2014. *Evaluation of changes of filtration properties of fractured rock under technogenic karst.* Mine research and information bulletin (scientific-research magazine), 3: 304–308.

Belikov, B. 1962. *Elastic properties of rocks.* Studia Geophysica et Geodaetica. Volume 6, Issue 1: 75–85.

Progressive Technologies of Coal, Coalbed Methane, and Ores Mining – Bondarenko, Kovalevs'ka & Ganushevych (eds)
© 2014 Taylor & Francis Group, London, ISBN: 978-1-138-02699-5

Hydrogeomechanical processes of occurrence in disturbed rock mass by mine workings

I. Sadovenko & V. Tymoshchuk
National Mining University, Dnipropetrovs'k, Ukraine

ABSTRACT: Features of hydrogeomechanical processes in the "underground workings-lining-saturated rock" considered in this paper. The dependence of deformation of the opening working lining on hydraulic pressure outside of lining was determined on the results of numerical modeling. Limit values for hydraulic loads on contour of lining were identified for the considered range of values of physical and mechanical properties of rocks and lining. The dependence of deformation of roof rocks on the hydraulic load and position loaded surfaces of lamination relative to stope was determined by analyzing the simulation results of the stress-strain state of a layered rock mass, which contains flooded rocks in stope roof.

1 INTRODUCTION

The problem of management of rock mass both when opening coal deposits and coal extraction due to the complexity of interaction between technical objects – opening and mine workings with geological environment under its heterogeneity and influence of hydraulic component.

Examples of violations of coal mining process, emergency situations and related resource and financial losses include violations of shaft lining when deeping in water-saturated rocks and caving powered roof support to "strictly", which are specific coal processing in geological conditions of Western Donbass mines.

When driving opening excavation through water-bearing strata the construction of waterproof layers when fixing sustainable water-bearing rocks may leads to a sharp water head increase outside of lining and weighing unload lining. Rock and lining strength at theirs' contact is inevitably lost in this case, so maybe its rapid violation that occurred in practice (Sadovenko 1986 & Rumin 1987).

2 RESEARCH RESULTS

It is based on the variant review of complex action of geomechanical and hydraulic components at the contact of the system "rock contour-lining of shaft". Finite element axially symmetric (radial) model of the rock mass with the placed therein mine opening is accepted as a base.

Interaction in modeled system "mine opening-concrete lining-rock mass" is controlled by using the contact surface which provides the possibility of autonomic displacement of calculation points along the lining plane under hydrostatic load. The hydrostatic load is controlled by hydraulic pressure value outside of lining provided that the gradual transition from the first phase (formation of a hydraulic depression while "deeping and casing" in water-bearing rocks, water head decline by the value of natural pressure) to the second one (smoothing the hydraulic depression when recovering level behind a constructed lining).

Obtained numerical solutions show that constructed lining deforms during the transition from the phase when a hydraulic depression forms to phase of its smoothing when water level restores. This leads to destruction of lining when a certain pressure threshold at contact "rock-concrete lining" reached (Figure 1).

Relative volume strains in rock mass and concrete lining shown in figure correspond to the characteristic value of their physical and mechanical properties: $E_\kappa = 1.0 \cdot 10^6$ kPa, $\nu_\kappa = 0.2$, $\gamma_k = 23.0$ kN/m³, $C_k = 2.0 \cdot 10^4$ kPa, $\varphi_k = 25°$, $E_n = 1.0 \cdot 10^7$, $\nu_n = 0.2$, $\gamma_n = 22.0$ kN/m³, $C_n = 2.5 \cdot 10^3$ kPa, $\varphi_n = 30°$. The concrete lining destroys when reaching a critical level of load, which in given conditions corresponds to the magnitude of hydraulic pressure on the lining contour about 15.0 m. The resulting solution corresponds to the variant of interaction lining and rocks, which is characterized by low-strain stiffness of the contact surface – within it the modulus of deformation is $0.1 E_n$.

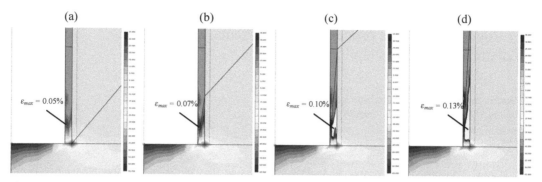

(a)   (b)   (c)   (d)

$\varepsilon_{max} = 0.05\%$ $\varepsilon_{max} = 0.07\%$ $\varepsilon_{max} = 0.10\%$ $\varepsilon_{max} = 0.13\%$

Figure 1. Strain development in the lining of mine opening when water level restores, volume strain ε, $[\cdot 10^{-3}]$: (a), (b), (c) and (d) – achieved strain level ε_{max} when water head is 0.0; 5.0; 10.0; 15.0 m at the outer contour of lining relatively to the mine face.

It should be noted that the development of behind limit strain on contour "rock-lining of opening working" under these conditions is characterized by damaging strains in the bottom of lining within its 5-meter interval. Moreover maximum deformation corresponds to the middle of the interval.

Graphs in Figure 2 illustrate the dependence of the horizontal displacements in the internal lining contour, as well as relative deformations within loaded contours on the value of hydraulic load when groundwater pressure is from 0 to 30 m on the outer contour of lining.

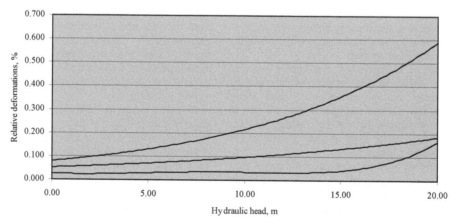

Figure 2. Diagrams of dependence volume deformations of concrete lining on hydraulic load on contour "rocks-lining" for deformation modulus of lining material: $1 - 0.5 \cdot 10^5$ kPa; $2 - 1.0 \cdot 10^6$ kPa; $3 - 5.0 \cdot 10^6$ kPa.

The sharp increase in displacements and strains in the graphs corresponds to the appearance of inelastic deformation zones in lining, which are accompanied by lining destruction under hydraulic load. As in the case shown in Figure 1, the maximum allowable stresses in lining form in the lower part of opening working, where hydraulic pressure values are maximum. Moreover the interval of destructive strain coincides with the radius of shaft when the size of its, deformation and strength properties of the lining material are given.

A somewhat different situation is observed when deformation properties of the lining material and rocks are similar within hydraulically loaded contour. Thus, when deformation modulus of lining and water-saturated rocks are $E_n = E_\kappa = 1.0 \cdot 10^6$ kPa destructive deformations in the concrete lining body are missing during hydraulic head is reduced from 0.0 to 25.0 m. In this case, as in the previous variant, the stiffness in the plane of contact surface corresponds to the deformation modulus $0.1 \, E_n$.

When system "lining-rock" reaches the contact strength, equivalent to deformation characteristics of lining material and rock around the mine working, destructive deformations are missing during all the phase of hydraulic depression smoothing outside the lining. For this variant, the hydraulic load on the lining contour corresponds to the reduction of water table to 90.0 m height.

Analytical calculations and numerical simulations show that the destructive deformations in the lining of mine working are connected with the failure of contact strength of rock and lining during the transition from the phase of a hydraulic depression formation to the phase of its smoothing with water head restoration.

In these conditions, hydrogeodynamics of "lining-water-bearing rocks" is defined by the deformation properties of shaft lining and water-saturated rocks, as well as their contact resistance, which depends on the construction technology of lining. Contact strength of hydraulically loaded contour is lost at a

significantly reduced the deformation properties of lining in comparison to the surrounding rocks.

Occurrences of hydrodynamic processes when there is a hydraulic load on the rock mass at stope roof is also examined at numerical geomechanical models that take into account effect of hydraulic component on the stress-strain state of undermined strata (Sadovenko & Tymoshchuk 2012).

Evaluation of hydrostatic stress action within the place of lamination of undermined rock mass on the conditions of excavation processing was to determine the vertical displacements of immediate roof within front and breaking rows of hydraulic props of powered roof support. The range of hydrostatic pressure within the lamination surfaces was limited by pressure value in Buchak horizon, which is about 90.0–100.0 meters in conditions of coal processing with Western Donbass mines

(a)

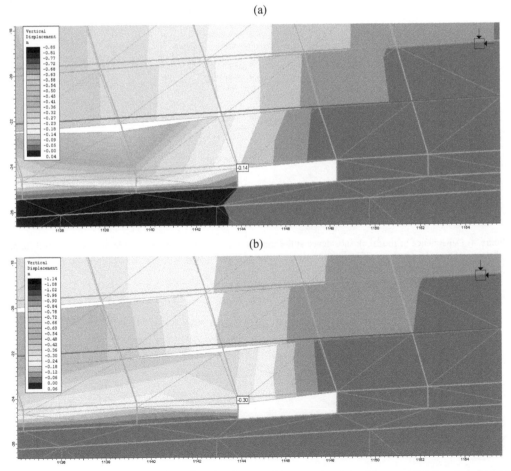

(b)

Figure 3. Character of the simulated deformation of rock mass subject to undermining it by room: (a) there is no hydraulic load; (b) hydraulic load p = 900 kPa.

In Figure 3 is a diagram of the roof rock deformation under conditions of its undermining by room. Vertical displacement of hydraulic props in the absence of hydraulic load within the existing lamination surfaces were within 0.20–0.30 meters. Subsidence of argillite layers within the interval of overhanging for given variant of solution corresponds to the beginning of shearing the immediate roof at stope.

Hydraulic load in the plane of lamination in base of main roof provides development of additional load on the hydraulic prop contour of powered support, resulting in additional subsidence at supporting contour reaches values of 0.45–0.70 m, found by the results of solutions.

In considered range of values of hydraulic load the dependence of subsidence of estimated contour on magnitude of hydrostatic pressure is described by graphs of roof point displacements at contour of powered roof support (Figure 4).

It should be noted that after reaching the maximum calculated load equal to 900 kPa at the base of main roof at a distance of $5m$, (where m – excavation capacity of the coal seam) in interaction zone of powered roof support with immediate roof, destructive deformations of roof rocks with loss of their strength are occurring.

Subsidences of props of powered roof support have similar character when forming hydraulically loaded contours within lamination planes that are at different distances from the roof of mined coal seam. However, reducing the impact of hydraulic load on the upper contour points subsidence of powered roof support occurs with increasing distance from the coal seam roof to lamination plane, and the maximum value of contour subsidence is in the range 0.30–0.50 m.

The dependence of interaction of roof rocks and powered roof support from the impact of spatial position of hydraulically loaded lamination planes is illustrated by the graphs in Figure 5.

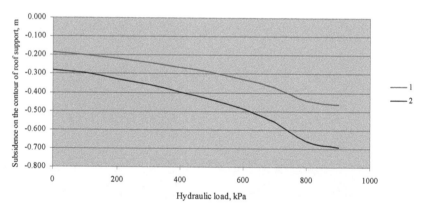

Figure. 4. Dependence of roof rock subsidence at the contour of powered roof support on hydraulic load level (in plane at a distance of $5m$): 1 – front row of hydraulic props; 2 – row of breaking props.

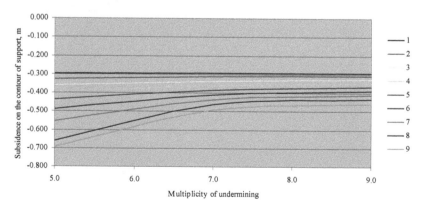

Figure 5. Dependence of roof rock subsidence on the contour of powered roof support from the relative distance to lamination plane: 1–9 – hydraulic load of 100 to 900 kPa (100 kPa increments).

It is seen that subsidence of points within contour propped up by powered roof support depends on the level of hydraulic load and the distance to the lamination plane; moreover, the intensity of subsidence increases with hydraulic load increasing within lamination contours, which are the most close to the stope.

3 CONCLUSIONS

The results of studies of the interaction of man-made objects with geological environment in conditions of development hydrogeomechanical processes are of great scientific and practical importance, as follows.

The established mechanism of hydraulic load on lining of opening working allows doing assessment of the geomechanical stability of lining under the action of a hydraulic component. Definition of maximum permissible level of hydraulic load for a filter lining in water-saturated rocks allows performing its parameterization for specific conditions of shaft sinking.

Determined regularities in hydrodynamic regime formation within undermined massif allow to define level of influence of hydraulic overload, created on contours of locally discrete surfaces of lamination within immediate and main roofs of stope, on character and distribution of load on the longwall powered roof support, and identify the main areas of technological action on the hydraulic component of the stress-strain state of rock roof.

Research in the formation of hydrogeomechanical state of hydraulically loaded massifs forms the basis for developing the management schemes for rock mass around opening and mine workings and substantiation their parameters that are important for trouble-free processing of coal seams in difficult geological conditions of Western Donbas mines.

REFERENCES

Sadovenko, I. 1986. *Calculated dependences for the definition of lining shaft loading.* Mine construction, 11: 15–17.

Rumin, A. 1987. *On calculation of concrete lining of shafts with the pressure of groundwater.* Mine construction, 9: 30–31.

Sadovenko, I. & Tymoshchuk, V. 2012. *Simulation of mechanism of hydrogeodynamic load of coal layer main roof above the powered roof support.* Dnipropetrovs'k: Collection of research papers National Mining University, 39: 5–11.

Computer-aided multy-object distributtion system for prompt project management

V. Nazimko
Donets'k National Technical University, Donets'k, Ukraine

M. Illiashov & E. Youshkov
PJSC "Donetsksteel", Donets'k, Ukraine

ABSTRACT: A prompt project management comes to action when limit the time that is given for the decision making is not sufficient to find actually optimal choice of the command. We developed an automated computer-aided multy-object distributed system as adviser for prompt project management that fills a gap between classic and agile project management. A criterion of prompt management optimality has been developed as a function of the rule relevance that used for decision making. We proposed a threshold that assists to develop new rule. An example of industrial testing has been provided to show the effectiveness of the proposed automated adviser for coal extraction project management.

1 INTRODUCTION

Application of project management (PM) often faces the problems of uncertainty, time and resources limits in real world. There are a good number of areas where these problems are typical and frequently posed. The industries operating in natural environment, experience serious problems that are consequence of the uncertainties, which have geologic, ecologic, biologic, human psychology nature. For example, mining industry suffers from geologic ambiguity and extremely complex problem of geophysical exploration. It worth to remark that coal inseam tomography costs up to $100 thousands per one kilometer of a longwall panel whereas its reliability does not exceed 70–75%. It is a typical situation when drilling of a well to the depth of 5 km may not provide sufficient oil extraction despite efforts and waste of several millions. Complex biological interrelations that complicated by uncertain factors and unforeseen events may convert positive expectations into ecologic catastrophe.

Very often, negative consequences of the uncertainty cause a time shortage and a project manager should generate decisions promptly, not having a time to find the actually optimal solution. Several techniques of PM were developed to combat the uncertainty troubles and to mitigate deficiency of time. However, the rate of accumulation of these problems is bigger than the pace of the therapy development. In spite of growing interest in project management, projects continue to fail at an astonishing rate (Thomas & Mengel 2008) due to the negative impact of complexity, chaos and uncertainty.

Practitioners, researchers and scientists have developed sophisticated means for automatic control of technological processes that are actually dynamic systems that operate in noisy (Levine 1996), turbulent (Crittenden & Crittenden 2012) environment and uncertain condition (Nguyen et al. 2013). However powerful tools developed for the automatic control cannot be applied in the project management process directly because technological processes use numerical inputs and are controlled by numerically measurable signals. They exploit a mathematical model of a dynamic system in a form of differential or algebraic equations using numerical parameters as inputs and outputs. Then a control system finds the relevant governing signal using a transfer function (for linear dynamic object) or specific mathematic transformations.

To the contrary, any project management process massively involves linguistic input information and verbal commands. As Cooke-Devies (2002) wrote "…it is fast becoming accepted wisdom that it is people who deliver projects, not processes and systems". However, people operate with fuzzy categories, managing the project and still make mistakes, taking subjective decisions, losses which increase considerably in the face of uncertainty, chaos and deficiency of time when prompt right decisions are critical. It is obvious that these losses can be reduced by the introduction of automated decision making. At present, the problem of automatic control of a project was not put. Therefore, development of method for automatic generation of decisions for prompt project management is an urgent task.

2 "PROMPT" DEFINITION

The term "prompt" had not sharp meaning and usually authors used it to express alert reaction. Instead, we prefer strict concept of this notion. Very often, projects are implemented in a shortage of resources and time. This is especially true for such industries as the urgent medical attention, aftermath of natural and man-made disasters, coal industry, which operates in a chronic lack of time and project resources such as financial, material, human, energy. In this regard, it is necessary to design PMS for flexible decision-making that would ensure it's maximum effectiveness within the allowable time and project resources under severe constraints (e.g. underground safety limitations).

Limit the time that is given for the decision, is the principal difference from the traditional automated control. Note that the allowable time limit for decision-making in real projects often depends on the current situation and can vary greatly in the course of the project (Guldemond et al. 2008). We can describe this limitation by the equation (Pupkov & Egupov 2004, Chiang et al. 2011)

$$T_c = f(t, k1, \dots kn) \le T_o, \qquad (1)$$

where T_c is a current time limit that is available for prompt resolving a recent problem of the project; f is a function that depends on the current time and the project parameters $k_1, \dots k_n$; T_o is the time, which is necessary for making optimal decision.

The problem of MODPPM design is relatively new because we should translate to mathematical framework such concepts as control of timing, cost, project quality, human resources, communications, risk management, project deliverables. This problem is close to decision support system (DSS) that has been developing for approximately 40 years (Power 2007).

To develop MODPPM, we will use best findings and practices from DSS, automatic control of technological processes, distributed control and fuzzy technique.

3 ARCHITECTURE OF MODPPM

We proposed to separate control and data flows that comply with physical sense of separating signals into thermodynamic forces (input and feedback information) and fluxes (commands) that allowed decentralizing the management of the project. Architecture of multy-object distributed prompt project management is actually organizational structure. It provides the organizational basis for a project, determines the role of every individual and indicates

how he or she interconnects to each other. The architecture of a team and various roles its members will depend on the nature of the project (Han & Hovav 2013). For instance in coal industry, depth of coal mining reached to 400-600m in USA and exceeded 1300m in Germany and Ukraine. High level of ground pressure causes roof falls and dangerous dynamic subsidence of rock mass. Poorly predictable bursts of coal and gas frequently lead to methane explosion and large scale disasters that claimed the lives of hundreds miners. Keeping in mind that underground coal extraction is extremely dangerous process we will select strong form of a project matrix where a project manager is primarily responsible for the project. Functional managers provide technical expertise and assign resources as needed.

Figure 1 demonstrates distributed structure of the project management system that was separated into two sections, namely control section that generate commands and informational part that analyses input and feedback data and prepares decisions. Executive committee is at the top of hierarchy (7th level) and represented by the Board of Directors that implements the overall executive authority of a coalmine. Decisions of the committee might be deliberated and weighed at meetings and not generated promptly. Steering committee is at the 6th level of the hierarchy and consists of a group of people charged with regular oversight of the project. Collectively they should represent all significant areas of participation in the project and have authority to take decisions on behalf of those areas. Members of the steering committee are heads of financial, logistic, safety services, Vice Presidents, or Directors, plus external representatives. The Project Director is a member but should report to the Steering Committee.

Prompt management of the project starts from Project Manager that is at the 4th level of the hierarchy. Part of PMS that is responsible for the prompt distributed management received the index Θ in connection with the external similarity. The project components outlined by dotted rectangular and are at the zero level of the hierarchy. Both control section and data section have four levels. Level 1 represents automated control and populated by the project team members (TM), namely electricians, foremen, team leader managers, inspectors, accountants, quality managers, internal auditors, facilitators, communication specialists, risk managers, marketing specialists. Level 2 implements the automated management and was formed from the deputies of project team leader who are actually sections chief deputies (SCD) at any coal mine. Project team leaders or section chiefs (SC) are coordinating control efforts at the level 3. Finally, project manager or technical director (TD) implements prompt decisions at the level 4.

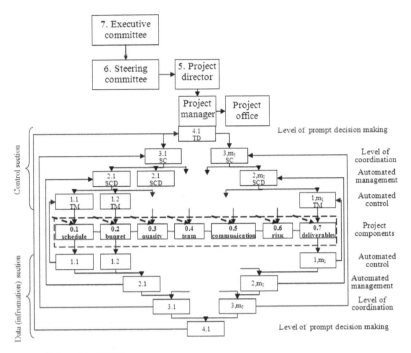

Figure 1. Θ-structure of hierarchical PMS.

Data or information section physically consists of database or knowledge base fields, access to which corresponds to the level. Dotted arrows indicate links to the rest of components of the project that not indicated in the Figure 1 for clarity. We would like to stress that dividing the structure into sections for control implementation and information treatment promotes self-organization of the PMS.

4 SELECTION OF A BASIS FOR COMMANDS GENERATION DUE TO PROJECT MANAGEMENT

Project management process is implemented through generation and issuing of the commands. It is natural for people, as the members of a project team, to use a rule-based reasoning for command generation (Ting et al. 2010). Majority instructions for the safe underground coal extraction use the rules, that accumulated severe experience and that underlies the lives of people. Here are typical rules that are used in everyday practice.

"If the uniaxial strength of the immediate roof reduced down to 20 MPa then the density of rock bolts should be increased twice".

"If the execution of the project is behind the schedule for two months, and the equipment is worn out by 50%, one should take reasonable loans".

Last rule is of financial nature and some rules use mixture of ambiguous logic and uniquely specific amounts. Therefore, we should involve fuzzy logic to cope with this problem (Kasabov 1998). Figure 2 depicts distribution of membership functions for rock stability characterization. Having such charts, we may facilitate transformation of fuzzy concept to numerical values and back. The most reliable way to accumulate membership functions and relevant rules is to use method of analogs according (Anonymous 2008). These valuable data may be placed to a relational knowledge base that is powerful tool for selection of the relevant rule and hence generation the command.

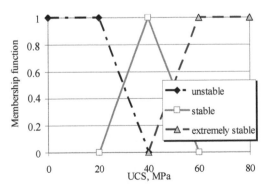

Figure 2. Membership functions for strength of the rock.

Figure 3 demonstrates the object oriented structure that accomplishes this task. The object consists of the rule components and an engine for selection of the relevant rule. We classified the components according the project sections. This structure is very compact and efficient because the components are the building blocks for the rules (see the dash-dotted downward arrows) and the rules control these components in return (indicated by upward dotted arrows). Furthermore, one rule (number ii) may affect several components. These bilateral relations were implemented by references (pointers) and linked lists.

The key problem is how to select the optimal rule.

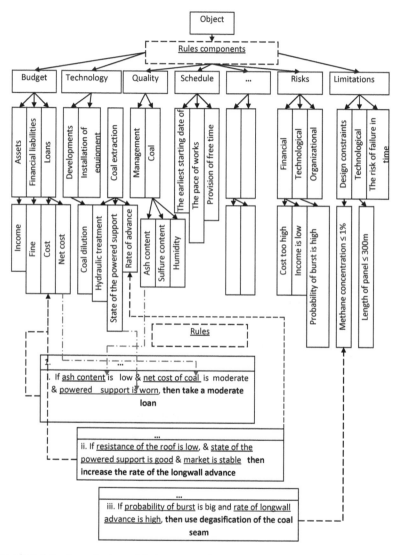

Figure 3. Object oriented structure of rules and their components.

5 DEVELOPMENT OF AN OPTIMALITY CRITERION

We chose combined structure of objective function according (Leyvand et al. 2010). Our function includes the absolute value of the discrepancy D between actual output of dynamic object and a setpoint that generated by control system and a penalty function. The reason is that plain discrepancy concords with the first low of thermodynamics and does not indicate how D has been reduced. However, it is important to know how many energy and efforts we dissipated to minimize D. The penalty function serves to clarify this issue and accounts the entropy production component. Therefore, we propose the next equation to calculate criterion of optimality

$$J(\theta) = w_d \sum (y(t) - s(t))^2 + w_p \rho (u(t-1))^2, \qquad (2)$$

where J is the optimality criterion or the objective function; θ is the indicates to input vector (for example, the amount of production, raw coal ash, its cost); w_d and w_p are the weight factors; y and s are the actual output and setpoint of the dynamic object correspondently; $\rho(u)$ is the penalty function that depends on control signal or command u; t is the denotes the discrete time elapsing during realization of a project as a dynamic system.

We proposed to define the current value of the penalty function by the relationship

$$\rho = 1/(-\sum ri(t) log_2 ri(t)), \qquad (3)$$

where $r(t)$ is the relevance of the rules in the database and symbol log_2 used to identify the binary logarithm. We find a current value of the penalty function as the sum through all previous discrete increments of time that has passed since the beginning of the project.

The relevance of a rule is much more informative than the frequency (probability) of the rule call. As we can see in the object structure (see Figure 3), any rule consists of the input vector and a command showing in bold. Thus, the relevance accounts a breadth of match. The more the vector matches the rule, the higher the weight of the relevance. Relevance decrypts an inverse rule frequency. Rare rules (within the entire knowledge base) receive a higher weight of the relevance. In addition, it is important to count the frequency, namely the number of times a rule call occurs in the process of management. Another advantage of the relevance is ability to count the density of a rule or the comparable length of retrieved rule.

6 MECHANISM OF THE OPTIMAL RULE SEARCH

We use Wagner–Fischer algorithm to measure the Levenshtein distance between two strings of characters (Navarro 2001). Computer matches an input vector string and component section of every rule in relevant project block. The rule that has minimum distance is a candidate to optimal rule. Finally, this rule may be refined or trimmed to input vector using fuzzy technique.

It should be mentioned that finally chosen rule is not optimal actually, because prompt management implies finding a command that approximately corresponds to really optimal control signal. There is not sufficient amount of time to find the actually optimum command (see Equation (1)).

We introduced a threshold to detect a situation when there is not relevant rule in the knowledge base. Normalizing all calculated distances, computer examines the minimum distance. If this distance is more than the threshold, say more than 30%, it does mean that current input vector is essentially new and has not any matching rule that would generate a command, which is close to the optimum. Therefore, the project experts create a new rule that reacts to new highlighted input vector as properly as possible. Actually, prompt project management fills a gap between classic and agile project management (Dalcher 2011).

7 IMPLEMENTATION OF MODPPM

Industrial testing of the developed computer-aided system for prompt project management in coal industry showed that the project risks block is exactly in form and content as the other units that are responsible for the current situation in staff management. So to handle the cases that create project risks and emergency situations, there is no need to develop special computer routine. There are some peculiarities related to the consideration of design constraints, but they governed by logical operators. Furthermore, the practice pointed to the need for substantial expansion the "design constraints" block. As a result, the effectiveness of automated control system design increased because the constraints use all parameters that characterize the current state of the project and its components.

For example, a mathematical formalization of the algorithms project management is greatly simplified and precision control increased, as long as we entered such design constraints as the current position of the longwall in space, present state material stocks, financial reserves, characteristics of the underground environment, etc. Thus, information about the situation when "the header is in the face" is significant because this machine interferes to accomplish the measures for reducing the risk of sudden burst of coal and gas in the face. Having 2.5 million on the current account of the company will not afford to buy a drill machine, which costs 2.7 million and so on. Importantly, the accuracy and efficiency of automated project management significantly increased after incorporating such restrictions, because even the most experienced member of the project team is unable to keep up with all the current components of the project and to respond to changing current situations properly.

The essential feature of the system is that it operates as a counsel or an adviser. This system generates a prompt command and supplies it by input signal vector and the circumstances in which it arose and which are important for current decision-making. However, a project manager or a team member formulates the final decision or command at their level of responsibility. Nevertheless, practice has shown that people increasingly agree with the management decisions, which the automated system provided, as far as knowledge base components and rules accumulate and become refined. This indicates a great future of developed computer-aided project management.

We tested the computer-aided prompt project management as an advisor in Pokrovs'ke Colliery. The software package has been installed on the local computer network. The package and knowledge base were on the server in Automatic Control and IT Department. Three remote local personal computers were connected into the network and used within 20 days during the testing this system in 4th South Panel from 162d to 182d day of Panel operation. Only three project blocks were presented in the knowledge base, namely schedule, risk and deliverables on the first stage of implementation. Foremen who were on duty had access to the 1st level of the knowledge base. They replenished their database fields with current information and called the rules when it was necessary.

Two section chief deputies had access to the level 2 whereas the section chief as the project team leader has been coordinating control efforts at the levels from 1 to 3. Technical director as the project manager has been supervising prompt decisions at the levels from 1 to 4. Training specialists were on duty every day and supervised the process of implementation. All selected and implemented decisions were recorded in the minutes. This facilitated analysis of management history because all used commands were combined with input information and design constraints.

Figure 4 depicts dynamics of the average daily value for the objective function, which has increased steadily. This indicates to accumulation of valuable information due to PMS implementation as a result of training. Assessment of the management performance may be done using daily output of coal from the Panel (Figure 5) during experimental (range b) and control (diapason a) periods. The average value of coal output during the control period was 2120 ± 753 tons per day, while it increased to the level of 2453 ± 986 tons per day during the experiment.

Since Zacharova and Nazimko (2012) found that the variation in coal production is consistent with the normal law, statistical comparisons were performed using t-test method. It was found that increasing of coal production from 4th South Panel associated with the new control method with a probability of 86.1%.

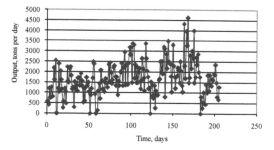

Figure 4. Dynamics of objective.

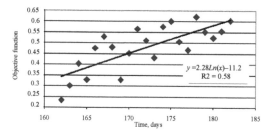

Figure 5. Coal output from 4th South panel.

8 CONCLUSIONS

The industries operating in natural environment, experience serious problems that are consequence of the uncertainties, which have different nature, cause a time shortage and project managers should generate decisions promptly, not having a time to find the actually optimal solution.

Sophisticated powerful means for automatic control of technological processes cannot assist to combat the uncertainty problem because they manipulate with numerical signals, information and commands whereas project management deals with chiefly linguistic input information and verbal commands.

We proposed to separate control and data flows that comply with physical sense of separating signals into thermodynamic forces (input and feedback information) and fluxes (commands) that allowed decentralizing the management of the project. The distribution of project management system is both spatial separation of objects management and allocation of control functions through building of the hierarchy of management. This allowed constructing a multy-object distributed prompt project management system (MODPPM) that differs essentially from industrial system analogs by domination of linguistic input information and verbal commands. We used the term "prompt" to emphasis that limit the time, which is given for the decision making is not sufficient to find actually optimal choice of command.

We developed four-level Θ-structure of MODPPM. First level fulfills automated control by team leader managers, level 2 implements the automated management that is responsibility of deputies of project team leader who coordinates this process and positioned at third level. Finally, a project manager implements prompt decisions at the level 4. All management staff uses automated computer-aided system of MODPPM as an adviser. PM commands are drawn from the rules that accumulate in a knowledge base during project activity, consist of component and command section and reside in the relevant blocks of the project. The object oriented structure of MODPPM is compact and efficient because the components are the building blocks for the rules and the rules control these components in return.

We proposed a criterion of prompt management optimality that consists of two components, namely deviation of actual output of a project from setpoint and penalty function that depends on the relevance of the rules in the knowledge base. Search of the relevant rule continues by comparison input vector of data and current information with components section of the rules in the block using Wagner-Fisher algorithm. We then use fuzzy technique to refine the most relevant rule that is the closest to optimal decision.

To detect essentially new input vector, a threshold has been introduced. If the shorter distance between the vector and a rule is more than the threshold, the project experts should create a new rule that reacts to new input vector as properly as possible.

Industrial testing of the developed computer-aided system for prompt project management in coal industry demonstrated its efficiency because average daily value for the objective function has increased steadily and output of the coal rose by 15.7% during testing period. In the future, we are planning additional verification of the developed system.

REFERENCES

A Guide to the Project Management Body of Knowledge (PMBOK® Guide, 2008) - Fourth Edition. Project Management Institute.

Al-Assadi, S.A.K., Patel, R.V., Zaheer-uddin, M., Verma, M.S. & Breitinger, J. 2004. *Robust decentralized control of HVAC systems using H∞H∞-performance measures.* Journal of the Franklin Institute, 341 (7): 543–567.

Bauerl, N.W., Donkers, M.C.F., Van de Wouw, N. & Heemels, W.P.M.H. 2013. *Decentralized observer-based control via networked communication.* Automatica, 49 (7): 2074–2086.

Chiang, C., Lin, C.-S. & Chin, S.-P. 2011. *Optimizing time limits for maximum sales response in Internet shopping promotions.* Expert Systems with Applications, 38 (1): 520–526.

Cooke-Davies, T. 2002. *The "real" success factors on projects.* International Journal of Project Management, 20 (3): 185–190.

Crittenden, V.L. & Crittenden, W.F. 2012. *Corporate governance in emerging economies: Understanding the game.* Business Horizons, 55 (6): 567–574.

Dalcher D. 2011. *Project management the agile way: Making it work in the enterprise.* Project Management Journal, 42: (1) 92–104.

Guldemond, T.A., Hurink, J.L, Paulus, J.J. & Schutten, J.M.J. 2008. *Time-Constrained Project Scheduling.* Journal of Scheduling, 11 (2): 137–148.

Han, J.Y. & Hovav, A. 2013. *To bridge or to bond? Diverse social connections in an IS project team.* International Journal of Project Management, 31 (3): 378–390.

Kasabov, N.K. 1998. *Foundations of Neural Network, Fuzzy Systems, and Knowledge Engineering.* MIT Press, London, 167–418.

Kleindorfer, D., Lindsell, C.J., Moomaw, C.J., Alwell, K., Woo, D., Flaherty, M.L., Adeoye, O., Zakaria, T., Broderick, J.P. & Kissela, B.M., 2010. *Which stroke symptoms prompt a 911 call? A population-based study.* The American Journal of Emergency Medicine, 28 (5): 607–612.

Levine, W.S., ed. 1996. *The Control Handbook.* CRC Press, New York.

Leyvand, Y., Shabtay, D. & Steiner, G. 2010. *A unified approach for scheduling with convex resource consumption functions using positional penalties.* European Journal of Operational Research, 206 (2): 301–312.

Navarro, G. 2001. *A guided tour to approximate string matching.* ACM Computing Surveys, 33 (1): 31–88.

Nguyen, H.-N., Gutman, P.-O., Olaru, S. & Hov, M. 2013. *Implicit improved vertex control for uncertain, time-varying linear discrete-time systems with state and control constraints.* Automatica, 49 (9): 2754–2759.

Power, D.J. 2007. *A Brief History of Decision Support Systems.* DSSResources.COM, World Wide Web, http://DSSResources.COM/history/dsshistory.html, version 4.0, March 10, 2007.

Pupkov, K.A. & Egupov, N.D. ed. 2004. *Methods of classical and modern theory of automatic management.* Moscow State Technical University Press, Moscow: 126–134 (in Russian).

Thomas, J. & Mengel, T. 2008. *Preparing project managers to deal with complexity - Advanced project management education.* International Journal of Project Management, 26 (3): 304–315.

Ting, S.L., Wang, W.M., Kwok, S.K., Tsang, H C.A. & Lee, W.B. 2010. *RACER: Rule-Associated CasE-based Reasoning for supporting General Practitioners in pre-* *scription making.* Expert Systems with Applications, 37 (12): 8079–8089.

Zacharova, L.N. & Nazimko, V.V. 2012. *Research of sensitivity of coal extraction project and its risks.* Radio-Electronics and Computer Systems, 53 (1): 157–164 (in Russian).

Progressive Technologies of Coal, Coalbed Methane, and Ores Mining – Bondarenko, Kovalevs'ka & Ganushevych (eds)
© 2014 Taylor & Francis Group, London, ISBN: 978-1-138-02699-5

Review of seismograms typical for an in-seam seismic technique in conditions of different coal basins

A. Antsiferov & O. Glukhov
Ukrainian State Research and Design Institute of Mining Geology, Rock Mechanics and Mine Surveying (UkrNIMI) National Academy of Sciences of Ukraine, Donets'k, Ukraine

ABSTRACT: The main factors that specify the structure and attributes of seismograms typical for in-seam seismic technique in conditions of different coal basins are considered. When employing this technique in conditions of Donbass the factor of generating wave trains by complex structure-low-velocity interbedded layers should be taken into account. Seismogram structures and attributes typical for conditions of Donbass are given.

1 INTRODUCTION

In-seam seismic prospecting (ISS) technique is one of the most efficient instruments to predict coal seam faults.

ISS is based on the results of T. Krey's research (Krey 1962, 1981). He described the mathematical solution for waves transmitted through low-velocity layers (the so called *channel waves*). T. Krey shows that faults can be detected by seismic reflection (RE) technique.

This method has two basic variations. The first one is based on searching waves reflected from geologic anomaly (reflection method). The second one is based on the analysis of wave trains transmitted through it (transmission method). According to the records of UkrNIMI, NAS of Ukraine, transmission method has been used only in one third of hundreds of the cases. In all other cases reflection method was employed at the areas scheduled for mining operations.

ISS is employed for searching practically all types of geologic anomalies (see Figure 1). More than half of them are tectonic faults. In 20% of all the cases employment of ISS was aimed at detecting the location of large faults; in the rest cases the aim was to determine parameters of low-amplitude faulting. Efficiency of seismic acoustic technique to detect tectonic faults is 80%.

ISS is used in coal basins all over the world where coal is produced by underground method. It is widely employed in Ukraine, Russia, Kazakhstan, Germany, India, China and other countries.

The main factors that specify the structure and attributes of seismograms typical for ISS technique in conditions of different coal basins are considered.

When employing this technique in conditions of Donbass the factor of generating wave trains by complex structure-low-velocity interbedded layers should be taken into account. Seismic data and statistical data used in this paper are taken from seismic recording archive of the UkrNIMI.

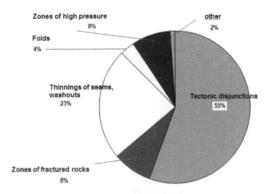

Figure 1. Employment of ISS technique.

2 STRUCTURES OF THE OBSERVED SEISMOGRAMS AND THEIR ATTRIBUTES

Each coal-production region has its own characteristic aspects. Yet structures of the observed seismograms are practically the same. They usually contain *P*-waves and *S*-waves. These waves propagate in wallrock and are almost dispersion-free. Coal seam in the adjacent strata is low-velocity (interbedded) layer. As a result, specific channel waves are observed (we shall call them *CH*-waves). These waves are generated due to the wave interference inside coal seam and are of two types: Rayleigh waves that oscillate in plane normal to the plane of coal seam and Love waves, which oscillate in the seam plane normal to the direction of their propagation. The wave energy of Love-type seam waves is more strongly tied to the seam than energy of Rayleigh waves. It is their important property and the main reason for a wide use of Love seam waves. *CH* wave frequencies depend on the seam thickness. Thicker seams generate lower-frequency *CH* waves.

CH waves are dispersive. Looking at dispersion curves (Figure 2) we can specify their peculiar properties. Phase and group velocities of CH waves are different. For Rayleigh waves the difference between phase and group velocities decreases with decreasing shear wave velocity in rocks. The minimums of group velocities yield special phase in the seam wave train. It is called the "Airy-phase" (Figure 2). *CH* wave "Airy-phase" envelope maximum propagates with velocity $\approx 0.9\ V_s$, where V_s is *S* wave propagation velocity in coal. These *S* waves are the basic instrument of in-seam seismic survey in its classic variant (Dresen & Ruter 1994, Schott & Uhl 1997).

It should be appreciated that the higher is the seismic frequency, the larger is its attenuation with the distance. Therefore, although in some experiments *CH* waves with the frequency of up to 1000 Hz were observed, in large majority of practical cases the frequency of qualitative recording did not exceed 600 Hz. 1000 Hz *CH* waves are observed only at comparatively small distances when using exploders that cannot be employed in gaseous mines.

The typical thickness of coal seams for deposits in Russia and Kazakhstan exceed 2.0 meters. This enables to observe the "Airy-phase" *CH* waves in most of coal seams. Moreover, in some cases (if the thickness is about 4.0 m and more) there is not only the fundamental but also the first mode of "Airy-phase" *CH* waves.

Coal seams in Ukraine are typically very thin. Most coal reserves are accumulated in seams with the thickness 0.6–1.0 m. And as the result, most commonly we are not able to record the "Airy-phase" *CH* waves. According to our statistics, *CH* waves can be observed only in 15% of all the cases. So, we deal with the "non-classic" situation.

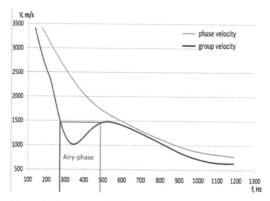

Figure 2. The pattern of CH wave dispersion curves.

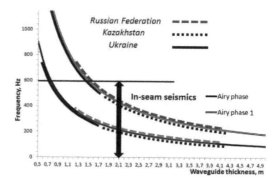

Figure 3. Frequency bands of the "Airy-phase" CH wave principal mode and the first mode for different coal basins.

In terms of physics, "Airy-phase" *CH* wave in direction normal to the plane of seam occurrence can be considered as a standing wave. Standing wave mode frequencies can be evaluated on the basis of the values of seam thickness and wave propagation velocity. Curves showing frequency bands of the "Airy-phase" *CH* wave fundamental mode and the first mode for several coal basins are given in Figure 3 where Ukraine is represented by Donets'k Coal Basin (Donbass), Russian Federation – by Pechora Coal Basin, Kazakhstan – by Karaganda Coal Basin. Although the curves do not reflect complete picture correctly, the trend is truthful.

The sharpness of coal-rock acoustic interface is also important factor specifying the part of the full wave field energy transferred by CH wave in coal seam.

Figure 4 shows distribution patterns for velocities of *P* (*Vp*) and *S* (*Vs*) waves in various rocks typical for coal basins in Ukraine, Russia and Kazakhstan.

Coals in Russia are relatively strong with relatively high velocities of propagation of seismic waves. The greatest difference between the wave propagation velocities in coal and in surrounding rocks is observed in Kazakhstan. Typical for Ukraine is the greatest difference between seismic wave velocities in coal and in rocks.

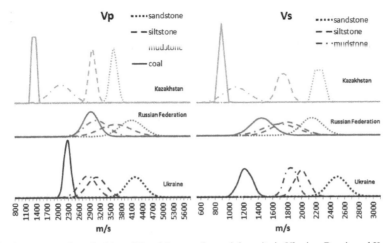

Figure 4. Distribution patterns for velocities of *P* and *S* waves for coal deposits in Ukraine, Russia and Kazakhstan.

Oscillations of seismic waves do not concentrate solely inside coal seam. According to theory, they penetrate into rocks at the distance of the order of half of wavelength. In conditions of thin coal seams of Donbass this distance can be 2–5 times larger than the seam thickness. This zone will be involved in generating classical channel wave. More over, under certain conditions it can generate a separate channel.

Taking into consideration thin coal seams of Donbass, we can divide them into two groups for our purpose. The first group is coal seam in tough rocks that forms classic low-velocity layer (waveguide). When loose rocks (for example, mudstone) adjoin the seam, and also when there are con-

tiguous seams, a waveguide of complex structure is formed. These are cases of the second group. For the waveguide of complex structure to be formed, it is necessary that the thickness of loose rock or the thickness of the band between contiguous seams does not exceed 3.0–4.0 m. More than half (55%) of coal seams belong to the second group.

Seismograms of Rayleigh waves obtained for coal seams of the first group with the thickness up to 1.5 m consist generally of two wave trains. The first wave train is *P* wave with frequency band in the range of 40–50 up to 250 Hz. The second wave train is S wave with frequency band of 80 to 300 Hz in the initial section and with a frequency up to 500 Hz in its tail part (Figure 5a).

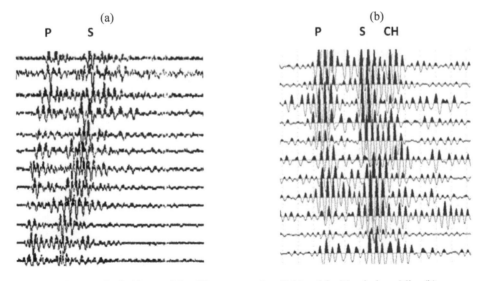

Figure 5 Real seismic traces for Stakhanova Mine (Krasnoarmeyskugol) (a) and for Mospinskaya Mine (b).

Seismograms obtained for coal seams of the second group consist of three wave trains. The first wave train is P wave with frequency band of 50 to 250 Hz. The second wave train is S wave with frequency band of 60 to 305 Hz, and the third wave train is CH wave formed by complex-structure waveguide with a frequency band 250 Hz and more (Figure 5b).

No matter what group coal seam belongs to, if its thickness is more than 1.5 m, "Airy-phase" CH waves with frequency band of 300 Hz and more are generated.

Lowe waves almost always contain only one wavetrain. First-break wavelet velocity equals to S-wave propagation velocity in wallrock. Initial wavetrain frequency is 70–150 Hz. Velocity and frequency of envelope maximum fits the "Airy-phase".

Information-bearing of any given wave trains depends on the employed technique.

With the employment of transmission method in coal seams of the first group P waves are information-bearing approximately in 60% of the cases; S waves are information-bearing in 90% of the cases. In 45 % of all the cases this is the only information-bearing wave. In less than 5% of the cases (if thickness of the seam exceeds 1.25 m) P waves are information-bearing approximately in 30% of the cases; S waves are information-bearing in 60% of the cases. CH waves that fit complex-structure low-velocity interbedded layer are information-bearing in 60% of the cases. Of real methodic interest is the fact that approximately in 30% of the cases this is the only information-bearing wave. And only in 5% of the cases the "Airy-phase" CH wave bears information on the fault.

Quality of waves with the use of reflection method depends essentially on particular technique. Recently on the basis of the classical reflection method UkrNIMI, NAS of Ukraine has developed seismic location method (SL). It differs in that seismic energy excitation, and recording is made directly in the mine roadway. The distance of seismic location is 25–50 m.

When employing RE method the most information-bearing waves in more than 70% of the cases are S waves. Second in quality are CH waves generated by low-velocity layer of complex structure. The fifth part is the cases when P waves are information-bearing. And only in each 10th case and only at distances less than or equal to 50–70 m "Airy-phase" CH waves are observed. With SL the most information-bearing are CH waves generated by low-velocity layer of complex structure; less informative are P waves, and "Airy-phase" CH waves are far less information-bearing.

The difference in distribution of the quality of wavetrains is due to the difference in prediction range of RE and SL. When RE is used at comparatively large distances, attenuation of relatively high-frequency classic CH waves is rather high.

3 CONCLUSIONS

As can be seen from the above, we have considered the main factors that specify the structure and attributes of seismograms typical for in-seam seismic technique in conditions of different coal basins. These are seam thickness, sharpness of coal–rock acoustic interface, structure of sole and roof rocks. The latter is of crucial importance when employing ISS technique in conditions of thin coal seams of Donbass. An additional wave guide can be formed by coal seam together with adjoining rocks. In this case non-classic channel wave can be observed, which is widely used to predict geological anomalies in conditions of Donbass.

REFERENCES

Krey, T.C. 1962. *Channel waves as a tool of applied geophysics in coal mining.* Geophysics, Vol. 28, #5: 701–714.
Krey, T.C. 1981. *Theoretical and practical aspects of absorption in the application of in-seam seismic coal exploration.* Geophysics, Vol. 47, #12: 1645–1656.
Buchanan, D.J., Davis, R., Jackson, P.J. & Taylor, P.M. 1981. *Fault location by channel wave seismology in United Kingdom coal seams.* Geophysics, Vol. 46, #7: 994–1002.
Dresen, L. & Rüter, H. 1994. *Seismic coal exploration. Part B: in-seam seismics.* Oxford: Pergamon.
Schott, W., Uhl, O. 1997. *Flözwellenseismische Untersuchungen auf dem Bergwerk Ensdorf der Saarbergwerke AG.* Glückauf 1997, 133 #7/8: 480–490.

The authors would like to thank Tatiana Skopich for her work on the manuscript and translation into English.

Progressive Technologies of Coal, Coalbed Methane, and Ores Mining – Bondarenko, Kovalevs'ka & Ganushevych (eds)
© 2014 Taylor & Francis Group, London, ISBN: 978-1-138-02699-5

Link of specific electric resistance with qualitative and strength characteristics of ores

V. Portnov, R. Kamarov & A. Mausymbaeva
Karaganda State Technical University, Karaganda, Kazakhstan

V. Yurov
Karaganda State University, Karaganda, Kazakhstan

ABSTRACT: Regularities of change in specific electric resistance of the ores from the concentration of sulphide minerals, as well as the hardness and durability of rocks on the compression.

On the basis of non-equilibrium statistical thermodynamics the link of conductivity of metals with thermodynamic parameters is received. The formula linking the strength criterion of rock with their thermodynamic parameters is deduced. The dependence for the calculation of the diameter of the piece of ore is obtained from the measurement of specific electric resistance. Thermodynamic criteria of minerals.mechanism of selective destruction of minerals and the dependence of fracture energy of the surface tension of the mineral and the thermodynamic parameters are presented.

INTRODUCTION

Rocks and ores are characterized by electrical properties. The electromagnetic field is a carrier of information about the nature of the geoelectric, and, therefore, the geological section. Any for electrical prospecting method can be viewed as a system designed to gather information about the nature of the geological section.

Artificial electromagnetic field is created, any source, and the source of natural field is the physical and chemical processes in the Earth, or Earth processes of interaction with cosmic radiation of different types.

Primary electromagnetic field (the field that is in the air), interacting with electrically non-uniform cross-section, acquires some of the features that can be characterized by the difference between the primary and the resulting fields, that is an anomaly. It is clear that it is part of the field of anomalous carries information about the geological aspect.

We have received an expression relating the dissipative interaction of electric dipoles in the containing environment with probability (Portnov 2003)

$$P = \frac{2\Delta S}{k\tau} exp\left\{-\frac{E_m - G^0 / N}{kT}\right\}, \qquad (1)$$

where ΔS – change in entropy in the dissipative process; E_m – average value of the ground state energy of electric dipoles; τ – time for relaxation; k – Boltz-

mann constant; T = absolute temperature; G^0 – potential Gibbs system thermostat (for rock) + system of electric dipoles (conductive ore); N – number of electric dipoles.

The majority of the dissipative processes in nature describe the Arrhenius equation

$$P = v\,exp\left(-E_a / kT\right), \qquad (2)$$

where E_a – activation energy; v – frequency factor.

Comparing the expression with (1), we find

$$\Delta S = \frac{vkT}{2} exp\left(-\frac{E_m + E_a - G^0 / N}{kT}\right). \qquad (3)$$

The frequency factor in most cases of practical importance $v = 1/\tau$ and the expression (3) will be overwritten in the form

$$\frac{2\Delta S}{k} = exp\left\{-\frac{E_m + E_a - G^0 / N}{kT}\right\}. \qquad (4)$$

Application of an external electric field on the system of electric dipoles, immersion in the thermostat will change its entropy

$$\Delta S = \frac{Ne(\Delta E)^2}{2kT^2}, \qquad (5)$$

where e – charge of the electron; E – tension of an external electric field; N – number of electric dipoles.

After simple transformation, given (4) we get

$$\Delta E = \frac{2N}{e}\sqrt{kTG^0}. \qquad (6)$$

It is known (Poplawski 1981) that the change of entropy object is inversely proportional to the number of ΔI information about it, i.e.

$$\Delta S = \frac{k\ln 2}{\Delta I}, \qquad (7)$$

where $k\ln 2$ – energy equivalent information.

On the other hand, information about the electrical properties of the object is the conductivity σ, i.e.

$$\Delta I = const \cdot \sigma. \qquad (8)$$

Subject to (3) and (8) for conduction are

$$\sigma = \frac{1}{\rho} = \frac{kT}{C_3}\frac{eN}{G^0}, \qquad (9)$$

where e – charge of the electron; N – number of electric dipoles; C_3 – constant electrical for this method.

Electrical anomalies are

$$\Delta E = \phi\sqrt{TG^0 c_p}. \qquad (10)$$

Equation (10) allows using the natural field for building out geotechnical mapping of minerals by studying electric fields local nature. Natural field study by measuring its potential or capacity for daily surface gradient, wells and mines.

The possibility of application of the method of natural field for specific geological problems is determined by a number of factors, the most important of which are the following:

– direct or indirect relationship of geological object or phenomenon of nature natural fields;

– the presence and level of natural or artificial fields-noise;

– the economic efficiency of the method (the cost of field works in comparison with other methods).

The most common method of natural field used for exploratory purposes, geological mapping and geological research. However, the development of ore deposits of this method is often used because of the intense industrial noise.

Equations (9) and (10) show the relationship of measured electrical quantities with the concentration of mineral components and electrical parameters through the energy of Gibbs. If we take to account that

$$G^0 = H - TS + PV, \qquad (11)$$

where H – enthalpy reflects the degree of metamorphism of minerals; S – entropy describes the degree of disorder; T – temperature characteristic education of the mineral; P – pressure proportional to the hardness of a mineral.

To establish the relationship between the volumetric content of ore minerals and bulk electrical resistivity (ρ) it is needed to carry out vertical electrical sounding (VES) with symmetric installation of Schlumberger to the depth to which the drilling of exploratory wells and explosive core and slime. These wells determine the volumetric concentration of ore minerals, and on passports of strength σ and hardness scale of M. Protodyakonov. The research shows correlation ρ with the content of the ore minerals (correlation $R^2 > 0.97$) (Figure 1) as well as with the hardness of rocks (Figure 2) regularities of change in strength of rocks on the compression of their density (Figure 3).

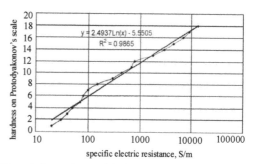

Figure 1. The dependence of resistivity (ρ) of the volume of sulphide minerals (m): 1 – medium-grained quartz diorites; 2 – granite-porphyries. 3 – oxidized ore.

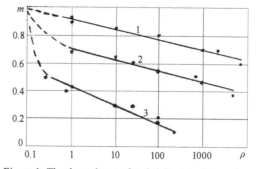

Figure 2. Dependence of the hardness of rocks and ores resistivity.

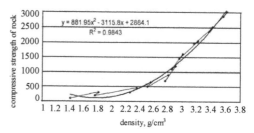

Figure 3. Dependence of strength of rocks and ores on the compression from density.

66

Concentration on 3-d definition, error of equations is ± 9.3% RH, and strength characteristics of ± 8.6% RH.

Change patterns of specific electric resistance of rocks and ores from their content of sulphide minerals form the basis of the technological mapping of the horizons quarries of a number of deposits of polymetallic ores (Mausymbaeva 2010). Evaluation of the quality of ore horizons plans is basedon not only allows you to significantly reduce the volume of geological exploration and explosive slurry sampling wells, and manage mining operations to stabilize the predetermined qualitative indicators (Mausymbaeva & Portnov 2010).

2 ESTIMATING THE SIZE OF CRUSHED ORE

To determine the probability of dissipative processes in the expression (1) denote: ΔS is the change of entropy in dissipative process; E_m – average value of the ground state energy of defects; τ is the relaxation time.

Define the probability of dissipative processes as the ratio of the energy of destruction $(e)_r$ to the crystal lattice energy $(e)_{of\ the\ CD}$. The numerator of the exponent we replace the $kT_{(m)}(c)$ taking into consideration that $\Delta S/k = \Delta H/kT_{(m)}$ (ΔH is the enthalpy or latent heat of fusion – T_m).

The subsystem is the phonon relaxation time

$$\frac{1}{\tau} = \sqrt{\frac{E(1-\nu)}{\gamma(1+\nu)(1-2\nu)}}, \qquad (12)$$

where E – Jung's module; ν – Poisson's coefficient; γ – density.

Test of strength will result in the form of

$$E_{st} = \frac{2L_m E_{cr}}{kT_m}\sqrt{\frac{E(1-\nu)}{\gamma(1+\nu)(1-2\nu)}}\left(1-\frac{T}{T_m}\right). \qquad (13)$$

In adiabatic approximation

$$E_{st} = \frac{2L_m E_{cr}}{kT_m}\sqrt{\frac{E(1-\nu)}{\sigma(1+\nu)(1-2\nu)}}. \qquad (14)$$

Equation (7) it can be concluded that the energy flow plays a role in the flow of information to which the surface layer structure self-organization process reacts.

Consider the process of crushing the ore in the Thermodynamics of information processes. Unlike the works of R.P. Poplavsky (Poplawski 1981, 1975), use the apparatus of non-equilibrium statistical thermodynamics, remaining, however, the ideology of the transition processes.

At each stage of elementary information interaction (crushing) increase in entropy (ore) ΔS lies within (Poplawski 1975)

$$1/\Delta^2 \geq \Delta S \geq \Delta S_{min} = 2/\Delta, \qquad (15)$$

here $\sqrt{\overline{\Delta r^2}/Dr} = \Delta r \leq \Delta << 1, \quad Dr = \left\langle\left(r-\langle r\rangle\right)^2\right\rangle,$

where $\overline{\Delta r^2}$ – average square of thermal fluctuations, Δ – error of installation of the operating parameter in the process of destruction can be a dislocation density.

Left border (15) corresponds to maximum irreversible implementation of the transition process, and the right side is the optimal delay it.

On the other hand, the non-entropy effect (ordering system $\Delta K = -\Delta S$) (Poplawski 1975, 1972)

$$\Delta K \approx ln(1/\Delta) \approx ln\sqrt{U/T} \approx \Delta I, \qquad (16)$$

where ΔI – obtained in the process of crushing amount of information.

Thus, the enthropy ore crushing process

$$\eta \leq \eta_{max} = \frac{1}{2}\frac{\Delta K}{\Delta S_{min}} = \frac{\Delta}{4}ln\frac{1}{\Delta} << 1, \quad \Delta << 1. \qquad (17)$$

If (17) substitute an expression that evaluates to a crushing η (Tursunbayeva et al. 2010)

$$\eta = \frac{kT}{C_1}\cdot\frac{A}{G^0}\cdot\overline{N}, \qquad (18)$$

where A – work of (energy) crush; T – temperature; G^0 – mineral Gibbs's potential of a massive sample; $\overline{N}$ – the average number of basic media destruction (proportional to the number of defects); C_1 – constant, then we obtain

$$\eta = \frac{kT}{C_1}\cdot\frac{A}{G^0}\cdot\overline{N} = B\frac{D^3}{G^0} \leq \eta_{max} = \frac{\Delta}{4}ln\frac{1}{\Delta}. \qquad (19)$$

From the formula (19) the thermodynamic constraints on size D crushed ore are derived that we must take into account both in the process of grinding and crushing in the design of machines and plants.

Comparing the expression (9) and (18) get the relationship between the work of crushing rocks and ores and their resistivity

$$A \approx C_{el}\rho_0. \qquad (20)$$

For the practical use of electrical measurements of rocks and ores (ρ_e) at their natural bedding, when determining work of their crushing (And) it is necessary to use standards with known values ρ_{et} and A_{et}

$$\frac{A}{A_{et}} = \frac{\rho_e}{\rho_{et}}. \tag{21}$$

From the formula (18) the more potential Gibbs energy, the more energy of mineral destruction should lead to its fragmentation.

If as a function of response to external electrical field

$$j = \sigma E = K_e \cdot \frac{A}{G^0}, \tag{22}$$

where E – electric field strength; A – work of the crushed mineral; K_e – then to limit the diameter of the crushed ore with $\eta = 1$ (Tursunbayeva et al. 2010)

$$D_{max} = \left(K_{EK} \frac{G^0}{\rho} \right)^{1/3}, \tag{23}$$

where K_{EK} – coefficient of proportionality.

For multi-component ore

$$\rho_l = X_1\rho_1 + X_2\rho_2 + ... + X_n\rho_n = \sum_{i=1}^{n} X_i\rho_i, \tag{24}$$

where X_i – concentration of i-th mineral with resistivity ρ_i.

Equation (23) and (24) showthe discrete distribution of crushed ore in size depending on relative resistance whose values for most minerals are known (Physical properties... 1984). From the formula (23) it is shown that the diameter of the crashed ore can be determined by G^0/ρ. The situation is different in the case of ore grinding, in which case all parameters, shown in the works (Yurov 2009, Portnov et al. 2009 & Yurov 2009): $a = \rho$ are functions of mineral particle size

$$A = A_0 \left(1 - \frac{d}{r} \right), \tag{25}$$

where A_0 – physical parameter of solid sample; r – size lump of ore, and the critical radius d is equal to

$$d = \frac{2\Omega\upsilon}{RT}, \tag{26}$$

where Ω – surface tension of the mineral; υ – molar volume; R – gas constant.

As can be seen from (22), decrease of the mentioned parameters leads to a dramatic increase of crushing. That is why the energy consumption for grinding ore is accounted for half of all energy consumption in the mining industry. In this case, the di-

ameter of crushed ore is equal to

$$D_{max} = \left(M_{EK} \frac{\Omega S}{\rho_0(1 - d/r)} \right)^{0.4}, \tag{27}$$

where S – surface area of the mineral; M_{EK} – proportionality constant; ρ_0 – resistivity of rocks and ores.

Product of ΩS represents the dispersion of a single pattern of increasing surface energy with increasing hardness of minerals on the Mohs scale and degree of dispersion.

3 CONCLUSIONS

Using measurements of specific electric resistance of the working area, and shoulder of the exploratory pits and boreholes, provide an opportunity to the primal slime control quality of extraction of polymetallic ores-the content of mineral components, the volume of ore, with a significant reduction of costs for geological sampling and plan expenses for crushing rocks and ores when developing the deposit.

Dependence of strength of rocks and minerals from the grinding environment (hardness reduction adsorption phenomenon) is known for a long time. Maximum work of destruction (and, therefore, the greatest change in specific surface energy) when dispersing in a vacuum, inert gas or liquid. Minimum job destruction (and thus minimum specific surface energy) was that when dispersing in liquid environments containing surfactants (surfactants). It does not follow, however, that the mineral substances dispersed in surfactant environment, are less chemical activity: reducing the value of σ is offset by a corresponding increase on the free surface, the product of the σ remains constant or varies slightly.

REFERENCES

Portnov, V. 2003. *A thermodynamic approach to the problems of geophysical testing of iron ore deposits*. Karaganda: KSTU: 178.
Poplawski, R. 1981. *Thermodynamics of information processes*. Moscow: Nauka: 255.
Mausymbaeva, A. 2010. *Development of geophysical methods to assess the quality of ores of precious and non-ferrous metals in exploitation*: Author. Diss. Candidate. Tehn. Sciences. Karaganda: 26.
Mausymbaeva, A. & Portnov, V. *Entropy model in exploration geophysics*. Bulletin of the University, Series "Physics", 2: 54–58.
Poplawski, R. 1975. *Thermodynamic models of information processes*. Volume 115, #3: 465–501.

Poplawski, R. 1972. *On thermodynamic limits of accuracy of physical measurement*: 562–565.

Tursunbayeva, A., Mausymbaeva, A., Portnov, V. & Yurov, V. 2010. *Thermodynamics crushing ore heap leaching of metals. Part 1 Nonequilibrium statistical thermodynamics of crushing*. Bulletin of ENU. LN Gumilev, #4 (77): 50–60.

Tursunbayeva, A., Mausymbaeva, A., Portnov, V. & Yurov, V. 2010. *Thermodynamics crushing of ore heap leaching of metals. Part 2 . Analogies method*. Bulletin of ENU. LN Gumilev, 4(77): 61–70.

Physical properties of rocks and minerals. 1984. Reference geophysics / Ed. by N. Dortman. Moscow: Nedra: 455.

Yurov, V. 2009. *Properties of small particles*. Bulletin of the University, Series "Physics", 2(54): 41–47.

Portnov, V., Yurov, V., Tursunbayeva, A. & Puzeeva, M. 2009. *Thermodynamics and thermal conductivity of minerals*. Regional Gazette East, 2: 14–18.

Yurov, V. 2009. *Some aspects of the physics of solid surfaces*. Bulletin of the University, Series "Physics", 1(53): 45–54.

Efficiency determination of using magnetic separation while processing of basalt raw material

Ye. Malanchuk
National University of Water Management and Nature Resources Use, Rivne, Ukraine

ABSTRACT: Basalts of Volyn' and Rivne Region attracted researchers for their unique properties in mineralogical and chemical composition. In the geological cross section basalts of the Volyn' series represent four retinue with layers of tuffs and breccia. Copperbelt of lavo-breccia rise to the surface in quarry of the vilagge Ivanchy and recorded in wells around the quarry. Patchy of lavo-breccia occur in tuffs in the form of horizons with thickness 1.0–0.7 m. Copper content in lavo-breccia ranges from (0.04%) to 5.0%. Copper is in the native state, which is especially valuable for mining. Copper is in the native state, which is especially valuable for mining. There are nuggets of dendritic form with weight 700–800 g. First finds of native copper in the Volyn' basalts were noted by geologists near Velykuy Midsk village in thirties of last century. It should be noted that deposit of native copper in the trap of basalts is a rarity that is why their study and pre-commercial preparations are on track of developing new, sometimes unique technologies. There are three global segments of native copper mineralization in basalts: 1) in the Great Lake area (USA); 2) Volyn' traps; 3) traps of Taimyr, Novaya Zemlya, Alaska, Yukon, British Columbia and China. From these fields only Michigan (USA) copper ore district (group of fields Keweenawan Peninsula) was exploited and is the most famous in the world reserves of native copper.

1 INTRODUCTION

Currently in open pit mining is only use basalt (mainly for the production of crushed stone). Accompanying it is tuff and breccia which are dump rock mass, which is stored and represents technogenic deposit with a high proportion of native copper, iron, titanium and other precious metals. Performed complex of research showed that basalt is a valuable mineral and requires complex processing for extracting useful components, contained in amounts representing the commercial interest and the possibility of their technical extraction.

The presence of impurities in the basalt array in the form of tuff and breccia does not reduce relevance of the ideas and values of complex processing, since these components contain same beneficial ingredients as basalt. The main ones are native copper, iron (titan magnetite) (Bulat et al. 2010). Spectral analysis showed that they contain copper oxides, rare and precious metals, extraction of which requires more advance techniques.

As the analysis of the available information on the use of magnetic separation in the processing cycle of ores and placers of ferrous and rare metals selection of nonferrous metal is primary, and iron concentrate – secondary. With the same position was explored issue of the magnetic separation of copper raw material in

basalt quarry. In the experiments to extract them were used standard methods of ore preparation before magnetic separation of raw materials: crushing, grinding and screening. The most effective classes of grain size for the magnetic separation were studied $(+0.63 \div 2.5 \text{ mm})$ $(+1.0 \div -0.63 \text{ mm})$.

The subject of research in this work was determination of degree of the magnetic susceptibility of all three components of basalt raw material from the Rafalovka basalt quarry – basalt, tuff and breccia.

2 MAGNETIC SEPARATION OF THE BASALT RAW MATERIAL

The objectives of the research was to consider how effective magnetic separators can be to allocate copper minerals in the tails of the magnetic separation, and at what size of copper extraction to tails of magnetic separation will be maximized.

Frankly speaking, the issue of application of magnetic separation to concentrate the copper minerals in the tailings of separation was investigated.

Preparation of samples for research was based on their preliminary crushing and grinding to class size less than 3 mm, in accordance with recommendations for dry magnetic separation of weakly magnetic ores (Nadutuy et al. 2012). Crushed rock mass was classified into four classes by their size, and in each class was defined the magnetic part (in two or three levels) and nonmagnetic part in the weight and percentage of sample's weight. Studies were conducted in laboratory conditions on the magnetic drum separator PBCU-0.5/0.2 in the dry magnetic separation. Mineralogical analysis was performed separately for the magnetic and nonmagnetic parts of the sample. Native copper content in each weighed portion was estimated.

Initial experimental data, obtained for further analysis, are shown in Tables 1–4.

Table 1. Indicators of dry magnetic separation of narrow basalt classes.

Grain size, mm	Product	Weight, 10^{-3} kg	Output in class, %	Output from initial, %	Content Cu, %	Extraction of Cu in class, %	Extraction of Cu from initial, %
−2.5 + 1.6	Concentrate 2	66	75.9	16.62	3.5	45.8	22.2
	Nonmagnetic	21	24.1	5.29	13.0	54.2	26.2
Total by the class		87	100.0	21.91	5.79	100.0	48.4
−1.6 + 0.8	Concentrate 1+2+3	87	79.8	21.91	0.37	9.0	3.1
	Nonmagnetic	22	20.2	5.54	15.0	91.0	31.7
Total by the class		109	100.0	27.46	3.33	100.0	34.8
−0.8 + 0.25	Concentrate 1+ 2	36	31.0	9.07	0.0014	0.04	0.0
	Nonmagnetic	80	69.0	20.15	1.5	99.96	11.5
Total by the class		116	100.0	29.22	1.03	100.0	11.5
−0.25	Concentrate 1+2	30	35.3	7.56	0	0	0
	Nonmagnetic	55	64.7	13.85	1.0	100.0	5.3
Total by the class		85	100.0	21.41	0.65	100.0	5.3
Total in sample		397		100.0	2.624		100.0

Table 2. Output materials. Mineralogical and particle size analysis of samples from Rafalovka quarry.

Class, mm	Product	Weight, g	Output, %	Mineral content	Copper content, %	Output	Extraction
2.5 + 1.6	-	–	–	–	–	–	–
	Magn. 2	66	75.86	Basalt – 96–97%. Native copper – 3–4%	3.5	16.62	0.582
	Nonmagn.	21	21.14	Basalt – 85%. Native copper – 15% (10% – opened and in aggregates – 5%)	13.0	5.29	0.688
−1.6 + 0.8	Magn. 1	19	17.43	Basalt – more than 99%. Native copper in aggregates – single grains. Malachite – single aggregates. Cuprite – single grains	0.01	4.786	0.0004
	Magn. 2	51	46.79	Basalt – more than 99%. Native copper in aggregates – single grains. Malachite – less than 1%	0.6	12.85	0.077
	Magn. 3	17	15.60	Basalt – more 99%. Native copper in aggregates – single grains	0.1	4.282	0.004
	Nonmagn.	22	20.18	Basalt – 75–80%. Native copper 10–15 in aggregates – 5–7%. Quartz – 5%. Malachite in aggregates 5–7%	15.0	5.542	0.831
−0.8 + 0.25	Magn. 1	5	4.31	Basalt – 100%. Malachite – single grains, in aggregates	0.01	1.259	0.004
	Magn. 2	31	26.72	Basalt – 100%	0	7.809	0
	Nonmagn.	80	68.97	Basalt – 94–96%. Quartz – 2–3%. Native copper in aggregates – 2–3%	1.5	20.15	0.302
−0.25	Magn. 1	8	9.41	Basalt – 100%	0	2.015	0
	Magn. 2	22	25.88	Basalt – 100%	0	5.542	0
	Nonmagn.	55	64.71	Basalt – 99–100%. Native copper up to 1%	1.0		0.139
Totall		397	100.0	Native copper in basalt sample 2.624%		100.0	2.624

72

Thus, the results of tuff and breccia study from the Rafalovka deposit showed feasibility of further study of magnetic separation operation, since all three of the most characteristic rocks have a high magnetic susceptibility: basalt magnetic product 55% from breccia – 33%, from tufa – 54%.

Dry magnetic separation of basalt. Sample was crushed and sieved to 4 narrow classes in the range –2.5...–0.25 mm. Basalt has the highest density of the three breeds studied $2.6 \cdot 10^{-3}$ kg/m^3 (breccia – 1.8, tuff – 1.3). Crushing was performed so that the output of each of the four classes was approximately the same – about 20–30%.

3 RESULTS OF CALCULATIONS

Under magnetic separation for narrow size grades of basalt it is characteristic that for the top two major classes there is a high yield of non-magnetic fraction. The content of copper in the non-magnetic fraction of large classes is very high: 13 and 15% (the last figure corresponds to the minimum condition of the finished copper concentrate).

Analysis of copper extraction (from the initial content) confirmed that the copper is mainly extracted from the major classes of 0.8 –2.5 mm. For them, the total extraction is 48.4 + 34.8 = 83.2 (%). The remaining share of extraction, 11.5 + 5.3 = 16.8 (%), accounted for smaller classes ... –0.8 –0.25 mm.

However, it is noteworthy that the largest class of 1.6 –2.5 mm is poor separated: yield of tails is very small (output in the non-magnetic fraction ~ 5% vs. 16.62% in magnetic), extraction of copper in both products differ not too much (22% in the magnetic and non-magnetic 26%), i.e, copper of upper class is equally divided between the magnetic and non-magnetic products.

All smaller classes –1.6 mm have another trend: copper extraction in non-magnetic product consistently higher than magnetic. Under the magnetic separation of basalt feed size should not be greater than 1.6 mm.

Picture of dry magnetic separation of basalt shows a summary Table 3, which shows that the magnetic product of small classes –1.6 mm does not have copper magnetic fraction is rich in copper only because of presence of large separation classes 1.6 mm.

Table 3. Summary data of magnetic separation of basalt.

Class size, mm	Output from starting material, %	Content of Cu, %	Extraction of Cu, %	Magnetic, %			Nonmagnetic,%		
				Output	Content	Extraction	Output	Content	Extraction
–2.5 + 1.6	21.91	5.79	48.39	16.62	3.5	22.2	5.29	13.0	26.21
–1.6 + 0.8	27.46	3.33	34.80	21.91	0.37	3.1	5.54	15.0	31.68
–0.8 + 0.25	29.22	1.03	11.53	9.07	0.0014	0.0	20.15	1.5	11.52
–0.25	21.41	0.65	5.28	7.56	0.0	0.0	13.85	1.0	5.28
Total	100.0	2.624	100.0	55.16	1.20	25.3	44.84	4.37	74.70

Question arises, how copper may get into the magnetic fraction (especially in large classes) if all of copper minerals are non-magnetic? There are two reasons. First, is the following. Although the mineralogy clearly shows that the deposit is rich in oxidized and native copper (non-sulfide) (Tables 1–4), basaltic materials still have sulphide copper, ie, chalcopyrite, and, more importantly, usually accompanying chalcopyrite such minerals as pyrrhotite and pyrite.

Although pyrite and chalcopyrite (copper pyrites, $CuFeS_2$) – nonmagnetic, but magnetic pyrrhotite or pyrite FeS_2 – strongly magnetic mineral. It removed during the separation, and with it, a splice copper minerals are extracted.

The second reason for copper extraction in the magnetic concentrate of large classes lies in the fact that splices of copper extracted with native iron,

magnetite, titanomagnetite, as well as iron and copper sulphides such as bornite Cu_3FeS_2. Mineralogical analysis showed presence in rocks of the Rafalovka quarry all these minerals. With them, in fact, is related high output of the magnetic fraction during the separation.

Major findings of basalt magnetic separation are that under the size 0 –2.5 mm, firstly, there is a high output of the magnetic fraction – 55 (16%), secondly, the amount of copper in the tailings of separation increases up to 1.7 times compared to in the initial copper (from 2.6 to 4.4%). However, both products have high copper content, indicating that there is a need to reduce its size.

Magnetic separation of basalt showed the possibility of copper concentration in tailings. However, from position of iron concentration, the resulting high output of magnetic fraction – more likely disadvantage than advantage, because it was obtained at a relatively coarse conditions, and, of course, along with a large mass of magnetic product and attracted a lot of splices, and gangue.

Due to this magnetic product is poor in iron (less than 50% Fe). It needed to be improved, i.e grind and re-separated. To increase the copper extraction in the tails of the magnetic separation, achieved growth rate in tailings (1.7 times) can be increased (up to $2 \div 3$ times), if you reduce feed size and crush basalt to the size 1.6 mm.

Table 4. Dependence of basalt raw material components from the magnetic field (in tesla) for two size classes of ore-preparation product.

# experiment	Induction, tesla	Tuff		Basalt		Breccia	
		+0.63–2.5 mm	+0.1–0.63 mm	+0.63–2.5 mm	+0.1–0.63 mm	+0.63–2.5 mm	+0.1–0.63 mm
1	0.08	63.2	30.5	73.6	29.0	51.3	21.0
2	0.16	59.5	37.0	76.4	35.5	60.5	26.2
3	0.3	51.7	37.7	68.0	40.5	64.4	31.6
4	0.44	49.7	31.9	63.5	38.5	68.5	36.4
5	0.58	44.8	32.5	60.5	30.4	63.1	28.5
6	1.3	6.2	5.8	10.6	7.6	20.2	12.9
7	Nonmagmetic	56.4	33.4	29.4	18.5	36.2	21.4
8	2	331.4	208.9	380.0	200.0	364.2	178.0
Total in each sample, g		331.4+208.9+104.5-645 (g)		380.0+200.0+65.0-645 (g)		364.2+178.0+102.8-645 (g)	

4 CONCLUSIONS

Experimentally established, the magnetic susceptibility of all three components of basaltic materials - tuff basalt and breccia. This indicates practicability of inclusion into technological scheme an integrated waste-free processing of raw basalt magnetic separation operation for separating titanomagnetit.

Dry magnetic classification of basalt, tuff, breccia and the narrow size classes showed that the most promising in outputs of the original amount and recovery of copper in each class (in %) are the classes: ($-2.5 \div 1.6$ mm) ($-1.6 \div 0.8$ mm), ($-0.8 \div 0.25$ mm) ($-0.25 \div 0.1$ mm).

Summary of the studies on the magnetic susceptibility of raw materials made in the form of experimental and regression output of tuff, basalt and breccia from the induction of separator's magnetic field.

Generalized regression model of outputs for all three components was set.

REFERENCES

Bulat, A.F., Nadutiy, V.P. & Malanchuk Z.R. 2010. *Prospects complex processing of raw basalt Volyn*. International collection of scientific works "Geotechnical Mechanics": Vol. 85: 3–7. Dnipropetrovs'k: IGTM NAN Ukraine.

Nadutuy, V.P. Erpert, A.M. & Malanchuk, Ye.Z. 2012. *Summary of the studies of the magnetic susceptibility of basalt raw material components*. Scientific-Technical Collection "Minerals processing": Issue 51 (92.): 144–149. Dnipropetrovs'k: National Mining University.

Progressive Technologies of Coal, Coalbed Methane, and Ores Mining – Bondarenko, Kovalevs'ka & Ganushevych (eds)
© 2014 Taylor & Francis Group, London, ISBN: 978-1-138-02699-5

Influence of stoping and coal extraction on gas release level from the underworked massif

Y. Kryzhanovskiy, M. Antoshchenko, R. Gasyuk & D. Shepelevich
Donbass State Technical University, Alchevs'k, Ukraine

ABSTRACT: Till now it has not been studied about cooperative influence of breakage works and coal excavation onto gas release in breakage faces. Carried out researches allowed determining the principles of achieving the planned load and dynamics of gas release on drawing breakage faces away from holing chute.

1 INTRODUCTION

It is determined, that gas release from breaking areas at other equal conditions depends both on coal production (Guidance on design... 1994), and development of second workings (Antoshchenko et al. 2013). Changing the load onto breakage face can be done at any stage of development of second working. In the early period of longwall face operation is achievement of planned volumes of coal production. Classically such a progress is achieved as a rule gradually after staring the operation of a breakage face. Along with production growth the restarts second working development caused by face's drawing away from a holing chute. During this period they observe the growth of gas release from underworking zones connected with shift processes in rocks including the initial squeezing of main roof.

At prognostication of gas release they offer its direct dependence on coal production (Guidance on design... 1994). The development level of second working is not considered in this case. Collective impact of two components like coal production level and development level of second working in the blocks has not been studied till now. So, investigation of this problem is rather important because its successful solving influences the safety of mine works on gas component.

2 IMPACT OF DEVELOPMENT LEVEL OF SECOND WORKING ONTO COAL PRODUCTION LEVEL AND GAS RELEASE DYNAMICS

Research technique considers analysis of coal production changes in the early operation period of blocks. Along with load growth and drawing faces away from holing chutes they carried out observations for changes of methane release into degassing wells and blocks.

Research task was to study principles of achieving planned volumes of coal production on drawing breakage faces away from holing chutes and impact of these processes onto gas release dynamics from underworking zones.

Experiments have been made in a mine "Sukhodolskaya–Vostochnaya" PC "Krasnodonugol" and a mine "Izvestiya" SH "Donbassantracite" at working off the seams i_3' and l_2^e.

Logically proceeding they supposed that coal production growing (A) after beginning of block operation on drawing a face away from a holing chute (L) is occurred on exponential dependence

$$A = A_m \left(1 - e^{-k_1 l}\right), \qquad (1)$$

where A_m – planned(maximum) coal production level, t/day; k_1 – empiric index characterizing cola production change for certain longwall face.

When choosing the ratio between gas release and distance from breakage face and holing chute (L) they proceed from fact that methane release into this working before second working starts is not great, and rarely exceeds 1 m³/min. It is explained by absence of coal breakage during assembly works. Their durability can be from some weeks to some months. For this period of time gas release from open surfaces of holing chute reduces to minimum. On this reason back ground value of gas release before second working starts is equal to 0 and current change of methane release (I_c) on increasing a distance (L) can be described by exponential curve, starting out from a source of coordinate grid

$$I_c = I_m \left(1 - e^{-k_2 L}\right), \qquad (2)$$

where I_m – gas release level, which corresponds planned coal production A_m, m³/min; k_2 – empiric index characterizing mine-geological conditions of certain longwall face operation.

On developing second working methane release is performed both into workings and degassing wells.

Processing results for experimental data according to Equations (1) and (2) are put into Table 1. As an example (Figure 1) there have been presented the diagrams of changes for coal production level and total gas release into workings and well at opening of 25th western longwall face in mine "Sukhodolskaya-Vostochnaya". You can see from diagrams that volumes of coal production were in creased at drawing breakage face away from holing chute for 64 meters. Total gas release, which corresponded

this level of coal production, occurred when distance between breakage face and holing chute was about 200 meters. Such ratio indicates that if distance is less than 200 meters the level of gas release is affected by two factors – coal production and development level of second workings. After drawing away for more than 200 meters gas release (40.6 m³/min) was determined by planned loading (1486 t/day) on breakage face of the 25th western longwall face (Table 1), and development of second workings (drawing face away from holing chute) did not affect the methane release processes (Figure 1).

(a)

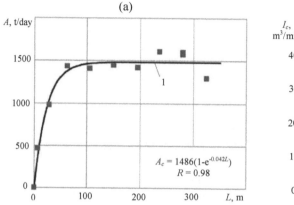

(b)

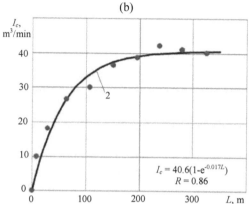

Figure 1. Example of coal production (A) change (a) and total gas release (I_c) in to workings and well (b) at working of 25th western longwall face of seam i'_3 at "Sukhodolskaya–Vostochnaya" mine: ■, ● – experimental data; 1, 2 – change curves for coal production and gas release respectively; R – correlation.

Table1. Information about empiric indexes of exponential equation and the parameters, which describe the correlation between coal production (A) and gas release (I_c) according to Equations (1) and (2) if drawing breakage face away from holing chutes.

Longwall face	Equation 1				Equation 2			
	Production A_m, t/day	Empiric index k_1	Correlation R	Standard deviation σ_A, t/day	Methane release I_m, m³/min	Empiric index k_2	Correlation R	Standard deviation σ_I, m³/day
"Sukhodolskaya-Vostochnaya" mine								
12th eastern	1039	0.124	0.96	77	21.9	0.050	0.89	3.2
24th eastern	2034	0.046	1.00	28	53.3	0.034	0.95	7.8
25th western	1486	0.042	0.98	102	40.6	0.017	0.86	2.5
34th eastern	687	0.093	0.95	65	12.8	0.018	0.97	1.0
37th western	954	0.018	0.95	113	16.4	0.021	0.99	0.9
Average value	1240	0.065	–	77	29.0	0.028	–	3.1
"Izvestiya" mine								
1st western	280	0.019	0.87	62	–	–	–	–
2nd western	1095	0.028	0.90	250	37.3	0.009	0.95	4.4
3rd western	1313	0.012	0.93	201	37.7	0.011	0.97	4.4
4th western	1722	0.002	0.64	397	–	–	–	–
5th western	1222	0.012	0.85	261	42.4	0.023	0.91	7.4
6th western	775	0.426	0.36	280	22.7	0.052	0.84	4.8

Continuation of Table 1.

Longwall face	Equation 1				Equation 2			
	Produc-tion A_m, t/day	Empiric index k_1	Corre-lation R	Standard deviation σ_A, t/day	Methane release I_m, m³/min	Empiric index k_2	Corre-lation R	Standard deviation σ_I, m³/day
7th western	1195	0.024	0.96	145	–	–	–	–
8th western	1342	0.019	0.99	71	65.1	0.025	0.96	9.2
9th western	872	0.073	0.99	74	13.9	0.032	0.98	1.6
Average value	1081.2	0.111	–	166	36.0	0.033	–	5.8

Such change in coal production and gas release was specific to all blocks of "Sukhodolskaya-Vostochnaya" mine. It is proved (Table 1) by high values of correlation (R) both for coal production (0.95÷1.00), and for methane release (0.86÷0.99).

For Conditions of "Izvestiya" mine coal production increase was going the same way (Table 1) at seven of nine blocks ($R = 0.85÷0.99$). At two blocks correlation links were much lower. At operation of 6th western longwall face such dependence was absolutely absent ($R = 0.36$). Atypical change of coal production (Figure 2) because of worsening of mining and geological conditions did not affect the nature of change for total gas release ($R = 0.84$). It testifies of significant impact on the level of gas release in the early operation period of blocks by processes of underworked seams squeezing if drawing breakage face away from holing chutes.

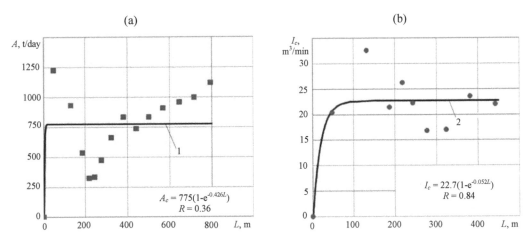

Figure 2. Example of coal production (A) change (a) and total gas release (I_c) in to workings and well (b) at working of 6th western longwall face of seam l_2^u at "Izvestiya" mine: ■, ● – experimental data; 1, 2 – change curves for coal production and gas release respectively; R – correlation.

Given examples for two mines indicated that in mostly cases (more than 85%) achievement of planned loads happens pursuant to exponential dependence. Then they analyzed empiric indexes of Equations (1) and (2), for which correlation ratio exceeded 0.84. There have been determined direct dependence I_m from A_m. They are rather individual for certain mining and geological and geotechnological conditions (Figure 3).

Empiric indexes k_1 and k_2 almost did not depend on A_m. Correlation indexes (r) are equal 0.30 and 0.14. In its turn k_1 and k_2 correlate to each other (Figure 4), that is stipulated obviously by similar squeezing processes in underworked seams in different geological conditions. The more absolute value k_1 has got, the more steep is a curve of coal production growth according to Equation (1). From ratio between $k_1 > k_2$ (Figure 4), gas release curve by Equation (2) will be sloping comparing to the coal production curve (1) if drawing breakage face away from holing chute. This indicates, that gas release growth retards in time and space from load growth onto breakage face.

For studied blocks (Table 1) there has been determined a correlation between standard derivation σ_A and value A_m ($r = 0.19$), that obviously is connected with different mechanical supply as well as technology and organization of production cycles in the blocks.

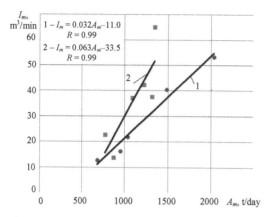

Figure 3. Ratio between maximum gas release (I_m) and planned load (A_m) onto breakage face: ●, ■ – experimental data; 1, 2 – averaged lines for blocks in "Sukhodolskaya-Vostochnaya" mine and "Izvestiya" mine respectively; R – correlation.

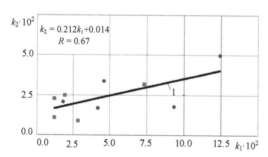

Figure 4. Ration between empiric indexes k_2 from k_1: ●, ■ – experimental data for blocks in "Sukhodolskaya-Vostochnaya" mine and "Izvestiya" mine respectively; 1 – averaged line; R – correlation.

Between standard derivation of gas release σ_I and component I_m they have observed direct dependence (Figure 5). Ration between σ_I and I_m testifies that change of σ_I is occurred in great degree comparing to I_m. It indicates that if I_m grows than unevenness of gas release grows as well but not reduces as it is stated in (Guidance on design... 1994).

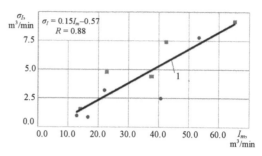

Figure 5. Ratio between standard derivation of methane release (σ_I) and its maximum value (I_m): ●, ■ – experimental data for blocks in "Sukhodolskaya-Vostochnaya" mine and "Izvestiya" mine respectively; 1 – averaged line.

3 CONCLUSIONS

Carried out researches allowed determining the principles of gas release dynamics in the early period of blocks operation depending on coal production level and development level of second workings. They are in the following:

– in most cases(more than 85%) coal production growth in the blocks is described by exponential ratio;

– dynamics in all cases corresponds to exponential curve if drawing breakage face away from holing chute;

– for certain mining and geological conditions there is its individual dependence between gas release and planned (maximum) coal production;

– gas release growth in the early period of block operation retards in time from coal production growth;

– there have been determined a direct ratio between indexes k_1 and k_2 for Equations 1 and 2;

– the big values of gas release correspond to greater unevenness of its release.

REFERENCES

Guidance on design of coal mines' ventilation. 1994 / ed. board: S.V. Yanko [and others]; under edition by S.V. Yanko. Kyiv: Osnova: 311.

Antoshchenko, N.I., Kulakova, S.I. & Filatiiev, M.V. 2013. *Prognostication of gas release from underworking coal seams*. Coal of Ukraine, 1: 44–49.

Progressive Technologies of Coal, Coalbed Methane, and Ores Mining – Bondarenko, Kovalevs'ka & Ganushevych (eds)
© 2014 Taylor & Francis Group, London, ISBN: 978-1-138-02699-5

Substantiation of cablebolts installation parameters in conditions of soft hoist rocks

Yu. Khalymendyk & A. Baryshnikov
National Mining University, Dnipropetrovs'k, Ukraine

V. Khalymendyk
Scientific production association "Mekhanik", Donets'k, Ukraine

ABSTRACT: A review of existing approaches to substantiation of cablebolts installation parameters is made; combining of geometrical levelling with observation at extensometers is proposed for monitoring over deformation of "support-rock massif" system; the peculiarities of massif geomechanical behavior in conditions of soft rocks are investigated; scheme of load on gateroad support behind longwall face is developed; cablebolts installation parameters in conditions of soft enclosing rocks are substantiated.

1 INTRODUCTION

World experience shows that the efficiency of coal-mining enterprise increases with the intensification of the coal extraction process. At the same time for the stable operation of the mining enterprises is required to maintain its constituent elements in the normal operating condition. One of the most "difficult" in this sense element is mine working. To achieve high rates of longwall face advance the usage of direct-flow ventilation schemes is required, which is provided by maintenance of gateroads both before and behind the longwall face (Khalymendyk 2011, Zborshchik 2011). With increasing depth of mining, the mechanism of rock deformation around mine working changes (Litvinsky 2009a), that leads to significant displacements of mine working rock contour, especially in soft fissile enclosing rocks. Under such conditions usage of the standing supports only does not always lead to the intended result. Effective solution is to combine standing support with reinforcement of the rock mass by means of bolting. This scheme has proven its effectiveness at mining enterprises worldwide (Junker 2013, Prusek 2011). However, in areas of abutment pressure, especially behind longwall face, zones of rock disintegration around mine workings reach a significant size, resulting in an increased load on the standing supports and deformation of bolted rock layers. This eventually entails significant losses of mine workings cross-section. To prevent such processes, a system of deep-laid bolts only (usually cablebolts 3–8 m in length and more) (Tadolini McDonnell 2010), or a system of two-level bolting with conventional roof bolts (first level with bolts 2–3 m in length) and deep-laid bolts (longer than bolts of first level – usually cablebolts) (Razumov et. al 2011, Sasaoka et. al 2013) are applied in the world mining practice.

System of gateroad support reinforcement in front of longwall face and at intersection by means of cablebolts has proven its effectiveness in conditions of soft enclosing rocks at mine of Western Donbass, Ukraine (Baryshnikov 2013, Voronin et. al 2013). Application of cablebolts for repair-free gateroad maintenance behind longwall face requires solving an actual problem of substantiation of their installation parameters in mentioned geological conditions.

2 REVIEW OF EXISTING APPROACHES

Operating principle of cablebolts consist in fastening of rocks in expected disintegration zone (before beginning of disintegration process) to overlying massif, that is less subjected to deformation, and realizing preliminary resistance to rock displacement in case of their pre-tensioning. The most important technological parameters are: length of cable anchors that provides their fixation into undisturbed rocks; load-bearing capacity and installation density.

These parameters of cable bolts installation depend on the following parameters of rock massif deforming:
– height (depth) of rock disintegration in the roof of mine working;
– volume of rock enclosed in the expected zone of rock disintegration.

To substantiate installation options of cable bolts

several approaches to forecasting and determination of the deformation parameters of rock massif are known in the world practice:

– roof fall analysis;
– theory of pressure arch (dome of natural equilibrium);
– numerical modelling;
– in-situ measurements.

Roof fall analysis is empirical-statistical theory (Junker 2006). It is based on registering the height of roof falls in gateroads on a particular seam, the mathematical analysis of obtained data and calculation of weighted average height of the fall. Bolting depth should be more than the weighted average height of the roof fall.

Pressure arch theory is used in foreign state standart for calculating the parameters of two-level roof bolting (Pozolotin et. al 2013, Federal rules... 2013). In accordance with this approach, bolting depth should exceed the height of the arch, and the mass of rock forming the load is determined from the volume of delaminated rock within the arch.

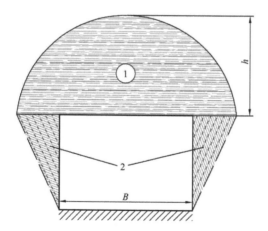

Figure 1. Scheme of arch pressure formation (Pozolotin et. al 2013). 1 – delaminated rocks in roof of gateroad; 2 – disrupted rocks in ribs of gateroad; h – height of pressure arch; B – width of gateroad.

The advantage of this approach is simplicity of calculations. However, the pressure arch theory fit to conditions of shallow depths and a single mine working. Results of researches (Zhdankin & Zhdankin 1990, Zharov 2007, Prusek 2008, Esterhuizen & Barczak 2006) show, that the spatial shape of rocks disintegration zone in the roof of a gateroad behind longwall face is different from the pressure arch.

In accordance with the approach, the height of the arch in the area of longwall influence is corrected by introducing the empirical coefficients. The change

in the shape of the rock disruption zone is not taken into account, which leads to an underestimation of the load on support behind longwall face.

Also, taking into account the variability and diversity of geological and mining conditions, the advisability of additional in-situ investigations and subsequent adjustment of cablebolts installation parameters is emphasized (Rogachkov et. al 2012).

Mathematical modeling provides a wide range of variation of geological conditions when solving the problems of identifying areas of increased stress in the simulated rock massif. It should be noted that the determination of rock disintegration zone is based on failure criterion. Consequently, the accuracy of the disintegration zone parameters determination depends on the choice of the failure criterion and its constituent parameters (Shashenko et. al 2008, Litvinsky 2009b). In addition, for an acceptable level of reliability of the results is necessary to calibrate the model carefully. Calibration is performed by comparison the results obtained in-situ (eg, at observation stations), with the results obtained on the model, and subsequent correction of parameters of the model and/or failure criterion (Prusek 2008, Prusek & Lubosik 2010). In abutment pressure zones and behind longwall face the deformation of rock massif around gateroad evolve beyond ultimate compressive strength. Realization of this process in the model requires additional studies of the rock samples strength to obtain complete curves "stress-strain" (Shashenko et. al 2008). If we ignore mentioned features of mathematical modeling, we obtain only a qualitative picture of the deformation process, rather than necessary quantitative.

In-situ observation over the "support-rock massif" system deforming are performed using extensometers and measurements of displacement of gateroad rock contour and support elements.

Let's consider how results of such observations can be used for substantiation of cablebolts installation parameters.

3 RESEARCH TECHNIQUE

The idea of geometric monitoring system "support-rock massif" is as follows. Multiple-position extensometers are used to monitor the rock displacement. Their design is based on placing anchors with predetermined spacing in borehole, which is drilled in roof of mine working (Figure 2). Each anchor connected to wire rope, which is stuck out of a borehole. When zone of disintegration forms, rock layers will be displaced, consequently anchors and ends of ropes will also be displaced. Observations at extensometers include measurement of anchors displacement with respect to the deepest-placed anchor, which, as a rule, is accepted as a reference.

placed (highest-placed) anchor is also measured with respect to the rib-embedded marks (Figure 3) (Novikov & Shestopalov 2012).

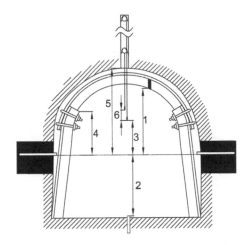

Figure 2. Measurement of rock layers displacement at extensometers. a – displacement of rock layer; l – spacing of anchors; n – number of anchor; 1– position of the anchor before rock layer displacement; 2 – position of the anchor after rock layer displacement.

Figure 3. Typical observations over displacement of "support-rock massif" system elements. 1–5 – Measurements from the level of rib-embedded marks to: 1 – arch support beam, 2 – gateroad floor, 3 – the deepest-placed anchor of the extensometer, 4 – arch support rack, 5 – collar of the borehole of extensometer; 6 – displacements of the anchors of extensometer with respect to the deepest (highest) placed anchor.

Displacements of the roof, floor and elements of support are measured with respect to the marks which are embedded in the ribs of a mine working (Prusek 2008, Prusek & Lubosik 2010). Usually at such observation stations the extensometers are mounted. In this case displacement of the deepest-

This approach is substantiated when the rib-embedded marks are immovable. In condition of soft rocks there is a phenomenon of gateroad rib deformation (Figure 4).

Figure 4. Condition of rock massif around gateroad at mine "Stepova" – phenomenon of rib deformation. 1 – rock layers position before deformation; 2 – rock layers position after deformation; 3 – coal seam.

Consequently, rib-embedded marks would be moved together with deformed rocks. It inevitably leads to decrease in reliability of gained results. That's why it is proposed to combine levelling with measurements at observation stations (Figure 5).

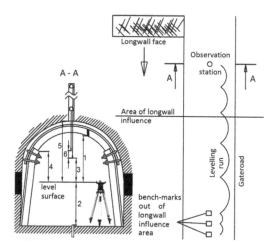

Figure 5. Levelling of "support – rock massif" system elements and anchors of extensometer. 1–5 reading from staff placed on: 1 – arch support beam, 2 – gateroad floor, 3 – the deepest placed anchor of the extensometer, 4 – arch support rack, 5 – collar of the borehole of extensometer; 6 – displacements of the anchors of extensometer with respect to the deepest (highest) placed anchor.

The bench-marks of levelling run are placed out of a zone of longwall influence that provides their immovability. And usage of modern levelling instrument provides sufficient accuracy of determination of the elements position at the observation station.

4 RESULTS OF RESEARCH

Such combined technique has been applied in the course of observations in #165 gateroad at mine "Stepova" (Western Donbass, Ukraine) during actual mining in conditions of soft rocks.

The gateroad had been driven from the roadways of level 300 m down the dip of the coal seam C_6 to the level of 490 m, with average inclination 4°. Coal seam C_6 is fractured, simple structured, has no cohesion with enclosing rocks. Extracting seam thickness is 1.04 m. Enclosing rocks are interstratified siltstones and mudstones with ultimate compressive strength up to 25 MPa and a weak cohesion.

#165 gateroad was supported by means of KShPU-17.7, with spacing 0.7 m (Figure 6). The roof of the gateroad is bolted on the depth of 2.2 m

with 5 bolts in row.

Maintenance of the #165 gateroad in the area of abutment pressure and at the intersection with longwall was performed as part of mine experiment (Baryshnikov 2013) by means of two rows 6 m long cable bolts with load bearing capacity of 210 kN (Figure 6, #4). Density of cable bolt installation was 0.3 pcs/m².

Two wooden props were installed under the each steel arch and protective construction was built up behind longwall face (Figure 6, # 7–10).

Geometric monitoring over "support – rock massif" system deforming was performed by means of the combined technique. Six observation stations were installed. Boreholes of extensometers were 8–9 m in length and 32 mm in diameter. Anchors were placed with spacing of 1.0 m in boreholes.

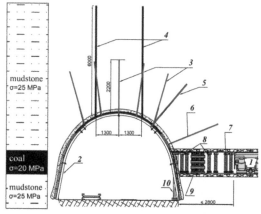

Figure 6. Support and reinforcing pattern of the #165 gateroad at intersection with 163 longwall: 1 – longwall set of equipment; 2 – steel arch support KShPU-17.7; 3 – rock bolts; 4 – cable bolts, length 6.0 м, paired installation, spacing 1.4 м; 5 – rock bolt for retaining of support beam; 6 – rock bolt for strengthening of roof above protective construction; elements of protective construction: 7 – breaker row; 8 – breaker props; 9 – chock; 10 – wooden prop between roof of the seam and floor of the gateroad.

Let's consider maximum displacements of anchors and strain of rock massif behind longwall face.

At the stations #1 and #3 stretching strain evolves to a height of 7.0 m in the top of the gateroad with the changeover from stretching to compression (Figure 7). At the station #2 stretching strain evolves to a depth of 6.5 m

The greatest displacements of anchors occur up to height of 4–5 m in the gateroad roof (Figure 7). Anchors placed higher than 7.0 m in roof are exposed to uniform sagging on magnitude of up to 0.2 m.

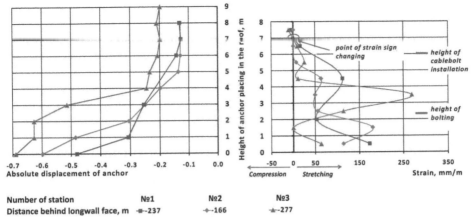

| Number of station | №1 | №2 | №3 |
| Distance behind longwall face, m | -237 | -166 | -277 |

Figure 7. Maximum displacements of anchors and strain of rock massif behind longwall face.

Installation density and load-bearing capacity of cable bolts was sufficient to maintain gateroad in area of abutment pressure in front of longwall face and at intersection (Baryshnikov 2013). However, behind longwall face the weight of rocks in disintegration zone exceeded the load-bearing capacity of cablebolts, which led to the breaking away of the anchor locks and bearing plates (Figure 8a, b).

(a)   (b)

Figure 8. Breaking away of the anchor locks and bearing plates as a result of an excessive load on cablebolt.

The difference in the behavior of massif deformation in stretching zone (0–7.0 m in the gateroad roof) obtained on three stations may be explained by the fact that:

– cablebolts were non-pretensioned. Consequently, between bearing plate and rock massif were gaps (different in size), that allowed initial rock displacement.

– the actual resistance of protective construction, built up on border "gateroad-goaf", varied. Consequently, sagging of roof over protective construction was allowed. Functional connection between sagging of roof over protective construction and value of vertical convergence in gateroad was highlighted in paper (Khalymendyk 2011). In such a manner sagging of roof over protective construction led to greater displacements of rocks in roof of gateroad.

In summary, at height of 7.0 m in roof of gateroad the presence of "neutral" rock layer which is not subjected to stretching strain is established. This layer and overlying rocks sag uniformly on magnitude of up to 0.2 m behind longwall face.

Using results of researches (Zhdankin & Zhdankin 1990, Zharov 2007, Prusek 2008, Esterhuizen & Barczak 2006) and obtained results, the simplified scheme of load on support is formed (Figure 9).

Weight of rocks Q enclosed in the running meter of area $acdef$ creates load on the gateroad support and protective construction. Now, knowing rebuff of protective construction P_p, support of gateroad P_g and load-bearing capacity of cablebolt P_{cb}, we can calculate the density of cable bolt installation per 1 m² of gateroad

$$n = \frac{Q - P_g - P_p}{P_{cb} \cdot B}, \qquad (1)$$

where B – width of the gateroad.

At the same load-bearing capacity of first-level roof bolts is ignored, because they operate only to reinforce the rocks of gateroad contour and work efficient in front of longwall face.

Working length of cablebolt should provide its fixation in a uniformly sagging rocks deeper than contour of rock disintegration zone, i.e. above the line bc in accordance with the scheme (Figure 9).

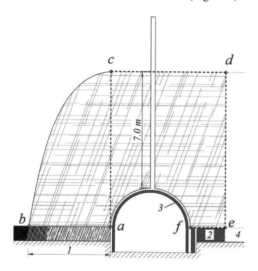

Figure 9. Scheme of load on gateroad support behind longwall face: 1 – depth of rock disruption in rib; 2 – protective construction; 3 – arch support; 4 – goaf; $abcdef$ – zone of rock disintegration; $acdef$ – disintegrated rocks that form load on support and protective construction.

It is impossible to resist uniform sagging of rock layers, as it occurs as a result of the elastic bending deflection of massif because of coal seam extraction. That's why, support of the gateroad and protective construction should compensate sagging of "neutral" layer. Providing of such yielding is realized due to: the deformation of rock massif in disintegration zone; deformation of protective construction; presence of gaps between arch support and contour rocks of gateroad.

As can be seen from the above the installation parameters of cablebolts and standing support of gateroad are substantiated reliably.

The proposed approach may seem labor-consuming and "belated", because it requires obser-

vations behind longwall face. However, engineers of mine surveying service quite cope with the implementation of these works. Presence of extensometers exceeding the bolting depth, with 2-3-anchors – so-called "tell-tales" – is required for the control of roof rock massif conditions and functioning of bolts and cablebolts. Periodic leveling of gateroad roof during actual mining – a common practice, part of the work routine of the district mine surveyor. Combining such kind of leveling with a bit "improved" tell-tales is not difficult and time-consuming for a qualified mine engineer-surveyor.

Of course, preliminary calculations of support installation parameters are necessary for mine workings support in different geological and technical conditions of other mine fields. But performing of field observations so far at extracting the first panel allows clarifying and reliable substantiation of the cablebolts and standing support installation parameters in order to maintain the roadway in front and behind longwall face effectively. Then obtained parameters can be extrapolated to adjacent districts with obligatory supervision over efficiency of bolting using the same approach.

4 CONCLUSIONS

Usage of in-situ observations over deforming of the "support-rock massif" system in conditions of soft enclosing rock allowed establishing of:

– zone of rocks stretching deformation that formed to a height of 7 m in the roof of gateroad;

– uniform sagging zone of rock layers above 7.0 m. Value of sagging is up to 0.2 m.

Basing on the obtained results, cablebolts and standing support installation parameters are substantiated in order to maintain gateroad behind longwall face.

REFERENCES

Baryshnikov, A.S. 2013. *Usage of cablebolts for gateroad maintenance*. Topical Issues of Rational Use of Natural Resources. Saint Petersburg, NMRU. Part 1: 88–90.

Esterhuizen, G.S. & Barczak, T.M. 2006. *Development of ground response curves for longwall tailgate support design*. Proceedings of the 41st U.S. Rock Mechanics Symposium, Colorado. http://www.cdc.gov/niosh/mining/UserFiles/works/pdfs/dogrc.pdf

Federal rules and regulations in the field of industrial safety. *Instructions for the calculation and application of roof bolting in coal mines*. Order of Federal Service for Environmental, Technological, and Nuclear Supervision 17.12.2013 #610.

Junker, M. 2006. *Gebirgsbeherrschung von Flözstrecken*. Verlag Glückauf, Essen. Germany.

Junker, M. 2013. *German mining expertise: meeting the future challenges of international mining.* Proceedings of 23rd World Mining Congress. Montreal, Canada http://www.cim.org/en/Publications-and-Technical-Resources/Publications/Proceedings/2013/8/23rd-World-Mining-Congress/WMCO-2013-08-191.aspx

Khalymendyk, Yu.M. 2011. *Providing of the reusing of gateroads.* Coal of Ukraine, 4: 51–54.

Litvinsky, G.G. 2009a. *The fundamental regularities and new classification of rock pressure in mining developments.* Proceedings of Donetsk National Technical University. Vol. 10 (151): 21–28.

Litvinsky, G.G. 2009b. *The analytical theory of strength of rocks and massifs.* DonSTU. Donets'k: Nord-Press http://sggs-donstu.ucoz.ru/Knigi/BOOK_Litvinskij.pdf.

Novikov, A.O. & Shestopalov, I.N. 2012. *The examination of recommendations for calculation of combined suppot parameters.* Research papers UkrNDMI NAS of Ukraine. Donets'k. Issue 10: 250–269.

Pozolotin, A.S., Rozenbaum, M.A., Renev, A.A., Razumov, Ye.A. & Chernyakhovskiy S.M. 2013. *Method of calculation of great depth bolting for supporting excavations in various mining & geological and mining & technical conditions of coal mines.* Coal, 4: 32–34. http://www.ugolinfo.ru/Free/042013.pdf

Prusek, S. & Lubosik, Z. 2006. *Monitoring of a longwall gate road maintained behind the caving extraction front.* Bergbau in Polen und Deutschland - Chancen für Innovationen und Kooperation: Freiberger Forschungsforum, 57. Berg- und Hüttenmännischer Tag C 519: 84–95.

Prusek, S. 2008. *Modification of parameters in the Hoek-Brown failure criterion for gate road deformation prediction by means of numerical modeling.* Glückauf # 9 (144): 529–534.

Prusek, S. & Lubosik, Z. 2010. *Geometrical description of gateroad roof sag.* Proceedings of the 29th International Conference on Ground Control in Mining, Morgantown, WV. http://icgcm.conferenceacademy.com/papers/detail.aspx?subdomain=icgcm&iid=303

Prusek, S., Lubosik, Z., Dvorsky, P. & Horak, P. 2011. *Gateroad support in the Czech and Polish coal mining industry – present state and future developments.* Proceedings of the 30th International Conference on Ground Control in Mining, Morgantown, WV. http://icgcm.conferenceacademy.com/papers/detail.aspx?subdomain=ICGCM&iid=920

Razumov, E.A., Hrechyshkyn, P.V., Samok, A.V. & Pozolotyn A.S. 2011. *Case history of cablebolts application for maintenance and reuse of the coal mine's roadways.* Coal #6: 26–27 http://rank42.ru/assets/files/public/06.12.pdf

Rogachkov, A.V., Pozolotin, A.S., Isambetov, V.F., Muravskiy, P.I. & Grechishkin P.V. 2012. *Use of modern technical monitoring means to ensure the bolting design parameters meet the changing underground working conditions.* Coal # 12: 38–40 http://www.ugolinfo.ru/Free/122012.pdf

Sasaoka, T., Shimada, H., Takamoto, H., Hamanaka, A., Matsui, K. & Oya, J. 2013. *Applicability of rock bolting system and ground control management under weak strata in Indonesia.* Coal International. Vol. 261 Issue 6: 32–37.

Shashenko A.N., Sdvizhkova Ye.A. & Gapeyev S.N. 2008. *Deformability and strength of rock massifs.* Dnipropetrovs'k: NMU.

Tadolini, S. & McDonnell, J. 2010. *Cable bolts – an effective primary support system.* Proceedings of the 29th International Conference on Ground Control in Mining, Morgantown, WV. http://icgcm.conferenceacademy.com/papers/detail.aspx?subdomain=icgcm&iid=314

Voronin, S.A., Efremov, A.V., Panchenko, V.V., Khalymendyk, Yu.M., Bruy, A.V. & Baryshnikov, A.S. 2013. *Usage of cablebolts in conditions of soft rocks.* Coal of Ukraine #6: 24–26.

Zborshchik, M.P. 2011. *Reusing of gateroads is the urgent task of coal mines.* Coal of Ukraine #1: 17–21.

Zharov A.I. 2007. *Objective laws of geomechanical processes at pillarless mine flowsheets.* Second edition. Moscow: MSMU.

Zhdankin, N.A. & Zhdankin, A.A. 1990. *Geomechanics of mine working: intersection longwall – gateroad.* Novosybirs'k: Nauka.

Plasma processing of steam coal and mine methane in controlled atmosphere for metallurgical production

L. Kholyavchenko & S. Davydov
M.S. Polyakov Institute of Geotechnical Mechanics, Dnipropetrovs'k, Ukraine

S. Oparin
Ukrainian State University of Chemical Technology, Dnipropetrovs'k, Ukraine

A. Maksakova
NGO "Doniks", Donets'k, Ukraine

ABSTRACT: It is submitted the usage efficiency of plasma-chemical processes for energy conversion mine methane-air mixtures and steam coal. It is shown that adjustments of output product can pro-usual coal-water mixtures. In this case, the whole range of safety and reducing atmospheres (final product) is generated by a plasma-chemical reactor when changing modes of its operation and composition of mixture.

1 INTRODUCTION

Continuously rising prices for oil and natural gas, inevitably reducing their inventories, serious environmental requirements for ambient environment negatively affect the competitiveness of the economy of many countries, including Ukraine, prompting experts and scientists from many countries to seek new and alternative sources their replacement. Intensively developed research and development aimed at exploiting and deeper processing of low-grade, steam coal, mine methane, restored natural resources, waste products. This is especially significant for the direction of Ukraine, where the volume of such materials are not particularly limited, and in the next 150–200 years may provide a solution to energy problems. Particular attention in a number of these problems should be paid to the production of controlled (protective and restorative) atmospheres in metallurgy background continuously declining production volumes and increased cost of coking coal and natural gas. Designed to prevent contact of the metal with oxygen environment during its thermal processing, they are chosen depending on the requirements for the metal, the temperature, duration and nature of the thermal effects, and cost of gas generation units for it (Estrin 1973). Range of gases such large volumes of its production. So, in modern metallurgical industry is continually evolving and improving technology of the twentieth century – no coke receipt of metallurgical raw materials (sponge iron), as an alternative to the production domain (Borodin 1998). The technology is based on the direct reduction of iron in a reducing gas medium, which is a high energy synthesis gas ($CO + H_2$) is traditionally produced carbon conversion of natural gas by gasification (Estrin 1973, Heat treatment... 1989).

However, a significant increase in natural gas prices leads to an increase in the cost of metal to reduce the growth and spread of the latest advanced technology for steel making. Retention rates for its distribution requires more available and cheap raw materials efficient and modern production methods of reducing gas. These can be of varying degrees of metamorphism of coal, mine methane and methane-air mixtures, renewable energy sources, waste. There are many schemes and designs of gas generators for production of controlled atmospheres. Their design principles and based on the controlled partial combustion reaction ($\alpha = 0.27$–0.3) of hydrocarbons in the presence of catalysts (plant- MIMI-SHZ) (Heat treatment... 1989). CARBOCAT generator, based on the method of producing a controlled atmosphere, wherein the gas-air mixture ($\alpha = 0.3$–0.4) fed to the working space, further mixing it with hydrocarbon additives (10–15%). It is known by many other generators series ENG (Heat treatment... 1989, Low-temperature... 1992).

However, the above installation apparatus and methods for producing a controlled atmosphere in itself bulky metal-costly, requiring purification of natural gas from the medium, wherein the catalysts used. This is further cumbersome expensive equipment. Moreover, the production of controlled atmospheres such ways and means, often limited range of the gases through strongly fixed temperature gasification processes based on direct combustion of carbon-containing media. Synthesis gas, including any quality-controlled atmosphere can be obtained from other carbon media by allothermic gasification. For example, varying degrees of metamorphism of coal, peat, coal mine methane, municipal solid waste are.

By this time in Ukraine, as in other countries, there is a problem processing CMM due to lack of optimal utilization of its technology, which is either vented to the atmosphere or burned inefficiently torching method, increasing environmental pollution. Produced from coal seams in the process of degassing and mining deposits, coalbed methane has a strong variability of the composition and, more often, lower its concentration in methane-air mixture. Stocks of coal mine methane in Ukraine are rather high 3.0–3.5 trillion. cu. m, which is much higher than natural gas reserves in existing fields of Ukraine. That is, the volume of such raw materials in Ukraine is significant and rational utilization of them can provide a solution to the problems of energy and raw materials for many years (Bulat et al. 2010).

2 OBJECTIVE

There is a need of an efficient way to process them, based on co-building an independent, high-controlled areas with large local energy densities. These properties are plasma torches, plasma chemical reactors. With their help, you can create an extreme degree of no equilibrium systems, which clearly appeared to catalytic effects of coal gasification with them, which would produce high-energy environmentally friendly, including restorative and protective gases, with a wide range of control of their properties (Bulat et al. 2010, Patent Ukraine #58359). Plasma processing of coals of varying degrees of metamorphism, methane-air mixtures of mine, various combinations of these carbon-containing media, does not require additional measures for their purification from sulfur and other contaminants that at high temperature become stable neutral compounds. Allothermic plasma method is selective with respect to the final product. Its composition is mainly dependent on the nature of the oxidizing medium. The protective atmosphere may be generated by using air as the oxidant. As part of the final product will prevail neutral nitrogen. Restoration with using steam in addition to the synthesis gas are absent or inclusion. The high intensity of reaction processes in the plasma flow, due to a temperature of 2000 to 6000 K, which provides an implementation of waste-free, eco-friendly technologies for complex processing any carbon media in a different combination. Plasma torches and plasma chemical reactors are compact construction, low metal content, high efficiency (70–90%) and in the flow capacity.

3 RESULTS

To assess the possibility of using plasma technology to create the means of production of controlled atmospheres, we performed studies of gasification processes of methane and methane-air mixtures thereof with energy coals with different ratios of incoming components and oxidizer. In the study of high temperature processes used universal software "Astra 4", based on the fundamental laws of thermodynamics. The studies used the equilibrium composition of methane-air mixtures at ratios (CH_4/air) 1/4; 2/3; 3/7 and atmospheric pressure, as well as mixtures thereof with a coal and water. Found that all the above methane-air mixture can be processed into protective or restorative atmosphere plasma method. However, their qualitative characteristics depend on feed composition, nature of the oxidant and the process temperature. Plasma method allows studies in a wide range of temperatures from 1000 to 6000 K, which is governed by the magnitude of the arc current and plasma gas flow rate (in the case of mixture). The research results are shown in Table 1.

Their analysis shows that the methane- air mixture number 2 (CH_4/air = 3/7), used as the plasma , gives the final product is close to balanced. At a temperature of carbon conversion is complete there is the largest amount of protective ($N_2 - 37\%$) and reductive ($CO + H_2 - 60\%$), constituting a controlled atmosphere. It minimum of oxidant (H_2O, O_2) (1%), environmentally problematic (CO_2) components and unoxidized compound N_2S. In this mixture the amount of the oxidant (air) corresponds to the stoichiometric, which ensures complete conversion of the carbon. The dependence on the temperature of the gasification process to mix number two is shown in Figure 1.

Dependence shows that the temperature greatly affects the quality and quantity of the end product, with its optimal value. It has been established that complete conversion to gaseous carbon and carbon-containing solid media occurs at a temperature from 1600 to 2000 K. At temperatures indicated below, the final product may be of carbon residues in the compounds of N_2S, water and CH_4. Increased temperatures above 2000K not affect on composition of the gas phase and leads to higher energy costs.

Table 1. Results of studies of plasma gasification of methane-air mixtures.

# of mix-tures	Ratio of mixture	The ratio of components	The total volume of gas m³/kg of mixture	The gas composition, % (vol.)							
				H_2	CO	H_2+CO	N_2	N_2C	H_2O	CO_2	other
1	$CH_4(20\%)$ + air (80%)	CH_4: air = 20:80 (vol.)	1.048	22.54	14.73	37.27	51.35	0.08	9.83	1.39	0.08
1-2	Mixture 1 + coal $AS1$ 10% (mass)	Mixture 1:coal = 100:10 (mass)	1.082	29.49	23.49	52.98	44.24	1.20	1.15	0.20	0.23
1-3	Mixture 1 + coal $AS1$ 50% + water (stoichiometry from C of coal)	Mixture 1: coal: H_2O = 100:50 Mixture 145 (mass)	1.212	39.33	33.64	72.97	22.83	0.49	2.78	0.52	0.41
2	CH_4 (30%) + air (70%)	CH_4: air = 30:70 (vol.)	1.282	40.75	19.41	60.16	37.30	1.38	0.97	0.10	0.09
3	CH_4 (40%) + air (60%)	CH_4: air = 40:60 (vol.)	1.438	52.17	16.47	68.64	21.46	9.68	0.09	0.01	0.12
3-1	Mixture 3 + water 10% (mass)	Mixture 3:water = 100:10 (mass)	1.512	52.04	20.00	72.04	24.18	3.30	0.34	0	0.02
3-2	Mixture 3 + coal $AS1$ 50% + water (stoichiometry from C of coal)	Mixture 3:coal $AS1$: H_2O = 100:50:45 (mass)	1.462	51.26	32.24	83.50	13.92	0.93	1.11	0.15	0.39
4	CH_4 + air (stoichiometry)	CH_4: air = 100:429.18 (mass) or 100:332.7 (vol.)	1.275	40.23	19.40	59.63	37.84	1.19	1.12	0.12	0.1
5	CH_4 + oxygen (stoichiometry)	CH_4: oxygen = 100:100 (mass) or 100:69.93 (vol.)	2.096	66.47	33.30	99.77	0	0	0.08	0.01	0.14
6	CH_4 + water (stoichiometry)	CH_4: water = 100:112.5 (mass)	2.59	73.74	24.46	98.2	0	0	1.55	0	0.25
7	Steam coal mine them. "Heroev Kosmosa" + water (stoichiometry)	Coal HK: water = 100:85.4 (mass)	1.55	55.97	43.09	99.06	0.13	0.19	0.16	0.03	0.43

Decline as an increase in CH_4 (a mixture of number 1 and 3 mixture, respectively) leads to disruption of oxidative balance and requires an adjustment to certain activities. Thus, in the plasma processing lean number 1 wherein CH_4/air = 1/4 (Table 1). In the final product is an excess of oxidant in the form of water (9.63%) and more than 1% CO_2. Protective nitrogen product more than 51%, and reducing the synthesis gas to 38%. Additive methane – air mixture 10% coal dust, fine adjusts the final product. It increases the regenerative component to 53% protective reduced to 44% and the harmful impurities (CO_2 and H_2O) are not more than 1.5%. Strict selection of coal in a mixture, it is possible to obtain a final product without any harmful components or adjust the desired composition of the protective and restorative components. Thus, an increase in the coal mixture (mixture # 1–3), and an oxidizer, as steam, controlled atmospheres can be significantly enriched (up to 73%) reduction component.

Protective component (N_2) thus drops to 23%. And as was noted above, strict selection of the original components can achieve minimum or zero turn harmful inclusions.

In mixtures where $CH_4 > 30\%$, there is a lack of oxidative environment, which manifests an increase in the composition of the compound N_2S final product. Corrections it can be produced by adding a mixture of plasma-forming oxidant. The best in this case, it is water vapor. In a mixture of CH_4 with > 30% is prevailing reducing environment ($CO + H_2$).

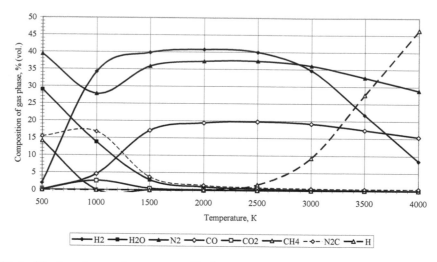

Figure 1. Nature of the dependence on temperature gasification process.

Increase of methane in the mixture up to 40% (mixture number 3, Table 1) leads to increasing in reducing component of controlled atmosphere, where it reaches 69%. Composition of the protective oxide is reduced to 21%. In the composition of the final product is observed up to 10% of unoxidized N_2S compound that indicates a lack of oxidant in the mixture. If the outcome of the mixture add 10% water (mixture # 3–2), then the replacement component controlled atmosphere will rise to 72%, while protecting only 3–4% due to the oxidation of the compound N_2S.

Contents of components of plasma processing with different concentrations CMM graphically represented in Figure 2. Mixture with the content of CH_4 and 30%, the processed products are in excess of oxidative environment in the form of H_2O, and require adjustment by the addition of carbon-containing components. In these mixtures, the predominant component is protective (N_2) controlled atmosphere.

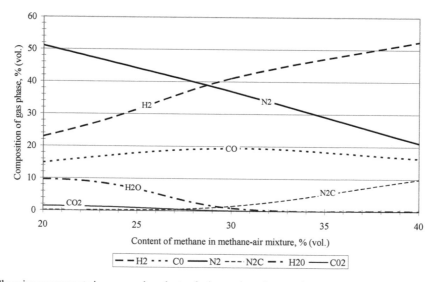

Figure 2. Changing components in processed products of mine methane from methane content in the methane-air mixture.

Gasification of pure methane (natural gas) in the air oxidizing environment (mixture number 4) is virtually identical qualitative and quantitative indicators from the gasification of methane-air mixture (mixture number 2). Obviously it is preferable to use passing mine methane, which is cheaper than natural gas.

Net reducing environment without any inclusions can be prepared for plasma processing of methane in oxygen (mixture number 5) or steam (a mixture of number 6) environment. In these cases, a reducing environment is presented by 98–99% synthesis gas. However, steam reforming of methane is preferable in terms of quantity of the final product. Thus, when the same amount of methane in the feed mixture the total volume of gas at steam conversion 23–24% more of the hydrogen due to degraded moisture oxidant.

Reducing environment without any additional inclusions can get steam-plasma gasification and steam coal (a mixture of number 7, Table 1). Controlled atmosphere in this case is represented by (98–99%) high-energy reducing synthesis gas ($CO + H_2$). Volumetric output it to 1 kg of the mixture was 1.55 m^3, and 1 kg of dry coal 2.5–3 m^3, depending on the concentration of carbon in the coal. In the gas phase of the final product is almost no environmentally problematic components.

It should be noted feature of the plasma energy converters carbonaceous environments. With high local energy density (103–104 W/cm^2), adjustable temperature from 2000 to 6000 K, they have a low metal content, small size and low capital costs, respectively. They are universal to the composition and quality of the feedstock and the selectivity of generating the final product, easily be automated in a production process, as carbonaceous energy conversion takes place in the continuous media stream. High and adjustable over a wide range (2000–6000) to the temperature of the process of energy transformations, clean oxidizing environments (water, oxygen mixtures thereof), allow to get the product to the appropriate environmental indicators, and in some respects exceeds the requirements of the European standard.

4 CONCLUSIONS

1. Continuous rise in prices for oil and natural gas production technology dictates necessity of perfection controlled atmospheres and synthesis gas from cheaper and more accessible, renewable carbon-containing media – steam coal, coal mine methane-air mixtures of household and industrial waste.

2. Steam-plasma gasification of carbonaceous environments, including mine methane-air mixtures, the adjustment of their coal and water make it possible to obtain all the necessary range of controlled atmospheres, satisfying the requirements of metal heat treatment including direct reduction of iron.

3. Found that the temperature of complete conversion of the carbon of 1700–2000 K. In this mode, there is a maximum energy conversion yield of the end product, and it includes no environmentally problematic components.

4. Plasma reactors easily adaptable to a wide range of carbon-containing media are selective in the final product of gasification, a little metal-easy handling and automatable in operational processes.

REFERENCES

Estrin, B. 1973. *Production and use of controlled atmospheres.* Moscow: Metallurgy: 392.

Borodin, V.I. 1998. *Low-temperature plasma processes for obtaining metals.* Moscow: Metallurgy: 273.

Heat treatment in mechanical engineering. 1989. Reference /ed. Y. Lakhtina / Moscow: Mechanical Engineering: 783.

Low-temperature plasma. 1992. Plasma metallurgy. Vol. 8. Novosibirsk: Nauka: 265.

Gasification of solid fuels. 1988. Chemical encyclopedia. Vol. 1. Moscow: 878.

Bulat, A., Alymov, B., Holyavchenko, L. & Davydov, S. 2010. *Complex processing of coal-water fuel plasma gasification.* Geotechnical Mechanics: Interagency. Sat Nauchn. Tr. IGTM NAS. MY. 81: 20–27.

Patent Ukraine #58359. S23S8/00, S21D 1/74. *Method of manufacturing a controlled atmosphere with natural gas heat treatment* / Maksakov A., Alimov V., Davydov S. (Electronic resource).

Progressive Technologies of Coal, Coalbed Methane, and Ores Mining – Bondarenko, Kovalevs'ka & Ganushevych (eds)
© 2014 Taylor & Francis Group, London, ISBN: 978-1-138-02699-5

The study of ecological state of waste disposal areas of coal mining enterprises

A. Gorova, A. Pavlychenko & S. Kulyna
National Mining University, Dnipropetrovs'k, Ukraine

O. Shkremetko
St. Petersburg Energy Institute, St. Petersburg, Russia

ABSTRACT: The features of the impact of the waste dumps of coal mines on the ecological state of the environmental objects have been analysed. The bioindicative assessment of the state of atmospheric air and soil on the waste disposal areas of coal mining enterprises have been carried out. A set of measures aimed at reducing environmental hazards of the wastes of mining enterprises have been proposed.

1 INTRODUCTION

The actual ecological situation in the mining regions of Ukraine shows that there are a growing environmental hazards and extreme anthropogenic overload. The coal industry in Ukraine belongs to the major environmental polluters. At the same time a significant impact on the environment is made not only directly in the process of coal mining, but for many years after its completion. One of the significant sources of contamination of the environment is the waste dumps occupying large areas of fertile lands. About 40 million m³ of dump rocks are stored in dumps every year. The majority of coal mining wastes potentially has toxic and mutagenic properties as contains a significant amount of heavy metals that are practically not biodegradable in the environment, and is therefore especially dangerous for the living organisms (Gorova 2013).

The wastes of coal production are in most cases large man-made sources of a stable negative impact on the environmental objects. At all stages of the waste dumps existence – from the rock mass winding to the extinction of internal and external physicochemical, mineral forming, biological and other processes – there is a gradual heating of rock masses, oxidation of volatile compounds, leaching of active components, acid inflow of a newly formed solutions, air and water erosion of the dump slopes, etc. In the study of internal and external features typical for rock masses there are problems connected with complex transformations in the dumps, because most reactions take place simultaneously, reducing or increasing the chemical activity of individual components (Krupskaya 2013 & Derbentseva 2013).

The issues concerning the peculiarities of migration of contaminants, depending on the dumps condition, namely the stages of internal and external physical and chemical transformations, natural or artificial intervention in the processes, violation of integrity of the body and the slopes of the previously created dump, etc. remain unresolved (Pavlychenko 2013).

The waste dumps of coal mines are a significant threat to natural and urban landscapes and health of population, which requires the development of efficient measures for their recording and ecological recovery, and choosing right directions of their further use (Kolesnik 2012). In addition, the waste dump areas are manmade deposits containing valuable components – coal (up to 30%), alumina (up to 15%), oxides of silicon and iron (up to 20%), rare-earth metals (up to 260 g/t) (Broyde 2004).

Chervonohrad Mining Region (ChMR) is one of the largest coal basins of Western Ukraine, where the hard coal has been extracted by underground mining for more than 50 years. On a relatively small area of 180 sq. km the twelve coal mines and Central Concentrating Mill (CCM) are located, and 266 ha of area are designated for waste heaps. About 80% of technogenic load falls on an area of 30 sq. km, where seven coal mines are located and most of the region's population resides. Such situation led to a local transformation of natural landscapes and their components, formation of a manmade and natural and technogenic landforms and withdrawal of vast areas of fertile lands for disposal of wastes of mining enterprises. In addition, the ecological situation in the region, as in general in Ukraine, is complicated by socio-economic problems (decline in the profitability of coal mining, mines closure, unemployment, economic crisis and social tensions), all of this lead to a high relevance of studies aimed at determination of the impact of wastes on the environmental objects and health of population of mining regions.

The control of the environmental quality as a rule is performed with the application of physical and chemical researches, which allow determination of contaminant concentration, but not always allow an efficient determination of their impact on the living organisms. That is why it is recommended to use not only physical and chemical methods for monitoring of the environment, but also biological methods which should complement them in an integral way. The application of a set of methods including highly sensitive bioindication methods allows taking into account the overall impact of the waste components on different biosystems and managing the environmental risks on the mining enterprises (Gorova 2013).

The solving of the problem of waste management of mining enterprises requires the development of scientific principles for the improvement of environmental safety of waste disposal areas, which will allow the assessment of actual environmental hazard of manmade objects and waste components, prediction of changes in the environment and social and ecological systems, and suggesting the ways of ecologically safe operation of mining enterprises.

The results of comprehensive studies will allow determination of the levels of environmental hazard of the waste dumps and prevention of contamination of soils, surface and underground waters, and atmospheric air on the waste disposal areas. The use of natural materials will reduce the intensity of migration of contaminants from the waste dumps and improve the ecological state of the environment in the coal mining regions (Pavlichenko 2013).

2 FORMULATING THE PROBLEM

The storage of coal mining wastes leads to the deterioration of ecological state of environmental ob-

jects and causes environmental risks. The situation is worsened by the lack of clear and reasonable criteria for the assessment of the environmental risks' levels. In order to solve the problem of improvement of environmental safety at waste dumping, the bioindicative method is proposed to be applied for the assessment of the waste toxicity level. Thus, the purpose of this study is a comprehensive assessment of the impact of waste dumps of mining enterprises on the ecological state of the environment of ChMR.

3 MATERIALS UNDER ANALYSIS

Most solid industrial wastes in ChMR are created as a result of operation of coal mines and Central Concentrating Mill (CCM). As a result of many years of coal mining in the region, there are 15 waste dumps on the area of 266 ha, 10 of which are operated. From 1.18 to 47.6 t/year of suspended solid particles get into the atmospheric air on the territories adjacent to the waste dumps. The deformation processes occurring on the waste heaps of mines are characterized by formation of gullies, the width and depth of which are ranging from 2–4 m and from 1–3.5 m, respectively, and in which the rock is moving for a distance reaching sometimes 6 m. The composition of rocks is represented by the mudstones, siltstones, coal duns, and hard coal. The rocks consist of the granular-fragmental components – red, grey and black rock fragments of 150–200 mm. In most cases the waste dumps consist of mudstones containing sorbed heavy metals: *Li, V, B, P, Zn, Pb, Bi, Co* (Table 1) and a significant amount of pyrite doped with arsenic and mercury. Under the influence of the weathering factors pyrite oxidizes and heavy metals are converted into soluble forms. In addition, the sulphuric acid is formed under the heaps, which contributes to thermochemical leaching of rocks and significant contamination of soils and underground waters.

Table 1. Content of gross forms of heavy metals, mg/kg of ash (Baranov 2008).

Type of rock	Zn	Cd	Ni	Co	Pb	Cu	Cr	Fe
Burned rock (red)	15.4±0.3	1.8 ±0.4	53.2±0.2	19.37±0.7	8.3±0.5	3.4±0.3	3.1±0.1	28.7±1.2
Unburned rock (black)	4.9 ±0.2	3.3±0.1	24.9±0.5	57.17 ±1.6	7.0 ±0.3	2.9 ±0.1	2.0±0.3	33.9±0.8

The level of radioactive contamination of the environment of ChMR exceeds the level of geochemical background by 3–10 times. The radioecological studies have revealed that the level of exposure dose of mining objects of the region exceeds the background level by 1.2–1.6 times. At the same time the

average parameter of radioactive contamination by Cs^{137} is higher than a background parameter by 1.3–1.7 times, and by Sr^{90} – by 2.0–5.0 times (Toxic-hygienic... 1992).

The spontaneous ignition and self heating of mine rocks is observed in the region, which is confirmed by the results of temperature surveys. On the surface of the heap of Chervonohrads'ka Mine the temperature of 800 °C was recorded on February 2010. On an inactive heap of Vizeis'ka Mine the areas with the temperature of 1280 °C were revealed, and on the heap of Vidrodzhennia Mine – of 1460 °C.

General characteristics of the waste dumps are shown in Table 2.

Table 2. Characteristics of the waste dumps of Chervonohrag Mining Region.

Mine	Form	Height, m	Deformations	Foundation area, thousand sq. m.	Annual volume, thousand m^3	Displacement angel	Rocks volume, thousand m^3
Velykomostivs'ka	herringbone	8.9	erosive scouring of slopes	1.4	–	30–60	3.6
	conical	24		57.8	–	40–42	395.5
	flat	7.8		21.5	21.5	28–36	81.8
Mezhyrichans'ka	flat	8.1	absent	36.7	–	30–35	296.8
	flat	10–12	absent	282	6.7	32–35	2210
Vidrodzhennia	conical	41.1	absent	63	–	50–80	1086.2
	flat	30.7	absent	152	40	29–32	2702.8
Lisova	conical	35.1	absent	46	–	30	834
	conical	34.1	absent	4.5	–	30	609
	flat	15.3	absent	5.1	47	35	732
Zarichna	flat	41	absent	148.3	20	29	3286
Stepova	flat	28.2	absent	165.4	52.3	37	3090
Bendiuzka (closed)	conical	32	absent	44	–	–	519
Chervonohrads'ka	flat	35	absent	23.3	50	37	2589
Vizeis'ka (closed)	flat	45	erosive scouring of slopes	99.9	–	37	2480.04
CCM	flat	68	erosive scouring of slopes	75	–	–	13000

The analysis of the state of the waste dumps in the region indicates a low efficiency and incompleteness of remediation. The works for preventing negative impact of waste heaps in the region have been carried out beginning from the 60-th years of XX century, but have been mainly directed towards the implementation of a set of measures preventing spontaneous ignition of the heaps, and remediation on the areas of land subsidence. The biological recultivation on the waste dumps is not carried out at all. Such situation is primarily connected with the insufficient financing of remediation and therefore the waste heaps are overgrowing. However, if the waste dump has a steep slant, there is almost no vegetation on it.

Taking into consideration the comprehensive negative impact of coal mining wastes on the environmental components the toxicity levels of atmospheric air and soils have been examined in the area of impact of the waste dumps.

The examinations have been carried out on the territory of ChMR, where the monitoring sites located at different distances from the waste dumps of coal mines were selected. The area of Volytsia Village of Sokal' District of L'viv Region was selected as a reference area as there were no mining enterprises and a minimal impact of other manmade factors was identified.

On the territory of each monitoring point the comprehensive bioindication studies of toxicity levels of atmospheric air and soils have been carried out. The toxicity of atmospheric air was determined by cytogenetic test – "Plant pollen sterility". The results of bioindication studies were reduced to a unified dimensionless system of conditional indices of damageability of biosystems (Guidelines 2007).

For determination of the impact of the waste dumps on the growth processes of hardy-shrub species the method consisting in measurement of biometric parameters of trees growing on the waste heaps was used. For this purpose the height of a tree and the thickness of a tree trunk were measured on the height of 1.3 m. The assessment of intensity of anthropogenic impact on the quality of the environment have been performed with the application of morphometric approach, which is based on the assessment of internal individual variability of morphological structures, namely, the degree of fluctuating asymmetry (FA) (Guidelines 2003).

The type of vegetation on the heaps depends on their location: the herbaceous plants prevail on the heap located on the area of agricultural lands, and forest plants prevail on the heaps located on the area of forests, respectively. The overgrowing of the waste dumps, as a rule, begins from the foot, where the growth processes are more or less even, while in the upper part of the dumps the growth is notably reduced. The vegetation of the waste dumps is represented by a number of grass, shrub and tree species, which take active part in the overgrowing processes. It should be noted that there is almost no vegetation on the heaps of CCM. The most popular of the total number of hardy-shrub species on the waste dumps are *Betula pendula* Roth, *Populus tremula* L., *Pinus sylvestris* L., *Robinia pseudoacacia* L., *Salix caprea* L.

Characteristics of spreading of hardy-shrub species and results of measuring of their morphometric parameters are shown in Table 3.

Table 3. Characteristics of hardy-shrub species of the waste dumps of Chervonohrad Mining Region.

Hardy-shrub species	Trunk diameter, m min-max average	Height, m min-max average	Conditions of growth on the waste dump
Betula pendula Roth	0.12–0.60 0.31	1.5-25 8	Mainly on the areas of deposits of burned rocks
Populus tremula L.	0.3–0.7 0.41	2.2-2.0 14.2	On the foot of the heaps
Pinus sylvestris L.	0.12–1.35 0.36	0.5-11 5	Foots and gentle slopes of the heaps
Robinia pseudoacacia L.	0.3–0.7 0.51	1.2-9 4.6	On the foot of the heaps
Salix caprea L.	0.42–1.05 0.67	1.3-4.5 2.8	On the foot of the heaps and on steep slopes

The analysis of data of Table 3 has revealed that the best growth processes of plants are observed on the foot of the waste dumps, while on the upper part of the heaps considerable blowing erosion processes occur.

The results of the assessment of FA indices of weeping birch (*Betula pendula* Roth) are shown in Table 4.

The analysis of data of Table 4 has revealed the negative impact of the waste dumps on the state and development of the leaves of weeping birch in comparison with the reference indices. FA indice (the average values) for the plants growing on the heap 2.3 times exceeded the reference values. According to the asymmetry indices the lowest one (I point) is observed on the reference area, where the quality of the environment is assessed as meeting the "conditional norm". As to the FA indices for certain parts of the waste heap, the numerical values were different, but despite this all of them have got the highest point (V), which has identified the quality of the environment in the area of the heap as "significant deviation from the norm".

The results of cytogenetic assessment of the state of air basin in the area of impact of the waste dumps are shown in Table 5.

Table 4. FA indices for weeping birch on the waste dumps of Velykomostivs'ka Mine.

Sampling area	FA indices	
	n	$X \pm m$
Foot of the heap (northern direction)	10	0.076±0.008
Foot of the heap (southern direction)	10	0.059±0.006
Foot of the heap (eastern direction)	10	0.064±0.007
Foot of the heap (western direction)	10	0.056±0.008
Slope of the heap (northern direction)	10	0.085±0.012
Slope of the heap (southern direction)	10	0.089±0.008
Slope of the heap (eastern direction)	10	0.075±0.012
Slope of the heap (western direction)	10	0.061±0.006
Upper part of the heap – plateau	10	0.086±0.007
Upper part of the heap – plateau	10	0.089±0.008
Average value for the heap		0.074±0.004
Average value for Volytsia Village (reference value)		0.031±0.002

Note: n – number of processed samples, X – value of asymmetry in the sample, m – non-sampling error.

Table 5. Bioindication assessment of the state of atmospheric air in the area of impact of the waste dump of Velyko-mostivs'ka Mine.

SN	Characteristics of monitoring point of the heap	Phytometer	PFD	Degree of biosystem damage[1]	Biosystem state[2]	Category of environmental safety of atmosphere[3]
1	Eastern part of slope	*Betula pendula* Roth	0.468	A	Th	MD
2	Western part of slope	*Betula pendula* Roth	0.435	A	Th	MD
3	Northern part of slope	*Betula pendula* Roth	0.450	A	Th	MD
4	North-West part of foot	*Matricaria chamomilla* L.	0.507	AA	Cr	D
5	North-West part of foot	*Achillea millefolium* L.	0.466	A	Th	MD
7	Western part of foot	*Robinia pseudoacacia* L.	0.262	A	C	MD
IPFD for the waste dump 0.431±0.032				A	Th	MD
reference indices of Volytsia Village 0.128±0.022				L	R	S

Notes: [1]degree of biosystem damage: L – low, A – average; AA – above average; [2]biosystem state: R – reference, C – conflict, Th – threatening, Cr – critical; [3]category of environmental safety of atmosphere: S – safe, MD – moderately dangerous; D – dangerous.

The analysis of average values of provisional figures of damage (*PFD*) of plant pollen cells allowed the assessment of the state of atmospheric air in the area of impact of the waste dumps of the mine by general toxicity as "moderately dangerous", and the degree of bioindicator damage as "average". The comparison of the results of assessment of pollen sterility of plants growing on the waste heaps with the results from the reference area allowed making the conclusion that the degree of pollen sterility for bioindicators selected from the territory free from the impact of mining enterprises and waste dumps is 2–15 times lower.

The performed studies have shown that the ecological situation for the reference area was determined as "reference", and for the waste dumps region as "moderately dangerous" as the numerical values of integral provisional figures of damage (*IPFD*) accounted for 0.128 and 0.431 standard units respectively. As the reference value of *IPFD* is 0.3 standard units, the 43% deviation from the norm is observed on the area of impact of the waste dump.

Thus, as a result of studies it was determined that the waste dumps are one of the main sources of contamination of environment in ChMR. The ecological situation on the waste disposal areas of coal mining enterprises is assessed as "moderately dangerous".

The basic scheme shown on Figure 1 is proposed for studying the state of the waste dumps and choosing the further directions of their management.

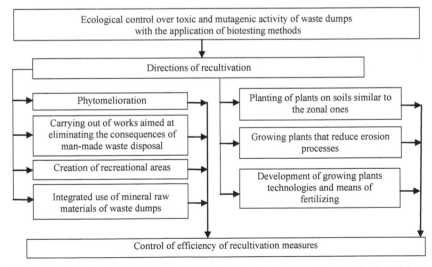

Figure 1. The scheme of application of biotesting methods for choosing the directions of recultivation of the waste dumps.

The performed studies allow asserting that the recultivation works on the waste heaps are carried out not in full; there is no stage of biological recultivation. The waste heaps are overgrowing. In addition, for prevention of negative impact of the rock forming processes on the environment of Chervonohrad Mining Region it is recommended to:

– implement modern technologies of technological process of coal mining, which allow partial or full filling of mine workings with the rocks;

– use the solid wastes of coal mining and concentrating for building of roads, artificial protective structures (dams), producing of building materials, etc.;

– conduct a detailed chemical analysis of the rocks that will be used for agricultural recultivation and other purposes;

– at closing of mines the re-vegetation including landscaping, afforestation and planting of perennial grasses should be compulsory, as this will stop deformation processes on the waste heaps and further contamination of the environment by toxic components of the waste dumps. The biological recultivation will enhance the biological productivity of disturbed lands and surrounding areas and will provide the improved sanitary conditions in the areas of impact of the waste heaps.

– use the biotesting methods for determination of directions of recultivation and respective express assessment of the efficiency of implemented measures.

The wastes of mining enterprises are the source of negative impact on the state of environmental objects, so there is a need for further implementation of bioindication methods for comprehensive assessment of the levels of their environmental hazard.

For the improvement of ecological state of areas adjacent to the waste dumps a set of environmental measures aimed at reducing of migration activity of the contaminants to the environmental components shall be implemented. The use of wastes of mining enterprises for filling of the mine workings will reduce the deformation of rock massif and land surface and respectively prevent the processes of land subsidence and underflooding of settlements and agricultural lands.

REFERENCES

Gorova, A., Pavlychenko, A. & Borysovs'ka, O. 2013. *The study of ecological state of waste disposal areas of energy and mining companies*, Annual Scientific-Technical Collection – Mining of Mineral Deposits, Leiden, The Netherlands : CRC Press / Balkema: 169–171.

Krupskaya, L.T., Zvereva, V.P. & Leonenko, A.V. 2013. Impact of technogenic systems on the environment and human health in the priamurye and primorye territories. Contemporary Problems of Ecology (March 2013), Volume 6, Issue 2: 223–227.

Derbentseva, A., Krupskaya, L. & Nazarkina, A. 2013. *Analysis and Assessment of the Environment in the Area of Abandoned Coal Mines in Primorsky Region.* Mechanics and Materials Vols. 260–261: 872–875.

Pavlychenko, A. & Kovalenko, A. 2013. *The investigation of rock dumps influence to the levels of heavy metals contamination of soil.* Annual Scientific-Technical Colletion – Mining of Mineral Deposits. Leiden, The Netherlands: CRC Press / Balkema: 237–238.

Broyde, Z.S., Makarov, E.A. & Broyde, H.Z. 2004. *Identification, registration, classification and certification in the field of waste management from their genesis to the certification of man-made deposits.* Scientific Papers of Donets'k National Technical University. Series of. Mining and Geology, 81: 44–48.

Kolesnik, V.Ye., Fedotov, V.V. & Buchavy, Yu.V. 2012. *Generalized algorithm of diversification of waste rock dump handling technologies in coal mines.* Scientific Bulletin of National Mining University, 4: 138–142.

Pavlichenko, A.V. & Kroik, A.A. 2013. *Geochemical assessment of the role of aeration zone rocks in pollution of ground waters by heavy metals.* Scientific Bulletin of National Mining University, 5: 93–99.

Baranov, V. 2008. *Ecological description of the waste dump of coal mines of CCM of CJSC "Lvivsystemenergo" as an object for landscaping.* Bulletin of L'viv University. Biological Issue, 46: 172–178.

Toxic-hygienic characteristics of the rocks of the waste heap of the mine #8 "Velykomostivska". 1992. LoD-NGMI, Ukrzapadugol: 8.

Guidelines 2.2.12-141-2007. Survey and Zoning by the Degree of Influence of Anthropogenic Factors on the State of Environmental Objects with the Application of Integrated Cytogenetic Assessment Methods / Gorova A.I., Ryzhenko S.A., Skvortsova T.V. Kyiv: "Polimed": 35.

Guidelines for the Assessment of the Environment Quality by the State of Living Beings (Estimation of Stability of Living Organisms in terms of the Asymmetry of Morphological Structures). 2003. Approved by the Resolution of RosEcology, 460-p: 25. Moscow.

Progressive Technologies of Coal, Coalbed Methane, and Ores Mining – Bondarenko, Kovalevs'ka & Ganushevych (eds)
© 2014 Taylor & Francis Group, London, ISBN: 978-1-138-02699-5

Experimental investigation of digging the organic-mineral sediments of the Black Sea

C. Drebenstedt
Technical University "Mining Academy Freiberg", Freiberg, Germany

V. Franchuk & T. Shepel
National Mining University, Dnipropetrovs'k, Ukraine

ABSTRACT: The results of experimental investigation of digging the deep-sea organic-mineral sediments with a bucket working tool are given. Data about physic-mechanical and rheological properties of sapropel and coccolith sediments, sampled in the Black Seaare given. The influence of cutting velocity, digging depth, bucket width and inclination on the bucket filling parameters is established. The diagrams show dependences of geometrical parameters of ground bulk into the bucket and the digging force from the digging process parameters. Thenotions of the ultimate bucket filling condition and the ultimate bucket filling parameters are introduced. Specific features of digging the coccolith and sapropel sediments in water are founded.

1 INTRODUCTION

Minerals of the Black Sea have a great potential for economical development of countries in the Black Sea region. Except polymetallic nodules, total stocks of which are about 5.6 million tons, gas-hydrates – 25 trillion m^3, placers of non-ferrous and precious metals, there were explored more than 320 billion m^3 of sapropel sediments in the abyssal see (Shnjukov 2004). According to the results of carried out investigations in Ukraine and Bulgaria, sapropeltogether with covering sediments – coccolith and diatom oozes – may represent a significant agronomic, industrial and ecological importance (Dimitrov2010, Dimitrov 1999, Degodjuk 2000).

Sapropel, coccolith and diatom sediments of the Black Sea belong to deep-water organic-mineral silts (DWOMS); their total thickness is 0.35–2.0 m, and they occur at the depth of 400-2,200 m from water surface (OMGOR 2010). In Ukraine investigation of DWOMS were being done from 1992 till 2.000 on the national programs for marine research. During that period, geological investigations, and tests on the application DWOMS in agriculture were conducted. However, that works were not finished because of insufficient level of financial support.

Beginning in 2012, within the scientific and technical cooperation between the Freiberg mining Academy, National mining University and the Department of Marine Geology and Sedimentary Mineralization of the National Academy of Sciences of Ukraine, it was performed a series of experiments to study the possibility the technical realization of the project on DWOMS deposits development. Studies have shown that the fine structure of the sediments does not allow applying effectively the known methods of dewatering. It limits the possibility of using the hydraulic mining methods. Therefore, with the current processing technologies the most promising method of mining is mechanical. It may be used bucket working tools, for example, as part of the cable-bucket mining systems(Franchuk 2011, Rybar 2011), which are the most simple from the viewpoint of design, manufacture, operation and maintenance. At designing such systems, it need to identify workloads on the working tools, taking into account physical, mechanical and rheological properties of excavated sediments, and also geological and mining-technical aspects of mining. For a substantiation of the bucket tools parameters and determining its workloads in DWOMS deposits development, on the basis of laboratory equipment of the Freiberg Mining Academy the studies of DWONS digging process have been conducted.

2 ANALYSIS OF RECENT RESEARCH

A lot of scientific works were devoted to determination of loads, which influence on working tools of the underwater excavating machines. To define digging force P_d, as a rule, it is used formula (1) (Patzold 2008) (Figure 1)

$$P_d = P_c + P_f + P_m ,\qquad(1)$$

where P_c – cutting force; P_f – friction force of the working tool on the ground; P_m – force of pushing the ground in front of the bucket, and moving cut soil into the bucket.

Applying to the development of underwater deposits, V.I. Balovnev proposed to account the influence of hydrostatic pressure as follows (Sukach 2004)

$$P_d = \frac{P_R}{P_{01}}\left(P_c + P_f + P_m\right), \qquad (2)$$

where P_R/P_{01} – ratio of digging force under hydrostatic pressure (P_R) to digging force in atmospheric conditions (P_{01}); it may be determined by the results of physical modeling in the pressure chamber.

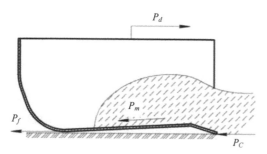

Figure 1. Scheme of influence the forces on the bucket working tool.

D.D. Turgumbaev investigated the process of developing the deep trenches under a layer of mud fluid in the construction of underground structures by the way "wall in the ground". According to the results of his research, the force of digging the ground with a plane dozer blade may be determined by the expression (Turgumbaev 1992)

$$P_d = P_c + P_f + P_R. \qquad (3)$$

Some of the recent studies on cutting the saturated soils with flat blade were held in Kyiv Institute of Construction and Architecture (now it is called Kyiv National University of Construction and Architecture), Kiev, Ukraine. The results of investigation on cutting process in pressure chamber at the pressure from 0 till 3 MPa showed that the force of cutting the fully water-saturated soils does not depend on pressure (Sukach 2004). It was also shown, that tests on cutting the water-saturated soils could be carried out in the atmospheric conditions without creating hydrostatic pressure.

According to the results of conducted by M.K. Sukach investigations, the force of cutting the water-saturated soils with a plane blade rises at cutting velocity from 0.02 till 0.10 m/s (Sukach 2004). The degree of influence the cutting velocity on the force depends on the cutting angle. For soft water-saturated grounds the minimal cutting force corresponds to the angles of 45–65°. Value of cutting force is in direct proportion to the blade width b. Increasing the depth of cutting h also leads to raising the cutting force. Optimal ratio of blade width and cutting depth, at which cutting force takes the minimal value, is in the range $b/h = 2...4$.

V.A. Lobanov (Lobanov 1983), V.G. Moiseenko (Moiseenko 1987), A.A. Karoshkin, S.P. Ogorodnikov, A.I. Koptelov and others scientists proposed different mathematical models for determining the force of cutting the ground under water, which are applied to different types and operation conditions of excavating machines. However, experience of onshore mining shows that for blade and bucket working tools (of dozers, scrappers, excavators) the part of components P_f and P_m in general balance of resistance can be from 30 to 70% depending on the soil strength (Nedorezov 2010). Furthermore, with the reduction of soil strength, share of these components increases. Therefore, at digging the soft water-saturated soils, share of the cutting force in general resistance balance should be much smaller than the force of moving the ground into the bucket.

In (Sukach 2005) it is given the condition, when the process of cutting the plasticity ground with a plant blade drops into the process of moving the ground to the sides. In this case filling of the bucket stops, and on the sides of the trench are formed the ground piles. The condition has the form

$$p_0 \geq 2\tau_0, \qquad (4)$$

where p_0 – pressure of cut ground in front of the bucket on the area ABC (Figure 2); τ_0 – shear strength.

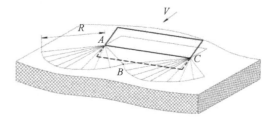

Figure 2. Soil stress state in front of the plane blade.

Size of the plasticity zone R in front of the blade (Figure 2), in which should be satisfied the condition (4), is (Sukach 2005)

$$R = \frac{b}{\sqrt{2}}, \qquad (5)$$

where b – blade width.

Condition (4) shows that if some geometrical parameters of ground, which is pushed in front of the bucket, are achieved, the process of bucket filling will be stopped. Therefore, the maximum volume of soil in the bucket should be limited. Hereinafter the state of bucket filling, at which cutting process drops into the process of moving the ground to the sides, is referred as the ultimate bucket filling condition; parameters that correspond to this state (volume of ground into the bucket, geometrical parameters of ground bulk, digging way, time of filling) are referred as ultimate filling parameters. If volume of the bucket exceeds the ultimate ground volume, efficiency of using the mining equipment will be reduced.

Thus, for choosing the rational geometric parameters of the bucket and determining loads on the working tools for the development of DWOMS, the process of digging the soft water-saturated soils should be studied more in details.

The purpose of this article is to determine the regularities of the process of digging the deep-sea organic-mineral sediments on the basis of laboratory research.

2 THE RESEARCH EQUIPMENT

Experimental investigations were carried out on the basis of three-dimensional cutting test machine in Technidsche Universiotät Bergakademie Freiberg (Freiberg, Germany).

General view of the machine is shown on Figure 3.

For the possibility to investigate the process of digging the water-saturated soils, for the cutting test machine were made supplementary instruments (Figure 4), which include: container for samples, mount for cutting tools and electronic equipment. Dimensions of container for samples are 200×200×800 mm.

Mount for cutting tools consists from: holder, floating chase, which is connected with holder by

elastic plates. Bucket models were fixed on floating chase with the bolt joint.

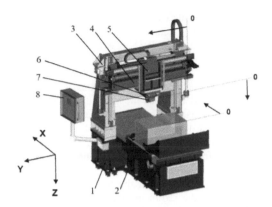

Figure 3. General view of the cutting test machine: 1 – bed-frame; 2 – movable worktable (X-axis); 3 – bridge; 4 – traverse; 5 – instrumental carriage; 6 –force measuring system; 7 – tool holder; 8 – control panel.

Electronic equipment includes force sensor, which is fixed on the holder, and signal data recorder. Way of the force sensor fastening provides measurement of digging force tangential component (the normal component of the digging force, according to (Sukach 2004), is insignificant). The results of measurements are recorded on the memory card with a frequency of 10 Hz. Maximum measurement error of used equipment in the range 0–50 N is not greater than 1 N.

U-shaped bucket models with the width of 0.04, 0.06, 0.08 and 0.10 m were made of stainless steel plate with the thickness of 0.5 mm. Bucket model with the width of 0.07 m was made with transparent side walls for research the bucket filling process with the video tools using.

To prepare the ground for experiments was developed profiling blade, and other auxiliary tools.

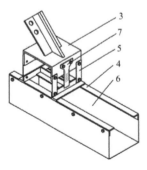

Figure 4. Instruments for digging the DWOMS: 1 – container for samples; 2 – mount for cutting tools; 3 – holder; 4 – floating chase; 5 – elastic plate; 6 – bucket model; 7 – force sensor.

3 CONDITIONS OF CARRYNG OUT THE EXPERIMENTS

The experiments were conducted with the field soil samples using, which were sampled in the Black Sea at the depth of 1935 m with a scrapper dredge (the 73th voyage of RV "Professor Vodyanitskiy", June 2013). The samples of coccolith and sapropel sediments were packed in hermetically sealed containers in order to keep the humidity of the samples. Packing, transportation and storage of samples were carried out in accordance with the requirements (Standard DSTU B V. 2.1-8-2001 2001).

Tools for studying the digging process were fastened on the cutting test machine. Samples preparation was carried out in the following order. After unpacking, the sediments were fitted into the container for samples on the movable worktable. Herewith, the presence of possible foreign solid inclusions was controlled (bone fish, vegetation residues, etc.). The surface of samples into the container was leveled by using the profiling blade. Part of the ground was selected to determine its physical and mechanical characteristics – humidity, density, viscosity and adhesion. Researches of digging process were done in atmospheric and underwater conditions. In the last case, the container with the prepared samples was filled by fresh water. In that state the samples were kept for at least 24 hours to equalize the pressure of water in the soil pores.

According to the results of 16 tests, which were conducted under the same conditions, it was found that the distribution of measured values (the ultimate length and height of ground bulk into the bucket, ground volume and digging force, Figure 5) is subject to the normal law. Herewith, variation coefficient v of the studied parameters did not exceed 10%. The required number of parallel experiments, in accordance with the recommendations (Moiseenko 1987), was calculated by the formula

$$ n = \frac{v^2}{\Delta_\alpha^2} t_{\alpha,k}^2 = \frac{0.1^2}{0.1^2} 1.64^2 = 2.69 \approx 3, $$

where $v = 0.1$ – coefficient of variation; $\Delta_\alpha = 0.1$ – maximal relative error; $t_{\alpha,k} = 1.64$ – fractile of a probability distribution t at probability belief 0.95 (is determined according to statistical Tables).

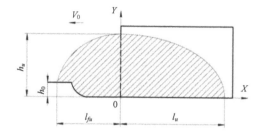

Figure 5. Geometrical parameters of ground bulk, which were measured at the tests.

During the digging of ground with bucket of 0.10 m width, at the digging depth of more than 0.035 m it was failed to achieve the ultimate bucket filling condition. Therefore, the maximal digging depth was set at value of 0.03 m. Digging was stopped at the attaining of the distance of $2R_{b=0.10} \approx 0.14$ between bucket digging edge and the opposite wall of the container (where R is defined by the formula (5) for the bucket with the maximal width). After every digging test, the ground on each side of the trench was removed to the required level with the profiling blade. When the thickness of ground in the container had been attained of $R_{b=0.10} \cdot cos\ 45° = 0.05$ m, the sample was removed from the container, and then prepared a new sample from remained in hermetically sealed package sediments.

During the tests, it was studied dependence of ultimate bucket filling parameters from the cutting velocity (v_0), digging depth (h_0) and the bucket width (b). Tests were conducted for coccolith and sapropel sediments in the atmospheric and underwater conditions.

Levels of the digging parameters variation are given in Table 1. Because the amount of soil samples for the research was limited, a variation of the single parameter was carried out by fixing the others on the level: $b = 0.06$ m, $h_0 = 0.01$ m, $v_0 = 0.1$ m/s, $\alpha = 1.5°$.

Table 1. Parameters of digging process.

Parameters	Symbol	Parameter value	
		in the atmosphere	in water
Bucket width, m	b	0.04; 0.06; 0.08; 0.10	0.07
Digging depth, m	h_0	0.01; 0.02; 0.03	0.01
Cutting velocity, m/s	v_0	0.05; 0.1; 0.15; 0.2; 0.5; 0.75; 1.00	0.01
Angle of bucket inclination, deg.	α	1.5; 15; 30; 45; 90	1.5

At digging in the atmosphere, the following parameters were defined (Figure 5): ultimate mass of ground into the bucket, ultimate height (h_u) and length (l_u) of ground bulk into the bucket, length (lfu) of pushed in front of the bucket ground, its mass, and digging force (F_{du}).The dimension of the linear parameters was performed with a metal ruler with an accuracy of 1 mm. Additional control of linear parameters was carried out by markers, which were plotted on the bucket surface. Measurement of mass characteristics accomplished with the aid analytical scales with an accuracy of 0.01 g.

At digging in the water, such parameters were measured as: ultimate mass of ground into the bucket, ultimate height and length of ground bulk into the bucket, and digging force. The analysis of the ground bulk form into the bucket, which had transparent walls, was made by using the camera and camcorder; processing of the received images was done on the PC.

To eliminate the ground friction, the bucket models were set with angle 1.5° to the horizontal surface; it were also used two plates with size 1×3 mm, set on both side walls outside of the bucket. Data from the memory card had been transferred to a PC for further processing and analysis after every test.

4 THE RESULTS OF EXPERIMENTAL INVESTIGATIONS

The properties of used samples. Studies of the grain-size have been done with the laser light scattering particle size analyzer HELOS (H0735) & QUIXEL, R3: 0.5/0.9–175 μm. The results showed, that in sapropel sediment the particles with size 0.5–1.0 μm were about 13%, 1.0–10 μm – 71%, 10–100 μm – 16%. Particles of size more than 125 μm were not revealed. Particles of coccolith sediment are distributed as follows: 0.5–1.0 μm – 12%, 1.0–10 μm – 68%, 10–100 μm – 20%. Particles of size more than 125 μm were not revealed too.

Physical properties were determined in accordance with the requirements (Standard DSTU B V. 2.1-17-2009).The average humidity of sapropel sediment was 222.68%; for coccolith sediment it was 188.43%. The densitieswere 1.219 kg/m^3 and 1.269 kg/m^3 respectively.

Critical shear stress was determined with automated device for rotation shearing "Fluegelsonde". Average critical shear stress for the sapropel samples was 156 Pa, for the coccolith samples – 187 Pa.

Viscosity of samples was defined with automated rotation viscosimeter "Thermo Scientific HAAKE MARS III". Plastic viscosity amounted 807.4 Pa·s for the sapropel sediment, and 226.8 Pa·s for the coccolith silt.

Digging process investigation in the atmosphere. At digging the sapropel and coccolith sediments in the atmosphere with the bucket model, it was observed characteristic for digging the saturated soils cutting of flow chip. Cut ground moved into the bucket and formed the ground bulk (Figure 6a). At ground deformation, it could be seen displacement zones, which were manifested in waviness of the ground bulk surface. Bucket filling mode was uniform (not periodic). During digging, the movement of soil into the bucket was being occurred simultaneously in the horizontal and vertical directions. Contour of ground bulk in parallel to the side walls plane had the form of two articulated quarters of an ellipse with the total vertical axis in the plane of the cutting edge. Following the bucket, the trench rectangular cross-section was formed (Figure 6b). When the ultimate bucket filling condition had been achieved, on the both sides of the trench were formed ground piles.

(a)   (b)

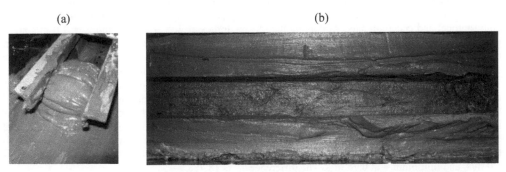

Figure 6. Photos of the ground bulk into the bucket (a) and trench (b) at digging DWOMS in the atmosphere.

The influence of cutting velocity on the bucket filling parameters. The diagrams on Figure 7 show that influence of cutting velocity on the bucket filling parameters in the range 0.05–1.00 m/s is complicated. At digging the sapropel sediment, function $l_u(v_0)$ rises in diapason $v_0 = 0.05$–0.5 m/s, and then it decreases (Figure 7a). The maximal value of ultimate length of filling corresponds to cutting velocity $v_0 = 0.5$ m/s, and minimal one is observed at $v_0 = 0.05$ m/s. At cutting velocity range 0.05–1.0 m/s the maximum value of the parameter l_u exceeds the minimum value by nearly 12%.

At digging the coccolith sediment, function $l_u(v_0)$ is decreasing monotonically in the entire range of the velocity variation (Figure 7a). The maximal value of the parameter l_u at $v_0 = 0.05$ m/s exceeds the minimum value at $v_0 = 1.0$ m/s by almost 18%.

Changing the function $h_u(v_0)$ depends little on the cutting velocity (Figure 7b). The maximal value of the ultimate height of filling is differs from the minimal value by 6.0% for sapropel sediment, and by 7.6 for coccolith sediment.

(a)	(b)

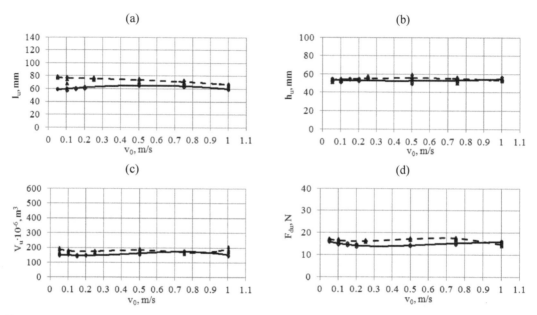

(c)	(d)

Figure 7. The graphs of dependencies of the ultimate bucket filling parameters from cutting velocity (dotted lines – for coccolith sediment, continuous line – for sapropel sediment).

The ultimate volume of ground into the bucket (V_u) was calculated as the ratio of the ground mass to its density. Function $V_u(v_0)$ has wave-like appearance for both sapropel and coccolith silts. At digging the sapropel sediments, function $V_u(v_0)$ decreases in the range 0.05–0.20 m/s, then rises in the cutting velocity diapason 0.20–0.75 m/s, and then it decreases again. The maximum value corresponds to cutting velocity of 0.75 m/s, the minimal one – at 0.15 m/s. The difference between the maximum and minimum values is 21.8% (concerning the minimum value).

At digging the cocolith sediment, function $V_u(v_0)$ takes the maximal value at a cutting velocity of 1.0 m/s, and minimum – at 0.75 m/s. The difference between the maximum and minimum values of ground volume into the bucket is 15.4%.

Function $F_{du}(v_0)$, at digging the sapropel silt, de-creases in the range of 0.05.–0.30 m/s, and then it gradually increases. The minimum value corresponds to the cutting velocity of 0.30 m/s. Digging forces at $v_0 = 0.05$ m/s and $v_0 = 1.0$ m/s are almost equal. The difference between the maximum and minimum values of digging force is 13.5% in the cutting velocity range of 0.05–1.0 m/s .

At digging the cocolith sediment, function $F_{du}(v_0)$ increases in the diapason of 0.25–0.75 m/s; in the velocity ranges of 0.05–0.25 m/s and 0.75–1.0 m/s the function is a decreasing. The minimal value of digging force is observed at $v_0 = 0.75$ m/s. The difference between the maximum and minimum values is 18.7%.

The complex character of dependences of the ul-timate bucket filling parameters from the cutting ve-locity can be explained as follows. Mainly, for such type of ground as DWOMS, the resistance of moving the sediments into the bucket depends from the value of viscous friction force. At increasing the cutting velocity the viscous friction forces increase, that resulting to raising the resistance of the movement the soil into the bucket. High resistance leads to coming less soil volume into the bucket. By increasing the resistance forces, less volume of soil comes into the bucket. At the same time, by increas-

ing the cutting velocity, the dynamic pressure of soil flow, which influences on the surface of cut ground in front of the bucket, raises. That contributes to filling the bucket. Thus, the type of the dependences the ultimate parameters of filling and the digging force from the cutting velocity, mainly, caused by rheological and mechanical characteristics of the developed soils.

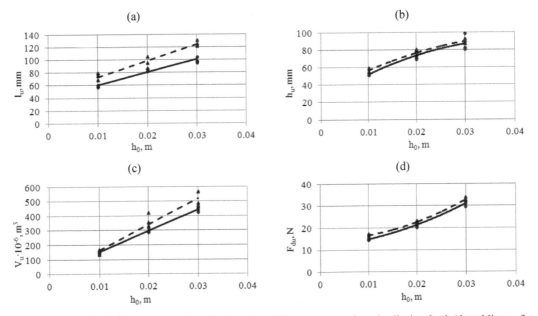

Figure 8. The graphs of dependencies of the ultimate bucket filling parameters from the digging depth (dotted lines – for coccolith sediment, continuous line – for sapropel sediment).

By increasing the digging depth and the width of the bucket, parameters l_u, h_u, V_u and F_{du} proportionally increase (Figures 8 and 9). At digging the coccolith sediment, functions $l_u(h_0)$, $l_u(b)$, $V_u(h_0)$ and $V_u(h_0)$ increase more intensively, than at digging the sapropel sediments. The ultimate length of filling for the coccolith silt in 1.1–1.25 times more, than for sapropel sediment (Figures 8a and 9a) due to higher strength and lower plastic viscosity of the coccolith samples. The intensities of increasing the ultimate height at digging the sapropel and coccolith silts are almost the same (Figures 8b and 9b). The difference between values of ultimate height for sapropel and coccolith sediments is insignificant (about 5%).

The intensities of increasing the function $F_{du}(h_0)$ are the same for both coccolith and sapropel silts (Figure 8d). Moreover, the digging force for the re-

spective digging depths differs slightly. At digging the sapropel sediments, function $F_{du}(b)$ increases a little more intensely, than at digging coccolith silt (Figure 9d), which is caused, apparently, by the lower viscosity of the latter.

Influence of bucket tool inclination on the parameters of its filling. The tests were carried out at digging the sapropel sediment in the atmospheric condition. At the Figure 10, a the superimposed photo of digging with inclined bucket of 1.5; 30; 45; 60 and 90° to the horizon is shown. According to the results of conducted experiments, the contour of the ground bulk remains virtually unchanged in the range of bucket inclination 1.5–90°. But with increasing the inclination of the bucket, the volume of soil into the bucket decreases. The bottom wall of the bucket "trims" the part of the main ellipse contour, which has semiaxis h_u and l_u.

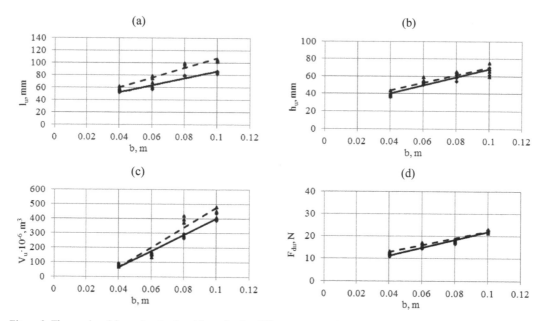

(a)　　　　　　　　　　　　　　(b)

(c)　　　　　　　　　　　　　　(d)

Figure 9. The graphs of dependencies the ultimate bucket filling parameters from bucket width (dotted lines – for coccolith sediment, continuous line – for sapropel sediment).

(a)

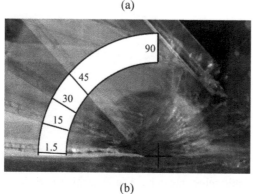

(b)

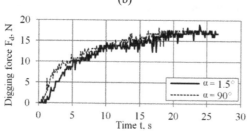

Figure 10. The superimposed photo of digging with inclined bucket (a) and oscillograms of digging force (b) ($b = 0.07$ m; $h_0 = 0.01$ m; $v_0 = 0.01$ m/s).

Digging force oscillograms at the angles of bucket inclination of 1.5 and 90° are given at the Figure 10(b). As it can be seen from the graphs, the

maximal digging force at these corners is on the same level. However, the digging force at the bucket inclination of 90° increases more intense than at 1.5°. This may be explained as follows. To achieve the ultimate bucket filling condition, the height of ground bulk should be equal h_u. At $\alpha = 90°$ the bucket works as a dozer blade, therefore, to achieve the ultimate height of filling it needed the less volume of ground, because all cut ground is spent on forming the ground bulk in front of the bucket. At $\alpha = 1.5°$ the main part of cut ground moves into the bucket, thus the way of digging should be more.

Investigation of digging the DWOMS in water. Experiments have shown that the nature of DWOMS digging process in the water is quite different from the digging in the atmospheric conditions. First of all, it is manifested in the formation of cracks at the deformation of a cut ground, which moves into the bucket. Cracks were observed even after keeping the samples in water for more than 36 hours. After draining the water from the container, cracks dulled and became barely visible.

At digging in water with 0.01 m/s velocity, at first the flow chip was separated, and after its deformation the ground bulk was formed into the bucket (Figure 11a). At the deformation it was observed the occurrence of cracks.

When the slope angle of pushed in front of the bucket ground had been achieved the value of 45–

106

50°, it was occurred rolling-down of the upper chip elements. At the ultimate bucket filling condition, the discontinuous chip was separated, the ground piles were formed from a lot of individual chip elements on both sides of the trench. Cracks were also been visible on the soil profile (Figure 11b).

(a)

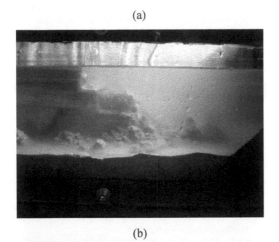

(b)

Figure 11. Photos of the ground bulk into the bucket (a) and the trench (b) at digging the DWOMS in water.

Cracking at digging the DWOMS in water can be explained as follows. The stress-strain state of the chip is being changed during digging process. If the maximal shear stress exceeds the critical shear stress, a shift of soil along glide lines is happen. In this case forming a gap between the shifted relative to each other elements of the soil depends on the ability of the particles to coagulate. For coagulation it is required the least possible distance between the particles. In atmospheric conditions the density of DWOMS is much greater than in water. Therefore, the convergence of soil particles is achieved by the weight of the overlying ground. Furthermore, the tackiness of DWOMS, which characterizing the ability of particles to stick together, in the air is in about 1.5 times more than in water. There fore, cracks on the ground surface converge badly in water. The influence of the cutting velocity on the process of cracking in water has not been investigated.

Oscillograms of digging force at digging sapropel and coccolith sediments in water are shown at the Figure 12. As it can be seen from the oscillograms, unlike digging in atmospheric conditions, in the water the maximal digging force for sapropel silt is higher than digging force for coccolith sediment. At the same time, at digging in atmospheric conditions with the bucket made of organic glass, digging force was on 10–15% lower than at digging in water with the same bucket. This is caused by features of digging the DWOMS in water. Expulsive force in water provides an increasing of ultimate height and length of the ground bulk into the bucket. The increasing of the geometrical sizes of the ground bulk into the bucket leads to rising of the frictional area that provides the increasing of the digging force. At digging the coccolith silt in water, the ultimate length of bucket filling exceeds on 18% the one in the atmospheric conditions. For sapropel sediment, that exceeding in water achieved more than 50%. The ultimate height of filling in water exceeds on 15–18% such parameter in the atmospheric conditions for the both types of sediments.

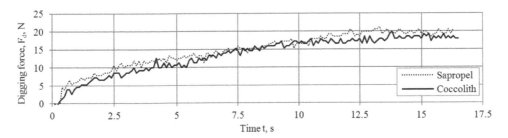

Figure 12. Oscillograms of digging forces at digging DWOMS in water ($b = 0.07$ m; $h_0 = 0.01$ m; $v_0 = 0.01$ m/s).

It should be also noticed that influence of texture and physicochemical properties of surface on ultimate bucket filling parameters at the digging in water and in air have not been investigated. However, this impact can be significant.

5 CONCLUSIONS

At digging the soft water-saturated grounds with the bucket working tools, takes place the ultimate bucket filling condition, at which the process of cutting the plasticity ground drops into the process of moving the ground to the sides. In this case filling of the bucket stops, and the ground piles are formed on the sides of the trench.

At digging the DWOMS in the atmosphere, the influence of cutting velocity on the bucket filling parameters in the range of 0.05–1.00 m/s is complicated. The difference between the maximal and the minimal values at digging the sapropel sediment made: $l_u(v_0) - 12\%$, $h_u(v_0) - 6\%$, $V_u(v_0) - 21.8\%$, $F_{du}(v_0) - 13.5\%$; at digging the coccolith silt: $l_u(v_0) - 18\%$, $h_u(v_0) - 7.6\%$, $V_u(v_0) - 15.4\%$, $F_{du}(v_0) - 18.7\%$. The ultimate bucket filling parameters and digging force rise proportionally by increasing of digging depth and bucket width.

By changing the angle of bucket inclination from 1.5 till 90°, the contour of ground bulk into the bucket remains unchanged at the ultimate bucket filling condition. At the angle of 90° the bucket works as a dozer blade. The digging force at 1.5 and 90° of bucket inclination has almost the same value.

The nature of DWOMS digging process in the water is quite different from the digging process in the atmospheric conditions. It is manifested in the formation of cracks at the deformation of the cut ground, which moves into the bucket. Because of the expulsive force, in water the ultimate height and length of the ground bulk into the bucket exceeds the appropriate parameters at digging in the atmosphere. The difference between the ultimate bucket filling parameters at digging in water and at digging in the atmosphere is: $l_u - 18$–50% and more, $h_u - 15$–18%, $F_{du} - 10$–15%.

REFERENCES

Shnjukov, E.F. & Ziborov, A.P. 2004. *Mineral wealth of the Black Sea.* Kyiv: Karbon-LTDPubl.: 290.

Dimitrov, D.P. 2010. *Geology and unconventional resources of the Black Sea.* Varna, OnglPubl.: 270.

Dimitrov, P.S. & Dimitrov, D.P. 1999. *Alternative energy resources from the bottom of the Black sea.* Geology and mineral resources of the Black Sea: 223–226.

Degodjuk, E.G., Kleshhenko, S.A. & Degodjuk, S.E. 2000. *Agronomical importance of sapropel and coccolith sediments of the Black Sea and technological safety issues.* Geology of the Black Sea and the Azov Sea: 164–174.

The problem of sapropel sediments in the Black Sea. 2010. Kyiv: MGOR Publ.

Franchuk, V.P., Shepel, T.V. & Ziborov, A.P. 2011. *Complex of the ship-based technological equipment "Sapropel" for sampling the deep-sea sediments.* Geology and Mineral Resources of the World Ocean, Vol. 1: 33–37.

Rybar, P., Hamrak, H. & Kosco, J. 2011. *Polymetalic kekonkrecie: bohatstvo na dne moria oceanov.* Kosice, Slovakrepublic: TUFBERG: 135.

Patzold, V., Gruhn, G. & Drebenstedt, C. 2008. *Der Nassabbau: Erkundung, Gewinnung, Aufbereitung, Bewertung.* Berlin: Springer-Verlag: 450.

Sukach, M.K. 2004. *Workflows deep machines.* Kyiv: Naukova Dumka: 364.

Turgumbaev, D.D. 1992. *Modeling the processes of underwater ground cutting by excavating machines to obtain the working loads.* Author's abstract. Moscow: 39.

Lobanov, V.A. 1983. *Offshore machines handbook.* Leningrad: Sudostroyenije: 288.

Moiseenko, V.G. 1987. *The operating loadings forecasting for earthmoving machines in the special operating conditions.* Kyiv: Vyshha shkola: 194.

Nedorezov, I.A. & Saveljev, A.G. 2010. *Building machines.* Moscow: MNTU named after Bauman N.E.: 119.

Sukach, M.K. & Magnushevskij, V.I. 2005. *The model of plastic waterlogged ground cutting with the earthmoving machine blade.* KhNAHU collection of proceedings. HNADU Publ., Vol. 29: 74–79.

Standard DSTU B V. 2.1-8-2001, 2001. *Grounds. Selection, packaging, transportation and storage of samples.* Kyiv: Derzhkommistbud of Ukraine: 16.

Standard DSTU B V. 2.1-17-2009, 2009. *Grounds. Laboratory methods for determining physical characteristics.* Kyiv: Minregionbud of Ukraine: 22.

Some aspects of technological processes control of an in-situ gasifier during coal seam gasification

V. Falshtynskyi, R. Dychkovskyi, P. Saik & V. Lozynskyi
National Mining University, Dnipropetrovs'k, Ukraine

ABSTRACT: The results of technological processes control research at coal seam gasification are resulted. It was received on the base of approbation the underground gasification technology on experimental in-situ gasifier which allows extending an application of this technology and conducting more high-quality process control of gasification due to the water inflow redistribution on length of gas outlet borehole. Conclusions are given according to conducted investigations. The purpose of the article is to explore ways to process control in the gasification of coal seam mining conditions with respect to geological conditions and construction of in-situ gasifier for the expansion of application of this technology.

1 INTRODUCTION

The coal in world power engineering makes more than 28%. In 2013 has been growth of fuel consumption – 2.5%. Ukrainian proved coal reserves ranked on seventh place in the world according World Energy Council report. In 2012 coal production in Ukraine amounted to 85.946 million tonnes, up 4.8% from 2011. In 2003, Ukraine produced 79.3 million tonnes.

Commercial, non-economic and anticipated resources make 117.5 billion tons which eliminate this source of energy in the first place in the energy sector (mpe.kmu.gov.ua). In coal seams with thickness ≥ 1.2 m concentrated 85% of coal reserves. Most of the proved reserves with thickness ≥ 0.8 m occurred in underworking and overworking of a coal, that make difficult or impossible exploration with traditional mining methods.

Reducing the coal demand in Ukrainian power engineering during 2012–2013, with extraction increasing associate with an imbalance of it usage in the energy and power sector, seasonally necessary, underdeveloped complex and industrial processing. Due to the unstable coal extraction become the growth of the cost and complexity of the technical and economic conditions and environmental safety in mines. These factors lead to losses, which in turn serve as one of the reasons for mine closure. Non-economic coal reserves make negative effect on the stable coal extraction. Closing down of an enterprise leads to negative social stress, which is associated with unemployment, lack of resources and degradation of human settlements.

2 STATE-OF-THE-ART-REVIEW

Exploration of non-economic and commercial resources in mines that closing down requires the introduction of low-cost, cost-effective, mobile, integrated, ecologically friendly technology based on the processes of borehole underground coal gasification (BUCG) that combines coal extraction and its energy and chemical usage.

Securing of environmental cleanliness of underground coal gasification processes associated to its controllability, impermeability of underground gasifier and integrated use of cogeneration technologies in closed purification and processing cycle of gasification products.

Energy and chemical complex on the BUCG base is a mobile modular enterprise that provides intensive productivity increase, quality and diversity of gasification products, enables fast and without lossless redirect output final product in the form of heat, electricity, chemicals and products due to the flexibility of technological parameters wit account of dynamic changes in mining conditions.

Price increase for oil, gas and coal associated with the expenses for extraction, transportation, processing, environment protection and depletion of commercial reserves.

The solution of this problem requires a complex comprehensive approach, which is fully consistent with the concept of technological schemes of borehole underground coal gasification in mine conditions was developed in the National mining university on the department of underground mining (Bondarenko 2010).

3 THE MAIN PART OF THE ARTICLE

At the design and operation of experimental in-situ gasifier (mine Barabara, Poland) (Falshtynskyi 2010) took into account the following geological, geotechnological and mining factors:

– water content of a deposit, presence of static (dynamic) groundwater in rock and coal mass;

– presence of tectonic and anthropogenic disturbances rock and coal strata;

– elemental composition of the coal and rocks;

– mechanical and thermal properties of the coal;

– geomechanical parameters of rocks of roof, ground, formation and changes at gasification (step of collapse, areas of technological fracturing, parameters of bearing pressure zone);

– geometric parameters of the coal seam, roof and ground.

Experimental area of a mine field delimits by mine workings and has a form close to the two perpendicular rectangles. The depth of the coal seam from 15.4 to 18.1 m. Lithological variety of roof is represented by sand and clay rocks. Aquifer represented by static water and is directly under the sediment at a depth of 0.3 meters. The next aquifer located in coal seam ground at a depth of 20–40 m.

Estimated average water inflow in the gasification zone is 2.55 m^3/min. Filtration rate of coal seam – $2.69 \cdot 10^{-5}$ m/min, the roof rocks – $6.79 \cdot 10^{-5}$ m/min, rock ground – $4.46 \cdot 10^{-5}$ m/min.

Coal seam has complex structure, sustained by the power with the angle of inclination 0–5°. Extracting seam thickness – 1.4 m, thickness of washout – 0.1 m. Coal humidity – 14.7%, ash content – 22.1%, the yield of volatile content – 32.4%, calorific value – 21.6 MJ/m^3.

Small depth of the whole coal seam attitude, its technological options and anthropogenic disturbance of rock massif causing a number of factors that are associated with the application process of coal seam gasification in combined regime of blast and forced removal of the BUCG products. These conditions were taken into account at flowsheets designing and development the technological parameters of the gasification process.

For the experiment proposed three flowsheets of in-situ gasifier was offered (Falshtynskyi 2010). Take into account the geological and mining conditions and the results of experimental and analytical studies were selected technological schemes of experimental in-situ gasifier that are shown on Figure 1.

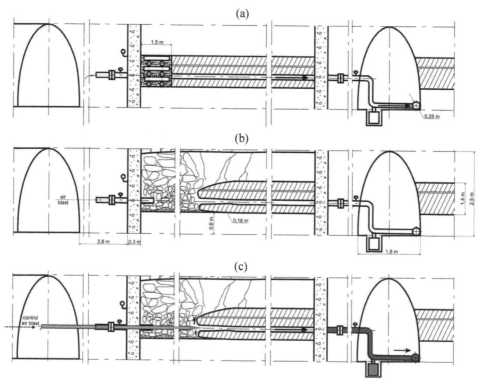

Figure 1. Technological schemes of experimental in-situ gasifier with air blast supply: (a) start of gasification (coal ignition), (b) coal seam gasification; (c) coal seam gasification by controlled retraction injection point.

4 STUDY KEY FINDING

Air blast oxygen enrichment took place in not a large number (O_2 less than 27%, $P = 0.165$–0.21 MPa), giving oxidant pulse, and the fate of ballast gases (CO_2, H_2S, N_2, O_2) in the gas generator was changed from 40.8 to 58.6%, depending on the fate of oxygen in the composition of the blast, time flow and pulse duration. Heat of combustion of generator gas made $Q = 3.89$–5.3 MJ/m^3.

Enrichment of oxygen blast steam happened pulse, containing a pair composed of the blast within 375–569 g/m^3, the pressure in the gas generator was within 0.219–0.34 MPa, while the yield of ballast gases was 19.8–26.2%. Heat of combustion of generator gas made $Q = 6.12$–8.4 MJ/m^3.

Technological parameters of coal seam gasification are shown on Table 1 and Figure 2.

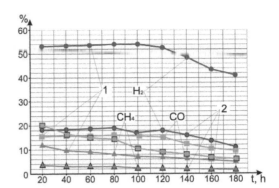

Figure 2. Diagram of gasification yield with account the time and blast composition: 1 – enriched air, 2 – vapor – saturated oxygen.

Table 1. Technological parameters of coal gasification.

Technological parameters	Values
Pressure in combustion channel, P (MPa)	1.65–2.3
Temperature of the process, T (°C)	1000–1200
Temperature of the gases in pipe ,T_o (°C)	309–516
Combustion face advance rate, v (m / day)	1.4–2.9
Calorific power of gasifier, Q_c (Gcal / h)	3.75–7.2
Electric power of gasifier, E_p (kWh)	4350–8362

During the experiment at coal seam gasification in mine conditions tested the physical and chemical methods of gasification process adaptation to change the parameters gasifier – rock massif.

Changing the pressure in the gasifier, injection blast on fire is shown on Figure 3. With the help of smoke exhaust, which was on the ground surface above gaifier in gas-pipeline formed estuary discharge point 3 (discharge), which could provide additional waterproofing of combustion channel, due to redistribution of static and dynamic groundwater (Silin-Bekchurin 1960) of overlaying rock stratum.

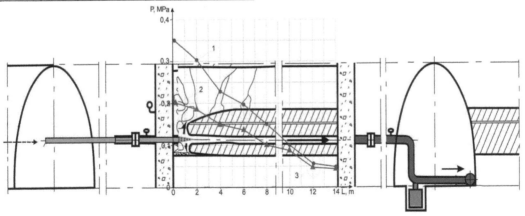

Figure 3. Control of hydro-field in reaction channel and gas outlet borehole; 1 – oxygen blast injection saturated by steam, 2 – air injection enriched by oxygen, 3 – point of discharge.

This model of inflows control in the combustion zone of gasifier include mining and geological, hydrogeological conditions, geomechanical parameters of rock massif (lithological variety and filtration properties of overlaying rock stratum) and technological parameters of the process (combined system of air blast and gas removal from gasifier) that provides a water removal from the reaction channel and water distribution along the length of the gasifier.

The proposed scheme of in-situ gasifier waterproofing was tested in laboratory and bench tests (Falshtynskyi 2013).

5 CONCLUSIONS

Technological scheme of in-situ gasification control are proposed. They allow to increase range of application the technology off BUCG and meet production and complex processing of solid fuels in place of its occurrence in an environmentally safe and closed cycle with chemical products, energy and heat reception.

Implementation the combined regime of air blast allowed spending more quality pressurization of combustion face and the reaction channel of in-situ gasifier by reallocating of water inflow along the borehole length.

The technology is effective at commercial and non-economic coal extraction in mine, do not requires a significant investment in deposit opening, mine development, utilization and waste storage, extends the long life operating of mines, solves a major social problem of employment.

Compact underground design of main equipment for in-situ gasifier reduces the surface area of the complex. Location of purification and processing equipment on-site mine do not need additional self-weaning earth.

REFERENCES

http://mpe.kmu.gov.ua/fuel/control/uk/doccatalog/list?curr Dir=50358

Bondarenko, V.I., Falshtynskyi, V.S., Dychkovskyi, R.O., Tabachenko, M.M., Medjanyk, V.Ju. & Ruskyh, V.V. 2010. *Method of underground gasification.* Patent #89278 UA from 11.01.2010.

Falshtynskyi, V., Dychkovskyi, R., Stanczyk, K. & Swiadrowski, J. 2010. *Analytical determination of parameters of material and thermal balance and physical parameters of coal seam work-out on mine «Barbara», Poland.* Dnipropetrovs'k: National mining university: 157–161.

Falshtynskyi, V., Dychkovskyi, R., Stanchyk, K., Svjadrovski, E. & Lozynskyi, V.G. 2010. *Justification of tehnological scheames of experimental in-situ gasifier.* Scientific Bulletin of NMU, 3: 34–38.

Silin-Bekchurin, A.I., & Bogoroditskii, K.F. 1960. *Role of ground water and other natural factors in the underground gasification of coal (with the Lisichansk facility and the one outside Moscow as examples),* 23: 17–66.

Falshtynskyi, V.S., Dychkovskyi, R.O., Lozynskyi, V.G. & Saik, P.B. 2013. *Determination of the technological parameters of borehole underground coal gasification for thin coal seams.* Journal of Sustainable Mining, 12: 8–16.

Progressive Technologies of Coal, Coalbed Methane, and Ores Mining – Bondarenko, Kovalevs'ka & Ganushevych (eds)
© 2014 Taylor & Francis Group, London, ISBN: 978-1-138-02699-5

Methane receiving from coal and technogenic deposits

R. Agaiev, V. Vlasenko & E. Kliuev
M.S. Polyakov Institute of Geotechnical Mechanics, Dnipropetrovs'k, Ukraine

ABSTRACT: In underground mining major problematic factor is methane. The goal of the study is to produce further information regarding the following question: what technology should be applied to obtain more substitute of natural gas (methane) for further commercial application. The results of theoretical and experimental investigations about hydrodynamic, pneumohydrodynamic and thermal impacts were given. Some prospects of methane receiving from coal and technogenic deposits were described.

1 INTRODUCTION

Over the last years coal deposit methane is considered as independent mineral resource. Experts had estimated the workable reserves of methane in Donets'k basin at depths to 3 km in quantity of 12–25 trillion m^3 (Kompaniec et al. 2007). On the one hand, methane is one of the most promising energy sources, on the other hand – is a danger to miners and can lead to different accidents, as well as one of the major environmental pollutants. It has been recognized that methane of coal-rock massif was discovered in cracks and pores in sorption and free states. As a result during mining, the integrity of rock massif is violated that leads to the methane releasing in goaf. For the prevention of further accidents in coal mines and in order to normalize of the workflow provides a number of regulatory steps for rock and coal massif degassing intensification.

Apart from, the significant methane amount may contain in wastes products of coal industry. In order to receive it, it's necessary to implement thermal impact on coal-contained material by means of temperature heating under atmospheric pressure. In such a way, it is possible to receive not only methane, but also other technological gases, such as hydrogen, and carbon monoxide, which are not considered in this paper.

2 FORMULATING THE PROBLEM

The method of hydrodynamic impact was developed by collaborators of the M.S. Polyakov Institute of Geotechnical Mechanics under the NASU. With this method, not only mining force and gas pressure force, but also vibration properties of the "coal seam – enclosing rocks" were more activated. As a result, the effective degassing of coal seams by means of hydrodynamic impact was achieved. Degassing process is accompanied by active methane emission through the technological borehole. If this borehole connects to the pipeline system, it will be able to ensure the methane supply to the further application.

Besides, one of the most effective ways of coal seam methane receiving is based on drilling surface borehole and active impact on gas-coal width. However, methane extraction from coal deposits is complicated by low massif permeability, pore space colmatation and other features of methane trap, which leads to lower production rate of surface boreholes and smaller service life. One of the main reasons for this is the strong absorption of drilling mud in zones of increasing fracturing and the necessity of special grouting mortar application. In Donets'k basin seam pressure conditions is, typically, below hydrostatic, that facilitates the penetration of drilling mud to productive levels.

It is common knowledge that one of the most perspective and up-to-date ways for obtaining of substitute of natural gas (methane) from coal and coal-contained material is the thermal impact in oxygen-free medium. Peat, lignite, bituminous shale were used often in such processes. High-ash coals and wash slurry wastes begin to use as a feedstock in thermal processing (Prihodchenko et al. 2010). Intensive investigations in this field of study predetermined by the undoubted advantages, such us low pressure, absence of oxidizing reagents or catalysts, simplicity of process organization and low capital intensity.

3 THE METHOD OF HYDRODYNAMIC IMPACT THROUGH THE UNDERGROUND BOREHOLE IN COAL SEAMS

Degassing process efficiency was confirmed by a number of mining-experimental works in conditions of Donets'k basin mines, which were carried out at "F.E. Dzerzhinsky Mine", "Severnaya" mine and "V.I. Lenin Mine". As a result, the procedure of industrial testing of the method for coal seams exploration by means of hydrodynamic impact through crosscuts was developed. In addition, the rules of hydrodynamic impact through the underground borehole in coal seams, exposed to gas-dynamic phenomena before its exploration was approved by "Bureau of the Central Commission for ventilation, degassing and gas-dynamic phenomena control in coal mines in Ukraine".

Technological borehole was drilled from faces of intermediate technology crosscuts according to the assembly diagram, shown on Figure 1.

Drilling process of technological borehole was carried out in two stages: 1) borehole drilling with diameter of 80 mm to a project length; 2) drilling-out of the borehole rock part to a diameter of 150–190 mm. Thus, the borehole was cased and plugged up by metal pipe with length of 2–2.5 m and outer diameter of 104–112 mm. The device for hydrodynamic impact was installed on the fixing flange of the casing pipe after setting of cement-sand mortar. The device was connected with two pumps (Patents of Ukraine... 1997).

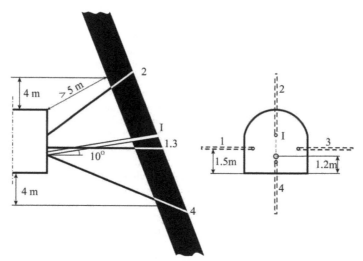

Figure 1. Location diagram of technological borehole: 1–4 – checking boreholes for measuring of gas seam pressure, I – technological borehole.

The coal seam hydrodynamic impact was carried out as follows. Water from a fire protection flight was fed in filtration mode at a speed of 30–40 l/min. Rapid depressurization was performed upon reaching the pressure of 5–7 MPa and a fluid from the borehole with destroyed coal with segregated gas were released. Pressure gauges installed on the remote-control station monitored the water pressure in boreholes.

The cycles of fluid supply and release had been repeated until calculated amount of coal was discontinued from the technological borehole (Bulat et al. 2003).

All process parameters were registered by the devices that installed at the remote-control station. The efficiency evaluation of hydrodynamic impact was implemented according to the amount of the ex-tracted coal from the technological borehole, degassing coefficient of the treated zone and changing of the methane yield kinetics (criterion B) using the DS-03 measuring instrument Shazhko (2011).

In order to determine the amount of the extracted coal the dividing wall was build up, it has the calibrated volume. The efficiency of coal release was defined by the volume that must be more than 2% of the treatment zone.

Degassing coefficient was calculated using the formula

$$k_d = \frac{V_a}{V_P},\qquad(1)$$

where V_a – actual gas amount, released through the

114

technological borehole; V_P – estimated gas amount in the treatment zone, defined by the formula

$$V_p = Sm\chi\gamma , \qquad (2)$$

where S – area of the treatment zone; m – seam thickness, m; χ – coal seam natural gas content.

The degassing efficiency was defined by the coefficient that must be at least 0.45.

According to the kinetics of methane yield the hydrodynamic impact will be effective if the inequality is satisfied

$$B = 0.44 P_k + 1.14 \cdot 10^8 D_{ef} \leq 4 , \qquad (3)$$

where P_k – pressure in a dish during the methane desorption from coal; D_{ef} – methane effective diffusion in the coal seam, determined by the formula

$$D_{ef} = \frac{R_1^2 \cdot \ln P_2 - R_2^2 \cdot \ln P_1}{6t \ln \frac{P_2}{P_1}} , \qquad (4)$$

where R_1, R_2 – sizes fractions of coal (0.4–0.5 mm and 1.0–1.6 mm); P_1, P_2 – relative methane content in the two fractions; t-time.

Estimation of coal seam gas-dynamic activity in time and space during hydrodynamic impact was made by the pressure changing in a dish and by the parameters of methane mass transfer in the coal seam. As an example, the results of hydrodynamic impact on the coal seam m_3-"Tolstyj" on depth 1146 m at "F.E. Dzerzhinsky Mine" were represented. 27 cycles had been implemented and 7 tons of coal was released from the borehole. At the end of the impact seam pressure in boreholes were as follows: borehole #1 – 0.20 MPa, borehole #2 – 0.22 MPa, borehole #3 – 0.25 MPa and borehole # 4 – 0.23 MPa. Coal seam was degassed on the area of 145 m². Estimated methane amount in the treatment zone was calculated according to the formula (2) and was 5.9 thousand m³. The actual volume of released methane from the treatment zone was 3.3 thousand m³. Degassing coefficient had reached the value of 0.56. Measurements results of the methane kinetics from coal in storage reservoir after hydrodynamic impact are shown in Figure 2.

Curves describing the character of methane yield from coal, represented in Figure 2, are logarithmic and can be determined by the following Equations: $y = 189.9 \ln(x) + 29.21$ with approximation reliability $R^2 = 0.94$ for measuring sizes fractions of coal 0.4–0.5 mm and $y = 149.1 \ln(x) + 19.98$ with ap-

proximation reliability $R^2 = 0.99$ for measuring sizes fractions of coal 1.0–1.6 mm.

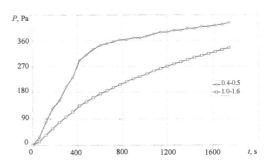

Figure 2. Change kinetics of pressure in the storage reservoir during the methane desorption from coal seam m_3-"Tolsty" after hydrodynamic impact.

Using Equation (4), the change dynamics of the methane mass transfer coefficient was shown in Figure 3.

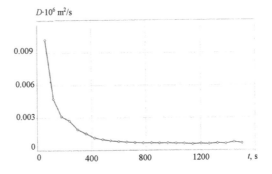

Figure 3. Change dynamics of the methane mass transfer coefficient from coal seam m_3-"Tolstyj" after hydrodynamic impact.

According to the Figure 3 the curve is, typically, hyperbolic, which means a steady decrease of the methane effective diffusion in coal seam. Thus, according to formula (3) the criterion B is equal to

$$B = 0.44 \cdot 7125 \cdot 10^{-3} + 1.14 \cdot 10^8 \cdot 0.4 \cdot 10^{-8} =$$

$$= 3.14 + 0.46 = 3.6 < 4.0$$

As a result, calculated value of the methane kinetics helps to conclude that the hydrodynamic impact was effective.

4 THE METHOD OF PNEUMOHYDRO-DYNAMIC IMPACT ON SURFACE DEGASSING BOREHOLES

The Department of Technologies for Underground Coal Mining of IGTM under the NAS of Ukraine had developed the method of pneumohydrodynamic impact (PHDI) on surface degassing boreholes (SDP) (Sofiiskyi et al. 2011).

The main objective of PHDI is to release filtration system of the near-borehole zone from colmation formations (pieces of clays, coals, rocks and others). The principle of impact lies in fluid (water) sign-reverse filtration in the massif treatment zone, created by alternating loads due to changes of air pressure on the water column located at the bottom of the borehole. Wherein, the fluid depressurization time in the massif depends on the output speed of the compressed air from the borehole.

As a result, the theoretical researches and the development of a mathematical model help to find out the permeability massif coefficient and the reverse filtration rate during PHDI, which are increasing due to forming of water-clay suspensions and carrying them out the filtration system into the borehole (Figure 4). It should be noted sufficiently big reverse filtration rate to wash out effectively colmation plug.

Presented calculations showed that the path length, traversed by the fluid, during depressurization is more than 1 m, water- clay slurry, moving in the direction of the borehole will be take out from coal-rock massif.

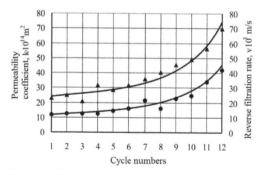

Figure 4. Changing the permeability coefficient and reverse filtration rate during PHDI: ▲ – permeability coefficient; ● – reverse filtration rate.

PHDI was carried as follows (Figure 5). SDP was drilling from the Earth's surface. The bottom of borehole is located at a distance at least of 5 coal bed thickness. The entire length of the well was cased by metal pipes with grouting of the borehole clearance. The perforation of casing pipes was made on the area of productive levels. SDP is filled with water approximately 30 m above the productive level. The tubing string was sank in the borehole. Borehole head was equipped by PHDI devices for process control as well through tubing string as through the intertubular space. Calculated air pressure was generated by high-pressure piston compressor in SDP. After that, the air discharge was implemented by means of PHDI device.

Cycles of pumping and depressurization form the sign-reverse loads on filtration volume whereby, due to the forward and reverse fluid filtration occur losses of colmatation material from pores and cracks in the SDP that facilitates connection between borehole and rock massif filtration volume.

Experimental investigations for improving of methane releasing and increasing of the operation period of SDP were made at "A.F. Zasyad'ko Mine". Before getting started 25 SDP were drilled on underworked rock mass of coal bed m_3 of which only 16 were functioning (Bulat 2002).

As a result, the experimental work carried out PHDI only on 7 SDP, which were detached from the pipeline or were not connected due to yield lack or low methane content. Thus, as a result, the boreholes average operation period had increased by 4.7 times, and methane yield – by 3.0 times, which expands the possibilities of methane utilization.

5 THE METHOD OF THERMAL IMPACT ON COAL AND COAL-CONTAINED MATERIALS

The physical essence of the thermal impact process of coal and coal-contained materials is as follows. According to the Fuks-Krevelen pyrolysis theory (Pomerantsev et al. 1986) with a sufficient degree of accuracy it can be described as a chain of chemical reactions, involving the decomposition of organic matter occurred in the adjacent layer and in the outer surface of the particles.

Physical-chemical transformations begin to appear at temperatures of about 200 °C. However, at 120 °C constitution moisture and adsorbed carbon gases, such us carbon dioxide, methane, air component were released. Thus, there is no decomposition of the organic matter, but probably occurs some certain changes in its internal structure.

At temperatures greater than 200 °C a certain quantity of water vapor and carbon dioxide are evolved, that is a result of complicated chemical reactions, involving, in general, in the external polar groups.

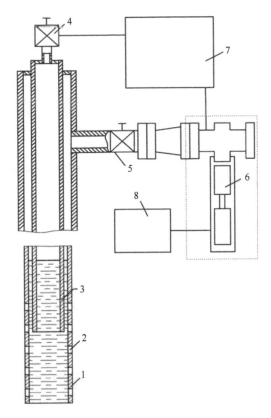

Figure 5. Equipment assembly diagram for PHDI: 1 – casing pipe; 2 – perforations; 3 – tubing string ; 4, 5 – latch; 6 – device of PHDI; 7 – compressor station; 8 – pumping station.

In the temperature range of 250–325 °C processes of coal structure decomposition are increased. The intensive emission of water vapor, carbon dioxide, some amounts of hydrogen sulfide and organic compounds of sulfur are occurred. At this stage, the oxygen content in the coal-contained material is markedly reduced. However, in this temperature range weak chemical bonds are splitting. Profound changes in the internal structure of the organic matter have not been observed yet.

At temperatures above 350 °C the organic matter are intensively decomposed with the formation of free and unstable substance groups. Occurred recombination processes were developed in two directions:

1) formation of condensed products characterized by a high content of carbon and low content of hydrogen;

2) formation of liquid and gaseous products enriched with hydrogen.

Deep decomposition of organic matter and emission of liquid products are completed generally at a temperature of about 550 °C. These processes are accompanied by releasing of hydrogen, methane, carbon monoxide and nitrogen.

During further temperature heating the yield of methane, nitrogen, oxygen, carbon dioxide were not observed, however, the hydrogen and carbon monoxide were formed in significant quantities.

But the temperature dependences of the methane yield from coal and coal-contained material during thermal impact were not defined. The materials used for research are substandard sapropelite and coal slime from L'viv – Volyn' basin.

The theoretical investigations were held by means of the computer program "ASTRA-3" and based on the fundamental laws of thermodynamics, which enables the calculation of the equilibrium composition of heterogeneous systems, depending on the component feedstock composition and the process temperature (Bulat et al. 2003). Initial data for the simulation are shown in Tables 1 and 2.

Table 1. Results of elemental and technical analysis.

Feedstock	Carbon, %	Hydrogen, %	Nitrogen, %	Oxigen, %	Sulphur, %	Moisture, %	Ash content, %	Volatiles yield, %
Sapropelite	34.8	2.5	0.7	3.2	0.3	1.8	56.7	52.2
Coal slime sample average	26.8	1.8	2.9	20.4	2.8	1.2	44.1	39.9
Coal slime sample #2	26.9	3.1	2.8	19.5	1.7	2.1	43.9	33.1
Coal slime sample #3	26.2	3.1	2.8	19.8	1.7	1.6	44.8	37.6
Coal slime sample #8	25.6	3.0	2.9	19.9	1.8	1.4	45.4	34.7

Table 2. Results of the ash analysis.

Feedstock	The sample content of chemical substances,%						
	SiO_2	Al_2O_3	Fe_2O_3	SO_3	CaO	MgO	K_2O++Na_2O
Sapropelite	47.3	26.7	20.0	0.5	2.3	2.2	1.0
Coal slime sample average	42.9	23.7	10.4	2.8	5.0	2.5	12.6

As we can see from these tables, sapropelite and coal slime may classify as the high-ash materials with significant sulfur content and volatile yield. However, these properties allow to consider them as the principal feedstock for thermal impact.

The calculation results are presented in Figures 6 and 7.

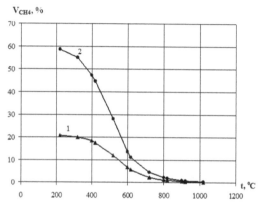

Figure 6. The methane yield during thermal impact on coal slime (1) and sapropelite (2).

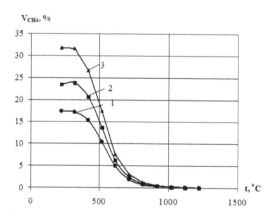

Figure 7. The methane yield during thermal impact on binary mixtures of coal slime with sapropelite in proportions of 2:1 (1), 1:1 (2), 1:2 (3).

As we can see from Figures 1 and 2, methane generates at the beginning of the process, at temperatures greater than 500 °C intensive methane emission was observed. After that amount is markedly reduced due to its decomposition at temperatures above 1000 °C and the yield of methane was not registered. It was determined that the greatest amount of methane is formed at a temperature of 200 °C during the thermal impact on sapropelite. Thus, the percentage methane content is 40% higher than in the gas from coal slurry.

6 CONCLUSIONS

1. Method of hydrodynamic impact helps to increase methane desorption from disintegrated zone of coal seam.

2. The hydrodynamic impact on the coal seam m_3 -"Tolstyj" promote to release 3.3 thousand m³ of methane from the treatment zone. In such a way, the borehole connection to a mine system pipeline allows to use released methane for commercial purposes.

4. Average methane extraction from one borehole by means of pneumohydrodynamic impact on surface degassing boreholes is 5.1 million m³ per year.

3. The perspective way for obtaining of substitute of natural gas (methane) is thermal impact on coal and coal-contained material in oxygen-free medium.

4. Generalization of the theoretical calculation results suggests to believe that the methane yield was observed during thermal impact on sapropelite, coal slime and their binary mixtures in the range of 200–1000 °C due to physical-chemical processes and the emission of adsorbed gases.

5. At temperatures above 1000 °C methane emission do not occurs due to its decomposition into carbon monoxide and molecular hydrogen.

6. The gas phase yield in an amount of about 200 m³ per one ton of feedstock was registered during thermal impact on coal-contained materials; thus expected gas resources amount is approximately 20 billion m³ per year, that equivalently to about 10 billion m³ of natural gas.

REFERENCES

Kompaniec, O.I., Anciferov, V.A. & Kryghovska, L.M. 2007. *Prediction of methane accumulation zones in the undisturbed mining coal-rock massif.* Coal of Ukraine, 1: 30–31.

Prihodchenko, V.L., Slashcheva, Ye.A., Osenniy, V.Ya., Koval, N.V. & Kliujev, E.S. 2010. *The influence of* heating conditions of low-grade coals and coal slimes on thermal degradation products. Geotechnical Mechanics, 89: 65–72.

Patents of Ukraine UA #19956. 1997. The device for hydrodynamic impact on the coal seam / Amelin, V.A., Baradulin, Ye.H., Demidov, I.P., Nechytailo, V.O., Portianko, V.P., Sylin, D.P., Sofiiskyi, K.K. & Shmelev, M.O. // Institute of Geotechnical Mechanics under NAS of Ukraine.

Bulat, A.F., Sofiiskyi, K.K. & Sylin, D.P. 2003. *The hydrodynamic impact on gas-bearing coal seams.* Dnipropetrovs'k: Poligrafist.

Shazhko, Ya.V. 2011. *Expres method of determining the pressure and the amount of methane in coal seams.* Physical-technical problems of mining, 14, pp. 60–67.

Sofiiskyij, K.K., Filimonov, P.E. & Guniya, D.P. 2011. *Methane extraction from coal deposits by applying of dynamic.* pneumohydro- and pneumodynamic impact through surface degassing boreholes on coal-rock massif. Geotechnical Mechanics, 92: 62–68.

Bulat, A.F. 2002. *Degassing of coal-rock massif at "A.F. Zasyad'ko Mine" through boreholes, drilled from surface.* Geotechnical Mechanics, 37: 49–57.

Pomerantsev, V.V., Alfredov, K.M. & Ahmedov, D.B. 1986. *Fundamentals of practical combustion theory /* Pomerantsev V.V. (ed.). Leningrad: Energoatomizdat.

Progressive Technologies of Coal, Coalbed Methane, and Ores Mining – Bondarenko, Kovalevs'ka & Ganushevych (eds)
© 2014 Taylor & Francis Group, London, ISBN: 978-1-138-02699-5

Application aspects of adsorption opening effect of solids pore space surface

V. Biletskyi
Donets'k National Technical University, Donets'k, Ukraine

ABSTRACT: The paper describes a new identified effect of adsorption opening of the surface of solids pore space, which accompanies well-known Rehbinder effect. Some application aspects of the new physical effect are reviewed, which are expedient to use in new technologies and for diagnostics of the state of rocks.

1 PROBLEM DEFINITION AND STATE-OF-THE-ART

Papers (Biletskyi, Sergeiev et al. 2012 & Biletskyi 2013) reported a new effect which we have detected for coal, specifically an adsorption opening effect of the solids pore space surface, which is a physical effect accompanying well-known Rehbinder effect and involving a dominant transition of the inner surface of pores into the external surface of grains when pre-wetted solids are crushed. Changes of surface characteristics of crushed material owing to this effect result in changes of rheological and other properties of crushed mass. The above is caused by a change of the zeta potential of crushed coal grains due to a dominant opening of oxidized pore surfaces, which brings about hydrophilization of coal substance – a shift of the hydrophilic-hydrophobic balance of the total external surface of crushed coal to the hydrophilic area.

The paper aims to review applied aspects of the adsorption opening effect of the solids pore space surface.

Main points. The identified effect of the shift of the hydrophilic-hydrophobic balance of the total solids external surface to the hydrophilic area after solids are crushed may take place, provided the following:

1. Porosity of solids and availability of open pores which surface gets oxidized in time;

2. Prior encroachment (wetting) of porous solids, which reduces their hardness (Rehbinder effect). The latter is explained firstly by a wedging action of molecules adsorbed in the tips of cracks, especially molecules of surface active agents, and secondly by cooling of microcracks sized several atoms of crystal lattice in water environment and therefore impossibility of their 'recovery'. Subsequent mechanical impacts expand these microcracks, which makes breakage of pre-wetted solids easier compared to dry solids.

The aggregate hydrophilic-hydrophobic balance of the external surface of coal grains after their crushing without pre-wetting will shift to the hydrophobic area, whereas in case of pre-wetting to the hydrophilic area (the "inverted" surface of pores is more hydrophilic than the freshly opened surface of coal substance).

The effect of the shift of the hydrophilic-hydrophobic balance of the total external surface of porous solids to the hydrophilic area after solids are crushed has the following consequences.

1. Changes of the grain size and hydrophilization of the surface of coal crushed with the use of Rehbinder effect and adsorption opening of the pore space surface (see Table 1 and Figure 1) improve rheological and stability characteristics of highly loaded coal-water slurry (coal-water fuel). The above is inter alia demonstrated in our paper (Biletskyi, Svitlyi & Krut 2012).

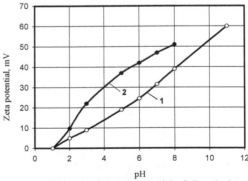

Figure 1. Change of the zeta potential of G-ranked coal (gas coal) used to produce coal-water slurry with the use of different techniques: 1 – without coal wetting prior to crushing; 2 – with coal wetting prior to crushing.

2. The effect of adsorption opening of the pore space surface during rock breakage impacts physical and chemical characteristics of the surface of crushed material and hence its floatability and dewaterability.

3. Manifestation of Rehbinder effect and adsorption opening of the pore space surface during drilling is limited by a number of factors, specifically by the fact that rock at the well bottom is often encroached with underground water which is usually mineralized. At a natural salt concentration of underground water above 5% introduction of additional substances into drill mud brings no sensible reduction of rock hardness in the well bottom (Yevseiev 2010). The above effects may be manifested to the maximum in a dry bottom with porous rock. In addition, it is proposed to use reagents which would promote the formation of adsorption layers on the surface of crushed material, i.e. hydrophilization reagents. With the latter in use, the grain surface gets hydrophilized, rock shearing resistance reduced, which correlates with the mechanism of the identified effect of adsorption opening of the pore space surface.

Table 1. Main parameters and properties of coal-water slurries[*].

Parameters	Option	Slurry retention time, hours					
		0	24	72	120	240	360
Shear stress τ (Pa) at shear rate $\dot{\varepsilon} = 9s^{-1}$	1	10.5	12.5	13.6	14.2	16.0	16.2
	2	13.5	13.4	13.5	13.6	13.6	13.5
Apparent viscosity η (Pa·s) at shear rate $\dot{\varepsilon} = 9s^{-1}$	1	1.17	1.39	1.51	1.58	1.78	1.80
	2	1.50	1.49	1.50	1.51	1,51	1.51
Static stability, %	1	100	96.8	90.8	85.0	74.6	66.7
	2	100	100	100	100	99.6	98.8

[*]1, 2 – crushing with and without pre-wetting of coal, respectively.

4. The effect of the shift of the hydrophilic-hydrophobic balance of the total external surface of porous solids to the hydrophilic area after solids are crushed may also manifest itself spontaneously during rock breaking which precedes, for example, sill flow. With the above in view, in areas prone to such breaking it is necessary to determine the degree of rock porosity and surface condition of pores. In case of high porosity and oxidation (hydrophilization) of the surface of rock pores the breaking of the latter will result in easily flowable and more stable slurries (suspensions).

2 CONCLUSIONS

1. The effect of adsorption opening of the solids pore space surface accompanies well-known Rehbinder effect and involves transition of the inner surface of pores into the external surface of grains when solids are crushed. A change of the hydrophobic-hydrophilic balance of the solid phase when crushed, resulting from the effect of adsorption opening of the solids pore space surface, takes place if the hydrophilic-hydrophobic balance of the inner surface of pores, which in case of Rehbinder effect in place becomes the external surface, is different from the hydrophobic-hydrophilic balance of the newly formed external surface of crushed dry material (if there is no Rehbinder effect).

2. The effect of adsorption opening of the solids pore space surface has a number of applied aspects. Due to the effect of a shift of the hydrophilic-hydrophobic balance of the coal total external surface to the hydrophilic area after coal is crushed, rheological and stability characteristics of highly loaded coal-water slurry (coal-water fuel) are improved. When the mineral is broken the above effect has an impact on physical and chemical properties of crushed material surface and hence its floatability and dewaterability. Hydrophilization of the crushed material surface reduces rock shearing resistance and therefore intensifies the drilling process.

REFERENCES

Biletskyi, V.S., Sergeiev, P.V., Krut, O.A., Svitlyi, Y.G. & Zubkova, Y.M. 2012. *Effects of adsorption decrease of hardness and solids pore space surface opening in production of coal-water fuel.* Mineral processing. Issue #48 (89): 54–60.

Biletskyi, V.S. 2013. *Effect of adsorption opening of solids pore space surface.* Miners' Forum 2013. Proceedings of the International Conference. 2–5 October 2013, Dnipropetrovs'k: National Mining University: 222–225.

Biletskyi, V.S., Svitlyi, Y.G. & Krut, O.A. 2012. *Use of Rehbinder effect in production of coal-water fuel.* Mining Herald. Kryvyi Rih State University, 95: 40–144.

Yevseiev, V.D. 2010. *On potentiality of using Rehbinder effect in drilling wells.* Izvestiya of Tomsk Polytechnic University. Issue #1. Vol. 317: 165–169.

Progressive Technologies of Coal, Coalbed Methane, and Ores Mining – Bondarenko, Kovalevs'ka & Ganushevych (eds)
© *2014 Taylor & Francis Group, London, ISBN: 978-1-138-02699-5*

Rotary drilling system efficiency reserve

M. Dudlya, V. Sirik, V. Rastsvetaev & T. Morozova
National Mining University, Dnipropetrovs'k, Ukraine

ABSTRACT: To make deep well drilling efficient, it is required to apply sinking increasers which 2.7 to 9.3 times speed up drilling bit rotation. Optimum drilling bit and composite bore bit rotations are 700 to 100 min^{-1}. Torque of drilling pipes becomes from 2 to 5 times less owing to low rotation of drilling pipes having thin walls. Mechanical drilling speed increases from 2 to 4 times; bore bit resource become 50 to 70% higher.

1 INTRODUCTION

Despite a number of disadvantages, rotary drilling rigs are still basic ones while drilling deep oil and gas wells. The wells depth is 5 km and more; the wells diameter is hundreds mm; and the drilling pipes weight is hundreds tons. While rotating, boring column consisting of a set of heavy-weight drilling pipes (HWDP), columns of drilling pipes, and their conjunctions consumes more power if the well depth is more than 2000 m.

Efficiency of face rocks destruction depends on the load applied to rock-breaking tool – boring bit – and on its rotation frequency. The drilling technology parameters determine capacity value transferred from rotating mechanism (rotor); they are limited by capacity of a drive and strength properties of the rotating mechanism. According to the State Standard (GOST 16293-89, Mysluk et al. 2002) current drilling rigs of 7–10 class (BU-4000, BU-5000, BU-6500, BU-8000 according to 26-02-807-76 Branch Standard) are equipped with a rotor drive which power is 370 to 440 kW. If rotor speed is 100 rotations per minute, then operating torque on a rotary table of U7-560-6 and U7-760 types is within 30–35 kNm (Epshtein et al. 1979).

Following drilling bits with 112 to 490-mm diameter are applied as rock-breaking tool while drilling deep wells:
– three-cutter (GOST 20692-750) finger bits with abrasion-resisting material facing and reinforced by carbide inserts;
– wing pike-shaped two-, three-, and six-bladed bits of crushing and cutting type;
– diamond bits reinforced by natural diamonds and artificial ones;
– bits by V.N. Bakul Institute of Superhard Materials reinforced by composite elements of slavutich type;

– milling bits reinforced by diamond inserts of stratapaks type.

Depending upon a bit type, its diameter, and abrasion-resisting material, optimum drilling parameters differ greatly, being:

On axial load:
– 0.05 to 0.15 kN/mm for bore bits;
– 0.4 to 1.0 kN/mm for rolling drilling bits; and
– 0.25 to 0.50 kN/mm for diamond bore bits, and diamond and hard-alloy drilling bits.

On peripheral velocity (rotation velocity for such bit diameters as 244.5 and 190.5 mm):
– 100 to 280 and 150 to 350 rotations per minute for bore bits;
– 40 to 120 and 50 to 200 rotations per minute for rolling drilling bits; and
– 700 to 1000 rotations per minute for diamond drilling bits and composite ones.

On torque:
– specific torque value for rolling drilling bits is 10 to 15 Nm/kN according to information by productive enterprises;
– specific torque is 1.5 to 2 times higher for milling bits to compare with rolling drilling bits; and
– specific torque for diamond drilling bits and composite ones is 7 to 12 Nm/kN if the diameter is 190.5 to 215.9 mm.

Torque for drilling column rotation in tens times more than the torque value required to destroy rocks of a face.

The National Mining University developed innovative equipment to provide optimum conditions for large-diameter wells drilling within hard rocks:
– bottom-hole bit feeding mechanism (BHBFM) for axial load developing;
– well increaser to develop high rotations of a bit under insignificant rotation velocity of drilling assembly.

2 BOTTOM-HOLE BIT FEEDING MECHANISM

Axial load on boring bit is the basic drilling parameter influencing on the quantity of energy transferred to a face, and consumed for the face rocks destruction. In the process of wells drilling, linear dependence is available between axial load and mechanical velocity; however, if a well is rather deep, load losses on a bit reaches 70% due to friction of drilling pipes on well walls. Figure 1 shows data concerning load transfer coefficient K depending upon the well depth L as a result of measuring pressure applied to drilling crown. The measurements were performed by face tracer (ZPON-73) while using diamond-drilling of wells which diameter is 76 mm, and interval of well depth is 296 to 638 m. Drilling assembly rotation velocity was 118 and 288 rotations per minute, and axial load according to MKN-1 device was 7.0 to 1.5 kN.

Test drilling using ZMP-73 was performed within similar modes as drilling by standard assembly and column set. Standard drilling assembly consisted of column set (diamond drilling crown, core-cutter, and single column pipe) and 50-mm RTJ drilling pipes (SBTM-50). Drilling assembly with UKN-1 consisted of diamond crown, core-cutter, ejector core equipment, hydraulic hammer GV-1, and SBTM-50 drilling pipes. Test assembly consisted of a column set, ZMP-73face feed unit, and SBTM-50 drilling pipes.

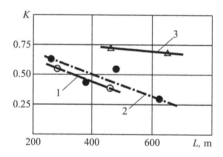

Figure 1. The dependence of axial load transfer coefficient on well depth and the type of drilling assembly: 1 – Standard drilling assembly; 2 – Versatile column assembly (UKN-1); 3 – Face feeding mechanism (ZMP-73).

ZMP-73 face feeding mechanism (Figure 2) consists of control section 1, several intermediate sections 2, and valve section 3. Sections encompass: control piston 4, intermediate pistons 5, valve piston 6, spinning shaft 7, adaptor 8, sleeve 9, spring 10, valve sleeve 11, valve 12, cylinder 13, spinning shaft 14, nozzle 15, nut 16, and sealing cups 17 and 18.

ZMP-73 operates as follows. When drilling assembly is placed on a face, spinning shaft 7, valve 12, and pistons 4, 5 and 6 are raised. Along central channel 15 and internal channels in pistons, fluid gets into restriction hole of valve 12. Pressure fail occurs in liquid acting on pistons. Total pressure on spinning shaft takes place; through adapter and column pipe it is transferred to a drilling crown (a bit). Weight of drilling pipes develops load on ZMP-73; the load is either similar or higher than pressure by ZMP-73. As a result, drilling assembly rotates. Such a drilling mode generates effective destruction of a face rocks.

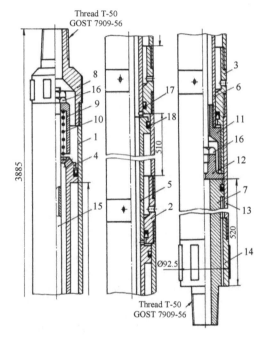

Figure 2. ZMP-73 face feeding mechanism (motor assembly): 1 – control section; 2 – intermediate section; 3 – valve section; 4 – control piston; 6 – valve piston; 7 – spinning shaft; 8 – adapter; 9 – sleeve; 10 – spring; 11 – valve sleeve; 12 –valve; 13 – cylinder; 14 – spinning shaft; 15 – nozzle; 16 – nut; 17 and 18 – sealing cups.

When a run is over, drilling pump is de-energized, liquid pressure drops, and drilling assembly is lifted from the well.

Less effective drilling mode is available; however it is simpler as for drilling process control. In this case after column assembly has been placed on a face, hydraulic rods of drilling machine (SKB-type machine with hydraulic feeding) are fixed in up position; in such way grease can not flow into hydraulic cylinders. When drilling pump is energized and a shaft of drilling machine starts rotating, drilling assembly penetration per stroke value of control piston 4 takes place. In this context, valve 12 is separated from a piston of spinning shaft 7 by the spring 10; pressure in ZMP-73 drops. Thus, drilling assembly together with pistons of drilling machine hydraulic feed moves down up to placing the valve on a piston of spinning shaft. The flow is stopped; pressure failure in a restriction arises, and drilling process takes place again.

In the process of ZMP operation up-directed reactive force equal to weight capacity arises due to action of liquid pressure on a container throats. The force is compensated by the weight of drilling pipes and forces of pipes friction on well walls; they are also down-directed being reaction of drilling pipes interaction in the process of their moving down. They depend on frequency of pipes rotation as well as on the length of compressed part of the pipes.

Hence, the length of compressed part of the pipes (weight of compressed part) can be reduced by the value of pipes friction forces on well wall. According to the results of measurements, axial load losses while drilling by ordinary assembly are 40 to 70%. Taking pressure reserve as 30 to 40%, axial load reduced by the value can be estimated while drilling a well with the help of ZMP-73 with constant movement of assembly. That is

$$C_{ass} = (0.6...0.7)C_{bit}, [\text{H}]$$

where C_{ass} – load on a drilling assembly developed by drilling pipes weight, H; C_{bit} – load on drilling crown (bit) developed by ZMP-73.

Results of test wells drilling using ZMP-73 are in Table 1.

Table 1. Results of test wells drilling using ZMP-73.

Expedition	Assembly type	Drilling volume, run	Distance per run		Drilling velocity	
			m	%	m per hour	%
L'viv	Standard	48	4.00	100	0.86	100
	ZMP-73	51	5.55	138	1.32	153
Transcarpathia	Standard	52	2.96	100	1.26	100
	ZMP-73	74	3.75	127	2.15	170
Kryvyi Rih	Standard	18	3.67	100	1.14	100
	ZMP-73	10	4.15	113	1.45	127

To increase efficiency of applying face feeding mechanisms and to avoid using heavy-weight drilling pipes in drilling column for developing reactive power especially while drilling deep oil, gas, and methane wells in Donbass, face feeding mechanisms are designed. The mechanisms consist of two systems: feeding device of ZMP-73 type tested and confirmed possible high technical and economic drilling ratios, and anchor device transferring reactive power on well walls. Table 2 shows technical characteristics of face mechanisms to drill wells for hard, liquid, and gaseous minerals.

Table 2. Technical characteristics of ZMP.

Ratio	Measurement unit	ZMP standard size		
		ZMP-89	ZMP-146	ZMP-219
Body diameter	mm	90	166	44.5
Piston diameter	mm	80	125	200
Piston stroke	mm	200	200	200
Pump feeding	l per minute	120	300	1020
Axial load	kN	27	120	200
Length	mm	4.5	4.0	3.5
Weight	kg	120	180	230

Today horizontal directional drilling of wells along mineral layer is widely used; that provides significant increase in well flow and its service life.

Drilling assembly consisting of a face feeding mechanism which develops optimum load on a drilling bit, and downhole drilling motor rotating drilling bit provide significant increase in cost-performance ratio of drilling along mineral layer – water, oil, natural gas, and methane.

Such a drilling approach need not rotation of a column of drilling pipes; heavy drilling pipes are not required which favours material supply cost saving, and, subsequently, cost of drilling.

Drilling assembly consisting of ZMP and screw

motor gives ability to change direction of well axis not only in vertical plane but also in horizontal one along mineral layer.

3 FACE INCREASERS

3.1 *Problem formulation*

To make destruction of rocks of well face effective, rotational frequency of rolling cutter bit should be 250 to 350 rotations per minute. Table 1 shows that required drive capacity to rotate column of drilling pipes within the well due to friction on well walls is 1023 to 4044 kW when rotational frequency is 240 to 360 rotations per minute and well depth is 6000 m. To provide the strength of drilling column at total stresses in pipes depending of pipes weight and torque, large-diameter drilling pipes with increased wall thickness are applied.

Weight of drilling pipes may be reduced owing to weight reduction of 1 meter of pipes (wall thickness), and decrease in rotational frequency of drilling column. Hence, if rotational frequency of drilling assembly is 30 rotations per minute instead of 240 rotations per minute, then torque value will be reduced by 4000–5000 Nm that is 4 to 5 times. Thus, 1 meter of drilling pipes can be reduced by this value, and it will decrease 30–50% decrease of tension loads.

Simultaneously with reducing rotational frequency of drilling pipes column up to 20–40 rotations per minute it is required to increase rotational frequency of rolling cutter drill bit up to 200–350 rotations per minute. Optimum rotational frequency for diamond bits and composite ones is 700–1000 rotations per minute. For this purpose it is required to apply submerged well increaser which 5 to 20 times raises rotational frequency.

3.2 *Power calculation as for drilling assembly rotation*

While drilling torque on a rotor of drill rig consists of following components

$$M = M_1 + M_2 + M_3 + M_4, \text{[Nm]} \qquad (1)$$

where M – total torque on rotor required to rotate drilling assembly and to destruct face rocks by a bit, Nm; M_1 – torque required to rotate drilling pipes in casing column, Nm; M_2 – torque required to rotate drilling pipe in uncased area of a well, Nm; M_3 – torque required to rotate heavy-weight drilling pipes, Nm; and M_4 – torque required to rotate drilling bit (to destruct face rock), Nm.

To analyze effect of drilling column rotational frequency on torque value (drilling power), assume the following:

– well can not have full-length casing column;

– heavy-weight drilling pipes are not available in drilling column; and

– power consumptions for face rocks destruction are similar under various drilling modes.

Such assumptions enable determining relative effect of rotational frequency and drilling column length according to power on idle rotation of pipes in a well. V.I. Matseichik formula (GOST 16293-89) for wells with 75–90° angle of slope to horizon for power consumption as for idle rotation of drilling pipes is

$$N = 0.5 L_p \gamma \mu k D \omega q f \left[\frac{\omega^2}{g} \left(\frac{k - \beta}{1 - \beta} \right)^2 \times \right.$$

$$\left. \times \left(0.43 - \frac{0.6}{1 + \sqrt{1 + H} + \sqrt{1 + \sqrt{1 + H}} \sqrt[4]{H}} \right) + \frac{sin\,\alpha}{f} \right], \quad (2)$$

where $\gamma = 1.0$ – coefficient taking into account radius of pipe flexure in a well; $\mu = 0.30$ – resistance coefficient when pipe column rotates; $k = 1.2$ – coefficient of well shaft development; $D = 0.250$ m – well diameter; ω – angular velocity of pipe axial rotation, rad per second; $q = 25.7$ kg – weight of 1 meter of a pipe; $f = 0.5(D - d) = 0.5(0.250 - 0.114) = 0.055$ m – maximum pipe flexure in a well; $g = 9.81$ meters per s^2 is gravity acceleration; $\beta = d/D = 114/250 = 0.45$ – coefficient taking into pipe diameter-well diameter ratio; L_p – length of expanded parts of pipes, m; we assume that the whole column of drilling pipes is expanded; $\alpha = 90°$ – angle of slope of a well to horizon;

$$H = \frac{10.72 \cdot E \cdot I \cdot \omega^2}{L_p^2 \cdot g \cdot q} \left(\frac{k - \beta}{1 - \beta} \right)^2 =$$

$$= \frac{10.72 \cdot 2 \cdot 10^6 \cdot 449 \cdot 10^{-4} \omega^2}{L_p^2 \cdot 9.81 \cdot 25.7} \times$$

$$\times \left(\frac{1.2 - 0.45}{1 - 0.45} \right)^2 \frac{\omega^2}{L_p^2} = 7110 \frac{\omega^2}{L_p^2}.$$

I.S. Kalinin formula obtained empirically according to the results of research while drilling wells which diameter is 76 to 93 mm, and which depth is up to 2000 m is

$$N = 4.6 \cdot 10^{-8} \cdot L \cdot n^2 \cdot \sqrt[3]{n}, \qquad (3)$$

where n – rotational frequency of pipes, rotations per minute.

G.M. Gevinian analytical formula taking into account diameters of wells and drilling pipes is

$$N = \beta \cdot 10^{-7} \cdot L \cdot n^2 \cdot \tau \cdot \frac{D^2 \cdot d^2}{D^2 - d^2}, \qquad (4)$$

where $\beta = 2.0\ldots3.5$ – coefficient taking into consideration effect of real conditions of a well; τ – shearing stress of clay mud, g per cm^2; and D and d – diameters of a well and drilling pipes, cm.

Assume following additional data to calculate power consumption:
– well depth $L = 1000, 3000, 5000$ and 6000 m;
– rotational frequency $n = 3.14; 6.28; 12.56; 25.12;$ and 37.68 rad per sec;

– angle of well slope to horizon $\theta = 90°$;
– axial load on a bit $P = 200$ кN;
– thickness of drilling pipes wall $\delta = 10$ mm;
– diameter of drilling pipes $d = 114$ mm;
– weight of 1 meter of a pipe $q = 290$ N;
– well diameter $D = 245$ mm;
– gaping between drilling pipes and well walls $\delta = 60$ mm.

Table 3 shows data of power calculation according to V.I. Matseichik formula required for idle rotation of drilling pipes.

Table 4 shows calculation values of power and torque at idle rotation of drilling pipes in a well according to G.M. Gevinian formula.

Table 3. Comparative data of power consumption for idle rotation of drilling pipes in a well according to V.I. Matseichik formula.

Well depth, m	Power consumption (kW) at rotational frequency, rad per sec (rotations per minute)				
	3.14 (30)	6.28 (60)	12.56 (120)	25.12 (240)	37.68 (360)
1000	0.34	2.47	29.2	272	989
3000	1.02	4.95	69.4	611	2359
5000	1.69	11.95	106.2	956	3499
6000	2.03	14.15	124.0	1023	4044

Table 4. Calculation values of power and torque at idle rotation of drilling pipes in a well according to G.M. Gevinian formula.

Well depth, m	Values at rotational frequency, rotations per minute									
	30		60		120		240		360	
	N, kW	M, Nm	N, kW	M, Nm	N, kW	M, Nm	N, kW	M, Nm	N, kW	M, Nm
1000	0.54	180	2.16	360	8.64	720	26.50	1100	77.80	2160
3000	1.62	540	6.48	1080	25.92	2160	79.50	3300	233.4	6480
5000	2.70	900	10.80	1800	43.20	3600	132.50	5500	389.0	10800
6000	3.14	1080	12.96	2160	51.84	4320	159.00	6600	466.8	12960

The data shows that up to depth of 2000 m a well drilling is possible at such rotational frequency of drilling assembly as 240–300 rotations per minute; increaser in not required. Deeper than 2000 m, increaser makes positive effect in the form of 3 to 5 times power drop on drilling assembly rotation; torque reduction is 10 to 20 times, electric power (diesel fuel) saving is 3 to 5 times, and 2 to 3 times increase in mechanical velocity of drilling and headway per bit.

3.3 Description of well increasers

Three variations of increases for different conditions of deep wells drilling have been developed according to current project:
– one-stage with 2.7 gear ratio; and
– two-stage with 4.4 gear ratio;
– three-stage with 9.3 gear ratio.
One-ratio increaser MZ-245/2.7 is designed to

transfer rotation from drilling pipes (heavy-weight drilling pipes) to drilling bit with 2.7 times increase of rotational frequency. Three-roller cutter bits with 244.5–290.5 mm diameter are used as rock distractive tool. MZ-245/2.7 increaser has a form of cylinder body with anchor runners which prevent a body from rotation, and one-stage planet increaser. Both assembling units are connected with threaded rods and screws.

Figure 3 shows one-stage increaser (without elements of anchor assembling unit).

Two-stage MZ-219/4.4 increaser differs from one-stage increaser by availability of extra planet mechanism. Thus, gear ratio is 4.4. applying two-stage increaser is rational while drilling by diamond drilling bits and composite ones which rotational frequency is 700 to 1000 rotations per minute at significant well depth when power of rotor drive provides application of optimum drilling modes.

Three-stage MZ-168/9.3 increaser's gear ratio is

9.3; that provides ability to develop optimum frequency of drilling bit rotation while drilling a well by diamond bits and composite ones at negligible (20 to 40 rotations per minute) frequency of rotor rotation. At such a drilling mode required power of rotor drive is quite lower than while drilling without increaser.

Table 5 shows designed specifications of face increasers.

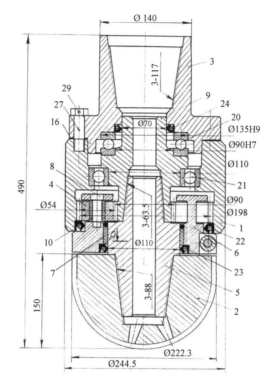

Figure 3. One-stage well increaser: 1 – body; 2 – drilling bit; 3 – adapter; 4 – satellite; 5 – nipple; 6 – carrier; 7 – bearing; 8 – axis; 9 – flange; 20 and 21 – bearings; 22, 23 and 24 – sealing rings; 27 – bolts.

Table 5. Specifications of face increasers.

Ratio	Measurement units	Value		
Increaser type	–	MZ-245/2.7	MZ-219/4.4	MZ-168/9.3
Body diameter	mm	244.5	219	168
The number of stages	–	1	2	3
Gear ratio	times	2.7	4.4	9.3
Allowable axial load	kN	200	150	100
Allowable torque	kNm	3.0	2.5	2.0
Frequency of pipes rotation	rot per min	77–154	32–101	32–77
Length	mm	800	1000	1200
Weight	kg	180	160	150

3.4 *Increaser operation while drilling a well*

Figure 4 shows a well drilling up to 6000 m depth applying well increasers. Drilling bit is either diamond bit or a bit reinforced by superhard materials. Optimum frequency of bit rotation is 700 to 1000 rotations per minute. Shaft predrilling according to a direction (a conductor) is performed at maximum frequency of drilling rig rotor rotation (245 rotations per minute) (Maslennikov & Matveev 1981). Drilling with the use of one-stage increaser with such frequency of drilling column rotation as 101 rotations per minute up to reaching power of rotor rotation of about 300 kW is carried out from 1100 to 2700 m of depth. A well drilling is continued up to designed depth with two-stage increaser application. Frequency of drilling pipes rotation is 77 rotations per minute, frequency of bit rotation is 920 rotations per minute, and power of rotor rotation is up to 270 kW.

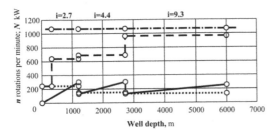

Figure 4. Diagram of a well drilling by diamond drilling bit with the use of increaser: ·····o····· – frequency of drilling pipes rotation; – -o- – – frequency of bit rotation; —o— – power of rotor drive; –··-o-··– – gear ratio.

Figure 5 shows a well rotor drilling by roller cutter bits with increaser. Frequency of roller cutter bit rotation up to the well bottom is 245–272 rotations per minute that is close to optimum. Drilling specifications will be much higher than while traditional rotor drilling of deep wells.

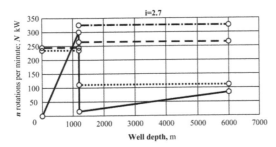

Figure 5. Diagram of a well drilling by roller cutter bit with the use of increaser: ·····o····· – frequency of drilling pipes rotation; – -o- – – frequency of bit rotation; —o— – power of rotor drive; –··-o-··– – gear ratio.

4 ADVANTAGES OF NEW METHOD OF WELLS DRILLING

4.1. Ability to apply drilling pipes with reduced wall thickness and drilling pipes with smaller diameter.

4.2. Ability to reduce rotor drive power and power intensity of face rocs destruction.

4.3. Reducing intensity of a well axis flexure owing to drilling assembly centering in a face area.

4.4. Ability to apply staged rock-destructive tool which provides use of various parameters of drilling mode.

4.5. Ability to design new effective drilling bits.

5 CONCLUSIONS

5.1. While developing axial load on drilling bit (crown) with drilling pipes weight, actual effort is 2–3 times less than given one.

5.2. Applying face feeding mechanism ZMP-73 enables to increase load on drilling bit (crown) and to lengthen 1.13–1.38 times distance per run when mechanical velocity of drilling becomes 1.27–1.70 times more.

5.3. Designed ZMP set in which anchor device is applied will help to avoid heavy-weight pipes application in drilling column, and to develop optimum load on a bit.

5.4. Projected growth of a bit resource will be 2–5 times, and increase in mechanical velocity of drilling is 1.5–2.5 times.

5.5. Drilling assembly consisting of ZMP and propelling screw gives ability to change direction of horizontal area of a well.

REFERENCES

GOST 16293-89. Drilling packaged plants for production and deep exploratory drilling. Basic parameters.
Mysluk, M.A., Rybchych, I.Y. & Yaremchiuk, R.S. 2002. Wells drilling. Book No 1. General bulletin. Drilling rigs. Equipment and tool. Kyiv: Interpress LTD: 332.
Epshtein, Ye.F., Matseichik, V.I., Ivahin, I.I. & Asaturin, Sh. 1979. Calculation of drilling pipes in geological prospecting drilling. Moscow: Nedra: 160.
Maslennikov, I.K. & Matveev, G.I. 1981. Tool for wells drilling. Reference aid. Moscow: Nedra: 335.

Impurity concentration test while moving highly dispersive systems

V. Golinko, D. Saveliev, Ya. Lebedev & T. Morozova
National Mining University, Dnipropetrovs'k, Ukraine

ABSTRACT: The results of study directed to improving sanitary and hygienic labor conditions within workings at rock fragmentation by drilling and blast method are given. According to the theory of aerodispersive system movement and on the basis of solution of differential equation of eddy diffusion of the dust particles suspended within ventilation flow, theoretical dependence which enables to determine fine dust concentration at any distance from the dust source along the length of the blind drift during the period of its ventilation taking into account air leak from the air duct was obtained.

1 INTRODUCTION

At steady-flow process and availability of momentary sources of impurities precipitation (explosions, sudden outburst, etc) dust and gaseous clouds on which free air current flowing from the duct affects for a very short period of time is injected into the working (Ushakov, Burchakov et al. 1987).

Impurities (dust, gas) located outside the boundaries of the free current penetrate into the air due to turbulent diffusion. Thus, mixing and intensive discharge of the dust and gaseous impurities from the face area takes place (Ushakov, Burchakov et al. 1987; Miletich, Yarovoy & Boiko 1972). Dust and product of explosion distributed along mine workings worsens sanitary and hygienic characteristic of the mine atmosphere at the working places. Various methods of lowering dust formation at blasting operations are used to reduce dust content of the mine atmosphere while driving workings by drilling-and-blasting method. While performing blasting operations it is required to determine dust concentration at any length of the blind drift during the whole period of its ventilation taking into account air leak from the air duct to determine efficiency of applied methods and ways of dust formation. It is possible to do that on the basis of solving differential equation of turbulent diffusion.

Nature of the motion of the dust particles and gas molecules is different, but to a certain degree. Dispersiveness of the dust particles is a limiting value.

2 FORMULATING THE PROBLEM

The works by Fuks N.A., Diakov V.V., Mednikov K.P., Belousov V.V., Ksenofontova A.I., Ushakov K.Z., Voronin V.N., Barenblatta G.I., Zhuravlev V.P., Klebanov F.S. and others are devoted to the motion of the solid particles within turbulent flow.

According to the theory of suspended particles within turbulent flow the nature of their motion within mine workings is determined by pulsation spectrum which is the characteristic feature of turbulent flows (Rsenofontova & Burchakov 1965).

In turbulent flows oscillatory modes which have various amplitudes and directions are heterodyned on the mean motion. Turbulent pulsations are characterized by the rate of speed and distances where they have noticeable changes. These distances are called motion scale. The quickest pulsations have the largest motion scale. For instance, when turbulent motion takes place within the channel the largest pulsation scale is the same as the channel diameter and the pulsation speed is the same as the speed in the center of the channel. Such large-scale pulsations make up the main portion of turbulent motion kinetic energy and perform substance transfer. Thus, there will be no low accuracy to admit that turbulent diffusion of the particles within the complex flow that is far from the walls is equal to the turbulent diffusion of the gas molecules (Rsenofontova & Burchakov 1965).

3 MATERIALS UNDER ANALYSIS

The task of particle vertical distribution within horizontal flow taking into account the action of the gravitational field is solved by N.A. Fuks. Taking into consideration the process of the balance of particle impaction under the action of the gravity force and their transition into the upper layer of the flow under the influence of turbulent diffusion he has got the following equation of distributing particles throughout the height

$$ln\frac{C_y}{C} = -\frac{\upsilon_e y}{D_t}, \qquad (1)$$

where C – particle concentration at the channel bottom, mg/m^3; C_y – concentration throughout the height of y, mg/m^3; y is the channel height, m; υ_e – rate of particles settling within still air, m/c; D_t – coefficient of turbulent diffusion, m^2/c.

$$D_t = 0.044\nu Re^{0.75}, \qquad (2)$$

where Re – Reynolds number; ν – kinematic air viscosity, m^2/c.

From equations (1) and (2) it can be inferred that

$$C_y = C\,exp\left(-\frac{\upsilon_e y}{0.0044\,Re^{0.75}}\right). \qquad (3)$$

Calculations performed according to the Equation (3) for the working with 8 m^2 cross-section area show that at 0.25–0.3 m/c speed of the air flow and diameter of the quartz particles up to 5 mkm, aerosol concentration throughout the height is practically the same. Experimental studies performed by V.V. Nedin and O.D. Neikov confirmed the truth of given calculations (Nedin & Neikov 1965).

Therefore, quartz particles with diameter up to 5 mkm are completely absorbed by the turbulent mass within such flows. Effect of gravity force which is far from the working walls is too low in comparison with the effect of aerodynamic forces.

Dust content of air is reduced in mines as removing from the sources of the dust release. This phenomenon is explained by the impaction of the dust particles on the walls, roof, and the working floor. Studies show that the dust is settled on the working surface at the low rates. Ventilation stream has also a great speed. Thus, in ore mines the accumulation of the dus within air workings where air velocity is 5–7 m/c takes place. Impaction of the dust particles within turbuletnt flow is possible under the influence of gravitational forces, diffusion processes and inertial forces. Analysis of theoretical and experi-

mental studies (Romanchenko, Rudenko & Kosterenko 2011) shows that the particles are completely absorbed by pulsations within turbulent flow at sufficient distance from the walls. Their movement along the flow and throughout the flow cross-section is mainly determined by the dynamic parameters of the flow. Gravitational forces do not influence on the movement of the fine particles. It was determined that the dust settling with particle size less than 10 mkm under the influence of gravitational forces is practically impossible in the drift at the speed of ventilation stream more than 0.2 m / c (Nedin & Neikov 1965). Thus, it is possible to confirm that the difference of the processes of turbulent movement of solid particles and gas molecules depends on the dust dispersiveness. For the fine-grained systems this difference can be neglected without accuracy loss of practical calculations. While comparing the motion pattern of suspended particles and gas molecules within turbulent flow it should be started from the general equations of mechanics of the turbulent motion of multicomponent media.

It is possible to obtain theoretical dependence which enables to determine concentration of the fine dust (< 10 mkm) at any distance from the source of the dust formation throughout the length of the blind drift during the period of its ventilation taking into account air leak from the air duct on the basis of differential equation of turbulent diffusion.

After explosion the process of ventilating the long blind workings while injecting the fresh air through the air duct is in continuous tension and deformation of the dust cloud along the working by the leakage of the additional air (Figure 1) (Miletich, Yarovoy & Boiko 1972).

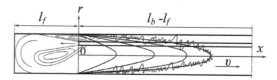

Figure 1. The chart of deformation and tension of the dust cloud within the blind working at nonstationary process of the dust inflow and forced ventilation.

While moving the dust cloud from the face dust liquefaction and consequently the change of its concentration throughout the length of the mine working takes place. Air leak from the air duct influences on the degree of liquefaction of the dust cloud. Besides, the dust is settled within the working. It also influences on the change of its concentration.

Thus, two basic processes, that is, deformation and tension of the dust cloud by the yield of the fresh air as well as dust settling from the cloud due to the forces of turbulent diffusion which have pulsation nature with continuous vector change and determined forces with constant direction and module (forces of gravitational settling, resistance forces, aerodynamic forces of ventilation airflow) influence on the change of the dust concentration along the length of the working. So, total change of the fine airborne dust concentration stipulated by the resultant actions of above mentioned forces will be determined by the equation

$$C(x) = C_T(x) + C_Q(x),\tag{4}$$

where C_T and C_Q – correspondingly, the change of the airborne dust concentration along the length of the working stipulated by the phenomenon of turbulent diffusion and increase of air consumption within the working, mg/m³.

Correlation of given forces is different for the particles of various mass determined by its size (diameter) and quasidensity (relationship of aerodynamic active volume of the particle with its mass). The speed of the air flow or the first derivates of coordinates with time influence on the correlation of forces (Romanchenko, Rudenko & Kosterenko 2011).

Within considered task of determining concentration of the fine airborne dust (≤ 10 mkm) it is possible to neglect the influence of determined forces without the loss of accuracy. It is assumed that given dust fractions are turbulent settled particles. For these particles weight force is completely compensated by aerodynamic forces and gravitational acceleration is less or equal to the acceleration given to the particle by aerodynamic forces.

Dependence of the change of the aerosol fractions concentration subjected to the processes of turbulent settling according to the performed studies has the following type

$$C_T(x) = C_0 \exp(-k\Delta x \upsilon_x),\tag{5}$$

where C_0 – dust concentration in the face zone, mg/m³; k – mass ratio of the airborne monodispersive dust within considered primitive volume of the working to the mass of settled dust per the area unit.

To determine dust concentration taking into account deformation of the dust cloud, mathematical task can be solved as follows.

It is assumed that at the initial instant ($t = 0$) dust concentration is the function of the working radius, i.e. $C = C_0\varphi(r)$ in the cylindrical coordinate system datum point of which is located within the flat surface of the discharged duct (Figure 1).

It is assumed that airborne dust doesn't settle to the surface of the working ($r = R$). The speed of

the air movement (υ) of considered section is taken equal to its average value. Dust concentration beyond the face zone ($x = 0$) depends on the radius of the working and is changed exponentially

$$C_Q = C_0 \exp\left(-\frac{\upsilon t}{l_3}\right) f(r).\tag{6}$$

Taking into account given assumption the change of airborne dust concentration can be described by the equation of turbulent diffusion:

$$\frac{\partial c_Q}{\partial t} + \upsilon\frac{\partial c_Q}{\partial x} + AC = D_x\left(\frac{\partial^2 c_Q}{\partial x^2}\right) +$$
$$+ D_r\left(\frac{1}{r}\frac{\partial c_Q}{\partial r} + \frac{\partial^2 c_Q}{\partial r^2}\right),\tag{7}$$

conditions of uniqueness

$$C_Q(0, r, x) = C_0\varphi(r).\tag{8}$$

$$C_Q(t, r, 0) = C_0\varphi r \exp(-\upsilon t / l_f).\tag{9}$$

$$\frac{\partial C_Q(t, R, x)}{\partial r} = 0.\tag{10}$$

The value taking into account the change of airborne dust concentration due to air leak from the air duct is determined by the equation

$$A = \frac{(Q_f - Q_i)\upsilon}{Q_i x},\tag{11}$$

where Q_i – amount of air which intakes into the face area, m³/c; Q_f – amount of air at the beginning of the working, m³/c.

Method of Hankel end integral transformation was used for solving given problem in the following type

$$\overline{C_Q}(t, \varepsilon, x) = \int_0^R r C_Q(t, \varepsilon, x) I_0(\varepsilon r) dr,\tag{12}$$

with the formula of inversion

$$C_Q(t, r, x) = \frac{r}{R^2}\overline{C_Q}(t, 0, x) +$$
$$+ \frac{r}{R^2}\sum_{i=1}^{\infty}\frac{I_0(\varepsilon_i r)}{I_0^2(\varepsilon_i R)}\overline{C_Q}(t, \varepsilon_i, x),\tag{13}$$

after that variable r was excluded from the quotation.

133

Furthermore, Laplas integral transformation according to the variable t was used in the form

$$C_Q(p,\varepsilon,x)=\int_0^\infty exp(-pt)\overline{C_Q}(t,\varepsilon,x)dt .\qquad (14)$$

By the help of the Table (Ditkin & Prudnikov 1965) transformation according to the variable t was performed.

After given transformation solution of the problem in dimensionless form was obtained.

$$\frac{C_Q}{C_0}=2\int_0^1 y\Phi(y)dy\cdot exp(-BFo)+\int_0^1 yF(y)dy\cdot exp\left(\frac{Pe}{2}x-\beta Fo\right)\cdot\left\{exp\left(-\sqrt{\frac{Pe^2}{4}+B-\beta}x\right)\times\right.$$

$$\times erfc\left[\frac{x}{2\sqrt{Fo}}-\sqrt{\left(\frac{Pe^2}{4}+B-\beta\right)Fo}\right]+exp\left(\frac{Pe^2}{4}+B-\beta x\right)\cdot erfc\left[\frac{x}{2\sqrt{Fo}}-\sqrt{\left(\frac{Pe^2}{4}+B-\beta\right)Fo}\right]\right\}$$

$$-\int_0^1 y\Phi(y)dy\cdot\int exp(-BFo)\left[erfc\left(\frac{x}{2\sqrt{Fo}}-\frac{Pe}{2}\sqrt{Fo}\right)+exp(Pe\cdot x)\cdot erfc\left(\frac{x}{2\sqrt{Fo}}+\frac{Pe}{2}\sqrt{Fo}\right)\right]+$$

$$+\sum_{i=1}^\infty\frac{I_0(\mu_i y)}{I_0^2(\mu_i)}\left[\int_0^1 yF(y)I_0(\mu_i y)dy\cdot exp\left(\frac{Pe}{2}x-\beta Fo\right)\right]\cdot\left\{exp\left(-\sqrt{\frac{Pe^2}{4}+B-\beta+k\mu_i^2}x\right)\times\right.$$

$$\times erfc\left[\frac{x}{2\sqrt{Fo}}-\sqrt{\left(\frac{Pe^2}{4}+B-\beta+k\mu_i^2\right)Fo}\right]+exp\left(\sqrt{\frac{Pe^2}{4}+B-\beta+k\mu_i^2}x\right)\times$$

$$\times erfc\left[\frac{x}{2\sqrt{Fo}}-\sqrt{\left(\frac{Pe^2}{4}+B-\beta+k\mu_i^2\right)Fo}\right]\right\}-exp\left[-\left(B+k\mu_i^2\right)Fo\right]\cdot\int_0^1 y\Phi(y)I_0(\mu_i y)dy\times$$

$$\times\left\{erfc\left(\frac{x}{2\sqrt{Fo}}-\frac{Pe}{2}\sqrt{Fo}\right)+exp(Pex)\cdot erfc\left(\frac{x}{2\sqrt{Fo}}+\frac{Pe}{2}\sqrt{Fo}\right)\right\}+$$

$$-\int_0^1 y\Phi(y)I_0(\mu_i y)dy\cdot exp\left[-\left(k\mu_i^2+B\right)Fo\right],\qquad (15)$$

where Fo and Pe – correspondingly Fourier and Peclet numbers: $Fo=\dfrac{D_x t}{R^2}$; $Pe=\dfrac{\upsilon R}{D_x}$; $k=\dfrac{D_r}{D_x}$;

$x=\dfrac{l_b-l_f}{K}$; $y=\dfrac{r}{R}$; $B=\dfrac{AR^2}{D_x}$; $\beta=\dfrac{\upsilon}{l_f}\dfrac{R^2}{D_x}$;

$\mu_i=\varepsilon_i R$; $\Phi(y)=\dfrac{\varphi(r)}{C_0}$; $F(y)=\dfrac{f(r)}{C_0}$.

In the absence of harmful impurities within development working before blasting operations $C(0,y,x)=\Phi(y)=0$.

Experimental studies (Nedin & Neikov 1965) show that the change of airborne dust concentration throughout the height of the working is practically the same. Therefore, it is possible to assume that the flow of turbulent diffusion is directed along the axis x. Then $F(x)=1$ and solution (15) can be represented in the following way

$$C_Q = C_0 \frac{1}{2} exp\left(\frac{Pe}{2}x \quad \beta F_o\right)\left\{exp\left(-\sqrt{\frac{Pe^2}{4}+B-\beta x}\right)erfc\left[\frac{x}{2\sqrt{Fo}}-\sqrt{\frac{Pe^2}{4}+B-\beta}Fo\right]+\right.$$

$$\left.+exp\left(\sqrt{\frac{Pe^2}{4}+B-\beta x}\right)erfc\left[\frac{x}{2\sqrt{Fo}}+\sqrt{\frac{Pe^2}{4}+B-\beta}Fo\right]\right\}. \qquad (16)$$

The value B in the Equations (15) and (16) is determined by the formula

$$B = \frac{AR^2}{D_x} = \frac{(Q_f - Q_i)\upsilon R^2}{Q_i x D_x}. \qquad (17)$$

Degree of the change of airborne dust concentration within the face area depends on the activity of the free jet stipulated by the value of the air consumption coming out of the end of the air duct. It is characterized by the value β determined

by equation

$$\beta = \frac{Q_i R^2}{S D_x l_3}. \qquad (18)$$

On the basis of Equation (4), (5), and (16) the change of the fine airborne dust concentration throughout the length of the ventilated mine working taking into account the influence of the effective forces is expressed by the following dependence

$$C(x) = C_0\,exp(-k\Delta x \upsilon_x) + C_0 \frac{1}{2}exp\left(\frac{Pe}{2}x - \beta Fo\right)\left\{exp\left(-\sqrt{\frac{Pe^2}{4}+B-\beta}_x\right)\times\right.$$

$$\left.\times erfc\left[\frac{x}{2\sqrt{Fo}}-\sqrt{\frac{Pe^2}{4}+B-\beta}Fo\right]+exp\left(\sqrt{\frac{Pe^2}{4}+B-\beta x}\right)\times erfc\left[\frac{x}{2\sqrt{Fo}}+\sqrt{\sqrt{\frac{Pe^2}{4}}+B-\beta}Fo\right]\right\}. \quad (19)$$

Initial dust content of the air C_0 within the zones of the explosion products waste is determined by the following formula (Yanov & Vaschenko 1977)

$$C_0 = K_p l_d f \frac{K_w}{K_i}, \qquad (20)$$

where K_p – proportionality factor ($K_p = 60$ for the initial maximum dust content and $K_p = 20$ is for the average dust content); K_w – coefficient depending on the blasting method (under electric blasting $K_w = 0.5$, under ignition blasting $K_w = 1.0$); K_i – coefficient depending on the rocks and water content within the working (for the dry face $K_i = 0.5$; for the face with the humid rocks or artificial irrigation $K_i = 1.0$; for the face with the weak inflow and small drip $K_i = 2.0$; for the watered face with the great drip or internal water

stemming $K_i = 3.0$); f – rock-hardness ratio by the scale of professor M.M. Protodiakonov; l_d – average length of the bore, m.

Under blasting operations along stratified and fractured rocks correlation 0.7–0.9 is making in the formula. Under blasting operations within vertical shafts, inclined workings and raises correlation equal to 0.1; 0.5 and 0.1 is making.

4 CONCLUSIONS

The process of distributing fine dust formed due to the blasting operations while driving development workings is considered. According to the theory of the movement of the aerodispersive systems and on the basis of solving differential equation of turbulent diffusion of the dust particles weighted within ventilation flow, theoretical dependence enabling to determine concentration of the fine dust at any distance from the dust source along the length of the blind drift at the moment of its ventilation taking into account air leak from the air duct is obtained.

REFERENCES

Ushakov, K.Z., Burchakov, A.S., Puchkov, L.A. & Medvedev, I.I. 1987. *Aerology of the mining enterprises*. Moscow: Nedra: 421.

Miletich, A.F., Yarovoy, I.M. & Boiko, V.A. 1972. *Mining and industrial aerology*. Moscow: Nedra: 248.

Nedin, V.V. & Neikov, O.D. 1965. *Dust control in the mines*. Moscow: Nedra: 216.

Ditkin, V.A. & Prudnikov, A.Y. 1965. *Operational calculus manual*. Moscow: Higher school: 346.

Romanchenko, S.B., Rudenko, Yu.F. & Kosterenko, V.N. 2011. *Dust dynamics in coal mines*. Moscow: Publishing House "Mining engineering" LLC "Cimmerian center": 256.

Yanov, A.P. & Vaschenko, V.S. 1977. *Preventing mine atmosphere against contamination*. Moscow: Nedra: 287.

Rsenofontova, A.I. & Burchakov, A.S. 1965. *Theory and practice of dust control in coal mines*. Moscow: Nedra: 227.

Progressive Technologies of Coal, Coalbed Methane, and Ores Mining – Bondarenko, Kovalevs'ka & Ganushevych (eds)
© 2014 Taylor & Francis Group, London, ISBN: 978-1-138-02699-5

Assessment of goaf degassing wells shear due to their longwall undermining

I. Diedich
PJSC "Donetsksteel", Donets'k, Ukraine

V. Nazimko
Donets'k National Technical University, Donets'k, Ukraine

ABSTRACT: Degassing of coal deposits through the wells that are drilled from the surface is efficient measure to increase underground safety and utilize valuable energetic resource. However longwalls that undermined the degassing wells corrupt them that reduces degassing efficiency. We investigated process of ground subsidence during its undermining and found the factors that impact stability of the well essentially. Special diagram has been constructed to predict shear deformations of stratified rock mass that are the most critical for the well stability.

1 INTRODUCTION

Ukrainian coal industry provides guaranty for its energy independence. Economic feasibility of this industry demands intensive coal extraction using longwalls that yield from 5000 to 7000 ton raw coal per day. High rate of longwall advance causes intensive methane emanation that exposes underground atmosphere to danger of gas explosion. Traditional technologies that used to eliminate the gas approached to limit of their capability. As usually, ventilation and degassing of rock mass through underground boreholes may seize no more than 7–10 m^3 of the gas from every ton of coal. However methane content increased up to 20–25 m^3/t due to depth of mining increase. To combat this problem, additional degassing technology expands implementation using so called goaf wells for degassing the rock mass from the surface.

Western coal industry extracts coal from a depth that typically deviates in the range from 200 to 400 m. Such a moderate depth and specific geologic conditions cause high rock mass permeability that makes possible to draw the essential share of methane from the virgin rock mass. Unfortunately, Ukrainian coal deposits have low permeability that can not provide the degassing of rock mass without special additional measures. According practice, intensive methane release starts after undermining of layered rock mass that accumulated the fossil gas in coal riders and disseminated organic matter.

The undermining causes disintegration of rock mass to separate layers and blocks that produces intensive structure of fractures. These fractures create the networks that increase rock mass permeability dramatically, by several orders. Therefore longwall undermining triggers degassing process and goaf well might seize the methane reducing danger of explosion. However ground movement produces deformation that may destruct goaf wells. So far, there is not sufficient knowledge concerning how the wells collapse and which deformations are the most harmful for the wells.

We investigated this problem and developed a method to predict dangerous deformations of rock mass that are the main reason of the well collapse.

2 EXPERIMENTAL DATA

Different productivity of degassing wells has been investigated in situ during undermining goaf wells in Pokrovske coal mine. The depth of coal seam extraction was 850 m and seam height varied in a range from 1.7 to 2.1 m. Degassing wells were drilled in advance before longwall approached the wells. The wells had diameter 150 mm and a casing were inserted to prevent early borehole collapse. Gas output from the wells was monitored using special flowmeter. As it was stressed before, active rock mass degassing started immediately after undermining wells. As a rule, initial rate of degassing was from 6 to 12 cubic meters per minute but the gas flow attenuated as the longwall retired from the undermined well.

Figure 1 demonstrates exponential slowdown of methane output from experimental well #14. Active degassing of the well lasted approximately for 2 months declining down to 0.5 m^3/min.

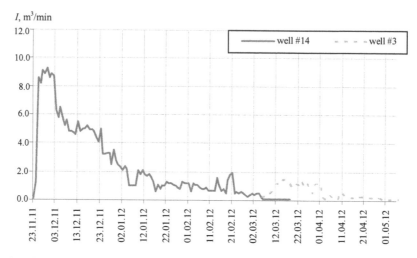

I, m³/min

Figure 1. Monitored gas output from different.

We examined experimental wells with special camera that was protected by durable casing (Illiashov et al. 2013). This case could withstand underground water pressure up to 6 MPa. The casing was downed into well using special winch. The camera registered all zones that were destructed or corrupted in the wells. It was detected that the major destruction of the well has the form of concentrated shear. This shear was oriented along stratification and occurred at the contacts of adjacent rock layers. Mostly every well had corrupted intervals but shear magnitude was different. Size of the residual diameter in the well #14 was not less then 100 mm. It does mean that the camera went through all experimental intervals of the well and residual aperture of the well was more then radius of this well.

It was found (Nazimko et al. 2010) that aerodynamic resistance of a borehole reduces slightly if residual diameter of this hole is more then initial radius. However the resistance abruptly increases as soon as hole aperture reduces down to initial radius. This effect was confirmed in the case of the well #3. This goaf well demonstrated poor output of methane because maximum rate of the flux was 1.5 m³/min. When we inserted the camera into the well, we discovered that it was actually cut completely at the depth 430 m.

Therefore experiments demonstrated that main reason of the goaf wells destruction is negative action of concentrated shear. Hence to assess well destruction, a method should be developed for prediction of layered rock mass shear.

3 NUMERICAL SIMULATION OF ROCK MASS STRESS STATE

Deformation of undermined rock mass has been investigated by FLAC3D model (ITASCA Consulting group 2008). Stratified rock mass was undermined by panel at the depth of 600 m. Panel was 200 m length and moved 200 m every month. Height of coal seam was 1.7 m. It has been extracted by 40 m stopes. As FLAC3D has the second law of Newton as the basic law, rate of advance was employed through explicit account of the time.

Distribution of shear strain generated in undermined rock mass depicted in Figure 2. Shear magnitude declines as the distance from the extracted seam increases on the normal direction to stratification. Process of shear magnitude increasing stabilized after panel withdrawal up to 200 m.

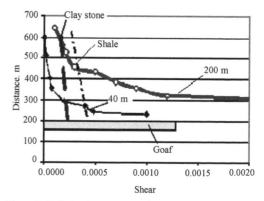

Figure 2. Relation between actual and critical share.

Two intermitted lines indicate critical shear strain for dry shale and clay stone according (Schwartz 1964). It should be mentioned that the more moisture content the more strength of the rock reduces. Actually, it does mean that the shear that generated after essential withdrawal of the panel may split the rock layers at any distance between extracted seam and surface excluding cover ground. The most probable spots where concentrated shear may occur situated at the contacts of clay or silt stones that covered by water saturated sandstones.

Another behavior of stratified rock mass has been detected during its undermining. Increase of longwall advance rate has positive impact because shear magnitude diminishes. This effect is practically important because the shear may be controlled by the rate advance regulation. Goaf well monitoring confirmed this positive effect.

All findings nave been concentrated in a diagram that may be used to predict magnitude of concentrated shear during undermining of stratified rock mass (Figure 3). Intermitted arrows indicate clue for practical implementation of the diagram. To forecast the spots of critical shear deformation, namely concentrated shear that is dangerous for goaf well stability, we need now all essential factors: geologic consequence in vertical direction, water content in rock layers (especially in sandstones), distance from the top of extracted coal seam up to the surface and rate of longwall advance.

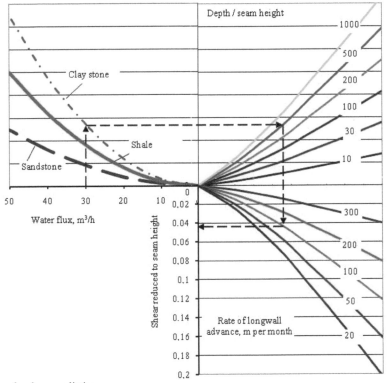

Figure 3. Diagram for shear prediction.

One should start from the water content and move to the curve that indicates the type of rock (for example silt or clay stone as indicated in Figure 3). Next, move to the undermining factor that calculated as ratio "Distance from the spot to the top of the extracted coal seam divided by the height of extracted coal seam". Finally, move to the curve that indicates rate of longwall advance and get reduced shear value on the vertical axis. To recover absolute magnitude of the shear, the reduced shear should be multiplied by the height of extracted coal seam.

Critical shear will more then internal radius of the goaf well. If the diagram predicts the concentrated shear that is more then critical value, we should choose an action to neutralize this negative effect. The straitest method to eliminate shear magnitude is to lift the rate of advance for longwall that will extract the coal seam under the goaf well if it is possible.

4 CONCLUSION

We found essential factors that regulate magnitude of concentrated shear during stratified rock mass undermining with longwall. Type of rock, water content, ratio "Distance from the spot to the top of the extracted coal seam divided by the height of extracted coal seam" and rate of longwall advance make the most important impact on the shear magnitude. Special diagram has been constructed to predict concentrated shear magnitude considering all these factors. The concentrated shear reduces aerodynamic resistance of the degassing well essentially if the shear magnitude is more then inner radius of the well.

REFERENCES

Illiashov, M., Halimendikov, E., Nazimko, V. & Diedich, I. 2013. *Concentrated shear forecasting for goaf wells that were drilled from the surface*. Mining report Gluckauf, 3: 62–68

ITASCA Consulting group. 2008. *FLAC3D manual.*

Nazimko, V., Demchenko, A. & Bruchanov, P. 2010. *Impact of the goal well deformation on its aerodynamic resistance*. Methods for safety maintenance at coal mines, 2(26). Makiivka: MakNII: 25–42.

Schwartz, A.E. 1964. *Failure Of Rock In The Triaxial Shear Test*. The 6th U.S Symposium on Rock Mechanics (USRMS), 28–30 October, Rolla, Missouri. American Rock Mechanics Association: 109–151.

Progressive Technologies of Coal, Coalbed Methane, and Ores Mining – Bondarenko, Kovalevs'ka & Ganushevych (eds)
© 2014 Taylor & Francis Group, London, ISBN: 978-1-138-02699-5

Parameters determination of the sudden coal and gas outbursts preventing method

V. Demin, N. Nemova & T. Demina
Karaganda State Technical University, Karaganda, Kazakhstan

ABSTRACT: One of the main ways to prevent sudden coal and gas outbursts in the mines of the Karaganda coal basin both when opening coal seams and driving is the drilling of degassing wells of 0.13–0.25 m in diameter. The effectiveness of its application depends on the correct determination of the main parameters. These include the diameter d, length l_c, the number n, minimum advance wells, their mutual location, size, protected (dangerous) areas in the circuit and the contour of the excavation. The value of minimum advance wells should make 5 meters, a two-meter safe (secure) area should be provided on flat and inclined seams in the roadway, and in particular outburst sites, its value is assumed to be 4 m on both sides of the excavation. On thick flat seams, the zone located 2–4 m above the mine working contour when having technocially disturbed coal.

1 INTRODUCTION

The length of leading holes of diameter 130–250 mm during the excavation on a layer should be 10–20 m and the number determined by the formula (Cai, Bondarenko & Isabek 2009; Cai, Issabekov et al. 2008)

$$n = \frac{S_{zo}}{S_{ef}} = \frac{(a+2b)m_n}{S_{ef}},\qquad(1)$$

where S_{ef} – area where the effective influence of drainage wells, m²; S_{zo} – protected area (dangerous) areas disturbed by a pack of coal, m², a – average width of the output on the bedding pack or packs of broken coal in the rough, m; b – width of the cavity of the treated wells array of disturbed coal the contour generation, m, m_n – power pack or packs of broken coal, m.

The effective area of influence S_{ef} well received in the form of a rectangle whose size is on the l_n lamination, and cross lamination l_k. For holes of diameter 200–250 mm $l_n = 2.6$ m, $l_k = 1.4$ m and a diameter 130–150 mm respectively 1.7 and 0.9 m (Instructions for the safe… 1995).

The design of drainage wells with flat seams outcrop crosscut or other reveals the workings defined as (Instructions for the safe… 1995).

$$n = \frac{(a+2b)\cdot(h+b)}{S_{ef}\cdot\sin\alpha},\qquad(2)$$

where a – rough crosscut width, m; h – height crosscut rough, m; b – bandwidth-treated wells array coal crosscut from the sides and below (above) it is normal, m, α – dip angle, deg.

When drilling diameter 130–250 mm value $S_{ef} = 6.8$ m² (dimensions of a rectangle $S_{ef} = 2.6$ m²) and a diameter 80–120 mm $S_{ef} = 2.6$ m² (dimensions of a rectangle 1.6×1.6 m).

From formulas (1) and (2) it follows that the number of leading holes is defined by the distance from its walls to the point in the massif at which the effective degasification of coal is carried out. There are several methods of determining the r_{ef}.

KazNIIBGP method, based on the registration of reducing the gas pressure (P) in the control boreholes drilled at different distances from the well. For value r_{ef} is taken away from it, which value decreased 24 hours not less than 50% of its initial value. The method is based on the registration of "burst" of gas emissions from the control holes when drilling the well (Horunji & Friedman 1973). For the effective radius of influence of pre-drilled wells is taken away from it to the hole, the rate of gas release q_n which increased by three measurements of not less than 10% compared to the last value q_n to its drilling.

r_{ef} parameter and is determined by the outburst of R (Horunji & Friedman 1973; Temporary instruction... 1961) equal

$$R = (S_{max} - 1.8)(q_{n\,max} - 4),\qquad (3)$$

where S_{max} and $q_{n\,max}$ respectively, the maximum output value of drilling culm (l/min) and the rate of gas emission (l/min·m) from one meter-interval hole.

For the Karaganda basin $R = 2$. When $R \geq 2$ formation zone is considered outburst and $R \vartriangleleft 2$ – non-hazardous. For the r_{ef} was taken away from the contour well to the middle of the interval between the adjacent control blast-holes, based on which the hazardous and non-hazardous values of R are established.

2 THE RESULTS OF THE EXPERIMENTAL OBSERVATIONS

Comparative tests at the mine Lenin in the Karaganda coal basin showed the following values of the effective radius of influence of a well diameter of 0.13 m: the first method – 1.25 m, the second –

1.5 m , and the third – 1.3 m (Antonov 1987). In this experiment were set numerical values of r_{ef} – up and down r_3, r_4 in cross layering can stretch assumed to be 0.5 m Analysis of the methods of finding the r_{ef} shows that the definition of this parameter by measuring the rate of gas makes up 4 and 20% higher values, compared with the first method (The technical operation... 1941–1946). Proved (Antonov 1987) that the method based on the registration of the gas pressure yielded satisfactory results only up to a depth of mining 400–500 m, and the method of recording "burst" of gassing allows you to install only the limit (max) values of r_{ef}.

Finding an effective radius of influence of a well in terms of the current forecast of outburst R in the control boreholes drilled at different distances from the well studied, is more rapid and objective and applicable even at depths greater than 500 m instructions (Instructions for the safe... 1995) shows the refined formula for the R, which has as follows

$$R = (S_{max} - 1.8)(q_{n\,max} - 4) - 6\qquad (4)$$

when $R \geq 0$ burst-formation zone is considered.

Table 1. The radii of the effective influence of leading (decontamination) wells r_{ef} at the opening seams crosscut.

Mine Saransk	H/d, 10^3	Plast	Depth H, m	Diameter hole d, mm	Elemental gas pressure in the reservoir, MPa	r_{ef}, m
Mine Saransk	4.6	K-10	460	100	3.45	1.0
	2.15	K-12	430	200	3.40	1.25
	2.08	K-12	520	250	4.30	1.20
Karaganda	3.92	K-12	490	125	3.97	1.00
Sokurskaya	0.87	K-13	340	390	2.40	1.40
	0.90	K-12	350	390	2.70	1.45
Toparskaya	2.4	K-18	240	100	1.56	1.25
	2.4	K-18	240	100	1.50	1.40
	2.0	K-18	340	170	2.55	1.80
	1.35	K-16	338	250	1.96	1.60
Them 50th Anniversary of the October Revolution	4.36	K-12	436	100	3.46	1.00
	5.0	K-10	500	100	4.12	1,00
	2.0	K-7	500	250	3.50	1.20
Lenin	1.42	D-6	400	280	2.10	1.20
	3.23	D-6	420	130	2.30	1.10
	3.3	D-8	428	130	2.35	1.25
Mean						1.25

Table 1 shows the results of determination KazNIIBGP ref before opening the coal seams (Antonov 1987). The assessment of this parameter was made to reduce the gas pressure to the outburst values in the control borehole drilling wells, and approved by the reaction layer in the implementation of the drilling tool into the formation at a distance approximately equal to the r_{ef} on its walls.

Tabular analysis shows that the depth of laying mines $H \geq 400$ numerical value of the effective radius of influence of a well in seven cases (43.75 %) less than the average of its value $r_{ef} = 1.25$ m sure, this may lead to an underestimation of the number of leading wells workings passed at a depth of 400 m, and indicates the need to consider when determining the number H of wells by the formulas (1) and (2). For ease of calculations, the effective circular area of influence is replaced by the well that is is approximately equal to rectangular , while on the bedding layer with a small margin ref is assumed to be 1.2 m , and this value is recommended in order to find the effective influence of the drainage hole in the current instructions (Instructions for the safe 1995).

The degree of degassing of adjacent to the well coal, of course, depends on the stress-strain state of the coal massif in front of the mine working face. It affects the fracture content of coal and its permeability in different points of the massif, the pressure of methane and filtration of gas from the formation into the wellbore. For this reason, the table data were processed according to r_{ef} H/d (Figure 1) allowing to consider this condition of coal massif. This gave the following empirical expression

$$r_{ef} = 1.59e^{0.0001(H/d)}. \tag{5}$$

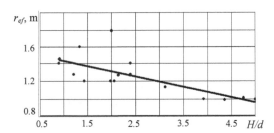

Figure 1. Dependence of r_{ef} H/d .

Correlation ratio $\eta = 0.76$, indicating, according to Cheddoku (Cai, Issabekov et al. 2008), the close relationship between the radius of the effective influence of pre-drilled wells and ref parameter (H/d). The coefficient of determination $\eta^2 = 0.58$. This means that a 58% variation due to the numerical value ref ratio (H/d), and 42% other relevant factors. To those include natural gas-bearing coal seam gas-dynamic and structural prop-

erties of coal, gas weathering depth H_0 , etc.

The analysis of the obtained relation (5) shows that for fixed values of d -diameter pre-drilled wells on the radius ref can have a significant impact depth of the excavation H .

Zone of effective influence of a well depends on the fracture zone forming in the vicinity of the well. The larger the zone the greater the surface of the carbon formed, from which methane is released.

To determine the fracture zone around the well, the laws of continuum mechanics are applied. Originally elastic problem is solved to determine the stress state of rock around the well, and then, on the basis of the criterion of destruction, determined into zones of inelastic deformations (UZND) within which rocks are considered destroyed (The technical operation... 1941–1946).

Figure 2 shows the propagation pattern of wells around the fracture zone according to the depth.

3 ANALYTICAL MODELING OF THE FRACTURE ZONE TAKING INTO ACCOUNT THE TIME FACTOR

This task was performed on a PC using the procedure developed by the department "Mining Karaganda State Technical University" (Cai, Bondarenko & Isabek 2009). The calculations were performed for the human cargo Walker, passed through the bed d_6 (Cai, Issabekov et al. 2008). In carrying forward-drilled production wells with a diameter of 250 mm. The required area of destruction is determined by taking into account the time factor. UZND formation time taken per day , as the effective area of influence of the well defined in terms of reduction of gas pressure near the control well not less than 50% throughout the day.

The data obtained are plotted in Figure 3, which shows the dependence UZND zone size, particularly diameter D_y (UZND diameter) of the depth of the generation.

However, the effective diameter of $D_y D_{ef}$ will be more, because eventually the expiry of methane gas into the well will occur not only from the area UZND, but also the adjacent areas $D_{ef}/D_y = \hat{e}$, respectively

$$\hat{e} = S_{ef}/S_y. \tag{6}$$

Assuming that the formation of the value of $\hat{e}$ is constant, perform calculation of the required number of wells at different depths. Considering the formula (2) it will power.

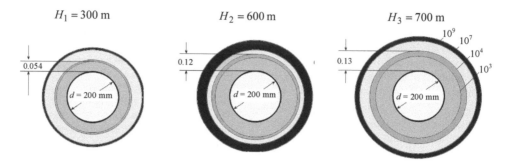

$H_1 = 300$ m $H_2 = 600$ m $H_3 = 700$ m

0.054   0.12   0.13

$d = 200$ mm $d = 200$ mm $d = 200$ mm

10^9 10^7
10^4
10^3

Figure 2. Effect of the depth of the hole (H) in the radius (r_{ef}): $H_1 \triangleleft H_2 \triangleleft H_3$.

The coefficient $\hat{e}$ is determined experimentally in mines.

Thus, the number of wells to reduce the degassing reservoir dangerous for release opening in the formation will change as the depth of the excavation.

$$n = \frac{(a + 2b) \cdot (h + b)}{k \cdot S_{ef} \sin\alpha}. \tag{7}$$

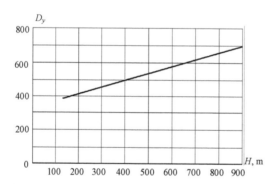

Figure 3. Graph of diameter D_y of depth h.

For example, it was calculated the number of project pre-drilled wells with a diameter of 200 mm to expose the outburst flat seams crosscut at a depth of 850 m (at a depth of mining operations are underway at the mine. Kostenko CD JSC "ArcelorMittal Temirtau") at $\alpha = 10°$, $b = 1.5m$, $f = 5.4m$ and $h = 3.3m$ under the current regulations and c obtained taking into account the dependence (5), determines the effect of H on the r_{ef}. In the first case, $n = 34$ wells. The numerical value of r_{ef} found from (5), equal to 0.96 m. Turning to the rectangular shape of the effective influence of the well, creating some stock, we will $r_{ef} = 0.92m$, then $n = 56$ wells. These data indicate the need for mandatory accounting reduce ref with increasing depth of the

workings of H, otherwise they can be dangerous due to the possibility of a VVUG even after drilling degasification wells.

Figure 4 shows the number of wells on the diameter wells.

In driving on the workings of the outburst seam deviation of the number of wells with and without dependence on H r_{ef} is less than at the opening of seams, as well the influence of the effective area in the formula (1) will only be changed by varying the l_H.

In (Instructions for the safe... 1995) proposed the following relationship between the pressure of gas in the reservoir P and depth H and H_0 – the depth of the gas zone of weathering:

at $H \triangleleft 600$ m

$$P = 0.1(H - H_0) + 1; \tag{8}$$

at $H \triangleleft 600$ m

$$P = 0.06(H - H_0)^{1.1}. \tag{9}$$

After determining the depth H of the expression (5) and substituting it into (9) one obtains that

$$P = 0.1\left[10^4 \cdot d \cdot (ln1.59 - ln r_{ef}) - H_0\right] + 1. \tag{10}$$

Similarly, when $H \triangleleft 600$ m

$$P = 0.1\left[10^4 \cdot d \cdot (ln1.59 - ln r_{ef}) - H_0\right]^{1.1}. \tag{11}$$

By processing tabular data revealed the following empirical relationship (Figure 5) with a maximum value of the coefficient of determination $r_d^2 = 0.32$

$$r_{ef} = 1.697 e^{-0.111P}. \tag{12}$$

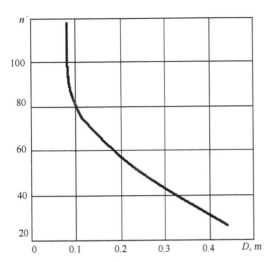

Figure 4. The relationship between the number of wells (n) from the borehole diameter (D).

The numerical value of $\eta = 0.57$, suggests a noticeable tightness between the parameters r_{ef} and P, and the resulting value r_d^2 indicates a significant impact on the value of r_{ef} other factors.

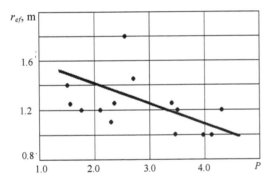

Figure 5. The dependence of r_{ef} P.

The natural gas-bearing strata x for mines in the Karaganda basin is defined as (Design Manual ventilation... 1997)

$$x = \frac{1.3 \cdot a \cdot (H - H_0)}{1 + b \cdot (H - H_0)} \cdot \frac{100 - A^c - W}{100}, \qquad (13)$$

where A^c – ash formation, %; a and b – coefficients depending on the properties of the reservoir; W ash content, %.

Determined from the expression (13) depth H, and the expansion of its value in (4), we find that

$$r_{ef} = 1.59 - exp\frac{10^4}{d} \cdot \frac{x}{\dfrac{1.3 \cdot a \cdot (1000 - A^c - W)}{100}} \cdot \qquad (15)$$

The resulting Equation (15) more fully reflects not only a quantitative but also a qualitative correlation radius of the effective influence of an advance well with a variety of factors.

Processing tabular data revealed a parametric relationship with the correlation ratio $\eta = 0.93$ between the ratio $k_r = r_{ef} / d$ and a diameter d degassing wells of the form (Figure 6)

$$k_r = 1.76 \cdot d^{-0.793}. \qquad (16)$$

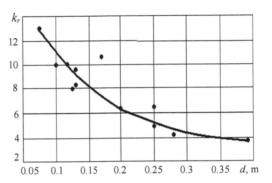

Figure 6. The dependence of k_r d.

4 CONCLUSIONS

1. With an increase in the d numerical value the parameter k_r rapidly decreases. This indicates that the change in diameter of degassing wells in a big way does not lead to significant changes in r_{ef}.

This conclusion is described by the fact that in tectonically disturbed coal seams the gas is released through the pre-drilled wells in the process of drilling and the capacity of all wells being of 0.08–0.39 m at current gas flows stays about the same (Antonov 1987).

2. When choosing a diameter of wells sit is necessary to consider the possibility to exclude sudden outbursts of coal and gas during drilling and the minimum required time for this process t_b and degassing t_d of coal in the developed area to a safe value indicators outburst. Significant amount of methane releases from the leading wells mainly during introduction of the drilling tool into the massif and coal after drilling the emission of gas practically ceases, i.e. $t_b \approx t_d$. It is found (Antonov 1987]) that

at screw drilling optimal values of d are in the range of 0.2 to 0.25 m, and during smooth drilling with washing – 0.13 m.

REFERENCES

Cai, B.N., Bondarenko, T.T. & Isabek, E.T. 2009. *To determine the parameters of a method for preventing sudden coal and gas drilling degasificationwells.* Almaty, Kazakhstan. Mining Journal, 2: 34–36.

Cai, B.N., Issabekov, E.T., Androsov, A.A. & Lebedev, G.V. 2008. *Influence of the depth of laying working by the number of degasification wells in mine workings.* Karaganda: MES, KSTU, Proc. Inter-University Student Conference "Student and scientific and technological progress", 1: 22–23.

Instructions for the safe conduct of mining operations at the seams prone to sudden outbursts of coal and gas. 1995. Karaganda KazNIIBGP, Approved. MEandUP RK 21.06.1995: 177.

Horunji, Y.T. & Friedman, I.S. 1973. *Ways of improving the methods of opening outburst layers of Donbass.* Ways and means to develop the outburst coal seams.

Moscow: Scientific message IGD A.A. Skochinskiy. Perf. 182: 41–43.

Temporary instruction to establish sites in Kuzbass mines reservoirs, hazardous and non-hazardous to sudden outbursts of coal and gas. 1961. Kemerovo Kemerovo CSTI People's Commissars: 25.

Antonov, A.A. 1987. *Forecast and prevention of sudden coal and gas at the opening of thick flat seams Karaganda basin.* Karaganda: Diss. on competition degree Cand. Tehn. Science: 30.

Cai, B.N., Isabekov, E.T., Androsov, A.A. & Lebedev, G.V. 2008. *Influence of the depth of laying production by the number of degasification wells in mines.* Karaganda: MES, KSTU, Proc. Inter-University Student Conference "Student and scientific and technological progress", 1: 22–23.

The technical operation of coal mines. 1941-1946. Moscow: Ugletekhizdat: 308.

Design Manual ventilation of coal mines. 1997. Almaty: 258.

Antonov, A.A. 1987. *Forecast and prevention of sudden coal and gas at the opening of thick flat seams Karaganda basin.* Diss. on competition degree Cand. Tehn. Science. Karaganda: KarPTI: 225.

Progressive Technologies of Coal, Coalbed Methane, and Ores Mining – Bondarenko, Kovalevs'ka & Ganushevych (eds)
© 2014 Taylor & Francis Group, London, ISBN: 978-1-138-02699-5

Investigations and technical decisions of the mining and ecological problems while closing coal mines

P. Dolzhikov & S. Semiriagin
Donbass State Technical University, Alchevs'k, Ukraine

ABSTRACT: The complex technical decisions of the ecological problems are proposed on the basis of the results of the investigations of the geomechanical and hydrogeological processes in the hydroactivated rocks while closing coal mines: liquidations of the excavations and gaps, utilization of the industrial wastes and forming of the artificial foundations. The examples of the developed technologies are given.

1 INTRODUCTION

Accumulated experience in implementation of the physical liquidation of excavations shows that there are mining and ecological problems and complications almost in each object, they require new technical solutions. These problems are always interrelated by time, causes and consequences, but they can be classified conditionally by the sources, and they can be divided into three groups by the consequences: geomechanical, hydrogeological, ecological. Analysis of the mining and ecological consequences while liquidation of the slopes showed close interconnection of three components – the state of the massif, excavation and technology of its liquidation. Investigation of this problem allowed working out the classification of the sources of the mining and ecological problems (Kipko 2004, 2005).

Concerning liquidation of the slopes and excavations it is necessary to note the following complications (Kipko 2005): changing of the hydrogeological state in the adjoined massif; activation of the geomechanical processes; reducing of the bearing capacity of the rocks in the result of the underground water steeping; deformation of the surface in the result of underflooding and waterlogging; destruction of the brick beam fillingup; the surface gaps after shrinkage of the material mine shafts filling up; gas contamination of the deleted part of the excavation and uncontrolled methane output on the surface; pollution of the adjacent territories with underground water streams; destruction of the adjacent surface buildings and service.

2 THE AIMS AND TASKS OF THE INVESTIGATIONS

This formed mining and ecological situation in the regions of coal mines closing requires complex approach to its solution which includes:

– studying of the new geological processes and general geomechanical and hydrogeological state of the closed mines territories with the help of geophysical and boreholed methods;

– application of the modern technologies of the guaranteed excavations filling up;

– industrial wastes utilization by means of their using in the grouting and fill suspensions;

– safe liquidation of the surface gaps, prevention of zones development of the massif's active deformation over the excavations, preservation of the soil and plant layer;

– forming of the artificial foundations and watertight cushions in the underflooding zones.

3 RESULTS OF THE INVESTIGATIONS AND DESIGNS

In the process of mines liquidation with goaf full flooding, strength properties of rocks (especially clay rocks) decrease, which may result to loss of steady state of the rock strata and to activation processes of displacement, whereby there may appear additional processes and phenomena in the massif, which directly impact on building and operated objects, life activity of the population.

It is known that by the moment of mine closure repeatedly undermined massif has got bundle and hangs zones, not fully compacted loose rocks, land parts with anthropogenic discontinuities. According to the basic positions of geomechanics the collapse zone, secondary fissility zone and beam deflection zone are formed in the goaf and its roof (Kipko 2005).

As a result of research and compilation of factual material it is set that the following effects develop in the watery rock strata (Dolzhikov 2013):

– the effect of the massif's block restructure;

– the effect of hydrodynamic stress state of the massif in the zone of deflection;

– local rock bursts (mountain earthquake effect).

To confirm this fact, there are numerous data of the hydrogeological observations and filtration works performed in the wells in the fields of active and closed mines of Dolzhano-Rovenets'kyi geological and industrial district, as well as data of Ministry of Emergency Situations (MES) about the manifestations of the rock bursts.

The object and regional development of the hydrogeomechanical stress and volumetric distribution of the hydrodynamic phenomena in the massif in case of flooding workings can cause local manifestations of rock bursts (mountain earthquake effect). Moreover, it should be noted that this phenomenon in most of their cases develops in the zones of tectonic disturbances, which has a strong development in Dolzhano-Rovenets'kyi coal-mining district.

Periodicity manifestation of the effect of mining earthquake is clearly seen in the graphs (Figure 1). This periodicity is explained by the new model of the viscoelastic deformation of the flooded rock mass (Maxwell-Kelvin-Voigt).

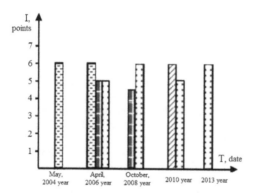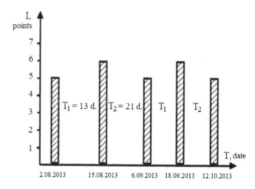

Figure 1. Date and intensities of manifestation of the local rock bursts: ⬚ – Rovenki; ▨ – village Kalinovka; ▤ – urban settlement Novodarievka; ▨ – village Maloriazantsevo.

The basic precondition for activation of geomechanical processes are preserved residual voids and secondary fissility in the underground space, and the cause of failures over the inclined shafts is poor cancelling of the crack mouths or its full absence, and also shrinkage of the packing material and the conditions for the bypass of the collapsed material to fall deep into production (Kipko 2005).

All the factors, actively influencing the development of the surface damages, can be divided into two main groups: provoking and attendant. There are three major groups of provoking factors: changes in rock strength; change in the stress state around the old extraction; dynamic and seismic effects.

We may include weathering, porosity, tectonic faults and the strength of layers, their physical and mechanical properties, bedding angles, composition and thickness of sediments, sizes of voids, etc. among the attendant factors. Number of the attendant factors is large, their impact on the development process of the failures is different, but they act together as they reflect the state of the rock mass and excavation at the moment. On the other hand, the hydrogeological factors have got predominant influence. This is because the thickness, located above the old mine workings at a shallow depth, is a zone of fracturing and weathering, it is broken by the secondary anthropogenic cracks caused by the influence of mining. This creates the conditions for intensive filtering into mines, both surface and water from aquifers in this column. Incoming water column wet and erode the available cracks, contribute the development of the process of collapse of rocks in the extractions, which leads to forming of collapse craters on the surface, they have got a significant size. It is set that the most favorable conditions for the ingress of water (surface and from the aquifer) into the subsided thickness are created by the presence of geological failures or developed network of the small-amplitude disturbance.

When closing the mines the basic way to eliminate courses is adopted – burnt rock filling up. The technological scheme of the shafts backfill includes the development of the waste dump, moving to the trunk of burned rock and rock gravity feed or conveyor in mines (Kipko 2005). However, experience shows that the way to eliminate the mining slopes does not guarantee their full filling up.

The analysis of mining and technical condition of the courses and forming of filling mass by the backfill method showed the necessity for their classification, division on the categories according to the degree of the environmental hazards. The result of such division must be the choice of the technological scheme of the courses liquidation. Here, obviously, there must be correspondence of the costs for termination of the course and the achievement of the necessary ecological safety in the post liquidation period. It is more rational, all the inclined slopes may be classified into three groups taking into account the degree of the reasonableness of the environmental safety. The basis of the proposed classification is based on two main criteria: the first and the main is ecological safety, the second is economy.

Summing up the results of studying of mining and technical state of the inclined workings and process of forming of filling mass by the method of filling the fused rock, the conclusion is made about need of the complex approach for the technical solutions. Such approach is based on the appropriate combination of ways of the burnt rock excavation backfill and plugging underlay residual voids and cavities in rocks filling (Kipko 2005) (Figure 2).

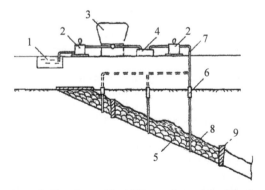

Figure 2. Flowsheet of backfilling and grouting of the inclined course through the wells: 1 – tank; 2 – pump; 3 – compositor; 4 – loading hopper; 5 – backfilling; 6 – well; 7 – pipeline; 8 – solution; 9 – barricade.

Technological schemes of voids plugging are based on the use of fine-grained materials put to the course in the nonpressure or pressure mode. For these works the most economically and technologically appropriate way is to use clay and cement materials with fillers of solutions which give the sufficient strength after structuring and are water-resistant and waterproof.

It is more appropriate to use uncemented backfill materials. One of promising application of fine slag and crushed overburden is to use them as curing activators and fillers in the production of the grouting and stowing mixtures. As industrial wastes used in the grouting and stowing mixtures, the burnt and unburnt rocks, sludge concentrators, TPP fly ash and dumping slag of the metallurgical industry are applied (Kipko 2008). Possibility of slag depleted solution using in blast furnace as the base materials is reasoned by its involving a wide range of minerals, including hydraulically active minerals, which allows using slag as a major component of filling mixture without loss of hydraulic activity (Dolzhikov 2013).

For the development of the grouting and stowing mixtures as the base raw material the wastes were used:

– milled rock dump (coal mine "Tsentralnaya");

– burned rocks of the dump (coal mine "Tsentralnaya");

– depleted slag (OJSC "AMP");

– ash plant (Lugans'k TPP).

As an additive bentonite gel powder was used, and Portland cement M400 was the binder.

Tables 1 and 2 show the research of the chemical composition of the industrial wastes, composition and properties of formulations of the grouting and stowing mixtures.

The deformations of the earth surface in the form of dips and uneven ground subsidence appeared and continue to appear in the areas of existing and abandoned mining enterprises, cities with developed underground infrastructure at different times.

The essence of the newly developed method is to eliminate failures by layering grouting and stowing massif with given strength and deformation parameters based on the use of burned rock dumps (Figure 3).

Specifications for strength and deformation characteristics of the grouting and stowing materials used in the elimination of gaps are formulated on the grounds of the researches:

– stabilization and preventing of the gaps development;

– durability, no need for monitoring;

– water-tightness and gas-tightness of the filling massif;

– given deformation to resistance and durability;

– cheapness and availability of the key components.

Table 1. Chemical composition of industrial wastes, %.

Name	lb	SiO$_2$	SO$_3$	Fe$_2$O$_3$	Al$_2$O$_3$	CaO	MgO	K$_2$O+Na$_2$O
Coal mine "Tsentralnaya"								
Rock dump	11.6	52.7	4.0	8.9	18.6	3.5	1.9	5.6
Burned rock dump	8.1	52.8	3.8	8.4	19.0	2.7	1.7	3.5
OJSC "AMP"								
Depleted ground slag	2.5	3	–	–	8.6	44.7	5.6	–
Lugans'k TPP								
Ash power station	7.3	50.0	1.0	13.5	24.0	2.6	1.6	1.0

Table 2. Composition and properties of formulations of the grouting and stowing mixtures.

#	Parameters of mixtures	Values			
		On the base of rock dumps	On the base of burnt rock	On the base of depleted slag	On the base of ash
1	Quantity of wastes, kg/m^3	818	816	840	825
2	Quantity of bentonite, kg/m^3	5.9	5.9	10	20
3	Quantity of water, kg/m^3	626	626	625	620
4	Density of the basic suspension, kg/m^3	1450	1450	1455	1450
5	Quantity of cement, kg/m^3	40	40	–	40
6	Density of solution, kg/m^3	1474	1474	1460	1460
7	Plastic strength, kPa in 10 days	296	286	580	865
8	Dynamic tension, Pa	60	47	42	20
9	Structural viscosity, 10^{-3} Pa/s	46	52	54	65
10	Permissible plastic strength, kPa	32.9	31.7	35.6	36.0
11	Shrinkage, %	0	0	0	1.0
12	Flowing, cm	9–10	10–11	10–11	8–9
13	Slope of repose, degree	15	12	17	16

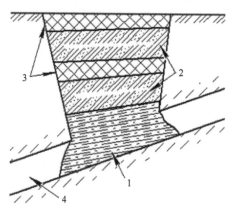

Figure 3. Scheme of elimination of gaps by layering plugging: 1 – supporting safety cement pad; 2 – gap filling with burned rock; 3 – stabilization layers of mixture of clay and cement solution with the addition of burned rock; 4 – underground cavity.

Thus, the proposed method is effective elimination technology of failed deformations on the sur-face of the earth, through a combination of burned rock and backfill resource injection grouting mortars, providing guaranteed filling voids, which allows solving the technical and environmental problems.

Analysis of geological sections in the foundations of buildings on the undermined areas of the mining towns allowed to divide them into three types, and to develop the technological schemes of their plugging with the help of the viscoplastic solutions.

The geological section of the first type is characterized by the overburden with the capacity of 50 m and more. They are usually introduced with the sandy-clay soils. As a result of the development of mineral resources in the ground massif, the decompression zones were formed which caused the reduction of strength and increase of the filtration characteristics of the soil.

The second type of geological section is introduced by the rock and semi rocks of Carboniferous age, reaching the surface of the earth or overlain by Quaternary sediments of low capacity. Geological cross-section of this type is introduced by the character for Donbass region rhythmic alternation of sandstone, shale and sandy clay, with subordinate power layers of limestone and coal.

The third type of the geological section is found in the areas of the closed coal mines with full flooding of the produced space, it is the main feature of the mining towns of Lugans'k region. The result is a change in the massif fissility and strength properties of rocks, as well as activation of geomechanical processes due to the hydrodynamic effects on the massif.

Therefore, the violation of geomechanical equilibrium in the massif takes place in the case of mines flooding, it causes displacement of rocks and development of deformations in the foundations. To stabilize the deformation process is reasonable to perform grouting work (Figure 4).

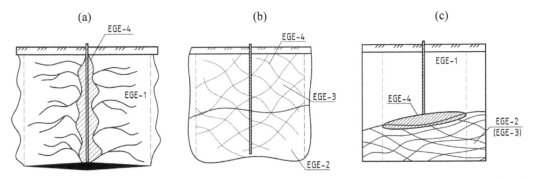

Figure 4. Grouting process model with viscoplastic solution (for one well): (a) decompression zone; (b) zones of cracks system; (c) scheme of stabilizing cushion forming.

4 CONCLUSIONS

Thus, the complex approach to the research and to the solution of environmental problems in the areas of closed mines allows us to offer a wide range of technical measures (research and grouting works), we can achieve almost equilibrium state in the rock massif, the surface is saved, utilization of industrial wastes is realized and the safety of industrial and civilian objects are ensured.

REFERENCES

Kipko, E. 2004. *Complex method of plugging while mine building.* 2nd ed. Dnipropetrovs'k: National Mining University: 367.

Kipko, E. 2005. *Complex technology of liquidation of mine slopes.* Donets'k: Nord-Press: 220.

Dolzhikov, P. 2013. *About necessity of engineering safety of the closed mines areas from the emergency situations by the methods of plugging.* Cases of emergency: safety and protection, 1(3): 112–115.

Kipko, E. 2008. *Projecting of the clay and cement plugging solutions in mining.* Dniprodzerzhyns'k: Publishing house "Andrei": 176.

Dolzhikov, P. 2013. *Projecting of slag and clayey plugging and filling suspensions for the ground gaps liquidation.* Collected scientific works of Donbass State Technical University, 40(1): 33–37.

Progressive Technologies of Coal, Coalbed Methane, and Ores Mining – Bondarenko, Kovalevs'ka & Ganushevych (eds)
© 2014 Taylor & Francis Group, London, ISBN: 978-1-138-02699-5

Development of truck-conveyor transport in open pit mining of Russia

A. Glebov, G. Karmaev & V. Bersenev
Institute of Mining of Ural Branch of RAS, Yekaterinburg, Russia

ABSTRACT: The necessity of truck-conveyor transport improvement is inseparably linked with the progress of priority direction of deep open pit levels development with employment flow-line technology (FLT). Its efficiency is proved by numerous scientific and design developments as well as by the practice of flow-line systems operation in Russian and foreign open pits.

1 INTRODUCTION

The progress of FLT in deep open pits should be based on advanced technological decisions: mobile crushing-transfer points application in the open pits, rock mass steep-inclined conveyor hoisting from an open pit to the receiving points on the surface; rational systems of opening up levels for arrangement crushing-conveyor complexes by multi-functional use of separate pit workings and transport communications (Yakovlev 2003).

The FLT progress depends to a great extent on truck-conveyor transport improvement stipulated by its separate kinds operation in the conditions of its preferable application. Thus its long-term progress should be defined by the decisions analogous to those defining the FLT progress. At the same time it is expedient to employ modernized efficient equipment with the parameters most completely conforming to the specific mining-technical conditions of mineral resources mining.

The effect from efficient truck equipment employment is conditioned by the dependences of variation the capacity and manufacturing cost of 1t·km transport operation against load-carrying capacity of a dump truck. A dump truck's technical capacity Q_a is in direct dependence from its load-carrying capacity and is defined from the expression

$$Q_a = \frac{60 \cdot q_a \cdot k_q}{t_p}, \text{ [t/h]}$$

where q_a – load-carrying capacity of a dump truck, t; k_q – utilization factor of a dump truck's load-carrying capacity; t_p – the time of a truck's run, min.

The 1 t·km cost of a transport operation $C_{t \cdot km}$ is inversely proportional to dump truck's capacity. It can be calculated from the formula

$$C_{t \cdot km} = \frac{O_{a\Sigma}}{Q_a \cdot T_a \cdot L_{ar}}, \text{ [rbl/t·km]}$$

or substituting Q_a value, we'll derive

$$C_{t \cdot km} = \frac{O_{a\Sigma} \cdot t_p}{60 q_a \cdot k_q \cdot T_a \cdot L_{ar}}, \text{ [rbl/t·km]}$$

where T_a – annual operation time fund of a truck, h; L_{ar} – annual truck's run with a load, km; $O_{a\Sigma}$ – operating costs for a truck, motor roads maintenance, garage facilities as well as for permanent mining operations extinction at this dump truck falling, roubles.

The behavior of these dependences variation is shown in Figure 1 (Vasiliev 1975). As is evident from Figure 1, by changing dump truck's load-carrying capacity from 40 to 160 tons its capacity increases 2.4–2.6 times. The manufacturing cost of 1 t·km transport operation, for this, decreases twice. According to foreign data, by increasing a truck's load-carrying capacity from 25 to 50 tons the trucking maintenance cost reduces to 0.9% per every ton of load-carrying capacity. Further load-carrying capacity raising from 50 to 100 tons brings to reduction the maintenance cost of transportation to 0.4% per every ton.

Therefore raising truck's load-carrying capacity can considerably improve technical-and-economic indices of truck-conveyor transport employment in the open pits. But however, on selection dump trucks the necessary motor roads broadening should be taken into account that takes place with trucks load-carrying capacity increase involving excavation of extra overburden volumes and the increase of open pit's dimensions in the plan.

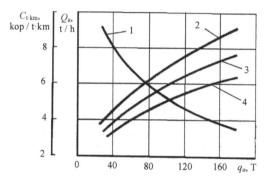

Figure 1. The variation of 1 t·km manufacturing cost $C_{t·km}$ (1) and a truck capacity Q_a (2, 3, 4) against its load-carrying capacity, the distance of transportation being 1.5 km: 2, 3, 4 – the speed of loaded trucks movement being 30, 18 and 12 km/h accordingly.

Certain rise in efficiency of truck-conveyor transport operation could be attained by employment conveyor equipment with optimum parameters.

2 THE CRUSHING-CONVEYOR SYSTEM PROGRESS

The researches display that the difference in the expenses on ancillary structures and permanent mining operations being slight (1–8%), there is essential reduction both of equipment expenditures (15–30%) and manufacturing cost of 1 ton of load transportation (16–28%) in the variant of employment conveyors with optimum parameters.

It is impossible to realize the advantages of truck-conveyor transport more completely if crushing-conveyor complexes are stationary arranged. Their application in such a pattern does not bring into line with both the dynamics of mining operations progress (lowering mining operations 7–15 meters per a year) and the conditions of forming technologic freight traffic. This has been spoken of at the Mining congress held in Chicago (the USA) in 1988. In the reports and messages it has been noted that in a number of the USA and SAR ore pits and in the pits of some other countries conveyor systems with stationary crushers as a rule "put in dead storage" a part of ore reserves under them and do not completely substitute truck transport. These systems, though being economically efficient than truck transport, their application with pit's depth rise becomes less profitable.

Table 1. The sites' dimensions for crushing-transfer points in the iron ore pits of CSC countries.

An open pit	The pattern of mining for a conveyor hoist	The elevation of the CTP's site, m	The dimension of the STP's site, m × m
Inguletzky	By an inclined shaft and cross heading	−60 −60 −120 −180	140×80 140×70 140×80 100×70
Annovsky	By a steep trench By an inclined shaft	±0 −30	360×260 140×75
Pervomaisky	By the inclined shaft and cross heading	−115	240×140
The SMIW pit	By an inclined shaft	−90	480×280
N3 of the NKMIW	By the inclined shaft and cross heading	−60	120×100
N2-bis of the NKMIW	By an inclined shaft	−45	180×160
Poltavsky: South-west flank South flank	By a steep trench By the inclined shaft and cross heading	−25 −120	210×100 150×70
Olenegorsky	By an inclined shaft	+75	120×90
Stoilensky	By a steep trench	+50	280×220
Kovdorsky	By a steep trench	+142 +10	280×90 260×90

The crushing-conveyor system progress takes place in two directions: making powerful mobile crushers with traditional conveyor installations and making steep-inclined conveyors. The stationarity of CCC' objects and irrationality of employing FLT mining development systems condition the large volume of mining-permanent and mining-development operations (up to 75% from the total complexes' cost). A substantial volume of mining-development operations is required to be performed only preparing sites for arrangement crushing-transfer points (CTP) in the open pit with dimensions in the plan from 70×100 up to 240×440 m (Table 1). The extra volumes of overburden rocks stripping are given in Table 2 for the sites with average dimensions to arrange a 200×100 m crushing-transfer point depending on the depth of its location on the pit edge and the pattern of its forming.

It is evident that forming the CTP's site with pre-spacing an open pit permanent edge the overburden volume will be substantially larger than if the site is arranged in the pillar of rocks. But however in the first case lower the CTP site the permanent pit edge agrees within the design contour; and in the second case the pillar of rocks reduces the designed pit depth with planned ore volumes being lost. The pe-

riod of CTP operation on one concentration floor reaches not less than 8–10 years.

This brings to unpractical application of assembly truck transport, the actual distance of transportation by which along the face-DTP section increases annually up to 150–250 meters.

The necessity of holding up rational haulage shoulder by truck transport is evident. Besides prime cost, the environment pollution is essentially decreased at the expense of restriction the motor transport operating zone by lower pit benches.

The CTP's stationarity during its carrying to another place in the pit space predetermines repeated application merely the mechanical equipment (crushers, a feeder); their cost makes up not more than 28% from the total expenditures on CTP. In connection with this there arise extra expenditures for civil engineering works on a new site of CTP sitting.

Table 2. The volumes of overburden stripping for a crushing-transfer point site.

The depth of laying the CTP site	The volume of overburden stripping for the CTP site, million. m^3	
	with open pit edge pre-spacing	on the pillar of rocks in the pit's contour
100	3.4	2.0
200	9.6	4.0
300	18.6	6.0
400	30.4	8.0
500	40.5	10.0

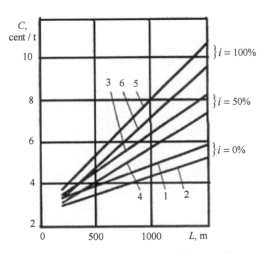

Figure 2. The variation prime cost C of 1 ton rock mass transportation against the distance L of transportation by dump trucks with 45 t load-carrying capacity (1, 3, 5) and 85 t (2, 4, 6) accordingly under different motor roads' grades.

Employing mobile (semi-portable) crushing-transfer installations (CTI) the flexibility of truck-conveyor transport increases essentially. The block-

module constructive design provides for their periodic movement in pit space as far as mining depth increases. Such CTI could be made on the basis of existing equipment having the capacity of 600, 1000 and 1350 m^3/h. the CTI type, periodicity and transfer step are set according to the results of optimization technological freight traffics derived with due regard for forming the pits' operating zone and rock mass volumes space and time distribution.

For the first time in the CSC countries territory a crushing-conveyor complex with semi-portable crushing installation has been assembled in the Poltavsky iron ore open pit in 1996. The investigations of the IM MFM USSR on grounding the efficiency of FLT employment in the open pits with semi-portable CTI and complex's equipment in open construction have forerun this. The realization of this solution has been put into practice on the basis of the equipment produced by the German firm Krupp Fordertechnik. The decision was made in terms of the indices analysis of two technology versions with truck (ground version) and truck-conveyor transport. It turned out that the pit having a 160 meters depth and ore capacity 2500 t/h, the application of truck-conveyor transport was preferable. Under roughly equal investments for both versions the cost price of 1 ore ton is almost two times less than in the ground version at the expense of lower operating costs on energy resources, tires, spare parts and wages (Figure 3) (Martinenko 1966).

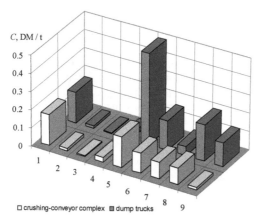

Figure 3. Operating costs according to accounting cost items: 1 - depreciation; 2 – percentage payment; 3 – insurance; 4 – taxes; 5 – energy resources; 6 – spare parts; 7 – materials and lubrication; 8 – tires; 9 – wages.

In the Kovdorsky MTW the crushing-conveyor-dumping complex has been operating since 1999; its

CTP is equipped by semi-portable crushing-transfer installations with SHTCDP-1500×2100 jaw crushers having the capacity up to 600 m³/h. Three semi-portable CTI are in operation; they are provided for carrying by large blocks (modules) to a new concentration level as far as the pit is deepened. The complex's conveyor hoist in open construction is mounted on the pit edge at an angle of 16°.

Technological potentialities of truck-conveyor transport in the FLT systems with mobile CTI are more completely realized with employment steep-inclined conveyor hoist. Steep-inclined conveyors, as compared with traditional ones, are able to move loose materials at a slope up to 50–60° and more. This permits to reduce essentially the volumes of permanent-mining operations while the route for conveyor hoist arrangement on the pit's edge is being prepared. The influence of a conveyor hoist route slope and rock mass hoisting height can be retraced by the data of Table 4 where the calculated values of unit capital outlays and operating costs for crushing-conveyor complexes with annual volume of haulage at 20 million tons are cited.

The costs' decrease for crusher-conveyor complexes with a 60° route slope of a conveyor hoist as compared with the complexes where a conveyor hoist is set at an 18° slope is cited below.

Table 4. Unit costs for crushing-conveyor complexes

The slope of a conveyor hoist, degree	The material's hoisting height by conveyor transport, m		
	100	300	600
	Unit costs		
18	2.6/0.60*⁾	6.0/1.33	13.8/2.54
25	2.5/0.61	5.7/1.31	13.0/2.58
35	2.5/0.61	5.4/1.22	11.0/2.29
45	2.4/0.58	5.0/1.14	9.9/2.0
60	2.4/0.58	4.5/1.11	8.6/1.91

*) above the line capital outlays' values are cited, under the line – the ones of operating costs.

Rock mass hoisting height	100	200	300	400	500	600
Unit costs decrease, %:						
capital outlays	8	20	25	32	37	39
operating costs	3	8	17	21	23	25

The information cited displays that steep-inclined conveyor hoists employment is expedient to consider under the rock mass hoisting height being more than 100–150 m. This concerns mining-technological conditions in which the application of both traditional and steep-inclined conveyors is possible. In other pit conditions, when it is rather difficult to arrange traditional conveyors without extra volumes of mining-permanent operations, the em-

ployment of steep-inclined conveyors is beyond doubt practically from the depth of transferring to combined truck-conveyor transport.

The investigations on FLT systems employment with truck-conveyor transport in the Kosto-mukshsky MIW performed by the IM UB RAS jointly with IW specialists are indicative of the prospects of semi-portable STI and steep-inclined conveyors employment. Two stages of arrangement the complexes are set forth to realization with due regard for the prospects of mining operations and pit transport progress in terms of comparison by variants. The first stage is the complex arrangement on a temporary section of a non-mining flank in the pit's northern face with in-pit transferring to railway transport; the second stage is stationary arrangement a complex on the finite contour in the southern face with conveyor line exit to upper levels. According to the second stage the FLT introduction is put into practice in 11–12 years after the first (ore) FLT line has been put into operation. At both stages of FLT introduction the construction of parallel operating ore and rock crushing-conveyor complexes is outlined with application uniform equipment and, in case of emergency, transferring freight traffics to any of them.

Technical and economic analysis has displayed high efficiency of employment FLT complexes with semi-portable CTI and steep-inclined conveyor hoists in the Kostomukshsky open pit. Thus, the FLT complex introduction at the first stage will permit to reduce the running overburden volumes to 5 million m^3. Arranging the FLT complex on the permanent flank (the second stage), the overburden volumes in the finite pit's contours will be reduced to 17.6–18 million m^3 as compared with the basic variant of truck and railway transport employment. On the whole, applying FLT with steep-inclined hoisting, the volumes of mining-permanent operations are reduced by 3–4.5 times; diesel fuel expenses – by 1.8–2.5 times as well as dust release and toxic agents ejection is reduced to 35–35% (Yakovlev 1997).

3 SYSTEM OF CONCENTRATION LEVELS DEVELOPMENT USING TRUCK-CONVEYOR TRANSPORT IN THE OPEN PITS

The studied above technical solutions on employment more perfect and mobile equipment do not completely determine the efficiency of truck-conveyor transport operation in the FLT systems. Here the system of concentration levels development for CTP is of great importance. The most practical of all known is the system of development with conveyor hoist arrangement in the inclined shaft and cross headings, located out of the finite pit contour. The procedure of development with these very shortages does not permit to realize the advantages of truck-conveyor transport. The most rational of the known is the system of development with CCC arrangement in open construction on the pit's flank with minimal (inessential) supplementary flank sep-

aration or its complete absence. The essence of such system of concentration levels development, worked out in the IM UB RAS consists of the following (Yakovlev 2003).

The stationary conveyor hoist is set on the pit's permanent flank parallel to the boundary between the working and permanent pit's flanks. CTP is arranged on the temporary pillars of rocks along this boundary. The transmitting belt conveyor is equipped in a semi-trench on the permanent pit's flank (Figure 4). The distance between the pillar of rocks for CTP and stationary conveyor hoist should provide safety of the latter, the pillar for CTP is being developed by blasting operations. This development system can be realized using uniflank pit mining system. Only in this case both the stationary and transmitting conveyors could be arranged on the permanent pit's flank and CTP – on the temporary pillar of rocks of the working pit's flank.

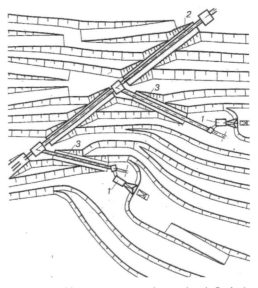

Figure 5. Crushing-conveyor complex on the pit flank: 1 – TCTI; 2 – temporary hoisting conveyor; 3 – stationary conveyor hoist; 4, 5 - transmitting conveyors; 6 – permanent truck crossovers.

The stationary conveyor hoist on the permanent pit flank can consist of several flights with transferring units between them. For arrangement transferring units and disposition conveyors' drive stations it is necessary to form appropriate sites and transport communications for entries to them. For excluding supplementary transport communications on the permanent pit flank both the conveyor hoist and truck crossover routes should be correlated with each other. It is expedient that the entry to the sites of transferring units should be carried out directly from the permanent truck crossover. For this purpose the truck crossover is performed in a loop form and its turning sites should be integrated with the sites of transferring units between conveyor flights. Constructing transferring units' sites safety berms of a pit flank should be used at most. For this purpose the development operations for each conveyor flight are performed so that the flight in its lower section should be arranged in the trench and in its upper section it should be arranged on the supports only (Figure 6).

In this case the conveyor mounting is carried out by a building crane from the building sites on safety berms, the entries to which are arranged from a crossover in the form of workings-embankments. Along this very crossover the necessary equipment and building structures are delivered on a tractor bogie. Safety berms application for between conveyors' transferring units as well as a building crane disposition and construction the road up to the tractor bogie reduce the volumes of mining-preparatory operations to the minimum and the construction procedure is applicable for both belt and steep-inclined conveyors.

4 CONCLUSIONS

Besides the above-stated technical and technological decisions orientated on the efficiency increase of truck-conveyor transport employment in the pits, there are other measures promoting this problem handling. Such measures are orientated on the rise of employed equipment operating period, the level of its service maintenance applying advanced procedures and facilities, as well on the improving production organization and others.

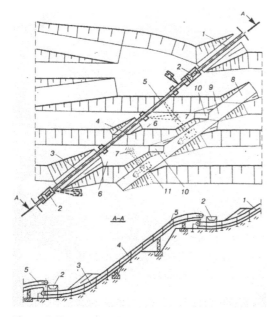

Figure 6. Construction the gallery with conveyor hoist on the permanent pit flank: 1 – conveyor gallery; 2 – transfer unit between conveyor flights; 3 – the lower section of conveyor flight disposed in a trench; 4 – the central section of conveyor flight disposed on the supports; 5 – the upper section of conveyor flight disposed on the supports; 6 – a building site; 7 – a building crane; 8 – a recess; 9 – an embankment; 10 – entries to the building sites; 11 – a tractor bogie.

REFERENCES

Yakovlev, V.L. 2003. *Prospect decisions in the field of flow-line technology of deep open pits*. Mining magazine. Vol. 4–5: 51–56.
Vasiliev, M.V. 1975. *Combined transport in the open pits*. Moscow: Nedra.
Martinenko, V. & Elios, K.H. 1966. *The first crushing-conveyor complex produced by KRUPP firm in the Poltavsky MTW*. Mining industry. Vol. 3: 27–29.
Yakovlev, V.L., Tulkin, A.P. & Karmaev, G.D. 1997. *The progress of FLT systems with steep-inclined conveyor hoisting is an efficient way of mining deep open pits*. Yekaterinburg: IM UB RAS: 194–205.
Yakovlev, V.L., Smirnov, V.P. & Bersenev, V.A. 2003. *The arrangement of crushing-conveyor complexes in deep open pits*. Yekaterinburg: IM UB RAS.

Progressive Technologies of Coal, Coalbed Methane, and Ores Mining – Bondarenko, Kovalevs'ka & Ganushevych (eds)
© 2014 Taylor & Francis Group, London, ISBN: 978-1-138-02699-5

Service of self-propelled mining equipment in underground conditions

M. Stupnik & V. Kalinichenko
Kryvyi Rih National University, Kryvyi Rih, Ukraine

I. Bah & V. Pozdnyakov
UGANC, Conakry, Republic of Guinea

D. Keita
ISMB, Boke, Conakry, Republic of Guinea

ABSTRACT: In this paper we developed and improved service principles of self-propelled diesel equipment in conditions of deep horizons of underground mines. The basic requirements for maintenance of mining equipment in mines are developed and fleshed out.

1 INTRODUCTION

One of the areas of technical re-equipment of underground mines in recent years is the use of self-propelled equipment. Complexity of designs, the high cost and stringent requirements for reliability of self-propelled machines create a situation where the task of improving the efficiency of their use is highly relevant (Korolenko et al. 2012, Kalinichenko et al. 2007, 1992 & Gireeev 1979).

Self-propelled equipment operation experience shows that machine downtime is mainly a consequence of the imperfection of maintenance services. In addition, the small efficiency of the maintenance services leads to high labor and material costs to maintain the equipment in good working condition.

Thus, the improvement of maintenance and repair of self-propelled mining equipment in underground conditions is a challenge that has important scientific and practical importance.

Research and publications analysis. In the mines, where vehicles with diesel drive used should get garages, warehouses, petroleum oil and lubricants (POL) and refueling points.

Garage should provide parking for all the scheduling of the park except for the machines that are to be repaired. Repair shop should be provided in the absence of the possibility of shipping the equipment to the surface without disassembling it to the basic units. Location of the repair shop is allowed in one passage with a garage subject to their division with fireproof wall with fire doors and the presence of independent exits (Kalnitsky 1990 & Kuleshov et al. 1997).

Garage cell and maintenance shop sizes should be determined taking into account the dimensions of the serviced machinery, necessary passages for people between the machines and the possibility of maneuvering into and out of a machine having the largest radius of rotation, as well as the device stationary jobs (Koval et al. 1987 & Monsini 1998).

In the initial period of application of self-propelled equipment it is necessary to solve complex issues on creation of a new system of its technical operation – maintenance and repair. It is necessary to develop a new structure of maintenance service, create or adapt to underground conditions stationary repair sites and items, mobile maintenance facilities, to provide specialized maintenance personnel, to resolve issues of implementing aggregate-node method of repair and supply of spare parts, design repair documentation.

In addition, a large depth of deposits development is characterized by the fact that the issue of self-propelled equipment, requiring repair on the surface, and descend into the mine of refurbished machines is quite labor-intensive process.

These requirements lead to the need to develop and improve new service principles of mining equipment in underground conditions.

Paper goal is to develop and improve the service principles of self-propelled diesel equipment in conditions of deep horizons of underground mines.

2 MATERIAL AND RESULTS PRESENTATION

Self-propelled diesel equipment manufacturers are foreign manufacturers, the main of which are companies such as Atlas Copco, Sandvik, Caterpillar, etc. Drilling rigs of Swedish companies Atlas Copco and Sandvik (formerly Temrok) have been proved theirselves in the underground environment for many decades.

This modern high-tech equipment should have proper care and timely repair in order to maintain and improve performance, longer service life, reliability and safety.

Corrective maintenance of mining equipment is a separate event in the framework of the mine activity. It ensures uninterrupted operation of the equipment, resulting in higher reliability and efficiency of mining equipment and thereby increases the efficiency of the mine as a whole.

Enterprises can solve the problem of machines repair in several ways:

• repair all mining equipment with own structures.

• repair and maintenance of mining equipment can be trusted to service centers of manufacturers.

• do not repair the equipment.

Refusal of equipment repair mainly affects the service life of equipment – it will be significantly reduced, causing the need for additional investment in fleet renewal. Ordered and uninterrupted operation of the mine will be endangered as well: all processes depend on the reliability and availability of equipment, which in turn will affect the implementation of the plan, and, as a consequence, the cost of production, revenue, and production efficiency.

Proper maintenance of underground mobile equipment requires a number of requirements for repair:

• qualified service personnel;

• appropriate maintenance equipmen;

• original parts, spare parts and components for cars;

• the frequency of inspection, maintenance and preventative maintenance.

Most competent in meeting the above requirements are service centers of equipment manufacturers, as accumulating experience in repair of equipment, training personnel of servicing this equipment and having access to spare parts, branded service centers approach to the task of repair or maintenance of underground self-propelled equipment most professionally.

Service of equipment by manufacturer service centers would be beneficial for the life cycle of the equipment; however, for the company it will bring additional costs: repair and service, the cost of production. For example, long-term delivery of parts, spare parts, and components increase period of equipment non-working condition, as equipment manufacturers prefer to use the original parts without having warehouses and reserves nearby.

Repair and service works with own company structures are complicated by several aspects of the repair work:

• the need for training and maintenance of personnel. Modern equipment of the mining industry is not only self-propelled machine, but the whole automated system with on-board computers, sensors and other necessary paraphernalia of modern production. Repair and adjustment of such systems is not limited to using a hammer and screwdriver, specially trained craftsmen who understand the whole mechanism of this particular machine and can cope with any problem are needed here. Accordingly, training of mechanics and engineers of specialized equipment service will have additional expenses for the enterprise;

• repairs should be carried out with specialized equipment in designated areas at the optimum time with minimal effort. The company needs to build and equip underground items for complex service of mining equipment;

• items required for service (spares, parts, components) should always be close at hand, in order to minimize the machines idle time. Elements must be performed on the spot, that will bear the costs of setting up an enterprise for the production of parts and the cost of manufacture of parts itself. Or you can buy spare parts and store them in their warehouses, the minus are additional expenses to provide warehousing and purchase rather than manufacture, elements.

In this regard, the authors propose improved system of self-propelled mining equipment maintenance and repair, which corresponds to developing in time the means of mechanization of mining operations and takes into account the operation of mobile equipment in underground conditions. Such a system is called adaptive, able to adapt to changing operating conditions. Planning of maintenance and repair in this system is based on computer processed information on the actual resource elements of machines installed by the computer groups , taking into account the stability of the individual elements and assemblies. Optimization of volume and timing of maintenance and repair of machines is determined based on an assessment of the actual condition of the equipment by means of technical diagnostics.

3 CONCLUSIONS

Developed by the authors adaptive system of maintenance and repair allows to eliminate drawbacks of existing systems. First, it does not require high material costs and organizational change when introducing it to the mining enterprise. Second, consideration of the effect of mining conditions on the durability and reliability of self-propelled equipment allows to determine the needs of the frequency of maintenance and repair (M and R). This requirement is especially true for the M and R system of loading and delivering equipment as the mixing of operations of loading and delivery of the rock mass determines the increase in the number of mining factors affecting their reliability. Third, a significant development in computer technology offers new opportunities for solving problems of organization, planning and management of M and R.

REFERENCES

Korolenko, M.K., Stupnik, M.I. Kalinichenko, V.A., Peregudov, V.V. & Protasov, V.P. 2012. *Expanding the resource base of Krivbas underground by involving the extraction of magnetite quartzite.* Kryvyi Rih: Dionis: 236 (978-866-2775-49-9).

Kalinichenko, V.A., Zhukov, S.A. & Kalinichenko, Y.V. 2007. *Development tendency of the mining sector and problems of technical re-equipment of underground mines.* Kryvyi Rih: Mineral: 163.

Gireeev, P.M. 1979. *Typical underground mobile equipment maintenance items.* Mining Journal, 3.

Kalinchenko, Y.P. 1992. *Installation, maintenance and repair of transport mining machinery.* Field Reference. Moscow: Nedra.

Kalnitsky, Y.B. 1990. *Safe operation of underground mobile equipment* / Worker's Guide Ed. by Tikhonov N.V. Moscow: Nedra.

Kuleshov, A.A., Popovich, A.E. & Trusov S.P. 1997. *Ways to improve mobile equipment at mine "North" ore-mining complex "Pechenganikel'".* Mining Journal, 11.

Koval, A.N., Gorlin, A.M. & Chekavsky, V.I. 1987. *Maintenance and repair of mining equipment.* Moscow: Nedra.

Monsini, K.R. 1998. Firm "Karerpillar" mining machines maintenance concept. Mining Journal: 11–12.

The use of information and communication technologies in the mathematical preparation of engineers-ecologists: problems and prospects

O. Grebonkina
Donets'k National University, Donets'k, Ukraine

ABSTRACT: In the article the question of application of modern information and communication technologies in the preparation of engineers – ecologists is given, are some of the problems and prospects for their implementation in the educational process are examples of the use of such technology in teaching mathematics to students of ecology specialty.

1 INTRODUCTION

In today's world are increase problems of ecology. In Ukraine, in particular, the acute problem of water pollution, air emissions of pollutants, disturbance of the Earth's surface due to mining, etc. The need for engineers – ecologists like is obvious. Hence, there is a need for training in appropriate skills. A special place in ecologists took the mathematical preparation. The course of higher mathematics students gets the knowledge necessary to understand the special items. Math classes develop skills of construction of mathematical models of processes and phenomena that occur in the environment, develop skills in the application of the mathematical device in addressing practical professional tasks.

2 OBJECT AND PROBLEMS

To address contemporary ecological issues are experts with several competencies. The main subject-line mathematical competences include (Rakov 2005):

– procedural competence (the ability to accomplish a range of common math problems);

– technological competence (mastery of modern information and communication technologies (ICTs));

– research and methodological competence (ability to assess the appropriateness of the use of and mathematical methods and ICTs tools for solving social individually meaningful tasks).

Such requirements for engineers-ecologists are forcing universities to explore new methods and forms of training. The changes are not only content, but also a form of educational material. This primarily refers to the fundamental disciplines: mathematics, physics, chemistry, biology.

In mathematics, it is important to teach students to solve applied math problems and use modern information technology.

Analysis of current research and publications showed that the relevant questions of the development of technologies and development tools learning software (Bashmakov 2003, Trius 2005); create conditions for the development of information and communication competence of students (Bukov 2008).

In high school, there are two main trends in the use of ICTs in the educational process. The first is learning by computer. To do this, you create e-learning program. The development of such programs on the course of higher mathematics deals with a number of native experts. Prepared electronic teaching packages, electronic multimedia complexes, electronic textbooks and manuals (Alekseeva et al. 2011, Borisov et al. 2004, Grebonkina et al. 2011, Selyakova & Yakusheva 2012).

The second trend is the computer support for traditional learning (Flegantov 2012, Grebonkina 2013, Harashehouk et al. 2013, Zahirnak & Chornyi 2013, Zuza & Tumko 2013). As a supporter of second track, believe that the use of ICTs in education is necessary, but should not completely replace the traditional form of learning material. To edit math expressions and work with them in any electronic program, it is necessary to understand the structure of expressions, the logic of their build. Skills to work with mathematical objects are formed and developed more effectively in class by the instructor live, in solving problems analytically without the use of ICTs No electronic program is a substitute for direct communication of the student and the teacher Moreover, when training ecologists math using modem ICTs arise specific difficulties.

The purpose of this article is to present the experience of ICTs use, specify the problems and prospects of application of such technology in the mathematical preparation of engineers-ecologists.

Note that students of ecological occupations are types of modern information resources (Kofanova 2010):

– multimedia presentations;

– Online encyclopedias that are analogues of traditional reference books;

– program simulators;

– software system of controlling knowledge, including different questionnaires and testing programs;

– online tutorials and training courses, which can combine educational and methodical complex for all or some of the above.

We share the view that when designing of educational process with the use of information technologies should respect the principle of diversity of species and forms of training (Dyachkin 2009). Show how this principle is implemented in teaching higher mathematics students of specialty "Ecology, environmental protection and balanced nature management" in Donets'k National Technical University (DonNTU).

In the course of higher mathematics in the specialty of 60% of the training time is devoted to the auditorium, 40% for independent work of students. Lessons are organized in the form of lectures and practical exercises. The theoretical part of the course is given at lectures. The training material is in the form of demonstration lectures. This allows you to save time, avoid mechanical errata, give more information, a shift in the statement. In traditional filing rate is not always possible to show examples of application of mathematical methods in solving applied problems of ecology. Because mathematics is abstract input concepts, algorithms and methods for problem solving, students might give the impression that the mathematical apparatus does not apply in their future work. On the demonstration lectures always have the opportunity to show applications and thus stress the need to study mathematics for success in professional life.

The main problems of using demonstration lectures feel relatively fast paced delivery of new material and the absence of a process step charts, graphs, drawings. Students at the same time watching the slides and listen to the teacher, Therefore, they may not have time to understand the conversion completed principle of construction schedule, etc.

We proposed for the student's lectures in the form of presentations created using PowerPoint. Select explain the ease of use and broad dissemination of it. The program is part of the "Office" and slides created with its help, can be viewed on virtually any computer. Of course a files *.pptx lake up a large amount of memory, but this lack is compensated by the opportunities of working with text, pictures, video clips.

Consider that in preparing the slides for a presentation you should follow these recommendations (Ignatenka et al. 2010):

– apply clear type fonts Arial, Times New Roman, Tahoma;

– definitions and highlights should be in bold or in color;

– font size and pictures on slides should be adapted to the size of the audience;

– the slides should be a bit of a lecture just 15–20 slides;

– don't overload the slides of text and pictures, to avoid simply moving lecture on slides;

– don't use music lectures.

Subject to these rules of use of presentations in educational process is promising. Demonstration lectures allow easy structuring training material, provide visibility into supply, speed and accuracy of graphics. All this contributes to a better absorption of the material at significant savings m training time.

We have developed a course in higher mathematics for students of ecology specialties (Grebonkina 2013, Grebonkina & Bondarenko 2013). Most lectures theoretical material is illustrated not only abstract examples, and applied tasks. Such tasks require students to deeper theoretical knowledge, the ability to establish a link between the different sections of the course of higher mathematics; it is necessary to solve must be solved in multiple ways, compare, analyses the results (Primakov & Razduj 2010). The professional orientation allows the skills of practical interpretation of the result, to develop the ability to make a prediction on the dynamics of an ecological process.

For clarity, here is giving a fragment of the lecture on "Differential equations of the first order" that has been prepared for students – ecologists (Figure 1).

The use of ICTs in practical training in higher mathematics allows you to create not only a mathematical competency, but also computer competency. It makes sense after a sufficient number of common tasks in a notebook check the correctness of their decision by means of ICTs. This gives mathematics helps contribute to training at lesson and indirectly promotes computer literacy students because they have to work with application programs. Test results can be done through programs such as Mathcad, Maple, Mathematica. We work with the ecologists use Mathcad (Pliš & Slivina 2000). Select explain the fact that, first, Mathcad is running in most widely used Windows operating system. Secondly, when you work with Mathcad all mathematical expressions are of such a kind as to work with them on paper pen.

(a)

Differential equations with separable variables

• *Definition.* Differential equations in which variables can be split by multiplying both sides of an equation by the same expression, called *differential equations with separable variables:*

$$\frac{dy}{dx} = \frac{f_2(x)}{f_1(y)},$$

where $f_1(y)$, $f_2(x)$ are the given function.

Multiply both sides of equation by $f_1(y)dx$

$$f_1(y)dy = f_2(x)dx$$

Integrating, we obtain the relationship between variables x and y, liberated from their differentials:

$$\int f_1(y)dy = \int f_2(x)dx \text{ – an integral equation}$$

(b)

Example. The system consists of water and dissolved organic waste in it. Find waste concentration at time t, if the initial time it was L_0.

Let $L_{(t)}$ concentration of waste at any time t. Then the change in concentration over time is described by differential equation with multiple variables with the initial condition:

$$\frac{dL(t)}{dt} = -kL(t), \ L(0) = L_0,$$

where k is the coefficient of oxygen consumption, 1/day. General solution of the equation: $L(t) = e^{-kt+c}$. Given the initial condition, we have the $L(t) = L_0 e^{-kt}$.

• Response $L(t) = L_0 e^{-kt}$.

Figure 1. Fragment of the demonstration lectures: (a) definition of equation; (b) example of applied task.

Next, here is a fragment of the task on "Derivative of function" analytical method and using Mathcad (Figure 2).

The results of the challenge are the same in both cases. Using Mathcad response received quicker and easier. But not exclude the practice of solving problems on paper. As you can see in Figure 2b – pro-

gram will give a definitive answer. For students – ecologists more effective is the first solution. When teaching mathematics to demonstrate step-by-step execution of the job. Detailed solution helps build students skills to build mathematical models of real-world processes and their subsequent decision. Electronic programs better use to monitor results. A specific problem in the work with the students – ecologists believe, also, their low level of development or inability to work with Mathcad at all.

(a)

Find the derivative of a function $y = \cos^3 2x$

$$\frac{dy}{dx} = (\cos^3 2x)';$$

$$\frac{dy}{dx} = 3\cos^2 2x \cdot (\cos 2x)'$$

$$\frac{dy}{dx} = 3\cos^2 2x \cdot (-\sin 2x) \cdot (2x)'$$

$$\frac{dy}{dx} = -6\cos^2 2x \cdot \sin 2x$$

(b)

$$y := (\cos(2x))^3$$

$$\frac{d}{dx} y \to -6 \cdot \cos(2 \cdot x)^2 \cdot \sin(2 \cdot x)$$

Figure 2. Fragment of the practical tasks of higher mathematics: (a) task analysis method; (b) task using Mathcad.

Great prospects of using the ICTs are in independent work of students. Of course, you need an electronic textbook. The DonNTU students – ecologists Handbook for higher mathematics (Grebonkina 2014), which is duplicated in electronic text document Word and PDF file. Also in electronic form are available for individual homework (Grebonkina et al. 2011). In addition, the uses of information technologies to make an independent mathematical preparation element of research work. You can have students find examples of tasks are modeled ecological processes. Traditional training for this type of work is not enough time. The Internet provides unlimited opportunities of finding problems at a relatively low cost. As a result, students in the study of mathematics are also ecological thinking.

Using ICTs is easily organized monitoring of student learning in mathematics (Grebonkina 2013). We asked ecologists tests in *"My Test"*. This system is easy to use. It frees the teacher from the mechanical level learning, enables you to receive new jobs in each of the next poll. Students go through in such a system, the more fun and easier than on paper. It is convenient tool independent check of their mathematical training

Despite these advantages and prospects of use of ICTs in the educational process, we have some technical problems. For the successful application of such technology requires a computer with relevant programs that in the real world has no place in all universities. The problem can be partly solved by using their own laptops for students.

3 CONCLUSIONS

The use of information technologies in educational process contributes to the intensive use of new teaching methods. The mathematical preparation of ecologists such technologies should be regarded not as a process and as a learning tool. Their use contributes to a better understanding of the course material, an ability to work with different kinds of information. Modern mathematical program is one of the main tools in the professional activities of engineer – ecologists. Therefore, the use of ICTs in learning mathematical disciplines creates conditions for increase of level of computer literacy and vocational training students. Thanks to ICTs, you can include in a course of mathematics objectives environmental content, which makes learning more effective. Despite some challenges, the introduction of modern means of information and communication technologies in the process of mathematical training conservationists, promotes the strengthening of independent work of students, enhancing their motivation and interest in teaming.

REFERENCES

Alekseeva, I.V., Gajdej, V.O., Duhovichnyi, V.O., Konovalov, N.R. & Fedorova, LB. 2011. *Fxperience creating and using teaching methods of higher mathematics, collection of scientific and methodical works*. Donets'k: DonNTU, Vol 7: 16–41.

Bashmakov, A. 2003. *Development of computer textbooks and instructional systems*. Moscow: Filin: 675.

Borisov, S.I., Dolmatov, A.V., Kruchinin, V.V. & Tomilenko, V.A. 2004. *Computer tutorial "TMCDO Higher mathematics-1"*. Open education, 3: 12–17.

Bukov, S.A. 2008. *Information's communication competence of students*, Higher education today, 12: 16–20.

Dyachkin, O.D. 2009. *Experience design methodology of computer teaching mathematics*. Open and distance learning, 4: 24–30.

Flegantov, L.O. 2012. *Study functions of one variable by using a web service. Wolfram Alpha Theory and methods of reaching mathematics, physics. Informatics*. Kryvyi Rig: NMetAU. Issue I. Vol. X: 326.

Grebonkina, O.S. 2013. *The issue of thematic control of students' knowledge of higher mathematics*. Teacher education theory and practice. Kamyanets-Podilsky. Vol. 13: 225–229.

Grebonkina, O.S., 2013. *Experience of establishing demonstration lectures in higher mathematics for students of the Faculty of ecology and chemical technology*. Collection of scientific and methodical works. Vol 8: 68–73.

Grebonkina, O.S. 2014. *The methods of higher mathematics in chemistry*. Donets'k: "VIC": 108.

Grebonkina, O.S. & Bondarenko, O.M. 2013. *The use of demonstration lectures in higher mathematics in training engineers, ecologists*. Materials VIII international scientific and technical conference "Problems of mining, geology and ecology of mining production". Donets'k: Mir knigi: 302–306.

Grebonkina, O.S., Evseeva, O.G., Klemina, S.I. & Savin, O.I. 2011. *Individual assignments on higher mathematics Electronic resource access mode* http //library don-ntu edu ua/hooks/met /ml837 zip

Harashchouk, O., Kutsenko, V. & Sodol, I. 2013. *Improving the qualm of training personnel fundamental mission of higher education*. High school, # 2: 22–36. Kyiv.

Ignatenko, V.M., Nefedchenko, V.F. & Opanasyuk, O.S., 2010. *Using multimedia presentations and electronic textbook in basic school subjects*. New teaching technologies, 63, Part 1: 67–71.

Kofanova, O. 2012. *Stimulating educational cognitive activities of the students-ecologists by means of information-communicative technologies*. High school, #8: 72–87. Kyiv.

Plis, A.I. & Slivina, N.A., 2000. *Math-cad 2000 Math workshop*. Moscow. Finance and Statistics: 656.

Primakov, A.V. & Razduj, O.M. 2010. *Some methodological peculiarities of mathematics in the context of requirements for presenting physics with innovation of educational policy*. New teaching technologies, 65: 43–48.

Rakov, S.A. 2005. *Mathematical education ICT-competence approach*. Kharkiv: "Fact": 360.

Selyakova, E. & Yakusheva, E. 2012. *Management independent work of students of chemistry department on example the theme "Elements of mathematical analysis"*. Didactics of mathematics Problems and Investigations, 3: 78–83. Donets'k.

Trius, Vu.V., 2005. *Computer-oriented methodical teaching of mathematics*. Cherkasy: Ukraine-Brama: 400.

Zahirniak, M. & Chornyi, O. 2013. *Information Communication technologies in training experts of technical specialties* High school, #1: 7–19. Kyiv.

Zuza, A.V. & Tumko, Yu., G. 2013. *The use of computer-based tutorials in the process of studying higher mathematics*. Collection of scientific and methodical works, Issue 8: 137–144. Donets'k.

Progressive Technologies of Coal, Coalbed Methane, and Ores Mining – Bondarenko, Kovalevs'ka & Ganushevych (eds)
© *2014 Taylor & Francis Group, London, ISBN: 978-1-138-02699-5*

Evaluation of changes in rock mass permeability due to long-time repeated mining

V. Driban & N. Dubrova
Ukrainian State Research and Design Institute of Mining Geology, Rock Mechanics and Mine Surveying (UkrNIMI) National Academy of Sciences of Ukraine, Donets'k, Ukraine

ABSTRACT: Conditions of wallrock exposed to long-time repeated undermining in the Central Donbass are evaluated. It is shown that long-term repeated undermining results in formation of quasi-regular network of vertically oriented man-induced reservoirs. Simulation modeling is made to evaluate permeability rate of wallrock due to the impact of ultimate deformation and deformation exceeding the maximum value. The rate of repeated undermining on permeability of wallrock is determined.

1 INTRODUCTION

It is known that for the whole period of development of Donets'k Coal Basin (Donbass) about 11.2 MT of coal have been extracted. As a result, rock mass has been deformed over the vast areas that was accompanied by ground subsidence and increase in permeability of rocks. These processes had extremely adverse impact on ecological safety of the region; in particular they increased exchange of ground – surface water that contained toxic substances – contaminants in inadmissible concentrations. Examples of this negative impact were accidents in mines due to air intoxication with lethal concentrations of harmful toxic pollutants (HTP) in Uglegorskaya and Aleksandr-Zapad coalmines in 1989–1999 (Gosk 2004).

Study and prediction of pollutant distribution in rock mass is rather complicated integrated task that requires accounting of many man-induced causes and endogenous factors, particularly in conditions of long-time repeated constant undermining of rock mass, complicated geologic structure and high tectonic faulting of the territory under consideration. We would like to emphasize that repeated undermining for more than 130 years in conditions of the Central Donbas is one of the most significant factors involved in profound changes in filtration properties of wallrock.

The goal of our work is to evaluate faulting of rock mass due to the repeated successive impact of alternating deformations arising in the process of field development in order to predict its permeability.

2 MATERIAL AND METHOD

The objects of our research (Figure 1a) were three mining allotments: Aleksandr-Zapad mine, where there was mine air intoxication, and two adjacent mines – Kalinina and Kondratievka of the total area of more than 45 sq. km. The mines had been operating since the beginning of the last century and are working up to 20 coal seams at the depths down to 750–1.000 m. As the testing model to obtain estimation characteristics of deformation fields in an arbitrary point of rock mass we chose three coal seams (52–56 degree dip): $k_5{}^2$, k_7, $k_7{}^1$ (Kalinina coal mine) that were being worked for 80 years from 1920's to 1990's of the last century.

Approach to calculations was as follows. All longwalls within each seam were joined together by time duration of mining, time interval being 10 years. The approximation of 1,000 longwalls gave 124 mining outlines in three coal seams which were divided into groups. Each group corresponds to time duration of mining: 1920's, 1930', 1940's, 1950's, 1960's, 1970's, 1980's, and 1990's (see Figure 1b). To obtain values of deformation in an arbitrary point of rock mass initial data on the depth of mining of all the approximated outlines of production workings at the zero ("0") level were reduced to the values corresponded to the appropriate levels: "-250", "-450" and "-650" meters. Contours above the specified level were deleted from the calculation.

A large number of seams being worked, steep pitch and small interbed space define geological conditions for mining the area under consideration and require the repeated undermining to be taken into account in order to obtain reliable estimation of rock mass stress-deformation conditions.

For that reason calculation of deformation parameters for a point in rock mass shall take into account at each subsequent stage of mining the impact of many overlying old mine works. For the calculation of maximum mining-related subsidence, taking into account old mine workings, the following recurrence equation is used (1)

$$\eta_{t_n}^{tot} = \eta_{t_n}^{max} + \frac{\Delta q}{q_0}\eta_{t_{n-1}}^{tot},$$ (1)

where t_n – calculation period; n is number of calculation periods; η_t^{max} – maximum first undermining-related subsidence; η_t^{tot} – total maximum subsidence, taking into account the impact of old mine workings; q_0 – relative magnitude of maximum subsidence determined from formula; Δq – difference between relative maximum subsidence due to the second undermining and the first undermining taken equal to 0.15 for the conditions in Donbass.

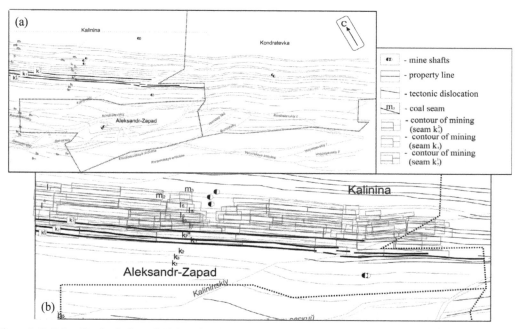

Figure 1. Kalinina, Kondratievka and Aleksandr-Zapad mining allotments (a) and Plans of time-approximated outlines of production workings (b).

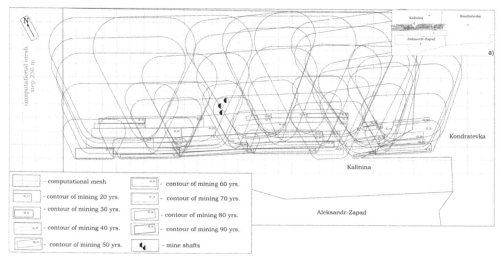

Figure 2. Affected areas of mining contours in seam k_5^2.

Computation and generation of subsidence troughs (Figure 2) were made using *"Undermining"* software (version 2.1) developed in UkrNIMI based on the methods set forth in the current State Standard of Ukraine GSTU 101.00159226.001-2003.

3 RESULTS AND DISCUSSION

Computation of rock mass deformation for each group of the contours enables to show the dynamics of variation of rock mass stress-deformation conditions for an arbitrary point in rock mass, the interval being 10 years. As an example Figure 3(a, b) presents

maps of distribution of horizontal deformations ε_p on the strike of rocks within the area of interest for decades. Distributions of horizontal deformations are characterized by alternate space-connected compression and tension zones traceable at all stages of mining. Compression and tension zones are nearly equal in area and are characterized by nearly equal by magnitude values and approximately equal space interval of alternation within each time interval under consideration. Distances between alternating maximum and minimum values are on the average 350 m for 1920's–1950' and about 1.000 m for 1960's–1990's.

(a)   (b)

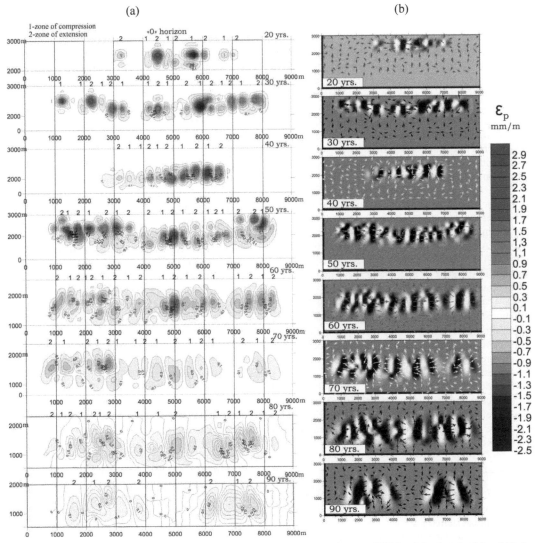

Figure 3. Variation dynamics of rock mass stress-deformation conditions at the zero ("0") level for decades (a) and Variation dynamics of rock mass stress-deformation conditions at the zero ("0") level for decades with vector component (b).

Figure 3(a) shows that areas of rock mass (for example, an interval of 4.000–5.000 m) have been affecting tensile deformation for two first decades of mining and transformed further into zones of compression observed in all areas of rock mass under survey. This goes to prove that rock mass being mined was influenced by horizontal alternating deformations, i. e. repeated step-by-step alternation of compression and tension zones. Such alternation of compression-tension zones was identified also at "-250", "-450" and "-650" meter horizons. So we can conclude that the whole rock mass was influenced by alternating differently directed deformations, which under the repeated alternation of compression-tension zones led to the multiplying of fractures and changes in filtration properties of rock mass in general, respectively.

Treatment of rock mass deformation around moving longwall on a decade basis makes it possible to trace transformation of the deformed rock mass in time. Figure 3(b) shows vector diagrams of decade-based alternation of rock mass deformation conditions.

Maps of the overall horizon-oriented distribution of horizontal deformations on the strike (ε_p) and transverse to the strike (ε_q) of the seam were generated with the account of undermining repetition in the computation of deformation. Distribution of overall horizontal deformations in rock mass is smoother than distribution on a decade basis because of the zeroing of alternating deformations due to the multiple subsequent compression and tension zones. At the same time we see a drive to the formation of the zones of increased permeability with increase in the step of alternating deformations up to 800 m on the average.

Increase in the values of horizontal tensile deformations, which are 5÷7 mm/m at some areas and are 20÷30 times of the limiting values, is indicative of profound changes in conditions of wallrock due to undermining.

To evaluate permeability of fractured formation we made simulation modeling based on the general analytical expression for permeability (2) as applied to the elastic medium with isotropic distribution of fracture systems obtained by E.S. Romm (Romm 1966, Romm 1967)

$$K_0 = Mb_0^3 / 18(b_0 + L), \qquad (2)$$

where b_0 – width of fracture opening; L – distance among walls of adjacent fractures; M – total number of fracture systems.

The aim of our simulation modeling is to determine behavior of permeability of rocks with different initial porosity under the impact of ultimate deformation and deformation exceeding the maximum value. Special emphasis should be made that in this experiment we intentionally consider the case of occurrence of large (ultimate and limiting) deformations accompanied by rock failure. The peculiarity of transformation of the fractured formations under the impact of ultimate deformation and deformation exceeding the maximum value is their discrete nature. It means that the main portion of deformation concentrates in general in isolated well-developed maximally opened fractures. Respectively large deformations will manifest themselves not in the formation of new fractures but in the opening of both already existing natural fractures and fractures arouse at the initial stages of fracture failure. Let us consider the limiting case when maximum deformation concentrates in the sorest spots of the sample – in isolated fractures and determine the impact of behavior of an isolated open fracture on permeability of formation due to the effect of large tensile deformations. Thus, Equation (2) will have the form (3)

$$K_0 = \frac{M}{18} \cdot \frac{b_0^3}{(b_0 + L)} + \frac{M}{18} \cdot \frac{b_{ed}^3}{(b_{ed} + L_{ed})}, \qquad (3)$$

where b_{ed} – width of the opening of the largest fracture; L_{ed} – distance among walls of the adjacent fractures (it describes the density of open fractures).

For our computations we take the following conditions: tensile deformations are 1–20 mm/m with different initial fracturing (number of fractures is 1, 25, 50, 100, 250, 500 and 1000), general porosity being preserved as 1, 5, and 30%; density of open fractures is 0.5 to 5.0 per one meter. In other words, the aim of our experiment is to recalculate the value of permeability of rocks of different fracturing after the impact of ultimate deformation and deformation exceeding the maximum value localized into an isolated fracture.

Based on the results of multiversion simulation the curves are plotted. They show changes in permeability of rock mass of different initial porosity versus the effect of horizontal tensile deformations and density of open fractures. Figure 4 shows as an example the permeability curves for rocks of different initial fracturing under the impact of ultimate and limiting tensile deformations ε_p with one open fracture for 1.0 meter, porosity of rockwall being 5%.

For universal representation of the obtained results and evaluation of relative changes in permeability with the occurrence of an isolated large fracture per 1.0 meter in conditions of limiting deformation we normalized the obtained values with respect to the permeability values in the initial conditions (see Figure 5).

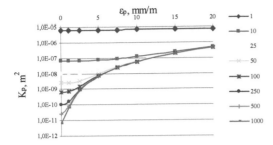

Figure 4. Permeability of rocks with different initial fracturing under the impact of ultimate and limiting tensile deformations ε_p with one open fracture per 1.0 meter, porosity being 5%.

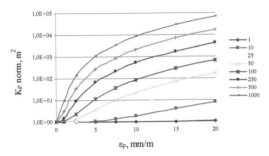

Figure 5. Relative changes in permeability of rocks under the impact of ultimate and limiting tensile deformations ε_p with one open fracture per 1.0 meter, porosity being 5%.

Let us group rocks in our experiment for permeability according to classification by N.A. Plotnikov (Maksimov 1967). For example, with distribution of initial porosity per 1000 fractures rocks are almost impermeable (VI). With distribution per 500 and 250 fractures rocks are rather weakly permeable; with distribution per 50 fractures rocks are weakly permeable and rather weakly permeable; with distribution per one fracture rocks are well permeable and very well permeable (extreme case).

The curves in Figure 4 and 5 show that rocks of equal porosity but with different density and fracture opening and hence with different initial permeability respond to ultimate and tensile deformations differently. For example, permeability of rocks of group I and group II (well permeable and very well permeable rocks respectively) (rows 1, 10 in Figure 4 and 5) described as extreme case with isolated open fracture practically does not change. Permeability of rocks of group III and group IV (permeable and weakly permeable rocks respectively) (rows 25, 50 in Figure 4 and 5) under the impact of ultimate deformations changes weakly (approximately half an order of magnitude). Under the impact of limiting deformations it increases from

$0.5 \div 1 \cdot 10^{-8}$ to $6 \cdot 10^{-7}$. Permeability of rocks of group IV and group V (weakly permeable and rather weakly permeable rocks respectively) (rows 100, 250 in Figure 4 and 5) under the impact of ultimate deformations also practically does not change. Under the impact of deformations exceeding the maximum value it increases already by two – three orders of magnitude from 10^{-9}–10^{-10} to 10^{-7}. However, under the impact of ultimate deformations rather weakly permeable and almost impermeable rocks of group V and group VI (rows 500, 1000 in Figure 4 and 5) become more permeable: by two orders of magnitude (from 10^{-12}–10^{-10}) and under the impact of deformations exceeding the maximum value – up to five orders of magnitude (from 10^{-12}–10^{-11} to 10^{-7}).

Detailed analysis of the obtained data enables us to make the following conclusion. Properties of well permeable rocks under the impact of ultimate deformation and deformation exceeding the maximum value do not change appreciably. However, weakly permeable and rather weakly permeable rocks are more sensitive to the effect of large tensile deformations, which change their permeability up to five orders of magnitude. More over, under the impact of ultimate deformation and deformation exceeding the maximum value there are smoothing of permeability of rocks, smoothing of differences in initial permeability and stabilization of values at the level $1 \div 6 \cdot 10^{-7}$. We emphasize distinctly different behavior of permeability with rock ultimate deformation and deformation exceeding the maximum value. If in case of deformation exceeding the maximum value there is uniform tension, then with arising of breaking tensile deformations they concentrate discretely on the existing and newly arose man-caused fractures. Just this very circumstance influences critically behavior of permeability.

Analysis of the obtained curves shows that with tensile deformations of the order of magnitude 3–5 mm/m permeabilities of rocks (of almost impermeable to weakly permeable) become almost equal. With deformations 15 mm/m the differences in properties of all rocks, including also very well permeable, disappear.

Selection of the range of the influencing deformations in the simulation model (1–20 mm/m) is specified by the correspondence to deformations arising with different degree of undermining. Our computations show the following. Undermining of only one seam results in the occurrence in rock mass of horizontal tensile deformations of about 2 mm/m. The model of rock mass deformation considered above shows that for the case of three-seam undermining horizontal deformations of the order of magnitude 3–5 mm/m arise. From dynamic analysis of rock mass deformations in periods of ten years (see Figure 3) we see the occurrence of overall tensile deformations of more than 8 mm/m (Driban 2013, Dubrova 2014). Comparison of the results of simulation enables us to determine the influence of the number of seams being worked on permeability of wallrock.

In order to determine the influence of density of large open fractures on the changes in wallrock permeability with deformation exceeding the maximum value we made computations for fracture densities 0.5, 1, 2 and 5. Example of changes in permeability with fracture density is shown in Figure 6.

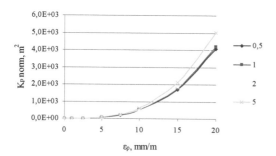

Figure 6. Relative changes in permeability of rocks for different density of open fractures in conditions of fracturing 250, density being 5%.

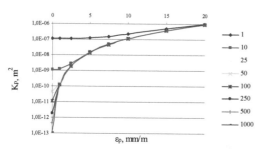

Figure 7. Changes in permeability of rocks with different initial fracturing under the impact of ultimate deformation and tensile deformation exceeding the maximum value ε_p with two open fractures per 1.0 meter, porosity being 1%.

The obtained results show insignificant changes in permeability of wallrock that experiences ultimate deformation on the density of open fractures. We can conclude that under the impact of deformation exceeding the maximum value, density of fracture distribution is not significant at all. In other words, there are such changes in overall wallrock permeability at which structural differences are of no importance and have no significant influence.

Similar simulation modeling was also made for wallrock with initial porosity 1 and 30%. Figure 7

shows the example of changes in wallrock permeability with porosity 1% under the impact of ultimate deformation and deformation exceeding the maximum value. Relative changes in wallrock permeability, porosity for different density of fractures being 1%, are shown in Figure 8. The example of changes in wallrock permeability with porosity 30% under the impact of ultimate deformation and tensile deformation exceeding the maximum value ε_p with two open fractures per 1 meter is shown in Figure 9.

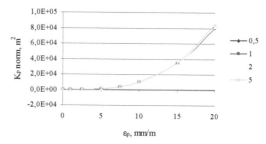

Figure 8. Relative changes in permeability of rocks for different density of fractures (100 fractures), density being 1%.

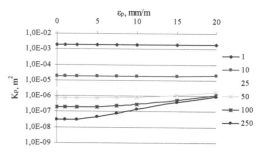

Figure 9. Changes in permeability of rocks with different initial fracturing under the impact of ultimate deformation and tensile deformation exceeding the maximum value ε_p with two open fractures per 1.0 meter, porosity being 30%.

Analysis of permeability of wallrock with different permeability enables us to make the following conclusions. Firstly, rocks with lower initial porosity are more prone to changes in permeability under the impact of deformation fields. Rocks with initial porosity 1 and 5% can change their permeability by 5–7 orders of magnitude under the impact of ultimate deformation and deformation exceeding the maximum value. Rocks with very large initial porosity can change their permeability by no more than 1–1.5 orders of magnitude with extremely high deformation of rock mass.

It should be mentioned that such insignificant changes in permeability of wallrock with high porosity are explained by their large initial permeability, which can't change significantly. It stabilizes at the level 10^{-7}-10^{-6} as does the permeability of practically monolithic rocks under the impact of deformation exceeding the maximum value.

Maximum changes in permeability in all cases fall on the range E_p = 3–5 mm/m. That is, in this deformation range there is a discontinuous increase in permeability by 3–5 orders of magnitude. With increase in deformation, permeability increases by 1–2 orders of magnitude and is characterized by weak stable growth.

As we already mentioned, one-seam undermining results in horizontal tensile deformations of the order of magnitude 2 mm/m arising in rock mass. The model considered above to determine deformations of the arbitrary point in rock mass shows maximum horizontal deformations of 3–5 orders of magnitude arising in conditions of three-seam undermining (see Figure 3).

We note that just such values of tensile deformations equalize permeability of rocks from almost impermeable (group VI) to very well permeable (group I).

As a result of the analysis of specific conditions for working coal in the Central Donbass (in particular approximately equal thickness of workable coal seams and close distribution of deformations in rock mass when working each of steeply dipping layers) we can conclude that our model is universal.

Three-seam undermining in conditions of the Central Donbass results in rock mass permeability of the order of magnitude 10^{-8} (10 darcy) for practically all type of rocks. Further increase in deformations from ultimate to exceeding the maximum value (up to 20 mm/m) has no longer any noticeable influence (full coalescence of permeability curves).

Thus, we can make conclusion that deformations arising with three-seam undermining have significant influence for the crucial changes in permeability of wallrock.

4 CONCLUSIONS

The long-time repeated undermining of coal seams in condition of steep dip results in formation within rock mass of alternating spatially connected zones of compression/tension substituting each other during mining and generating quasi-regular network of vertically oriented man-induced reservoirs.

Detailed analysis of the obtained data enables us to make the following conclusion. Properties of well permeable rocks under the impact of ultimate deformation and deformation exceeding the maximum value do not change appreciably. However, weakly permeable and rather weakly permeable rocks are more sensitive to the effect of large tensile deformations, which change their permeability up to seven orders of magnitude. More over, under the impact of ultimate deformation and deformation exceeding the maximum value there are smoothing of permeability of rocks, smoothing of differences in initial permeability and stabilization of values at the level $1 \div 6 \cdot 10^{-7}$.

The obtained results show insignificant changes in permeability of wallrock that experiences ultimate deformation on the density of open fractures. We can conclude that under the impact of deformation exceeding the maximum value, density of fracture distribution is not significant at all. In other words, there are such changes in overall wallrock permeability at which structural differences are of no importance and have no significant influence.

We make conclusion that deformations arising with three-seam undermining have significant influence for the crucial changes in permeability of wallrock.

In our opinion, the incremental approach of step-by-step evaluation of stress field transformations, deformations and permeability of rock mass in time can be the basis to solve problems related to prediction of contamination due to repeated constant undermining of rock mass (Driban 2012).

REFERENCES

Driban, V.A. & Dubrova, N.A. 2012. *Studying of quantity-related and dimensional distributed parameters of hazardous pollutants in a man-made faulted rock mass of complex structure.* Transactions of UkrNIMI NAN Ukraine, 11: 306–318.

Driban, V.A., Grischenkov, N.N., Hodyirev, E.D. & Dubrova, N.A. 2013. *Special considerations relating to formation of man-induced reservoirs in the process of working coal seams in conditions of the central region of the Donets'k coal basin.* Transactions of UkrNIMI NAN Ukraine, Part 1, # 13: 220–237.

Dubrova, N.A. 2014. *Measurement of change of rock mass filtration characteristics in consequence of long-time repeated mining.* Transactions of Kremenchuk Mykhailo Ostrohradskyi State Polytechnic University, 1 (84): 96–106.

Gosk, E., Slyadnev, V.A., Yurkova, N.A. & Yakovlev, E.A. 2004. *Forward estimate of eco-geological risk of Gorlovka urban complex mine flooding.* Ecotechnologies and resource saving, 3: 60–65.

Maksimov, V.M. 1967. *Hydrogeologist reference tool.* Leningrad: Nedra.

Romm, E.S., 1966. *Filtration properties of faulted rock.* Moscow: Nedra.

Romm, E.S. 1985. *Structural models of rock nonsolid space.* Leningrad: Nedra.

Impulse technologies of borehole dilling – technologies of the XXI century

A. Kozhevnykov
National Mining University, Dnipropetrovs'k, Ukraine

ABSTRACT: The authors examine the well drilling impulse technologies classification that includes such drilling techniques as monoparametric, biparametric and threeparametric. Theoretical foundation of monoparametric technique of long well drilling with the use of high-frequent hydropercussion machinery with hydraulic wave reflectors were designed. Suggested criteria of rotary-percussion drilling efficiency are confirmed by boring conducting practice.

1 INTRODUCTION

The mineral and raw material base development of every country requires the exploration and development of coal, iron ore, polymetallic and other mineral deposits; this includes big depths. In connection with this a row of problems emerges. One of these problems is the drilling of deep wells (to 2000 thousand and deeper) of different purposes: geological exploration, engineering, etc. Resource charges are particularly big by the drilling of deep interval wells (100–2000 and deeper) in fast rocks.

Under these conditions rotation drilling technique is traditionally-used. It has tendency to penetration speed reduction while the increasing of a well depth. This tendency is caused by the natural pattern of face capacity reduction by drilling using a rock-destroying tool in connection with the drop of drilling practice parameters (rotating frequency, axial load, washer fluid amount) while the increasing of a well depth. Therefore deep well drilling efficiency increase is the most important science and technical problem.

The rotation drilling technique of wells is carried out with stationary, time-constant drilling practice parameters – axial load $F = const$, rotating frequency n=const, washer fluid amount $Q = const$.

By rotation drilling the rock destruction can be intensified by load impact of definite frequency and force on the rock-destroying tool (crown, chisel) imposition; i.e. drilling impulse technique implementation when $F = var$.

The article objective is to develop methods and technologies of well drilling with varied combinations of drilling practice time-constant and time-variable parameters.

Theoretical analysis of the technologies of well drilling with variant combinations of drilling practice time-constant and time-variable parameters shows that the drilling with variable axial load is the first variant of impulse technology.

It is suggested to classify all variants of impulse technologies into:

- monoparametric (one of the three parameters is variable) with three variations;
- biparametric (two parameters are pairwise variable) with three variations;
- threeparametric (three parameters are variable) in one variation (Kozhevnykov 1992, 1998, 2003).

The classification of drilling impulse technologies is shown at the table 1.

Currently practical implementation has one drilling impulse technology – those one with variable axial load. Another 6 variations of impulse technologies exist; they require their researches, development and implementation. We have wide perspectives to implement either monoparametric or biparametric or threeparametric impulse technologies.

In this article the investigation results of the development of monoparametric well drilling impulse technique with one practice parameter change – axial load ($F = var$), washer fluid amount ($Q = var$) and rotating frequency ($n = var$) are presented.

2 WELL DRILLING IMPULSE TECHNOLOGY WITH VARIABLE AXIAL LOAD

By the implementation of these impulse technologies into practical well drilling the application of percussion mechanisms (hydraulic, pneumatic, electrical, vibration, magnetostrictors – both face and surface) makes the creation of load impacts possible.

Depending on energy and stroke frequency magnitudes there are percussive-rotary and rotary-percussion drilling.

Percussive-rotary drilling is characterized by the high magnitudes of stroke energy and the low magnitudes of stroke frequency. It requires the application of crown bits and chisels of ruggedness.

Rotary-percussion drilling is characterized by the low magnitudes of stroke energy and high magnitudes of stroke frequency. Its practical implementation is possible due to the application of drilling hard metal and diamond crown bits which are used in rotation drilling.

Percussive-rotary drilling is well-known and widely-used in the drilling of blast wells and wells of different purposes.

Table 1. The classification of drilling impulse technologies.

Stationary	Technologies of drilling						
	Impulse						
	Monoparametric			Biparametric			Threeparametric
$F = const$ $n = const$ $Q = const$	$F = var$ $n = const$ $Q = const$	$F = const$ $n = var$ $Q = const$	$F = const$ $n = const$ $Q = var$	$F = var$ $n = var$ $Q = const$	$F = var$ $n = const$ $Q = var$	$F = const$ $n = var$ $Q = var$	$F = var$ $n = var$ $Q = var$
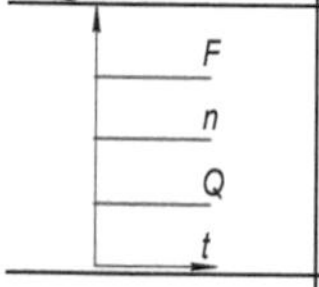 (F, n, Q, t graphs)	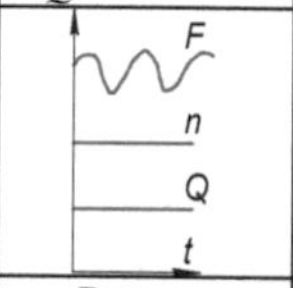 (F, n, Q, t graphs)	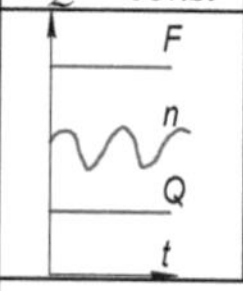 (F, n, Q, t graphs)	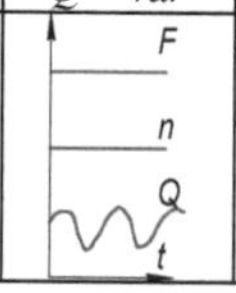 (F, n, Q, t graphs)	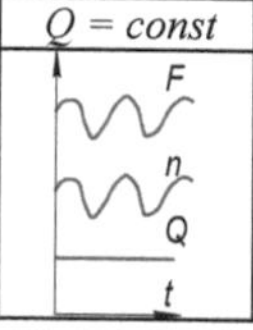 (F, n, Q, t graphs)	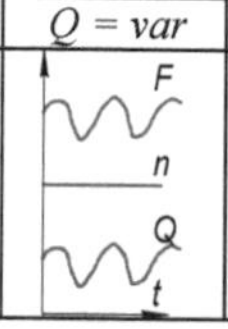 (F, n, Q, t graphs)	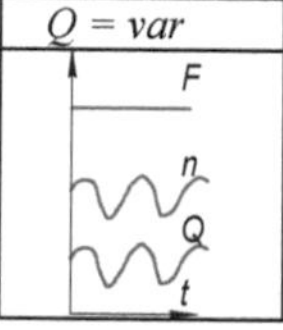 (F, n, Q, t graphs)	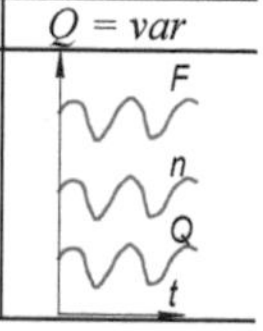 (F, n, Q, t graphs)
Rotation drilling	Rotary-percussion drilling						

Rotary-percussion drilling emerged in 70s of the XX century. It was connected with the development of high-frequent hydraulic hammers. The definition of the rational combination of drilling practice parameters (axial load, rotating frequency, washer fluid amount, stroke energy and stroke frequency) had required the application of research complex, including theoretical research, experimental bench investigations and industrial well experiment.

Theoretical and bench researches were carried out at the National Mining University (Dnipropetrovs'k, Ukraine); field experimental studies were carrying out while exploration drilling work in Ukraine and Kazakhstan.

In comparison with the rotation drilling the efficiency (i.e. the penetration speed raise) of rotary-percussion drilling can be defined with the help of further formula

$$K_{ef} = \left(1 + \frac{N_{si}}{N_{rd}}\right) \cdot \frac{A_{rd}}{A_{rd-si}}, \qquad (1)$$

where K_{ef} – efficiency factor of rotary-percussion drilling; N_{si} – capacity of imposed stroke impulses; N_{rd} – face capacity by rotation drilling; A_{rd} – energy intensity of rock destruction process by rotation drilling; A_{rd-si} – energy intensity of rock destruction process by percussion drilling.

Thus, formula (1) is the energy efficiency measure of rotary-percussion drilling. This measure includes:

• on the one hand: energy level of additional stroke impulses in the form of percussion capacity N_{si} and energy level of rotation drilling; by this one stroke impulses impose additionally in the form of drilling capacity N_{rd};

• on the other hand: the energy intensity (specific volume activity) of the rock destruction process by both rotation drilling and rotary-percussion drilling – A_{ed} and A_{rd-si}.

According to (1) the efficiency of rotary-percussion drilling rises in direct proportion to the ratio of percussion capacity and rotation drilling capacity and the ratio of the energy intensity of rock destruction process by both rotation drilling and rotary-percussion drilling.

So as, on the one hand

$$N_{si} = A_{si} \cdot f, \qquad (2)$$

where A_{si} – stroke energy, f – stroke frequency and on the other hand, having taken N_{rd} for the diamond rotation drilling according to the well-known dependence (Shamshev 1973)

$$N_{rd} = 2 \cdot 10^{-7} \cdot F \cdot d_{cd} \cdot n$$

and for hard metal drilling

$$N_{rd} = 5.3 \cdot 10^{-7} \cdot (0.137 + \mu) \cdot F \cdot d_{cd} \cdot n,$$

where A – forward force; d_{cd} – the average crown diameter; n – rotating frequency; μ – friction coefficient, we will get energy efficiency measure of rotary-percussion drilling with diamond crowns

$$K_{ef} = \left(1 + \frac{A_{si} \cdot f}{2 \cdot 10^{-7} \cdot F \cdot d_{cd} \cdot n}\right) \cdot \frac{A_{rd}}{A_{rd-si}} \qquad (3)$$

and hard metal crowns

$$K_{ef} = \left(1 + \frac{A_{si} \cdot f}{5.3 \cdot 10^{-7} \cdot (0.137 + \mu) \cdot F \cdot d_{cd} \cdot n}\right) \times$$

$$\times \frac{A_{rd}}{A_{rd-si}}. \qquad (4)$$

The K_{ef} defining formulas (1), (3) and (4) let us determine important theoretical and practical recommendations:

• the efficiency of rotary-percussion drilling depends on:
• stroke impulses parameters (energy A_{si} and stroke frequency f);
• rotary drilling practice parameters (forward force F and rotating frequency n);
• crown design parameters (d_{cd});
• crown and rock interaction parameters (μ, A_{rd}, A_{rd-si});
• if energy and stroke frequency increase the efficiency of rotary-percussion drilling increases either;
• the efficiency of rotary-percussion drilling reduces if forward force and rotating frequency increase and percussion capacity is constant (if F and n are increasing to keep K_{ef} it is necessary to enhance A_{si} and f);
• it is possible to obtain the same penetration speed both by rotation and rotary-percussion drilling, but rotating frequency measure and axial load are less by rotary-percussion drilling than by rotation drilling;
• when the creation of high axial loads and rotating frequency is impossible, there are geological, technical and technological reasons why in complicated drilling conditions hummers should be applied. Thus, besides the main purpose of rock destruction process intensification when rotation drilling on additionally imposing percussion impulses, hummers should be considered as the mean which compensates lacking face capacity on rock-destroying tool that is created by drill column rotation and statistical axial load operation.

Thus, energy approach to the rotary-percussion drilling efficiency evaluation enabled us to get the energy measure of its efficiency that makes it possible:

• to predict the magnitude K_{ef} with given percussion capacity levels and known N_{rd} for the concrete conditions;
• to define necessary percussion capacity level for given magnitude K_{ef} for the concrete geotechnical drilling conditions.

At the Figure 1 the change of rotary-percussion drilling efficiency actual factor K_{ef}^{fact} by rotary drilling face capacity calculated value increase N_{rd}^{calc} resulted from wells drilling of 59 mm diameter with the help of high-frequent hydraulic machinery with hydraulic wave reflectors in Kryvyi Rig geological exploration expedition in Ukraine is presented.

Experimental data from Figure 1 confirm theoretical dependence $K_{ef} = f(N_{rd})$ pattern.

The change of rotary-percussion drilling efficiency with well depth growth is the one of particular scientific and practical interest.

The analysis of energy measure rotary-percussion drilling efficiency theoretical dependence (1), (3), (4) on the drilling practice and rotary percussion face capacity parameters leads to the new theoretical proposition of the great practical importance. This proposition says that as the rotating frequency and, therefore, face capacity decreases being the depth function, so the efficiency factor being the well depth function will increase.

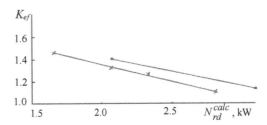

Figure 1. The dependence $K_{ef} = f(N_{rd})$ by industrial testing: • – preliminary testing; x – approval testing.

At the Table 2 and Figure 2 there are virtual experimental data according to the change of axial load, rotating frequency, penetration speed and rotary-percussion drilling efficiency coefficient by well depth increasing in industrial conditions. The data have resulted from wells drilling in Kryvyi Rih geological exploration expedition.

Table 2. The change of rotary-percussion drilling efficiency by well depth increasing.

Well depth interval, m	Average drilling depth, m	Axial load, daN	Rotating frequency, min⁻¹	Penetration grilling speed, m/h		$K_{ef} = \dfrac{V_{rd-si}}{V_{rd}}$
				V_{rd}	V_{rd-si}	
668–927	797.6	1200	231	1.10	1.20	1.09
1264–1365	1314.8	1500	136	0.61	0.80	1.31
1484–1497	1490	800	203	0.46	0.66	1.43
2102–2280	2191	1700	136	0.47	0.60	1.28

As follows from the Table 2 when a well depth increases the ratio of the velocities of rotary-percussion drilling and rotation drilling increases either.

Thus, rotary-percussion drilling is the only one of existing drilling technologies the efficiency of which does not decrease or stay at the stable level relatively to the bases of comparison but, on the contrary, rises.

There have been designed and adopted hydraulic complexes to implement the impulse technologies with variable axial load by wells drilling on the big depths; these complexes include high-frequent hydropercussion machinery and hydraulic wave reflectors. Those reflectors allowed to carry out the activity of hydropercussion machinery with the washer liquid charge equaled to the rated consumption by diamond rotation drilling. That has provided with the application of hydropercussion drilling on the depth to 2000 m and more. Furthermore, besides the cycle one, transannular and internal phase installations of hydraulic wave reflectors in complex with hydraulic hammer were suggested.

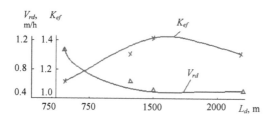

Figure 2. The change of K_{ef} and V_{rf} by well depth increasing.

3 WELL DRILLING IMPULSE TECHNIQUE WITH VARIABLE WASHER LIQUID CHARGE

At present it is known the row of technological processes to construct a well by washing the face with the variable washer liquid charge is to be carried out. They are the follow:

1. Gasoline pump drilling. This drilling technique is to apply to raise core output by rock drilling when rocks are easily washed away by core drilling with washing occurs.

2. The wells drilling with the use of drowned pulsing pumps. This technique is used to drill wells in the presence of washer liquid absorption creating well bottom washing.

3. Diamond drilling under conditions of drilling crowns polishing. By rocks drilling that cause matrix grinding or crown diamonds polishing the technology of diamond crown sharpening is used directly at the well face. Sharpening technology consists in drilling on limited period of time with having drill pump switched on and off periodically.

4. Coreless drilling with impulse washing. Pulsing spurt allows to improve face and well cleaning from slime.

5. Well drilling with the use of hydraulic hammers and hydrovibrators. Because of the pressure pulsation that accompanies with the activity of this equipment there is also washer liquid charge.

6. The investigation and processing of liquid minerals productive strata with the help of hydrodynamic methods.

7. Breakdown elimination by hydrovibration well drilling method.

At the National Mining University drilling laboratory experimental bench studies of the diamond drilling with washer liquid impulse delivery were carried out. The single-layered diamond crowns with the diameter 59 and 76 mm were used for drilling which was conducting through the granite blocks.

Diamond drilling impulse technology with $Q = var$ has provided us with the penetration speed growth in 1.5–2.5 times because in this mode the rock destruction process at the well face is intensified through the thermo-mechanical effect (Kozhevnykov 1998).

4 IMPULSE DRILLING TECHNOLOGY WITH VARIABLE ROTATING FREQUENCY

Figure 3 shows the drilling methods with an impulse tool rotation depending on different symmetric impulses forming approach.

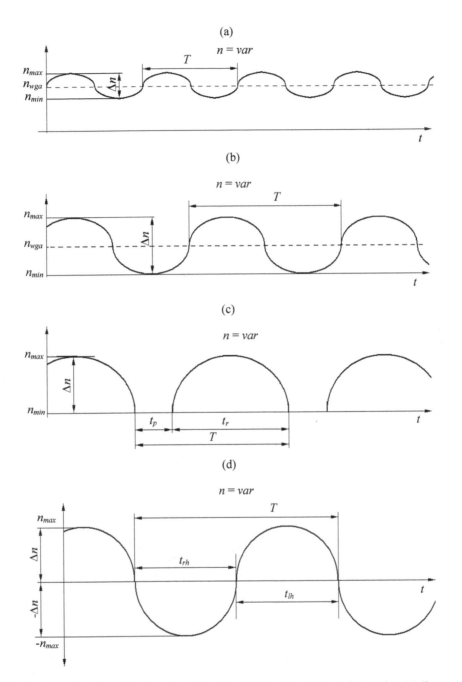

Figure 3. Drilling methods with an impulse tool rotation: (a) variable rotation; (b) pulsating rotation; (c) discrete rotation; (d) reversible rotation.

It is proposed to distinguish the drilling methods with an impulse tool rotation the amplitude of tool's rotation frequency change into a drilling with a: variable (Figure 3a), pulsating (Figure 3b), discrete (Figure 3c) and reversible (Figure 3d) tool rotation.

Pulse drilling technology with variable rotation of the drilling tool (Figure 3a), characterized by the following parameters: maximum rotation speed n_{max}; minimum rotation speed n_{min}; weight-average speed n_{wgav}; amplitude of rotation frequency change

Δn; period(duration of rotation frequency change cycle) T; rate of rotation frequency change f.

Impulse drilling technology with pulsating tool rotation (Figure 3b) is characterized by the same parameters as the technology with variable rotation, but the change in rotation frequency is derived from the maximum speed n_{max} to zero.

By using pulsating tool rotation modepulsations of rotation frequency follow each other without pauses between them.

The main feature of the impulse drilling technology with discrete rotation (Figure 3c) is the presence of pauses between impulses while rotation frequency changes from zero to n_{max}. In this case, just as in the pulsating mode $\Delta n = n_{max}$ and the period is equal to the sum of durations of two phases: rotation phase and pause phase

$$T = t_r + t_p,$$

where t_r – duration of tool rotation phase, t_p – duration of rotation pause phase, i.e. when tool stays without rotation.

In this mode of tool rotation $t_p = t_r$. If t_p becomes equal to zero pause will not happen ($t_p = 0$) and rotation mode would transform from discrete to pulsating.

Impulse drilling technology with reversible rotation is characterized by the presence of the two rotation directions in the same cycle: forward and reverse, i.e. right-hand and left-hand rotation. Drilling tool rotates in the opposite direction after its rotation frequency reached zero. Thus, the period of tool's rotational frequency change in its reverse motion is equal to the sum of the durations of the two phases

$$T = t_{rh} + t_{lh},$$

where t_{rh} – duration of the forward rotation phase, for example right-hand rotation; t_{lh} – duration of the reverse rotation phase, left-hand rotation accordingly.

1. Deep wells core drilling. By the core drilling the rock-destroying tool with the variable rotating frequency is attained by the resilient element installation above the core equipment.

According to (Vozdvijenskiy, Sidorenko & Skornyakov 1978) the use of coil spring as a resilient element in Kryvyi Rig basin by a drilling assembly testing the penetration speed growth in 25–30% has been taken.

The drilling assembly which has the resilient element in the form of grommets was designed at the Dnipropetrovs'k Mining Institute (which now is National Mining University) (Epshtein, Shepel & Kozhevnykov 1979). Production test by the drilling of the wells of the depth to 1200 m had demonstrated that new equipment application allowed to improve:

– penetration speed: by hard metal crowns – in 25%, by diamond crowns – in 30%;

– drifting in one turn: by hard metal crowns – in 15%, by diamond crowns – in 12%.

2. Short wells and blast holes drilling. At the Kuzbas State Polytechnic University (Yakunin 1989) the foundation of boring machinery for blast holes and short wells with impulse and interrupted rotation drilling construction diagrams was made. By the drilling of wells with the diameter 135 mm from a day surface on the depth 3,5 m in rocks with the competence $f = 8$–10 the drilling technique with the use of pulse rotating tool had been tested. Technical and economical effect has been taken.

3. Oil and gas deep wells drilling. At the Poltava department of the Ukrainian State of Geological Exploration several variants of the constructions of the device named bottom drill collar which has the resilient element in the form of the longitudinal elastic nail were developed. The bottom drill collars were industrially tested by the drilling of deep wells (to 5 km) with the diameter 215.9 mm and 295.3 mm in Ukraine. The use of bottom drill collars provides the increase of cone chisel debugging (Svitalka 2001):

– penetration speed increase in 71.1%;

– drifting in one chisel in 126.4%;

4. Wells drilling with the use of hydraulic hammers. By rotary-percussion drilling (Voskresenskiy 1961, Kogan 1964, Graf & Kogan 1972, Smirnov 1971, Sokolinskiy 1982) under the action of longitudinal stroke impulse the rock-destroying tool slows down till the full stop, i.e. its angular velocity equals zero. The tool stop conforms to the torque leap, which is named torque shock. Torque shock might be aroused by drill pipe core curling angle increase and by the conversion of rotating core kinetic energy to retarded core curling potential energy.

Thus, by percussive-rotary drilling rock-destroying tool angular velocity changes from nominal (for the given rotating frequency) to minimal, up to zero. Therefore, percussive-rotary drilling might be referred to impulse technology when $n = var$. To implement percussive-rotary drilling they use percussion mechanisms – hydraulic, pneumonic and mechanical. Hydraulic hammers are the drilling mechanisms with variable axial load which is the impulse monoparametric technology ($F = var$).

Besides that acting hydraulic hammers create pressure and washer liquid charge impulses, i.e. they are equipment to implement the biparametric impulse technique ($n = var, F = var$).

On the ground of facts above it follows that hydropercussion mechanisms are the devices for threeparametric impulse technology implementation, when all three drilling practice parameters are variable ($F = var$, $Q = var$, $n = var$).

5 CONCLUSIONS

Hereby the impulse drilling technologies classification was suggested. It includes 7 impulse technologies: three monoparametric technologies, three biparametric technologies and one threeparametric technology.

All impulse technologies have great technical and economical effect. All technologies are at the same stage of developing, investigation and implementation.

The most developed and implemented in drilling activity is impulse monoparametric drilling technology with variable axial load ($F = var$).

The rest of impulse drilling technologies are at the stage of theoretical, experimental, laboratory investigation or production testing. The carrying out of the big set of research and experiment works will allow us to get and implement into well drilling activity of different purposes by mineral deposit exploration highly-effective mono-, bi- and threeparametric impulse technologies that allow to reduce considerably well construction terms and cost when exploration and production work conducting.

REFERENCES

Kozhevnykov, A.A. 1992. *Well Drilling Impulse Technology. Interstate conference reports theses*. Rocks drilling mechanic. Grozny.

Kozhevnykov, A.A. 1998. *The Scientific Base of Deep Geological Exploration Wells Rotary-percussion Drilling with the Use of High-frequent Hydropercussion Ma-chinery Equipped with Hydraulic Wave Reflectors*. Thesis. 05.15.10. 365. Dnipropetrovs'k.

Kozhevnykov, A.A., Goshovsky, S.V. & Martinenko, I.I. 2003. *Geological Exploration Wells Drilling Impulse Technology*. Kyiv: UkrGRII: 208.

Shamshev, F.A. Tarakanov, S.N., Kudryashev, B.B. and others. 1973. *Exploration Drilling Technology and Technique*. Moscow: Nedra: 496.

Vozdvijenskiy, B.I., Sidorenko, A.K. & Skornyakov A.L. 1978. *The Modern Methods of Wells Drilling*. Moscow: Nedra: 342.

Epshtein, E.F., Shepel, A.I., Kozhevnykov, A.A. and others. 1979. *The Results of Diamond and Hard metal Crown Bit Debugging with the Use of Rubber-metal Damping Sub*. The theses of National scientific and technical conference "The ways of hard metal and diamond drilling tool production improvement and the expansion of its application fields". Samarkand: 184–186.

Yakunin, M.K. 1989. *The Theory and Rocks Drilling Method with Pulsing Tool Advance Designing*. The abstract of the doctoral candidate's thesis. The Kuzbas State Polytechnic Institute, KuzPI typography.

Svitalka, P.I., Duda, Z.N., Nevejin, V.V. & Sloviev, V.V. 2001. *The Influence of Bottom Drill Collar Application on the Techno-economic Indices of Well Drilling*. Scientific and technological compilation IFTUNG, Ivano-Frankivs'k: Issue 38 (volume 2): 29–34.

Voskresenskiy, F.F. Kichigin, A.V. and others. 1961. *Vibration and Rotary-percussion Drilling*. Moscow: Gostopizdat: 244.

Kogan, D.I. 1964. *The Investigation and Development of Rock-destroying Bit Effective Constructions for Exploration Wells Drilling in Fast Rocks with the Use of Hydropercussion Method*: Technical doctoral candidate's thesis. Toms'k: 306.

Graf, L.E. & Kogan, D.I. 1972. *Hydropercussion Machinery and Tool*. Moscow: Nedra: 207.

Smirnov, O.V. 1971. *The Investigation of Hydropercussion Face Machine Hammer Parameters Influence on Drilling Indices*: Technical doctoral candidate's thesis: 04.138: 183. Moscow.

Sokolinskiy, V.B. 1982. *Percussion Destruction Machinery*. Moscow: Engineering.

Progressive Technologies of Coal, Coalbed Methane, and Ores Mining – Bondarenko, Kovalevs'ka & Ganushevych (eds)
© 2014 Taylor & Francis Group, London, ISBN: 978-1-138-02699-5

Research of vibrational activation effect on cement-loess mixture in jet grouting technology

S. Vlasov, N. Maksymova-Gulyaeva & E. Tymchenko
National Mining University, Dnipropetrovs'k, Ukraine

S. Vlasov
Vyatka State University, Kirov, Russia

ABSTRACT: In article application of vibrational activation of soil mixtures in order to increase of their strength at jet technology of loess material consolidation is substantiated.

1 INTRODUCTION

Jet technology allows to solve a complex of different problems in building for foundation bases strengthening of existing buildings and constructions, increasing bearing capacity of piles, in foundation pits digging, shallow underground tunneling, ground water cutoffs and landslide protection constructions. Development and improvement of this technology is especially important for usage at the loess strata spreading territories.

2 FORMULATING THE PROBLEM

High degree of dispersion and, consequently, a large specific surface of loess clay sand and loam particles explain the qualitative difference between the processes of their interaction with fixing substances and the processes occurring in concrete. It consists that the cement in concrete is exposed to hydrolysis and hydration in the low-activity environment with low specific surface of the filler, i.e. hardens in own hydrolysis products. The usual course of hardening cement varies in cement-loess mixture. Hydration and hydrolysis occurs in the conditions of active physical and chemical interaction with the finely dispersed part of soils.

However, it is not possible to create a perfectly homogeneous mixture with a complete lack of pore space in practice. Heterogeneity of cement-loess mixtures, like the ordinary concrete is explained primarily by different dispersion of components.

In loess soils pulverescent content (0.05–0.005 mm) is 70–80%, and it is dominated by small pulverescent fraction (0.01–0.005 mm). Clay particles (less than 0.005 mm) – 10–15%, but proportion of sand fraction (2–0.05 mm) is completely insignifi-

cant. The particle size of alite-and-belite cement composition is 0.1–0.01 mm (Maksymov & Evtushenko 1978).

During intensive formation of crystallization structure in one microvolume of finely dispersed particles, this process is happens with less activity in another microvolume with larger particles simultaneously. In this way, take place formation of disordered spatial crystal structure with nonuniform distribution of internal stresses.

Another reason of mixture heterogeneity is specific characteristics of soils. Loess formations as natural substance are genetically heterogeneous by poriness and mineralogical composition, even if they represent one engineering-geological element. Minerals of fine fraction display various activities during interaction with products of cement hydrolysis and hydration. Along with this, during destruction of soil with high-pressure jets of cement slurry, part of gases including air enclosed in the macro- and micropores of loess soils, remains in the mixture forming its own porous space.

Cement stone is also heterogeneous, with changing parameters of strength and deformability depending on mineralogical composition and hardening conditions. The value of strength with taking into account weakening of structure by inner-gel and capillary pores is depends on the water-cement ratio (W/C), the degree of compaction, etc. According to necessary conditions for jet technology W/C = 1 (Ginsburg 1979) is overflow. It exceeds the amount of water needed for chemical reaction with cement that ultimately leads to the formation of capillaries in the cement stone.

It should be noted that macropores have larger concentration of tensile stresses than micropores. In addition, they saturated with water easier. This implies that total mixture heterogeneity associated with nonuniform water saturation, poriness and different dispersion of binder and soil causes nonuniform distribution of internal stresses, which reduces the final strength of the mixture.

From all presented arguments happens necessity of jet technology improvement in the direction of maximally uniform cement-loess mixture creation. The list of operations of discussed technology should be supplemented by vibrational activation of constructions to solve this problem.

3 PRESENTATION OF MATERIALS AND RESULTS

The essence of vibrational activation consists in the fact that internal friction force between particles of the mixture is largely overcome by the transmitted vibration energy. As a result, it acquires increased mobility and self-packing, releasing gases contained in it (Markov & Cherkasov 1963).

At vibrational activation primarily weak crystalline splices are destroyed, mixture components are milled and uniformly distributed over the volume as a result of particle collisions. Physical and chemical processes are intensified, contacts between particles are ordered and as a result the setting time reduces. Totally, microporous structure with a uniform distribution of internal stresses is formed.

In cement-loess dispersions due to the ability of the particle surface loess soils to interaction (adsorptive, chemical, adhesive) with cement hydration products primary spatial structure develops rather quickly. Vibrational solidification should be carried out during the period of coagulation structure formation – setting period. This is due to the fact that the next step of hardening, the crystallization structure which is formed under mechanical impact is irreversibly destroyed, which adversely affects the final strength of the final material. In other words, we are talking about hereditary irreversibility of the process. Its heredity lies in the fact that all prehistory has a big impact on the process, i.e. component properties, processing methods of the mixture preparation and compacting, initial structure, strength, etc. "Defects" are put in the system during structure formation, as a result will affect on the strength and other properties of hardened mixtures.

We investigated the effect of vibrational activation on the kinetics of changes of loess-cement mixtures strength characteristics that are hardening in normal conditions. The experiments used a constant composition of the mixture with a weight ratio of the components – Portland cement (M400) – 50%; hydrated lime – 5% from the weight of soil. Water-cement ratio – 1, as required to ensure the normal

operation of the monitor in jet technology (Vlasov, Maksymova-Gulyaeva & Maksymova 2013).

Cubic forms with a size of facets 50, 100 mm and prism-forms $40 \times 40 \times 160$ mm in size were filled with the mixture. The processing lasting 60 and 120 s, was performed with a frequency of 50 Hz. This vibration frequency is selected based on the research of the concrete vibrational processing.

Analysis of results showed that strength of concrete naturally grows with increasing of frequency in the range of 10–50 Hz.

Use of low frequency processing can lead to demixing. Most researchers believe that increasing of vibrator frequency grows the efficiency of vibration both for reducing the time required to achieve a sufficient solidification, and for obtaining the maximum density. However, experiments of cement mortar compaction by audible and ultrasonic frequency performed by L.P. Petrunkina and Y.P. Petrunkin have shown that these frequencies are not effective: first of them are fading extremely fast, their visible range is only 5–6 cm. Range of the second was even less significant, it is limited with radius only few millimeters.

Activation time of 60 and 120 s is taken on the results of an effective concrete processing. Increasing the duration of processing can lead to demixing. Vibrational processing increases the density of the non-activated mixture from 1.47 to 1.88 g/cm^3 after retreatment.

Obviously, this leads to activation of the physicochemical structure formation processes that reduce setting time. Researches have shown that the initial setting time reduced to 3 h 40 minutes after the primary mixture processing and up to two hours after the secondary. The end of setting consequently happens after 6 h 30 m and 3 h 50 m (Table 1).

Table 1. Results of vibrational processing effect research on cement-loess mixture setting.

Duration of vibrational processing, s	Terms of the mixture setting, h-min		Mixture density, g/cm^3
	begin	end	
60	3–40	6–30	1.63
120	2–30	4–30	1.74
60 (retreatment)	2–00	3–50	1.88

The released water of coagulation structure remains in the hardening system at vibratory compacting. It has a negative impact on the setting process. However, as noted above, the value of water-mixture factor decreases when constructing piles in insufficiently moist loess soils. In this regard, greater effectiveness can be expected from the vibrational processing in natural conditions. Thus, the issue of vibrational activation mode should be decided in each particular case.

Further series of tests were performed to determine compressive and flexural strength of the mixture. The main seal quality criterion was samples-cubes compressive strength. Analysis of the results shows that strength decreasing of samples associated with intense formation of finely crystalline rigid structure was eliminated after vibrational processing (Figure 1).

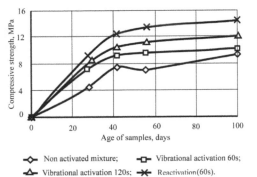

Non activated mixture; ◻ Vibrational activation 60s;
△ Vibrational activation 120s; ✕ Reactivation(60s).

Figure 1. The dependence between compressive strength and duration of mixture vibrational activation.

Strength change under uniaxial compression of cement-loess mixtures after 60 s and 120 s vibrational activation and after retreatment over time is described by the exponential dependence with a high degree of reliability (Table 2).

In addition, experiments have shown that the activated mixture gains strength in the early stages of hardening. Then gain of strength gradually decreases (Table 3).

Thus, after a 60 s processing compressive strength of 28-day old sample increased by 56% and after 120 s – by 78% relative to non-activated samples. As for flexural strength, its value after the 60s activation

increased by 34%.

The effectiveness (K_v) of thickening action is determined by ratio of mixture strength index after the vibrational activation (R^v) to the same indicator of ordinary prepared mixture (R^o).

The greatest effect of vibrational processing, as already noted, is reached in the initial period of hardening (Figure 2). The graph shows that the effect of vibrational processing is reduced at the enhanced crystallization structure formation in ordinary prepared mixture.

At the end of the second stage of structure formation when crystallization frame based on the spatial coagulation frame begins to form, but the mixture is still able to show thixotropic properties with more complete solvation of cement particles, there is a positive effect during re-vibrocompaction. Repeated vibrations allows to internal stresses to relax and heal the structural defects that formed.

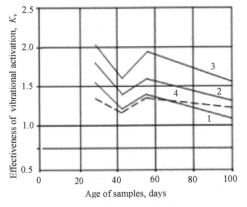

Figure 2. Dependence of between of vibrational processing efficiency from duration of exposure and setting time: 1 – 60 s; 2 – 120 s; 3 – repeated impact (compression); 4 – 60 s (flexion).

Table 2. Results of cement-loess mixture compressive strength value approximation after vibrational activation, depending on the hardening period.

Duration of vibrational activation, s	Equation of dependence of the compressive strength from the timing of hardening	The accuracy of approximation R^2
60	$R_{comp} = 11.53(1 - exp^{-0.04t})$	0.93
120	$R_{comp} = 12.46(1 - exp^{-0.034t})$	0.94
retreatment	$R_{comp} = 14.83(1 - exp^{-0.34t})$	0.94

In an experimental investigation, repeated vibration with 60 s duration increased the samples strength up to 104% in an early period of hardening decreasing gradually to 55% in the later period.

According to experimental researches we can assume that the maximum effect of vibrational activation is achieved if the primary and secondary proc-

essing occurs during the setting period. This conclusion is substantiated by the fact that our task consists in the elimination of defects in the coagulation frame – basis of the crystallizational structure and also, by the fact that we simulate natural technological process conditions when the structure of cement-loess structure has completed.

Table 3. Results of research of vibrational activation effect on mixtures strength.

Age of sample, days	Compressive strength, MPa	Flexural strength, MPa	Vibrational activation		Compressive strength, MPa	Flexural strength, MPa	Compression		Flexion	
			Frequency, Hz	Duration, s			Efficiency	Strength gain, %	Efficiency	Strength gain, %
28	4.5	1.64	50	60	7.0	2.20	1.56	55.6	1.34	34.1
42	7.5	2.30	50	60	9.2	2.81	1.23	22.7	1.22	22.2
56	7.0	2.20	50	60	9.7	2.97	1.39	38.6	1.35	35.0
100	9.3	2.58	50	60	10.0	3.13	1.08	7.5	1.21	21.3
28	4.5	–	50	120	8.0	–	1.78	77.8	–	–
42	7.5	–	50	120	10.5	–	1.40	40.0	–	–
56	7.0	–	50	120	12.2	–	1.31	31.2	–	–
Retreatment										
28	4.5	–	50	60	9.2	–	2.04	104.4	–	–
42	7.5	–	50	60	12.4	–	1.65	65.3	–	–
56	7.0	–	50	60	13.5	–	1.93	92.8	–	–
100	9.3	–	50	60	14.4	–	1.55	54.8	–	–

4 CONCLUSIONS

Dependence between the strength of cement-loess mixture and its vibrational processing time of 60 s and 120 s with 50 Hz frequency was obtained for the first time. It can be expressed with exponential dependence $y = a(1 - e^{-bx})$ with high precision.

The effectiveness of vibrational activation of mixture containing 5% of lime and 50% of cement in a 42 days of age increased by 1.23 times after 60 s processing and by 1.4 times after 120 s processing. The maximal values of compressive strength are 9.2 and 10.5 MPa respectively.

Since the problem in question is multifaceted, the authors believe that further researches in finding ways to improve manufacturing operations and equipment are necessary. Therefore, undoubtedly, works will be continued by the authors and other researchers.

REFERENCES

Maksymov, A. & Evtushenko, V. 1978. *Rocks plugging.* Moscow: Nedra: 180.

Ginsburg, L. 1979. *Landslide protection retaining structures.* Moscow: Stroyizdat: 81.

Markov, L. & Cherkasov, I. 1963. *Improvement of soils characteristics by surfactants and structure-forming agents.* Moscow: Auto-transizdat: 176.

Vlasov, S., Maksymova-Gulyaeva, N. & Maksymova, E. 2013. *Modification of cement-loess mixtures in jet technology during mastering underground area.* Mining of mineral deposits: Collection of scientific papers. CRC Press/Balkema. Dnipropetrovs'k: LizunovPress Ltd.

Mathematical modeling of hydraulic treatment of anisotropic coal seam

V. Pavlysh, O. Grebonkina & S. Grebyonkin
Donets'k National Technical University, Donets'k, Ukraine

ABSTRACT: The problem of theoretical bases development of hydraulic impact on anisotropic coal seam for study and calculation improvement of technological schemes parameters is considered.

1 INTRODUCTION

Hydraulic impact on coal seam is an important way of reducing manifestations of major hazards during underground coal mining. The essence of the hydraulic impact is in implementation under pressure into the project area of coal seam substances with fluidity (liquids, gases, aerosols, suspensions), i.e. the discharge fluid into coal seam with the purpose of its condition manifestation.

Using the technology of liquid injection through single holes, to achieve a uniform high-quality processing of the massif is not possible due to the anisotropy of seepage of reservoir properties. To overcome the negative influence of structure formation for quality improvement and increasing of quality impact was developed cascade hydraulic method (Pavlysh & Grebyonkin 2006).

Effective method of process investigation is mathematical modeling.

The purpose of work – development of theoretical base for investigation and technology parameters calculation of hydraulic impact on anisotropic coal seam.

Main part. According to the physical essence of cascade process and specifics of fluid flow in coal massif filter equation should satisfy following conditions:

– implicit computation of fluid flow coordinates;

– possibility of investigation hard, elastic and elastic nonlinear filtering modes;

– presence of effective numerical solution method.

Velocity of front fluid flow can be determined from the system

$$\frac{k}{\mu}\frac{\partial^2 P}{\partial x^2} = \frac{\partial W}{\partial t} , \qquad (1)$$

$$\frac{\partial W}{\partial t} = \alpha P(W_0 - W) , \qquad (2)$$

$$P(0,t) = P_C , \qquad (3)$$

$$P(x \geq l,t) = 0; W(x \geq l,t) = 0 , \qquad (4)$$

$$l = l(t); l(0) = 0 ,$$

$$\frac{dl}{dt} = -\frac{k}{\mu m}\frac{\partial P}{\partial x}\Big|_{x=l} , \qquad (5)$$

where $l(t)$ – position of the front; W – increase of water saturation; W_0 – maximum increase of water saturation.

From theory of nonlinear parabolic equations it is borehole-known that solution of equations

$$\frac{\partial P}{\partial t} = div[\lambda(P)gradP]; \qquad (6)$$

$$\lambda(P) = aP^n; \quad n > 0. \qquad (7)$$

Has a zero initial condition of target velocity perturbations (Pavlysh 2007). Obviously, that Equation (6) up to the coefficient $\lambda(P)$ is the equation of elastic fluid filtration in porous medium. Assuming that dependency (7) is accept, with help of special selection coefficient a and an exponent n find out that this velocity value is accepted to calculated by Equation (5). Without limiting of generality to simplify assumptions we will consider task in one-dimensional setting.

Characteristic value of time is calculated from equation

$$t_{x1} = \frac{x_x^2}{\xi P_x} , \qquad (8)$$

where $\xi = \dfrac{k}{\mu n_e}$; n_e – effective porosity; P_x – characteristic value of pressure (the pressure at some point near the filtration front edge); x_x – distance from this point to surface front.

It was considered that Equation (6) is rightly near the contour of area that occupied by fluid, we get an expression for t characteristic value of the time

$$t_{x2} = \frac{x_x^2}{aP_x^n} .$$ (9)

For equality of characteristic values is enough execution two of conditions

$$a = \xi; \quad n = 1.$$

Then Equation (6) near the contour takes the form (in a one-dimensional setting)

$$\frac{\partial P}{\partial t} = \xi \frac{\partial}{\partial x} \left(P \frac{\partial P}{\partial x} \right)$$ (10)

on condition $\xi = const.$

Let's show that velocity of movement of free moving surface of liquid accepted from Equation (10) is accepted to Equation (5). Write down it (10) in the form of

$$\frac{\partial P}{\partial t} = \frac{\xi}{2} \frac{\partial^2 P^2}{\partial x^2} .$$ (11)

Equation (11) on the form is similar to equation of polytrope gas filtration

$$\frac{\partial \rho}{\partial t} = a^2 \frac{\partial^2 \rho^{n+1}}{\partial x^2} .$$ (12)

At $n = 1$ for which (Barenblatt 1953) were received right solutions like "traveling wave"

$$\rho = \begin{cases} \left\{ \frac{c}{a^2} \left[(ct-x)\frac{n}{n+1} + B\frac{n}{n+1} \right] \right\}^{1/n}, & 0 \le x \le ct + B; \\ 0, & x \ge ct + B, \end{cases}$$ (13)

where c –distribution velocity of gas leading edge; B – constant that defined by the equality

$$\frac{n}{n+1} \rho^n (0,0) = \frac{c}{a^2} B .$$ (14)

For Equation (13), accepting condition $P(0,t) = P_x$ is following

$$B = \frac{P_x \xi}{c} - ct ;$$ (15)

$$P = \begin{cases} \frac{c}{\xi} \left(\frac{P_x \xi}{c} - x \right), & 0 \le x \le \frac{P_x \xi}{c}; \\ 0, & x \ge \frac{P_x \xi}{c}. \end{cases}$$ (16)

Differentiating (16) on x, we accept an expression for velocity of front edge

$$c = -\xi \frac{\partial P}{\partial x} .$$ (17)

A similar Equation (10), has been previously received by V.V. Liventsev (Liventsev 1964) for isotropic seam, however, was not solved and investigated due to the lack of exact methods. Equation (10) cannot be used in the research of interaction of counter of fluid flow, and in the presence of water wall as pressure transmission velocity defined by Equation (10), is equal to the velocity of fluid. Therefore, for correct calculation of pressure distribution in the filled area of Equation (10) has to be combined with the equation of piezoconduction (Shchelkachev 1946). Inasmuch the Equation is differing from (10) only by coefficient type, it will not impact significantly on algorithm complexity.

Thus, turning to n – dimensional setting, write down the equation of elastic fluid filtration in anisotropic reservoir in the form of:

$$\frac{\partial P}{\partial t} = div[\lambda(P)gradP];$$ (18)

$$\lambda(P) = \begin{cases} \xi P, & x_i \ge l_i - \Delta x_i; \\ \chi, & x_i < l_i - \Delta x_i, \end{cases}$$ (19)

where i – number of coordinate; Δx_i – some small size; χ – coefficient of piezoconduction conductivity that defined according to Shchelkachev (1946).

During necessity of nonlinear elastic mode investigation, coefficient χ is replaced by the product of $\chi(1 + \alpha P)$.

Equation (18) is solved numerically on a computer with using the method of finite differences. Application of difference schemes for solving boundary value tasks for Equations (6) and (7) at $n \ge 2$ described in Samarskiy (1963). During usage of this method for equations solution (18)–(19), value Δx_i is equal to the step of grid area.

Attaching to Equation (18) initial and boundary conditions that corresponding to different technological schemes and operation interactions, we accept a mathematical model of hydraulic impact on coal seam in filtration mode.

Take in future $P = P - P_0$.

Then, as the initial conditions can be taken following:

$$P(x,y,z,0) = 0.$$ (20)

The most common types of boundary conditions are Pavlysh (2007).

Set discharge pressure (condition of I type)

$$P(G_c, t) = P_H(t), \quad (21)$$

where G_c – coordinates of boreholes in considered area.

Usually on borehole is constant pressure mode

$$P_H = const,$$

or injection rate (condition of II type)

$$\frac{k}{\mu} S_c \frac{\partial P(G_c, t)}{\partial n} = q(t), \quad (22)$$

where S_c –surface area of borehole filtration part.

On the borehole can be a relation between pressure and the injection rate (condition of III type)

$$P(G_c, t) = f[q(t)], \quad (23)$$

where f – function that determines a specific correlation.

On outcome boreholes or contoured workings pressure is equal to zero

$$P(G_1, t) = 0, \quad (24)$$

where G_1 – coordinates of outcome boreholes or mine working.

In contact with side rocks (during assuming of their impenetrability) usually applies condition of flow absence

$$\frac{\partial P(G_2, t)}{\partial n} = 0, \quad (25)$$

where G_2 –corresponding boundary of the area.

More difficult is to set boundary conditions when filtering toward stretch untouched massif. In such cases, for parabolic equations setting condition that suggested the presence of outcome borehole far from injection, G_{max}

$$P(G_{maa}, t) = 0, \quad (26)$$

where G_{max} – contour that marked by borehole or absence of flow on random, quite remote from the borehole boundary area G

$$\frac{\partial P(G, t)}{\partial n} = 0. \quad (27)$$

2 ANALYSIS OF RESULTS AND RECOMMENDS FOR ITS APPLICATION

The condition (26), in principle, makes it possible to accept analytical solutions (Barenblatt 1953), but does not allow to use numerical algorithm. Therefore, application of this condition leads to unduly inflate the size of solution, or to a substantial error, because of untouched area on the border of massif is actually replace the naked surface. Condition (27) corresponds for the presence of water at the border, leading to inflated values of pressure: liquid. The most correct would be set on the border of this flow that corresponds to a real drain fluid into the massif outside the area. With the minor mistake we accepted that flow value at the edge of researched area is a constant

$$\frac{\partial P(G, t)}{\partial n} = const. \quad (28)$$

Equation (18) with the specific area of the edge set of conditions (20)–(25), (28) is a mathematical model of pressure filtration of liquids in a coal seam.

3 CONCLUSIONS

Mathematical model of hydraulic impact on anisotropic coal seam with a moving boundary front of filtered fluid is based on non-linear parabolic equation of elastic fluid of filtration mode in highly fractured-porous solid environment, which is supplemented by boundary conditions that reflect technological schemes of seam preparation.

Application of proposed model provides possibility of theoretical investigation processes with advanced calculations of technology options, but also construction of the automated management systems and design of preparation schemes.

REFERENCES

Pavlysh, V.N. & Grebyonkin, S.S. 2006. *Physic-technical basis of hydraulic impact on coal seams*: monograph. Donets'k: "Vik": 269.

Pavlysh, V.N. & Shtern, Y.M. 2007. Basic *theory and technology parameters of the process of hydropneumatic impact on coal seams*: monograph. Donets'k: "Vik": 409.

Barenblatt, G.I. 1953. *One class of exact solutions of one-dimensional unsteady tasks flat gas filtration in porous medium*. Journal of applied mathematics and mechanics, 17. Issue 6: 739–742.

Liventsev, V.V. 1964. *Experimental and analytical method of calculation and forecast parameters of moisture of coal seams*: Dis. of Cand. Tech. of Sciences: 186. Moscow.

Shchelkachev, V.N. 1946. *Basic equations of motion of an elastic liquid in elastic porous medium:* bstracts of reports. USSR, Vol. 52, TN 2: 10–12.

Samarskiy, A.A. & Sobol, I.M. 1963. *Examples of numerical calculation of temperature waves.* Journal of computational mathematics and mathematical physics, 4. Vol. 3: 702–719.

Economic efficiency of heat pump technology for geothermal heat recovery from mine water

Y. Oksen & O. Samusia
National Mining University, Dnipropetrovs'k, Ukraine

ABSTRACT: The analysis of the influence of the duration of the daily operation cycle and thermal capacity of the heat pump hot water system that is uniquely associated with it on the economic performance of the technology under the action of under the conditions of three-time-zone electrical energy rate is conducted. It has been shown that duration of the daily operation cycle increases and the heat pump system electrical power consumption decreases, the payback period of the capital costs for its construction decreases.

1 INTRODUCTION

The heat pump technology utilization for the geothermal heat waste heat recovery from mining water and ventilation air is one of the most promising solutions to the problem of coal and iron mines heat supply. The heat pump technology distinguishes its high energy efficiency and ecological friendliness but requires essential investments for its realization. Because of that the decision as for a heat pump unit assembling must be made on the bases of economic criteria (Samusia 2013).

Capital expenditures for the construction of heat pump units (HPU) are determined mainly by the power of the assembled equipment and operational expenditures by the cost of electricity consumed. In case of heat pump technology utilization for hot water supply system when it is required to provide the preparation of the required daily volume of the hot water under the conditions tree-zone time electrical rate. These indicators are closely related to the duration of the unit's daily operation cycle. Moreover, as the duration of the operation cycle increases the electrical power and capital costs decrease, and the cost of electricity consumed and operating costs increase.

The purpose of the current research is to estimate the influence of energy datum such as heat and electrical power generated on the economic efficiency of heat pump technology for hot water supply for a mine by recovering mine water waste heat.

2 METHOLOGY AND RESEARCH RESULTS

The main technical index that must be provided by the technology is daily needed in hot water V_{day}, m³.

The required volume of the hot water can be prepared by the equipment of different heat rate (heat power) Q_h, kW, that will depend on the daily operating cycle duration τ_{day}, h, and determined as

$$Q_h = c_w \rho_w \frac{V_{day}}{3600\,\tau_{day}}(t_{h2} - t_{h1}), \qquad (1)$$

where $c_w = 4.19$ kJ/(kg·°C) и $\rho_w = 1000$ kg/m³ – heat capacity and the density of the water heated for the hot water system; t_{h1} and t_{h2} its initial and final temperature, °C.

Heat power of the heat pumps determines substantially their costs and capital investments. The capital investments can be calculated on the basis of the data about the existing units that increase linearly as heat power rises.

$$K = A + k_q Q_h, \qquad (2)$$

where A – the permanent part of the capital investments; k_q – specific capital investments (the capital investment and unit's heat rate ration).

It has been shown from (1) and (2) that as the daily operating cycle duration increase τ_{day} results in the decrease in the power of installed equipment Q_h and heat pump's capital investments K. At the condition of the operation of tree-zone electrical energy rate, the increase of τ_{day} results in the increase of the portion of heat pump unit's operation time and consequently the increase of electrical energy cost.

As a result in the current research three variants of the heat pump unit operation have been investigated. In those variants the daily operational cycle duration corresponds to the ultimate duration of the zones of tree-zones rate: 7 hours (nightly period), 18 hours (nightly and semi-peak period) and 24 hours (round-the-clock period). It has been stated that at the first and second variants the equipment operates at 100% load, and at the third period the 18-hours period operates at 100% and 6-hours period at 50% load.

The electrical power consumed by heat pump unit relates to the heat power generated as (Moroziuk 2006)

$$N_{el} = \frac{Q_h}{COP}, \tag{3}$$

where COP – heat pump performance efficiency.

As the heat pump diagram the diagram with the intermediate water circuits has been considered. The diagram provides the heat transfer from the mine water to the heat pump's refrigerant and from the refrigerant to the heated water. In the intermediate water circuits the water is purified from hardness salts (Samusia, Oksen, Komissarov & Radiuk 2012).

If to neglect by electrical consumption of the support equipment (water pumps), the COP will be defined by the properties of the heat pump thermodynamic cycle. As the thermodynamic cycle, the cycle with vapor superheating in the evaporator and subcooling in the condenser with the cold heating water (Samusia, Oksen & Komissarov 2012). The value of COP depends primarily on the refrigerant boiling and condensing temperature. And those temperatures are determined by temperatures of the low- and high temperature of the heat sources (mine and heating water) and the restrictions on temperature differences in the heat exchanges (Samusia 2010). The calculation of the heat pump thermodynamic cycle and the heat pump's energy parameters made for the following conditions: daily hot water flow rate $V_{day} = 140$ м³; the initial clean water temperature heated for the system of hot water supply, at winter time $t_{h1win} = 8\,°C$, at summer time $t_{h1sum} = 12\,°C$; final temperature $t_{h2} = 45\,°C$; mine water temperature $t_{x1} = 12\,°C$; minimal temperature differences on the cold and hot ends of the heat exchanges of the mine and clean water, the heat pump evaporator $5\,°C$; on the cold and the section that corresponds to the dew point of the refrigerator and on the hot end of the heat pump condenser the temperature is 5, 7 and 15 °C, the superheating of the refrigerant vapor on the end of the evaporator is $6\,°C$, isoentropic efficiency of the heat pump compressor is 0.7. The heat pump working fluid is the Freon R407C (Goman et al. 2012 & Kyrychenko et al. 2012). According to the calculation of the heat pump's COP at the given conditions is $COP = 4.3$. The main results of the calculation of the heat pump

energy parameters and hot water flow rate while operating with 100% load on 7, 18 and 24-hours cycles have been show in the Table 1.

Table 1. Heat pump's energy and technical parameters.

Heat pump's operation cycle duration τ_{day}, hours	Q_h, kW	Q_x, kW	N_{el}, kW	V_h, m³/h
7	861	661	200	20.00
18	335	257	78	7.78
24	287	220	67	6.67

The annual economy from the heat pump technology utilization

$$E = C^{bp} - C^{hp}, \tag{4}$$

where C^{bp} and C^{hp} – annual operating costs for the hot water production using traditional (by using coal boiler plant) and heat pump technologies, thousands of UAH/year.

In case of heat pump technology the operating costs determined by the electrical energy consumption costs C_{el}^{hp}

$$C^{hp} = C_{el}^{hp}. \tag{5}$$

The electrical energy costs consumed by the heat pump unit per year are calculated subject to the difference in the load at winter and summer period and the three-zone time electrical energy rate

$$C_{el}^{hp} = \left(N_{el1}\tau_1 c_{el1} + N_{el2}\tau_2 c_{el2} + N_{el3}\tau_3 c_{el3} \right) \times$$
$$\times \left(n_{win} + n_{sum} k_{sum} \right), \tag{6}$$

where N_{el1}, N_{el2} and N_{el3} – electrical power consumed by the heat pump unit during the night, semi-peak and peak electrical energy rate at winter time, kW; τ_1, τ_2 and τ_3 – heat pump operating time at night, semi-peak and peak periods, hours/day; c_{el1}, c_{el2}, c_{el3} – electrical energy rate at night, semi-peak and peak periods; k_{sum} – coefficient that takes into account the decrease in heat pump load at summer period compared to winter calculated as

$$k_{sum} = \frac{t_{h2} - t_{h1sum}}{t_{h2} - t_{h1win}}. \tag{7}$$

In case of the usage of the traditional technology of the water heating using coal boiler plant, the operating casts determined as

$$C^{bp} = c_q Q_y, \tag{8}$$

where c_q – the cost of 1 Gcal of the heat generated in the coal boiler plant, UAH/Gcal.

The amount of heat required for the annual volume of hot water, Gcal/year,

$$Q_y = 0.86 \cdot 10^{-3} \rho_w c_w V_{day} \times$$

$$\times \left[n_{win}(t_{h2} - t_{h1win}) + n_{sum}(t_{h2} - t_{h1\,sum}) \right], \quad (9)$$

where n_{win} and n_{sum} – number of unit's operating days at winter and summer time.

While calculating the costs parameters the following has been stated: the constant of the heat pump's operating costs is $A = 480$ thousand of UAH, specific operating costs on the heat pump unit assembling is $k_q = 4.65$ thousands of AUH/kW; the number of the heat pump's operating days in winter and summer time $n_{win} = n_{sum} = 180$ days; electrical power rate at night, semi-peak and peak periods corresponding 0.257; 0.749 and 1.233 UAH/kW·h; the cost of 1Gcal of the heat generated in the coal boiler plants is $c_q = 530$ UAH/Gcal.

The results of the costs parameters calculation for heat pump and traditional technologies have been shown in Table 2.

Table 2. Costs parameters of the compared technologies.

τ_{day}, hours	K, $10^3 \cdot$UAH/year	C^{hp}, $10^3 \cdot$UAH/year	C^{bp}, $10^3 \cdot$UAH/year	E, $10^3 \cdot$UAH/year
7	4484	118	942	824
18	2037	256	942	686
24	1815	300	942	642

The Tables 1 and 2 show that as daily operating cycle duration increases the heat pump heat power and capital assembling costs decreases; and the operating costs also increases that leads to the annual economy decrease. Let us find out how the payback time changes.

The calculation of the amortization and heat pump capital costs payback time has been made by using book cost reduction technique. According to that method, the amortization rate is calculated as

$$n_{am} = 1 - T\sqrt{\frac{L}{I}}, \quad (10)$$

where L and I – liquidating and initial cost of the object; T – equipment useful life.

The useful life of the heat pump and heat exchange equipment have been considerate to be $T = 15$ years. The value of liquidating cost has been taken 3.5% from the initial cost. Then according to (10), the amortization rate is $n_{am} = 0.2$.

The payback time of the capital costs is determined on the basis of the data about discount net efficiency income (DNEI). The DNEI is obtained as a result of the unit operation.

The calculation of DNEI is made as follows.

The initial book cost of the heat pump BC_0 is considered to be equal the capital costs that are required for its creation

$$BC_0 = K. \quad (11)$$

The book cost of the unit at the end of each year of the operation is determined as

$$BC_t = BC_{t-1} - AR_t, \quad (12)$$

where BC_t – book cost of the unit at the end of the previous year; AR_t – amortization rate at the current year.

The amortization rate

$$AR_t = n_{am}BC_{t-1}. \quad (13)$$

The annual income money flow that is calculated as the sum of amortization rate and annual economy

$$MF_t = AR_t + E. \quad (14)$$

The annual discount inlet money flow is calculated as

$$DMF_t = KD_t \cdot MF_t, \quad (15)$$

where KD_t – coefficient of discounting

The coefficient of discounting

$$KD_t = \frac{1}{\left[(1+r)(1+\alpha)\right]^t}, \quad (16)$$

where r – rate of inflation (given 11.2% per year, $r = 0.122$); α – discount rate (given $\alpha = 0.165$).

The cumulative discount inlet money flow and DMF for T years of unit operating

$$KDMF_T = \sum_1^T DMF_t, \quad (17)$$

$$DNEI_T = KDDP_T - BC_0. \quad (18)$$

On the basis of the calculations results the graphs of the DNEI changes depending on heat pump operating period have been created. The period at which DNEI equals zero is payback time of the capital costs of the heat pump.

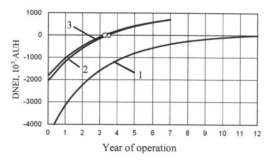

Figure 1. DNEI changes with the operating years of the heat pump operating at 7-hours (line 1), 18-hours (2 line) and 24-hours (line 3) operating cycle.

Figure 1 shows that the minimal operation life of the capital costs of the heat pump is reached when the heat pump has the longest operation cycle duration and corresponds to the minimal electrical power of the heat pump equipment. It indicates that the decrease of capital costs plays more important role than operation costs increase.

3 CONCLUSIONS

The research shows that in the case of heat pump technology utilization for the preparation of the hot water for hot water supply system when it is required to provide the preparation of the daily volume of hot water under the conditions of three-zone time electricity pay rate, the heat pump unit that is characterized by lower thermal capacity and lower capital costs are more economically efficient.

REFERENCES

Samusia, V., Oksen, Y. & Radiuk, M. 2013. *Heat pumps for mine water waste heat recovery*. Proceedings of the international scientific and technical conference "Mining of mineral deposits", Taylor & Francis group, London, UK: 153–157.

Moroziuk, T. 2006. *The theory of chillers and heat pumps*: 712. Odesa.

Samusia, V., Oksen, Y., Komissarov, Y. & Radiuk, M. 2012. *Heat pumps for mine water waste heat recovery. Hot water supply unit*. State higher educational Institution National Mining University. Dnipropetrovs'k: Scientific Bulletin of National Mining University, 4: 143–144.

Samusia, V., Oksen, Y. & Komissarov, Y. 2012. *Evaluation of the energy efficiency of the heat pump system, using a complex heat outgoing air flow and waste water*. Mining Electrical Mechanics and Automatics: Proceedings of scientific works. Issue 88: 131–136.

Samusia, V., Oksen, Y. & Radiuk M. 2010. *Efficiency estimation of heat pump waste heat recovery*. State higher educational Institution National Mining University. Dnipropetrovs'k: Scientific Bulletin of National Mining University, 6: 78–82.

Goman, Kyrychenko, Y., Samusia, V. & Kyrychenko, V. 2012. *Experimental investigation of aeroelastic and hydroelastic instability parameters of a marine pipeline*. Geomechanical processes during underground mining, Taylor & Francis group, – London, UK, – 2012, ISBN: 978-0-415-66174-4: 163–167.

Kyrychenko, Y., Samusia, V. & Kyrychenko, V. 2012. *Software development for the automatic control system of deep water hydrohoist*. Geomechanical processes during underground mining, Taylor & Francis group, – London, UK, – 2012, ISBN: 978-0-415-66174-4: 81–86.

Methodological basics of gas-drainage pipeline engineering for transporting wet firedamp in winter time

V. Alabiev
PJSC "Zasyadko Mine", Donets'k, Ukraine

S. Alekseenko & I. Shaykhlislamova
National Mining University, Dnipropetrovs'k, Ukraine

ABSTRACT: The work presents the developed mathematical model of heat-and-mass transfer processes of the atmospheric air with humid firedamp transported through surface pipelines in winter seasons. In terms of the integral of a differential equation of the convection-diffusion heat transfer in a pipeline, methodological fundamentals have been worked out to calculate the critical length of mine surface pipelines when their inside surface freezing process do not occur. Engineering methods of calculating thermodynamic properties of the methane-air mixture are provided.

1 INPRODUCTION

In view of energy independence support, the use of degassed coalmine methane has good prospects in Ukraine. This conditions the development of pipeline systems working all year round on the coal mine surface. The firedamp, or methane-air mixture (further MAM), at the output of vacuum-air pumps of mine-gas-drainage plants contains suspended moisture and has one hundred percent relative humidity. On cooling MAM, there occurs condensation of water vapour which MAM contains. With negative values of the free-air temperature, the condensation product transforms directly into ice which narrows the inner dimension till their complete clogging. As freeze protection, thermal insulation of gas-drainage pipelines is applied. However, the issue of using thermal insulation in order to avoid overspending must be solved with reference to a thermal design considering the MAM thermodynamic properties and the environment, remoteness

of vacuum-pump stations from consumers, and other factors. Nevertheless, at the present time there are no methods which allow doing such calculations.

The purpose of the work is to develop methodological fundamentals to calculate the preassembled length of the pipelines transporting the degassed coalmine methane with negative values of the free-air temperature.

2 THE MAJOR PART

The pipeline design scheme of a line element is depicted in Figure 1 being x a longitudinal coordinate, m; r a transverse coordinate, m; T is the MAM temperature, K; T_0 is the free-air temperature, K; R is the pipeline radius, m; S is the cross-section area of the pipeline, m^2.

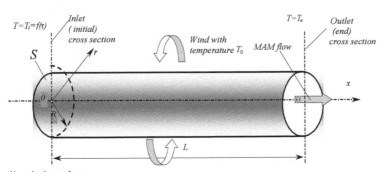

Figure 1. The pipeline design scheme.

Let's accept that up to the point of feeding MAM into the pipeline, the temperature inside the pipeline was equal to the temperature in the ambient atmosphere T_0, and the MAM initial temperature T_i increases from level T_0 to critical values T^* within specific time frames. The heat flux rate on the pipeline surface into the ambient atmosphere is defined with Fourier's law. Subsequently, in differential calculus the mathematical formulation of the heat-and-mass transfer problem with the MAM moving in the pipe takes the form of equation of convection-diffusion heat transfer in the pipe (Tsoy 1984)

$$\frac{\partial T}{\partial \tau} + u \frac{\partial T}{\partial x} = a \frac{1}{r} \frac{\partial}{\partial r} \left(r \frac{\partial T}{\partial r} \right), \qquad (1)$$

with the initial condition

$$T(x, r, 0) = T_0, \qquad (2)$$

and the boundary of third kind

$$T(0, r, \tau) = f(\tau), \qquad (3)$$

$$-\lambda \frac{\partial T}{\partial r} \bigg|_{r=R} = k \cdot \left(T \big|_{r=R} - T_0 \right), \qquad (4)$$

where τ – time, s; u – MAM velocity in the pipeline, mps; a – MAM temperature conductivity coefficient, m^2ps; λ – MAM thermal conductivity coefficient, W/m·K; k – heat transfer coefficient, W/m^2·K.

In expressions (1), (4) the values of the temperature conductivity coefficient and the heat transfer coefficient are estimated according to (Mikheev & Mikheeva 1973)

$$a = \frac{\lambda}{\rho \cdot c}, \qquad (5)$$

$$k = \frac{1}{\dfrac{1}{\alpha_g} + \dfrac{\delta}{\lambda_T} + \dfrac{1}{\alpha_a}}, \qquad (6)$$

where ρ – MAM density, kg/m^3; c – MAM heat capacity, J/(kg·K), α_g – coefficient of heat-exchange between the MAM and the inner surface of the gas pipeline , W/(m^2·K); δ – gas pipe wall thickness, m; λ_T – thermal conductivity coefficient of gas pipeline material, W/(m·K); α_a – coefficient of heat-exchange from the gas pipeline outer surface to the ambient air, W/(m^2·K).

The MAM thermophysical properties which are included in (1)–(6) depend on the dynamics of MAM temperature and pressure changes over a distance and time. Let's assume that the MAM pressure drop along the pipeline length as a result of friction force is inconsiderable compared to absolute pressure while the MAM enters the gas pipeline. In this case according to the law of gas mass conservation with steady-state motion, the MAM density can be considered constant (Baskakov 1982 & Loytsianskiy 1970). Consequently, while doing practical calculation we will take the values of the MAM thermophysical properties in the context of a steady mode of the MAM average temperature and constant pressure.

With regard to (5) we will write Equation (1) as follows

$$c\rho \frac{\partial T}{\partial \tau} + c\rho u \frac{\partial T}{\partial x} = \lambda \frac{1}{r} \frac{\partial}{\partial r} \left(r \frac{\partial T}{\partial r} \right). \qquad (7)$$

Equation (7) accounts for both longitudinal and radial heat transfer. The pipeline radius is considerably smaller than its length. Consequently, transverse heat flow is considerably smaller than the longitudinal one. That is why while modelling the MAM heat and mass transfer process in the gas pipeline, it is worth using the MAM averaged temperature in the cross section (Bobrovskiy 1972). To do this, we multiply both parts of Equation (7) by r, integrate over this coordinate within the limits of 0 to R and divide by the cross-section area of the pipeline S. As a result, Equation (7) takes on the following form

$$c\rho \frac{\partial \overline{T}}{\partial \tau} + c\rho u \frac{\partial \overline{T}}{\partial x} = k \frac{\Omega}{S} \left(T_0 - \overline{T} \right), \qquad (8)$$

$$\overline{T} = \frac{1}{S} \int_0^R r \cdot T \cdot dr, \qquad (9)$$

where Ω – pipeline perimeter, m.

After solving Equation (8) under the boundary condition (3) with the initial condition $T(x, 0) = T_0$ we get the formula for the length estimation of the gas pipeline whose inner surface is ice free (Alabiev 2006)

$$L = \frac{1}{4} d \frac{1}{St} \ln \frac{T_i - T_0}{T_e - T_0}, \qquad (10)$$

where d – gas pipeline diameter, m; T_i, T_e – MAM temperature at the beginning and the end of the pipeline, °C; T_0 – atmospheric air temperature, °C; St – Stanton number.

As (10) shows, the estimation of the critical length of the gas pipeline resolves itself into estimation of the Stanton number St which is connected to the Nusselt number (Nu) and the Peclet number (Pe) with relation (Mikheev & Mikheeva 1973)

$$St = \frac{Nu}{Pe} . \tag{11}$$

Peclet number is defined as (Mikheev & Mikheeva 1973)

$$Pe = \frac{u \cdot d}{a} , \tag{12}$$

in which the MAM movement velocity in the pipeline is estimated with formula

$$u = \frac{Q}{S} , \text{[m/s]} \tag{13}$$

where Q – MAM consumption in the pipeline, m^3/s; S – stands for the pipeline cross section.

The MAM thermal conductivity coefficient in formula (5) to estimate the MAM temperature conductivity coefficient can be defined as average weighted for the thermal conductivity coefficients of the air and methane

$$\lambda = (1 - \psi) \cdot \lambda_a + \psi \cdot \lambda_m , \text{W/(m·°K)}, \tag{14}$$

where λ_a – air thermal conductivity coefficient; λ_m – methane thermal conductivity coefficient; ψ – methane concentration in the MAM, unit fraction.

The air thermal conductivity coefficient λ_a depends on the temperature and is accepted according to Table 1 or is estimated with a high degree of accuracy according to empirical formula

$$\lambda_a = \frac{2.44 + 0.0078 \cdot T}{100} , \text{[W/(m·°K)]} \tag{15}$$

Table 1. The air thermal conductivity coefficient.

Temperature, °K	243	263	283	303	323	333
$\lambda_a \cdot 10^2$, W/(m·°K)	2.20	2.36	2.51	2.67	2.83	2.90

The methane thermal conductivity coefficient λ_m also depends on the temperature and is taken according to Table 2 or is estimated according to empirical formula

$$\lambda_m = \frac{3.06 + 0.0139 \cdot T}{100} \text{[W/(m·°K)]} \tag{16}$$

Table 2. The methane thermal conductivity coefficient.

Temperature, °K	240	260	280	300	320	340
$\lambda_m \cdot 10^2$, W/(m·°K)	2.64	2.88	3.13	3.42	3.72	4.02

In expressions (15) and (16) T is the MAM logarithmic mean temperature which is defined according to formula (Alabiev 2006)

$$T = T_0 + \frac{T_i - T_e}{\ln \dfrac{T_i - T_0}{T_e - T_0}} , \text{[°C]} \tag{17}$$

Taking into consideration that the MAM consists of a mixture of dry air, water vapour and methane, the MAM density in formula (5) for the MAM temperature conductivity coefficient estimation can be defined according to the following recommendations (Chernichenko & Podgorniy 2003)

$$\rho = (1 - \psi) \cdot (\rho_a + \rho_v) + \psi \rho_m , \text{[kg/m}^3\text{]} \tag{18}$$

where ρ_a – dry air density; ρ_v – water vapour density; ρ_m – methane density.

The densities of dry air, water vapour and methane are estimated according to formulas

$$\rho_a = 3.488 \cdot \frac{P - \varphi \cdot P_p}{T + 273} , \text{[kg/m}^3\text{]} \tag{19}$$

$$\rho_v = 2.168 \cdot \varphi \cdot \frac{P_p}{T + 273} , \text{[kg/m}^3\text{]} \tag{20}$$

$$\rho_m = 1.928 \cdot \frac{P}{T + 273} , \text{[kg/m}^3\text{]} \tag{21}$$

where φ – MAM relative humidity, unit fraction; P – MAM absolute pressure in the gas pipeline, kPa; P_p – partial pressure of saturated steams at MAM average temperature which is defined based on the reference literature or empirical dependence

$$P_p = 0.516 \cdot E^{0.0591 \cdot T} , \text{[kPa]} \tag{22}$$

MAM absolute pressure in the gas pipeline can be calculated from the formula

$$P = P_0 + P_m , \text{[kPa]} \tag{23}$$

where P_0 – atmospheric pressure, kPa; P_m – MAM pressure in the gas pipeline, kPa.

The estimation of MAM specific heat per unit mass in expression (5) is done according to formula, J/(kg·°K) (Federal Standard of Ukraine 2002)

$$c = \frac{(1-\psi)\cdot\rho_a\cdot c_a + \psi\cdot\rho_m\cdot c_m}{\rho}, \tag{24}$$

where c_a – specific heat per unit mass of moist air, J/(kg·°K); c_m – specific heat per unit mass of humid air of methane, J/(kg·°K).

The specific heat per unit mass of moist air is defined

$$c_a = 1005 + 1880 \cdot d_a, \tag{25}$$

$$d_a = 0.622 \cdot \varepsilon \cdot \frac{\varphi \cdot P_p}{P - \varphi \cdot P_p}, \tag{26}$$

where d_a – humidity of moist air, kg/kg; ε – correction factor of methane concentration evaluation in MAM taken from Table 3 (Chernichenko & Podgorniy 2003).

Table 3. Correction factor of methane concentration evaluation in MAM.

Methane content in MAM, %	Correction factor	Methane content in MAM, %	Correction factor
25	1.20	60	1.48
30	1.24	65	1.52
35	1.28	70	1.56
40	1.32	75	1.61
45	1.36	80	1.65
50	1.40	85	1.69
55	1.44	90	1.73

For engineering evaluation an empirical relationship has been obtained to estimate the correction factor which accounts for the methane content in MAM

$$\varepsilon = 0.99 + 0.82 \cdot \psi. \tag{27}$$

The specific heat per unit mass of methane in formula (16) is defined according to Table 4 (Zacheruchenko & Zhuravliov 1969) or is estimated using the empirical formula

$$c_m = 2170 + 2.8 \cdot T. \tag{28}$$

Table 4. The specific heat per unit mass of methane.

Temperature, °K	255	273	298	300	323	373
c_m, kJ/(kg·°K)	2.14	2.17	2.23	2.23	2.29	2.44

The equivalent Nusselt number in formula (11) is estimated by formula (Mikheev & Mikheeva 1973)

$$Nu = \frac{k}{\lambda}d. \tag{29}$$

According to (Mikheev & Mikheeva 1973), in the process of heat interchange between the gas flow and the inner surface of the pipe, the Nusselt number is equal to

$$Nu_g = 0.021 \cdot Re^{0.80} \cdot Pr^{0.43} \cdot \left(\frac{Pr}{Pr_T}\right)^{0.25}, \tag{30}$$

where Pr_T – Prandtl number at the MAM temperature equal to the pipe inner surface temperature.

Taking into account the fact that $Pr/Pr_T \approx 1$ for the air, formula (30) assumes the form

$$Nu_g = 0.021 \cdot Re^{0.80} \cdot Pr^{0.43}. \tag{31}$$

During the pipeline filling with atmospheric air, the Nusselt number is (Mikheev & Mikheeva 1973)

$$Nu_a = 0.245 \cdot Re_a^{0.60}, \tag{32}$$

$$Re_a = \frac{u_a \cdot d}{v_a}, \tag{33}$$

where u_a and v_a – velocity and kinetic viscosity of the atmospheric air (wind).

Formula (33) refers to cases of the strongest heat exchange when the wind blows crosswise. Using (31) and (32) we define the value of the equivalent Nusselt number. It follows from (29) that

$$\alpha_g = \frac{\lambda}{d}Nu_g; \quad \alpha_a = \frac{\lambda_a}{d}Nu_a, \tag{34}$$

where λ_a – thermal conductivity coefficient of the atmospheric air, W/(m·K).

After the substitution of (34) for (6) we have

$$k = \frac{1}{\dfrac{d}{\lambda}\dfrac{1}{Nu_g} + \dfrac{\delta}{\lambda_T} + \dfrac{d}{\lambda_a}\dfrac{1}{Nu_a}} =$$

$$\frac{\lambda}{d}\frac{Nu_g}{1 + \dfrac{\lambda}{\lambda_T}\dfrac{\delta}{d}Nu_g + \dfrac{\lambda}{\lambda_a}\dfrac{Nu_g}{Nu_a}}, \tag{35}$$

and according to (29)

$$Nu = \frac{Nu_g}{1 + \dfrac{\lambda}{\lambda_t}\cdot\dfrac{\delta}{d}Nu_g + \dfrac{\lambda}{\lambda_a}\cdot\dfrac{Nu_g}{Nu_a}}, \tag{36}$$

where Nu_g – Nusselt number for MAM; λ_t equivalent thermal conductivity coefficient of the pipeline, W/(m·°K); δ – equivalent wall thickness of the pipe, m; λ_a – thermal conductivity coefficient of the atmospheric air, W/(m ·°K); Nu_a – Nusselt number for the atmospheric air.

The equivalent wall thickness of the pipe is defined using the formula

$$\delta = \delta_0 + \delta_i \text{, [m]} \tag{37}$$

where δ_0 – wall thickness of the pipe, m; δ_i – insulation thickness, m.

The equivalent of thermal conductivity coefficient of the pipeline is estimated by the formula

$$\lambda_t = \frac{\lambda_0 \cdot \delta_0 + \lambda_i \cdot \delta_i}{\delta_0 + \delta_i} \text{, [W/(m·°K)]} \tag{38}$$

where λ_0 – thermal conductivity coefficient of the pipeline, W/(m·°K); λ_i – insulation thermal conductivity coefficient, W/(m·K).

The thermal conductivity coefficient of the atmospheric air is taken from Table 1 or is estimated from the formula

$$\lambda_a = \frac{2.44 + 0.0078 \cdot T_a}{100} \text{ [W/(m·°K)]} \tag{39}$$

The Nusselt number for MAM in expression (36) is calculated by the formula (Alabiev 2006)

$$Nu_g = 0.0237 \cdot Re^{0.80} , \tag{40}$$

where Re – Reynolds number for MAM.

The estimation of the Reynolds number is done using the formula

$$Re = \frac{u \cdot d}{v} , \tag{41}$$

where v – kinematic viscosity coefficient of MAM. It is taken from Table 5 (Federal Standard of Ukraine 2002) or is estimated from the empirical formula

$$v = (17.44 + 0.06 \cdot T - 0.038 \cdot P) \cdot 10^{-6} . \tag{42}$$

The Nusselt number for the atmospheric air in expression (36) is estimated using the formula

$$Nu_a = 0.245 \cdot Re_a^{0.6} , \tag{43}$$

where Re_a – Reynolds number for the atmospheric air defined from the formula

$$Re_a = \frac{\omega \cdot d}{v_a} , \tag{44}$$

where ω – atmospheric air velocity, m/s; v_a – kinematic viscosity of the atmospheric air. It is taken from Table 6 (Federal Standard of Ukraine 2002) or is estimated from the empirical formula

$$v_a = (13.36 + 0.092 \cdot T_a) \cdot 10^{-6} \text{ [m}^2\text{/s]} \tag{45}$$

Table 5. MAM kinematic viscosity coefficient, $v \cdot 10^6$.

Tempera-	Pressure, kPa			
ture, °K	100	200	400	600
240	11.43	5.73	2.87	1.92
250	12.35	6.19	3.10	2.08
260	13.29	6.66	3.34	2.23
270	14.26	7.14	3.58	2.39
280	15.26	7.65	3.83	2.56
290	16.29	8.15	4.08	2.73
300	17.33	8.67	4.34	2.90
310	18.41	9.21	4.61	3.08
320	19.54	9.77	4.89	3.26
330	20.65	10.35	5.17	3.46
340	21.80	10.91	5.46	3.65
350	22.96	11.49	5.75	3.84

Table 6. The kinematic viscosity coefficient of the atmospheric air.

Temperature, °C	$v_a \cdot 10^6$	Temperature, °C	$v_a \cdot 10^6$
−30	10.8	30	16.00
−20	12.79	40	16.96
−10	12.43	50	17.95
0	13.28	60	18.97
10	14.16	70	20.02
20	15.06	80	21.09

3 CONCLUSIONS

Methods of estimating the preassembled length of the gas pipeline whose inner surface is ice free while transporting wet firedamp extracted with mine-gas-drainage systems have been worked out. The methods can be used by engineering technicians while designing mine gas pipeline systems which will allow increasing the operation and maintenance safety in winter time.

REFERENCES

Tsoy, P.V. 1984. *Methods of heat and mass transfer estimation*, Moscow: Energoatomizdat: 416.
Mikheev, M.A. & Mikheeva, I.M. 1973. *Heat transfer basics*. Moscow: Energia: 343.

199

Baskakov, A.P., Berg, B.V. & Vitt, O.K. 1982. *Heat engineering*. Moscow: Energoizdat: 264.

Loytsianskiy, L.G.1970. *Fluid mechanics*. Moscow: Nedra: 904.

Bobrovskiy, S.A., Shcherbakov, S.G. & Guseynzade, M.A. 1972. *Gas movement in gas pipelines with track selection*. Moscow: Nauka: 192

Alabiev, V.P. 2006. *Analytical solution of heat and mass exchange while transporting wet firedamp in winter time*. Bulletin of Volodymyr Dahl East Ukrainian National University: Scientific journal, 6 (100). P. 2: 44–53.

Chernichenko, V.K. & Podgorniy, N.Ye. 2003. *Methods of wet firedamp thermodynamic parameters estimation. Ways and means of developing safe and healthy labour conditions in coal mines*: Collection of research papers. Makeevka: MakNII: 200–206.

Develop (Federal Standard of Ukraine) *"Manual on downcast ventilating shaft and hole heating on base of direct fired air heaters which use coalmine methane as fuel"*. 2002. The Research Work Report (interim). Makeevka: MakNII: 138.

Zacheruchenko, V.A. & Zhuravliov, A.M. 1969. *Thermalphysical properties of gaseous and liquid methane*. Moscow: Standards Publishing House: 236.

Progressive Technologies of Coal, Coalbed Methane, and Ores Mining – Bondarenko, Kovalevs'ka & Ganushevych (eds)
© 2014 Taylor & Francis Group, London, ISBN: 978-1-138-02699-5

Formation dynamics of coal seams morphostructure of the deep-seated horizons of the L'viv-Volyn' basin

K. Bezruchko
M.S. Polyakov Institute of Geotechnical Mechanics, Dnipropetrovs'k, Ukraine

M. Matrofailo
Institute of Geology and Geochemistry of Combustible Minerals, L'viv, Ukraine

ABSTRACT: The peculiarities of morphology and genesis of the commercial Visean coal seam $v_0{}^3$, which occurs in deep horizons of the extreme north-west of the L'viv-Volyn' basin (Kovel prospect) are shown. The confinement of accumulation of the source matter of the seam to the inherited tectonic valley-like depression of latitudinal stretch is noted. For the first time the dependence between the peculiarities of the Pre-Carboniferous paleorelief, which becomes apparent due to splitting of the coal seam and complication of its structure towards the valley-like depression is traced. The new type of coal accumulation, which is characteristic of the platform coal-bearing formations, located just on the erosional surface of underlying deposits, is established.

1 INTRODUCTION

More and more attention is paid to the problems of evolution of the earth's crust and its separate structural elements of all ranks and geological bodies of different scale, to their formation and subsequent changes from the position of geodynamics. Scales of manifestation of dynamic processes are different in time and in each region separately. A definite character of such processes is characteristic of the formation of the coal-bearing associations on the whole and at different stages of their development. These processes are similar on the whole, but they have their own distinctive features. The formation of the coal-bearing rock association in the territory of the Kovel coal area of the L'viv-Volyn' Coal Basin (LVB) is such characteristic example.

At the same time, it should be noted that constant systematic working out of commercial reserves by the majority of coal mines of the LVB, the closing of unprofitable coal-producing enterprises cause the necessity of further quest for new areas and deep horizons with commercial coal presence and favourable conditions for the development of coal seams (Shulga et. al. 1995, Kostyk et. al. 2012). Its northern continuation – the Kovel' prospecting area, the field of the distribution of the Carboniferous deposits along the state border on Poland from the town of Vladimir-Volyns'ky to the border on Belarus – belongs to that kind excepting the south-western part of basin. The boundary of the area is a conventional line which in the north passes through the Belarus frontier, in the

east through the village of Kamyanka, the village of Zabuzhzhie, the village of Mosyr, the village of Nikitychy, in the south up to the town of Ustilng, in the west – through the Polish frontier. It stretches in a form of a strip, 15–20 km wide, covering a distance of t about 100 km along the state frontier. Its total area comes to 420 km².

Limited facts obtained as a result of drilling of the few prospecting boreholes in this territory have not allowed earlier to determine special features of the structure and the potential of coal presence in the Carboniferous deposits developed here. Preceding investigations, in particular (Shulga & Shpakova 1958, Pomyanovskaya & Zavialova 1959, Bobrovnyk et. al. 1962 etc.) have allowed us to determine the lithological composition of formations of Carboniferous, their belonging to the upper part of the Visean stage as well as their position with angle and stratigraphic unconformity on different-age deposits of the Lower Paleozoic.

Over the last years, while carrying out prospecting works for coal (Lviv GRE) and geological surveying works (Rivne GRE) in the Kovel' prospecting area we have obtained new facts. Their complex studies have was useful to revealing of peculiarities of the geological structure, coal presence, coal bed morphology and to estimate in a new way the influence of dynamic processes upon the formation of the morphostructure of the coal seam $v_0{}^3$ that belongs to deep horizons of LVB. On the whole, that determines the necessity of other approach to carry out explorative-prospecting works and contributes to predictive appraisal of resources of the northern part the basin.

2 GOAL OF INVESTIGATIONS

To analyze special features of the environment and dynamics of formation of morphology of the coal seam v_0^3 and change in its basic parameters in the Kovel' perspective area, to determine its morphostructural and morphogenetic peculiarities, to compile a map of morphology of the seam in which to indicate its commercial value at individual areas for expansion of commercial coal-bearing potential of the L'viv-Volyn' Coal Basin.

3 GENERAL PRINCIPLES

In the structural plan, the Kovel' prospecting perspective area is situated in the northern part of LVB located within the limits of the L'viv Paleozoic deep of the south-western edge of the East European Platform and is contiguous to the Volyn' field (Figure 1). It should be also noted that the prospecting area is located within the limits of the Kovel' tectonic protrusion which is a part of the Kovel'-Hrubeshuv transverse uplift separating the L'viv Paleozoic deep from the Lyublin one (Znamenskaya & Chebanenko 1985).Horst-thrust dislocations in Pre-Visean time are characteristic of the uplift (Bretonian phase of the Hercynian tectogenesis) that caused a wave-like character of the boundary of distribution of Carboniferous and subhorizontal occurrence of coal-bearing deposits on the eroded surface of strongly dislocated rocks of the Lower Paleozoic. According to (Vlasov 1990) the thickness of eroded Pre-Carboniferous deposits within the limits of the Kovel' protrusion comes to 0.5–1.7 km.

Rocks of the Carboniferous system occur on the eroded different-age deposits of the Lower Paleozoic and are represented by the Visean and Serpukhovian stages (Figure 2). The thickness of Carboniferous in the area reaches no more than 224 m.

Figure 1. Map of morphology of coal seam v_0^3 of the Kovel' coal-bearing area of the L'viv-Volyn' Basin: 1 – boundary of epigenetic washout of coal-bearing formation; 2 – isopaches of coal seam, m; 3 – isohypses of coalbed foot, m; 4 – contour of mainly epipeaty washouts of coal seam; 5 – coal seam of composite structure (two and more coal units); 6 – line of splitting of coal seam at different stratigraphic levels; 7 – tectonic dislocations with a break of continuity; 8 – borehole and its number; 9 – structure of seam and thickness of coal units and rock interlayers; 10 – location of detailed cross-sections; 11 – alphabetical index of coal seam; 12 – state frontier; 13 – boggy seaside lowland; 14 – delta; 15 – directions of paleostreams; 16 – Kovel' Carboniferous hydrographic system; 17 – boundary of paleogeographic zones.

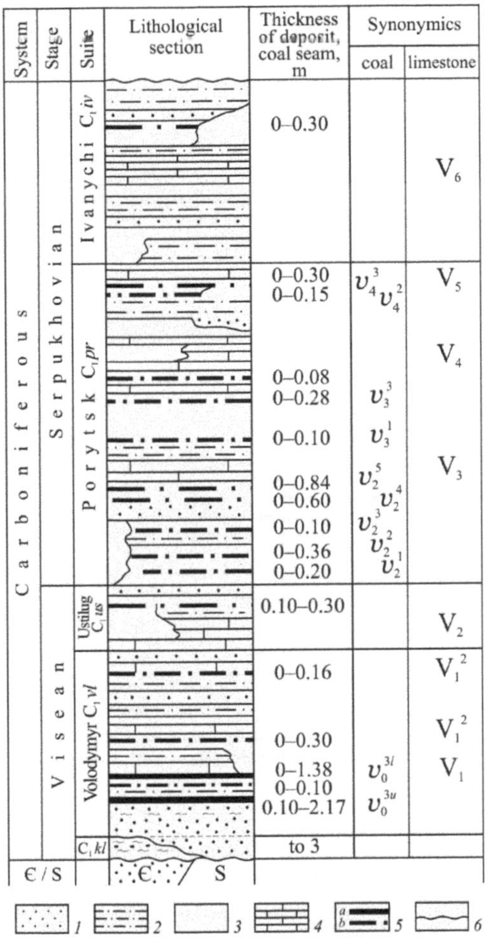

System	Stage	Suite	Lithological section	Thickness of deposit, coal seam, m	Synonymics coal	Synonymics limestone
Carboniferous	Serpukhovian	Ivanychi C_1iv		0–0.30		V_6
		Porytsk C_1pr		0–0.30 / 0–0.15	v_4^3 v_4^2	V_5
						V_4
				0–0.08 / 0–0.28	v_3^3	
				0–0.10	v_3^1	V_3
				0–0.84 / 0–0.60	v_2^5 v_2^4	
				0–0.10	v_2^3	
				0–0.36	v_2^2	
				0–0.20	v_2^1	
	Visean	Ustilug C_1us		0.10–0.30		V_2
		Volodymyr C_1vl		0–0.16		V_1^2
				0–0.30		V_1^2
				0–1.38 / 0–0.10	v_0^{3l}	V_1
				0.10–2.17	v_0^{3u}	
Є / S		C_1kl		to 3		S

Figure 2. Lithological-stratigraphical section of coal-bearing deposits of the Kovel' perspective area: 1 – sandstone; 2 – siltstone; 3 – argillite; 4 – limestone; 5 – coal seams of workable (*a*) and unworkable (*b*) thickness; 6 – stratigraphic unconformities.

4 METHODS OF INVESTIGATIONS

To study the geological structure, morphology, forming conditions of the coal seams and coal-bearing deposits of the Kovel' coal-bearing area we have used a complex of investigations based on the formational analysis (Shulga et. al. 2007). In addition, we have carried out a lithological description of the sections and have given a lithological-facies characteristic of rocks, have correlated coal-bearing deposits of LVB and its northern continuation, have characterized the coal-bearing potential of the thickness, have conducted a morphological analysis of the coal seams, have considered conditions of the peat accumulation and have carried out a dynamic analysis of formation of the Carboniferous deposits.

In particular, the morphological analysis of coal seams was conducted using the methods developed and used by us during investigations in the basin near Moscow as well as in Donets and L'viv-Volyn' Basin (Shulga 1981; Shulga et. al. 1992, 2010). Also we have used methods of mapping and geological-industrial typification of basic morphological parameters of the coal seams, lithological-stratigraphic analysis and construction of detailed morphological sections.

Reconstruction of dynamics of formation of carboniferous deposits and, the great thing, of morphology of the coal seam v_0^3 has carried out by famous methods proceeding from notions of block tectonics, conditions and mechanism of the coal seam formation (Vasiliev 1950, Lomashov 1958, Shulga 1962, Sergeev 1976, Volkov 1976, Znamenskaya & Chebanenko 1985).

5 MORPHOLOGICAL FEATURES OF THE SEAM

The seam v_0^3 is distributed throughout the whole territory of the Kovel' coal-bearing area: through the Shatsk, Lyuboml and Novyny areas. It consists of two workable coal seams: v_0^{3l} (lower) and v_0^{3u} (upper) (Figures 1 and 3).

The lower coal seam v_0^{3l} occurs at depths of 319.6 to 551.2 m and is distributed in all sites of the area. Its thickness varies from 0.10 to 2.17 m. In the Novyny and Lyuboml area it's conditional and varies from 0.59 to 2.17 m. Farther north (in the Shatsk area) the seam has a highly thin unworkable thickness reaching 0.30–0.35 m. Changeability of thickness is low, only in the west of the Lyuboml area it is strong and very strong. The seam structure varies from simple to composite. Without rock interlayers, the seam is found in the Novyny, Shatsk areas and in the east of the Lyuboml area. With one or two rock interlayers of 0.10–0.38 m thick it is developed in the western, the deeply buried part of the Lyuboml area. Interlayers are mainly represented by argillites and rarely by sandstones. Coaly argillite of 0.15–0.25 m trick is developed in the basement, in the middle part and in the cover of the seam, and in the borehole 2944 it substitutes coal. The coal seam v_0^{3l} is splitted. This local splitting (borehole 7005) belongs to bifurcation and is distributed across the area of 15.2 km². The thickness of rock interlayer represented by argillite comes to 1.9 m.

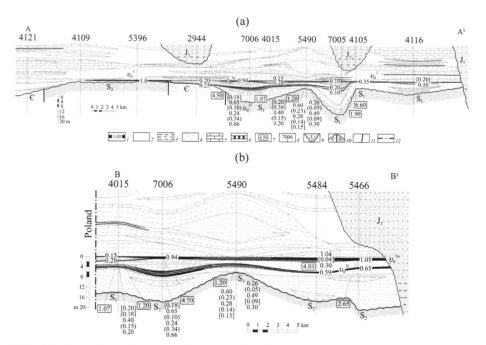

Figure 3. Morphological sections of coal seam v_0^3 along A–A[1] (a) and B–B[1] (b) (lines of section are indicated in Figure 1): 1 – coal seam and its thickness; 2 – argillite; 3 – siltstone; 4 – sandstone; 5 – limestone; 6 – coaly argillite; 7 – thickness of rock interlayer of splitting; 8 – borehole and its number; 9 – Jurassic washout of coal-bearing formation; 10 – washed surface of Cambrian and Silurian deposits underlying coal-bearing deposits; 11 – tectonic dislocations with a break of continuity; 12 – state frontier.

The upper coal seam v_0^{3u} occurs at depth from 319.6 to 546.7 m under limestone V_1. The thickness of the seam varies from 0.1 to 1.38 m. With conditional thickness ranging from 0.94 to 1.38 m, it is found in the greater part of the Lyuboml area, further west its thickness is very thin and comes up to 0.10–0.15 m. Changeability of the thickness is wear. The seam belongs to relatively sustained. The seam structure is changed from simple to medium complexity and composite. Rock interlayers are composed of coaly mudstone and argillite 0.04–0.09 m thick. In the borehole 2944 in the south of the Lyuboml area, the coaly mudstone 0.20 m thick fully substitutes coal.

Similar to the lower seam the coal seam v_0^{3u} is locally splitted too. The splitting belongs to bifurcation and is widespread throughout the area of 7.9 km². The thickness of the rock interlayer, represented by siltstone and partly by sandstone, comes to 1.2 m. Local splittings of coal seam v_0^{3l} and v_0^{3u} are located in the tectonically the most active central Lyuboml site of the Kovel' area.

As it was noted before, on the whole the coal seam v_0^3 in the Kovel' area is of composite structure, and it is splitted into two conditional coal seams v_0^{3l} and v_0^{3u} reaching the maximum thickness in the Lyuboml area. Its genetic peculiarities are concluded in that the thickness of merged compact part of the coal seam before splitting is comparable to the total thickness of coal packs composing it (in neighbouring boreholes), and complication of their structure and gradual increase of the thickness of the rock interlayer occurs in the direction of the centre of the study area. Splitting rock interlayer is composed of sandstones, mudstone and siltstones. The thickness of dividing deposits reaches 6–8 m. It is also characteristic of changes in thickness of the whole coal-bearing series that reaches 25–30 m in the centre of the study area and decreases in the northern and the southern direction. Gradient of splitting in the meridian direction at different sections changes from 0.16 to 0.45 m/km, on average it is 0.28 m/km. The splitting of the seam v_0^3, contours of which stretch sublatitudinally, occurs in the direction from the north to the south and from the south to the north throughout the area of 202.7 km² forming a composite stage-by-stage (three-phase) bifurcation. It should be noted that its contours are broken by the state frontier in the west, and in the east with the whole coal series – by washing-out of the coal-bearing formation.

Morphology of the coal seam v_0^3 was changed under the influence of intraformational and epigenetic wash-out of the Carboniferous deposits. The fragment of intraformational wash-out that belongs mainly to epipeaty wash-outs of the coal seam is located in the Novyny area (Figure 1). It is obvious that it is caused by abrasion activity, because the occurrence of limestone in the seam cover was found. Its contour passes beyond the area limits.

Wash-out of the coal-bearing series in the study area (epigenetic wash-out of the coal-bearing formation) has decreased the area of distribution of the coal seams to a great extent, the coal presence of commercial Value and mainly has formed modern outlines of their morphology that reflect only a part of Carboniferous deposits that occupied much greater territory before Post-Carboniferous wash-out. The modern configuration of the boundary of epigenetic wash-out is secondary: denudational. It stretches submeridionally and covers the whole Kovel area. Proceeding from correlation of the surface, the contour of wash-out and morphology of coal seams (their thickness, composition, stretching of the boundaries of splitting and etc.) one can come to a conclusion that at least a part of the seam is destroyed, and the coal-bearing thickness comparable to preserved one by the area of development.

On the whole, the modern configuration of the contour of distribution of the Carboniferous deposits of LVB, is due to shows of Asturian tectonic movements and later deep Pre-Upper Jurassic and Pre-Upper Cretaceous erosion and abrasion levels (Shulga et. al. 2007). The L'viv-Volyn' Basin, and Kovel' coal-bearing area in particular, compose the most uplifted closing outlying part of the vast L'viv-Lyublin deep where the Post-Carboniferous denudational processes occurred especially intensively. That has caused the absence in the stratigraphic section of deposits of Carboniferous younger Late Bashkirian (Westfalian A) in the basin's central part and Late Serpukhovian (Post-Ivanychi) in the territory of the Kovel' area.

6 DYNAMICS OF THE FORMATION OF THE SEAM MORPHOSTRUCTURE

Formation of coal-bearing potential of deep horizons of LVB is caused by composite interaction both primary and secondary factors. Formation of commercial coal seams of the lower subformation of different parts of the basin occurred in unequal conditions. In the northern edge of the basin within the limits of the Kovel' coal-bearing area, the commercial coal seam v_0^3 is located close to pre-coal basement represented by deposits of Early Paleozoic.

Essential changeability of its morphology is determined by specific conditions of coal forming, and is caused by typical stage-by-stage dynamics of the formation of the Carboniferous deposits of the given area (Figure 4).

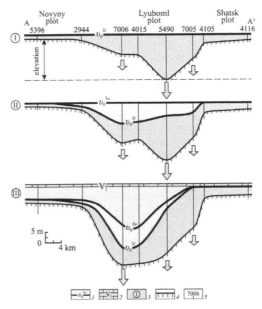

Figure 4. Dynamics of formation of accumulative-tectonic splitting (bifurcation) of coal seam v_0^3 and surface of pre-Carboniferous deposits along the line A–A[1] in the Kovel' area of the L'viv-Volyn' Basin (line of section is indicated in Figure 1): 1 – coal seam and its synonymics; 2 – seam of limestone and its synonymics; 3 – formation stages of splitting: I – formation of lower coal seam, basement of splitting; II – formation of rock interlayer of splitting and upper coal seam; III – completion of splitting formation and its overburdening with rock deposits and limestone V_1; 4 – surface of pre-Carboniferous deposits; 5 – borehole and its number.

The first, initial stage – the formation of the lower seam v_0^{3l}, the basement of splitting. It is characterized by that to the beginning of the formation of coal-bearing deposits the studied area has represented itself as a marshy seaside lowland with strongly divided erosion-tectonic relief. Conducted paleoreconstruction using bottom of the coal seam v_0^{3l} as conventional (zero) horizon, and the principle that a depth range of stagnant waters while peat forming comes to 1–2 m (Volkov 1976) has shown the presence of a great (over 20 km wide) latitudinal valley-like reduction in the central part of the territory. As gradual reduction occurred, channel and floodplain alluvium accumulated in the Valley. There occurred smoothing of paleorelief that became a substrate for the formation of this seam throughout the large area. At the same time, the upper reaches of the paleoriver were located in the east – in the region of the Ukrainian Shield (Shulga & Znamenskaya 1995).

The second stage – formation of rock interlayer of splitting and upper coal seam v_0^{3u}. This stage is characterized by more intensive inherited subsidence of the valley-like reduction in comparison with marginal more elevated parts of the area that has led to the formation of the rock interlayer. Then the processes of deeping of the central part of the area are delayed, accumulation of terrigenous sediments smoothes the relief and as a result – a stable regime of peat-forming becomes settled throughout the whole area. In addition, it should be also noted that to the north and to the south a junction of the lower and the upper seams occurs and the single seam v_0^3 is formed.

The third stage – completion of the formation of splitting and its overlapping by terrigenous sediments and limestone V_1. It is characterized by replacement of a stable regime of peat forming for the regime of increase in the thickness of terrigenous formations in the central part of the area which is caused by the following inherited subsidence of the valley-like reduction. In addition, increase in the amplitude of subsidence of the paleovalley (blocks of the basement of Carboniferous deposits) is displaced in the southern direction; it is obvious that migration of the paleochannel occurred. On the whole, this was expressed by increased thickness of alluvium and coal-bearing deposits underlying limestone V_1 in the region of the paleovalley, as well as by accumulative-tectonic splitting of the coal seam v_0^3. This stage ends with the seam of limestone V_1 which is the upper boundary of the regressive macrocycle $v_0^3–V_1$ (cycle of the 3rd order) (Shulga et. al. 1992).

Such dependence between the coal presence and peculiarities of pre-carboniferous paleorelief within the limits of the northern continuation of LVB was revealed for the first time. It shows itself in bifurcation of the coal seam and complication of its structure in the direction of the valley-like subsidence. Thus, a new type of coal accumulation was established in LVB characteristic of the coal-bearing formations of the ancient platforms located directly on the erosional surfaces of underlying formations (Ivanov 1967). Formations near-Moscow basin, Dnieper basin and other coal basins belong to such kind (Korzhenevskaya et. al. 1962; Nagirny 1977; Radzivill et. al. 1987). Similarity of the forming conditions of the Visean coal-bearing deposits of the Moscow Basin and the lower coal-bearing subformation of the northern continuation of LVB becomes still more obvious, taking practically equal

material composition of both regions into consideration.

Taking into account the latitudinal stretching of the paleovalley, tracing it up to the state frontier with Poland, drilling data from the borehole Savin IG-1, that exposed the seam 2.0 m thick occurring almost directly of the denuded surface of the Late Paleozoic (Musial & Tabor 1988), there are reasons to believe that indicated peculiarities of the formation of the lower coal-bearing subformation of the northern continuation of LVB also existed in the frontier territory of the Lyubbin basin (Shulga et. al. 2007).

6 CONCLUSIONS

Sufficient changeability of morphology of the coal seam v_0^3 and peculiarities of the formation of the Carboniferous deposits within the limits of the Kovel coal-bearing are of LVB are determined by specific paleotectonic conditions of the coal forming, depending upon tectonically divided erosional relief of the basement represented by deposits of different age of Early Paleozoic and inherited dynamics of movements of its tectonic blocks in the central part of the territory. Such dependence between coal presence and peculiarities of the Pre-Carboniferous paleorelief within the limits of the northern edge of LVB was revealed for the first time.

Conducted paleoreconstructions of the formation of the Carboniferous deposits indicated the existence of the coarse (over 20 km wide) longitudinal valley-like subsidence in the central part of the territory where accumulated channel and lowland alluvium. During the period of formation of the coals seams v_0^{3l} and v_0^{3u} as well as of limestone V_1, it underwent the most inherited subsidence in comparison with the neighbouring northern and southern elevated areas. It was shown that increase in the amplitude of subsidence of the paleovalley was displaced in the southern direction that confirms migration of the paleochannel.

The morphostructure of the coal seam v_0^3 is characterized by a composite structure and accumulative-tectonic splitting into two conditional coal seams v_0^{3l} and v_0^{3u} forming a composite stage-by-stage bifurcation.

Executed investigations of dynamics of the formation of the coal seam v_0^3 make our notions of morphology coal seams more accurate and broaden our notions of the coal presence of the deep horizons of the L'viv-Volyn' Basin. Material stated in the paper is of scientific and applied value and testifies to prospects of the Kovel' coal-bearing area for prospecting.

REFERENCES

Shulga, V.F., Karavayev, V.Ya., Lelyk, B.I. et. al. 1995. *Mining-geological characteristic of the Lviv-Volyn Coal Basin*. Coal of Ukraine, 12. Kyiv.

Kostyk, I.E., Matrofailo, M.N., Shulga, V.F. & Korol, N.D. 2012. *On coal-bearing potential of deep horizons of Lviv-Volyn Basin*. Coal of Ukraine, 8. Kyiv.

Shulga, P.L. & Shpakova, V.B. 1958. *New geological section of Paleozoic deposits in north-western part of Volyn area*. Reports of Acad. of Sci. of Ukraine, 5. Kyiv.

Pomyanovska, G.M. & Zavialova, E.A. 1959. *New data on Carboniferous in north-western part of the Volyn Region*. Problems of stratigraphy, lithology and paleogeography of oil- and gas-bearing regions of Ukraine. Gostoptechizdat. Moscow.

Bobrovnyk, D.P., Boldyreva, T.O., Ishchenko, A.M. et. al. 1962. *Lviv-Volyv Coal Basin*. Publishing House of Acad. of Sci. of UkrSSR. Kyiv.

Znamenskaya, T.A. & Chebanenko, I.I. 1985. *Block tektonics of Volyn-Podillya area*. Naukova dumka. Kyiv.

Vlasov, B.I. 1990. *Paleozoic uplift*. Geotektonics of Volyn-Podillya. Naukova dumka. Kyiv.

Shulga, V.F., Zdanovski, A., Zaitseva, L.B. et. al. 2007. *Correlation of Carboniferous coal-bearing formations of Lviv-Volyn and Lyublin basins*. Vatra. Kyiv.

Shulga, V.F. 1981. *Lower Carboniferous coal-bearing formation of Donets basin*. Nauka. Moscow.

Shulga, V.F., Lelyk, B.I., Garun, V.I. et. al. 1992. *Atlas of lithogenetic types and formation conditions of coal-bearing deposits of Lviv-Volyn Basin*. Naukova dumka. Kyiv.

Shulga, V.F., Matrofailo, M.N., Kostyk, I.E. & Korol, N.D. 2010. *Study of morphology of coal seams in Ukraine. Modern state. Trends of further development*. Book of scientific works of IGN NASU. Issue 3. Kyiv.

Vasiliev, P.V. 1950. *Paleogeographical conditions of formation of coal-bearing deposits of Lower Carboniferous of the western slope of the Ural Mts*. Ugletechizdat. Moscow, Leningrad.

Lomashov, I.P. 1958. *On relief of limestone basement in the basin near-Moscov*. Proceedings of Academy of Science of the USSR. Series geol., 3. Moscow.

Shulga, V.F. 1962. *On facies study of coal-bearing deposits of the southern edge of the basin near-Moscow*. Proceedings of Academy of Science of the USSR. Series geol., 6. Moscow.

Sergeev, V.V. 1976. *Splitting of coal seams as indicator of shows of consedimentional tectonics*. Sov. Geology, 8. Moscow.

Volkov, V.N. 1976. *Conditions and mechanism of formation of coal seams*. Bulletin of LSU, 6. Leningrad.

Shulga, V.F. & Znamenskaya, T.A. 1995. *Carboniferous rivers of Lviv-Volyn Basin and their connection with tectonics*. Geol. journ., 2. Kyiv.

Ivanov, G.A. 1967. *Coal-bearing formations*. Nauka. Leningrad.

Korzhenevskaya, A.S., Shulga, V.F., Vinogradov, B.G. et. al. 1962. *Lithological characteristic of Visean coal-bearing thickness*. Geology of fields of coal and combustible shales in the USSR. Gosgeoltechizdat, Vol. 2. Moscow.

Nagirny, V.M. 1977. *Paleogeographic conditions of formation of Cainozoic brown coal deposits of Ukraine*. Naukova dumka. Kyiv.

Radzivill, A. Ya., Guridov, S.A., Samarin, M.A. et. al. 1987. *Dnieper brown coal basin*. Naukova dumka. Kyiv.

Musial, L. & Tabor, M. 1988. *Stratygrafia karbonu na podstawie makrofauny*. Karbon Lubelskiego Zaglebia Weglowego. Prace Inst. Geol, T. 122. Warszawa.

Progressive Technologies of Coal, Coalbed Methane, and Ores Mining – Bondarenko, Kovalevs'ka & Ganushevych (eds)
© 2014 Taylor & Francis Group, London, ISBN: 978-1-138-02699-5

Enhancement of coal seams and mined-out areas degassing productivity

V. Demin, Y. Steflyuk & T. Demina
Karaganda State Technical University, Karaganda, Kazakhstan

ABSTRACT: The article deals with importance of the necessity to expand the field of application on timely degassing preparation of coal seams reserves with the increase of the degassing volume and degassing efficiency.

1 INTRODUCTION

1.1 Gas and rock pressure control on coal mines of JSC "ArcelorMittal Temirtau"

Coal mines of JSC "ArcelorMittal Temirtau" are the best show of gas among world coal producers. The gas content of coal mines is 18–25 m³/t, and the gas release at the working area at times can reach 150 m³/min, due to the big coal saturation of multiple coal layers.

At active deep mining (500–700 m) the coal seams have very low gas permeability ($3 \div 5 \cdot 10^{-3}$ millidarsy).

Applied methods of timely massive coal seams degassing have not had the expected results.

The investigations of coal seams permeability in 2008 made by the firm *Ensafe-Petron* as well as by the *Australian scientist Yan Grey (Sigra firm)* confirmed the data about very low coal seams permeability and made it necessary to study more closely the properties of Karaganda basin coal seams for designing the appropriate timely degassing technology.

The preliminary embedded degassing of coal seams from the underground mine opening is widely used as well as mined out area degassing in the way of drilling the vertical holes from the earth surface.

However at present time its effectiveness is limited by using drill machines with poor productiveness and old fashioned drill technology. Up to 50 vertical holes are drilled from the surface per year (it takes 1 or 2 months to drill each hole) and up to 350 km from the underground working with maximum 120–130 m in length and if necessary they can be 200–250 m.

To provide the underworking of overlying highly gassy seams with the appropriate degassing level by underlying less gassy seams is widely used at the mines of Karaganda basin.

But not all seams have the appropriate underlying seams. That is why the above mentioned technology is used for the seams the mining of which is not more than 40% from total.

The underworking of overlying seams by underlying ones allows to reduce the gas content of the overlying seam to 90%, accelerates the pace of mine working 3–4 times as well as it discharges the working from the rock pressure, which allows to reduce 2–3 times labor costs for mine working maintenance as well as it allows to increase output per face.

The critical issue was to mine especially outburst-prone seam D_6, where due to preventive actions on coal and gas outburst prevention the speed of excavation was not more than 25–40 m/month that slowed down the timely preparation of extraction front.

The Coal Division has offered the technology of primary tunnel driving which is 10–12 m under the seam so that it allowed to discharge the seam from the rock pressure and to increase the gas permeability of the coal seam D_6 for the following degassing.

The given technology allowed increasing the speed of coal roadway construction up to 150 m/month, accelerating the preparation of working area on 5–6 months and repaying expenditures connected with the development of additional stone working which is not required in mining technology.

For 4 years of using this technology there was no case of sudden gas and coal outbursts and the amount of hazardous indicator values on outburst hazard of coal seam D_6 has been reduced 90%.

To provide degassing and maintenance work there is a department "SpecialMineMountageDegassing" (USShMD) working as well as for research and practice implementation of specialized work parameters which help ensure safe working conditions of miners. The basic of them are: degassing work complex; gas methane utilization; timely degassing preparation of coal seams; underground fire prevention and extinguishment; chemical strengthening of coal rock mass; control on comfortable work place, the evaluation of its dust content, noise, vibration, illumination, release into the atmosphere of flame products and quality of dropping water from the mine.

The work of the department specialized on the complex solutions of safety issues has affected positively the work of the basin mines. Within the 40 year work period of USShMD department the methane capture by means of degassing has exceeded 4 milliard m^3. Practically all working faces were working with high force with application of degassing facilities and its effectiveness was 60–80%.

Exclusive standards to degassing were accompanied by reconstruction of mine degassing systems, increase of work volume on drilling the holes and fixing the gas line of big diameter.

At present time the gas recovered by degassing facilities is being used in boiler houses of 5 mines: Lenina mine, Kostenko mine, "Shakhtinskaya", "Abayskaya", "Saranskaya" and "Kazakhstanskaya" – for heating the surface complexes and colorific installations.

For the period of time since 1996 (since the foundation of Coal Division JSC "ArcelorMittal Temir-tau") till 2012 632 mln. m^3 of methane were recovered by means of degassing, and 144 mln. m^3 were utilized in boiler houses of the mines that is 23% from methane capture.

Gathered experience of hole drilling, optimal application of any given ways of degassing, depending on certain conditions allowed to cover the growing requirement of the mines in methane recovery due to load increase on working faces and its use.

The problem of safe and effective mining work at the mines of Coal Division has become more challenging recently in connection with the intensification of production process. It has resulted in the load increase on the working face as well as in its advancing speed. With that the gas content at extraction area and mines has strongly increased. The following ways of methane remove are applied at the extraction area: vertical holes from the surface; preparatory seam degassing; advanced seam degassing; holes into rock cavity; drainage holes from pair to air roadway; insulated methane drainage due to the barrier by degassing; insulated methane drainage due to the barrier by general shaft depression.

The approximate gas balance of the working area of the face is given in the Table 1.

Table 1. The approximate gas balance of the working area of the face.

Recovery method	Methane pick off,			Efficiency of gas drainage,%
	Qcm, m^3/min	K,%	QCH_4, m^3/min	
1. By ventilation facilities	2910	1.0	29.1	–
2. Vertical holes	62.8	85	53.3	42.6
3. Embedded degassing	48.8	8	3.9	3.1
4. Holes in the rock cavity	10.2	53	5.4	4.3
5. Drainage wells	10.6	35	3.7	3.0
6. Insulated methane drainage due to the barrier (degassing)	46.0	25	11.5	9.2
7. Gas drainage by general shaft depression	120	15	18.0	
Total:			124.9	62.2

The basic ratio in gas balance refers to vertical holes from the surface – 53.3 m^3/min which is 42.6% from total methane pick off. The methane recovered by the vertical holes with high concentration up to 70–80% and quite high debit (production) 35–40 m^3/min can be utilized in gas boiler house by the way of direct combustion to make the heat and hot water for the mine needs.

For degassing of the mined out area the holes are drilled into the rock cavity as well as gas drainage holes are made from the pair air roadway which are connected with gas line. The total pick off of the given way goes to 10 m^3/min.

To reduce the gas content of working face the preliminary methane drainage of the seam is used.

The application of different schemes of gas release management allows to have high coefficient of degassing (0.6–0.8) as well as to provide with high load on the working face and safety when mining at high gas content work area.

The inventory of methane from coal production by the underground way at the mines of the Coal Division in JSC "ArcelorMittal Temirtau" is given in the Table 2.

One of the main problems for the mines of the Coal Division was and is the degassing of coal seams. Aiming to guarantee the deep coal seam degassing the contract was concluded with the firm "Ensf Petron" in order to determine the gas permeability of the seam D_6 and the reasonability of implementing the technology of degassing the seam from the surface (so called the feather degassing scheme). The work on equipping the mines of the Coal Division with automated systems of aerogas control continues according to the contract made with the English firm "*Devis Derby*". Due to the investment programme they will be implemented at two mines – "Shakhtinskaya" and "Kazakhstanskaya".

Table 2.The inventory of methane from coal production by the underground way at the mines of the coal division in JSC "ArcelorMittalTemirtau".

Mine	Actual coal production, mln. t	Methane pick off, mln. m³				Volume of utilized methane, mln. m³	Volume of methane burned as torchlight, mln. m³	Methane emission, mln. m³			
		de-gas-sing	venti-lation	from holes of hydrau-lic seam fractu-ring	total			de-gas-sing	venti-lation	from holes of hydrau-lic seam fractu-ring	total
Kostenko	1.81	11.88	32.84	–	44.72	6.93	–	4.95	32.84	–	37.79
Kuzembaeva	1.443	2.68	44.05	–	46.73	–	–	2.68	44.05	–	46.73
"Saranskaya"	1.515	15.61	48.6	–	64.21	0.43	–	15.18	48.6	–	63.78
"Abayskaya"	0.91	18.96	21.77	–	40.73	-	–	18.96	21.77	–	40.73
"Kazakhstan-skaya"	1.312	2.36	14.52	1.43	18.31	–	1.09	2.36	14.52	0.34	17.22
Lenina	1.297	8.93	42.0	0.36	51.29	5.22	0.27	3.71	42.0	0.09	45.8
"Shakhtin-skaya"	1.073	12.4	40.67	–	53.07	0.04	–	12.36	40.67	–	53.03
"Tentekskaya"	1.199	1.73	13.55	–	15.28	–	–	1.73	13.55	–	15.28
Total in CoalDiv:	10.559	74.55	258.0	1.79	334.3	12.62	1.36	61.93	258.0	0.43	320.3

1.2 The volume of methane recovered from the mines of the coal division and the effectiveness of degassing

The industrial resources of methane within the exploited depths by the mines i.e. the methane which can be recovered and used is 10.12 bln. m³ (see the Table 3). Total prognostic resources of methane are evaluated as 3.5 trillion. m³, industrial ones are 1.8–1.9 trillion. m³.

Karaganda Coal Basin is one of the most complicated basins concerning the methane demonstration due to high coal bearing capacity; closeness of methane content seams which have the average and big capacity; high natural methane content in coal seams due to the presence of the gas and coal quaternary deposits etc.

Certain success has been achieved while conducting the degassing work in the basin both in preliminary technology and current degassing of coal seams.

Table 3. Industrial resources of methane within the exploited depths by the mines.

Area	Depth limits when calculating methane resources, m	Methane content of coal, m³/t	Total methane resources, bln. m3				Methane resources of industrial importance, mlrd. m³
			In work coal seams	In coal interlayers	In adjacent strata	total	
Industrial	270–835	10–22	6.8	3.1	14.9	24.8	1.98
Abaysky	140–800	10–25	6.1	2.1	12.3	20.5	1.68
Saransky	264–830	10–25	6,.2	2.8	13.6	22.6	1.81
Shakhtinsky	100–685	10–25	7.4	8.1	23.2	38.7	4.65
Total:			26.5	16.1	64.0	106.6	10.12

The experience of working the seams with high speed of advancing the working faces has showed that the mined out area is a prevailing gas emission source in overall gas balance structure of working area share of which is 80–90% of escaping gas. At that the gas content of working area reaches 100–150 m³/min. In these conditions the reduction of gas release was made by using almost all known ways of degassing and schemes of insulated methane drainage.

The ways of mined out area degassing are widely developed where the share of methane release is 80%.

The volume of methane recovered when conducting the mining work was reduced from 1100 mln. m^3/year at the end of 80s to 285 mln. m^3/year in 2001 and that is connected with sharp decrease of coal mine volume. Out of this volume 234 mln. m^3 is removed by ventilation means and 51 mln. m^3 of methane is removed by degassing. At present mining work is being done at the depth of 500–600 m and at these depths the coal seams have quite low gas permeability which made us search new technologies on coal seams permeability increase, increase of methane recovery aiming to conduct safe mining work.

The use of degassing methods allows having its sustainable effectiveness which is 46–55% for the working faces. Degassing work is cost efficient and provides with 3–4 mln. dollars of economic benefit per year. The gas recovered by means of degassing is used in the boiler plants at the mines Lenina, "Saranskaya" and Kostenko to heat the surface complexes and colorific installations in the volume of 14.7 mln. m^3 per year.

When processing the coal seams through the holes drilled from the surface when analyzing the various process technologies evaluated by the gas recovery factor it is determined that the best of them are caving formation with the use of cavitation effect.

Work productive trends taken in the Coal Division on utilization and recovery of methane are aimed at both coal mining improvement and methane emission reduction into the atmosphere. At that two main trends have been determined the methane outburst reduction by ventilation due to its parameter optimization and efficient use of degassing; volume growth of utilized methane removed by means of degassing and timely gas drainage preparation of specially treated seams.

For the work period of Coal Division JSC "ArcelorMittal Temirtau" from 2007 to 2012 certain results were reached using these trends: methane emission into the atmosphere was reduced 54.9 mln. m^3 for the period of 5 years; the volume of harmful emissions into the atmosphere of coal combustion gases has been cut down which was substituted by methane in the volume of about 5.92 thou. t; long term work plan is being performed on determining the parameters of timely gas drainage preparation with the purpose of large scale methane mining at all mines of Shakhtinsk region.

At the mines of Coal Division JSC "ArcelorMittal Temirtau" there is a coal seam mining of Karaganda and Dolinskaya series at the depth of 450–600 m and deeper, natural gas content of which reaches 17.0–19.0 m^3/t. At taken technology of coal getting

with the load up to 3500–4800 t per day the volume of gas in mining faces is 50–120 m^3/min. To carry out the planned coal mining and safe conditions of mining on the gas factor at high expected volume of gas at the extraction area is quite a difficult task.

Reduction of gas volume at extraction area at the mine is made both by means of ventilation and by using complex degassing methods including gas recovery from the working seam and mined out space.

With the purpose of efficiency evaluation of the ways which are used at the mines to manage gas release the factual data analysis was made on gas volume at extraction area and the quantity of gas eliminated beyond area limits by means of ventilation and degassing for the work period of faces 2007 through 2012. Gas volume at extraction area for the analyzed period changed in quite wide range from 5.1 to 107.3 m^3/min. At that the average daily load on the face was 1865–4647 t, with the increase of the load on the face there is an evident tendency of the face gas volume increase at extraction area. The face gas volume increase at high speed of advancing the mining faces is seen basically due to gas release increase from mined out area, the part of which in total gas balance of extraction area reaches 80–90%.

The gas volume reduction in the faces is made with the help of: preliminary-advanced degassing of working seam; degassing of mined out area by vertical holes, drilled from the surface; methane removal from the mined out area by perforated pipes, going behind the curtain wall, constructed in air gate; gas drainage heading connected through the curtain wall to vacuum gas line; underground holes, drilled above the rock cavity, holes across the strike of the coal seam.

It is necessary to expand the field of application of timely degassing preparation of coal seams reserves with the degassing volume increase and degassing efficiency increase (Figure 1).

1.3 The necessity of designing the technology of intensive recovery and complex use of methane in Karaganda basin

Methane mining realization is of special importance in the item on the agenda. Several variants are supposed to be here including:

– use of methane at Karaganda Metallurgical Complex instead of propane and black mineral oil. At economically viable cost of methane 0.05 dollar per 1 m^3 planned for yearly mining 125 mln. m^3 of methane with total cost 6.25 mln. Dollars which can replace 39.2 thou. t of propane and 54.0 thou. T of black mineral oil, that will ensure yearly profit 8.55 mln. Dollars. From the concernment point of investor this profit can be divided in the ratio 50/50%, with debt service payment of risk period during three – four years;

– methane can be used as fuel for auto transport that will lead to certain cost increase connected with the need in 6.5 mln. dollars, it will replace 53.8 thou. t of petroleum with the cost 14.3 mln. dollars and will make sure profitability 8 mln. dollars per year;

– liquefied or gassy methane can be used for everyday use (Alaygas) or in other branches of heavy and light industry (confectionery plant, PO "Carbide", margarine factory, plant on Mining Machines Repair etc.).

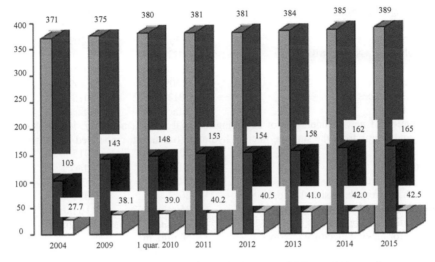

Figure 1. The volume of recovered gas by degassing and degassing efficiency: ■ - volume of methane recovered, mln m³; □ - degassing efficiency, %.

Gas volume increase in working faces and the necessity of using high efficiency management of gas release at working area with high load have predetermined advisability of making the analysis of gas situation in them and efficiency evaluation of different methods how to control gas, with the aim of designing the technological schemes how to manage the gas release at extraction area with high paces of faces advancement.

Technological advancement and identification of parameters on mining methane from the coal seams for industrial recovery and use of gas methane for industrial and everyday use which include the following innovation approaches: computer monitoring of forecast system on methane content in coal seams and calculation of gas reserves; making the ecologically pure technology how to utilize coal methane producing pure carbon and other chemical raw materials when forming the unit on mining and transporting gas methane.

To realize the project the following is necessary: development of the project for making the technology of industrial methane gas mining; work programmes on earthquake forecast of gas content evaluation, gas transmission and gas recovery and determining the regions of natural intensive rock jointing of natural gas collectors; hydrodynamic machining of coal seams from the holes drilled from the surface. Cost of a project: 10 mln. USD.

Economical efficiency on timely methane recovery by the example of Lenina mine was – 46 mln. KZT, with expenditures for 14 holes of hydraulic seam fracturing – 6 mln. KZT:

– by means of degassing 1 mln. m³ of methane makes additional mining 22.5 thou. t of coal. Timely capitation of 20 mln. m³ methane will ensure additional mining 450 thou. t, which even at 25% contribution of degassing gives 18 mln. KZT;

– from the holes of hydraulic seams fracturing the recovered methane was utilized on heating the shaft with cage winding which made it possible to burn 1600 thou. m³ of gas for three heating seasons, which is equivalent to 3 thou. t of coal with cost of more than 4 mln. KZT.

The wide realization of technology on intensive recovery and complex methane use will provide with methane safety at the mines; receiving a new cheap energy carrier for Heat Energy Complex of RK; ecological situation improvement; increase of coal mining due to its gas volume reduction with methane mining up to 3–4 bln. m³ per year.

In view of Resolution of the Government RK

from 5.06.97 #42 on change of license provisions for the Coal Division JSC "ArcelorMittal Temirtau" it is considered both to do the stone coal mining and gas methane recovery in the mine fields assigned by the Contract. That is why this issue is paid more and more attention in Coal Division.

The gas transmission at the mines of Coal Division is the lowest among the coal mining basins and differs – below in 1000 times that requires the implementation of new and improvement of existing technologies of coal seams degassing.

Factors determining the efficiency of degassing and fractional analysis of recovered methane air mixture are the following:

	2012	2015
1. Underworking of a seam	2012	2015
Number of faces	1 face	2 faces
Degassing efficiency	$\geq 70\%$	$\geq 70\%$
2. Drilling holes "across" the strike	2012	2015
Number of faces	2 faces	8 faces
Degassing efficiency	$\geq 40\%$	$\geq 40\%$
3. Remaining interfaces blocks	2012	2015
Number of faces	1 face	4 faces

4. Implementation of new technologies of degassing the coal seams (including: long holes drill, hydraulic partition of coal seams from the surface, interval hydraulic fracturing, embedded holes drill with regard to cleavage)

	2012	2015
Number of faces	2 faces	4 faces
Degassing efficiency	$\geq 70\%$	$\geq 70\%$
5. Use of methane from degassing the coal mines	2012	2015
Volume of recovered CH_4 more than 30%	25 mln. m^3	116 mln. m^3
Use of CH_4 for making heat	25 mln. m^3	60 mln. m^3
Use of CH_4 for making electrical energy –	56 mln. m^3	

The techniques of degassing work fulfillment are shown in the Table 4. The volumes, the variety and the efficiency of degassing work in the Coal Division are without equals in the world practice.

Table 4. The techniques of degassing work fulfillment in coal division.

#	Degassing method	Existing	In perspective
I	Timely degassing from the surface Holes of hydraulic seam fracturing Turnaround time Methane removal from one hole Methane removal from one ton of reserves Degassing reserves for one hole	6–8 years 0.1–0.3 m^3/min 3–5 m^3/t 200 thou. t	2–3 years 0.5 m^3/min 4–5 m^3/t
II	Timely degassing (underground) Embedded holes with regard of cleavage (L – 230 m) Embedded holes of control angle drill (L – 1000 m) Degassing deadline Distance between holes Average methane removal (265 holes) Methane removal from one ton of reserves Methane concentration in gas pipeline	1 year 4 m 8–10 m^3/min 2.3 m^3/t 3–10 %	0.5 year 8–10 m 14–16 m^3/min 4 m^3/t > 25%
III	Hole drill from drainage field workings into the contour of future workings on coal Work time The length of the hole along the rock On coal Number of holes in a bunch Number of bunches in the block Average methane removal Methane removal from one ton of reserves	> 3 months 10–12 m 5–7 m 5 pieces 10 pieces 03 m^3/min 8–10 m^3/t	Additional research is required due to outburst hazard
IV	Current degassing Vertical holes Gas drainage roadway Holes across the strike	14–17 m^3/min 23–25 m^3/min 15–18 m^3/min	not required not required up to 30 m^3/min

In Coal Division the realization of the following actions is offered on time reduction to fulfill the forecast and on blowout preventions.

1. Implementing the acoustic method of forecast: 2012 – at one mine; 2015 – at four mines.

2. Implementing the method of forecast on desorption coal index: 2012 – at one mine; 2015 – at eight mines.

3. Approbation of the forecast method at 25–30 m (by the German mines experience): 2012 – at one mine; 2015 – at four mines.

2 CONCLUSIONS

Implementing new methods of forecast on outburst hazard will allow:

– increasing the safety of mining work;

– reducing the time on forecast fulfillment ≈ in 2 times;

– reducing the time to fulfill blowout preventions in Coal Division on 40–50% (in % from work time).

REFERENCE

Instructions for the safe conduction of mining operations at the seams prone to sudden outbursts of coal and gas. 1995. Karaganda: KazNIIBGP: 177.

Antonov, A.A. 1987. *Forecast and prevention of sudden coal and gas at the opening of thick flat seams Karaganda basin.* Diss. on competition degree Cand. Tehn. Science. Karaganda: 30.

The technical operation of coal mines. 1941–1946. Moscow: Ugletekhizdat: 308.

Design Manual ventilation of coal mines. 1997. Almaty: 258.

Progressive Technologies of Coal, Coalbed Methane, and Ores Mining – Bondarenko, Kovalevs'ka & Ganushevych (eds)
© 2014 Taylor & Francis Group, London, ISBN: 978-1-138-02699-5

Research of deformations of the "base-fundament" system on underworked and hydroactivated territories by finite element method

P. Dolzhikov, K. Kiriyak & E. Ivlieva
Donbass State Technical University, Alchevs'k, Ukraine

ABSTRACT: The article investigates a deformation behavior of foundation bases in conditions of underworking and flooding. By the finite element method has been analyzed vertical dislocations of the foundation and the ground mass in hydroactivisation state and under its injection stabilization. It has been established the appropriatenesses of change of the absolute deformations on three types of engineering-geological sections which is characteristic for the conditions of Donbass.

1 INTRODUCTION

The full flooding of coal mines' workings in mining towns and villages has led to a change in physical and mechanical properties of rocks that caused of the weakening and activation of uneven deformation of the foundation bases. Buildings construction and facilities have exposed to damages or destructions as a result.

To ensure the stabilization of deformation processes in the ground massif and to protect against inadmissible subsidence of foundations are applied various engineering activities, rather prospective of which is a construction of artificial base.

To study the deformation process in the ground as a result of the formation of an artificial foundation on the underworked and flooded areas is applied the finite element method, which is implemented on the engineering program Phase2.

2 FORMULATING THE PROBLEM

The aim of the research is an estimation of the vertical dislocations of the "foundation-basis" system with injecting fixing of hydroactivization ground massif by the finite element method.

To achieve the aim was put the following objectives:

1. To analyze vertical dislocations of the foundation and the ground massif in natural state and on its hydroactivization by the method of finite element analysis in engineering program Phase2 considering of geomorphological features of rock mass and the

Coulomb-Mohr's strength criterion.

2. To determine values of the foundation subsidence obtained by injecting fixing of hydroactivization ground massif on three kinds of engineering-geological sections, which are specific to the conditions of Donbass.

3 MATERIALS UNDER ANALYSIS

The major factor of critical deformations of building constructions is a ground mass hydroactivization on three characteristic types of engineering and geological sections (Figure 1) at the construction and operation of buildings and facilities in conditions of the underworked and flooded areas of Donbass (Ivlieva & Furdey 2013).

The physical and mechanical characteristics of top engineering and geological elements (EGE) are shown in Table 1.

To estimate the geomechanical processes and to determine the values of absolute deformations it was used an engineering program of the finite element analysis Phase2 by version 8.0 of Rocscience Inc. (www.rocscience.com). The initial task is to determine the values of possible subsidence of the foundation bases of constructions in the natural state of ground mass for the three type models. For this purpose, based on obtained data and engineering and geological surveys, which are defined of the geometrical parameters of considered type sections of ground massif.

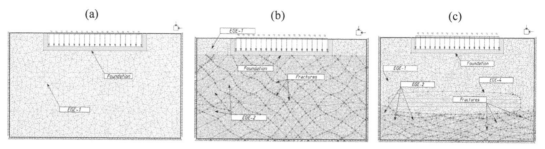

Figure 1. The characteristic types of engineering and geological sections in conditions of Donbass: (a) decompressed loam; (b) fracture-block system; (c) contact of cover and bed-rocks.

Table 1. The physical and mechanical characteristics of ground massif.

Type of sections, #	Name of engineering and geological elements	Density γ, kN/m^3	Cohesion C, kPa	Angle of internal friction φ, degree	Strains module E, kPa	The Poisson coefficient μ
1	EGE-1 – loam	20.0	14	20	9400	0.35
	EGE-2 – stabilized loam	21.0	16	23	28000	0.32
2	EGE-1 – loam	20.0	14	20	5400/9400	0.35
	EGE-2 – argillite	22.5	16	25	36000/88000	0.30
	EGE-2* – stabilized argillite	23.6	18	38	88000	0.28
	EGE-4 – clay and cement mortar	21.0	34	25	36000	0.27
	EGE-5 – filling	22.6	0	17	1500/36000	0.00
3	EGE-1 – loam	20.0	14	20	9400/25000	0.35
	EGE-2 – argillite	22.5	18	38	88000	0.30
	EGE-4 – clay and cement mortar	21.0	34	25	36000	0.27
	EGE-5 – filling	22.6	0	17	1500/36000	0.00

The values as a fraction: the numerator – in water-saturated condition; the denominator – in natural state.

The technogenic load on a model – a five-storey typical apartment building with semi-basement, represented as a cross section. The vertical load acting on the basis equals to 72 kN/m^2.

Area of geometric models was divided into finite elements, and for engineering geological layers were assigned physical and mechanical characteristics of ground mass according to the data shown in Table 1. As a criterion of strength for modeling of ground mass is selected the Coulomb-Mohr criterion often used to estimation of the strength of ground and soft rocks.

The model #1 is characterized by the general spread loam, which is modeled as a continuous isotropic medium. As a result of modeling was determined the vertical subsidences of foundations construction in natural conditions. The maximum values of vertical subsidences are observed at the base of the foundation in its central part and are $U_{y\,abs}$ = 0.044 m. The obtained values are within acceptable limits according to the requirements (BN&R 2.01.09-91 1991).

Operational complexity of the constructions in

conditions of undermined areas of Donbass is that the ground mass, which is above the workings, is experienced the intensive watering due to the rising of groundwater, this leads to a change of its deformation properties (Dolzhikov & Ivlieva 2013). The hydroactivisation of loam leads to a significant reduction in modulus of deformation and, as a result, increasing of the subsidences.

To determine the values of absolute dislocations has been carried out a modeling of ground mass in full water saturation. The results are shown in Figure 2(a).

As seen on the data, the result of loam watering and decrease its modulus of deformation is an increase of subsidences $U_{y\,abs}$ = 0.115 m, which is unacceptable for normal operation of the facility.

The next geotechnical model type is represented with the model #2, which is characterized by weak spread EGE-2 and argillite, which is broken with systems of fractures (EGE-5), with its going out to the surface. The ground massif is modeled as an isotropic medium broken with the systems of transversely isotropic fractures, which are appointed by physical and mechanical characteristics of Table 1. As a result of modeling was determined the vertical subsidences of foundations construction in natural conditions., the maximum absolute vertical dislocations are increased to mid of the foundation ($U_{y\,abs}$ = 0.046 m), but the character of display of dislocations izopoles indicates uneven subsidences due to structural weakening through a system of fissures.

As a result of modeling the ground mass in a natural state the maximum vertical dislocations of the foundation are $U_{y\,abs}$ = 0.038 m.

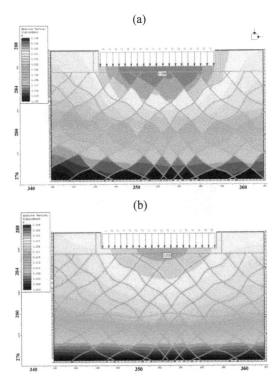

(a)

(b)

Figure 3. The izopols of vertical dislocations of the massif for the model # 2: (a) at the hydroactivisation, (b) at the injection stabilization.

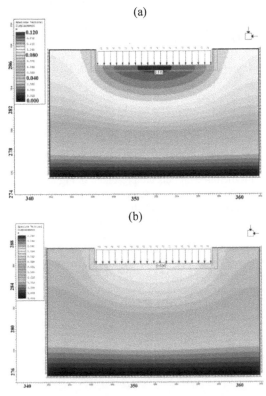

(a)

(b)

Figure 2. The izopols of vertical dislocations of the massif for the model #1: (a) at the hydroactivisation, (b) at the injection stabilization.

Also, the reduction of argillite strength occurs as a result of watering, which leads to significant reduction of its deformation properties. The results of loading modeling of hydroactivisation argillite is shown in Figure 3(a). The character of dislocation izopole is reflected to an increase of absolute value compared with a natural state of the massif, $U_{y\,abs}$ = 0.099 m and shows an increase of uneven values of subsidences.

The model #3 is also quite widespread. The ground massif is submitted with argillite (EGE-2), which is broken by a system of fissures (EGE-5) and is blocked by insignificant loam layer on thickness (EGE-1).

The hydroactivisation mechanism of loam is implemented by raising of the groundwater with a systems of fissures passing at EGE-2, thereby is caused wetting of EGE-1. The results of ground mass' modeling in wetting state are presented in Figure 4(a).

The maximum vertical dislocation of foundation is $U_{y\,abs}$ = 0.092 m, which is unacceptable. The attainment to critical values of vertical dislocations, causing uncontrolled redistribution of stresses and strains in exploited constructions, causing progressive destruction, thereby the building is led to emergency state, and the impossibility of its further use.

To prevent or stabilize of arising uncontrolled subsidence of foundations base for each of the standard models it was developed a scheme of injecting fixing of the unstable ground structures.

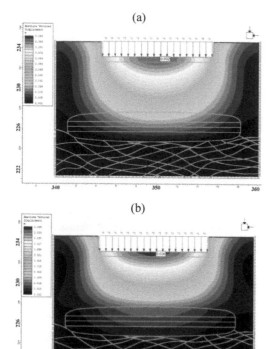

Figure 4. The izopols of vertical dislocations of the massif for the model #3: (a) at the hydroactivisation, (b) at the injection stabilization.

On the first model, the scheme is implemented as a hydrofracturing of water-saturated clay loam by the outflow of the liquid phase and formation of consolidation zones, and as a result the increase of density of the ground framework. The second model is characterized by the creation of artificial clay-cement-block system by pumping the mortar through vertical boring wells. The mortar for the first fills the fracture systems, forming a "monolithic" massif. The feature of the model #3 is the pumping with a clay-cement mortar on a contact of zones EGE-1 and EGE-2 thereby forming an injection stabilizing cushion which prevents the development of strains in the foundations base.

The physic-mechanical and deformation characteristics to the stabilized models are shown in Table 1, and the results of modeling are shown in Figures 2(b), 3(b) and 4(b).

As a result of numerical modeling was obtained the ground areas exposed to the maximum deformation of the values of absolute dislocations for each model: #1 – $U_{y\,abs} = 0.036$ m; #2 – $U_{y\,abs} = 0.030$ m; #3 – $U_{y\,abs} = 0.036$ m. The appropriatenesses of subsidence distribution of the system "foundation-base" for the three models are presented in Figure 5.

The analysis of getting appropriatenesses shows that for the model №1 the distribution of the subsidence is submitted to a hyperbolic function, for the model #2 is stipulated for a piecewise-broken dependence at the hydroactivization and a smooth degree dependence after the grouting, for the model #3 a dependence is almost linear with a fracture at a depth of stabilization cushion's setting. These results are demonstrated the efficiency of the foundation base grouting.

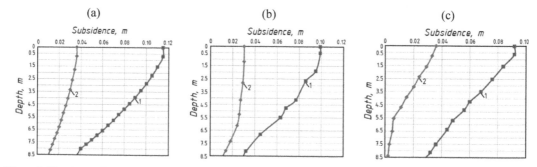

Figure 5. The appropriatenesses of distribution of the basis subsidence to (1) and after (2) the formation of an artificial base: (a) in a decompressed loam; (b) in an argillite broken with system of fissures; (c) with on a contact cover and bedrocks.

4 CONCLUSIONS

Based on the results it had been determined the appropriatenesses of changes in absolute deformations for each model. So for the model #1 the injection stabilization of water-saturated loam reduces the subsidence 3.2 times as much; the injection fixing of fissuring argillite and the formation of cement-block system in the model #2 shows a decrease in the absolute dislocation 3.3 times less; the formation of stabilizing cushion on a contact of the rocks in the implementation of the model #3 reduces the subsidence 2.6 times less.

REFERENCES

Ivlieva, E. & P. Furdey, P. 2013. *Formation of artificial base of the foundations at undermined areas.* Proceedings of the conference "Prospects of building technologies": 34–37.

BN&R 2,01.09-91. 1991. *Buildings and facilities on undermined areas and subsiding grounds.* Stroyizdat. 44. Moscow.

Dolzhikov, P. & Ivlieva, E. 2013. *Influence of watering and fracturing on the deformation properties of foundation bases.* Collection of scientific works DonSTU, 40: 168–172.

Kipko, E., Dolzhikov, P., Dudlya, N., Kipko, A., etc. 2004. *The complex method of tamping in the mines construction:* tutorial: 367. Dnipropetrovs'k.

Progressive Technologies of Coal, Coalbed Methane, and Ores Mining – Bondarenko, Kovalevs'ka & Ganushevych (eds)
© 2014 Taylor & Francis Group, London, ISBN: 978-1-138-02699-5

Submersible electromechanical transformers for energy efficient technologies of oil extraction

N. Zablodskiy, V. Pliugin & V. Gritsyuk
Donbass State Technical University, Alchevs'k, Ukraine

ABSTRACT: In paper by finite element method performed mathematical modeling of interconnected electromagnetic and thermal processes in the submersible electromechanical converter. The results indicate a high level of thermal stress in the active part of the submersible PEMC. The temperature on the surface of the rotor submersible converter reaches very high values, which suggests the possibility of providing rapid heating and high speed pumping oil and gas-liquid component of shale through the cavity of inner rotor of submersible PEMC.

1 INTRODUCTION

Thermal methods are widely used in well oil recovery method in the United States, Russia, Venezuela, Canada, France, with a heavy viscous oil developments, and in the absence of in-situ pressure. Wherein high temperature agents are used (steam, hot water, air, oil and combustion gas) and required huge costs for the heating of the fuel and energy transporting in the well. Recovery rate with this method is 0.4–0.6 for 2.5–2.7 tons of specific steam consumption per ton of product. Researches have shown that a big amount of heat during discharge of coolant from the field of drilling wells through the horizon is lost in the rocks of the roof as well, together with the formation of recoverable oil and hot water (Minaev 2001). Technologies and devices for electric heating of wells' bottom zone to ignite the oil reservoir near the wellbore during the process of in-situ combustion front moving and removing of paraffin-bitumen deposits from the walls of oil and gas wells are known (Mironov 2001). However, the reliability of used tubular electric heaters is very low. At the same time the efficiency of bottom zone warming is low because of weak turbulence in gas-liquid mixture and heat loss by convection in the upper layers of the liquid in the well, or by heat transfer to the roof rocks by direct electrical heating of pump-compressing tubes.

New EPICC technology (electricity production with in situ carbon capture), proposed by scientists of Stanford University (Brandt & Mulchandani 2011) and combines production of electricity with capture and storage of carbon dioxide underground can make available huge (up to 3 trillion barrels of oil) energy reserves in the form of oil shale. Oil shale is inherently difficult technological material. Now removing of chemical-technological component due to thermal processing of oil shale, specifically heating of raw materials to temperatures when a complete destruction of kerogen to the volatile substances, which by physical condensation are separated and used separately, is occurs (Osipov 2013). Oil shale is a promising resource for organic row materials for Ukraine having large deposits in Boltysh, Novo-Dmitrov and Kornetsk.

In given technology in order to maintain the thermal cracking processes of liquid hydrocarbons and their transportation to the separation sites saving electromechanical devices are required.

Market demands for increasing the commercial value of fuel and increasing the production rate of oil and gas wells, the introduction of new energy technologies in the oil shale can be solved by creating a series of submersible electromechanical converters (Zablodskiy 2008).

2 FORMULATING THE PROBLEM

Submersible electromechanical converters are belonging to a new class of electromechanical devices – polyfunctional electromechanical converters (PEMC) of technological purpose. This is the series of electromechanical converters modifications for various applications with a common technical framework and creation ideology.

Principles of PEMC creation can be divided into the following groups:

 – structural and functional integration;
 – integration of thermal processes;
 – self-regulation for the separation into compo-

nents of useful power;

– gearless ensure of low speed and torque fold division;

– noncriticality to the quality of energy supply harmonic composition.

Principles of PEMC structural and functional integration are in combine of technical system involving electromechanical, mechanical, and thermal subsystems to implement an integrated objective function set by functional features of the drive motor, actuator and heater. The aim of integration of thermal processes is the formation of heat removal capabilities from sources losses in PEMC, coordination of heat flows generated at the target converting of electricity into the heat and its canalization in the technological chain links, which requires heating work surfaces and volumes. The principle of self-regulation by the division into components of useful power is defines the mechanism of converter electromagnetic power share distribution on two useful output streams: stream of mechanical power output and stream of thermal power.

Overall efficiency of PEMC is very high, which provides an effective implementation of the principles of energy saving. Significant advantages of based technologies PEMC are not only the high value of efficiency, but also significant reductions in the production area, the number of units and the payback time (Zablodskiy 2008).

On Figure 1 shows an apparatus for removing deposits from the walls oil and gas wells (Patent #49409… 2004).

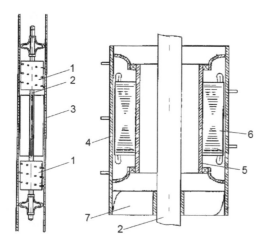

Figure 1. Apparatus for removing deposits from the walls oil and gas wells: 1 – rotating turbine; 2 – shell; 3 – pipeline; 4 – outer cylinder of rotating turbine; 5 – internal cylinder of rotating turbine; 6 – cylindrical inductor with windings; 7 – impeller of reverse rotation.

Some modifications of submersible PEMC can be performed with ring windings. This opens the possibility of using end parts not only for forming the resultant electromagnetic torque, but also to provide additional power to the heating in the end parts of the converter. The ring windings stacked in the annular grooves along the entire length of the zone, thus in the frontal portion of this winding are absent, and the length of the active portion increases. Application of ring windings stator – a new technical solution, for followed by a significant increase in the utilization of active materials and increased technical and economic indicators.

On Figure 2 shows a general view, as well active part of the submersible converter with a ring winding, intended for processing, production and transportation of viscous oil (Patent #39226… 2001). In this device uses a hollow ferromagnetic coaxial rotor that performs the functions of both rotor induction motor, heater actuator and protective housing. Material to be acting in two capacities: a mechanical load to the rotor created by viscous friction material; cooling fluid that bathes the rotor and take of his heat.

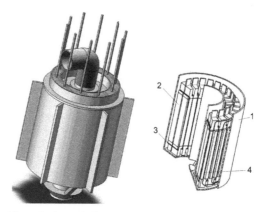

Figure 2. General view and active part of the submersible converter with a ring winding: 1 – hollow ferromagnetic coaxial rotor; 2 – a stator core with windings; 3 – winding ring type; 4 – outer winding drum type.

The basis for the study of processes in the new electromechanical devices is the mathematical modeling of electromagnetic and thermal fields. Obtained using a mathematical model of the distribution of the electromagnetic and thermal fields allows a preliminary step to determine the optimum geometrical dimensions of the active part of the converter and to assess its performance properties and characteristics.

The objective of this work is to conduct mathematical modeling interconnected electromagnetic and thermal processes in the submersible PEMC and determine optimal electromagnetic characteristics, as well maximum turbulence-forming and thermal effect on the gas-liquid and viscous oil shale components.

One of the most efficient numerical methods for solving problems is the finite element method. The advantage of this method is the relative ease and accuracy of description rather complex configurations, the possibility to consider nonlinearity of material properties, etc.

3 MATERIALS UNDER ANALYSIS

In general, the differential equation of the electromagnetic field in the partial derivatives with respect to the magnetic vector potential $\vec{A}$ has the following form

$$\frac{1}{\mu}\Delta\vec{A} - \gamma\frac{\partial\vec{A}}{\partial t} + \gamma\left(\vec{v}\times rot\vec{A}\right) = -\vec{J}_{ext}, \qquad (1)$$

where μ and γ – magnetic permeability and electrical conductivity portions of the medium within the computational domain; $\vec{J}_{ext}$ – density of external currents; $\vec{v}$ – velocity of the conductive body relative to the magnetic field source.

Equations of electromagnetic field are supplemented by boundary conditions – zero value of the vector magnetic potential and its normal derivative at the outer boundaries of the computational domain, which displays the active zone PEMC. Such boundaries are for immersion PEMC – massive external surface of coaxial rotor. On these surfaces are given homogeneous boundary conditions of the first kind

$$A(x,y,t) = 0. \qquad (2)$$

Setting of boundary condition (2) is equivalent to the adoption assuming no leakage flux into outer space through the border.

In solving the unsteady field equation must specify initial conditions – the values of unknown function inside the area in the initial settlement time t_0.

$$A(x,y,z,t)\big|_{t=t_0} = A_0(x,y,t_0). \qquad (3)$$

When analyzing the characteristic for submersible PEMC dynamic processes, such as start, is typically define the homogeneous initial condition

$$A_0(x,y,t_0) = 0.$$

To determine the induced current density in the rotor can be used the following expression, which follows from the first expression of Maxwell's equations

$$J_z = \frac{1}{\mu}\left(\frac{\partial B_y}{\partial x} - \frac{\partial B_x}{\partial y}\right). \qquad (4)$$

After the calculation of the field vector magnetic potential is the total magnetic flux linkage of each phase of the stator winding by the following expression

$$\Psi_k = \frac{2l_a w_c}{S_c}\int_{S_A} A_z \cdot ds, \qquad (5)$$

where l_a – active length of the stator; w_c – number of conductors in the slot; S_c – slot area; S_A – integration area, consisting of a total cross-sectional area of all the sides of phase coils are connected in series with a current direction;

At each point of the rotor specific the losses calculated by the expression

$$Q = J_z^2/\gamma(T), \qquad (6)$$

where $\gamma(T)$ – conductivity of the material of the rotor, the temperature-dependent T

$$\gamma(T) = \gamma_0/(1+\alpha(T-T_0)), \qquad (7)$$

where γ_0 – conductivity of "cold" rotor; T_0, T – ambient temperature and the temperature of the rotor material; α – temperature coefficient.

The differential equation of the thermal field in the partial derivatives with respect to temperature T has the following form

$$\lambda\Delta T - c\rho\frac{\partial T}{\partial t} = Q, \qquad (8)$$

where λ, c, ρ – accordingly thermal conductivity, specific heat and density of the material; Q – specific heat loss calculated by the expression (6).

In Cartesian coordinates for the two-dimensional field Equation (8) can be rewritten as follows

$$\lambda\frac{\partial^2 T}{\partial x^2} + \lambda\frac{\partial^2 T}{\partial y^2} - c\rho\frac{\partial T}{\partial t} = Q. \qquad (9)$$

Feedback equations of the electromagnetic and thermal fields manifested in mutual influence of temperature, conductivity, density, eddy currents and specific heat losses. As well as for the electromagnetic field equations for Equation (9) are given by the boundary and initial conditions. Choice of boundary conditions is due to features of rotor cooling PEMC. When constructing a mathematical model assumes that the heat transfer into the main load-coolant released in active elements PEMC thermal energy carried by convective heat transfer between the heated surface and the load-coolant liquid is described by (10). Such heat transfer takes place in accordance with equation Newton-Richman

$$\frac{\partial T}{\partial n} = -\frac{k}{\lambda}(T - T_0), \tag{10}$$

where k – heat transfer coefficient; T_0 – cooling air temperature.

The boundary condition (10) is defined by the outer surfaces of the rotor. From the physical point of view, the most appropriate is the assumption that the transfer of all the separated rotor Joule losses in the liquid material in the form of heat flow through the surface.

To improve the accuracy of the mathematical model according to the functionality made of temperature coefficients of the active materials on the temperature dependence of the magnetic permeability electrical steel of the magnetic induction, and the thermal conductivity of steel, the air load-cooling medium depending on the temperature.

Model geometry submersible PEMC represents a quarter cross-sectional of real design. Finite element mesh of model is shown in Figure 3. Figures 4 and 5

shows the results of simulation of the electromagnetic field of submersible PEMC operating in short-circuit mode (fixed rotor).

Figure 3. Finite element mesh of model submersible PEMC.

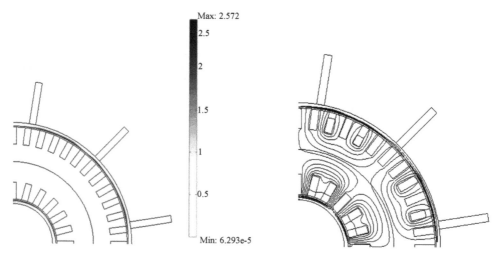

Figure 4. Distribution pattern of the normal component of the magnetic induction and the vector magnetic potential in the active part of the submersible PEMC.

These results indicate that in this mode, the magnetic field penetrates to a small depth of the array (3.2 mm) from the inner surface of the outer and inner cylinders. Induction in the rotor reaches values of 2.1–2.2 T.

The radial line along which the distribution plotted normal component of the magnetic induction (Figure 5) passes through the ferromagnetic portion of the inner rotor, the air gap, the stator teeth and the yoke and also through the gap and the outer rotor portion.

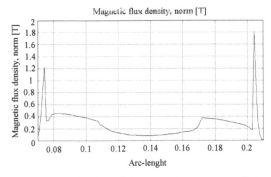

Figure 5. Distribution of the normal component of the magnetic induction along the radius of the submersible PEMC.

The result of the calculation of electromagnetic problem is the amount of power of thermal losses used to calculate paintings of thermal field PEMC, as well as data on the distribution of magnetic induction and eddy currents in the rotor, used to calculate the electromagnetic torque. Distribution of thermal field submersible PEMC shown in Figure 6. The results indicate a high level of thermal stress in the active part of the submersible PEMC. Thus, the main sources of heat release due to eddy currents are flowing outer and inner cylinders ferromagnetic rotor.

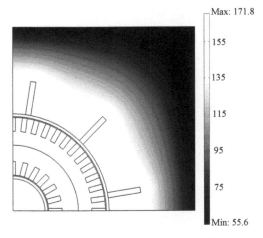

Figure 6. Distribution of thermal field submersible PEMC.

The temperature on the surface of rotor in short-circuit mode reaches 160–170 °C, which suggests the possibility of providing rapid heating and high speed pumping oil and gas-liquid component of

shale through the cavity inner ferromagnetic cylinder. Temperature coolant is near the surface of the rotor – with 145–155 °C. The calculated data showed that the resulting value of electromagnetic torque of two cylindrical portions of rotor PEMC is 143·Nm. This in turn indicates on satisfactory moment performance and turbulence-forming properties of submersible converter.

4 CONCLUSIONS

By finite element method performed mathematical modeling of interconnected electromagnetic and thermal processes in the submersible electromechanical converter. The results indicate a high level of thermal stress in the active part of the submersible PEMC. The temperature on the surface of the rotor submersible converter reaches very high values, which suggests the possibility of providing rapid heating and high speed pumping oil and gas-liquid component of shale through the cavity of the inner rotor of submersible PEMC.

REFERENCES

Minaev, Y. 2001. *Mine oil production:* tutorial. St. Petersburg mining institute.
Mironov, Yu., Ivanov, A. & Arzemasov, V. 2001. *Optimization of electrical heating equipments for oil boreholes clearance from paraffin.*Electricity, 6.
Brandt, A. & Mulchandani, H. 2011. *Oil shale as an energy resource in a CO₂ constrained world: the concept of electricity production with in situ carbon capture.* Energy & Fuels, 25.
Osipov, A., Shendrik, T. & Grischuk, S. 2013. *Oil shale – perspective resource of natural organic raw material in Ukraine.* Electrotechnology and resource saving, 6.
Zablodskiy, N. 2008. *Polyfunctional electromechanical transformers of technologic assignment:* monograph. Alchevsk: Donbass state technical university.
Patent #49409 Ukraine, MPK E21V 37/00. 2004. Advice for paraffin removal from oil-gas boreholes walls / Zablodskiy, N., Dorofieiev, V., Zakharchenko, P. & Shynkarenko, V. Applicant and patentee of Donbass state technical university. # 2001128246; declared 03.12.2001, Bul. # 10.
Patent #39226 Ukraine, MPK N05V 6/10. In-depth electric heater / Zablodskiy, N., Veremeenko, V.& Bondarev, V. Applicant and patentee of Donbass state technical university. – # 98031637; declared 31.03.1998; published 15.06.2001, Bul. # 5.

Progressive Technologies of Coal, Coalbed Methane, and Ores Mining – Bondarenko, Kovalevs'ka & Ganushevych (eds)
© 2014 Taylor & Francis Group, London, ISBN: 978-1-138-02699-5

Evaluation of surface subsidence during mining thin and very thin coal seams

O. Koshka, A. Yavors'kyy & D. Malashkevych
National Mining University, Dnipropetrovs'k, Ukraine

ABSTRACT: The state and relevance issue of leaving wall rock undercutting extracted from longwall and development faces in the goaf are shown. The most appropriate ways of decision in this issue have been substantiated. The mathematical apparatus developed at the National Mining University has been used for the evaluation of surface subsidence and construction subsidence trough Earth surface at different stowing goaf. The design model includes a stratum of sediment, carbon, coal seam, longwall face and mined-out space filled by stowing material and caved rock. The calculation results of surface subsidence, depending on the completeness of stowing and total roof convergence in the goaf have been obtained.

1 INTRODUCTION

The main factors deteriorating significantly the environmental situation in the coal-mining regions are: undermining Earth's surface and hoisting of large volumes mine waste rock, polluting areas of land, water and air basin.

For the environmental protection in the liquidation consequences the activity of coal enterprises is spent a considerable amount of money. However, none of the means let do away completely all negative consequences, as most of them are irreversible. Especially this problem stays at developing thin and very thin coal seams, balance reserves of which on Ukrainian mines, according to various sources, exceed 80% of the total coal reserves. As a result, coal mines develop a large number of seams up to 1 m, which are mined basically with internally wall rock undercutting.

Currently, on singular Ukrainian mines drawing rock volume on the surface exceeds the amount of mined coal. Waste rock from wall rock undercutting in longwall and development faces is drawn from mines together with coal forming the so-called "rock mass". In this case, use of electricity, underground and surface transport, hoisting, human and material resources is irrational. That's why, the prime cost of mined coal and dispatched rock mass increases correspondingly. The growth of ash content at the expense of clogging by waste rocks impairs its quality, and it means the reduce of the price, required the additional costs for transport and enrichment of the rock mass, storage waste on the surface. Therefore, for each specific mine and industry in general the question of the efficient mining

of thin and very thin coal seams without drawing of substantial quantities of waste rock on the surface – it is a question not only and not so much environmental as economic and technological.

One of the most realistic ways that allow to reduce significantly the negative effects of mining activities on the environment, reduce the cost of mined coal, improve its quality and price, is a wide use of the selective mining thin and very thin coal seams technology, providing leaving wall rock undercutting from longwall and development faces in minedout spaces (Buzilo 2012).

Leaving rocks in the mine will not only improve the quality of mined coal, reduce the pollution of the environment, but at the same time the complex of engineering problems such as the creation of safe working conditions of coal extraction under protected objects can also be solved. The stowing of goaf by undercutting rock will favorably influence on the minimizing process of rock mass displacement, and maintenance of buildings and structures.

2 STATE-OF-THE-ART SUMMARY

Today the most use-proved variant of such technology is the one variant with stowing undercutting rock from longwall face in the goaf by pneumatic stowing complexes "Titan-1" and "Titan-1M", the crushing and stowing stationary complexes (DZK) of domestic production and Czech stowing machine ZP-200, ZS-240 (Dzhvarsheishvili 1978).

The crushing and stowing complex "Titan" in due time was quite applied widely on Ukrainian mines for stowing and leaving waste rock in the goaf from development workings. The scale of application of this complex accounted about 50 units (Dzhvarshe-ishvili 1978). However, in 90-s it has been taken out from production by the reason of low efficiency.

Czech stowing machines ZP-200 and ZS-240 were used on "M. Gorky" mine for stowing under-cutting rock working with coal seams selective ex-traction. The stowing machines used compressed air from the all-mine pneumatic system. The material was prepared by one-stage scheme using two paral-lel operating crushers DO1. The prepared stowing material was moved by the belt conveyor to the stowing machine and was stowed in mined-out space.

On the flat seams the scraper stowing installation ZK-02 and ZK-03 types got a wide application. They were also used for waste rock stowing goaf and packs during development workings (over 400 installations). The other facilities of mechanical stowing were also used, particularly the centrifugal stowing machines (belt, drum and disc) which, however, have not been used widely.

The most common way in the coal mining indus-tries of former USSR was the hydraulic stowing and its unit weight accounted about 70–80%. The spe-cial extraction mechanized complexes were created (KMGZ, KVZ, etc.) that allowed to mechanize the process of seams development and stowing mine-out space. For stowing they used: sand, gravel, tails concentrators and in some cases rocks from waste dumps. So on "Red October" mine the hydraulic stowing complex GSK, which stowed goaf waste rock specially prepared on the surface, was in opera-tion.

On Donbass mines, that develop steep, steeply in-clined and inclined strata, was widely spread the gravity stowing method, under which the stowing material by gravity on seam floor, metal sheets or pans went to the goaf. This was used for full and partial stowing and packing the ventilation drift.

And the solid stowing has found its application; currently, this is used on a number of mines, mainly for cast strips building on connection longwall and drift ("Pokrovskaya" mine administration etc.).

3 BASIC UNIT

The analysis of stowing methods, that are used on Ukrainian mines in different years, showed that for the assigned task in the work, notably – leaving waste mined rock in mine-out space in the devel-opment and longwall faces, the pneumatic and grav-ity stowing methods are the most appropriate. Al-though they have a number of drawbacks, but allow to use the mine waste as stowing material in large

volumes.

Take the estimation of stowing methods influence (pneumatic and gravity) on the convergence of wall rock in the goaf at selective coal seam extraction and leaving wall rock undercutting in the goaf. The convergence of the wall rock in the goaf can be rep-resented as the total size of roof rock displacement during two periods – before and after filling mass building.

Figure 1 shows a principle scheme for size calcu-lation of roof convergence and residual height of stowing.

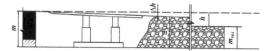

Figure 1. Scheme for calculation size of roof convergence.

Figure 1 shows that size roof convergence subject to rock shrinkage in mine-out space will be

$$h = m - m_{res} \text{ , [m]} \qquad (1)$$

where m – extraction hight, m; m_s – stowing height, m; m_{res} – residual stowing height after shrinkage, m; k_{res} – coefficient taking into accounts the residual height of stowing mass; $k_{res} = 0.80$ – for pneumatic stowing; 0.78 – for gravity stowing.

In case of full stowing goaf (at height)

$$(m - \Delta h) \cdot k_{res} \text{ [m]} \qquad (2)$$

For partial stowing

$$m_{res} = m_s \cdot k_{res} \text{ ,} \qquad (3)$$

where Δh – size of wall rock convergence in the stope, m.

$$\Delta h = \alpha m R \text{ , [m]} \qquad (4)$$

where α – coefficient taking into accounts the spe-cific side rock convergence, m^{-1}; R – distance from stope to stowing mass, m.

Input data for calculations:
– geological seam height: 0.8; 0.7; 0.6; 0.5, m;
– extraction seam height: 1.0 m;
– under-cut side rock height: 0.2; 0.3; 0.4; 0.5, m;
– under-cut side rock: mudstone;
– immediate roof: siltstone.

In terms of expression (2), under the full stowing condition (at height) and taking into account stow-ing shrinkage under extraction height 1m, the roof convergence will be: 0.38 m – for gravity stowing, 0.36 m-pneumatic stowing.

Table 1 shows the size of wall rock convergence for gravity and pneumatic stowing, and the residual height of stowing mass, when stowing goaf aims only leaving wall rock undercutting in the goaf.

Table 1. The size of roof convergence for gravity and pneumatic stowing and residual filling mass.

Goaf filling (at height), %	The size of total roof convergence in goaf h, m	Residual height of filling mass, m_{res}, m
90	0.44 / 0.43	0.56 / 0.57
80	0.50 / 0.49	0.50 / 0.51
70	0.57 / 0.56	0.43 / 0.44
60	0.63 / 0.62	0.37 / 0.38
50	0.69 / 0.68	0.31 / 0.32
40	0.75 / 0.74	0.25 / 0.26

Numerator – for gravity stowing; denominator – for pneumatic stowing.

Table 1 shows that for gravity method the total value of roof convergence is 0.01 m more than pneumatic one.

Thus, improving gravity stowing technology can be reached the minimum size of wall rock convergence, that are close approximate to pneumatic method of stowing mass.

The assess subsidence and construction subsidence troughs of Earth's surface at different filling the goaf by stowing material using mathematical apparatus developed at the National Mining University (Yavors'kyy 2010). Design model is shown in Figure 2. It includes a stratum of sediment, carbon (homogeneous stratum with the average properties of the bearing strata), coal seam, a longwall face and mined-out space, filled by stowing materials and stowing caving. The current loading is weight of rock.

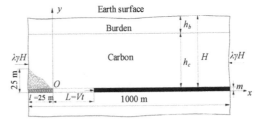

Figure 2. Design model.

The calculations are performed for the geological conditions of Donbass mines with the following initial data: $h_b = 140$ m; $H = 400$ m; seam thickness $m = 1$ m; average speed of moving longwall face $= 60$ m/month. Coal properties and bearing strata are given in Table 2.

The data in Table 2 for carbon are defined as the average values of the relevant characteristics of the bearing strata (mudstone, siltstone, sandstone and limestone).

Table 2. Properties of coal and bearing strata.

Rock	Volume weight, γ, t / m^3	Modulus of elasticity, $E \cdot 10^{-3}$, MPa	Poisson ratio, v	Creepage parameters $\delta \cdot 10^3$, c$^{\alpha \text{-} 1}$	α
Sediment	1.5	0.07	0.3	13.9	0.862
Carbon	2.76	3.18	0.3	6.84	0.7
Collapsed rock	1.5	0.01	0.499	–	–
Coal	1.47	20	0.35	2.32	0.7

4 THE RESULT OF THE STUDY

The subsidence trough construction that is formed above the moving front of second working was made by the superposition principle. This means that the shifting this point on the surface was determined as the sum of the individual effects of "elementary" remove elements. The length of the element was assumed to being equal to step of the main roof collapse. The step is 10–22 m in these conditions.

The subsidence troughs that were constructed by the results of calculations are shown in Figure 3.

The analysis of calculation results presented in Figure 3 shows, that the subsidence surface depends directly proportional on values of the total roof convergence in the goaf, which in turn is defined as the difference between the extracted height and the residual height of stowing mass. Thus, with the increasing the total roof convergence by two or three times, respectively to 0.5 and 0.75, the subsidence surface also increases by two, three times from

200 to 440 and 680 mm at a distance of 500 to 700 m from the face line.

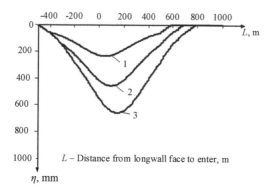

Figure 3. Subsidence Earth's surface at different values of the total convergence: 1 – $h/m = 0.25$; 2 – $h/m = 0.5$; 3 – $h/m = 0.75$.

The curves of calculated subsidence and horizontal displacements in the main section of trough on the surface at full undermining under concerned conditions are shown in Figure 4(a, b). The solid lines – the results of calculations performed without rock creeping, and the dashed lines – taking into account rock creeping.

(a)

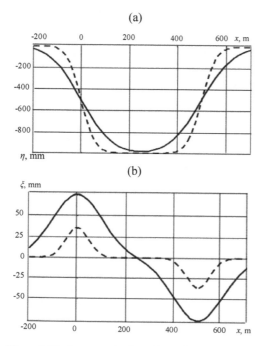

(b)

Figure 4. Displacements in the main section troughs of the Earth's surface: (a) subsidence; (b) horizontal displacements.

Figure 4(a) presents follow that the subsidence curve with the lapse of time flattens, wherein offset values, as compared with the calculation results nonmetering the rheological rock properties, increases by about 10–25%, and in the marginal parts of the trough decrease about the same.

In point of the horizontal displacements the following can be noted. The shape of the curve $\xi(x)$ is not changed with the lapse of time. Only the scale is changed: values $\xi(x)$ are reduced by 2.1 times.

Based on a statistical analysis of the calculations data it is set the correlation dependence of maximum relative subsidence of the parameter $z = \dfrac{h}{m}$ characterizing the total roof convergence in the goaf

$$\frac{\eta_{max}(z)}{\eta_0} = 5.021z^3 - 9.271z^2 + 97.72z - 0.2284. \quad (5)$$

where η_0 – maximum deflection of seam roof at full undermining; $z = \dfrac{h}{m} \in [0.05...1.0]$.

The correlation coefficient of this dependence $R = 0.99$.

Figure 5 shows the curve $\dfrac{\eta_{max}(z)}{\eta_0}$ constructed from the relation (2). The dots indicate the data of the respective calculation variants.

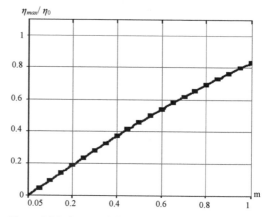

Figure 5. Maximum subsidence troughs at different values of the total convergence.

The analysis of calculation results showed that the values of subsidence troughs are 82–90% of the total roof convergence in the goaf.

5 CONCLUSIONS

At full stowing gravity the stowing can be used as effective measure for reducing the earth's surface subsidence almost as the pneumatic one. But the disadvantages of gravity stowing concede essentially. That is why it allows toassume such kind of goaf stowing the most rational for the conditions of majority mines in Donbass.

However, questions of gravity stowing technology in flat and inclined strata are characterized by the insufficient knowledge and the lack of scientific evidence and the main parameters of the application. These circumstances as well as a number of organizational and technical factors do not give the possibility of using this type of stowing for the leaving undercutting rock from longwall and development faces in the goaf yet. The solution of these issues would make the gravity stowing the efficient and cost-justifying method of leaving rock in the mine.

Using derived in NMU analytical relations for the geological conditions of Donbass mines the indicators of surface deformation in full and partial undermining have been identified. The following relationships have been found:

– the surface subsidence is directly proportional to value of the total roof convergence subsidence in the goaf, with increasing the total roof convergence subsidence from 0.25 to 0.5 and 0.75 of extracted height, the maximum values of the subsidence trough at a distance of 500 to 700 m from the face line, are increased respectively from 220 to 440 and 670 mm.

– maximum values of subsidence trough that is under full undermining by rock creeping are increased by 1.1–1.25 times, the horizontal displacement – by1.5–1.53 times.

REFERENCES

Buzilo, V.I., Koshka A.G., Serduk, V.P. etc. 2012. *The technology of selective mining of thin coal seams.* Monography: 138. Dnipropetrovs'k.

Buzilo, V.I., Sulaev, V.I., Koshka, A.G. etc. 2013. *Technology mining thin seams with stowing goaf.* Monography: 124. Dnipropetrovs'k.

Dzhvarsheishvili, A.G., Silagadze, V.A., Inashvili, A.K. & Shavgulidze, Sh.V. 1978. *Waste stowing mines.* Moscow: Niedra: 280.

Yavors'kyy, A.V., Koshka, O.G., Serduk, V.P. & Yavors'ka, O.O. 2010. *Stress-strain state of rock massif during mining flat coal seam under the protected objects.* Monography: 121. Dnipropetrovs'k.

Progressive Technologies of Coal, Coalbed Methane, and Ores Mining – Bondarenko, Kovalevs'ka & Ganushevych (eds)
© 2014 Taylor & Francis Group, London, ISBN: 978-1-138-02699-5

Research of influence of the unloading cavity on stress-strain state of the preparatory working bottom

N. Klishin, K. Sklepovich & P. Pron
Donbass State Technical University, Alchevs'k, Ukraine

ABSTRACT: Based on finite element modeling and mine measurements it was found a reduction of the bottom development working, influenced by the discharge cavity, carried by one slaughter with it. A simple tech way to ensure the sustainability of the excavation adjacent to the longwall face is created, with little material cost.

1 INTRODUCTION

There are many ways to prevent or reduce swelling bottom development working. As a rule, they have significant limitations. Some of them have poor efficiency due to the short period of their validity, due to the dangers related to the demolition work, due to the deformation of frame support during the conduct of such works (A.p. USSR #1506126... 1989). Others are costly to drill holes and install anchors, a significant increase in the laboriousness of tunnel works, limiting the functional properties of excavation due to the lower cavity width as compared with the distance between racks support (Pat. 2007577 Russian Federation... 1994). Therefore, there is a need to create a simple tech way to ensure the sustainability of the excavation adjacent to the lava.

The authors proposed a method to ensure sustainability of the excavation at swelling soil (Pat. #80703 Ukraine... 2013), according to which the excavation and a discharge cavity for excavation of swelling rocks are carried out by one slaughter and the constant arch support is set in the excavation.

The purpose of this article is to explore the influence of the discharge cavity on a stress-strain state of bottom and a value of its raise in the excavation adjacent to a stope.

2 RESULTS OF THE RESEARCH

At the mine "XIX Congress of the CPSU SE "Luganskugol" in the vent bias of layer l_1, mine studies of swelling soil's properties of the excavation were conducted: measurements of a value of bottom's raise, seismic-acoustic location for drilling holes in the soil and the air sampling of cracking (ASC).

Geological conditions of occurrence of the vent bias of layer l_1, adjacent to the first western longwall face of 700 m mine "XIX Congress of the CPSU SE "Luganskugol", are following.

Coal layer l_1 has a capacity of 1.0 m. Immediate roof of the layer is mudstone with a capacity of 5.95 m, which has the limit of the compressive strength 40 MPa. Immediate soil of the layer is siltstone with a capacity of 0.5 m and sandstone with a capacity of 2.9 m, which have the limit of the compressive strength respectively 40 and 60 MPa.

Mining conditions: width of the cource is 4.93 m, height is 3.14 m, rocks bomb is upper, length of longwall face is 252 m, method of working is pillar with reuse of development working.

The value of bottom's raise of a bias in the axial part of a bias was 780 mm in front of longwall face, at a distance of 20 m, and 950 mm at a distance of 1.4 m.

Scheme of drilling holes in the cource at a distance of 20 m in front of longwall face is shown in Figure 1.

The intensity of seismic-acoustic waves was measured during drilling hole by electric manual drill EMD-19. The transceiver of the downhole location device DLD-2M was disposed on both sides of the hole in turn at a distance of 0.5 m, designed by Institute of Mining named A.A. Skochinskiy (Nikitchenko 1973). The intensity of seismic-acoustic waves was measured by the depth of the hole. Feed force of drill is constant and equal 200 H; drilling was stopped during the transition period for the measurement on the other side of the hole. The results of seismic-acoustic location, conducted in the vent bias of layer l_1, are shown in Table 1.

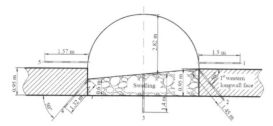

Figure 1. Scheme of drilling holes in the cource at a distance of 20 m in front of the longwall face.

Table 1. The results of measurements of the intensity of seismic-acoustic waves.

| # of hole | Length of hole (m) | Statements of DLD-2M, mkA | | |
		left of hole A_l, mkA	right of hole A_r, mkA	average A_{av}, mkA
1	1.5	400	800	600
2	1.45	500	600	550
3	1.4	400	200	300
4	1.32	300	250	275
5	1.57	200	300	250

To determine the filtration properties of the bottom in the underground conditions the method of air sampling of wells or the method of the air sampling of cracking (ASC) were used.

The ASC method is the following: the shot is divided into plots of 0.2 m and the outflow duration of specific volume of air is measured through each sealed plot of the hole (Efimenko 1977). On the basis of these measurements we calculated the permeability and the equivalent hole, which was used to evaluate the cracking and the stress state.

Permeability of bottom's rocks V_t is calculated by the formula

$$V_t = \frac{600\left(P_i - P_f\right)}{\Delta lt}, \text{Asc,} \qquad (1)$$

where P_i, P_f – initial and final pressure, MPa; t – time of the air outflow, s; Δl – length of the sealed plot of the hole, m.

Equivalent hole is obtained by calibration of the device ASC by means of air outflow through holes of different diameters in the smooth-walled tube; it is determined by the formula

$$S = 0.1V_t, \text{mm}^2. \qquad (2)$$

Measurement results of ASC are presented in Table 2.

Examination of Table 2 shows, that the value of equivalent hole, therefore the cracking, is higher in boreholes drilled near the longwall face. In the

borehole #2 at a distance of 0.5 m from its mouth, the equivalent hole is bigger more than 13 times compared with the borehole #4. This confirms the asymmetry of bottom's raise with a maximum from longwall face's side. In boreholes #1 and 5, drilled above the coal layer, values of equivalent hole are not different significantly.

For a detailed investigation of SSC it was used the numerical simulation of rock mass – the finite elements method (FEM).

For mine conditions of the vent bias of layer l_1 two dimensional FEM models are developed. In the first model, the excavation of the mining passport-section is carried out. In the second model, the discharge cavity for the entire width of the excavation in bottom is conducted by one slaughter with the excavation. Depth of the cavity is equal to the capacity of the unstable layer of bottom's rocks – 0.5 m.

Table 2. Measurement results of ASC.

# of hole	P_i, MPa	P_f, MPa	L, m	t, s	V_b Asc	S, mm^2
1	0.2	0.001	1.3	127	4.7	0.47
	0.2	0.001	0.8	61	9.8	0.98
	0.2	0.001	0.4	31	19.3	1.93
2	0.2	0.001	0.5	8	74.6	7.46
4	0.2	0.001	0.8	105	5.7	0.57
	0.2	0.001	0.6	108	5.5	0.55
	0.2	0.001	0.4	15	39.8	3.98
5	0.2	0.001	1.2	108	5.5	0.55
	0.2	0.001	1	80	7.5	0.75
	0.2	0.001	0.8	53	11.3	1.13
	0.2	0.001	0.6	63	9.5	0.95
	0.2	0.001	0.4	38	15.7	1.57

Model dimensions: length along the development working is 960 m; width is 1064 m; height is 933 m; they are taken with the influence of the stope. The model consists of 45274 iso-parametric hexagonal octagonal universal finite elements, minimum sizes of which are 0.2×0.15×0.25 m at coarees and up to 100×200×100 m at boundaries of the model. Connections are imposed on nodes of side surfaces of the model, limiting their movement in the horizontal direction. Connections are imposed on nodes of the bottom surface of the model, limiting their movement in horizontal and vertical directions (Komissarov 1983).

To estimate the stress-strain condition of rock mass the equivalent stress is calculated by the strength theory of Mor. These are simple stresses, action of which is equivalent to the stress action in the element, which is in the complex stress condition.

Figure 2 shows the distribution of equivalent stresses around the vent bias down grade without the discharge cavity (Figure 2 a, c) and with the discharge cavity (Figure 2b, d), in front of the longwall face in the distance of 19.6 and 1.4 m.

Outside the influence of longwall face (Figure 2a, b) in the first task, weak layer of siltstone with the capacity of 0.5 m is unable to perceive the load of the soil, so there are stresses of 1.14 MPa; therefore it will move to excavation's cavity without much opposition. In the underlying sandstone at a depth of up to 0.75 m across all width of the excavation, an area of increased stretching stresses is formed, reaching 3.96 MPa.

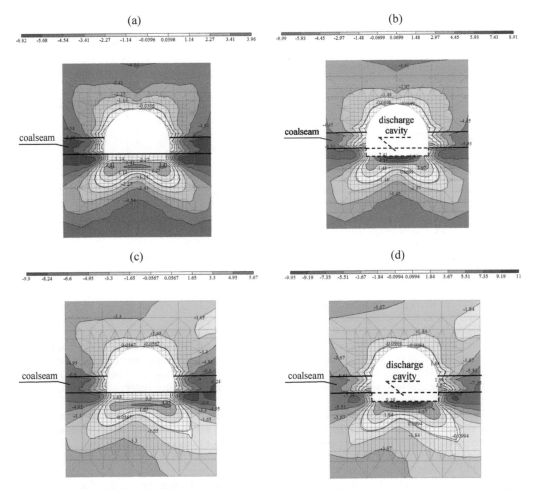

Figure 2. Equivalent stresses distribution around the vent bias dawn grade without the discharge cavity (a, c) and with the discharge cavity (b, d) at 19.6 and 1.4 m in front of the longwall face.

In the second task the swelling layer is absent. As a result, in sandstone stretching stresses are increased by 87% compared with the previous version. However, an area of increased stretching stresses with a maximum of 4.45 MPa is less than 0.35 m. Maximum stretching equivalent stresses of 8.91–7.41 MPa spread deep into the sandstone by an amount, not exceeding 0.25 m. They centered around the axis of excavation at a distance up to 1.5 m. In this area ultimate strength of sandstone is exceeded on strength, therefore it will be destroyed here. The overall picture of the distribution of equivalent stresses around the excavation with the discharge cavity and without it shows a decrease of compressive stresses around it up to 14%, which explains the decrease of swelling in front of longwall face.

At a distance of 1.4 m in front of the longwall face there is a redistribution of stresses. Equivalent stresses contour plots are shown in Figure 2 (c, d). In the siltstone layer stretching stresses reach 1.65 MPa. An area of maximum stretching stresses of 4.95 MPa is at a depth of 1.0 m in the underlying sandstone and it is shifted relative excavation's axis aside lava on 0.75 m. There is asymmetrical stress distribution around the excavation, smaller stresses focus aside massif, resulting in uneven soil raise with a maximum, shifted towards the longwall face.

In the second task in sandstone stretching stresses are bigger at 46% than the in first task. Maximum stretching equivalent stresses of 9.19–11.0 MPa spread deep into sandstone by 0.25 m. Compressive

stresses around the excavation with the discharge cavity is 0.5% bigger.

Figure 3 shows soil raise across the width of the slope, resulting by simulation at distances of 19.6 m (a) and 1.4 m (b) before longwall face.

Behind the front area of the reference pressure at a distance of 19.6 m in front of the longwall face, bottom displacements of the slope without discharge cavity are 6.8 mm, which is over 3.9 mm or 57% more than in the second task.

(a)

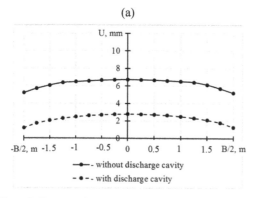

(b)

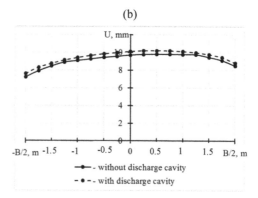

Figure 3. Bottom raise U (mm) across the width of the slope B (m) at a distance of 19.6 m (a) and 1.4 m (b) before the longwall face.

In the area of maximum reference pressure at a distance of 1.4 m in front of the longwall face, bottom displacements are equal 8.8 and 9.0 mm in the 1st and 2nd task. I.e. absolute bottom raise of the slope in the model with the discharge cavity is more over 0.2 mm or 2.3%, which is less than 10%, therefore, they are comparable to each other and the difference is within the calculation error.

At a distance of 21 m behind the longwall face, the character of bottom raise is similar to Figure 3(b). Displacements are 14.5 and 14.8 mm respectively.

Analyzing bottom displacements of the slope in nature and on models, we see that they differ by two points, due to elastic task setting in the simulation. Thus, simulation results match mine measurements with a tolerance.

Height of the excavation without a discharge cavity at distances of 19.6 and 1.4 m in front of the longwall face is 2.46 and 2.26 m respectively, and with a discharge cavity it was 3.15 and 2.64 m.

With the application of the proposed method of ensuring the sustainability of the development working, its height becomes bigger over 28% behind the reference pressure area and over 18% in it.

3 CONCLUSIONS

1. After the discharge cavity there is decrease of compressive stresses around the development working; there is growth of stretching stresses in the bottom while reducing areas of their distribution deep in the massif, which reduces the swelling of the bottom.

2. Application of the proposed method of ensuring the sustainability of the development working leads to a complete filling by rock of the discharge cavity ahead the longwall face. Residual height of the development working is at least 80% of its nameplate of value, i.e. the development working is sustainable and in good condition.

REFERENCES

A.p. USSR #1506126, Cl. E 21 D 11/00, 9/00. 1989. *Method of protection mining* / V.N. Reva, L.K. Nejman, S.I. Melnikov, A.V. Shmigol , V.Y. Kirichenko, S.M. Buchatsky / #4341765/23-03; appl. 14.12.87; publ. 09.07.89; Bull. #33.

Pat. #2007577 Russian Federation, IPC E 21 D 11/14. 1994. *Method of protecting mining in swelling bottom rocks* / K.I. Rut'kov, V.I. Fomichev. # u4885878/03; appl. 26.11.1990; publ. 15.02.1994.

Pat. #80703 Ukraine, IPC E 21 D 9/10. 2013. *Method of protecting mining in swelling soil* / M.K. Klishin, K.Z. Sklepovich, P.O. Pron'/ # u201214148; appl. 11.12.2012; publ. 10.06.2013; Bull. # 11.

Nikitchenko, R.F. 1973. *Sound sensing in mines*. Physical and mechanical properties of rocks: the scientific community. Moscow: Institute of Mining named A.A. Skochinskiy: 43-46.

Efimenko, A.A. 1977. *Measuring the permeability of rocks in the massif*. Underground coal mining, 5: 17-19.

Komissarov, S.N. 1983. *Manage of rock mass around cleaning development working*. Moscow: Nedra: 237.

Progressive Technologies of Coal, Coalbed Methane, and Ores Mining – Bondarenko, Kovalevs'ka & Ganushevych (eds)
© 2014 Taylor & Francis Group, London, ISBN: 978-1-138-02699-5

Investigation of stress-strain state of rock massif around the secondary chambers

O. Khomenko, M. Kononenko & M. Petlyovanyy
National Mining University, Dnipropetrovs'k, Ukraine

ABSTRACT: The analysis of stress-strain state of rock massif around the chambers of the second stage mining with using thermodynamic method are performed. The empirical equation dependencies of radial stresses on the distance from the chambers of the second stage mining are given. The regions of destructive deformations and their dimensions in the array filling massif chamber surrounding the second stage of mining level 940–1040 m are established.

1 THE PROBLEM AND ITS CONNECTION WITH SCIENTIFIC AND PRACTICAL TASKS

Main priority and strategic task of metal mining industry in the fierce competition in the global market of crude ore is increasing in the content of valuable component in its production. Characterization of mined ore materials quality is possible in terms of dilution. In developing ore deposits chamber mining with hardening backfill depending on the staged development of object except waste rock dilution is also backfilling material. With decreasing depth of mining operations and increase tensions of rock massif, special attention should be paid to the sustainability of rock and artificial massif that will provide a stable and efficient mining of ore deposits.

In this regard, considerable scientific interest in the study of stress state around the secondary chambers, working out which directly affects on the stability of the surrounding chamber filling massif and the amount of dilution of mined ore.

2 ANALYSIS OF RESEARCHES AND PUBLICATIONS

Questions of the study of stress-strain state of rocks chamber mining with hardening backfill were engaged Borisenko S.G. Voloschenko V.P. Kuz'menko A.M. Perepelitsa V.G. Tsarikovsky V.V. Chmarsky V.V. Chistyakov, E.P., Shvidko P.V., Usatyi V.Y., Khomenko O.E., Kononenko M.N. (Perepelitsa et al. 2006, Kaplenko & Tsarikovsky 2005, Kuz'menko, Usatyi et al. 2004, Lyashenko & Golik 2004, Kuz'menko & Petlyovanyy 2013, Kuz'menko, Ulanova et al. 2004, Russkikh et al. 2013).

The results of these researches have made a sig-

nificant contribution to improving the development of ore deposits chamber mining with hardening backfill, which consisted in the fullness of reserve recovery, improving the sustainability of massif outcrops and artificial massif, establishment of rational forms of chambers and improving the system design parameters. However, despite numerous researches, the resistance to expose an artificial massif is constantly pressing issue that requires clarification depending on the order mining treatment chambers, physical and mechanical properties of filling massif and enclosing rocks. Geomechanical processes that occur in the filling massif, when developing secondary chambers have been insufficiently studied.

3 PROBLEM STATEMENT

The purpose of research is the study carried out tension of the rock massif and backfill surrounding secondary chambers. The main objectives of analytical modeling are: a study of stress areas around the chamber and reveal patterns of change in radial stresses in a filling massif around the secondary chamber.

4 THE MAIN MATERIAL OF RESEARCH

Research of stress-strain state of rock massif secondary chambers was conducted by analytical method, which is based on the use of a synergistic approach (Knyazev & Kurdyumov 1992). Analysis of theoretical research methods allowed to choose the most appropriate method – thermodynamic (Lavrynenko & Lysak 1993), developed in Kryvyi Rih National University.

Analytical researches were conducted on the base of developed computational scheme modeling of thermodynamic processes in rock mass around the secondary chamber (Figure 1). The purpose of research is to study the stress of the rock massif and backfill surrounding secondary chamber.

Analytical researches performed for the geological conditions of the South -Belozersky deposit according to the developed calculation modeling scheme of thermodynamic processes in the rock massif around the secondary chamber (Figure 1). Rich iron ore deposits of South-Belozersky "Zaporizhzhya iron ore" (CJC "ZZRK") fulfill by chamber mining with hardening backfill. It is characterized by an altered form of chambers across the strike of the deposit. With this technology deposit develops by primary and secondary chambers. Primary chambers are situating in hanging wall, but secondary in footwall. Height of chambers 100–120 m, width 15–30 m, length 40–60 m.

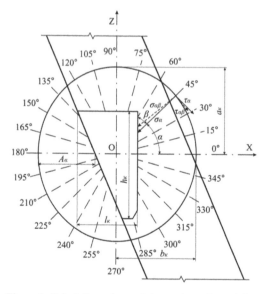

Figure 1. Calculation schem of thermodynamic modeling of the processes occurring in the vicinity of the secondary chamber: α – the angle between the ground and the calculated axis OX, hail; β – the angle between σ_a and the tangent at the intersection of calculated area with a contour chambers, hail; x – coordinate of the current position of the point on estimated area; A_a – distance from the site boundary along the unloading area to the contour of chamber; h_k – vertical span exposure massif, m; l_k – horizontal span exposure massif; O – center of the chamber, the beginning of the coordinate system XYZ; a_k – the vertical axis of unload zone, directed along the axis OZ; b_k – horizontal axis of the unload zone, directed along the axis OX; σ_a and τ_a – radial and tangential stresses in pristine massif; $\sigma_{a\beta}$ and $\tau_{a\beta}$ – radial and tangential residual potential stress.

As an example, consider the formation of the stress areas in a massif around the secondary chamber, which in turn, is the average chamber level 940–1040 m. To the modeling took the chamber with the average geological conditions (depth organize cleaning chamber $H = 1040$ m, the angle of incidence deposits $\alpha = 67°$, horizontal power fulfills deposits $m = 90$ m , the strength of the hanging-wall rock uniaxial compression 120 MPa , the strength of the rocks of the footwall uniaxial compression 90 MPa, strength backfill uniaxial compression 60 MPa) .

Massif unloading area surrounding the chamber is characterized by major areas of stress concentration. Located in the footwall rocks, massif ore and backfill, they are areas of expansion. As in the footwall rocks and filling massif area expansion partially "envelop" the chamber (Figure 2).

Main areas of stress concentration are located in the footwall rocks, massif ore and backfill. In all areas, there are tensile stresses, which vary in a linear relationship. Stress values in massif increase from border zone to the unload direction of the massif exposure chamber.

Tensile stress area is located in the footwall rocks in the center of chamber (Figure 2a). Its dimensions are 47 m. The magnitude of maximum tensile stress is equal to 4.8 MPa or about $1.2\gamma H$. Contour shape – ellipsoid. Tensile stress area is located in the footwall rocks and ore massif directly in the base of chamber (Figure 2a). Its dimensions are 25 m. The magnitude of maximum tensile stress is equal to 2.7 MPa or about $1.1\gamma H$. Contour shape – ellipsoid. Tensile stress area is located in a filling massif primary chamber (Figure 2a). Its dimensions are 46 m. The magnitude of maximum tensile stress is equal to 5 MPa or about $1.2\gamma H$. Contour shape – ellipsoid. Tensile stress area is located in filling massif at the top of the chamber (Figure 2a). Its dimensions are 31 m. The magnitude of maximum tensile stress is equal to 3.3 MPa or about $1.13\gamma H$. Contour shape – elliptic. Tensile stress area is located in the footwall rocks directly in the bottom chamber (Figure 2b). Its dimensions are 25 m. The magnitude of maximum tensile stress is equal to 2.7 MPa or about $1.1\gamma H$. Contour shape – ellipsoid, which is adjacent to the sides of the chamber. Tensile stress area is located on the sides of filling massif of secondary chambers (Figure 2b). Their dimensions are 37 m. The magnitude of the maximum tensile stress is 4.5 MPa or about $1.17\gamma H$. Contour shape – ellipsoid. Tensile stress area is located in filling massif at the top of the chamber (Figure 2b). Its dimensions are 31 m. The magnitude of maximum tensile stress is 3.3 MPa or about $1.13\gamma H$. Contour shape – elliptic.

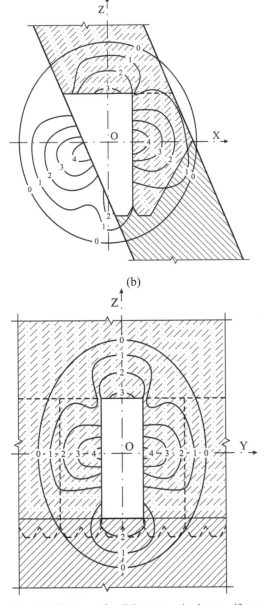

(a)

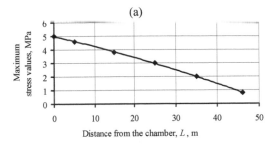

(a)

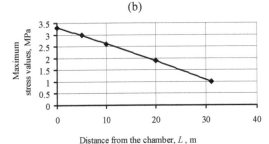

(b)

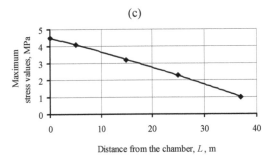

(c)

Figure 3. Maximum radial stress σ_r in filling massif of primary chambers (a), roof of chamber (b) and sides of chamber (c).

Figure 2. Contours of radial stress σ_r in the massif surrounding the secondary chamber in planes: ZOX (a) and ZOY (b), MPa.

Overall picture of stress- strain state of backfill with increasing distance from the secondary chambers can be observed by changing the maximum radial stresses (Figure 3). Analysis of tension revealed that secondary chambers are impact on filling massif in 1.1–1.5 times more than the ore and footwall rocks.

After conduction of approximation maximum values with help of program Microsoft Excel 2003 obtained empirical equations dependencies of radial stresses on distance to secondary chambers L. For a filling massif surrounding secondary chamber, empirical correlations have the form:
– the maximum tensile stress in the primary filling massif chambers

$$\sigma_r = 5 - 0.0005 \cdot L^2 - 0.069 \cdot L ; \qquad (1)$$

where L – distance from the chamber, m.
– the maximum tensile stress in filling massif at the top of secondary chamber

$$\sigma_r = 3.3 - 0.0003 \cdot L^2 - 0.065 \cdot L ; \qquad (2)$$

– the maximum tensile stress in filling massif in the sides of secondary chamber

$$\sigma_r = 4.5 - 0.0004 \cdot L^2 - 0.078 \cdot L \ . \tag{3}$$

Investigation of stress-strain state of containing chamber massif revealed patterns of development of stress fields in massif of unload zone of second stage of chambers mining; to determine changes in level of tension in core areas of unload zone of chambers on filling massif; to establish that change in tangential stress τ_r is similar to radial σ_r at smaller stresses. In addition, the quadratic dependence revealed stress changes in the filling massif with increasing distance from the chamber. Therefore, during determining of influence degree on stability of filling massif is necessary accurate account of stress-strain state distribution by chambers of second stage mining.

Physical and mechanical properties of the ore, rock and backfill, enclosing secondary chambers are presented in a wide range. Durability ores ranges from 60 to 80 MPa and the host rocks – from 80 to 140 MPa, backfill – 60 MPa. Bulk weight ores varies from 0.39 to 0.4 MN/m^3 and host rocks – from 0.21 to 0.29 MN/m^3. The above research results obtained without deformation properties of a filling backfill, the host chamber. This is due to the fact that change in the strength of massif has small effect on change of stress fields. Calculations showed that strength of rock has significant influence on deformation of containing chamber massif. Breaking strain occurs in places where the real stresses in massif of unload zone, exceed the maximum allowable tensile or shear. The difference between current and limiting stresses reflects safety factor n. As an example, considering formation of destructive strain-filling massif surrounding secondary chambers level 940–1040 m (Figure 4).

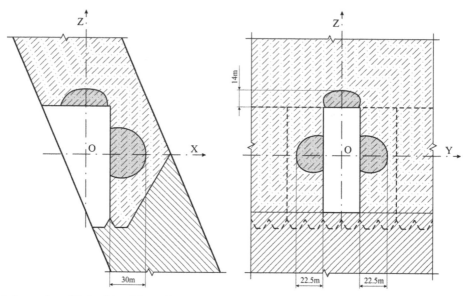

Figure 4. Areas of possible breaking deformations.

From figure clearly observes that deformation area of filling massif chamber around tabs become elongated shape resembling an ellipse and tabs are arranged in an array of the primary chamber, top and sides of the chamber. Massif of footwall rock and ore are in stable condition.

2/3 of chamber height filling massif primary chamber, deformation region extends from the side of the rock outcrop to the hanging wall at a distance of 30 m. Width in the area of deformation ranges 38–42 m. In roof of chamber filling massif overlying chamber level deformation region extends to a height of 14 m. Width of this region is 27–31 m. In sides of chamber filling massif secondary chambers along the strike of the deformation area extends to a distance of 22.5 m. Width of this region varies from 28 to 31 m. Executed researches confirm the importance of conducting filling operations in areas of deformation, which must be accompanied with an accurate account deformation filling massif around secondary chambers.

In established areas of deformation possible cracking, dumped filling massif. This is due to the fact that stability of filling massif is reduced upon exposure to tensile stress. According to the results of research, the maximum value of tensile stresses deep filling massif chamber of the first stage is 5 MPa. Strength tensile hardening backfill is 0.06 -

0.1 times the compressive strength. When exceeding the values of tensile stresses on the tensile strength tensile backfill artificial massif is destroyed. This means that in addition to the characteristic strength, taking into account the weight of overlying rocks and necessary depth to develop certain areas of chamber to fill a backfill with strength above normative as filling massif is in zone of influence of mining the secondary chamber.

Backfill massif steadiness decreases resistance during impact of tensile stresses. Researches show that maximum value of tensile stresses acting deep filling massif chamber of first stage is 5 MPa or about $1.2\gamma H$. Strength of tensile hardening backfill is $(0.06-0.1)$ σ_{szh}. During exceeding of tensile stresses values under limit of backfill artificial massif is destroying. During mining of ore reserves of 940 – 1040 m level filling chambers will be produce by composition of filling mixture of granulated blast furnace slag (20%), waste production flux (39%), rock (23%), water (18%), which provides strength to backfill with age of 6 months 8 – 9 MPa. Filling massif of composition is equivalent to the stability of rock mass $f = 5.5$ on a scale prof. Protodjakonov and is equal to compressive strength of 55–60 MPa.

Researches showed that impact of mining in area of secondary chamber, roof and filling massif chamber of the first stage working out at height of 2/3 chamber is mismatch to strength condition. Increasing of filling massif steadiness can be achieved by building layers of reinforcement in areas of its destruction and formation of different strength filling massif. The foregoing states that in construction of filling massif of primary chamber should be guided by its state of stress arising from the effects of secondary chamber mining. Scientific results will form the basis for development further recommendations for the formation of filling massif during the influence of second stage chamber mining.

5 CONCLUSIONS

1. Modeling computation scheme of thermodynamic processes in rock massif and backfill around the level of secondary chambers 940–1040 m. As result of researches, empirical equations of radial stress dependencies from the secondary chambers deep filling mass of the primary chamber.

2. It was revealed that main areas of stress concentration are located in the footwall rocks, ore and

filling massif. Contour shape – ellipsoid. Analysis of tension allowed determining that secondary chambers impact on masses in 1.1–1.5 times more than ore and on footwall rocks.

3. It was stated that formation of areas of destructive deformations happens only in surrounding backfill massif of secondary chamber. Massif of rocks and footwall ore is in stable condition.

4. It was established that on height of 2/3 of primary chamber filling massif does not match to strength and prone to fracture. From this it follows that in construction of filling massif of primary chamber it is necessary to construct strengthening layer of hardening backfill for resistance to increased tensile stresses and reaching steadiness of filling massif.

REFERENCES

Perepelitsa, V.G., Kolomiec, A.N. & Shmatovsky, L.D. 2006. *Task solution about stress- strain state of a rock during its fracturing.* Reports of NAS of Ukraine, Vol. 12: 44–51.

Kaplenko, Y.P. & Tsarikovsky, V.V. 2005. *Impact of stress state of rock massif of mine-geological conditions on parameters of outcrops and chambers form.* Development of ore deposits, Vol. 88: 21–24.

Kuz'menko, A.M., Usatyi, V.U. & Usatyi, V.V. 2004. *Investigation of the stress - strain state of inclined surface around chambers with high hardening backfill.* Geotechnical Mechanics, Vol. 49: 129–133.

Lyashenko, V.I. & Golik, V.I. 2004. *Scientific bases of state management of rock massifs of ore bodies preparation to extraction.* Sustainable development of the mining and metallurgical industry, Vol. 1: 39–44.

Kuz'menko, A.M. & Petlyovanyy, M.V. 2013. *Condition and perspectives of filling operations on underground ore mines of Ukraine.* Geotechnical mechanics, Vol. 110: 89–98.

Kuz'menko, A.M. Ulanova, N.P., Prikhodko, V.V. & Usatyi V.U. 2004. *High steadiness of secondary chambers surrounding multimodulus massif during development of iron ore deposit.* Geotechnical mechanics: 82–86.

Russkikh, V.V., Yavors'kyy, A.V., Chistyakov, E.P., & Zubko, S.A. 2013. *Study of rock geomechanical processes while mining two-level inter chamber pillars.* CRC Press/Balkema Taylor & Francis Group, London, United Kingdom: 149–152.

Knyazev, E.N. & Kurdyumov, S.P. 1992. *Synergetics as a new world outlook*: Dialogue with Prigogine. I. Problems of Philosophy, Vol. 12: 23–28.

Lavrynenko, V.F. & Lysak, V.I. 1993. *Physical processes in rock massif at imbalance.* Mining Journal, Vol 1: 1–6.

Progressive Technologies of Coal, Coalbed Methane, and Ores Mining – Bondarenko, Kovalevs'ka & Ganushevych (eds)
© 2014 Taylor & Francis Group, London, ISBN: 978-1-138-02699-5

Dilatancy at rock fracture

T. Isabek, A. Bahtybaeva, A. Imashev & N. Nemova
Karaganda State Technical University, Karaganda, Kazakhstan

A. Sudarikov
National University of mineral resources "Mine", Saint-Petersburg, Russia

ABSTRACT: The article discusses the change in volume of rocks during their destruction. It is known that the increase in the tested samples of rocks during their destruction is connected with the formation of micro-cracks in them. When solving geomechanical problems related to the assessment of stability of mine workings, it is necessary to clarify what causes the volume change. The paper presents the dilaton-phonon model of rock destruction, based on the kinetic theory of strength of solid materials.

1 INTRODUCTION

It is considered (Stavrogin & Protosyenya 1979, 1992 & Glushko 1982), that dilatancy which represents a volume increase of a tested specimen at rock fracture, is related to microcracks initiation in rock. It is important to study this phenomenon, since this is the reason of rock displacement in the outer boundary of mine workings.

Dilatancy mechanism in isotropic material of inhomogeneous structure is shown in Figure 1(b) (Stavrogin & Protosyenya 1979, 1992).

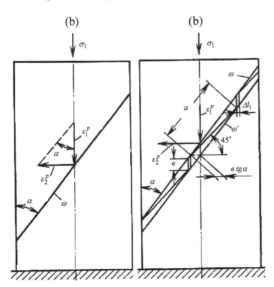

Figure 1. Rock breaking scheme: (a) with no discontinuity; (b) including disintegration in micro areas *в*.

Macroscopic shear cut ω, which is located at an angle of α to the vertical, is formed at specimen fracture in the class of stressed state $\sigma_1 > \sigma_2 = \sigma_3$.

Shear cut ω consists of a large number of benches formed of elements "a" and "$в$". The element "a" is located at 45° to the axis of the specimen, and the element "$в$" – vertical or close to vertical. Specimen fracture shifts the element "a" and fissures the element "$в$", forming microcracks of width "Δl" and "$в$". Separate step with areas of "a" and "$в$" in Figure 1(b) is called deformation or sliding element, the size of which depends on the rock structure. Dimension of "a" is equated to the average grain size.

2 RESEARCH RESULTS

The areas mutual orientation scheme of "a" slip and "$в$" fissure, as shown in Figure 1 (Stavrogin & Protosyenya 1979), depends on the type of stressed state C and is qualitatively similar to the microcrack initiation mechanism, illustrated in (Stavrogin, Protosyenya 1979). The number of "a" and "$в$" elements, involved in the deformation, and the angle α depends on the ratio $C = \sigma_2 / \sigma_1$. For instance, if C increases, the number of such elements will be increased also, so angle $\alpha \to 45°$ (Stavrogin & Protosyenya 1979).

In general, assuming that $x = в / a$, the micro sites number of slip in volume unit N of the body can be expressed as (Stavrogin & Protosyenya 1979, 1992)

$$N = \frac{1}{(a+в)^3} = \frac{1}{[a(1+x)]^3} . \qquad (1)$$

For $x = \infty$, the angle of $\alpha = 0$, and for $x = 0$, the angle of $\alpha = 0$ and macroscopic plane ω coincides with the area of the maximum shearing stresses τ. The first option ($x = \infty$) states that elements of slip "a" in the deformation process are missing, in the second – the elements of fissure "$в$". All intermediate values of x characterize the mixed deformation band and fracture involving both the elements of "a" and the elements of "$в$". Definitely, that statistical model of fracture, shown in Figure 1(b), corresponds to the values $0 < x < \infty$ and sufficiently explains the mechanism of specimen fracture volume change upwards due to the initiation of microcracks with dimensions $\Delta l \times в$. The boundary values of x correspond to the material fracture by pure slip and fissure. The given microcrack initiation mechanism does not work in these conditions.

Relative residual changes in volume θ^p, equal to (Glushko 1982)

$$\theta^p = \varepsilon_1^p + 2\varepsilon_2^p = \varepsilon_1^p\left(1 + 2\mu_p\right), \tag{2}$$

$\varepsilon_1^p \neq \varepsilon_2^p = \varepsilon_3^p$ – principle irreversible deformations;

$\mu_p = \dfrac{\varepsilon_2^p}{\varepsilon_1^p}$ – coefficient of relative permanent lateral deformations.

Using the rock fracture statistical model of part (Stavrogin & Protosyenya 1992), the following formula was obtained to determine θ^p

$$\theta^p = \left(\frac{\varepsilon_1^0}{N_0^{\sigma/M}}\right)\left\{\frac{1}{[a(1+x)]^В}\right\}^{\sigma/M} \times$$
$$\times \left(1 - 2\mu_0 e^{-GC}\right), \tag{3}$$

where ε_1^0, N_0, B, M, G – constant values, which depend on the material properties.

The Equation (3) shows, that rock structure, expressed by the parameter x, and stressed states, expressed by the ratio $C = \sigma_2/\sigma_1$, significantly influences value θ^p. If C decreases, the value θ^p will be increased. The value of the latter element increases with decreasing grain size that forms the rock.

To determine θ^p in formula (3), it is required to test a large number of specimens to find the empirical coefficients. In addition, dilatancy physical nature is not fully shown in Figure 1(b). For example, in part (Baklashov 1988) it is noted that the internal mechanism of out-of-limited rock deformation can

be also determined by the boundaries interaction of initiated macrocracks and mutual movement of fractured rocks separate parts. Hence, the change in its volumes can be caused not only by the microcracks initiation in the loaded rock, but also by the above-mentioned reasons.

Interesting conclusions can be obtained from modern ideas of dilaton-phonon model of the rock fracture strength, based on the kinetic (thermoactivation) theory of the solids strength (Zhurkov 1983 & Petrov 1983). According to this, rock disintegration is an integral result of thermoactivation processes occurring in the loaded material on micro- and mesoscopic levels, characterized by the initiation and subsequent merging of submicroscopic cracks. The process of microcracks initiation occur through the formation of critical dilatons – material density adverse fluctuations inside which provides a density of elastic energy, where the atomic bonding deformations reach the ultimate value ε_{cr} due to the athermal component of the pressure. Afterwards, this irreversible pumping of phonons from the environment begins into the dilaton area, leading to further thermal expansion of ties up to its complete failure and the formation of submicroscopic crack.

Based on the above statistics and the dilaton-phonon fracture model, it follows that the relative permanent volumetric deformations are due to the submicroscopic cracks formation in the loaded rock, which leads to relative deformations increase in the direction perpendicular to a maximal principal stress.

The Equation (2) provides that coefficient of relative permanent deformation μ_p significantly influences the value of θ^p. From the statistic model we have derived that the value of μ can be described by two Equations (Stavrogin & Protosyenya 1979):
– at phenomena as discontinuity (recording of microdiscontinuities in model with steps (Figure 1b)

$$\mu = \frac{1}{2tg\alpha}, \tag{4}$$

– for the case of slip in the plane ω without microdiscontinuities on the areas of "$в$" (Figure 1a)

$$\mu_1 = \frac{tg\alpha}{2}. \tag{5}$$

Table 1 shows the values of the relative permanent lateral deformation coefficients, derived for different rocks by an experimental approach μ_{exp}, and according to Equations (4) μ_{p1} and (5) μ_{p2} (Stavrogin & Protosyenya 1979).

Table 1. Values of relative permanent deformation coefficients.

Rock	σ_{comp}, MPa	μ_1	μ_2	μ_e	μ_e / μ_1
Marble 1	51.0	0.83	0.30	2.1	2.53
Marble 2	38.0	0.98	0.26	1.8	1.84
Talcohlorit	49.5	0.98	0.26	1.6	1.63
Diabase	116.7	1.12	0.22	2.0	1.79
Sandstone P-0	120.0	1.17	0.21	2.4	2.05
Sandstone P-01	125.5	1.22	0.20	2.4	1.97
Sandstone P-02	159.0	1.15	0.22	1.9	1.65
Sandstone P-03	143.5	1.15	0.22	3.0	2.61
Sandstone P-026	74.1	1.45	0.17	3.6	2.48
Outburst-prone sandstone	64.8	1.30	0.19	3.3	2.54
Sandstone not subjected to outbursts	72.5	1.50	0.17	3.2	2.13
Sandstone 8D	66.0	1.63	0.15	3.8	2.46
Quartz diorite D_2	125.0	1.86	0.13	1.8	0.97

Tabular analysis shows that, for all rock types, except quartz diorite D_2, the experimental values of the relative permanent lateral deformations exceed the design values μ_{p1}, determined in Equation (4), by 1.63–2.65 times. Numerical values μ_{p2}, calculated in Equation (5), are less μ_{pexp}.

The process of rock fracture consists of two stages: the preparatory and catastrophic. In the first stage, plastic deformations are implemented, preparing conditions for the critical dilaton formation and accordingly sub-microscopic cracks. The fracture is characterized by the stable accumulation and consolidation of microcracks. It is derived (Tsai 2007), that the stress deviator is responsible for the deformation processes at the plastic flow of material and the spherical stress tensor at the formation and disintegration of critical dilatons. This phase corresponds to the scheme in Figure 1(b), which characterizes the slip in the plane ω with the microdiscontinuity initiation on the areas of "ε", which can be identified with the critical dilatons, fracture of which form the microcracks of $\Delta l \times \varepsilon$ size. The final objective is to prepare the structure of the loading material to the catastrophic stage, so that the further rock fracture, occurred without the previous deformation, but due to the concurrent rotational deformation. This stage is characterized by the massive initiation of submicroscopic cracks and the intermittent growth of the relative lateral and volumetric deformations.

Indeed, the schemes shown in Figure 1 do not illustrate the changes in the coefficient μ, and thus, increase of fractured rock volume on the catastrophic fracture stage. It explains the conservative design values of μ_p in comparison to μ_{exp}. In addition, the analysis of the test conditions stated in (Stavrogin & Protosyenya 1979) and in Table 1

shows that the values μ_{exp} are derived for the supercritical (pre-limit) stage of microcracks development, i.e. the change of rock fractured volume is not taken into account in the pre-limit part of the diagram.

Big volume of experimental works on different rock deformation and fracture is shown in (Stavrogin & Protosyenya 1979, 1992). Its advantage is that tests on specimen fracture were performed for different values of $C = \sigma_2 / \sigma_1$ and time intervals of test and its disadvantage is that pre-limit deformation and strength characteristics of rocks were missing in most cases.

Figure 2 shows a plot of the maximum values of the relative permanent volumetric deformations, depending on the rock strength on the simple compression derived at different values of C (Stavrogin & Protosyenya 1979).

Maximum permanent volumetric deformations are substantially independent of σ_{comp}. Numerical values θ^p_{max} range within the limits $(1 \div 11) \cdot 10^{-3}$. The percentage increase of dilatancy in prelimi part will be $0.1 \div 1.1\%$. As the test data shows it is impossible to find any common mechanism of change θ^p from C and rock properties. Similar results were derived in (Glushko 1982), where θ^p for different rock varies from $0.13 \cdot 10^{-3}$ to $5 \cdot 10^{-3}$. For instance, if the inelastic deformation zone of 1.0 m size is formed in the mine working roof, the displacement will be 0.1–1.1 cm. These data contradict the measured values of the rock disintegration coefficient K_p, which ranges within the limits $1.04 \div 1.1$ (Glushko 1982, Chernyak & Burchakov 1984).

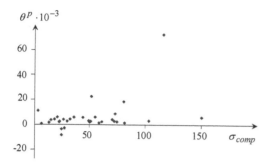

$\theta^P \cdot 10^{-3}$

Figure 2. Plot of θ^P against (σ_{comp}).

Table 2 shows the data of the rock deformation and strength characteristics at its fracture by simple compression (Stavrogin & Protosyenya 1992) (where $\theta^P, \cdot \sum \theta^P$ is volumetric deformation of disintegration on the shear strength and total accordingly and μ_{p1} and μ_{p2} are the lateral deformation coefficients in near-the-contour and out-of-limit zones accordingly).

It shows that the relative permanent volumetric deformation on the last stage of fracture exceeds the value of the same parameter at the point of maximum rock strength by 23–285 times. Tabular data show the independence of the lateral deformation coefficients from the time factor in near-the-contour μ_{p1} and out-of-limit μ_{p2} zones.

Table 2. Deformation and strength characteristics of rock.

Rock	ε_1, c^{-1}	σ_{comp}, MPa	θ^P 10^{-3}	$\sum \theta^P \cdot 10^{-3}$	μ_{p1}	μ_{p2}
Marble	$2 \cdot 10^{-6}$	74.5	0.55	55	1	2.5
	$2 \cdot 10^{-5}$	79	0.13	38	1	2.5
	$2 \cdot 10^{-4}$	81	0.6	55	1	2.5
	$2 \cdot 10^{-3}$	82.5	0.45	38	1	2.5
	$2 \cdot 10^{-2}$	90	1.25	45	1	2.5
	$2 \cdot 10^{-1}$	89	1.6	38	1	2.5
Solid sulphide ore	$1 \cdot 10^{-5}$	148	3.5	75	5	7
	$5 \cdot 10^{-2}$	173	1.5	48	5	7
	$2 \cdot 10^{-1}$	193	1.7	74	5	7
Biotitgranit	$1 \cdot 10^{-5}$	203	1.5	60	2	7
	$2 \cdot 10^{-4}$	200	0.8	105	2	7
	$5 \cdot 10^{-2}$	231	1	100	2	7
	$2 \cdot 10^{-1}$	247	1.1	70	2	7
Sandstone not subjected to outbursts	$3 \cdot 10^{-6}$	142	3	100	4	6
	$1 \cdot 10^{-1}$	156	5	120	4	6

Due to low values of θ^P in the pre-limit zone of the diagram $\sigma - \varepsilon$ compared to the same parameter in the out-of-limit zone and the above values of K_p, we can conclude that microcrack initiation is not the only reason for the rock volume increase in the mine working, which indicates the need for a detailed study of the dilatancy mechanism after the shear strength is exceeded. Disintegration coefficient after specimen fracture, as it is shown in the table, does not depend on the deformation velocity. Its average values are 1.05 for marble, 1.07 for massive sulphide ore, 1.1 for outburst prone sandstone and 1.1 for biotite granite and close to the disintegration coefficient, measured in mine workings. To identify the reason of significant change in rock dilatancy in the out-of-limit zone we consider the dynamics of volumetric deformations and fracturing process development, which is illustrated in Figure 3.

Maximum shearing stresses – τ; volumetric deformations change rate – $\dot{\varepsilon}_v = d\varepsilon_v / dt$; fracturing rate – $\dot{N}$, which is defined as the registered acoustic signals-registry time ratio are shown in Figure 3. Similar plots were derived after change of the test conditions. Analysis of plots $\dot{\varepsilon}_v = d\varepsilon_v / dt$ and $\dot{N}(t)$ shows, that compaction processes take place largely at the beginning of loading inside rock, although microfracturing processes occur simultaneously in this area. When exceeding the maximum value of τ, relative volumetric deformations change and fracturing rate intensively increases. Figure 3 shows that the intensive microfracturing process ends at maximum strength of rock, it results that a significant permanent volumetric deformation in out-of-limit zone is caused by other reasons.

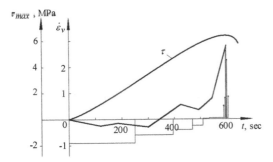

Figure 3. Plot of $\tau_{max}(t)$, $\dot{\varepsilon}_v(t)$ and $\dot{N}(t)$.

Let's consider the power consumption at rock fracture, using the stress-strain diagram (Figure 4).

The area, which is shaded and designated by the A_V^P, presents the process of dissipation, i.e. the work applied on the permanent deformations associated with the submicroscopic cracks formation and its subsequent merging into larger cracks.

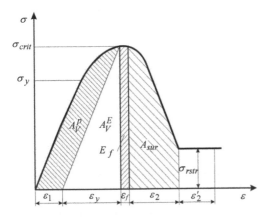

Figure 4. Stress-strain diagram.

According to the dilaton-phonon model of solids fracture (Zhurkov 1983 & Petrov 1983) these processes occur under the mechanical stress and thermal fluctuations inside the loaded material in the preparatory fracture stage. As a result of structural changes the elastic energy A_V^E is accumulated in the specimen by the time of maximum strength. In this case, the tested material is characterized by such a stress-strain state, where the critical dilatons, which are disintegrated by thermal fluctuations and led to mass initiation of the submicroscopic cracks with the subsequent merge into macro fissure, are formed on the main crack planned trajectory. It corresponds to the relative permanent deformation ε_f caused by the critical dilaton fracture under the

thermal fluctuations and the micro-cracks formation nearby σ_{crit}. Further, the costs of elastic energy A_V^E are spent on the concurrent rotational deformation and movement of the macrocracks A_{sur} adjoining surfaces. As the spherical stress tensor is responsible for fracture in the of cracks initiation area, we can conclude that cracks countersurfaces moving occur due to the elastic restitution of atomic and molecular links.

The cracks boundary layer presents the rough surface with microelevations and cavities alteration, so that any movement of the adjoining surfaces relative to each other leads to a decrease in the contact area between them and the increase of dilatancy.

If there is no contact between adjoining surfaces of cracks, than $\sigma_{rstr} = 0$, otherwise it is determined by the frictional force of one fracture surface of the other.

As provided by the above-mentioned, the costs of elastic energy A_V^E in out-of-limit area $\sigma - \varepsilon$ are calculated from the power consumption for thermal fluctuations E_f and the main cracks boundaries movement Asur. In addition, as provided by the test data (Baklashov 1988 & Olovyanny 2000), a large part of dilatancy is caused by the work. The numerical values of the permanent deformations in out-of-limits area are determined by the energy A_V^E, where its density inside the loaded rock can be defined as $a_y = 2^{-1}\varepsilon_y\sigma = 2^{-1}\sigma^2 E^{-1}$.

Let's determine the number of accumulated rock elastic energy required to transfer it to the out-of-limit state at loading in the mode $\sigma_1 > \sigma_2 = \sigma_3$. Density of elastic energy accumulated in the rock, will be equal to

$$a_y = \frac{\sigma_1^2}{2E_y} = \frac{2\tau + \sigma_2}{2E_y}. \tag{6}$$

It is known (Tsai 2007) that τ and the coefficient of the structure γ can be expressed as

$$\tau = 2kT\left[(ln\dot{\varepsilon} - \varepsilon_0) + U_0\right]\gamma^{-1}, \tag{7}$$

where k – Boltzmann's constant; T – absolute temperature of the tested body; $\dot{\varepsilon}$ – relative deformation rate; $\dot{\varepsilon}_0$ – barrierless relative deformation rate; U_0 – initial activation barrier; γ – structural parameter.

Rock structure coefficient including σ_2 is equal to (Tsai 2007, Tsai, Bondarenko & Bakhtybayev

2007, Tsai, Malakhov & Bakhtybayev 2007)

$$\gamma_\tau = \frac{\gamma_\sigma}{1.0 + A \cdot \sigma_3^{0.67}}, \tag{8}$$

where A – empirical coefficient depending on the rock strength on the simple compression σ_{comp}; γ_σ – rock structure coefficient.

Parameter γ can be determined in the empirical formula (Tsai 2007)

$$\gamma_\sigma = 3.55 \cdot 10^6 \, \sigma_{comp}^{-1.4} \; [\text{J/mole} \cdot \text{mm}^2/\text{kg}] \tag{9}$$

Presenting the right part of (7) in Equation (6) we derive

$$a_y = \frac{2kT\left[\left(ln\,\dot\varepsilon - ln\,\dot\varepsilon_0\right) + U_0\right]\gamma_\tau^{-1} + \sigma_2}{2E_y}. \tag{10}$$

The Equation (10) provides that due to increase of σ_2 at a given value of $\dot\varepsilon$, more rock has accumulated elastic energy is required to move it to the out-of-limit state. If the actual value a_y is less than the designed, determined by the formula (10), the permanent volumetric and lateral deformations will be determined only by microcrack initiation process, but not by the movement of the adjoining surfaces of macrocracks relative to each other. For the criteria the rock dilatancy for the given rock can be expressed as

$$\theta = f\left(T, \dot\varepsilon, \sigma_2, \sigma_1, \gamma_\tau\right) = f\left(T, \dot\varepsilon, c, \gamma_\tau\right). \tag{11}$$

Thus, dilatancy at rock fracture is caused by two factors: the submicroscopic cracks initiation and movement of the adjoining surfaces of macro fissures continuity relative to each other. In addition, the second mechanism of rock disintegration is dominant; its implementation is determined by the possibility of the elastic energy accumulation in the near-the-contour part of the diagram $\sigma - \varepsilon$. The disintegration coefficient may be equal to 1.001–1.005 at practical calculations when it is needed to determine the displacement of rock outer boundary except when the rock in out-of-limit. Otherwise, – $K_p = 1.04$–1.1.

3 CONCLUSIONS

1. *Temperature T.* With its decrease, durability of rocks raises, therefore at its comparison with tension isolines in the near-the-contour part of the mine working, the destruction may not be realized and if it is realized, the sizes of a zone of destruction will be smaller, therefore the shifts will be smaller as well. Decrease in the temperature of rocks leads to increase of its fragility and possibility of bigger accumulation of elastic energy at pre-limit deformation. It, in its turn, can lead to increase of coefficient of a loosening at the expense of movement of adjacent surfaces of cracks. For this reason cooling of the rock massif near excavations will be effective at decrease of dilatancy only when strength of the cooled rocks will be more than tensions occurring in the near-the-contour massif.

2. *Deformation speed $\dot\varepsilon$.* Increase leads $\dot\varepsilon$ to increase of strength of rock and the elastic energy, accumulated in the pre-limit part of the chart $\sigma - \varepsilon$. The first circumstance is favorable from the point of view of the decrease in probability of formation of zones of destruction near the development, the second leads to increase in a loosening of rocks if settlement tension is more than strength of rocks. In case the destruction zones are formed, and it depends on properties of rocks in which the development takes place, for decrease in a dilatancy at the expense of movement of walls of cracks it is necessary to reduce the speed of deformation of rocks $\dot\varepsilon$. It can be reached by installation of mine support right after an exposure of rocks. At application of supports the thorough backfill of the support area is necessary;

3. *Parameter $C = \sigma_2/\sigma_1$.* This parameter is defined by a mining depth. With removal from a development contour in depth of the massif parameter $C = \sigma_2/\sigma_1$ comes nearer to 1 and the probability of a dilatancy because of movement of adjacent surfaces of cracks decreases. The loosening of rocks can be reduced at the expense of application of a form of cross section of the developments excluding concentration of tension and low values of C.

4. *Coefficient γ_τ of structure of rocks.* This characteristic can change to a greater value at a weakening of rocks and to the smaller when hardening. In both cases there is an influence on strength of the rock. Weakening of rocks can be implemented by its preliminary moistening which leads to decrease in strength and increase in plastic properties. Decrease of strength leads to increase in zones of destruction, and increase in plastic properties to decrease in the chart of elastic energy accumulated in the pre-limit part $\sigma - \varepsilon$ and probability of a dilatancy because of movement of opposite surfaces of a crack. It is established (Tsai & Sudarikov 2007) that increase of humidity of rocks leads to decrease in durability and to increase in a zone of destruction by 1.5–3 times, then, in ultra boundary part of the chart a dilatancy

in 23–285 times more, than in the ultra boundary. Therefore, moistening of rocks close to the mine working can be an effective action for reduction of shift of rocks near the mine working. Besides, this measure increases safety of works as the dust emission decreases during breaking of rocks weight and the probability of emergence of the dynamic and gas dynamic phenomena decreases.

Near-the-contour rocks hardening (cementation, chemical hardening and so on) leads to strength limit increase of rocks and exclusion or decrease of destruction zones near the mine workings.

In case the strength of the developed rock massif is greater than tensions occurring in the near-the-contour massif, destructions will not occur and the dilatancy is excluded.

REFERENCES

Stavrogin, A.N. & Protosyenya, A.G. 1979. *Rock plasticity.* Moscow: Subsoil: 301.

Glushko, V.T. & Vinogradov, V.V. 1982. *Rock fracture and rock pressure prediction.* Moscow: Subsoil: 192.

Stavrogin, A.N. & Protosyenya, A.G. 1992. *Rock fracture and deformation mechanics.* Moscow: Subsoil: 224.

Baklashov, I.V. 1988. *Rock massive deformation and fracture.* Moscow: Subsoil: 277.

Zhurkov, S.N. 1983. *Dilaton mechanism of solid strength.* Physics of solid. t. 25, edition 10: 3119–3123.

Petrov, V.A. 1983. *Dilaton model of thermo fluctuated crack initiation.* Physics of solid, t. 26, edition 11: 3124–3127.

Tsai, B.N. 2007. *Thermoactive nature of rock strength.* Karaganda: KarGTU: 204.

Chernyak, I.L. & Burchakov, Y.I. 1984. *Control of rock pressure in deep mine roadways.* Moscow: Subsoil: 304.

Mark, C., Molinda, G.M., Schissler, A.P. & Wuest, W.J. 1994. *Evaluating Roof Control in Underground Coal Mines with the Coal Mine Roof Rating.* In: Peng, S.S., ed. Proceedings, 13th International Conference on Ground Control in Mining, Morgantown, WV, University of West Virginia: 252–260.

Gale, W.J., Fabjanczyk, M.W. & Guy, R.J. 1992. *Optimization of Reinforcement Design of Coal Mine Roadways.* In: Aziz, N.I., Peng, S., eds. Proceedings, 11th International Conference on Ground Control in Mining, Wollongong, New South Wales, Australia: University of Wollongong: 272–279.

Olovyanny, A.G. 2000. *The numerical model of deformation and fracture of rocks.* Surveying Bulletin, 2.

Scott, J.J. 1992. *Roof Bolting in Mining-USA.* Proceeding. Norsk Jord-og Fjellteknisk Forbuns, Oslo, Norway, November 27: 13.

Molinda, G.M. & Mark, C. 1994. *Coal Mine Roof Rating (CMRR): A Practical Rock Mass Rating for Coal Mines.* Pittsburgh, PA: U.S. Department of the Interior, Bureau of Mines, IC 9387: 83.

Tsai, B.N., Bondarenko, T.T. & Bakhtybayev, N.B. 2007. *The calculation of anchors with the development of conditioned zones of inelastic deformations.* Bulletin of ore, 3 (Peoples' Friendship University of Russia).

Tsai, B.N., Malakhov, A.A. & Bakhtybayev, N.B. 2007. *Justification of parameters including fixing workings of their service life.* Mining Journal Kazakhstan, 2.

Tsai, B.N. & Sudarikov, A.E. 2007. *Mechanics of underground structures.* Textbook. Karaganda. KSTU: 159.

Progressive Technologies of Coal, Coalbed Methane, and Ores Mining – Bondarenko, Kovalevs'ka & Ganushevych (eds)
© 2014 Taylor & Francis Group, London, ISBN: 978-1-138-02699-5

Ecological safety of emulsion explosives use at mining enterprises

T. Kholodenko, Ye. Ustimenko & L. Pidkamenna
State Enterprise Research-Industrial Complex "Pavlograd Chemical Plant", Ukraine

A. Pavlychenko
National Mining University, Dnipropetrovs'k, Ukraine

ABSTRACT: The features of influence of drilling and blasting operations at mining enterprises on the state of environment components have been analyzed. The integrated research of the state of natural environments in the territories adjacent to the mining enterprises using ERA emulsion explosives has been carried out. The determined high process and ecological safety of ERA emulsion explosives allows recommending them for use in surface and underground mining.

1 INTRODUCTION

The mining industry is the main source of raw materials for most sectors of the economy of Ukraine. At the same time, the exploitation of mineral deposits alters the natural groundwater dynamics, affects the atmosphere (dust, aerosol and gas pollution), land resources, as well as it negatively affects flora and fauna. The degree of such influence much depends on the method of mining and the effectiveness of environmental and resource-saving technologies applied at the enterprise (Kolesnik 2014, Yurchenko 2010, Gorova 2013, Saksin 2005).

The exploitation of most mineral deposits is carried out by means of drilling and blasting operations, for which TNT and TNT-based explosives have been used for decades. Along with emissions of toxic explosion gases, the environment is polluted with toxic components of explosives that pose a hazard to health of those people who live in mining regions (Kozlovskaya 2010 & Doludareva 2012). Furthermore, during blasting operations the ground vibrations occur, as well as shock and air waves are formed.

Large-scale blast environmental hazard during mining operations primarily depends on degree of ground level concentrations of pollutants outside the sanitary protection area of mining enterprises. Moreover, the concentrations and the range of pollutant dispersion depend on large-scale blast parameters, as well as the type of explosive used. The consumption of industrial explosives by mining enterprises increases annually in Ukraine (Yefremov 2010 & Zakharenkov 2010).

The change-over of the mining enterprises of Ukraine to the use of TNT-free environmentally-friendly explosives plays an important role in enhancing the ecological safety of mining processes. Replacement of TNT-containing explosives with TNT-free emulsion explosives allows reducing the content of harmful substances in detonation products and thus decreasing their release into the environment (Stupnik 2013 & Khomenko 2013). However, at present the ecological state of natural environments in the territories adjacent to the mining enterprises using emulsion explosives is understudied.

2 FORMULATING THE PROBLEM

Based on global experience and current trends of use of industrial explosives, Ukraine has developed blasting operations in the course of mining using emulsion explosives instead of TNT-containing and nitrate ester explosives at present. The implementation of TNT-free explosives in pits, quarries and mines was due to several economic factors, as well as severization of requirements for technical and ecological safety both during manufacture of industrial explosives an their use for blasting operations.

The development of modern technologies of drilling and blasting operations allows minimizing the degree of influence of a particular factor on the environment, and primarily due to optimization of characteristics of used explosives, initiating devices, parameters and conditions of blast, depending on geological conditions.

Considering the need for exploration and development of new deposits, especially near residential areas located in the immediate vicinity of the blast site, the appropriate reduction of seismic and shock action of explosions under current conditions is required. In addition, when carrying out blasting operations near residential areas it is required to eliminate the use of unbalanced oxygen explosives or explosives having unstable composition components that vary depending on the geological conditions of application of explosives in boreholes (jointing, water-bearing nature, chemical reaction with the environment, etc.).

Therefore, the objective of this work is the environmental and technological justification of the possibility of using emulsion explosives at mining enterprises.

3 GENERAL PART

At the present time most mining enterprises use emulsion explosives when carrying out blasting operations. The reason for this is that they do not contain materials classified as explosives and they acquire explosive properties only at the final stage of preparation. Together with environmental cleanness, emulsion explosives have a number of operational and technological advantages as compared to conventional explosives (Table 1).

As the analysis of Table 1 showed, Ammonites, Grammonites and Granulites have low water resistance, high and moderate sensitivity to mechanical and electrostatic charge impact. Furthermore, the use of TNT-containing explosives is accompanied by formation of significant amount of harmful gases. The above data touches upon the subject regarding the need to replace TNT with other explosives that are safer, environmentally friendly and cost-effective.

Production of emulsion explosives has changed the organization of blasting operations carried out at mining enterprises, and contributed to manufacture of new equipment. Such emulsion explosives as Ukrainite, Anemix, Emonite, Granemite and ERA explosives have been the most widely spread.

For emulsion explosives it is required to evaluate physico-chemical parameters, with which it has the properties predetermined for the explosive and to evaluate parameters, with which the substance of emulsion explosives becomes "inert", insensitive to initiating effects. Knowledge of these parameters allows managing processes of carrying out blasting operations ensuring the highest efficiency of their use, as well as safety in case of the need of borehole charge destruction.

Table 1. Comparative characteristics of explosives.

Characteristics	Ammonite			Grammonite		Granulite			
	A	AV	6ZhV	79/21	A	AS-4	AS-6	AS-8	M
Explosion heat, J/kg	4187	4020	4100	4100	3810	3810			3850
Volume of gaseous products of detonation, l/kg	890	870	910	880	895	940			974
Density, g/cm³	0.96			0.9		0.9			0.86
Water resistance	Low			Low		Low			
Sensitivity to mechanical impact	High			Moderate		Moderate			
Sensitivity to electrostatic charge impact	High			Moderate		Moderate			
Volume of toxic explosion gases, l/kg of explosive	40–60			68.3–76		25–27.3			
Toxicity of explosives as per GOST 12.1.007	2			2		3			

Physico-chemical characteristics of bulk and packaged emulsion explosives are shown in Tables 2 and 3.

As the analysis of Tables 2 and 3 showed, when using emulsion explosives the minimum amount of toxic pollutants is produced. Emulsion explosives have low sensitivity to mechanical impacts, can be applied in boreholes with any water content, as well as are highly effective for breaking of rocks of different hardness.

Emulsion explosives do not contain expensive and environmentally-dangerous TNT. All these results contribute to reduction of cost of blasting operations and speed up change-over of mining enterprises to widespread use of highly effective, environmentally friendly, water-resistant emulsion explosives. Use of emulsion explosives can reduce the cost of blasting operations due to improved water resistance and stability of properties of explosives, as well as their personnel safety and ecological safety.

Table 2. Physico-chemical characteristics of bulk emulsion explosives (Division 1.5).

Characteristics	Anemix	Ukrainite	ERA emulsion explosives (TU U 24.6-14310112-026:2007)		
			ERA-A	ERA-AM	ERA-Al
Density of explosive, g/cm^3	1.25–1.35	1.4–1.5	0.95-1.2	1.0–1.2	1.05–1.2
Explosive consumption rate, kg/m^3	1.09–1.18	1.13–1.2	0.9-1.0	0.86–0.94	0.9–1.0
Explosion heat, kJ/kg	3316	3100	3575	3940	3595
Volume of gaseous products of detonation, l/kg	1009	721–750	820	868	880
Volume of toxic explosion gases, l/kg of explosive	66	21.4	19.6	6.96	21.1
Volume of toxic explosion gases, l/m^3 g of weight	75.2	24.8	17.6	6.6	21.1
Detonation velocity, km/s	4.55	4.35	5.5	4.7	4.35
Critical diameter of detonation, mm	76	135	75	70	65

Table 3. Physico-chemical characteristics of packaged emulsion explosives (Division 1.1).

Characteristics	Standard value for cartridges of ERA-R emulsion explosives TU U24.6-14310112-022:2007		
	ERA-R1	ERA-R2	ERA-R3
Explosion heat, kJ/kg	3400	3800	3950
Volume of gaseous products of detonation, l/kg	925	952	890
Explosion temperature, °C	2282	2261	2908
TNT equivalent according to explosion heat	0.8	0.9	0.9
Detonation velocity, km/s, minimum	5	4.5	4.1
Impact sensitivity, J	over 50	over 50	over 50
Friction sensitivity, J	over 360	over 360	over 360
Electrostatic charge sensitivity, J, minimum	0.1	0.1	0.1
Volume of toxic explosion gases, l/kg of explosive	14.3	8.2	20.4
Critical diameter of unconfined charge detonation, mm	32	25	25
Brisance, mm, minimum	20	20	20
Toxicity	Moderately hazardous	Moderately hazardous	Moderately hazardous
Sensitivity to No.8 detonator	Sensitive	Sensitive	Sensitive

Production of Ukrainites does not need sophisticated equipment. When producing Ukrainite, only domestic products are used. The process of production and loading of Ukrainite into boreholes is carried out on the basis of already existing domestic production facilities (Krysin 2004, 2006).

The main components of ERA emulsion explosives making up a base of water-oil emulsion are: ammonium nitrate – 40–45%, calcium nitrate – 15–30%, industrial oil (diesel oil) – 4–5%, water – 8–12%. As energy additives increasing propellant effect of explosive emulsion, the mixture of ammonium nitrate and industrial oil in a stoichiometric ratio and /or aluminum powder, or products of solid propellant processing as viscous liquid slurry in the quantity of up to 10% are added to some compositions. (Shyman 2009 & Ustimenko 2009).

Currently used technologies of drilling and blasting operations on extraction of non-metallic construction materials usually involve use of non-water-resistant cheap explosives. They can include Granulites or Igdanites (ANFO) produced on site, as well as plant-produced TNT-containing explosives as Grammonite. The use of such explosives in quarries, especially in boreholes with a diameter of more than 160 mm, during charges initiation with detonating cords (DC) do not allow achieving an effective reduction of seismic impact, decreasing harmful gaseous air pollutants and increasing the efficiency of mining.

The data in Table 4 shows that a wide range of detonation velocity for composite explosives as ANFO and heavy ANFO (H-ANFO) and emulsion explosives mixtures allows changing the time of load application during blast, and thus affecting the deformation velocity of rock mass. On the one hand, this is due to compositional inhomogeneity of cheap explosives produced on site when forming charge column height. Such inhomogeneity is accompanied by formation of zones with strongly positive or strongly negative oxygen balance. On the other hand, insufficient time intervals of borehole charges initiation with DC and its destructive effect on TNT-free explosives increase the adverse impact on the environment.

It was found during comparison of blast results obtained in the areas of rock with variable rock characteristics along the bench height as far as its continuity decreases and jointing increases, that the higher is the rock breaking intensity with equal explosion heat of explosives, the lower is the velocity of detonation. The presence of air and water spaces in borehole charge may cause detonation interruption or deflagration process development when the energy of detonation of explosives is not accomplished in full. Solution to this problem can be provided by selecting the optimum performance parameters of mix-pump trucks, as well as selecting type and design of downhole explosive charge, which can be formed using a mix-pump truck. Process equipment mounted on UMS and RPS mix-pump trucks allows producing not only conventional emulsion explosives, but also so called heavy ANFO, composite emulsion explosives with emulsion content of 25 to 75%, and can produce cheap explosives as Igdanites or ANFO.

Table 4. Characteristics of composite explosives.

Characteristics	ANFO	H-ANFO 25/75	H-ANFO 50/50	Emulsion explosives mixtures 75/25	ERA-R2 emulsion explosives
Volumetric energy, MJ/dm^3	3.3	4.2–4.3	4.0–4.1	3.8–4.0	4.2–4.3
Volume of gaseous products of detonation, l/kg	974	941	908	841	921±31
Explosion temperature, °C	2870	2873	2892	2931	2585±323
Detonation velocity, km/s	2.0	<3.4	<3.8	~4.3	~4.5

In order to optimize the process parameters of the mix-pump trucks in relation to the existing non-metallic quarry conditions, the five processes of emulsion explosive mixtures production and loading into the boreholes (Figure 1) have been worked out. The first process is implemented by mixing of oxidizer and fuel phase in a stoichiometric proportion to produce emulsion explosive. The second one – by mixing of emulsion matrix and granulated oxidizer to produce emulsion explosive mixtures containing more than 50% of emulsion The third one – by mixing of oxidizer, fuel phase and energy additives in form of aluminum powder or products of processed explosives and rocket propellants generated as a result of their disposal to produce emulsion explosive mixtures containing more than 75% of emulsion. The fourth one – by mixing of granulated nitrate and industrial oil to produce granulites (ANFO). The fifth one – by mixing of emulsion explosive mixtures and granulites to produce so-called "heavy" ANFO mixtures. Such technologies allow producing explosives in a wide range of energetic characteristics with density of 0.9 to 1.25 g/cm^3, as well as granulated explosives (ANFO) and mixtures of emulsion explosives and ANFO (heavy ANFO) in a ratio from 50:50 to 25:75. Initiation of borehole charges and wiring-up of a surface circuit was performed using Prima-ERA non-electric initiation systems (NEIS) and ERA-R3 primed cartridges or T-800 (T-400H) detonator blocks.

Results of blasting operations using ERA emulsion explosives in non-metallic quarries having different geological structures and physical and chemical properties of rocks show that when blasting hard rocks and strongly coherent rocks having different water level, the required quality of rock crushing is reached.

The use of combined explosive charges (Figure 1) allows to control the energy saturation of borehole charge, eliminate spreading of emulsion explosives over fractures and simultaneously control the specific consumption through the low density of cheap explosives. When applying emulsion explosive cartridges or bags, their consumption may be also controlled through the formation of an annular space between the borehole walls and the side surface of explosive cartridges.

Analysis of blasting operations using ERA emulsion explosives allowed determining the further development of trends in mechanized production technology of ERA emulsion explosives. Herewith the major issue in the mechanized production and mechanical loading of ERA emulsion explosives is the need of purposeful activity in relation to optimizing of formation of the explosive borehole charge column.

In case of large-scale underground mining operations the drifting works are to be carried out. The main method used during drifting is the rock breakage using explosive blasthole charges. In order to form blasthole charges in nongaseous-and-dusty mines and quarries, the Class 2 explosives are used.

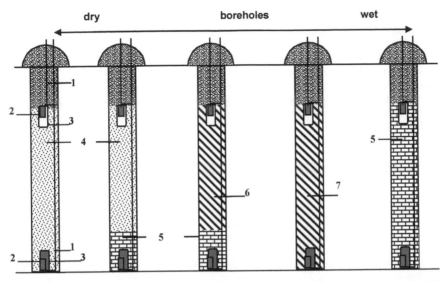

Figure 1. Design of borehole charges in boreholes with different water level: 1 – NEIS shock tubes; 2 – NEIS blasting cap; 3 – Primed cartridges (of ERA – R3 or T – 800 type); 4 – Granulite (ANFO); 5 – Emulsion explosives (mixtures) (emulsion > 75%); 6 – H-ANFO (emulsion<50%); 7 – emulsion explosive mixtures (emulsion > 50%).

Among these explosives the ammonites and detonits in form of screw-filled cartridges or bulk cartridges, and also bulk grammonites and granulites, which are loaded by pneumatic conveying, have obtained a wide-spread application in formation of blasthole charges. Lately, with the development of slurry explosive technologies, the forming of blasthole charges also uses the water-gel and water-emulsion explosives both in cartridges and pumped liquid (bulk) compositions (Kirichenko 2012).

Experience in application of packaged emulsion explosives as blasthole charges during underground workings development showed that with increasing density of blasthole distribution in the mine opening the effectiveness of blasting operations decreases due to reduction of energy output of the emulsion explosives explosion through the detonation attenuation and its transition to deflagration. This results in increasing the amount of harmful products of detonation when exploding. Derating of blasthole charges function may be caused by higher density of emulsion explosives due to compression by the elastic wave, propagating from the charge explosion of previous delay series. Efficiency of emulsion explosive blasthole charges, the density of which is controlled through the chemical gasification, when the distances between the blastholes are fixed, increase with increasing the time intervals of initiation delay between groups of charges (Kirichenko 2012).

Given the fact that many underground mines and quarries use traditional technologies and explosives

to form blasthole charges, the changeover to the use of emulsion explosives requires changing conditions of formation and initiation of blasthole charges to ensure effectiveness of their use and reduction of the amount of harmful gaseous explosion products.

To assess the impact of the explosion products (when using ERA emulsion explosives in quarries) on the ecological state of surrounding area the comprehensive studies on levels of environmental contamination were carried out. Sampling and relevant measurements were performed for different blasting conditions of ERA emulsion explosives in 19 regions of Ukraine, including usage in the quarries with different mining and geological conditions and blasting conditions. Here, the initiation of borehole charges and wiring up of surface circuit are carried out using Prima-ERA non-electric initiation system.

The results of the surveillance studies in the period from 2006 to 2014 showed that no hydrogen chloride and organochlorine contaminants are present at the explosion sites in the atmospheric environment when blasting ERA emulsion explosives. There were no ammonium perchlorate and organochlorine contaminants in soil and groundwater either, and the content of nitrate ions and chloride ions remained at background concentration levels formed in some or other quarry and was below maximum allowable concentrations. The concentrations of controlled pollutants were virtually unchanged after a major blast and met the current health-based exposure limits.

The study results have found out that the use of ERA emulsion explosives in mines does not change the state of natural environments both on the border of the sanitary protection zone and the residential area. Application of ERA emulsion explosives allows reducing the amount of harmful substances released into the environment during major blasting and thereby improving the living conditions of the population in the mining areas.

4 CONCLUSIONS

Ukrainian best practices in production of blasting operations in mining enterprises using emulsion explosives show that the relatively low cost and high efficiency of such explosives allow arranging their application instead of TNT-containing explosives for different mining conditions. At the same time, their use allows to achieve a significant improvement in parameters of process and ecological safety of blasting operations and reduce the output of overground and oversized fractions of extracted non-metallic minerals.

Explosion products resulting from the use of ERA emulsion explosives in the quarries do not cause the deterioration in quality of the natural environments both within quarries and beyond.

Application of emulsion explosives enables to eliminate some disadvantages that are typical of TNT-containing explosives such as high costs and a high risks during production and use as well as ecological hazard (toxicity). The explosion of emulsion explosives releases 10 to 40 times less toxic gases than that of TNT-containing explosives; that ensures reduction of gas clearing time in quarries, pits and mines, as well as improving sanitary-hygiene environment and ecological situation in mining regions.

REFERENCES

Kolesnik, V.Ye, Yurchenko, A.A., Litvinenko, A.A. & Pavlychenko, A.V. 2014. *Ways and means to enhance the environmental safety of massive explosions in quarries for iron dust factor.* Dnipropetrovs'k: Litograf: 112.

Yurchenko, A.A. 2010. *Physical processes in the emission of dust and gas clouds in the mass explosions in quarries.* Scientific Bulletin of National Mining University, 2: 85–88.

Gorova, A., Pavlychenko, A., Borysovs'ka, O. & Krups'ka, L. 2013. *The development of methodology for assessment of environmental risk degree in mining regions.* Annual Scientific-Technical Collection – Mining of Mineral Deposits. Leiden, The Netherlands: CRC Press / Balkema: 207–209.

Saksin, B.G. & Krupskaya, L.T. 2005. *Regional estimation of the effect of mining production on the environment.* Mining Journal, 2: 82–86.

Kozlovskaya, T.F. & Chebenko, B.N. *Ways to reduce the level of environmental danger in regions of the open pit mining.* Transactions of Kremenchuk Mykhailo Ostrohradskyi National University, 6(65). Ch. 1: 163–168.

Doludareva, Ya.S., Kozlovskaya, T.F., Lemizhanskaya, V.D. & Komir, A.I. 2012. *Influence of the surface-active substances implementation in the rock failure area on the intensity of rock crushing by means of the pulse loads.* Scientific Bulletin of National Mining University, 4: 93–97.

Yefremov, E.I. 2010. *Experience of use of simple explosives in open-cast mines of Ukraine.* Ukrainian Union of Mining Engineers. Information Bulletin, 4: 9–11.

Zakharenkov, Ye.I. 2010. *State of blasting in Ukraine. State supervision in the treatment of industrial explosives.* Ukrainian Union of Mining Engineers. Information Bulletin, 4: 3–8.

Stupnik, N.I., Kalinichenko, V.A., Fedko, M.B. & Mirchenko, Ye.G. 2013. *Prospects of application of TNT-free explosives in ore deposites developed by underground mining.* Scientific Bulletin of National Mining University, 1: 44–48.

Khomenko, O., Kononenko, M. & Myronova, I. 2013. *Blasting works technology to decrease an emission of harmful matters into the mine atmosphere.* Annual Scientific-Technical Collection – Mining of Mineral Deposits. Leiden, The Netherlands: CRC Press / Balkema: 231–235.

Krysin, R.S., Ishchenko, N.I., Klimenko, V.A., Piven', V.A. & Kuprin, V.P. 2004. *Explosive ukranit-PM-1: Equipment and fabrication technology,* Mining Journal, 8: 32–37.

Krysin, R.S. & Novinskij, V.V. 2006. *Model of the explosive rock crushing: monograph.* Dnipropetrovs'k, Art-Press: 144.

Shyman, L. & Ustimenko, Y. 2009. *Disposal and destruction processes of ammunition, missiles and explosives, which constitute danger when storing,* NATO Security through Science Series C: Environmental Security: 147–152.

Ustimenko, Ye.B., Shyman, L.N. & Kirichenko, A.L. 2009. *Peculiarities of water-emulsion explosive features for safe use during blasting.* Transactions of Kremenchuk Mykhailo Ostrohradskyi National University, 2(55), Part. 1: 86–89.

Kirichenko, A.L., Ustimenko, Ye.B., Shyman, L.N. & Politov, V.V. 2012. *Study of detonation characteristics of blast-hole charges of packaged emulsion explosives.* Scientific Bulletin of National Mining University, 6: 37–41.

Kirichenko, A.L., Ustimenko, Ye.B., Shyman, L.N., Podkamennaya, L.I. & Politov, V.V. 2012. *Method optimization of loading and initiation of blast-hole charges of packaged emulsion "ERA" explosives during the drifting operations in coal-bearing massifs.* Transactions of Kremenchuk Mykhailo Ostrohradskyi National University, 2(73), Part 2: 84–87.

Ways of soils detoxication that are contaminated by heavy metals using nature sorbents

O. Demenko & O. Borysovs'ka
National Mining University, Dnipropetrovs'k, Ukraine

ABSTRACT: The problem of anthropogenic soil contamination with heavy metals is considered. An overview of soil detoxication methods is done, based on the use of natural mineral sorbents, their advantages and disadvantages are discussed. Promising directions for further researches in this area are marked.

1 INTRODUCTION

Ecological situation on the territory of industrial agglomerations requires the implementation of priority measures aimed at ensuring the minimization and prevention of heavy metal pollution of the environmental objects. Soil pollution occupies a special place; the levels of this type of contamination in some cases reach dangerous values. In the surroundings of large industrial enterprises that are emitting large amounts of harmful substances into the atmosphere ecologically unfavorable areas with heavily contaminated soils are formed. Soil being peculiar biochemical filter is capable to retain heavy metals and inactivate them for a long time. However, with the increasing pollution of environment the protective capacity of the soil are exhausted and heavy metals come in excessive amounts into natural waters, plants and further along the trophic chain into animals and humans (Ubugunov 2004; Gorova et al. 2013; Pavlichenko & Kroik 2013).

Many heavy metals possess high affinity for the physiologically important organic compounds (eg, ferments) and are capable to inactivate them. Excess input of heavy metals in living organisms interferes with the metabolism processes; slows down growth and development. In agriculture, it is reflected in the quality of products and in reducing of its output. In vegetable and fodder crops accumulation of heavy metals often reaches dangerous level for humans and animals without any significant external manifestations. Heavy metals that entered into the human body are removed very slowly, and even small their receipts with food can cause a cumulative effect.

Besides human and animal health damage is done to the ecosystem of whole region. Heavy metals that are accumulated in the soil in large quantities are able to change many of its properties. First of all, the biological properties of the soil change: total

number of microorganisms reduces, their species composition shrinks, intensity of basic microbiological processes falls and the activity of soil enzymes falls too, etc. Severe pollution of heavy metals leads to a change of even more conservative soil features such as humus status, structure, *pH*, etc. It results in partial, and in some cases, in complete loss of soil fertility (Valkov 2004).

2 FORMULATING THE PROBLEM

In connection with the aforesaid, the issues of reducing the negative influence of heavy metals on groundwater ecosystems are very relevant today. Detoxification of contaminated soils and territories is one of effective ways of solving this problem. Detoxification means the set of techniques and methods, leading to a weakening or complete exemption from the toxic effect of heavy metals, as well as the creation of favorable conditions for soil self-purification.

3 GENERAL PART

Measures of detoxification of technologically-contaminated soils are represented by the following methods:

– *physical* – removal of contaminated layer of soil and its disposal;

– *chemical* – inactivation or reduction of the toxic effect of pollutants by ion exchange resins and organic substances that form chelates; liming, addition of organic materials and fertilizers, which adsorb pollutants or reduce their receipts to the plants; application of mineral substances and fertilizers;

– *biological* – growing crops that are tolerant to pollution and capable to take out toxic substances from soils (Kireycheva & Glazunova 1995;

Makarenko & Vakal 2011).

Review of the literature and patent sources on the use of effective methods of technogenic contaminated soils detoxification showed that methods based on the use of natural materials with sorption properties currently especially distinguished.

The increasing application of aluminosilicate mineral adsorbents is found in the world; these minerals are characterized by high absorbency, resistance to environmental influences and can serve as excellent carriers for fixing on the surface of various compounds. Zeolites, glauconites and various clay minerals are attracted particular attention of scientists (Volodin & Chelishchev 1993; Baydina 1995; Vezentsev et al. 2008; Borisovskaya & Fedotov 2011). Their widespread occurrence in nature, low cost and simple technology of application along with sufficiently high sorption properties make use of these natural materials as adsorbents for decontamination of soils very promising.

The results summary of a number of existing studies on the possible use of the above-mentioned natural materials for soils detoxification contaminated with heavy metals is given below.

Zeolites – natural hydroaluminosilicates with frame structure, which includes cavities and channels of molecular size, occupied by mobile cations and water molecules.

Photography of the zeolite surface with a magnification of 10.000 times shows the presence in its structure of input "windows" of pores and channels (Figure 1).

These minerals act on the principle of molecular sieves, separating a mixture of substances, depending on the size of the atoms and molecules. Structural features of the zeolites determine the participation of only cations in ion-exchange process, mainly cations of heavy metals. Na-type zeolite has the highest capacity. The equilibrium exchange capacity of such a typical zeolite as clinoptilolite is (mg-eq/100 g): lead – 96–196, cadmium – 125, mercury – 237, copper – 95–107, zinc – 109, cobalt – 44, nickel – 17 (Sizov et al. 1990).

In a comparative study of lead adsorption on the clinoptilolite and alfisol in the *pH* range from 3 to 5 is proved that the zeolite absorbs 20–30 times more lead than soil minerals. The outcome of the study was the recommendation to use clinoptilolite for remediation of soils contaminated with lead with a wide range of *pH*-factor. The experiment on the use of Iranian clinoptilolite on contaminated soils of Gilan province (North of Iran) showed that zeolite introduction stabilized content of *Cd* at the maximum permissible concentration. Moreover, adsorption increased with increasing doses of zeolite. Also a pore volume of the zeolite has a value. Thus, the

zeolite introduction with pore volume of 15 and 75% reduced *Cd* content by 12 and 35%, respectively (Bodnya 2008).

Figure 1. Photography of zeolite surface with magnification (GC "Zeolite technology", 2010).

Application of clinoptilolite to the soil in the dose of 15 t/ha increases the absorption capacity of the soil by 15–20%, the aftereffect lasts from 7 to 20 years (Grigora 1985).

A number of special studies of the zeolites effectiveness in specific areas and soils show the selective ability of this sorbent in relation to certain heavy metals. For example, the introduction of zeolites in the amount of 100 t/ha in the technogenic pollution chernozems of Belovo town (Kemerovo region) significantly reduced the toxicity of zinc and lead, and the content of mobile cadmium remained steady (Baydina 1994). Use of zeolites on model contaminated by heavy metals sod-podzolic soil in the dose of 300 t/ha significantly reduced the content of mobile zinc in the soil, whereas the amount of cadmium and lead remained almost unchanged (Mineev et al. 1989). Contradictions associated with detoxification of lead by zeolite may be due to the difference of soil and climatic conditions, specificity of soil contamination by heavy metals and physic and chemical features of zeolites of different deposits. When studying the immobilization of zinc ions using double-natural zeolites (Gordes and Bigadiç area, Western Turkey) proved that that at *pH* <4 zeolites was ineffective because their crystal structure broke down. Ion-exchange is a basic mechanism in the range of pH from 4 to 6. In general, it is preferable to use zeolite Gordes of the two zeolites, as it adsorbs zinc ions twice as efficient (Bodnya 2008).

Because the development of techniques to detoxify soils contaminated with heavy metals is a necessary activity when growing crops, it is important to take into account their reaction to sorbents introduction in the soils.

Thus, there is evidence that, although the zeolite introduction does not reduce the mobile lead content under the use on sod-podzolic soils, however, it reduces almost twice and over the metal accumulation of barley phytomass (Obukhov & Plekhanova 1995).

When studying the effect of the use of zeolite, compost and hydrated lime on content of *Pb, Cd, Zn* in soils of southwestern Italy and assessing the impact of toxic heavy metals on the growth and accumulation in seedlings of Lupinusalbus L. in terms of vegetative test, zeolite was the most effective in cadmium ions, and also contributed to the increase of the roots and the aboveground biomass of white lupine. (Castaldi 2005).

Inactivating effect of zeolite tuffs containing mordenite on the mobility of heavy metals in soils of the Transbaikalia and on the level of their receipts into potato plants and annual grasses is set. The Mukhor-Tala deposit of zeolite tuffs is located in the distance of 100 km from the city of Ulan-Ude (Ubugunov and others 2000; Doroshkevich et al. 2002).

In vegetation experiments with artificially contaminated soils by heavy metals in the comparative study of clinoptilolite, hydroxyapatite and mixed iron oxide is shown that zeolite dose 42 kg / ha, in most cases, reduces the absorption of *Cd* and *Pb* by corn and barley more efficiently than iron oxide and hydroxyapatite. Using zeolite tuffs of Metaksades region (Thrace, Greece) in relation to soils contaminated with mercury, was reflected in reducing its accumulation in seedlings and roots of Lucerne (Bodnja 2008).

Glauconites – hydrate aluminum silicates that contain potassium, minerals from group of hydromicas from subclass of phyllosilicates of intermittent and complex composition. Glauconites are characterized by high ion-exchange capacity (up to 15–20 mg-eq/100 g per 100 g of rock) and specific surface area (up to 100–115 m/g), and as a consequence – a very large absorption capacity.

At the electron microscopic photos of glauconite grain surface with a magnification of 10.000 and 20.000 times the typical for glauconites nanostruc-

ture is visible with randomly distributed illite-smectite flakes, size 1–2 microns, having dissected and scalloped edges (Figure 2).

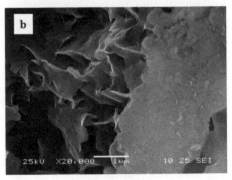

Figure 2. Electron microscopic image of the surface of glauconite grains with magnification of 10.000 (a) and 20000 (b) (Afanasjeva et al. 2013).

The main difference between glauconite and zeolites is that it has not carcass but layered structure. Thus the part of the intermolecular forces is not balanced by interaction of ions of the chemical elements located in one cavity of such layer. These forces can interact with the ions of chemical substances contained in the solutions or in the air as a result they accumulate on the surfaces of active plates that constituent common crystal. Active surface area increases significantly and this is the main difference between glauconite and zeolite (Kurbani-yazov & Abdimutalip 2012).

Glauconite sands attract attention, primarily by low cost of a product, high potassium content (9%), their ability to release potassium as readily digestible compounds for a long time, ability to absorb nuclides, heavy metals and toxicants.

Modeling of anthropogenic soil contamination has shown high efficiency of glauconite use in combination with bird droppings and phosphogypsum in detoxification of soil from heavy metals (Yurkova 2012). So, at application of a compost consisting of bird droppings and glauconite (1:1) in the first year of the experiment, the Cu content decreased by 28%, Ni – 26%, Pb – 21,5%, Zn – 35%. The concentration of Cd in the soil has not changed. In the third year of research the content of Zn decreased by 72%, Ni – 58%, Pb – 51%, Cd – 30%, Cu – 55%.

With a combination of poultry manure, phosphogypsum and glauconite in the ratio 1:1:1 absorption of heavy metals in the third year after compost application ranged from 49.0 to 82.0%, including Ni and Zn – 69.0 and 82.0% respectively. Proposed fertilizing and ameliorating composts besides reducing the negative impact of pollutants on soils, contribute to the improvement of state of irrigated chernozems and restore their fertility.

Clay – a fine-grained sedimentary rock that composed of one or more minerals of kaolinite, montmorillonite or other layered aluminosilicates (clay minerals), but may contain sand and carbonate particles.

Application of clays containing minerals with expanding crystal lattice (montmorillonite, illite, vermiculite) allows increasing considerably cation exchange capacity of soils. The bond strength of heavy metals with clay minerals depends on their structure and increases from kaolinite to montmorillonite. The bond strength also depends on the pH and on the content of organic substances. Is noted that soil clay minerals adsorb heavy metals the stronger the more organic matter are contained in the soil (Sizov et al. 1990). And properties of metals themselves are important. It iss established that the strength of fixation of clay minerals decreases in the next order: $Pb^{2+} > Zn^{2+} > Cd^{2+}$ (Titova 2005).

The possibility of clay use for the adsorption of heavy metals in soils in the example of the bentonite clay use of Kudrinsk deposit of Ukraine is shown (Ablaeva & Borisovskaya 2011).

To investigate the sorption capacity of clays on the territory experimental studies on the use of bentonite suspensions of different concentrations (30, 20 and 10%) were conducted for remediation of territory of Sevastopol department of JSC "Krymvtormet" that contaminated with heavy metals which contents exceeded the maximum permissible concentration in 2–83 times.

Stages of the study consisted of the application of suspensions on a certain area, keeping them up to dry, independently peeling (1–2 days) and collecting the dried layer. Solution consumption was 2.0 m^3 per 1 m^2. The results showed general regularity: the highest amount of contaminated particles is captured by 10% suspension of bentonite (Table 1).

Table 1. Heavy metal content in the bentonite clay before and after remediation of the territory of JSC "Krymvtormet", mg/kg (Ablaeva & Borisovskaya 2011).

Sample	Zn	Cd	Pb	Cu	Ni	Co	Cr
Soil	1500	15	2500	320	–	–	–
Sector #1							
10% of clay	1200	12	320	250	150	10	100
20% of clay	1000	12	200	220	80	8	80
30% of clay	900	8	140	150	40	8	40
Sector #2							
10% of clay	630	10	1000	200	63	12	150
20% of clay	600	8	800	80	40	10	120
30% of clay	500	8	630	120	32	6	100
Sector #3							
10% of clay	1000	10	1500	250	120	20	120
20% of clay	800	6.3	1000	180	80	15	100
30% of clay	450	5	700	50	60	8	100

Note. Sector #1 – slide-ways territory; sector #2 – area of non-ferrous metals cutting; sector #3 – department of cutting with installation of hydraulic shears.

It is established that a bentonite clay provides sorption of metals: zinc up to 80%, copper – 78%, cadmium – 80%, lead – 50%. Under the conducted investigations conclusions are drawn about the feasibility of using bentonite clay for urbanized areas.

Authors of the "Chistozem" preparation which is a mixture of such clay minerals as montmorillonite and palygorskite, reported that this preparation effectively absorbs all heavy metals under the joint presence (Chernyakhovskiy 2003). Under the joint presence of equal concentrations of Zn, Cu, Cd, Pb and Ni in a solution the greatest fraction of elements absorbed by the preparation was the fraction of Zn and Cu (25 and 27%, respectively).

The method of agricultural land purification consists in adding of montmorillonite and palygorskite in the amount more than 2 tons per 1 ha of land, followed by stirring. Montmorillonite is primarily used in those cases where the problem of remediation of highly contaminated soils is solved. Simultaneously, the improvement of water-physical and agronomic properties of soil is observed, nutrient reserves are increasing, the mass of symbiotic apparatus of nitrogen fixers is growing. Palygorskite slightly weaker absorbs heavy metals, but its effect as meliorant on light and depleted soils, is much stronger due to the high cation exchange capacity (250 mmol-ekv/100 g).

Calculations based on the results of laboratory studies have shown that the application of 2–3 tons of the "Chistozem" preparation on 1 hectare of arable land reduces the content of heavy metals of first danger class from average to low level. Analysis of the total harvest of four vegetation experiments and analysis of Cd content in soil showed that a dose of 3 t/ ha is quite sufficient to reduce the possibility of heavy metals intake into plants in 2 times.

The use of montmorillonite and palygorskite at application rates of 2–3 t / ha also increases the harvest by 15–20%. But the special value of these improvers lies in their ability to increase the aftereffect of organic and mineral fertilizers for 5–7 years, that significantly reduces the cost of agricultural production and allows obtaining environmentally friendly products. Unlike the other sorbents used in agroecological purposes, in particular zeolites or zeolite mixtures, proposed meliorants are stable in soil. It is connected with the structure of the crystal lattice and is confirmed by the presence of long-term existing (geologic time) natural smectite (montmorillonite) and palygorskite weathering crusts and soil formations.

4 CONCLUSIONS

Summarizing a brief review of natural sorbents application for detoxification of contaminated soils, it may be noted that their use is advantageous because they are relatively affordable raw materials, have a high degree of sorption, stay active for a long time, prolong action of fertilizers, protecting the soluble components from washing out of the soil, reduce the content of toxic elements in the vegetation.

However, despite all the advantages described above, it should be noted that considered natural mineral sorbents effectively adsorb the elements belonging to the group of heavy metals which are toxic to plants and at the same time are microcomponents of mineral nutrition (e.g., Zn, Cu). Therefore deterioration of conditions of plant nutrition with biogenic microelements is possible.

Thus, further studies are required to develop ways to detoxify soils that contaminated with heavy metals, based on the use of natural sorbents.

REFERENCES

Ubugunov, V.L. 2004. *Detoxication of soils of Baikal region that are contaminated by heavy metals*. Advances in current natural sciences, 2: 129–129.

Gorova, A., Pavlychenko, A. & Borysovs'ka, O. 2013. *The study of ecological state of waste disposal areas of energy and mining companies*. Annual Scientific-Technical Colletion - Mining of Mineral Deposits. The Netherlands: CRC Press / Balkema: 169–171.

Pavlichenko, A.V. & Kroik, A.A. 2013. *Geochemical assessment of the role of aeration zone rocks in pollution of ground waters by heavy metals*. Scientific Bulletin of National Mining University, 5: 93–99.

Valkov, V.F., Kazeev, K.S. & Kolesnikov, S.I. 2004. *Ecology of soils*. Part 3. Soil pollution. Rostov-on-Don, Russia.

Kireycheva, L.V. & Glazunova, I.V. 1995. *Detoxification methods of soils contaminated with heavy metals*. Eurasian Soil Science. Issue 7: 892–896.

Makarenko, N.O. & Vakal, S.V. 2011. *Experimental methods for the research of environmentally friendly mineral fertilizers efficiency*. Transactions of Kremenchuk Mykhailo Ostrohradskyi National University. Issue 6, Part 1: 142–144.

Volodin, V.F. & Chelishchev, N.F. 1989. *Zeolites are new types of mineralraw materials*. Moscow: Nedra.

Baydina, N.L. 1991. *About the use of zeolites as scavengers of heavy metals in technogenically contaminated soil*. Siberian Biological Journal, 6: 32–37.

Vesentsev, A.I., Troubitsin, M.A., Goldovskaya-Peristaya, L.F. & Volovicheva, N.A. 2008. *Sorption purification of soils from heavy metals*. Scientific statements of Belgorod State University, 3: 172–175.

Borisovskaya, Ye.A. and Fedotov, V.V. 2011. *The use of natural and synthetic materials as component of artificial soil substratums (review)*. Scientific Bulletin of National Mining University, 1: 84–88.

The site of GC "Zeolite technology". 2010. *Natural Zeolites - Sokirnity. Experience inusein the purification of water and waste water, available at:* www.zeomix.ru/aboutproject/usefularticles/67-experience-in-use-in-the-purificationof-water-and-wastewater.html (accessed April 25, 2014).

Sizov, A.P., Homyakov, D.M. & Homyakov, P.M. 1990. *Problems of pollution of soil and crop products*. Moscow: MGU.

Bodnya, M.S. 2008. *Application of zeolite-containing minerals for remediation of polluted soils. Problems of contemporary science and practice*. Vernadsky university, Vol. 2, #1: 142–149.

Grigora, T.I. 1985. *Action and aftereffect of zeolite-clinoptillolite on fertility of sod-podzolic soil*. Agronomy, 60: 31–35.

Mineev, V.G., Kochetavkin, A.B. & Nguyen, V.B. 1989. *Use of natural zeolites to prevent contamination of soil and plants by heavy metals*, Agrochemistry, 8: 89–95.

Obukhov, A.I. & Plekhanova, I.O. 1995. Detoxification sod-podzolic soils contaminated with heavy metals: theoretical and practical aspects. Agrochemistry, 2: 108–116.

Castaldi, P & Santona, P. 2005. *Heavy metal immobilization by chemical amendments in apolluted soil and influence on white lupin growth.* Chemosphere, Vol. 60, Issue. 3: 365–371.

Afanasjeva, N. I., Zorina, S. O., Gubaidullina, A. M., Naumkina, N. I. & Suchkova, G.G. 2013. *Crystal chemistry and genesis of glauconite from "Melovatka" section (Cenomanian, of South-Eastern Russian plate).* Litosfera, 2: 65–75.

Kurbaniyazov, S.K. & Abdimutalip, N.A. 2012. *Wide range of applications of glauconites and their role in modern society.* Scientific & practical journal "Researches in Science", 5 (available at: http://science.snauka.ru/2012/05/359 (accessed April 25, 2014).

Yurkova, R.E. 2012. *Methods for inactivation of heavy metals and soil fertility restoration of irrigated lands.* Scientific journal of the Russian Scientific Research Institute for Land Reclamation, 1: 1–12.

Titova, V.I., Dabakhov, M.V., Dabakhova, E.V. & Krasnov, D.G. 2005. *Recommendations for environmental assessment and measures to reduce soil pollution and contamination of environmental components adjacent to farmlands.* NGSKHA, Izd VVAGS, N. Novgorod, Russia.

Ablaeva, L.A. & Borisovskaya, E.A. 2011. *Perspective directions of natural clays for urban areas cleaning.* Reports of the National Academy of Sciences of Ukraine, 3, 187–192.

Chernyakhovskiy, D.A. 2003. *"Clay farming" – the future of clay sorbents in agriculture.* Agroecologycal journal, 4: 34–40.

Prognosis for free methane traps of structural and tectonic type in Donbas

K. Bezruchko, O. Prykhodchenko
M.S. Polyakov Institute of Geotechnical Mechanics, Dnipropetrovs'k, Ukraine

L. Tokar
National Mining University, Dnipropetrovs'k, Ukraine

ABSTRACT: The paper concerns prognosis evaluation mine fields and geological prospecting sites prospectivity for free methane output in traps of structural and tectonic type (local anticlinal structures) of coalbearing strata in terms of "Butovskaia" mine and "Kalmiuskii Rudnik" site in Donbas. Such traps form as a reservoir in decompactification zone occurring in local anticlinal upstructure portion as a result of fracturing in the process of brittle tensile deformations which exceed critical ruptural deformations. Availability of gas showing in expendable wells and mine workings in the process of their advance actually proves presence of free methane in the structures.

1 INTRODUCTION

A problem of coal deposits methane comprises the three basic aspects: providing safe coal mining in the process of roadway and stope development; using coal methane as essential energy feedstock and chemical raw materials; and improving environment through reduce of coal methane hazardous emissions in atmosphere.

Contrary to natural gas deposits, coal-bearing stratum mainly contains dispersed methane which is either sluggish or immobile. Low values of rock mass porosity being in the mean hundredths and thousandths of millidarcy prevent gas and water migration; as a result, much gas is accumulated in the form of deposits favourable to industrial deployment (Lukinov 2007). A process of rock mass decompactification and fracturing as a result of tectonic processes may contribute to gas and water redistribution within coal-rock massif and free methane absorbing in the form of concentrations. Hence, fracturing should be considered as one of basic factors of gas deposits formation in poor-porous coalbearing stratum (Lukinov 2007).

In broad terms, such types of traps as stratigraphical, lithological, structural, structural and tectonic, tectonic, technogenic, hydrodynamic, and those connected with magmatic processes are distinguished in coal-bearing strata. Also, there are combined traps formed by the joint action of two or several abovementioned geological factors.

For carbonic rocks of late katagenesis (effective porosity coefficient is up to 9%) including sand-stones, a mechanism for gas accumulations formation is considered. The accumulations are of structural and tectonic type; they are considered in accordance with an approach proposed by V.V. Lukinov (Lukinov 2007) and developed in papers (Bezruchko 2011, Lukinov & Bezruchko 2012). The approach main point is that methane traps in poorporous sandstones of coal-bearing stratum form as a reservoir in decompactification zone occurring in local anticlinal upstructure portion as a result of fracturing in the process of brittle tensile deformations which exceed critical ruptural deformations. Above-laying rocks having improved ductile properties are cap rock; as a result, they stay to be unbroken in the process of fold mashing. Impermeable seams of those very sandstones are its screen upstructure; their tensile deformation can not achieve its maximum for discontinuity. Incomplete anticlines (hemianticlines) can belong to such structures. They complicate monoclinal rock bedding; differ in a seam hypsometry deviation from approximating surface being classified as structural noses, structural terraces, and uniclinal flexures.

2 PROGNOSIS EVALUATION CONCERNING AVAILABILITY OF FREE METHANE ACCUMULATIONS WITHIN "BUTOVSKAIA" MINE

Such traps are possible by reason of a technique developed by IGTM of NAS of Ukraine (Lukinov & Bezruchko 2012). The technique involves generating local first-order structures maps, determining their basic parameters (amplitude, length, and width), calculating coefficients of rock mass linear and volumetric decompactification in upstructure portion. Then, filtration and capacity characteristics (i.e. poriness, water content, gas content, and porosity) of rocks in potential zone of free methane accumulation within local anticlinal structure are determined. Potential zone of methane accumulation is identified according to Sectoral standard of Ukraine (Free methane accumulation... 2005).

That very technique was applied to make prognosis evaluation concerning availability of free methane accumulations within "Butovskaia" mine field in Donets'k-Makeievka geological and industrial district (Lukinov, Bezruchko, Prykhodchenko & Shpak 2012).

Results of tectonic analysis on coal seam n_1 structural contours within the mine field determine two tectonic structures of "nose" type. Deviation of the seam hypsometry helps to prepare a map of coal seam n_1 local structures (Figure 1). Both local anticlinal structures (eastern and western) are traced well. The former is located along the eastern extension of a mine field (east of Oktiabrskii overthrusting No. 1) between Oktiabrskii overthrusting No. 1 and Panteleimonovskii overthrusting; the latter is west of Oktiabrskii overthrusting No. 1 in the neighbourhood of Щ-1040, 3998, and 3908 wells (Figure 1). Their major axes are stretched out along Oktiabrskii overthrusting No. 1 being oriented on the rock strike for western structure, and at the angle of about 45° to rock strike for eastern one. Respectively, minor axes of the structures, are perpendicular to Oktiabrskii overthrusting No. 1.

Highest level (maximum deviation from approximating surface) for eastern structure is between Щ-1009 and 3939 wells being +75.0 m. Dome of western structure is characterized by saddle shape having two peaks with +114.0 m level in the neighbourhood of well No. 3998, and +120.6 m in the neighbourhood of well No. Щ-1040 (Figure 1). Dimensions of eastern local anticlinal fold within +70 m contour line are 486 m×250 m. Dimensions of western local anticlinal fold within +100 m contour line are 1250 m × 330 m. As eastern part (upslope) of western anticlinal fold is faulted by overthrusting, plan determines its dimensions in the rough.

Values of amplitude and width of anticlinal folds are used to calculate following basic parameters of each structure: coefficient of a knee of a fold; coefficient of linear deformation along the length and across the width of a fold; coefficient of volumetric deformation; and critical thickness of sandstone. Critical thickness of sandstone is thickness of sand-

stone substrata not involved in fracturing process, and within which fractures can not occur as a result of less crenulation and tension stress insuffient for rock discontinuity. Critical thickness was calculated according to a procedure stated in Sectoral standard of Ukraine (Antsiferov & Bulat etc., 2005).

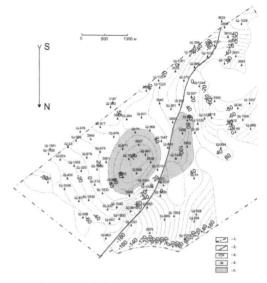

Figure 1. A map of First-Order Local Structures of Coal Seam n_1 within "Butovskaia" Mine Field: 1 – site boundaries; 2 – faults; 3 – expendable wells; 4 – contour lines of seam n_1 local structures; 5 – perspective zone.

According to the calculations, critical thickness of eastern structure seams which are not fractured, is 13.0–15.1 m within the structure (contour lines are +40 – +50 m), and 6.3 m within upstructure portion. Areas within which total thickness of each sandstone exceeds calculated critical thickness are zones to develop sandstone layers with improved capacity and filtration properties owing to fracturing.

Within eastern structure (well No. 3857) three sandstones are located over coal seam n_1. Inside the structure, thickness of two of them, namely $n_1 Sn_1^{1}$ and $n_1 Sn_2^{3}$ (upper part) exceeds calculated critical thickness. That is they have efficient thickness to be in the central part of the structure 25.5–27.2 and 11.9–13.6 m respectively.

Within western structure (well No. 3998) those very sandstones are located over coal seam n_1. Each of them has efficient thickness inside the structure; total thickness of each of them exceeds calculated critical thickness. In the mean (within contour line +80 m) their efficient thickness is as follows: 29.3 m for $n_1 Sn_1^{1}$ sandstone; 5.7 m for $n_1 Sn_2^{3}$ sandstone (lower part); and 27.2 m for $n_1 Sn_2^{3}$ sandstone (upper part).

Hence, availability of fissure and porous reservoirs is possible in the local anticlinal structures at depths down to 1000–1100 m. Of them, sandstones $n_1 Sn_1^1$ and $n_1^1 Sn_2^3$ (upper part) are the most prospective for eastern structure; $n_1 Sn_1^1$ and $n_1^1 Sn_2^3$ (lower and upper parts) are the most prospective for western structure. The sandstones within local anticlinal structures are prospective for free methane production.

Gas show which took place on the 23rd of December 2008 can confirm availability of gas accumulations. The gas show occurred in the process of eastern auxiliary slope mining within $n_1 Sn_1^1$ sandstone. The mine working is inside local eastern structure near its dome next to well No. Щ-991. Free methane flow was more than 7.5 m^3 per minute. Methane emission was observed for long; during two months total output exceeded 650.000 m^3.

Applying surface wells is recommended for examined conditions to make preliminary methane drainage of rock mass with related methane utilization. In future such surface wells may also be used for ongoing methane drainage while coal seam n_1 mining.

Rock mass undermining should be followed by extra volumes of free methane; it will depend on both its redistribution in sandstones and fixed methane release directly from coal seam. Hence, follow-up free gas production is possible in the process of preliminary and ongoing degassing. Moreover, preliminary degassing by means of pilot wells drilling directly from mine workings is also possible.

Prognosis evaluation of eastern local structure as for free methane accumulations availability is confirmed by 4th western discharging longwall mining which took place in 2012. According to individual project, preliminary degassing was performed; the process was implemented through pilot wells drilling from mine workings in the neighbourhood of 4th western discharging longwall. Early preparation and drilling of gas drainage wells make it possible to provide admissible concentration of gas-air mixture while mining within chosen local anticlinal structure. When 4th western discharging longwall fell outside the structure, gas emissions terminated.

3 PROGNOSIS EVALUATION CONCERNING AVAILABILITY OF FREE METHANE ACCUMULATIONS WITHIN "KALMIUSKII RUDNIK" SITE (O.F.ZASIADKO MINE)

Similar approach was applied to forecast prospectivity of "Kalmiuskii Rudnik" site (O.F. Zasiadko mine) as for free methane (Lukinov, Bezruchko, Prykhodchenko & Gunya 2012).

The mine is located in Donets'k-Makeievka district (Donbas) too. The district is in south zone of small size folds and faults within vast southern part of Kalmius-Torets basin (south-west of Donetsk coal field). O.F. Zasiadko mine field is within tectonic block limited by great various types: sublatitudinal Musketivskyi overthrusting is in the south; Vetkivska and Chaikivska flexures are in the west and in the east, respectively; and Kalmius-Torets basin is in the north. Tectonically, the mine field structure is rather complex. Strike is close to sublatitudinal; its azimuth is 85–100°. Close to monoclinal common rock bedding (on thickness strike and dipping) is someway complicated due to flat-lying bends, local folds, and rupture dislocations.

Based on the results of tectonic analysis, tectonic structure of "structural nose" is identified in the northern part of "Kalmiuskii Rudnik" site owing to contour lines of coal seam m_3. Using hypsometric plan of coal seam m_3 and variation of order 1, a map of regional slope – approximating surface of seam m_3 – is prepared. Deviations of the seam hypsometry from approximating surface help to prepare a map of local structures (Figure 2).

The structure is seen on maps of local structures drawn in terms of coal seam m_3, sandstone roof $m_4^0 Sm_4^1$ and sandstone roof $m_5^1 Sm_6^1$ as an open anticlinal fold. Structural recurrence on various stratigraphic levels indicates the fact that considerable rock thickness takes part in its formation. There is flattening of fold limb in the direction from the levels occurring lower towards the ones being higher than the section; it shows the increase of fold steepness along with the depth growth, i.e. close anticlinal fold can be formed on dipper levels lower than seam m_3.

Map of local structures (Figure 2) shows that the structure closes along +30 m isoline and have the following dimensions in plan (within these isoline): 3275 by 2000 m. Long axis of local anticlinal fold oriented along the rock extent being parallel to Vetkivsk flexural fold.

Consequently short axis is directed along m_3 seam inclination. Structural dimensions in the upstructure part (+80 m isoline) are 800 by 425 m. Maximum mark (maximum exceeding over approximating surface) is within the area of Щ-9 well being equal to +88.7 m. Basic parameters of the structure were calculated along the values of amplitude and width of anticlinal fold, i.e. coefficient of fold curvature (stepfold), coefficient of linear deformation according to fold length and width, coefficient of volume deformation and critical sandstone thickness.

Critical sandstone thickness is the thickness of the lower sandstone layers which are not effected by fissure formation process; moreover fissures are not formed within this thickness as the result of minor curvature and low stretching to disturb rock continuity. According to the calculation performed thickness of the layers without fissures is about 34 m on structure contour (+30 m isoline), 10–11 m within the upstructure part, and 21–23 m within middle part (within +40 − + 60 m isolines). Areas within which general thickness of each sandstone is more than the calculated critical one are the areas of development of sandstone layers with better capacitory and filtering properties at the expense of fissility.

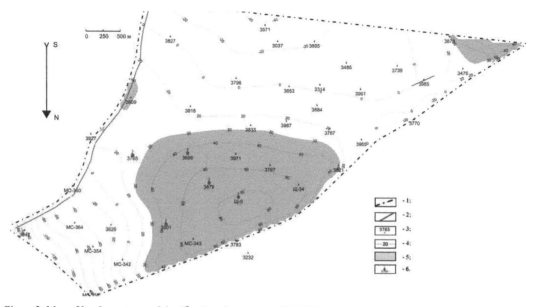

Figure 2. Map of local structures of the 1st order of m_3 seam within "Kalmius mine" area of O.F. Zsyadko mine: 1 – area boundaries; 2 – discontinuous fault; 3 – extension wells; 4 – isolines of local structures of m$_3$ seam; 5 – perspective zone; 6 – gas showing wells.

Upper sandstone layers develop effective thickness; their stretching while structure formation has exceeded critical boundary resulting in the development of crumbling deformations as well as the development of fracture zones. Effective sandstone thickness within the middle part of the structure (within +40 − +60 isolines) is taken as average value of effective sandstone thickness. Sandstones of M-3 level ($m_4^0Sm_4^1$ and $m_5^1Sm_6^1$) as well as sandstone of M-6 level (l_1Sl_2, L_1Sl_1) are stated to be the most perspective gas-bearing objects. Their average thickness is as follows: sandstone $m_5^1Sm_6^1$ – 34 m, $m_4^0Sm_4$ – 17 m, l_1Sl_2, and L_1Sl_1 (7 m each).

Thus, depths up to 1400 m shows possible interstitial collectors among which sandstones of M-3 level ($m_4^0Sm_4^1$ and m$_5^1Sm_6^1$) are the most prospective ones. At the depth of more than 1400 m perspective can be connected with the possible fissure collectors in the sandstones of Moskovskiy and Bashkirskiy tiers that may occur due to the increase of stratigraphic depth which can result in the growth of effective sandstone thickness. It means that sandstone of minor thickness can be used for collectors at deeper levels.

Gas showing in the form of gas and water available while drilling expendable wells within the area of interest. All the five wells within the specified structure are gas showing ones (see Figure 2). Thus, area located between Щ-9, 3879, 3971, 3797 wells (see Figure 2) is the most optimal territory for possible prospecting wells within the specified local anticlinal structure within "Kalmius mine" object. Advanced technologies for gas recovery of coal rock mass, in particular well structure with horizontal column part into productive levels and numerous hydraulic fracturing, can be used for commercial production of methane.

4 CONCLUSIONS

Prognosis evaluation of perspective mine fields and geological prospecting areas in terms of "Butovskaia" mine and area of "Kalmius mine" in Donbas demonstrates the perspectives of available free methane in traps of structural and tectonic types (local anticlinal structures) in low-porous sandstones of coal-bearing deposits. Traps of such type are formed as reservoirs within the decompactification zone that occurs in the upstructure part of local anticlinal structure at the expense of fissure formation at brittle stretch deformations exceeding critical ones as for tensile factor. Free methane occurrence in these structures is confirmed by the available gas loated in geological prospecting wells as well as in mine workings while their driving.

REFERENCES

Lukinov, V.V. 2007. *Mining and geological conditions of free methane accumulation within deposits.* Scientific Herald of National Mining University, 4.

Bezruchko, K.A. 2011. *Conditions for formation and preservation of methane accumulation within poor-porous coal-bearing carboniferous deposits.* Scientific Papers of UkrNDMI NAS of Ukraine: Collection of scientific papers / UkrNDMI. Issue 9, Vol. 2.

Lukinov, V.V. & Bezruchko, K.A. 2012 *Prospectivity assessment of local anticlinal structures in terms of available free methane accumulation.* Geotekhnichna mekhanika: Interdepartmental collection of scientific papers / IGTM NAS of Ukraine. Issue 102.

Free methane accumulation in non-disturbed coal massif. Technique to predict zones and determine their parameters: SSU 10.1.05411357.004:2005. Official edition: approved by Minugleprom of Ukraine: being in force since (29 November 2005) / V.A. Antsiferov, V.A. Baranov, A.F. Bulat, D.P. Gunya, M.E. Kplanets, etc. Kyiv: 12.

Lukinov, V.V., Bezruchko, K.A., Prykhodchenko, O.V. & Shpak, V.Y. 2012. *Prospectivity assessment of the areas to prospect free methane accumulation (in terms of "Butovska" mine).* Scientific Herald of the National Mining University. 2.

Lukinov, V.V., Bezruchko K.A., Prykhodchenko, O.V. & Gunya, D.P. 2012. *Prospectivity assessment of "Kalmius mine" area (Zasyadko mine) as for the available free methane accumulation.* Geoinformatika, 1(41).

Progressive Technologies of Coal, Coalbed Methane, and Ores Mining – Bondarenko, Kovalevs'ka & Ganushevych (eds)
© 2014 Taylor & Francis Group, London, ISBN: 978-1-138-02699-5

Improvement of methods of environmental hazard control of industrial waste

O. Borysovs'ka & A. Pavlychenko
National Mining University, Dnipropetrovs'k, Ukraine

ABSTRACT: The problem of Definition of waste hazard class is considered. Disadvantages of State Sanitary Rules and Regulations "Determination of class of danger of industrial waste to public health" are given. Directions of improving methods of environmental hazard control of industrial waste are considered. An overview of alternative methodology of classification of waste by the calculation method is done. Advantages and disadvantages of the existing methods have been studied on the example of waste of Dnipropetrovs'k incinerator.

1 INTRODUCTION

The negative side of growth of industrial production and the use of natural resources are high volumes of annual generation and accumulation of production and consumption waste, whereby solid waste currently occupied large areas of land. The difficult situation in the sphere of industrial waste management, developed in Ukraine now, is compounded by the lack of adequate methods for assessment of their class of danger. Currently, the only approved methods of definitions of waste hazard class are The State Sanitary Rules and Regulations "Hygienic requirements for behavior with industrial waste and determination of their class of danger to public health" (SSR&R 2.2.7-98 1998).

2 FORMULATING THE PROBLEM

According to this standard document, if the physical and chemical composition of the waste is not set, the class of hazard is determined by the manufacturer of these waste (or on his behalf) experimentally using experimental animals in laboratories accredited for this activity. If the physical and chemical composition of the waste is known, the class of hazard is determined by calculation.

Definition of waste hazard class by computational method is as follows. Toxicity index K_i is determined for each chemical ingredient that is included in waste composition, according to the formula

$$K_i = \frac{lg(LD_{50})_i}{S_i + 0.1 F_i + C_i^s},\qquad(1)$$

where LD_{50} – average lethal dose of the chemical ingredient, when introduced into the stomach, mg/kg; S – coefficient reflecting the solubility of the chemical ingredient in the water; it is obtained by dividing by 100 the solubility of substances in water in gram per 100 grams of water at a temperature not higher than 25 °C; F – coefficient of volatility of chemical ingredient, which is obtained by dividing by 760 the saturation of vapor pressure in mm of mercury column at a temperature of 25 °C; C^s – content of the substance in the total mass of waste, t/t; i – ordinal number of ingredient.

In the absence of LD_{50} for ingredients of waste, but in the presence of the hazard class of these ingredients in the occupational air, conditional values of LD_{50} are substituted in the formula 1, roughly defined in terms of a hazard class of substances in workplace air (Table 1).

Table 1. Hazard classes of substances in the air of the working area and the corresponding conditional LD50 values.

Hazard class in the air of working area	The equivalent of LD_{50}, mg/kg	$lg(LD_{50})$
I	15	1.176
II	150	2.176
III	5000	3.699
IV	>5000	3.778

After calculating the K_i for each ingredient of wastes no more than three, but at least two toxicity indexes which have the lowest K_i are selected, wherein $K_1 < K_2 < K_3$, and moreover, the next condition should be satisfied: $2K_1 \geq K_3$. If this condition is not met, according to the method, the third index does not take into account and the definition of the hazard class is carried out by two leading indexes of toxicity.

Then total hazard index K_Σ is calculated by the formula

$$K_\Sigma = \frac{1}{n^2} \sum_{i=1}^{n} K_i, \qquad (2)$$

where n – number of waste toxicity indexes of ingredients ($n = 2$ or $n = 3$). Thereafter, the hazard class of the waste is defined according to Table 2.

Considering the fact that a significant part of hazardous industrial waste has no implemented schemes of recycling, disposal or processing and is removed by burial or used as impurities or layers on landfills of solid industrial waste, that is, can have direct contact with the objects of the environment, their maximum permissible concentrations (MPC) in the soil are taken into account to determine their danger (Krupskaya 2012, 2013, Kolesnik 2012, Derbentseva 2013, Gorova 2013, Pavlychenko 2013).

Table 2. Classification of waste hazard by LD_{50}.

Total hazard index K_Σ	Hazard class	Degree of toxicity
Less than 1.3	I	Extremely hazardous
From 1.3 to 3.3	II	Highly hazardous
From 3.4 to 10.0	III	Moderately hazardous
10.0 or more	IV	Low hazardous

In this case, the toxicity index of each ingredient is determined by the formula

$$K_i = \frac{MPC_i}{S_i + 0,1F_i + C_i^s}, \qquad (3)$$

where MPC_i – maximum permissible concentration of toxic chemicals in the soil, mg/kg.

After calculating the K_i for each ingredient of wastes no more than three, but at least two toxicity indexes which have the lowest K_i are selected too, wherein $K_1 < K_2 < K_3$, and moreover, the next condition should be satisfied also: $2K_1 \geq K_3$. If this condition is not met, the third index does not take into account and the definition of the hazard class is carried out by two leading indexes of toxicity. Then the total hazard index K_Σ is determined by formula (2), then the hazard class and the degree of toxicity of the waste are determined according to Table 3.

This methodology, in our opinion, has several disadvantages:

– LD_{50}, hazard classes are in air of the working area and the MPC in the soil are established not for all substances that can be found in the waste;

– the danger of all waste ingredients are not taken into account when calculating, but only two or three priorities;

– a lack of information on the primary indicators of danger of waste components is not considered;

– environmental indicators of hazard of waste components are not considered, such as carcinogenicity, mutagenicity, etc.;

– composition of the waste is Not always accurately known;

– some types of industrial waste can have unstable qualitative composition that varies in time depending upon various conditions;

– qualitative and quantitative analysis requires a significant investment of time and money.

Table 3. Classification of waste hazard by MPC in the soil.

Total hazard index K_Σ	Hazard class	Degree of toxicity
Less than 2.0	I	Extremely hazardous
From 2.1 to 16.0	II	Highly hazardous
From 16.1 to 30.0	III	Moderately hazardous
30.1 or more	IV	Low hazardous

formulas used to determine the hazard class of waste by this method are also imperfect and criticized with mathematical and hygiene positions (Antamonov 1999, Nazarov 1994):

– values of the coefficients of solubility and volatility S and F in the formulas (1) and (3) differ by ten times and more; with the addition of such quite different values of S, F and C^s the their sum can take essential value even at extremely low content of substances C^s, but significant quantities of its coefficient of solubility S (for example, for antimony fluoride, which $S = 4.45$; or zinc chloride, which $S = 3.75$);

– change of toxicity index K_i has illogical nature: the smaller the index the higher the toxicity, and vice versa;

– the index of ingredient toxicity K_i itself is incomplete: This technique involves his account only when calculating the total hazard index K_Σ, that is itself the value of K_i is not self-sufficient characteristic of the mixture component and is not related to its class of toxicity;

– quadratic dependence of the total danger index K_Σ from n in the formula (2) is unreasonable, and consequently, this relationship is hypertrophied of in boundary situations: if $K_3 = 2K_1$ the K_Σ value depending on the choice of $n = 2$ or $n = 3$ may vary over twice.

Furthermore, in terms of environmental safety for determination of hazard characterization of wastes is not enough to use such indicators as the average lethal dose of the chemical ingredient LD_{50}, coefficient reflecting the solubility of the chemical ingredient in the water S, the coefficient of volatility of chemical ingredient F, its hazard class in the air of working area and the maximum permissible concentration of toxic chemicals in the soil. All these indicators take into account the impact of the waste or its components only on the human body, without taking into account their risk to other living organisms and the environment.

Thus, the imperfection of the methodology for determining the hazard class of waste causes differences in the interpretation of its provisions, the differences in the results of the calculation for the same waste and, as a consequence, the violation of hygienic requirements for the behavior of hazardous waste.

Therefore, the directions of improving methods of environmental hazard control of industrial waste are considered in this paper. Advantages and disadvantages of the existing methods have been studied on the example of waste of Dnipropetrovs'k incinerator.

3 GENERAL PART

Wastes of this industrial facility are presented by slag remaining after combustion of municipal solid waste in boilers, and fly ash, which is retained in the electric filters. In accordance with The State Sanitary Rules and Regulations "Hygienic requirements for behavior with industrial waste and determination of their class of danger to public health", slag refers to the fourth class of danger – low hazardous waste and fly ash refers to the third class – moderately hazardous waste.

According to the results of our research, the products of combustion of municipal solid waste (MSW) represent a significant hazard to the environment. High concentrations of heavy metals in these waste are found, and many of are presented by water-soluble and mobile forms.

Thus, Figures 1–2 show results of the determination the content of heavy metals extracted from the slag and fly ash by aqueous solution and ammonium acetate buffer ($pH = 4.8$).

Water soluble forms of heavy metals were found in the slag, such as Zn, Co, Ni, Cu, Cr and Mn, as well as in the fly ash – Pb, Zn, Co, Ni, Cu and Cr. Fraction of heavy metals, moving from these wastes to an aqueous solution is relatively low, while the proportion of metals extracted by acidic buffer solution indicates a significant risk of products of municipal solid waste combustion to the environment. Thus, practically all of the zinc and about 50% of lead (of the gross) passed from slag to the buffer so-

lution, in the experiment with ash the same parameters were respectively 40 and 70%. Therefore, in case of contact between waste and atmospheric precipitation with acid reaction substances of first class of danger – Zn and Pb – will leach out from these wastes in significant quantities. As for heavy metals of second and third hazard class – Co, Ni, Cu, Cr and Mn, they are also leached out of products of MSW combustion, although to a lesser extent. According to the intensity of leaching from waste heavy metals can be arranged in the following series in order of decreasing: for slag – $Zn > Pb > Mn > Ni > Co > Cr > Cu$; for fly ash – $Pb > Zn > Ni > Co > Mn > Cr > Cu$. From these series can be seen that heavy metals of first class of danger will primarily come to the environment from the products of combustion of solid waste in the case of leaching by acidic precipitation.

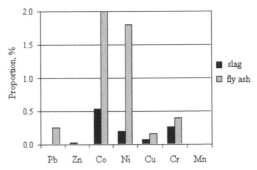

Figure 1. Proportion of water-soluble metals in the studied waste.

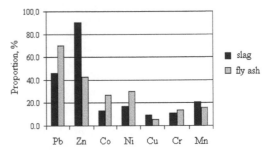

Figure 2. Proportion of mobile forms of metals in the studied waste.

It is obvious that the calculation method for determining the hazard class of waste adopted in our country is not quite correct, since this method does not consider the form of ingredient while calculating of hazard of waste ingredients.

"Sanitary rules on the definition of the hazard class of toxic waste of production and consumption", taken in Russia (SR 2.1.7.1386-03 2003) could be an alternative to domestic method. According to Russian standards, classification of waste by the calculation method performed on the basis of value of the total hazard index K, calculated by the sum of the hazards indexes of substances constituting waste (K_i). List of waste components and their quantitative content are established through qualitative and quantitative chemical analysis or composition of raw materials and the technology of its processing.

Index of danger of waste components K_i is calculated as the ratio of the waste components concentration C_i (mg/kg) and to coefficient of the hazard degree of component W_i

$$K_i = \frac{C_i}{W_i},\qquad(4)$$

$$lg\,W_i = 1,2(X_i - 1),\qquad(5)$$

where X_i – averaged parameter of hazard of waste components.

An information search of toxicological, hygienic and physical and chemical parameters of danger for each component is performed on the basis of the qualitative composition of waste. Herewith indicators such as carcinogenicity, mutagenicity, persistence, bioaccumulation, biological dissimilation and other are taken into account. On the value of danger index the last one is scored from 1 to 4. The first twelve indicators are used in the calculation.

The information index I is taken into account in calculating X_i, it depends on the used number of hazard indicators n. That is, the lack of information on the primary indicators of danger of waste components is reflected in the hazard index of ingredient K_i. Form of an ingredient in which it is present in the waste – fixed or mobile – also is taken into account in determining K_i.

Averaged parameter of hazard of waste components X_i is calculated by dividing the sum of scores for all indicators, including information index I on the total number of indicators.

Waste components, consisting of chemical elements such as silicon, titanium, sodium, potassium, calcium, carbon, phosphorus, sulfur in concentrations not exceeding their content in the main soil types belong to practically non hazardous components with averaged parameter of danger of component X_i at 4 .

If there are substances in waste with proven carcinogenicity for human averaged parameter is set $X_i = 1$ for these ingredients, hence $W_i = 1$, all other indications of danger are not considered, i.e. $K_i = C_i / 1 = C_i$.

The total hazard index K equals the sum of K_i of all components of waste

$$K = K_1 + K_2 + ... + K_n = \sum_{i=1}^{n} K_i .\qquad(6)$$

Ranking waste by hazard class in magnitude of K is carried out in accordance with Table 4.

Table 4. Hazard classification of waste on human health and the environment.

Hazard class	1 class	2 class	3 class	4 class
K	>50000	50000–1000	999–100	< 100

The advantages of this method of determining the hazard class:
– the danger all substances that enter the waste stream is taken into account without exception;
– not only hygienic but also environmental indicators of danger are used to determine the toxicity of each component;
– lack of information on the primary indicators of danger of each component is reflected on the indicator of information support I and, consequently, on the averaged parameter of hazard of waste components X_i etc.

The comparative analysis (Table 5) of native methods of the definition of waste hazard class and its foreign analogue showed that the most significant drawback of domestic sanitary rules is no need to confirm experimentally the hazard class, obtained by calculation. The exception is when the composition of the waste is unknown or when hazard class of waste obtained by the calculation method does not satisfy the manufacturer. Is more logical to add to these requirements the need to confirm experimentally the calculated hazard class in all cases, because it is obvious that even the most profound analysis of toxic, hygienic and physical and chemical properties of the waste components in isolation does not give a complete picture of how dangerous these components are in the complex.

By calculating the hazard class of products of MSW incineration by method (The official website of the Ministry of Natural Resources and Ecology... 2014), taking into account the mobile forms of heavy metals, it has been established that the wastes of Dnipropetrovs'k incinerator belong to the second class of hazard to humans and the environment – highly hazardous waste.

To confirm the hazard class, obtained by calculation, experimental studies of environmental hazard of the waste were carried out in accordance with the criteria given in Table 6.

Table 5. Comparative analysis of methods for determining the hazard class.

Method detail	Methods for determining the hazard class	
	SSR&R 2.2.7-98 (Ukraine)	SR 2.1.7.1386-03 (Russian Federation)
Rules allow establishing the hazard class of waste on human health	+	+
Rules allow establishing the hazard class of waste on environment	–	+
Calculation method applies, if qualitative and quantitative composition of waste is known	+	+
The calculation may be made by the manufacturer of waste	+	–
The danger all components of the waste are taken into account when calculating	–	+
The lack of information about each component is taken into account in determining its danger	–	+
Not only hygienic but also environmental indicators of danger of each component are used for the calculation	–	+
The need to confirm the lowest hazard class by experimental method	–	+

Phytotoxicity was determined by testing the water extract of the studied waste on the culture waste *Triticum aestivum L.* (common wheat) in the growth test. Mutagenic activity was determined on the culture of *Allium cepa L.* (onion) in Allium-test (Gorovaya 2004, Gorova 2013, Guidelines... 2007).

Calculation of indicative water-migration rate was conducted on the results of the chemical analysis of the buffer extract of waste ($pH = 4.8$) and the aqueous extract from the formulas:

$$IWMR_b = \sum \frac{C_i^b}{MPC_i^w}, \qquad (7)$$

$$IWMR_a = \sum \frac{C_i^a}{MPC_i^w}, \qquad (8)$$

where C_i^b and C_i^a – actual concentration of the each component in the buffer and aqueous extracts, respectively, mg/l; MPC_i^w – maximum permissible concentration of the component in water bodies mg/l.

Hazard assessment of waste affecting the biological activity of the soil was conducted by determining the shift of the redox potential over 100 mV the by adding the studied waste to the soil at a concentration from 10 to 50%.

Table 6. Environmental and hygienic characteristics and criteria for the classification of waste.

Danger characteristics	Hazard class			
	I – extremely hazardous	II – highly hazardous	III – moderately hazardous	IV – low hazardous
Phytotoxicity(ER_{50})	> 100	> 10–100	> 1–10	0.1–1
Mutagenic activity (multiplicity of excess test/control)	> 15	11–15	4–10	2–3
IWMR$_b$	> 1000	> 100–1000	> 10–100	≤ 10
IWMR$_a$	> 100	> 50–100	> 10–50	> 3–10
The redox potential of the soil (shift of ORP, mV)	> 250	> 200–250	> 100–200	100–15

Note: 1. ER_{50} – the average effective dilution of waste extract, causing inhibition of growth of sprouts roots of seeds by 50%.

2. IWMR$_b$, IWMR$_a$ – the indicative water-migration rate of buffer extract and an aqueous extract of waste, correspondingly.

Table 7 summarizes the results of the definition of hazard class of waste studied by calculation and experiment.

It was established that investigated waste possesses phytotoxic and mutagenic properties, pose a threat to the soil due to the high water migration ability of components and significant negative effect on the redox potential of the soil.

Table 8 shows a comparison of the results ob-tained experimentally and theoretically: maximum, minimum and average values are highlighted for the hazard classes of slag and fly ash. It has been established that experimental method has a wider range of hazard class, i.e. it characterize the danger of waste fuller and more detailed than the calculation method. In this case, there were five hazard indicators of waste, defined experimentally. By increasing the number of these parameters by studying air-migration danger of waste; morphological, immunological and other changes in animals and other factors, you can get an even better understanding of all the negative properties of the investigated waste.

Table 7. The results of definition of hazard class of waste.

Methods for determining the hazard class	Hazard class	
	slag	fly ash
Calculation method		
SSR&R 2.2.7-98 (UA)	IV– low hazardous	III – moderately hazardous
SR 2.1.7.1386-03 (RU)	II – highly hazardous	II – highly hazardous
Experimental method		
Phytotoxicity	IV – low hazardous	III – moderately hazardous
Mutagenic activity	III – moderately hazardous	II – highly hazardous
IWMR$_b$	II – highly hazardous	I – extremely hazardous
IWMR$_a$	III – moderately hazardous	I – extremely hazardous
ORPof the soil (shift of ORP, mV)	II – highly hazardous	III – moderately hazardous

Table 8. Comparison of results of definition of hazard classes.

Waste	Value of hazard class defined						Final hazard class
	by calculation method			by the experimental method			
	min	max	average	min	max	average	
Slag	2	3	2.3	1	4	2.6	III – moderately hazardous
Fly ash	2	3	2.3	1	3	2.0	II – highly hazardous

Averaging the results obtained for the slag and fly ash by experienced and by calculation showed a higher sensitivity of the experimental method in comparison with the calculated values. So, with the very significant differences in the chemical composition and properties of slag and fly ash of Dnipropetrovs'k incinerator for both types of waste in average the second hazard class was established by the calculation method, while experimentally the third hazard class was identified for slag and the second hazard class was confirmed for fly ash.

The final hazard class of products of MSW incineration has been revised and set based on the comparison the calculation results with experimental results.

4 CONCLUSIONS

Thus, it was found that the experimental method of waste hazard control enables more fully and accurately assess the degree of danger of various properties of waste to the environment components. Therefore, hazard class of waste defined by the standard calculation method must be confirmed or corrected by experimentation to assess adequately the environmental hazard. The implementation of this decision in practice would eliminate the discrepancies in the results of the calculation for the same waste and prevent violations of hygienic requirements for the behavior of hazardous waste.

REFERENCES

The State Environmental Inspectorate of Ukraine – Normative base. 2014. Available at: http://dei.gov.ua/ index.php?option=com_content&view=article&id=1298:sanpin-gigiyena&catid=23&Itemid=204 (accessed May 1).

The official website of the Ministry of Natural Resources and Ecology of the Russian Federation – Documents. 2014. available at: www.mnr.gov.ru/regulatory/ (accessed May 1)

Krupskaya, L.T., Bubnova, M.B., Zvereva, V.P. & Krupskiy, A.V. 2012. *Characteristics of mining-ecological monitoring of environmental objects changing under the influence of toxic waste tailing dump ("Solnechny GOK" Company).* Environmental Monitoring and Assessment. Vol. 184, Issue 5: 2775–2781.

Krupskaya, L.T., Zvereva, V.P. & Leonenko, A.V. 2013. *Impact of technogenic systems on the environment and human health in the priamurye and primorye territories.* Contemporary Problems of Ecology. Vol. 6, Issue 2: 223–227.

Kolesnik, V.Ye., Fedotov, V.V. & Buchavy, Yu.V. 2012. *Generalized algorithm of diversification of waste rock dump handling technologies in coal mines.* Scientific Bulletin of National Mining University, 4: 138–142.

Gorova, A., Pavlychenko, A. & Borysovs'ka, O. 2013. *The study of ecological state of waste disposal areas of energy and mining companies.* Annual Scientific-Technical Colletion – Mining of Mineral Deposits, Leiden, The Netherlands: CRC Press / Balkema: 169–171.

Gorova, A., Pavlychenko, A., Borysovs'ka, O. & Krups'ka, L. 2013. *The development of methodology for assessment of environmental risk degree in mining regions.* Annual Scientific-Technical Colletion – Mining of Mineral Deposits, Leiden, The Netherlands: CRC Press/ Balkema: 207–209.

Derbentseva, A., Krupskaya, L. & Nazarkina, A. 2013. *Analysis and Assessment of the Environment in the Area of Abandoned Coal Mines in Primorsky Region*. Mechanics and Materials, Vols. 260–261: 872–875.

Pavlychenko, A. & Kovalenko, A. 2013. *The investigation of rock dumps influence to the levels of heavy metals contamination of soil*. Annual Scientific-Technical Colletion – Mining of Mineral Deposits, Leiden, The Netherlands: CRC Press / Balkema: 237–238.

Antamonov, M. 1999. *An alternative approach to classification of industrial waste*. Hygiene and Sanitation, 4: 62–64.

Nazarov, A.A. & Pavlov, V.N. 1994. *Proposals for adjustment the formulas for calculating the toxic index of industrial waste to their classification*. Hygiene and Sanitation, 4, 16–18.

Gorovaya, A.I., Lapitskiy, V.N., Borisovskaya, E.A. & Pavlichenko, A.V. 2004. *Bioindication of toxicity and mutagenicity of slag of Dnepropetrovsk incinerator*. Scientific Bulletin of National Mining University, 6: 79–81.

Guidelines 2.2.12-141-2007. Survey and Zoning by the Degree of Influence of Anthropogenic Factors on the State of Environmental Objects with the Application of Integrated Cytogenetic Assessment Methods / Gorova A.I., Ryzhenko S.A., Skvortsova T.V. Kyiv: "Polimed": 35.

Progressive Technologies of Coal, Coalbed Methane, and Ores Mining – Bondarenko, Kovalevs'ka & Ganushevych (eds)
© 2014 Taylor & Francis Group, London, ISBN: 978-1-138-02699-5

Improvement of operational properties of two-layer filtering elements of dust respirators for mining operations

S. Cheberyacko, O. Yavors'ka & V. Tykhonenko
National Mining University, Dnipropetrovs'k, Ukraine

ABSTRACT: Calculation methods for two-layer filters for dust respirators under mining conditions are given. Basic mechanisms to determine their parameters ensuring minimum pressure differential under maximum dust holding capacity are defined. Conditions for optimal radius of preliminary filter fibers to ensure uniform growth of pressure differential on both filtering layers of a filter are determined. It is proved that to ensure maximum amount of filter-deposited dust it is necessary to increase fiber diameter along with the growth of aerosol size; on the other hand radius of preliminary filter fibers should be more than 6 mcm.

1 TOPICALITY

High-efficiency aerosol filter is the device in which air being cleaned goes through the layer of filtering material. The layer catches aerosol particles with the help of several catching mechanisms. Particles are accumulated gradually in filter volume. Here their pressure differential grows and operational properties become worse especially in case of high dust concentration within operation area. Consequently, increase of respiratory-protection period under severe mining conditions is quite a topical issue.

2 FORMULATING OF THE PROBLEM

There are the following structural, functional, and operational filter characteristics (Filatov 1997): coefficient of protection K; configuration and overall dimensions: filter height – H_f; width (or corrugation length) – W_f; length – L_f; initial pressure gradient – Δp; respiratory-protection period– t_k; filtering area – F; properties of filtering layer effecting coefficient of penetration: fiber radius – a, density of fiber packing – β; thickness of filtering layer – H; airflow resistance – R.

First four characteristics are selected according to the aim and conditions of filter use. Last two ones are determined according to the specified conditions. There are numerous variants meeting initial data. Following equation system determines functional relations (Stechkina & Kirsh 2003)

$$K = \frac{100}{exp\left(-\frac{2\beta H\eta}{\pi a}\right)}, \qquad (1)$$

$$E = 1 - K - exp\left(-2\beta H\eta/\pi a\right), \qquad (2)$$

$$\Delta p = \mu v R_0 L, \qquad (3)$$

where η – total coefficient of aerosol particles catching; μ – dynamic air viscosity, Pa·s; v – filtering rate, m/s; R_0 – dimensionless fiber force of airflow resistance; L – general fiber length in a filter, m^{-1}

$$L = \frac{\beta H}{\pi a^2}.$$

One more equation is required to complete this system. The best one is the determining respiratory-protection period of a dust respirator. This characteristic depends on coefficient of protection, breathing resistance, and dust holding capability of a filter as well. Respiratory-protection period of a respiratory protective device (RPD) is determined according to boundary allowable pressure differential. Its value is specified by normative documents: DSTU GOST 12.4.041:2006 – not more than 100 Pa at 30 l/min air loss while according to DSTU EN149-2003 for P1 class it is 400 Pa; for P2 class it is 500 Pa; for P3 class it is 700 Pa at 95 l/min air loss.

Development of the first respiratory protection devices has shown the necessity to increase filter dust capability and correspondingly their respiratory-protection period. It is clear that the performance improvement of high-efficiency filters made of ultrathin fibers characterized by better breathing resistance is of high interest. Solution of this problem is to make filters with alternating packing density as for their thickness allowing minimizing growth of pressure differential while aerosol catching. If packing density is uniform as for the thickness than dust deposit will be distributed nonuniformly. Its major amount is concentrated on the external filter layer with the considerable reduction of its operational life. It is possible to increase considerably dust holding capacity and operating life with variable density or fiber radii.

Thus there is the problem to improve operational characteristics of multistage system of air purification system under conditions of dust accumulation.

3 BASIC DATA

Consider two-layer filter (Figure 1). The problem is formulated as follows. It is necessary to determine such parameters of filtering layers for which the amount of particles deposited in the system will be maximum one at the moment of time when pressure differential reaches boundary allowable value if coefficient of filter protection is specified.

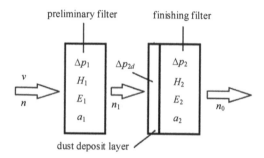

Figure 1. Scheme of two-layer filter.

We assume that the areas of two filters as well as their filtering rates are the same. The parameters that can be changed to ensure maximum respiratory-protection period are as follows: radii of clean fibers, their packing density, and thickness of filtering layer.

As internal layer is characterized by higher protective efficiency E_2 than external one, dust deposit layer increases fast. There is no volume filtering process under such conditions. It allows determining pressure differential on finishing filter as the total of initial pressure differential Δp_2 and pressure differential of dust deposit formed on its surface Δp_{2d}. Then it is possible to determine total pressure differential on two-layer filter according to the formula (Kirsh 1998)

$$\Delta p_f = \begin{cases} \Delta p_1 + \Delta p_2 + \Delta p_d & t < t' \\ \Delta p_1 + \Delta p_{1d} + \Delta p_{2d} & t > t' \end{cases}, \quad (4)$$

where Δp_1 – pressure differential on external filter, Pa; Δp_{1d} – pressure differential of dust deposit layer generated on the surface of external layer after finishing volume dust contamination, Pa; t' – time

when dust deposit appears on the surface of external filter, s. Pressure differential on internal filter is calculated according to the formula (Kirsh 1998)

$$\Delta p_2 = \frac{4\mu\beta Hv}{a^2 k_0(\beta)}, \quad (5)$$

where $k_0(\beta)$ – coefficient of filtering fiber resistance.

Pressure differential on dust deposit formed while filtering is determined by (Kirsh1998)

$$\Delta p_{2d} = \frac{3\mu v k_0(k_0)}{2\pi p r_p^2}, \quad (6)$$

where $k_0(\beta_2)$ – coefficient of dust deposit resistance that depends on dust deposit density β_2; ρ_p – density of an aerosol particle, kg/m^3; r_p – average radius of deposited particles, m;

Pressure differential on external layer taking into account dust deposit accumulation will be determined according to the formula

$$\Delta p(x,t) = \frac{4\mu Hv\cos(\omega t)}{a^2}\int_0^H \frac{\beta_1(x,t)dx}{F(\beta_1(x,t))}, \quad (7)$$

where

$$F(\beta_1) = \frac{4\pi}{-0.5\ln\beta_1 - 1.15} + 0.5\omega\sin\omega t\sqrt{\omega\mu p}\cos\omega t ;$$

ρ – air density; ω – pulsation frequency, min^{-1}; β_1 – density of dust-deposited fibers.

Since protective efficiency of the whole filter is determined mostly by internal layer, it is possible to determine its fiber radius using following equation

$$\frac{\eta_2(\alpha)}{\pi\alpha_2} = \frac{-\ln K}{2a_2 H},$$

and external filter parameters are possible to determine using value of pressure differential irrespective of protective efficiency value.

Total volume of dust particles deposited on definite filter area is possible to determine according to the formula (Kalnin 1977)

$$V = v_p n_0 vt = \pi\beta_1\alpha_1 t . \quad (8)$$

Probable dust amount deposited on respirator filter at known air dustiness is calculated according to the formula

$$M_d = \frac{CQt}{F}, \quad [\text{mg/m}^2] \quad (9)$$

where M_d – dust holding capacity, mg/m^2; C – general dust concentration, mg/m^3; t – working shift period, min; Q – lung ventilation volume of , m^3/min; F – filter area.

This formula makes it possible to determine parameters of filtering material ensuring allowable pressure differential until operating shift is over.

4 RESULTS AND DISCUSSION

Figure 2 represents curves $\Delta p(M_d)$ calculated for conditions of coal dust filtering with 10 mcm average particle size at $v = 1.5$ m/s airflow rate with fiber packing density being $\gamma = 0.05$ and thickness being $H = 6$ mm for various fiber diameters $d = 2$ mcm (1); 3.5 (2); 5 (3). Horizontal lines correspond to the specified pressure differential in the system. Curves areas (1, 2, 3) up to their cross with horizontal lines characterize volume filtering. If these curves are increased by value Δp_{2d} it is possible to obtain complete pressure differential of the system.

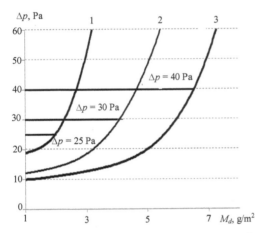

Figure 2. Dependence of pressure differential Δp on the amount of filter-deposited dust; curves are calculated at $v = 1.5$ m/s; $\gamma = 0.05$; $H = 6$ mm; and fiber diameter $d = 2$ mcm (1); 3.5 (2); 5 (3).

As Figure 2 shows there is optimal value of fiber radius for the specified pressure differential at which dust amount will be higher than in any other system. Cross point of the curves coincides roughly with the beginning of dust layer formation on front surface of a filter. It means that preliminary filter operates only in the mode of volume filtering. If fiber diameter in external layer is more than optimal one, it will result in rapid growth of pressure differential on finishing layer at the expense of increased particle penetration (Figure 3). On the contrary, if fiber diameter of preliminary layer is less than optimal one then pressure differential will grow sharply at the expense of dust deposit formation on its surface (Figure 4). Thus, it is clear that there is optimal fiber radius of preliminary filter to ensure

time-uniform growth of pressure differential on both filtering layers.

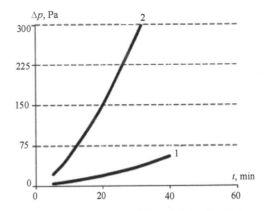

Figure 3. Change of pressure differential Δp of preliminary (1) and finishing (2) filters along with time t at dust concentration $C = 120$ mg/m³, filtering rate $v = 1.5$ m/s.

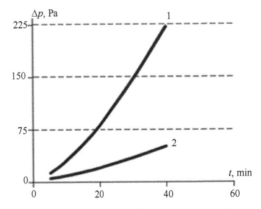

Figure 4. Change of pressure differential Δp of preliminary (1) and finishing (2) filters along with time t at dust concentration $C = 120$ mg/m³, filtering rate $v = 1.5$ m/s.

After determining optimal radius of preliminary filter it is necessary to estimate its coefficient of penetration using formula (2). Then, to calculate dust holding capacity of finishing filter it is necessary to determine dust aerosol concentration after preliminary C_1 using the formula

$$C_1 = KC \text{ , } [\text{mg/m}^3] \tag{10}$$

where K – coefficient of aerosol penetration though preliminary filter.

It is necessary to stress that dust concentration influences considerably fiber diameter of filtering material. That is why the calculations are performed for typical distribution of disperse coal dust composition (Golinko 2001). The problem is to determine optimal fiber size for catching first of all coal dust and ensuring maximum filter life under mining conditions.

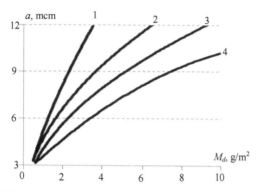

Figure 5. Dependence of fiber radius a of preliminary filter on its dust holding capacity Π with different size of deposited aerosol particles: $r_\text{n} = 5$ mcm (1); $r_\text{n} = 2$ mcm (2); $r_\text{n} = 1$ mcm (3); $r_\text{n} = 0.5$ mcm (4).

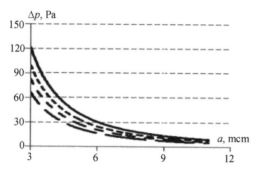

Figure 6. Dependence of pressure differential Δp of preliminary filter on fiber radius at aerosol particle sedimentation: —— diameter of particles is 5 mcm; - - - diameter of particles is 2 mcm; — — diameter of particles is 1 mcm; —— – diameter of particles is 0.5 mcm.

Thus, it is possible to determine dust holding capacity of preliminary layer per time t on the basis of dust concentration in the air of operating area. Use Figures 5 and 6 to determine fiber radius for the most efficient dust holding capacity. Then, having determined packing density and thickness of filtering layer ($\beta = 0.05$; $H = 6$ mm for polypropylene layers) calculate both filtering efficiency according to formula (1) and dust aerosol concentration after preliminary layer (10). Formula (7) is used to determine change of pressure differential on external layer per specified time t.

Fiber radius of finishing filter is calculated according to the specified efficiency of respirator filtration using formula (2) or Figure 8. Then, we cal-

culate dust holding capacity of finishing filter per the same time t (5). It is possible to clarify its radius using formula (1) and determine pressure differential increment generating dust layer deposited on it. Having compared its value with pressure differential on preliminary filter, it is possible to make conclusion as for the calculation accuracy. There should be such fiber radii of both preliminary and finishing layers so that pressure differential on both filters would be practically similar (Figure 8).

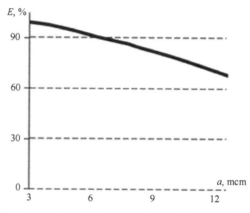

Figure 7. Dependence of filtration efficiency E of layered filtering material on fiber diameter d.

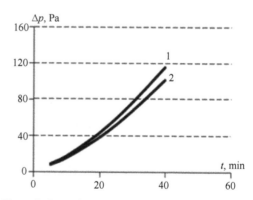

Figure 8. Dependence of pressure differential Δp increment upon time t in two-layer filter: 1 – preliminary filter, 2 – main filter.

The Table represents some calculation results of two-layer polypropylene filters with general filtration areas of 50 cm^2 at coal dust concentration of 500 mg/m^3 for specified filtration efficiency of $E = 99.99\%$.

Table 1. Calculation results for two-layer filters.

Filter operation Period t, min	Fiber radius of preliminary filter a, mcm	Dust holding capacity of preliminary filter, M_d, g/m²	Pressure differential on preliminary filter, Pa	Finishing filter radius, a, mcm	Dust holding capacity of finishing filter, M_d, g/m²	Pressure differential on finishing filter, Pa	Total filter dust holding capacity, M_d, g/m²	Total pressure differential on filter, Pa
5		0.01	13.2		0.001	5.67	0.011	18.9
10	6	0.03	30.9	2.5	0.002	9.53	0.032	40.4
20		0.06	81.0		0.005	20.39	0.065	101.4
30		0.09	145.8		0.008	34.43	0.098	180.2
5		0.01	8.5		0.002	7.9	0.012	16.4
10	7.5	0.03	17.5	2.5	0.004	15.9	0.034	33.4
20		0.06	43.2		0.009	38.3	0.069	81.5
30		0.09	76.4		0.013	67.2	0.103	143.6
5		0.01	7.6		0.002	8.8	0.012	16.4
10	8	0.03	15.0	2.5	0.005	18.5	0.035	33.5
20		0.06	36.1		0.001	45.6	0.061	81.7
30		0.09	63.5		0.015	80.7	0.105	144.2
5		0.01	5.6		0.004	22.1	0.014	27.7
10	9	0.03	9.4	2.5	0.007	55.8	0.037	65.2
20		0.06	20.2		0.014	151.2	0.074	171.4
30		0.09	34.2		0.021	274.7	0.111	308.9

As a result of calculations for the specified conditions optimal radius of preliminary and finishing filter fibers are obtained being $a = 6.5$ mcm and $a = 2.5$ mcm correspondingly that will provide the best filter characteristics as for dust holding capacity and filtration efficiency under the specified conditions.

5 CONCLUSIONS

1. Basic regularities to determine parameters of two-layer filter ensuring its minimum pressure differential at maximum dust holding capacity are determined.

2. Conditions for optimal radius of preliminary filter fibers to ensure uniform pressure differential increment on both filtering layers are defined.

3. To ensure maximum amount of filter-deposited dust along with the growth of aerosol particles it is necessary to increase fiber diameter; on the other hand radius of the preliminary filter fibers should not be less than 6 mcm.

4. It is established that typical distribution of coal dust disperse composition with concentration of 500 mg/m³ requires such optimal radii of preliminary and finishing filter fibers as $a = 6.5$ mcm and $a = 2.5$ mcm correspondingly.

REFERENCES

Filatov, Y.N. 1997. *Electric formation of fiber materials (EF-process)*. Moscow: L.Y. Karpov GNTs RFNIFKhI: 297.

Stechkina, V.A. & Kirsh, V.A. 2003. *Optimization of filter parameters in multistage system of fine gas cleaning*. Theoretical basis of chemical technology, 3. Vol. 37: 238–345.

Kirsh, V.A. 1998. *Method to calculate growth of pressure differential in aerosol filter in terms of filter blockage with solid particles*, 6. Vol. 60: 480–484.

Kalnin Y.V., Kanusik, N.S. & Illarionov, A.V. 1977. *Study of multi-layer fiber filters*. Improvement of working conditions on nonferrous-metal industry, 7: 112–117.

Golinko, V.I., Kolesnik, V.Y., Ishchenko, A.S. & Cheberyachko, S.I. 2001. *Development of plant for dust-proofing devices testing and dustiness control*. Scientific Bulletin of National Mining University, 3: 51–55.

Progressive Technologies of Coal, Coalbed Methane, and Ores Mining – Bondarenko, Kovalevs'ka & Ganushevych (eds)
© 2014 Taylor & Francis Group, London, ISBN: 978-1-138-02699-5

Analysis of longwall equipment for thin seams mining in conditions of Polish and Ukrainian mines

D. Astafiev & Y. Shapovalov
National Mining University, Dnipropetrovs'k, Ukraine

P. Kaminski & L. Herezy
AGH University of Science and Technology, Krakow, Poland

ABSTRACT: The analysis of current situation in coal mining industry of Ukraine and Poland is scrutinized. Features of current stoping equipment are reviewed. Technical characteristics of modern high-productive stoping equipment are given. The reasons about implementation of new shearers for increase load on longwall are substantiated.

1 INTRODUCTION

Starting from the middle of previous age coal energy that stayed "a motor" of world's industrial revolution became deadlock. In energetic balance gas, oil and nuclear fuel became more and more popular. However, in last year's situation changed in root and branch.

Journal "The Economist" (http: // www. economist.com/) recently turned attention that in Europe comes back hard coal and is possible its renaissance. World's energy report 2013 shown that in 2012 consumption of hard coal in Europe increased in comparison with previous years on 6.3%. Besides, in some countries of European Union quantity of produced energy with help of coal industry was grown on 50%.

Everybody knows that coal is one of the dirtiest sources of energy but it also the most available in all ways, from extraction (60% of total world stocks of organic fuel) to cost, because coal is still the cheapest energy resource. Also it is a single source of fuel which stocks in subsurface resources of Poland and Ukraine is enough for satisfaction of all economic sectors needs during several ages (Figure 1). At once, coal deposits are characterized by extremely hard mine-and-geological conditions of exploitation, so that is why most part of coal enterprises has small productive capacity. To improve the results of production it is necessary to apply the technology of the highest technical advancement. An example of this technique can be Caterpillar plow complex and a drum shearer made by Famur.

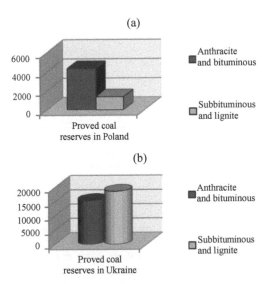

Figure 1. Comparison diagrams of proved coal stocks in Poland and Ukraine (according to BP statistical review of world energy).

According to BP statistical review of world energy by coal extraction level Ukraine and Poland take 12th and 9th places suitably (Table 1). That is why increasing of extraction volume is still priority of coal enterprises of these countries. The main ways how to solve this issue is implementation of new high-productivity stoping equipment and increase of load on longwalls.

Table 1. World's top countries in coal production. (http: //www.bp.com/en/global/corporate/about-bp/energyeconomics /statistical-review-of-world-energy-2013/review-by-energy-type/coal.html).

Country	Production in 2012, million tonnes	Changing of production in comparison with 2011, %	Share of total production, %
China	3650.0	+3.5	46.41
USA	922.1	-7.5	11.72
India	605.8	+6.0	7.70
Australia	431.2	+4.2	5.48
Indonesia	386.0	+9.0	4.91
Russia	354.8	+6.1	4.51
RSA	260.0	+3.1	3.31
Germany	196.2	+2.0	2.49
Poland	144.1	+3.6	1.83
Kazakhstan	116.4	+4.2	1.48
Columbia	89.2	+3.7	1.13
Ukraine	88.2	+4.0	1.12
Turkey	72.0	-5.6	0.92

2 SEAMS CLASSIFICATION AND COAL STOCKS

Coal seams classification by thickness in Ukraine and Poland is rather different. According to the operational regulations in Ukraine coal seams are divided into 4 groups (Table 2).

Table 2. Seams classification in Ukraine (Bondarenko 2003).

Groups of seams	Seam thickness, m
Very thin	< 0.7
Thin	0.71 – 1.2
Average thickness	1.21 – 3.5
Thick	>3.5

The main Ukrainian coal stocks are concentrated in flat (0–18°) seams with thickness from 0.7 to 2.6 m. More than 85% of proved stocks of Donbass are represented by seams with thickness to 1.2 m and level of extraction from these seams at present time consists only 54%. Current stoping equipment make coal-cutting with stone (in some cases more than 30% of seam thickness). As a result increase energy demands in longwall and cost of output product, sharply decreases quality of coal because of ash content growth (Shraiber 2008). In Poland, the seams are divided into:

- thin – 1.6 m;
- medium – from 1.6 to 2.5 m (or 3.5 m);
- thick – over 2.5 m (or 3.5 m).
- longwall excavations are divided into:
- low – from 1.2 to 1.5 m;
- medium – from 1.5 to 2.5 m;
- high – over 2.5 m.

A Currently, thin layers are operated in Lublin Węgiel Bogdanka, JSW and Katowice Coal Holding. The first longwall thin seams also purchased Coal Company (KW). Planned start of operation in

KW is planned for the second half of 2014.

Back to the exploitation of thin seams was come in 2009 after several years of absence. The reason was reducing of resources in the provided medium and thick seams. Resource data contained in thin seams is shown in Figure 2 and 3. For "Bogdanka" mine part of resources in thin seams is about 35% (Stopa 2008).

Figure 2. Distribution of stocks by seam thickness on mine "Bogdanka", Poland.

According to the data in the fields of mines JSW (JSW) documented a total of 40.3 million tonnes (14.1%) of operational reserves defaulting on seams with thickness up to 1.5 m. In addition, large amounts of coal remain in the seams thicknesses of 1.5–2.0 m, which can also be exploited profitably using plow technology. Figure 3 shows the structure of industrial resources in the mines of JSW SA. Perspective resource base for operation using the plowing are seams that remain within the areas Bzie-Debina 1 – West and Pawłowice 1. Volume of documented industrial resources in these areas is estimated at about 190 million tonnes, and the vast majority is hard coal.

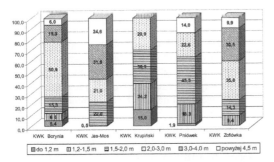

Figure 3. The structure of industrial resources of coal in the mines of JSW SA, along with thickness of the layers of compartments.

3 CURRENT POSITION OF STOPING EQUIPMENT

On the whole term of coal deposit exploitation always appeared question about equipment for coal extraction improvement. The main direction of stoping complex mechanization is implementation of modern shearers and expansion of its application area. Efficient operation of shearer on seams with thickness near 1 m is achieved during load on longwall more than 1500 t/day. A wide range of previous Ukrainian production shearers didn't reach required loads, as we can see proceeding from technical characteristics (Table 3).

Table 3. Technical characteristics of USSR-manufactured shearers.

Shearer type	1K101	K103	KA90	MK67
Production capacity, t / min	1.2–1.8	2.0–3.2	2.2–3.3	2.0–2.5
Extracting seam thickness, m	0.75–1.2	0.7–1.4	0.8-1.25	0.7–1.3
Nominal web width, m	0.63; 0.8	0.8	0.8	0.8
Maximal intake velocity, m / min	4.4	5.0	5.0	1.27
Weight, t	10.4	17.5	12.8	8.4

Nowadays on Western Donbass mines operate shearers of Corum Croup (in the past Scientific-Industrial Complex "Mining machinery"), and also complexes OSTROJ, OSTROJ 75/120 (with shearers MB 410E, MB 444 P) of "T Machinery", The Czech Republic-manufactured and plow Bucyrus DBT Europe GmbH-manufactured (Table 4).

Table 4. Technical characteristics of Corum Group and "T Machinery"– manufactured modern shearers.

Shearer type	UKD 200	UKD 200-250	UKD 300	UKD 400	KA 200	MB 410E	MB 444 P
Production capacity, t / min	3.0–5.0	3.3–5.5	4.0–8.0	5.5–12.0	5.0	3.0–9.0	6.0–7.0
Extracting seam thickness, m	0.8–1.3	0.85–1.3	0.85–1.3	0.85–1.5	0.8–1.25	0.9–1.8	0.88–1.3
Nominal web width, m	0.63; 0.8	0.63; 0.8	0.7; 0.8	0.7	0.8	0.8	0.7; 0.9
Maximal intake velocity, m / min	5.0	5.0	12.0	12.0	5.0	12.0	11.5
Weight, t	14.5	14.4	18.2	21.0	13.0	18.0	22.0

On Figure 4 you can see quantity of Ukrainian and foreign shearers that apply in mine-and-geological conditions of Western Donbass mines.

This figure shows that shearers of Ukraine-manufactured are almost equal to foreign counterparts in production capacity. However, the main problem is huge expenses during exploitation period because of short life of details and mechanisms. No doubts, foreign shearers are much expensive than Ukrainian, but exploitation expenses are less and all modern foreign equipment can use in few longwalls without hoisting on surface for overhaul. The level of extraction from flat thin seams by different shearers is represented on Figure 5.

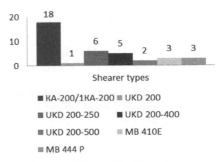

Figure 4. Shearers of Western Donbass mines.

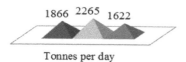

■ MB 410E   ■ MB 444 P   ■ UKD 200

1866 2265 1622

Tonnes per day

Figure 5. Level of coal extraction per day by Ukrainian and Czech shearers on example of "Zapadno-Donbasskaya" mine PJSC DTEK Pavlogradugol.

As for Poland, calculated economic viability of the planned project was the basis for the start drafting designs for usage plow technique under Coal Mine "Zofiówka" (Figure 6). The starting point for determining the selection of technical parameters of the complex plow was to define the basic data mining and geological seams planned for operation. Adopted one very important condition – the plow should also allow the operation of medium-sized seams (Kicki 2011).

Taken geological and mining details:
- the thickness of seams: 1.0–2.2 m.
- the incline of seams: longitudinal up to 10° and transvers up to 11°.
- geomechanical parameters of the rock mass: compressive strength RC of coal up to 10 MPa; compressive strength RC of roof up to 50 MPa; compressive strength RC of floor up to do 40 MPa.
- pollution of excavated rock up to 40%.

Next, were defined the basic parameters for complex devices:
- voltage: 3300 V;
- require the ability to work on longitudinal slopes up to 18°;
- require the ability to work on transverse slopes up to 20°;
- power bus pressure: 25–30 MPa;
- the length of the conveyor (from the axis of the star drive spout to the axis of the star power return): minimal 225 m;
- plow complex will be equipped with a set of components necessary for building both in the wall of the left and right;
- hazards which will be conducted in the wall: of water – grade I; of coal dust explosion – class B; of methane – IV category; of rock burst – none.

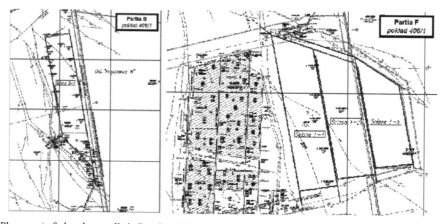

Figure 6. Placement of plow longwalls in Part B and F seam 406/1 on KWK "Zofiówka".

4 PERSPECTIVE EQUIPMENT FOR THIN COAL SEAMS DEVELOPMENT

Extraction using plow systems in Poland is carried out in recent years in several longwalls. Under the conditions of the mines JSW SA uses two complexes plow. Their performance in the process of coal exploitation in the recent period is limited by the presence of methane, which forces the output concentration limit, as rapid progress further releases of methane, hence the production results may not be comparable with the results of other mines. In such a situation the industry is also faced with a shearer complex used to exploit low coal seams in the Katowice Coal Holding SA (Zorychta 2008).

For the exploitation low seams, Katowice Coal Holding had acquired shearer complex type FL 12/18 (for the lower seams) production Famur SA (Figure 7). The technical equipment of the complex shearer for the low seams: shearer FS-200 – is a machine working "alongside" the AFC equipped with a no-linkage type Eicotrack ride system. Due to the design, the drum must be guided to the bottom of the body to find the space for the shearer.

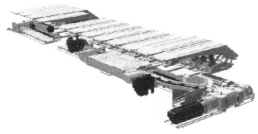

Figure 7. General view of mechanized longwall complex FL 12/18.

Technical characteristics:
- Mining height range: 1.2–1.8 m.
- Diameter of cutting drums: 1200 mm.
- Cutting depth: 900 mm.
- Power transferred to the drum: 200 kW.
- Installed capacity: 467.5 kW.

Mechanized roof support type FAZOS-10/19-2×2340 – is machinery with possibility of support-covering designed to select seams endangered and non-hazardous rock burst, mined by longwall with the collapse of the roof.

To work in seams impact rock mass housing is adapted through the use of relief valves. Each housing section is equipped with a pressure gauge in the space under the piston rack and with its own power source.

AFC type FFC 750 – equipped with a system of moving type Eicotrack adapted to cooperate with a combine type FS 200 and mechanized roof support type FAZOS 10/19-2×2340. The conveyor is equipped with three engines with gearbox size 25 and PL25 FKPL25 type complying with standard RAGN 335000.

Longwall conveyor – in this case applied PPZ-850/1×65/200 Nowomag type with crusher type KD-1600/1×160 Nowomag. It is adapted to cooperate with the drive cross the AFC FFC 750.

For the first time the complex FL 12/18 was used to operate the longwall No 703 in seam 407/1 in the mine Murcki-Staszic part Staszic. Wall was started in April 2011.

Katowice Coal Holding SA before buying the above longwall mining with minimal assumed the longwall of 3000 t/day (for comparison: "Bogdanka" SA assumes production of 10 000 t/day, and mining at such a level in practice by actually that is chosen and applied technology implements). Katowice Coal Holding SA in the initial phase of longwall extraction 703 was obtained at the level of 1.900 t/day average. In August, the average daily production was 3977

tons at the monthly progress of the longwall of at 197 m. The other hand the maximum daily production has been achieved on this wall has exceeded 6200 tonnes (Gach 2011).

The reason for lower production in the initial period of operation of the longwall was technical difficulties and the use of new
technologies, with which the crew have not yet had to deal with. After eliminating the causes of shutdowns and as a result of changes in work organization increased production, which in September was 4.700 tons per day of average.

Analysis of Ukrainian mining industry shows that in next years for increasing level of extraction it is necessary to implement high-productivity shearers of modern generation. That is why we choose foreign counterparts.

Shearer MB 12 version MB 320 (Figure 8) "T Machinery" – manufactured (http://www.tmachinery.cz/) designed for two-sided coal extraction and fitted by chainless haulage with frequency automatic control depending on hardness of extracted coal. Shearer safety movement along chain conveyor materialized because of two feed-motions with magnetic brake. Shearer manipulation can execute from panel on body frame, or remotely with help of wireless control system. The main advantages of these shearers are follows: short length, simple mounting/dismounting and increased rigidity because of units on single-block body frame.

Figure 8. General view of shearer MB12 "T Machinery"–manufactured.

Another variant are shearers (Figure 9) MG350/811-WD and MG180/420-BWD "Joy Global" – manufactured (China) that have wide potential of installed power requirement with application of electrohydraulic feed type (http: //gjmjru. icm.cn/products/caimeiji.htm).

Main technical and constructive specialties are following:
- increased power available per worker of shearer allows to make a cut in longwalls with high coal hardness coefficient with interlayers presence;

MG350/811-WD

MG180/420-BWD

Figure 9. Visual representation of shearers MG350/811-WD and MG180/420-BWD "Joy Global"-manufactured.

• feasibility of mechanized support and conveyor automatic control;

feasibility of simultaneous automatic regulation of two augers height;

• body frame of shearer executed as monolith force frame;

• main components of shearer are made as block construction;

• joining hydraulic jacks and bolts of block constructions are executed from high-tension steel;

• applies frequency converter of Joy-manufactured;

• presence of gear box cooling system of cutting part and electrical equipment;

• shearer has automatic operating system that provides supporting of sufficient speed and it automatic decrease or shearer tripping during increase of loads more than acceptable;

• control system provides diagnostic of shearer's work and operates in local mode or remotely, condition of main components is shown on technical data display;

• in electrical control system applies interface with protocol CAN and TFT monitor 4.7 inches.

5 CONCLUSIONS

1. Transition to renewable energy sources is very expensive and illusive. That is why to reach higher level of coal extraction, our countries have to increase load on longwall by means of modern stoping equipment application.

2. At present the question about implementation of new high-productive equipment is very actually and demands further development. Nowadays foreign shearers are much better than native counterpart, that's why it is better to use such equipment as: "Joy Global" and "T Machinery".

3. World's progress improves from day to day, so our native engineers and constructors have to find another ways to solve problems that always happened with current equipment.

4. There is no currently shearer complexes for thin seams, which manufacturers guaranteed achieve production level receive from the "Bogdanka" SA using technology plow (average production declared by the Polish producer of the shearer complex combine to thin seams of only 3.500 t/day, which is comparable with the results of 10 000 t/day achieved by "Bogdanka" SA techno-logy plow).

5. Proposed shearers and plows can operate without repair in 1.5–2 times longer that current equipment. That is why expenses will repay in short terms.

REFERENCES

Bondarenko, V., Kuzmenko, A., Gryadushii, Yu., etc. 2003. *Technology of underground mining of layered deposits.* Dnipropetrovs'k: National Mining University: 708.

Shraiber, A. & Redkin, V. 2008. *Modern and perspective technologies of coal extraction.* Kiev: Institute of Total Energetics. Issues of total energetics: 7–13.

Stopa, Z. 2008. *View of Thin Hard Coal Seam Mining in the L W "Bogdanka" Grou.* Poland: Materiały Szkoły Eksploatacji Podziemnej.

Zorychta, A., Tor, A.& Plutecki, J. 2008. *Possibilities of thin seams mining in Polish collieries, on example of Jastrzębska Spółka Węglowa SA.* Polska Akademia Nauk: Mineral Resources Management.

Gach, Z., Skrabaka, M., Nosal, T. & Holewa, R. 2011. *Mining of Thin Hard Coal Seams in the Mines Belonging to Katowicki Holding Węglowy Using Shearer Longwall System.* New Techniques and Technologies in Thin Coal Seam Exploitation. International Mining Forum.

Kicki, J. & Dyczko, A. 2011. *Plow technology – the practice of implementation.* Krakow. Monografia. Wydawca: Fundacja dla AGH. ISBN-978-83-62079-12-4.

BP Statistical review of world energy: *http://www.bp.com/en/global/corporate/about-bp/energy-economics/statistical-review-of-world-energy-2013/review-by-energy-type/coal.html*

Joy Global: *http://gjmjru.icm.cn/products/caimeiji.htm*

T Machinery: *http://www.tmachinery.cz/*

The Economist: *http://www.economist.com/*

Progressive Technologies of Coal, Coalbed Methane, and Ores Mining – Bondarenko, Kovalevs'ka & Ganushevych (eds)
© 2014 Taylor & Francis Group, London, ISBN: 978-1-138-02699-5

Perspective ways of mine locomotives autonomy increase

S. Bartashevskii, O. Koptovets, O. Novitskii & D. Afonin
National Mining University, Dnipropetrovs'k, Ukraine

ABSTRACT: The paper concerns methods of studying possible ways to improve autonomy of mine locomotives operating in Ukrainian coal mines being gas-and-dust-dangerous; characteristics of fuel cell basic systems, their fuel types as well as their storage conditions are analyzed; advanced systems of fuel cells are determined. Their application enables increasing travel of locomotive improving locomotive transport basic performance indicators.

1 INTRODUCTION

Electric locomotives are the main type of locomotives that used in coal mines; they are more than 95% of total amount. That depends on the fact that the type of locomotives does not contaminate mine air having very high efficiency (up to 80–85%) and hyperbolical traction performance. Capability of motor to withstand sudden overloads simplifies overcoming steep gradients of the way. For reasons of safety, gas-and-dust dangerous mines apply accumulator locomotives by MI and MS.

Steady growth of mean route extent is common practice today. That depends on the fact that mining operations approach the boundaries of forecasted mine fields sometimes leaving them. Moreover, battery capacity restricts travel of accumulator locomotives. Analysis of specification of available electrochemical systems of accumulators shows that in the near future increase in battery capacity will become increase in their mass and dimensions parameters (Debelyi et al. 2006).In the foreseeable future increase in mine locomotives travel owing to batteries of increased specific power capacity seems to be improbable (Bartashevskii 2011).

Attempts to replace accumulator electric locomotives with mine diesel locomotives using high-energy fuel identify a number of problems connected with their operation. Among them are:

– unbalance in standards of harmful substances content in diesel engine exhaust between equipment produced in EU countries and Ukraine (Mokhelnik & Kovarzch 2002);

– problems connected with exhaust gas dilution; especially it concerns locomotive performance in blind drifts;

– unbalance between maintenance personnel training level and requirements for maintenance of the locomotives motors, transmission systems, and exhaust control systems.

2 MAIN PART

Application of fuel cells (FCs) allowing avoidance of toxic exhaust while using high-energy fuel is trade-off decision.

FCs are electrochemical generators which can run continuously owing to steady supply of new portions of reagents and product takeout (Varypaiev et al. 1990). Practically, flashless, catalytic fuel oxidation is continuous process in FCs.

Advantages of fuel cells are:

– easy maintenance (practically they need not any maintenance);

– durability (working life is up to 10−15 years);

– use of high-energy fuel (hydrogen, methanol);

– efficiency is higher to compare with accumulator batteries (AB) and diesels (up to 80%);

– $2H_2O$ or $2H_2O + CO_2$ is basic coproduct (depending upon cell type).

Following types of fuel cells classification are available today:

– according to electrolyte type;

– according to a type of fuel and oxidizing compound;

– according to operational temperature which depends heavily on a type of electrolyte and fuel.

According to oxidizing compound, fuel cells are conditionally divided into oxygenous and aerial (the latter use atmospheric oxygen as oxidizing compound).

According to fuel type hydrogenous, methanol, and natural gas-based fuel cells are singled out; however, from viewpoint of chemistry, the latter ones should be regarded as hydrogenous as previously natural gas experiences conversion.

As for operational temperature they are conventionally divided into:

– high-temperature (up to 1000 °C);

– medium-temperature (up to 250 °C);

– low-temperature (up to 100 °C).

For locomotives operating in mine workings with restricted dimensions, in dust and gassed environment, overall dimensions of fuel cells, their operational temperature, fuel type being applied, and necessity to use extra equipment (fuel converters, cooling systems etc.) are of fundamental importance. Analysis of FCs specification is based on classification of FCs according to operational temperature.

Table 1 demonstrates specification of high- and medium-temperature FCs.

Table 1. High- and medium-temperature fuel cells at the stage of industrial and pilot production.

Type of fuel cell	Operational temperature	Energy generation efficiency	Fuel type	Application area
MCFC	550–700 °C	50–70%	Majority of hydrocarbon fuel types	Large and mid-sized units
SOFC	450–1000 °C	45–70%	Majority of hydrocarbon fuel types	Small, mid-sized, and large units
PAFC	100–220 °C	35–40%	Pure hydrogen	Large units

Fuel Cells on Carbonate Melt (MCFC). High operational temperature makes it possible to use natural gas without fuel processor. The cells apply electrolyte from a mixture of molten carbonate salts. The two types of mixtures are currently used: lithium carbonate and potassium carbonate, or lithium carbonate and sodium carbonate. To melt carbonate salts and reach high degree of ion mobility in electrolyte, up to 650 °C temperature is used for action of fuel cells with molt carbonate electrolyte.

Solid Oxide Fuel Cells (SOFC). Operational temperature can vary from 600 to 1000 °C; that makes it possible to use various fuel types without special pretreatment. Useable electrolyte is thin solid ceramic-based metallic oxide. Frequently it is ittrium-zirconium alloy being conductor material for oxygen ions (O_2^-).

Fuel Cells on Carbonates Melt (PAFC). Phosphoric acid-based (orthophosphoric acid-based) fuel cells is using orthophosphoric acid-based electrolyte (H_3PO_4) with up to 100% concentration. Ionic conductivity of orthophosphoric acid is low at low temperature; that is why the fuel cells are used at up to 150–220 °C temperatures.

High operational temperature of above-mentioned FCs keeps them from using for mine locomotives.

In this context, Table 2 demonstrates low-temperature FCs as the most suitable for mine electric locomotives.

Table 2. Low-temperature fuel cells at the stage of industrial and pilot production.

Type of fuel cell	Operational temperature	Energy generation efficiency	Fuel type	Application area
PEMFC	30–100 °C	35–50%	Pure hydrogen	Small units
DMFC	20–90 °C	20–30%	Methanol	Small units
AFC	50–200 °C	40–65%	Pure hydrogen	Small units

Fuel Cells with Proton Exchange Membrane (PEMFC). Solid polymeric membrane (thin plastic film) is electrolyte in the fuel cells. While water soaking, the polymer transmits protons rather than electrons.

Oxidizing reaction takes place within anode

$$H_2 \rightarrow 2H^+ + 2e^- . \tag{1}$$

Proton reduction reaction with water formation takes place within cathode.

$$\tfrac{1}{2}O_2 + 2H^+ + 2e^- \rightarrow H_2O^- . \tag{2}$$

Water is the reaction by-product.

Fuel Cells with Direct Methanol Oxidation (DMFC). Polymer is used as electrolyte; hydrogen ion (proton) is used as charge carrier. Within anode a process of methanol oxidation takes place; within catalyzator it transforms into carbon dioxide

$$CH_3OH + H_2O \rightarrow CO_2 + 6H^+ + 6e^- . \tag{3}$$

Within cathode, protons, transmitting through proton-exchange membrane, react with oxygen with water formation

$$1.5O_2 + 6H^+ + 6e^- \rightarrow 3H_2O . \tag{4}$$

Water and carbon dioxide are the reaction by-products.

Alkaline Fuel Cells (AFC). Electrolyte is used in alkaline fuel cells; the electrolyte is water solution of potassium hydroxide contained in stabilized porous matrix.

Within anode, hydroxide ion reacts with hydrogen with water formation

$$2H_2 + 4OH^- \rightarrow 4H_2O + 4e^-. \tag{5}$$

Within cathode, water transforms into hydroxylic cell

$$O_2 + 4e^- + 2H_2O \rightarrow 4OH^-. \tag{6}$$

Water is the reaction by-product.

As the Table explains, either hydrogen, or methanol is fuel for the cells.

Problems connected with hydrogen storage are the key ones while using it (Physicochemical Properties... 2013):

– pressurized hydrogen storage needs high-strength multilayered steel balloons with internal liners made of austenite steel. Hydrogen corrosion of metal results in leakage; in turn it invites danger of detonating gas explosion;

– pressurized hydrogen storage in Dewar flasks involves heavy expenses for hydrogen cooling down to −253 °C. Cryogenic storage facilities have not ideal thermal covering; that is why they are equipped with blowing systems inviting danger of detonating gas explosion;

– storage of hydrogen in metal hydrides. Hydrogen is very soluble in many metals; that allows for storing it in tanks having spongeous mass or nanotubes made of noble metals. Disadvantage of such a storage technique is in high cost and necessity to heat metal hydrides up for stimulation of gaseous hydrogen discharge; within boundary layer of filling compound/liner, hydrogen corrosion of tank metal results in leakage inviting detonating gas explosion.

Thus, no one of available storage technique can provide safe and efficient hydrogen storage.

Methanol is burnable, toxic, and explosive; it steams actively at room temperature (Druzhinin & Korichev 2009).

Hence, a type of fuel applied plays a key part while selecting FE to provide its safe use in terms of coal mines.

Analysis of related research makes it possible to identify two types of low-temperature fuel cells using pilot explosion- and flameproof and low-toxic fuel cells.

Formic acid and zinc are respectively their fuel.

Direct formic acid-based fuel cells (DFAFC) are a subtype of PEMFC– devices with direct supply of formic acid (Rice et al. 2002).

Formic (methanoic) acid within FC anode experiences oxidation with CO_2 formation

$$HCOOH \rightarrow CO_2 + 2H^+ + 2e^-. \tag{7}$$

Moreover, oxygen O_2 recovery takes place within cathode with H_2O water formation

$$O_2 + 2H^+ + 2e^- \rightarrow H_2O. \tag{8}$$

Thus, chemically pure water and carbon dioxide are by-products of fuel cell operation. They may be utilized in proper tanks to be later used in a mine.

At standard temperature formic acid is liquid. It is nonflammable without explosive vapor. Its handling is similar to that with acid electrolytes. In contrast to methanol, formic acid cannot penetrate membrane. By definition, DFAFC efficiency should be higher. Until recently many scientists have not regarded the technique as that having practical future. For years researchers rejected formic acid due to high electrochemical overvoltage resulting in drastic electric loss. However, recent results show that such inefficiency results from use of platinum as catalyzer which was common practice. After researchers of Illinois University ran a number of experiments with other materials, it emerged that if palladium is applied as a catalyzer, DFAFC efficiency is higher to compare with direct equivalent methanol fuel cells. Today American Company Tekion holds rights to the technique proposing own product line Formira Power Pack for microelectronics devices.

Necessity to use expensive platinum or palladic catalyzers to cover graphite nanotubes is among disadvantages of formic acid-based cells. Fuel (formic acid is the most dangerous of fatty acids!) is of serious hazard to personnel. In contrast to such inorganic acids as sulphuric one, it easily penetrates through fatty layer of skin. Even small its amount induces severe pain; invaded tissues turn into crust with up to several millimeters thickness. Evaporation of even few drops of formic acid may cause strong irritation of eyes and irritation of respiratory system.

In view of the above, another type of fuel cells is more potential for mine locomotives.

Zinc-Air Fuel Cells (ZAFC) (Electrochemistry Handbook 1981). Zinc oxidation with the help of atmospheric oxygen is applied for electrical energy generation. Figure 1 demonstrates configuration of zinc-air fuel cell.

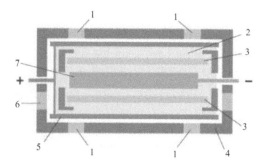

Figure 1. Zinc-Air Fuel Cell: 1 – air channels, 2 – electrolyte, 3– separator, 4 – capsule, 5 –cathode, 6 – air hole, 7 – Zinc anode.

Alongside of anode, a mixture of Zinc particles and electrolyte is located and oxidizing reaction takes place

$$Zn + 2H_2O^- \rightarrow Zn(OH)_2 + 2e^-. \qquad (9)$$

Alongside of cathode, water and atmospheric oxygen interreact forming hydroxyl (its molecule is oxygen atom and hydrogen atom with covalent forces)

$$O_2 + 2H_2O + 4e^- \rightarrow 4OH^-. \qquad (10)$$

As a result of hydroxyl-zinc mixture reaction, electrons moving to cathode are released. Zinc oxide and water are end products of the electrochemical reaction.

Power density of zinc-air fuel cells is 2 to 2.5 higher to compare with rechargeable lithium-ion cells being today energy thickest.

In the process of consumption, on the part of anode, cell becomes clear once more filling in with Zincy compound or zincky granules. Indisputability advantage of ZAFC is a capability to control reaction just correcting air feed into fuel cell. As for current zinc-air cells, air design humidity may reach 60%; if it increases up to 90%, then FC life experiences 1.5 times drop. However, from engineer viewpoint it is quite simple to dehumidify air and extract dust from it. It is possible to use a combination of air filters made from available nonwoven fabrics using silica aerogel SiO_2 it can absorb moisture content reaching up to 30% of dead weight.

Zinc can be regarded as optimum fuel type for mine locomotives. Zinc and its oxide are non-toxic, incombustible, and inexplosive; they need not specific protective measures for personnel. Zinc can be charged within any point of locomotive route; no special facilities for charge cases, fuel storages, their separate ventilation etc. are required.

3 CONCLUSIONS

Subsequent activities connected with the type of fuel cells make it possible to update available accumulator locomotives replacing battery cases by containers with fuel packing's, proper facilities to store and supply fuel and oxidizer, and to change controllers for adequate control systems.

Inturn, fuel cells will help to lift restrictions on route of locomotive using one charge improving basic performance indicators of coalmine locomotive transport.

REFERENCES

Debelyi, V., Debelyi, L. & Melnikov, S. 2006. *Key Trends in Mine Locomotive Transport Development.* Kyiv: Coal of Ukraine, 6: 30–31.

Bartashevskii, S. 2011. *Technological Principles to Transfer Mine Locomotive Transport to Unconventional Power.* Problems of Mining and Environment in Mining Practice (Collected Papers) Proceedings of the 6th Research and Practice Conference (13–14 May 2011). Donets'k: Donbass: 136–140.

Mokhelnik, P & Kovarzch, P. 2002. *Explosion-Protected Czech Mine Diesel Locomotives.* Glückauf, 1. 50–52.

Varypaev, V.N., Daosian, M.A. & Nikolskii, V.A. 1990. *Chemical Cells.* Moscow: High school: 240.

Physicochemical Properties of Matters. 2013. Chemistry Handbook [under the Editorship of Keeper, R.A.]. Khabarovsk: Prondo: 250.

Druzhinin, P.V. & Korichev, A.A. 2009. *On the Problem of Hydrogen Storage.* Engineering and Technological Problems of Service, 3(9): 51–53.

Rice, C., HaS., Masel, R.., Waszczuk, P., Wieckowski, A. & Tom Barnard. 2002. *Journal of Power Sources.* Volume 111: 83–89.

Electrochemistry Handbook [under the Editorship of Sukhotin, A.]. 1981. Leningrad: Chemistry: 400.

Progressive Technologies of Coal, Coalbed Methane, and Ores Mining – Bondarenko, Kovalevs'ka & Ganushevych (eds)
© 2014 Taylor & Francis Group, London, ISBN: 978-1-138-02699-5

The geomechanical substantiation of efficiency of ground dikes grouting of mine hydroheaders

P. Dolzhikov, N. Paleychuk & O. Ryzhikova
Donbass State Technical University, Alchevs'k, Ukraine

ABSTRACT: The article considers research results of dike body's stress state of mine hydroheader on the formation of ground deconsolidation zone and after fixing clay and cement grouting mortar. The expediency and efficiency of the dike grouting in the relevant geotechnical conditions is substantiated.

1 INTRODUCTION

Among variety of hydrotechnical constructions a special place is occupied by ground hydroheaders providing an accumulation and use of hydro resources of mining and ore processing fields. In Ukraine there are more than 500 hydroheaders differing by class of responsibility, capacity, height, length, relief of base, method of construction and other characteristics.

Currently, coal and mining industries widely use following kinds of hydrostorages as slime tanks, tailing dikes, mine water dikes, etc. Most of them are being built with local bulk materials with high water resistance on compaction, such as clay or loam.

The operational state of hydroheaders is determined by geotechnical, technical and technological factors, among which should be distinguished geological structure and relief of ground, quality of dike construction, geological and granulometric composition of storage dike, existence and magnitude of filtration leaks through the dike body and the base of hydrotechnical construction.

As at any industrial enterprises, at the hydroheaders there are occurred emergencies and accidents, the most typical causes of which are shown at Figure 1 (Arustamov 2006), that the most valid reason of emergencies at hydrotechnical objects is destruction of base and dikes.

After analyzing statistical data of accidents at the hydro constructions of Ukraine on the constructive features it was noted that destruction of dikes occurs 1.7 times more often than the destruction of the base (Kostyukov 2005, Dolzhikov 2006). Thus, the state of the dike of mine hydroheader is one of key factors that determining the economic feasibility of object operation as a violation of its stability can lead to significant economic and environmental damage connected with leaks of polluted water, which is usually found in operation period of the hydrotechnical construction.

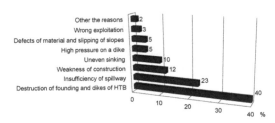

Figure 1. The reasons of emergencies and accidents at hydrotechnical objects.

By grouping accidents for the reasons, the most characteristic for mine hydroheader of Donbass, as a result of statistical analysis and technical documentation it was found that over 70% of accidents on the ground headers are connected with the appearance of local, actively developing filtration streams through the body and the base of the dike, and therefore its resistance is decreased, hydrogeological mode of header is changed and zones of ground deconsolidation are formed (Dolzhikov 2004).

At present there are many ways to eliminate an emergency condition of groundwater dikes, among them clay and cement grouting is recognized as one of the most efficient, allowing to consolidate decompressed ground and to eliminate filtration streams through the body and the base of the hydroheader dike (Dolzhikov 2004, 2006).

The advantages of this method are rather low cost, durability and possibility of using of clay material directly or nearby the object of grouting. However, practical recommendations to implement of grouting under specific conditions are related to variability of geotechnical information, state of the dike body and storage base and also a variation in the recipe of components in grouting mortar.

2 FORMULATING THE PROBLEM

Due to differences in geotechnical, technical, technological and operational conditions of the mine hydroheaders the use of clay and cement grouting mortar at a particular object requires an appropriate substantiation.

The article aims of geomechanical substantiation of feasibility and efficiency of grouting with clay and cement mortar, which is realizable with using of the finite element method (FEM) that allows to measure the stress-strain state (SSS) of grounds, rocks and grouting mortar, it is necessary to predict the scenarios of development of emergency situations on the HTS.

The objectives of the research are:
– to select a recipe of clay and cement mortar with appropriate strain-strength characteristics;
– to make a geomechanical stages of operating of the HTS' dikes;
– to research and to analyze the stress-strain state of hydroheader;
– to determine the efficiency of grouting with clay and cement mortar.

3 THE SUBSTANTIATION OF GROUTING EFFICIENCY OF THE GROUND DIKES BY THE METHOD OF EVENTUAL ELEMENTS

To implement a numerical simulation it had been used a software package "Lira", version 9.6, developed by "Lira-SAPR" Ltd. (Kyiv). Carrying out the numerical experiment it had been used triangular, rectangular and square eventual elements of ground deformation. The model is characterized by the following parameters: length on the horizontal axis (x) – 95.70 m, height (z) – 22.70 m; width of the dike at the base is $L = 15.50$ m; width of the dike at the top is $L_1 = 5.65$ m; dike height is $h = 11.95$ m; inclination angles of generatrix surface of the dike body α and β – 68° and 67° accordingly; inclination angle of hydroheader base δ to the horizontal axis x is 10°. The width of a water surface on the axis x – 63.55 m. The scheme indication – six degrees of freedom. Step of mesh triangulation FE – 10.00 × 10.00 m for EGE1, 1.00×1.00 m for EGE2, 0.25 × 0.25 m for a dike and 0.10 × 0.10 m for a deconsolidation zone, that due to the interest in study of stress-strain state base and dike while the stress distribution deep into rock massif for this problem less important. The load from the weight of water was set equals to 0.01 MN/m². Limit conditions – connections in all directions.

In the model has been accepted following materials: limestone (EGE1), loam (EGE2), decompressed loam (EGE3) and clay-cement grouting mortar (EGE4).

The rated scheme of the model is presented at Figure 2.

The stress-strain and physic-mechanical properties of model materials are shown in Table 1.

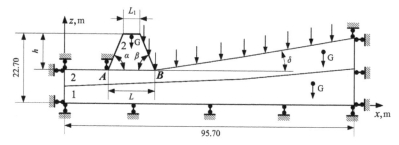

Figure 2. Chart to research of the stress-strain state of dike body and basis of mine hydrotechnical construction: 1 – engineer-geological element #1 (limestone); 2 – engineer-geological element #2 (loam).

Table 1. Physic-mechanical properties of model materials.

Engineer-geological element	E, MPa	v	H, m	γ, MN/m³	C, MPa	R_t, MPa	φ, degrees	k_e
1 (EGE1)	$17.4 \cdot 10^3$	0.17	0.10	$2.66 \cdot 10^{-3}$	23.5800	12.6000	33	3
2 (EGE 2)	14.71	0.35	0.10	$1.96 \cdot 10^{-3}$	0.0049	0.0005	16	3
3 (EGE 3)	12.85	0.40	0.10	$1.50 \cdot 10^{-3}$	0.0035	0.0002	16	3
4 (EGE 4)	24.51	0.27	0.10	$2.10 \cdot 10^{-3}$	0.0340	0.0035	25	3

The designations: E – the module of ground deformation on branche of the primary load; v – the Poisson coefficient; H – the calculation thickness of the element; γ – the unit weight of the ground; K_e – the conversion coefficient to the module of ground deformation on the secondary loading branche $E_e = k_e E$; C – the structural cohesion; R_t – the ultimate stress at the strain; φ – the angle of internal friction.

Since a process of ground decompression with the eventual element method is not possible to model, it was selected the most typical geomechanical situations to be studied such as an initial state of ground dike (Figure 3a), an initial stage of filtration to form a decompression zone (Figure 3b), an outcrop of deconsolidation zone on one of the sides of dike (Figure 3c), a formation of hydrospaltung cavity in the area of ground deconsolidation (Figure 3d) and a consolidation of reduced zone with a clay and cement grouting mortar (Figure 3e).

The geometric parameters and shapes of decompression zone and of hydrofracturing canal were accepted according to the recommendations to create a process model of the grouting of deconsolidated ground with viscoplastic solutions (Dolzhikov 2004, 2006), and also based on the actual data about the power of deconsolidated zone in different grounds (Dolzhikov 2006, Kostyukov 2005). At the final stage of numerical experiment in canal the presence of hydrofracturing (Figure 3d) of grouting mortar was modeled with assignment this area of the model of reduced stress-strength characteristics. Thus, in particular, the deformation module was taken equals to 19.6 MPa, which corresponds to the coefficient of a module reduction of $k = 0.8$, the Poisson coefficient was taken as the value of 0.32. This method allows to model a situation where a cavity is filled with a material that is a prerequisite for an adequate description of the contact conditions in continuum mechanics (Alexsandrov 1986).

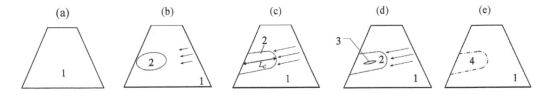

Figure 3. Design stages: (a) initial stage of exploitation of dike; (b) initial stage of filtration (formation of soil's deconsolidation zone); (c) development of soil's deconsolidation zone; (d) formation of the hydro-break channel; (e) – fixing of soil; 1 – initial state of soil; 2 – deconsolidation soil; 3 – cavity of hydro-break channel; 4 – envisaged soil; ← – direction of filtration, L_c – width of deconsolidation zone.

For modeling it was applied nonlinear uploading of a ground mass using the stepwise-iterative method. The minimum number of iterations – 300, the number of steps – 25, the coefficient to the load by steps – 0.04; were used uniform steps with a total coefficient of 0.99. Also, it was set the impact from a hitting of waves on the dike with a coefficient of inelastic base of 0.1 and a frequency of external influence 0.05 rad. per a sec. To determine the principal and equivalent stresses was used a calculated processor "Litera", that was selected a theory of the Coulomb-Mohr strength for grounds.

The analysis of nature distribution of the isofield of stress is implemented on an example of equivalent efforts, which distribution at various stages of modeling is illustraited in Figure 4.

The analysis of isofield distribution of the equivalent stresses in the dike body of header in the initial state (Figure 4a) showed the following: on the right side of the dike's base are maximum equivalent efforts equal to 8 kPa at point B, decreasing up to a value of 1 kPa at point A. By the changes of a distance from the characteristic points to study area of

the dike body there are perturbations of stresses σ_E. The perturbations reach the values of 4.45–6.75 kPa at a distance accordingly 0.6L and 0.3L from point B, that is caused by the influence of horizontal component of the load given by a weight of water.

It's rather characteristic the presence of boundary effect at point B. This is caused the external load on the dike body and the base of settling tank was applied only from point B. The distribution pattern of equivalent stresses is similar to the configuration horizontal σ_x and vertical σ_z efforts that is reflected to lack of an external load, especially – horizontal, on the lateral boundaries of the model.

The configuration of equivalent stresses in the initial stage of ground decompression (Figure 4b), as in the initial state, due to the influence of the vertical stresses and according to accepted the Coulomb-Mohr's theory of strength – the major efforts σ_3. Maximum values σ_E increased to 30 kPa (baseline σ_{Emax} = 8 kPa) and were centered around the point B. As in the case of majoir efforts σ_3, the distribution of equivalent stresses in the dike body due to the presence of weakened zone.

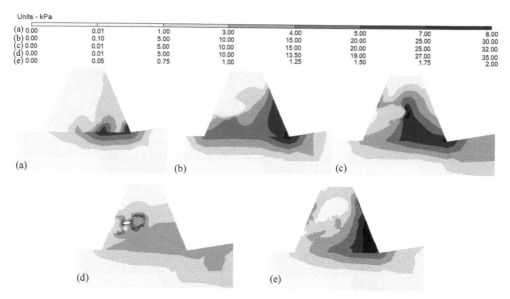

Figure 4. The distribution of the fields of equivalent stresses σ_E in a body and founding of hydrotechnical building's dike: (a) the initial stage of exploitation of the dike; (b) the initial stage of filtration (formation of ground's deconsolidation zone); (c) the development of ground's deconsolidation zone; (d) the formation of the hydro-break canal; (e) the fixing of ground.

Analyzing the izofield distribution of equivalent stresses under development of zone of ground deconsolidation in the dike body of hydroheader (Figure 4c) it is clearly distinguished two zones of maximum. The first, as in the previous configurations of model, is concentrated around point B, and the second is located in the central part of the dike body and due to the influence of locally weakened area of ground massif. The greatest values of equivalent stresses σ_E are 32 kPa. This value is 2 kPa more than at the forming of the decompression zone and 4 times more than in initial state.

The maximum values of the equivalent stresses in the forming of hydrofracturing canal are 35 kPa, which is 4 times more than the greatest values σ_E in initial state, 5 kPa more than in initial stage of decompression and 3 kPa more than at the development of decompression zone. The configuration of izofields of equivalent stresses is similar to the form of efforts σ_3 and σ_z, but extending of values to 10 kPa at the base of settling tank at a distance of 1.9L from point B (Figure 4d).

The greatest equivalent stresses after a fixing of deconsolidation zone with clay and cement mortar is 2kPa, which is 4 times less the efforts σ_E in the initial state, at 28 kPa less than stresses in the initial stage of decompression, at 30 kPa less than σ_E in developing of decompression zone of ground and at 33 kPa less than equivalent efforts at forming of hydrofracturing canal. The configuration of izofields

σ_E is equivalent to distribution of majoir stresses σ_3, and maximum efforts σ_E are observed around point B on the side of the dike body (Figure 4e).

For a presentation the efficiency of grouting of the decompression zone and the forming canal of hydrofracturing with clay and cement mortar with strain-strength characteristics, which are shown in Table 1, is performed an analysis of the equivalent stresses in plane of hydrofracturing canal across the width of decompression zone L_c, which are presented in Figure 5.

At initial state (Figure 5, curve 1) the equivalent stresses are smoothly evolved from the edge of the dike ($0.00L_c$) to its central part ($1.00L_c$) on a value from 0.7 to 2.2 kPa accordingly. The regularity of equivalent stresses' changes across the width of the dike is almost linear.

Forming a zone of ground deconsolidation (Figure 5, curve 2) the character of equivalent stresses in the decompression zone was changed from almost linear to exponential. The equivalent efforts increase from 1.2 kPa on the side of the dike to 10.6 kPa on a contact zone of the decompression ground with the ground in an undisturbed condition.

Developing a zone of ground deconsolidation (Figure 5, curve 3) it is clearly distinguished the maximum area, which is confined to a contact of deconsolidated and undisturbed zones. The efforts σ_E exponentially increase from 2.5 kPa on surface of the dike to 32 kPa on a contact of deconsolidated and undisturbed zones.

The formation of hydrofracturing canal determines a significant change of the equivalent stress in the whole area of decompression (Figure 5, curve 4). From the side surface of the dike the stresses σ_E increase exponentially from 4.7 to 35 kPa on a contact with forming canal of hydrofracturing. In direction from the edge of forming cavity of hydrofracturing to the central part of the dike the character of change of equivalent efforts close to exponential with maximum 34.7 kPa on a contact of a cavity of hydrofracturing and decompression zone and minimum 16.7 kPa – on a contact of deconsolidated and undisturbed zones.

About the effectiveness of grouting works can be seen by comparing value and character of the equivalent stresses in decompression ground at various stages of dike operation. So, after a fixing of deconsolidation zone of stress σ_E is 0.3 kPa on a side surface of the dike, linearly increasing up to 1 kPa on contact of unbroken and deconsolidation zones (Figure 5, curve 5).

Compare the maximum values of equivalent stresses in decompression zone at different stages of exploitation of the dike. Take as a basis an initial stress state for which the equivalent efforts in the dike body reach the values of 2.2 kPa. At formation the decompression zone of stress σ_E were increased 4.8 times as much and reached the values of 10.6 kPa. The development of deconsolidation zone contributes to increase the equivalent stresses 14.5 (32 kPa) times as much compared with an initial state and 3 times as much compared with initial step of decompression. At the formation of hydrofracturing canal the stresses increase 16 times as much compared with initial state and is 35 kPa. The application of clay and cement mortar as a material to make grouting works and to connect a deconsolidation disperse ground helps to stabilize the stress state and to reduce the equivalent stresses 2.2 times compared with initial state, 10.6 times compared with stress state in formation of deconsolidation zone and 35 times compared with equivalent stresses in the formation of hydrofracturing canal.

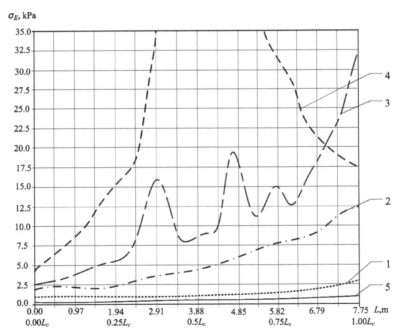

Figure 5. Analysis of equivalent stresses in plane of hydro-break canal in different configurations of a model: 1 – initial stage of a dike exploitation; 2 – initial stage of filtration (a formation of ground's deconsolidation zone); 3 – development of ground's deconsolidation zone; 4 – formation of the hydro-break canal; 5 – fixing of ground.

Thus, the use of grouting clay and cement mortar with shown in Table 1 strain-strength characteristics allows to increase a reliability of ground dike and to reduce equivalent stresses in its body compared with initial state, that will improve the operational safety of the mine hydroheader and will reduce the probability of an emergency in the dike body.

4 CONCLUSIONS

The analysis of stresses distribution in the dike body of mine hydroheader showed the following:

1. The presence in the dike body of a zone of ground deconsolidation causes stresses redistribution in a ground massif and their increase. The maximum values of equivalent stresses are concentrated around the decompression area that caused by a difference of deformation and strength properties.

2. At the formation of decompression zone the stress σ_E in the dike body of hydrostorage increased 4.8 times as much compared with initial state. The development of deconsolidation zone contributes to increase the equivalent stresses 14.5 times as much compared with initial state and 3 times as much compared with initial step of decompression. At the formation of hydrofracturing canal the stresses increase 16 times as much compared with initial state and is 35 kPa.

3. The application of clay and cement mortar helps to stabilize the stress state and to reduce the equivalent stresses 2.2 times compared with initial state, 10.6 times compared with stress state in formation of deconsolidation zone and 35 times compared with equivalent stresses in the formation of hydrofracturing canal.

4. The application of clay and cement mortar with the above characteristics is effective and allows to increase a reliability of ground dike and to provide a safety of hydroheader operation.

REFERENCES

Arustamov, E.A. 2006. *Safety of vital functions*. Textbook: 473. Moscow.

Kostyukov, E.V. 2005. *Research of bodily condition and estimation of stability of the ground dikes of hydrotechnical building of mountain enterprises by a geoelectric method*. Dissertation on the competition of graduate degree of candidate of engineering sciences on speciality 25.00.16 – Mining and gaz-petroleum industrial' geology, geophysics, surveyor business and geometry of bowels of the earth: 147. Kyemyerovo.

Dolzhikov, P.N. 2006. *New technical decisions at building of making, plug-back and fixing of mountain breeds*. Monograph: 256. Donets'k.

Dolzhikov, P.N. 2004. *Analytical researches of parameters of deconsolidation zones and processes of their filling. Mining industry of Ukraine and Poland*: issues of the *day and prospects*: Materials of the Ukrainian-Poland forum of miners: 635. Dnipropetrovs'k.

Alexsandrov, A.M. 1986. *Tasks of mechanics of continuous environments with the mixed border conditions*: 336. Moscow.

Progressive Technologies of Coal, Coalbed Methane, and Ores Mining – Bondarenko, Kovalevs'ka & Ganushevych (eds)
© 2014 Taylor & Francis Group, London, ISBN: 978-1-138-02699-5

Investigation of regularities of trapped gas recovery from watered macro heterogeneous gas fields

O. Kondrat & R. Kondrat
Ivano-Frankivs'k National Technical University of Oil and Gas, Ivano-Frankivs'k, Ukraine

ABSTRACT: Most of natural gas fields in Ukraine are associated with formation of water systems and are developed in a water drive mechanism. As a result of edge water inflow into the field production wells are being gradually watered and up to 15–30% of initial gas reservoirs become trapped. In conditions of the acute shortage of natural gas in Ukraine trapped gas removal from water flooded reservoirs has both scientific and practical importance. The known methods of trapped gas removing from flooded reservoirs were described. The existing studies of the trapped gas recovery were carried out on homogeneous reservoir models. This paper presents the experimental unit and the results of laboratory tests of trapped gas recovery by pressure reduction from high and low permeable homogeneous models and heterogeneous model from two different permeable layers. Qualitative and quantitative characteristics of the pressure reducing process in flooded homogeneous and heterogeneous formations and trapped gas recovery were established. Guidelines for the practical use of research results were specified.

1 INTRODUCTION

A large number of natural gas fields is associated with formation of water systems. During gas production from the field formation pressure decreases and creates pressure differential between the aquifer and gas-saturated zones, which allows water to flow into the field. Water drive mechanism of field development appears. Water entering the field, gradually waters production wells and some parts of gas in the reservoir becomes trapped. According to the field data of the worldwide gas fields development final gas recovery factor in conditions of water drive ranges from 0.4 to 0.98 and in average equals 0.7–0.85 (Zakirov 1988, Kondrat 1992 & Improvement of oil... 2000). Coefficient of residual gas saturation varies between 0.10–0.50 with average value 0.20–0.30. In the conditions of the world's hydrocarbon resources gradual depletion the issue of trapped gas removal from water flooded fields is paid great attention to. This question is particularly relevant for Ukraine, which imports most part of its consumed gas. As a result the problem of trapped gas recovery from flooded fields has both scientific and practical importance.

2 ORIGIN OF PROBLEM

In the gas fields flooded zone there is micro and macro trapped gas. Gas micro trapping is due to the heterogeneity of the pore space structure of productive deposits. Micro trapped gas is in a porous medium in a dispersed state in the form of separate, isolated bubbles and is concentrated in one-side open pores, small pores with a high resistance to water movement and some large pores that water bypassed. To remove micro trapped gas it is necessary to reduce pressure in flooded reservoir zones, leading to expansion of micro trapped gas volumes and creating interconnected grid of channels for its movement. Another possible method for gas extraction is pinched injection of gaseous agents into the flooded formations, such as non-hydrocarbon gases, particularly nitrogen, which can penetrate into the small pores, and create cross-cutting channels for gas flow. One more direction of micro trapped gas recovery is liquid injection with low and less then water surface tension into the flooded formations. It includes alcohols, solvents and other surfactants. These solutions support the creation of a solid displacement front and penetrate the pore channels that water bypasses because of large movement resistance or low capillary pressure.

Macro trapped gas is in the form of "pillars" in separate low permeable slightly drained areas of the reservoir with the initial gas saturation, which water bypasses due to the inhomogeneous structure of the productive deposits and uneven placement of production wells. The possible way to extract macro trapped gas is additional drilling of production wells in gas-saturated areas in the flooded zone with macro trapped gas. Additional wells drilling needs to reliably know the location of gas-saturated areas, but it is not always possible. Besides, significant financial costs are needed for drilling additional wells that may not pay off for minor gas reserves in gas-saturated areas or well can strike flooded or unproductive zones. Other types of macro trapped gas recovery is to reduce the pressure in the watered formations by forced gas and water production from flooded wells and using various gas displacement agents (Zakirov 1988, Kondrat 1992 & Improvement of oil... 2000).

The method of gas recovery from the flooded reservoirs was first patented in the U.S. (Patent USA #3134434... 1964). In its basis there is the assumption that all trapped gas expands as a result of reservoir pressure reduction, becomes moveable, flows to the top part of the structure and is extracted through roof wells which after a short period of two phase mixture gushing operates without water. However, the results of further laboratory studies by other authors show that while reducing the pressure in flooded reservoirs trapped gas firstly expands and only after that it begins to move (Zakirov 1988, Kondrat 1992 & Improvement of oil... 2000).

A method of gas recovery from the field in terms of water drive development mode by sampling gas through wells located in a "dry" zone, and simultaneous sampling of water from flooded wells in an amount which reduces the pressure in the reservoir to a value less than 25% of water pressure testing (Patent USA #4040487... 1977). The main disadvantage of this method is that by decreasing pressure to the specified value trapped gas expands and can hardly move. Therefore a low trapped gas recovery coefficient is achieved.

In Ivano-Frankivs'k National Technical University of Oil and Gas and the State University of Oil and Gas named after Gubkin (Moscow) according to the results of performed laboratory tests on formation models secondary gas extraction method was proposed which is based on the reduction of pressure in the flooded area by water production through the flooded wells (A.c. 991785 SSSR... 1981). Trapped gas is extracted together with water. In order to increase trapped gas recovery factor water production from flooded wells is carried out at the pressure which is equal to or less then the pressure, which is consistent with the achievement of maximum gas-water factor and the highest rate of change in time of gas recovery factor of the flooded area.

The implementation of the above method of gas extraction is connected with the necessity of water production from the flooded fields and utilization of large volumes of water.

Production of water from flooded wells can be significantly reduced if in the profile of structure,

which is confined to flooded reservoirs there are depleted gas reservoirs. According to the second method of gas extraction, that was proposed by IFNTUOG and UkrNDIgaz, in order to reduce the amount of water produced to the surface from the flooded gas fields and shorten the period of deposit further development, simultaneously with the production of water through flooded wells water cross-flow is performed in to above or below the section of depleted gas reservoirs through the wells located in the peripheral parts of the deposits and the residual gas extraction is carried out through the wells situated in the roof of watered and depleted gas reservoirs (A.c. #1314757... 1985). The implementation of this method will also increase the final gas recovery factor of depleted gas fields by water displacement of the residual gas.

A number of ways to intensify the process of trapped gas recovery from flooded reservoirs by pressure reduction was presented (Patent #2047742... 1995 & Patent #2379490... 2010). These methods allow increasing gas mobility in watered formations by periodic pressure pulses actions.

One more method for trapped gas removal from flooded reservoirs is to displace it by non-hydrocarbon gases injection. On the possibility of trapped gas displacement from the flooded formation model by nitrogen gas injection shows the results of laboratory tests which were performed in VNDIhazi (Ter-Sarkisov 1999).

These studies of the trapped gas recovery from the flooded fields by reducing the pressure in them and displacement (replacement) by non-hydrocarbon gases were made for homogeneous formations. Real gas-bearing formations are mainly homogeneous ones and are presented by alternating formations and areas of different permeability. Both qualitative and quantitative values characterising the process of trapped gas extraction from flooded reservoirs by pressure reduction are also of grate interest. The above-mentioned justifies the urgency of conducting further researches on the problem of trapped gas extraction from homogeneous and inhomogeneous formations.

3 EXPERIMENTAL UNIT AND
   METHODOLOGY OF INVESTIGATION

Laboratory studies were performed on linear models of the reservoir, which are thick-walled steel column with the length of 1.44 m and an inner diameter of 0.078 m. To prevent the slippage of water and gas along the walls the inner surface of the columns was made rough by attaching the layer of coarse sand with the help of epoxy. As porous medium a mixture of silica sand with a grain size of 0.5; 1; 2;

3 mm from the quarry in the area of the Bystrica Nadvirnianska (v. Pasichna) and marshalit (quartz flour) were used. A porous medium model with absolute permeability coefficients of 0.29 and 1.35 D and open porosity of 0.356 and 0.354 respectively was used in experiments. The scheme of the laboratory unit is shown in Figure 1. During preparation for the experiment the model in upright position was filled with portions of quartz sand. Every portion of sand was consolidated. After filling the model with sand absolute permeability factor of the porous medium was measured by using gas standard method. To determine the initial gas-saturated pore volume and the factor of open porosity the model was filled with gas to a certain pressure level, which afterwards was released from the model through the flow meter. By the amount of gas, the initial gas saturated pore volume of the model was determined and in relation of it to the internal volume of the model - the factor of open porosity. Then the model for a long time (6 hours) was vacuumized, filled with water and the permeability coefficient of model for water was determined. According to the amount of water entered the model open porosity factor was additionally controlled. To create residual water saturation the model in the upright position for a long time

was purged by gas from the top to the bottom and after that gas relative permeability was measured. Volume of gas-saturated pores of the model and the open porosity factor were determined by the amount of gas that is in at a given pressure level (method of formation pressure decrease).

In experiments the displacement of gas (nitrogen) by fresh water in horizontal models was carried out. During the experiments the pressure at the model inlet was maintained to be constant and equal to 4 MPa. The water into the reservoir model 5 entered from gravity cell 4. Required water pressure was created by gas cylinder 1 and was controlled by pressure reduction valve 2. Gas from the model was extracted at a constant flow rate which was achieved by using micro valve 7 of capillary tube on its way outlet. Gas flow rate was measured using a gas meter 8. Experiments were performed in self-similarity zone of gas by water displacement coefficient of similarity criterion $\pi 1$ and $\pi 2$. For model evacuation before filling it with water a separator 10, a moisture separator 11 and a vacuum pump 12 were used. For certain volume of gas input during the experiments to the model inlet the additional cell 6 was used.

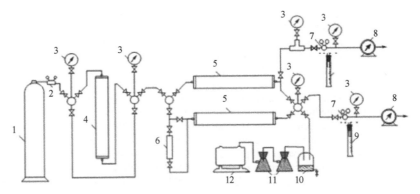

Figure 1. Experimental unit for the study of high-pressure gas displacement by water from the heterogeneous reservoir model: 1 – gas cylinder; 2 – reduction pressure valve; 3 – pressure gauges; 4 – gravity cell; 5 – models of the formation; 6 – additional cell; 7 – micro valve; 8– gas flow meter; 9 – measuring separator; 10 – separator; 11 – moisture separators; 12 – vacuum pump.

Experiments were performed separately with high and low permeable models and with two models simultaneously which have been set up in parallel and horizontal position. The latter one is modeling the case of joint development by a single well pattern of two formations with different permeability separated by an impermeable membrane (macrononuniform model) with common aquifer basin. After presence of water in the separator the model was washed with water until the termination of gas bub-

bles removal, and then the coefficients of gas by water displacement, residual gas saturation and water phase permeability were determined.

After that the inlet valve of the model was closed and with the help of the valve at the outlet pressure in the model was gradually reduced. During this process the pressure difference at the ends of the model did not exceed the maximum pressure drop during gas displacement by water. This option of reservoir pressure reduction simulates the process of trapped gas recovery from the flooded gas and gas-condensate fields after stoping artificial flooding. In some experiments, after flooding the model inlet was connected to a water cell with the gas bag. During the pressure drop water from water cell continued to flow into the model under the gas bag pressure. This option of pressure reduction simulates further development of flooded gas and gas-condensate fields with trapped gas in the presence of extra-contour aquifer basin from which water continues to enter the field.

Preparing reservoir model for these experiments was in blowing it by gas in a vertical position to create residual water saturation, and then the parameters of the model were measured.

During the experiments on heterogeneous model after the appearance of water at the outlet from high permeable reservoir the water supply continued until watering the low permeable reservoir. The amount of water which flows from high permeable reservoir during the period of time between watering high and low permeable reservoirs was measured.

4 STUDY RESULTS

According to the experimental data for high-permeability formation model initial gas saturation is 0.46–0.47; residual gas saturation – 0.23–0.24; water displacement factor – 0.5–0.521; for low-permeability formation – 0.218–0.219; 0.113–0.118; 0.458–0.470; for two formations macro heterogeneous model – 0.324–0.352; 0.197–0.209; 0.364–0.406. For studied porous medium with the increase of permeability rate increases primary and residual gas saturation and water displacement factors. For macro heterogeneous model the value of primary and residual gas saturation are between their values of high-permeability and low-permeability formations models but the water displacement rate is lower as for the low-permeability formation.

Water displacement from formation models allowed making residual gas saturation in porous medium. The basic in this paper is the study of the peculiarities of the trapped gas recovery process from flooded formations by decreasing pressure in them. While conducting the experiment it was fixed the time change of the inlet P_1 and outlet P_2 pressure of the model and gas V_g and water V_w volumes taken from the flooded model. According to the experimental data it was defined the average pressure in the model $\widetilde{P}$, gas recovery factor with residual (trapped) gas β_{res}, general gas recovery factor β_{gen}, gas Q_g and water Q_w rates and gas-water ratio G.

The results of the study were presented with the curves of change in dimensionless time $\bar{t}$ ($\bar{t} = \dfrac{t}{T}$, where t is the period of time from the beginning of pressure decrease in the flooded zone; T is a total duration of the pressure decrease process in the flooded model), peculiarities of the trapped gas production process from the flooded formations and dependences of the separate characteristics of the studied process on the dimensionless pressure $\dfrac{\overline{P}}{\overline{P}_{in}}$, characterizing the pressure decrease level in the flooded formation model compared to the flooded pressure ($\overline{P}_{in} = \dfrac{P_{in}}{Z_{in}}$, $\overline{P} = \dfrac{P}{Z}$, where P_{in}, P – gas trapped pressure and current pressure; Z_{in}, Z – gas loading rate under P_{in} and P and temperature 20 °C). In the Figure 2 it is given the peculiarities of the trapped gas recovery process by pressure decreasing from the high-permeable formation model (Figure 2a), low-permeable formation model (Figure 2b) and macro heterogeneous model (Figure 2c).

Figure 3 shows the dependences of the residual gas saturation of the flooded model α_{res}, gas recovery factor for residual (trapped) gas β_{tr} and total gas recovery factor β_{tot} on the level of pressure decrease for the high-permeable formation model (Figure 3a), low-permeable formation model (Figure 3b) and macro heterogeneous model (Figure 3c).

The analysis of the studied data shows the same character of the parameters change of trapped gas recovery processes from the high-permeability, low-permeability and macro heterogeneous flooded models of porous medium (Figures 2 and 3). In the initial period of the pressure decrease in general it is recovered water from a model but a water rate has a maximal value Q_w. While pressure decreasing in the flooded model, expansion of trapped gas, it movement and recovery take place. At first separate gas bubbles are recovered which are carried by water flow. Under pressure decreasing gas bubbles unites together and gas cross channels are made in the model. Gas being in nonwettable phase takes the central part of the pores which is characterized by lower filtration resistance rather than peripheral parts. Gas is characterized by a lower viscosity compared to water. That is why it starts to move faster than water. Gas rate increases dramatically and the water rate decreases. Simultaneously gas-water ratio increases dramatically which represents relation of the gas rate to the water rate. The rate of gas-water ratio increase is higher than gas rate as the water rate decreases.

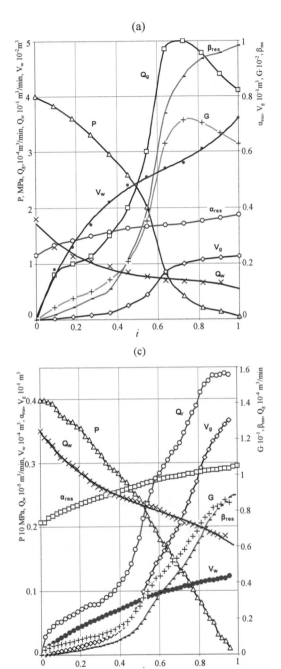

(a)

(b)

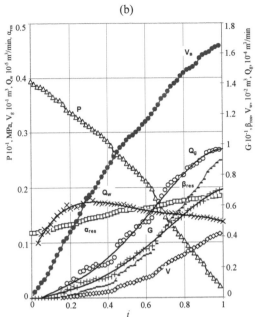

(c)

later decreases because of general trapped gas reservoir depletion. For the low-permeability formation model a continuous gas rate and gas-water ratio is observed because of low permeability of porous medium and significant filtration resistance to gas movement in small pores. For macro heterogeneous model, which consists of high-permeability and low-permeability formations, it is observed the dependences flattening for gas rate and gas-water ratio in the final period of pressure decrease or their maximum is displaced into the lower value of the relative pressure in the model compared to the high-permeability formation.

(a)

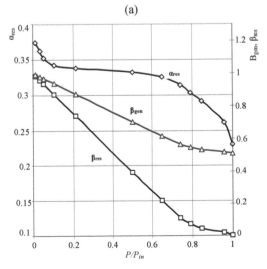

Figure 2. The peculiarities of the trapped gas recovery process from flooded gas field by pressure decreasing from the high-permeable formation model (a), low-permeable formation model (b) and macro heterogeneous formation model (c).

For the high-permeability formation model gas rate and gas-water ratio reach the maximum value,

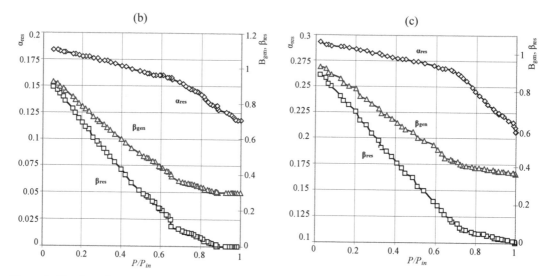

Figure 3. The dependences of the residual gas saturation of the flooded model α_{res}, gas recovery factor for residual (trapped) gas β_{tr} and total gas recovery factor β_{tot} on the level of pressure decrease for the high-permeable formation model (Figure 3a), low-permeable formation model (Figure 3b) and macroheterogeneous model (Figure 3c).

The analysis of the study results on the trapped gas recovery from the flooded formation models allows making the following conclusions. The graphical dependence of the time change of pressure in the high-permeability formation model has a curved character to the x-axis, the time of pressure reduction gradually decreasing in time. For the low-permeability formation models and macro heterogeneous model with two different-permeability formations it is observed a rapid pressure reduction over time, such dependences are convex to the x-axis and only in the final period of the trapped gas recovery – becomes flatten. Graphical dependences of the total water recovery have a curve character to the x-axis and gradually become flatten. Thus, under the pressure decrease in the high-permeability, low-permeability and macro heterogeneous formations models of 10% from the initial pressure (flooding pressure) is recovered 36.11, 19.39 and 20.46% of water from its total rate, under the pressure decrease from 10 up to 20% is recovered 11.11, 16.97 and 18.18% of water, and from the beginning of the process – 47.22, 36.36 and 38.64% of water, under pressure decrease from 20% up to 30% is recovered 13.89, 15.76 and 26.14% of water and from the beginning of process – 61.11, 52.12 and 64.78% of water, under pressure decrease from 30% up to 40% is recovered 6.94, 10.30 and 7.95% of water and from the beginning of the process 68.05, 62.42 and 72.73% of water, under the pressure decrease from 40 up to 50% is recovered 4.17, 9.09 and 4.55% and from the beginning of the process – 72.22, 71.51

and 77.28% of water. The presented data show that the main problems with trapped gas recovery will take place in the initial period of the process implementation because of the necessity of the water volumes production. Subsequently current water production decreases.

Graphical dependence for the gas-water ratio has a curved character to the x-axis certifying it rapid decrease over time and only in the final period this dependence becomes flatten.

According to the results of the conducted experimental data the affect of the pressure decrease level in the flooded model compared to the flooded pressure on the trapped gas movement is estimated and its recovery from the formation models. According to the study data while pressure is decreasing in the flooded model the residual gas saturation increases dramatically but later on stabilizes or increases slowly.

In some experiments in the final period of pressure reduction in the formation model it was observed a rapid increase of residual gas saturation which can be connected with the water displacement by gas from small pores into larger pores under its expansion and further movement of water with gas through large pores to the model outlet. At the time of 5% recovery of trapped gas the relative pressure in the formation model is the following: high-permeability formation – 0.82–0.87; low-permeability – 0.78–0.83 and macro heterogeneous formation – 0.76–0.801. To recover 5% of trapped gas it is necessary to produce the following amount of water which is recovered from the model: high-permeability formation – 0.262–0.388; low-permeability – 0.315–0.428 and macro heterogeneous formation – 0.333–0.464.

According to the experimental data after reaching some relative pressure in the formation model the dependence between gas recovery factor over trapped gas and relative pressure in the model becomes almost a straight line. A mentioned dependence is the straight line under the following: high-permeability formation – 0.694–0.74; low-permeability 0.655–0.76 and macro heterogeneous formation – 0.684–0.713. For the given values of relative pressure gas recovery factor over trapped gas reaches the following: high-permeability formation – 0.095–0.108; low-permeability 0.107–0.109 and macro heterogeneous formation – 0.107–0.111. To achieve the given values of gas recovery factor over trapped gas it is necessary to take the following amount of water (the proportion of total water recovery in the process of pressure decrease in the flooded formations): high-permeability formation – 0.573–0.605; low-permeability 0.573–0.607 and macro heterogeneous formation – 0.433–0.464.

Thus, to recover 5% of trapped gas it is necessary to decrease pressure in the flooded model on 13–24% from the initial pressure (flooded pressure) that in real conditions will require large amount of water production from flooded formation. That's why practically it is not possible to allow a complete flooding of the field and after it proceed to recover trapped gas. After the first flooding in the producing wells it is necessary to introduce the methods of gas and water recovery that will promote water movement slowing down into the field, to reduce the volume of flooded zone and to recover trapped gas from the flooded formations because of pressure decrease in them under flooded wells operation.

5 CONCLUSIONS

One of the ways of increasing the efficiency of own hydrocarbon resources usage is trapped gas recovery from flooded fields with water drive mechanism. The existing studies of the trapped gas recovery were carried out on homogeneous reservoir models.

Based on the results of experimental studies at homogeneous low and high permeable models and heterogeneous model, which consist of two models with different permeability, for the first time it was established qualitative and quantitative characteristics of gas recovery process from the homogeneous and heterogeneous reservoirs by pressure reduction. They have proved the effectiveness of technological method of pressure reduction in flooded formations. However, its implementation requires extraction from the deposit significant amount of water. Therefore, in the real field we can not allow a full fields flooding and only after that begin extracting trapped gas. After the first production well flooding it is necessary to perform joint extraction of gas and water from them, which enables to reduce the rate of field flooding and the amount of trapped gas.

REFERENCES

Zakirov, S.N. 1988. *Development of gas and gas-condensated deposits*. Moscow: Struna: 628.

Kondrat, R.M. 1992. *Gas-condensate release of coal seams*. Moscow: Nedra: 255.

Improvement of oil and gas deposits recovery technologies. 2000 / under edition by S.N. Zakirova. Moscow: Graal'.

Patent USA #3134434. 1964. *Increasing ultimate recovery from gas reservoirs* / Woody, L.D. Apl. date 19.06.1961, #118134: Publ. date 26.05.1964.

Patent USA 4040487. 1977. *Method for increasing the recovery of natural gas from a geo-pressured aquifer* / Cook, H.L., Geer, E.C. & Katz, D.L. Apl. date 24.05.1976, # 689621: Pub. date 9.08.1977.

A.c. 991785 SSSR, MKI E 21 V 43/00. 1981. *Method of gas repeat recovery* / Zakirov S.N., Kondrat R.M. & Kravtsov N.A. – #1218300292/22-03; Reported 11.06.1981. DSP.

A.c. 1314757 SSSR, MKI E 21 B 43/00. 1985. *Method of gas repeat recovery* / Kondrat R.M., Tokoy I.N. Fyk I.M. [and others] #3956225/22-03; Apl. date 24.09.1985. DSP.

Patent #2047742 Russia, MPK E 21 B 43/00. 1995. *Method of gas recovery from water-bearing seams* / Belonenko V.N. – #5030902/03; Apl. date 06.03.1992; Pub. date 10.11.1995, Bul. #31.

Patent #2379490 Russia, MPK E 21 B 43/18. 2010. *Recovery method of trapped by water gas* / Ulyashev V.E., Burakov Yu.G. – #2008133957/03; Apl. date 18.08.2008, Pub. date 20.01.2010.

Ter-Sarkisov, R.M. 1999. *Natural Gas Fields Development*. Moscow: Nedra: 659.

Progressive Technologies of Coal, Coalbed Methane, and Ores Mining – Bondarenko, Kovalevs'ka & Ganushevych (eds)
© *2014 Taylor & Francis Group, London, ISBN: 978-1-138-02699-5*

Perspectives of mine methane extraction in conditions of Donets'k gas-coal basin

Ye. Korovyaka, V. Astakhov & E. Manykian
National Mining University, Dnipropetrovs'k, Ukraine

ABSTRACT: The analysis of perspectives of mine methane extraction in conditions of Donets'k gas-coal basin is executed. Intensification of surface degasification method of gas-bearing coal seams is given.

1 INTRODUCTION

Continuously growing volumes of energy deposits set a new task before science and production divisions that consists in complex developing of alternative energy sources. One of the most perspective directions for fuel and energy complex of Ukraine is mine methane. At that technology of its extraction is based on application of traditional degasification methods.

However, in spite of perspective of such direction, up to date developing of mine methane deposits in industry is realized only by individual coal extraction enterprises. Analysis of researches in given direction showed that current degasification technology doesn't allow to receive enough quantity gas. Absence of effective methods of low-quality mine methane utilization and complex approach to its extraction is lead to atmospheric injection. It is caused because of traditionally using technological schemes of underground coal extraction do not directed on accompanying methane extraction.

Expert judgment of industrial situations showed that only complex approach to given problem that allow to connect technology of coal and methane extraction in united system will increase profitability of coal enterprises, labour safety, environmental protection and decrease energy dependence of our country.

Methane deposits in coal-bearing series are estimated in 1345 billion m^3, including 1181 billion m^3, absorbed by coal and 164 billion m^3 of free gas in porous layers, particularly, in sandstones (Karp 1993 & Pavlov 2005). These stocks are enormous, but scattered along numerous coal seams and sandstone layers in coal-bearing series.

It should be noted that, thrown in atmosphere methane by ventilation and degasification boreholes considerably damage environment and promotes to hothouse effect formation. Thus, in 2013 coal enterprises of Donbass extracted near 2000 million m^3 of methane, 1900 million m^3 was throw in the atmosphere. Mine ventilation systems were extract and throw in atmosphere 1750 million m^3, but from 250 million m^3 of methane that was captured by mine degasification systems and used only 100 million m^3. In that way, only 95% of extracted coal practically annually is lost irreparably, at this intensify hothouse effect and contaminate environment.

It is caused by serial complexes of equipment for underground gas pumping-out, near 50% is working in industrial procedure. On the most of degasification complexes methane concentration that reaches to surface is so minimal that it's not enough for further usage. In this connection on some mines gas is using only for auxiliaries.

Total reserves of methane in coal seams and sublayers of Donbass coal-bearing deposits are estimated in 1.2 trillion m^3, but with taking into account methane content in rocks – 25.4 trillion m^3. Analyzing official data of M.S. Polyakov Institute of Geotechnical Mechanics under the NASU (Bulat 1998) about gas emission and quantity of extracted methane on mines of Donets'k basin coal enterprises (Table 1) we can state that average value of gas content coals of Donets'k basin are equal to 8–15 m^3/tons, but gas specific emission during coal extraction (gas content) – 23–36 m^3/tons.

2 MAIN PART

It should be noted that specific emission often exceeds gas content of coal more than in three times. In some sources noted that maximum values reach to 100 m^3/tons of extracted coal and more. At this, with increasing of coalification, value gas content increase from 10–15 m^3/tons in gas-coal to 40 m^3/tons in anthracites. During comparison of gas content geological conditions of coal deposits was established that high gas concentration in side rocks differ Donets'k basin from others coal basins. It is supposed that only 10% of methane situates in coal seams, most quantity of gas situates in adjacent rocks. Dependence of gas content from tectonic situation also promotes changing of share in total quantity of gas.

The main source of mine methane extraction in Ukraine is degasification systems of active mines. Methane extraction from coal-bearing massif immediately connected with processes of preliminary, current and post-exploitation degasification (Table 2).

Table 1. Methane deposits on areas of Donbass active mines.

Coal enterprise	Methane content, m³/t	Specific emission, m³/t	Methane deposits (billion m³) profitable reserves		
			on all mines	active mines	industrial reserves
Donetskugol	12.9	37.1	30.2–86.9	19.9–57.3	14.4–41.5
Makeevugol	21.1	53.5	40.8–103.5	12.6–32.0	9.7–24.5
Oktyabrugol	18.2	55.8	11.1–34.1	7.2–22.0	5.8–17.9
Krasnoarmeiskugol	12.8	34.5	12.6–33.9	12.6–33.9	5.2–14.0
Krasnodonugol	21.9	38.4	38.2–66.9	11.5–20.2	8.7–15.3
Schakhterskantracyt	29.9	47.8	18.7–29.8	9.5–15.1	7.6–12.2
Donbassantracyt	15.9	38.0	25.3–60.4	5.8–13.8	4.2–9.9
Luganskugol	15.1	23.3	55.0–84.9	11.9–18.4	8.4–12.9
Artemugol	17.4	28.5	9.1–14.9	6.0–9.9	4.8–7.9
Stakhanovugol	14.2	23.7	27.1–45.3	6.5–10.8	4.0–6.7
Torezantracyt	6.7	16.5	7.8–18.7	1.7–4.3	1.4–3.4
Dobropolieugol	11.0	21.6	37.0–72.3	8.3–1.3	5.9–11.5
Ordzhonikidzeugol	15.4	25.2	7.5–12.3	3.6–5.9	2.9–4.7
Pavlogradugol	7.9	8.9	39.9–44.9	10.0–11.3	7.9–8.9
Pervomaiskugol	14.4	25.2	24.2–43.3	5.2–9.3	3.9–6.9
Dzerzhynskugol	16.8	26.8	5.0–7.9	2.5–4.0	1.8–2.9
Lisichanskugol	10.9	17.5	7.7–12.3	2.6–4.2	1.9–3.1
Average/Total	15.4	29.0	417.2–752.2	137.5–265.7	103.7–199.1

Table 2. Short characteristics of methane extraction processes from coal-bearing strata.

Degasification type	Borehole-drilling type	Rocks characteristics	Method of methane emission stimulation	Methane extraction method
Preliminary (before beginning of active mining operations)	from surface	coal	hydraulic fracturing	water pumping with incidental gas emission
			without stimulation	intensive self-expiration
		Sandstone (in dome structures)	without stimulation	intensive self-expiration
	underground	coal	without stimulation	intensive self-expiration
		coal-rock massif	without stimulation	intensive self-expiration
Current (in process of mining)	from surface	coal	mining operations (unload)	self-expiration (compaction by vacuum)
		coal-rock massif	mining operations (unload)	self-expiration (compaction by vacuum)
	underground	coal	mining operations (unload)	compaction by vacuum
		coal-rock massif	mining operations (unload)	compaction by vacuum
Post-exploitation	from surface	coal-rock massif	old mining operations (unload))	compaction by vacuum

As we can see from Table 2, preliminary degasification is single method of decreasing methane content of developed coal seam before beginning of mining operations. In some cases it is necessary to conduct preliminary degasification for decreasing danger of sudden outburst. Because degasification conducts before beginning of mining operations, probability of gas collection system disfunction in consequence of rock faulting is absence, at this usually extracted gas has enough high quality.

Ukrainian mines use practically all degasification methods that well-known in world practice. The most expanded degasification methods of nearby undermining seams that situated in stoping influence zone. The share of it is more than 80% of capture gas. However, effectiveness of degasification is still insufficient in many cases.

Gas extraction of conducted with help of underground degasification boreholes that drilled from mine workings in roof and bottom of working seams and connect with degasification system. Gas intake from massif in borehole is beginning from rock massif breaking moment in the process of coal extraction. Along degasification system, fresh air delivery on surface and at conditional methane composition (more than 25%) using as high-caloric fuel or as chemical raw material.

This method is the method of incidental methane extraction from coal-bearing cut during coal extraction and called mine. In native methodology and technology the main object of gas extraction as incidental mineral deposit is mine field, but source of extraction is massif, more exactly – rocks of roof and bottom of working coal seams together with accompanying seams that hit in unloading zone from rock pressure. These zones in conditions of Donbass expand in roof of seams on distance to 150–180 m, in bottom – 50–60 m.

Created by foreign companies "Amoco Production Company", "Enron", "Keln" methodology is designed for gas extraction as individual mineral deposit from coal seams that didn't involve in exploitation. At this, objects of extraction are not mines, but in detail prospected areas, source – gas-bearing coal, but not coal-contain rocks and accompanying seams. Gas extraction from coal is conducted with help of boreholes that were drilled from surface. Intensification of gas recovery is conducted by method coal seams hydraulic fracturing method.

We can call this method mineless, in comparison with first one – mining.

Especially perspective direction in development of methane extraction technology is the method of free hydrocarbon gases extraction from natural traps with help of vertical boreholes that were drilled from surface. This method is referring to mineless and traditional gas-field method of methane extraction.

In this context object of extraction is sandstone horizons with good collector properties. In rocks of low stage of katagenesis – collectors of grainy type, high stages – rock-fracture and fracture. In cases of close bedding to developed coal seams similar concentrations and gas traps can be sources of gas hazard of mine workings. At the same time at great gas reserves it can serve as sources of local gas supply.

Therefore, Donbass coal deposits present complex collector type of hydrocarbon gases, extraction of which is possible from three different sources:

- coal seams of prospected districts that situated outside mine field. Hydrocarbon gases are mainly represented by sorb form, gas extraction is conducted by mineless with help of drilled from surface boreholes;
- host rocks with accompanying seams of roof and bottom of working seams, active degasification and gas extraction is conducted only in process of coal extraction. Hydrocarbon gases in interlayers are mainly represented by fixed gas, but in rocks – dispersively-spreaded, free and solved in water. Gas extraction is conducted with help of underground degasification systems;
- small gas deposits, concentrations and tramps of free gas. Hydrocarbon gases situated here in mobile phase that allows extracting gas by gas-field method.

Intensive gas emission points at high level of seam penetration and presence of possibilities for conduction effective preliminary degasification and gas utilization. Factors that determine possibility of preliminary degasification in conditions of specified object are available time for achievement the intended effect of methane removal and expenses on drilling and borehole completion.

Different methods of preliminary degasification are using in the whole world. For borehole drilling from mine workings in seams on a depth from 100 to 200 m is using everywhere rotatory drilling. Boreholes with depth 1000 m and more can lay out with usage of directed drilling methods from underground mine workings by means of increasing of degasification effectiveness.

Besides, the possibility exists for seam degasification conduction on big areas immediately from surface. Methods of seam directed drilling from surface proved its effectiveness during conduction of preliminary degasification of coal seams with penetration range from 0.5 to 10 milidarcy (mD). In Australia, where totals mine volume of gas can reaches to $8 \text{ m}^3/\text{s}$, but effectiveness of gas capture on longwall is required on a level of 80%, it is planning combined appliance of preliminary and accompanying degasification with usage of advanced methods of directed drilling from surface (von Schoenfeldt 2008).

Experience of Australia and USA showed that during presence of possibility of seam degasification from surface appliance of this method is more effective in comparison with seam degasification from mine workings, because surface boreholes can be drilled long before beginning of mining operations. In this regard there is a less expectation of time reduction that can be used for effective degasification by reasons, connected with main technological process (Black & Aziz 2009).

Advantages of "surface" methods are in the fact that degasification can be conducted independently of mining operations, however possibility of its usage depends upon drilling depth, coal continuity and penetration, but also from different limits that caused by topographic factor or presence of surface constructions.

Given method is effective both technologically and economically during development of high-methane coal seams on the depth to 1000–1100 m from surface (at current drilling equipment). At further improvement of drilling operations and consequently decreasing of 1linear meter cost of degasification boreholes this degasification method will be economically viable even during mining on the depth more than 1000 m.

Vertical degasifications boreholes use when gas emission from under- and overworked seams in total gas balance of unit is equal to 60–70% at absolute methane emission on unit more than 10 m³/min. However, during usage for gas extraction with further utilization, appliance conditions will be determined by economic viability on the basis of calculations. Main parameters of calculations during degasification are following: depth and site of degasification boreholes; orientation in relative to preparatory and stoping workings that are degasified; distance between boreholes; diameter and construction of boreholes casing strings and also market cost of extracted gas.

During drilling of first degasification borehole it is necessary to cross the developed seam, in upper $\frac{1}{4} - \frac{1}{3}$ part of roof height and enter on 5–10 m in its bottom for sump creation which is necessary for drilling cuttings settlement and water exit from borehole.

Distance between boreholes depends upon degree of gas content of degasification massif, i.e. volume of methane deposits, mining depth and advance rate of stoping face and equal to 100–120 m. At very thick rocks boreholes not drill ahead on 10–30 m to seam that developing in order to borehole longwall after underworking miss collapse zone of exhausted rocks.

On the basis of proposed method set a task of degasification method improvement in which with help of including new technological parameters reaches possibility to increase unloading's in the range of given unit, decrease of labour intensity production and mounting of casing string and suction tube in borehole, increase of degasification efficiency and decrease of specific economic expenses.

The task is solved by well-known degasification method of gas-content deposits that including drill-ing and hermetic sealing of degasification borehole collar, casing string by perforated pipe with suction pipe that plug through degasification pipeline to vacuum air pump. According to invention, in borehole exploitation process, in casing pipe add suction pipe that consists of connection units of less diameter pipes, first one is perforated and has solidification from both sides.

The method is conducted in following way (Figure 1). Firstly preparing stoping space for borehole drilling (1), then driving degasification borehole (1) by drilling method. After mine working formation realizes casing by perforated pipes (3) and collar hermetic sealing (7) for borehole destruction exception (1) in exploitation process. In perforated casing pipe (3) firstly setting perforated unit (4) of less diameter that customized by elastic collar (2) in the beginning of section that provides its hermetic sealing and centering. Further with help of thread connection (5) to first perforated unit (4) joining next unit with elastic collar. Then, join solid units with advancing to need length in casing pipe (Bulat 1998).

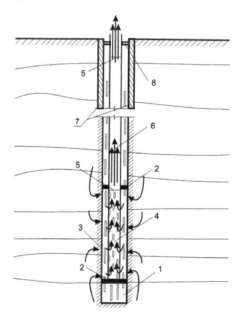

Figure 1. Scheme of degasification method of gas-bearing deposits: 1– borehole, 2 – elastic collar, 3 – perforated casing pipe, 4 – perforated unit of suction pipes, 5 – thread connection of suction pipes, 6 – typeset units of suction pipes, 7 – hermetic sealing borehole collar 8 – elastic solidification.

Whereat, space between perforated casing pipe (3) and collecting suction pipe (6) is pressurized by elastic solidification. Suction pipe (6) from typeset units connects to vacuum air pump.

If it is necessary, to system consistently or alternatively to typeset units can be added few perforated units (4) for unloading creation in concrete districts of degasification borehole (1).

Usage of given degasification method in spite of increasing of metal consumption process have to decrease prime cost and increase quality of extracted gas. Leak-in minimization near borehole collar and also increasing of unloading in zones with the biggest gas-penetration of degasified rocks give opportunity to receive mixture with methane concentration not less than 45% and also promote increasing of labour protection and increasing effectiveness of mineral deposits extraction.

Degasification of coal-bearing strata's by drilled from surface boreholes using practically on all main coal extraction world's countries (USA, China, Germany, Russia, Poland and The Czech Republic).

During usage of surface degasification of underworked strata's reached the highest average effectiveness of methane emission decreasing in mine workings, at this value of captive gas is equal to 70–90% of total emission.

3 CONCLUSIONS

Potential variants of mine methane usage in the range of concentrations 30–100% exists in extremely different areas, including following:

- usage as fuel in metallurgical furnaces, industrial furnaces and steam generating units;
- burning in internal-combustion engines or turbines for energy production;
- usage for natural gas flooding in pipelines;
- appliance as raw material for industry for fertilizers production;
- burning as motor fuel (liquefied natural gas LNG or compressed natural gas CNG).

Experience of industrially-developed countries shows that capital investments in advanced degasification technologies significantly increase economic efficiency of coal mines by means of fault time decreasing that caused by exceeding of methane maximum concentration limit in stoping and preparatory longwalls. Also create possibilities for utilization greater volume of gas and reduce methane emissions in the atmosphere.

REFERENCES

Karp, I.N. 1993. *About mine methane*. Ecotechnologies and resource-safety, 6: 5–7.

Pavlov, S.D. 2005. *The ways of coal deposits gas development*. Monograph: Koloryt: 336.

Bulat, A.F. 1998. *Creation of mine methane industry in fuel-energy complex of Ukraine*. Dnipropetrovs'k: Geotechnical mechanics: Interdepartmental collection of scientific papers, 10: 3–12.

von Schoenfeldt, H. 2008. *Advanced CMM and CBM Extraction Technologies*. Singapore: CBM conference.

Black, D. & Aziz, N. 2009. *Reducing Coal Mine GHG Emissions Through Effective Gas Drainage and Utilization*. Australian Institute of Mining and Metallurgy: Illawarra Branch: 217–214.

Progressive Technologies of Coal, Coalbed Methane, and Ores Mining – Bondarenko, Kovalevs'ka & Ganushevych (eds)
© 2014 Taylor & Francis Group, London, ISBN: 978-1-138-02699-5

Mining and geological conditions of methane redistribution within the undermined coal-rock massif

V. Lukinov, V. Prykhodchenko & L. Tokar
National Mining University, Dnipropetrovs'k, Ukraine

O. Prykhodchenko
M.S. Polyakov Institute of Geotechnical Mechanics, Dnipropetrovs'k, Ukraine

ABSTRACT: The paper considers problem of mining and geological conditions effecting methane redistribution within undermined coal-rock massif as well as determining rules for methane technogenic reserves accumulation according to coal-bearing strata geologic structure. Research results help to specify that increase of methane technogenic reserves accumulation within undermined coal-rock massif south-west northeast is typical for Donets'k-Makiivka district in Donbas. It is determined that increase in stratification depth in north-east sandstones is basic reason for increase in methane technogenic reserves accumulation; as for the coal grists, it is increase in their quantity and thickness.

1 INTRODUCTION

Today Ukraine experiences steady tendency for reduction of coal mines number. Moreover, there is also a great number of old closed mines; hence, the problem of methane recovery from coal-rock massif and mine workings is actual both economically and environmentally. About 120 mines have been closed at the territory of Ukrainian Donbas; similar number of mines is regarded as futureless being prepared to be removed from operation and phased out. Closing process is followed up by methane recovery to gob workings, isolated workings, and worked-out space of out-of-work mine workings. In this context methane eventually migrates through technogenic fractions up to earth surface inviting danger to population and objected located within a territory of closed mine take. In addition much methane is wasted despite it is environmentally friendly fuel. Determination of mining and geological conditions for methane redistribution within undermined coal-rock massif makes it possible to implement ideas concerning reliable decision of methane technogenic reserves expansion; that will give ability to attract extra power resources improving environment in coal mining regions by means of extraction and utilization of methane from closed mines and worked-out sites of productive mines.

2 METHODIC OF CALCULATION DENSITY OF METHANE TECHNOGENIC RESERVES

Technogenic methane is accumulated as a result of roof subsidence, fragmentation, and free and liberated gas recovery to mine workings as well as to fractured zones of coal-rock massif. Undermined coal strata fragmentation is followed by changes in the massif density (Lukinov & Klets 2002, Lukinov & Bezruchko 2010, Ayruny & Halazov 1990). Partially gas emission to mine workings takes place; fluid pressure in the massif falls. Analysis of rules concerning changes in pressure and rock density after mining makes it possible to evaluate accumulated technogenic methane reserves in undermined rock mass.

Direct measurements using seam tester in the process of well drilling give the most accurate information on massif pressure changes; however, such practice is very uncommon. Results of pressure measurements in wells within undermined rock mass (O.F. Zasiadko mine and "Chaikino" mine) help to conclude that there is a tendency for fluid pressure increase; the fluids saturate coal-rock massif. Linear dependence concerning increase of fluid pressure to upper boundary of underworking massif is proposed; within it pressure corresponds to a value within virgin rock mass. The linear dependence for fluid pressure change within undermined coal-rock massif P_{um} (Pa) may be shown either as a distance function behind a normal from a roof of undermined coal seam to rock stratum in which pressure is determined or through the dependence

$$P_{u.m.} = ah\rho_w g,$$ (1)

where a – coefficient involving height zone of undermining effect as well as undermined coal seam depth; h – distance behind a normal from undermined coal seam roof to rock stratum in which pressure is determined; ρ_w – water-mass density being equal to 1000 kg/m³; g – gravity sector, m/s². Coefficient a – determined for each well depending upon undermined massif height and fluid pressure within a zone where mass softening starts; that is

$$a = \frac{0.0085 \cdot H_u}{h_m},$$ (2)

where H_n – depth of undermining zone upper boundary (from surface), m; and h_u – height of undermining zone effect (from undermined seam), m. The dependence of pressure change makes it possible to determine its values everywhere in coal-rock massif.

Comparison of calculated pressure values in undermined coal-rock massif with virgin massif values demonstrates its heavy drop; that depends on increase in the massif porosity. As paper (Bulat 2000) determines, undermining can not give rise to apparent increase of efficient porosity. Fractures are the most important factor for changes in undermined coal-rock massif porosity. Determination of coal-rock massif destruction zone boundary can be conditionally divided into the two areas: former is an area disturbed by fractures where gas drains to a longwall; latter is a zone with increased porosity values (as for virgin state); from it gas gradually drains to area one, accumulating there and forming technogenic methane layerings. Hence, changes in porosity depend on fractures (fracture porosity $k_{e.p.}$). Efficient porosity of sandstones and efficient fracture porosity $k_{f.p.}$ form integral porosity $k_{i.e.p.}$ (Lukinov & Klets 2002).

In terms of calculations (Lukinov & Klets 2002), integral porosity $k_{i.e.p.}$ can be expressed as a ratio between pressure in a virgin massif P_s and pressure in softened massif $P_{u.m.}$ multiplied by a confident of efficient porosity $k_{e.p.}$

$$k_{i.e.p.} = \frac{P_s \cdot k_{e.p.}}{P_{u.m}}.$$ (3)

Following papers (Lidin & Ayruny 1969, Zabihaylo & Shyrokov 1974) determine logarithmic dependence of gas porosity changes on open porosity. Results of measurements concerning open porosity, efficient porosity, and gas porosity of sandstones helped to determine (Lukinov & Klets 2002) exponential dependence between values of gas po-

rosity coefficient k_p and efficient porosity $k_{e.p.}$; its corrected version is

$$k_p = e^{k_{e.p.} - 5}.$$ (4)

With undermined coal-rock massif, a coefficient of efficient porosity should be replaced by a coefficient of integral efficient porosity as the change in porosity depends of technogenic fracturing; then, integral porosity coefficient of undermined coal-rock massif is calculated. The calculation data help to identify a zone disturbed by fractures dividing undermined massif into zones of "slow" and "fast" gas. It is expedient draw boundaries between the zones according to a value of integral porosity $1 \cdot 10^{-15}$ m², lower which rocks become nonporosive. That makes it possible to refer coal-rock massif within a zone of "fast" gas to industrial collectors of 5th grade (according to classification (Khanin 1969)).

To check figures determining zones of "slow" and "fast" gas a method basing upon comparison of static pressure values is applied. The values have been determined according to water level in a well drilled on undermined coal-rock massif with hydrostatic pressure values. The results are used to determine coefficient of static pressure drop in a well $k_{s.p.d}$ demonstrating relative pressure deviation from hydrostatic one. In a virgin coal-rock massif, values of coefficient $k_{s.p.d}$ should be close to 0.8–0.9; that is 80–90% of hydrostatic pressure. So, the coefficient deviation will help to make accurate determination of boundary between zones where pressure drop takes place as a result of undermining and fluid migration to mine workings. Boundaries of "fast" and "slow" gas zones calculated with the help of two techniques coincide, which makes it possible to apply only a technique basing upon changes in integral porosity if water level measurements in a well are not available.

To calculate density of methane technogenic reserves it is required to determine interval disturbed as a result of mining operations; zones of "fast" and "slow" gas should be identified in it. As porosity in a zone of "fast" gas reaches important values (over $100 \cdot 10^{-15}$ m²) in the process of mining gas gets to active longwall being ventilated and degassed. Thus, they can not participate in methane technogenic accumulations formation. From a zone of "slow" gas, methane gradually gets to a zone of "fast" gas disturbed by fractures, and old mine workings structuring technogenic methane formations. Hence, technogenic collector is formed in a zone of "fast" gas and old mine workings; however, methane form a zone of "slow" gas fills it. Therefore, to evaluate density of methane technogenic reserves, it is required to calculate methane reserves in coal seams and sandstones located in a zone of "slow" gas. Within interval of "slow" gas zone, density of methane reserves is calculated for sandstones which thickness is over 5 m, and for coal grists which thickness is 0.2 m and more. Reserve density in sandstones ($P_{s.r.}$) is calculated as follows

$$P_{s.r.} = \sum_{i=1} x_{si} \cdot m_{si} , \qquad (5)$$

where x_{si} – gas content of sandstones, m³/m³; m_{si} – sandstone stratum thickness, m.

Following formula is applied to calculate methane density in coal seams and scuds

$$P_{c.r.} = \sum_{i=1} \left(x_{ci} - x_{coo} \right) \cdot m_c \cdot \rho_c , \qquad (6)$$

where x_{ci} – gas content of coal grist, m³/t; x_{coi} – residual gas content of coal grist, m³/t; m_c – coal grist thickness, m; and ρ_c – bed density of coal grist, t/m³.

Methane technogenic density is calculated for each well being a sum of methane density values in sandstones and grists located in a zone of "slow" gas

$$P_r = P_{s.r.} + P_{c.r.} . \qquad (7)$$

3 EVALUATION OF INFLUENCE OF REGIONAL DEPENDENCES VARIATION DENSITY OF TECHNOGENIC METHANE RESERVES

To forecast and identify zones potential from the viewpoint of technogenic methane reserves, it is required to evaluate regional dependences of technogenic methane reserves density variation. To solve the problem, it is necessary to track spatial changes in accumulated methane reserves. Donets'k-Makiivka geological and industrial area in Donbas is chosen to make regional evaluation of geological factors. A number of mine fields are chosen for local evaluations: O.F. Zasiadko mine, "Chaikino" mine, and V.M. Bazhanov mine. On the data of expendable wells, above technique is applied to calculate density of methane technogenic reserves for the mine fields.

Calculation results concerning methane technogenic reserves density for mines in Donets'k-Makiivka area help to determine that the reserves increase north-east. As paper (Alekeseiev 1969) confirms, successive and regular south-west in-

crease in coal-bearing formation thickness where total coal-bearing formation thickness is 1800–1900 m north-east where it is 3400–3450 m is one of typical features of medium coal formations. However, geological structure analyzed on well columns identifies that within mine field change in coal formation thickness can only effect change in total density of coal grists methane technogenic reserves; it can not exercise a significant influence on the factor for sandstones as increase in formation thickness results in increase of coal seams thickness and their number; that very time thickness of sandstones remains practically invariant within a mine field.

Paper (Lukinov & Zhykalyak 2005) determines increase in potential gas content of sandstones depending upon depth; maximums are achieved at the depth of 1500 m. Then their decrease takes place. As worked-out mine workings are located up to a depth of 1500 m, it is reputed that increase in potential gas content may be considered as a factor of increased density of accumulated methane technogenic reserves in sandstones. coal seam mined out in lowering directions with a direction of increase in accumulated technogenic methane reserves in sandstones were correlated. Hypsometric maps and maps of approximate surfaces of accumulated technogenic methane reserves in sandstones were generated to that end.

In O.F.Zasiadko mine (Figure 1a) a course of l_1 seam within a site under consideration is north-west with 15 to 40° dip azimuth. Bedding, close to monoclonal, is complicated by flat plicatures on the strike. Level of a seam floor varies from -400 m south-west to -1100 m along the northeast extension. In sandstones (Figure 1b) density of accumulated methane technogenic reserves increases north-east with 12° azimuth.

Such wells as ДМ-1924, ДМ-1896, ДМ-1139 and a number of others have minimum of accumulated technogenic reserves in sandstones. In those wells estimated values are near zero as sandstones within affected zone of undermining are close to worked-out seam; besides, they were degassed while mining. Maximums obtained in the neighbourhood of ДМ-1134 and ДМ-1937 wells being 23.6 m³/m², and 21.7 m³/m² respectively. Hence, comparison of the maps shows that azimuth of accumulated methane technogenic reserves density increase is close to azimuth of coal-bearing strata lowering.

(a)

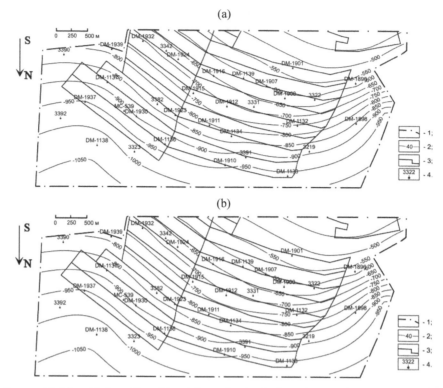

(b)

Figure 1. Hypsometric Map of l_1 Seam Floor (a) and Approximating Surface of Accumulated Methane Technogenic Reserves in Sandstones (b) over l_1 Seam within O.F. Zasiadko Mine Field: 1 – Mine field boundaries; 2 – Contour lines of coal seam foot level, m (a) and contour lines of accumulated methane technogenic reserves in sandstones, m³/m² (b). 3 – Boundaries of worked-out area as for coal seam; 4 – Expendable wells.

In "Chaikino" mine, coal seam m_3 is of monoclinal bedding complicated with turns; it lowers north-west with 310–350° azimuth (Figure 2a). Minimum levels of coal seam floor are −500 m along the southwest area of mine field near MC-261 well. Maximums (-1170 m) are in the neighbourhood of well No. 3927 near north-east border of a mine field. Comparison with a map of approximating surface of accumulated technogenic methane reserves density in sandstones (Figure 2b) makes it possible to conclude that in broad terms direction of density increase coincides with azimuth of coal-bearing strata lowering; its azimuth is 345°. Minimum density of accumulated methane technogenic reserves in sandstones is obtained for well No. 3707 being 11.7 m³/m². Maximums are obtained in the neighbourhood of wells No. Щ-59 (78.7 m³/m²), and No. 3600 (73.3 m³/m²). Hence, there is dependence between values of accumulated technogenic methane reserves in sandstones and stratification depth of coal-bearing strata in which technogenic deposits are being formed.

Within V.M.Bazhanov mine field, m_3 coal seam is also of monoclinal bedding complicated with insignificant turns behind bearing. Seam lowers northeast with 30–45° azimuth. The coal seam floor levels are −550 m in the neighbourhood of wells Nos.3867 and 3944 to - 1180 m in the neighbourhood of wells Nos. Щ-852 and Щ-814. Expendable wells data are used to prepare hypsometric map of coal seam m_3 floor (Figure 3a). A map of approximating surface is prepared using estimated results for accumulated methane technogenic reserves density in sandstones (Figure 3b). Such wells as Щ-624, Щ-474 and No. 3551 give minimum density of accumulated methane technogenic reserves in sandstones; it is about 5 m³/m². Maximums are in the neighbourhood of Щ-207 and Щ-817 wells being 66.8 and 62.7 m³/m². Increase in density of accumulated methane technogenic reserves takes place north-east; its azimuth is 45°. Thus, comparing the two maps we reason that azimuths of coal-bearing strata lowering and increase in accumulated methane technogenic reserves density in sandstones coincide.

(a)   (b)

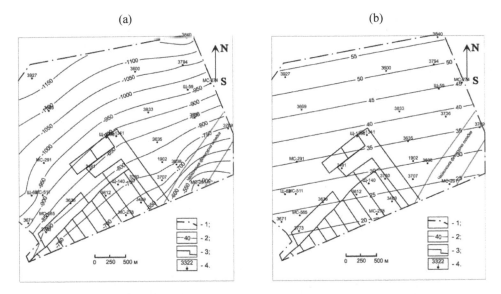

Figure 2. Hypsometric Map of m³ Seam Floor m_3 (a) and Approximating Surface of Accumulated Methane Technogenic Reserves in Sandstones (b) over m_3 Seam within "Chaikino" Mine Field: 1 – Boundaries of a Mine Field; 2 – Contour lines of coal seam floor, m (a) and contour lines of accumulated methane technogenic reserves, m³/m² (b); 3 – Boundaries of worked-out area in terms of coal seam; 4 – Expendable wells.

(a)   (b)

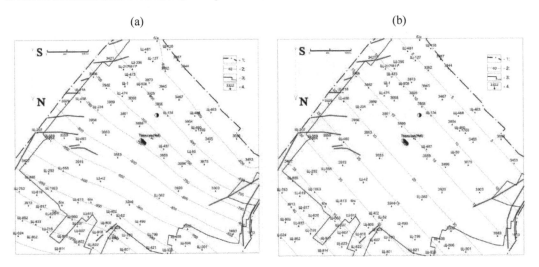

Figure 3. Hypsometric Map of m_3 Coal Seam Floor (a) and Approximating Surface of Accumulated Methane Technogenic Reserves Density in Sandstones (b) over m_3 Seam within V.M.Bazhanov Mine Field: 1 – Boundaries of a mine field; 2 – Contour lines of coal seam floor, m (a) and contour lines of accumulated methane technogenic reserves density, m³/m² (b). 3 – Boundaries of worked-out area in terms of coal seam; 4 – Expendable wells; 5 – Tectonic fault.

To examine assumption that increase in coal formation thickness is basic reason effecting increase in density of accumulated methane technogenic in north-east coal grists, a coefficient of undermined massif for $k_{b.m.c.}$ coal seams is introduced. The coefficient characterizes structure of undermined coal-rock strata. It also involves the number, thickness, and location of coal grists within a zone of "slow" gas being calculated in terms of each well. A coefficient of coal-rock strata structure for coal grists is calculated according the data of expendable wells for each seam; it is a total output of each grist thickness for its distance to worked-out coal seam floor

321

$$k_{b.m.c.} = \sum \left(m_{c1} M_{c1} + m_{c2} M_{c2} + ... + m_{ci} M_{ci} \right), \quad (8)$$

where $m_{c1,\ c2,...,ci}$ – coal seam thickness, m, and $M_{c1,\ c2,...,ci}$ – distance from worked-out coal seam floor to coal seam or to interlayer under study, m.

In O.F. Zasiadko mine calculations of coal-rock massif structure for coal grists over l_1 seam involved data of 22 expendable wells (Figure 4). Two to four coal seams and interlayers among which l_4, l_5, l_7^1, and l_8 are basic ones are located above worked-out coal seam within a zone of "slow" gas. Minimum of

coal-rock massif structure coefficient for coal grists is in the area of ДМ-1898 (199.7 m·m) and No. 3391 (228.7 m·m) wells. ДМ-1898 well drilled out following four coal grists in a zone of "slow" gas: l_4 (thickness (m) is 0.4 m; distance from coal seam l_1 floor is (M) 63.8 м), l_5 ($m = 0.4$ m, $M = 97.6$ m), l_7^1 ($m = 0.5$ m, $M = 101.2$ m), and l_8 ($m = 0.4$ m, $M = 211.4$ m). Well No. 3391 drilled out two coal seams l_4 ($m = 1.15$ m, $M = 78.2$ m) and l_8 ($m = 0.64$ m, $M = 216.85$ m).

(a)

(b)

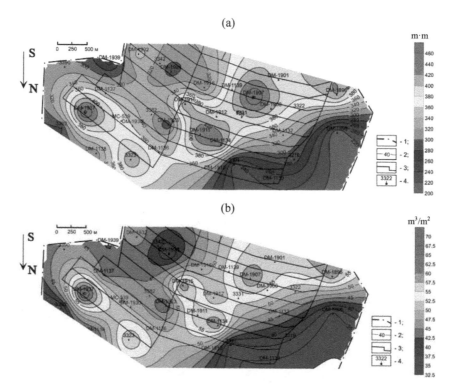

Figure 4. Contour Lines of Coal-Rock Massif Structure for Coal Grists, m·m (a) and Contour Lines of Accumulated Methane Technogenic Reserves Density in Coal Grists m³/m² (b) over a Seam l_1 in O.F.Zasiadko Mine Field: 1 – Boundaries of a mine field; 2 – Contour lines of coal-rock massif structure (a) and contour lines of accumulated methane technogenic reserves density (b); 3 – Boundaries of worked-out area in terms coal seam; 4 – Expendable wells.

Maximums of a coefficient are obtained for ДМ-1937 well located along the northeast extension of a mine field site under study. The coefficient value reaches 543.4 m·m. Such a value depends on availability of four rather thick coal grists. In addition, distance is optimum to form technogenic methane deposits. Those are seams l_4 (thickness is (m) 1.1 m, distance from coal seam l_1 (M) is 72.9 m); l_5 ($m = 0.35$ m, $M = 110.6$ m); l_7^1 ($m = 1.3$ m, $M = 180.0$ m); and l_8 ($m = 0.6$ m, $M = 197.0$ m). For comparison, Figure 4(b) demonstrates a map of con-

tour lines for density of accumulated methane technogenic reserves in coal grists above l_1 seam. Minimums of accumulated methane technogenic reserves density in coal grists are in the neighbourhood of ДМ-1898 and No. 3392 wells; they are 34.0 m³/m² and 35.5 m³/m² respectively. Maximums are in the neighbourhood of ДМ-1937 well (67.2 m³/m²). Comparison of the two maps makes it possible to conclude that a mine field sites with high coefficients of coal-rock massif structure for coal grists coincide with zones of high values of accumulated methane technogenic reserves density in coal grists.

In "Chaikino" mine, coefficient of coal-rock massif structure for coal grists above m_3 seam involved data of 18 expendable wells. The calculations results were used for a map in Figure 5 a. Two to eight coal grists are located above coal seam m_3 in a zone of "slow" gas. m_4, m_4^4, m_5'', m_5^6, and m_5^1 are the thickest. Results demonstrate that minimums of coal-rock massif structure coefficient for coal grists are in the neighbourhood of wells Nos. 3638 and 3780. The former is located along the southeast extension of a mine field site. The coefficient value is 91.3 m·m. Such a low value depends on availability of the only grist m_5^1 which thickness is 0.55 m. It is located at the distance of 166.0 m from coal seam m_3 in a zone of "slow" gas. The latter is in the neighbourhood of eastern block 17 near western part of a mine field. The coefficient value is 154.1 m·m. A well drilled out two coal seams (m_4^3 and m_5^1). Thickness of m_4^3 seam is 0.4 m; distance from worked-out coal seam is 109.4 m; as for m_5^1 seam, its thickness is 0.7 m, and distance is 157.7 m. Coefficient maximums are obtained for wells Nos. 3669 and 3840. Well No. 3669 is located along the

northwest extension of a mine field. Coefficient of coal-rock massif structure for coal grists reaches 327.1 m·m. The value depends on increase in number and thickness of coal grists in a zone of "slow" gas; those are m_4^1, m_4^2, m_4^3, m_5^1, and m_5^3 seams. Well No. 3840 is located in the neighbourhood of a mine field eastern border. The coefficient value is 286.8 m·m. A well in a zone of "slow" gas drilled out the four coal grists: m_4^1, m_5'', m_5^6, and m_5^1. Calculation results of coal-rock massif structure for coal grists were compared with values of accumulated methane technogenic reserves density in coal grists (Figure 5b). As the Figure demonstrates, zones of accumulated methane technogenic reserves density values coincides with zones of the coefficient minimums. Thus, in the neighbourhood of wells Nos. 3638 and 3780 minimum density of accumulated methane technogenic reserves in coal grists is 13.0 and 24.0 m³/m² respectively; wells Nos. 3669 and 3840 demonstrated maximums being 47.9 and 41.8 m³/m². Hence, strict association between density of accumulated methane technogenic reserves in coal grists and values of coal-rock massif structural coefficient values for coal grists.

(a) (b)

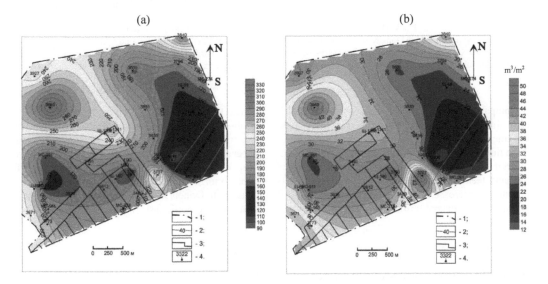

Figure 5. A Map of coal-rock massif structural coefficient for coal grists (a) and contour lines of accumulated methane technogenic reserves density in coal grists (b) above m_3 seam within "Chaikino" mine field: 1 – Borders of a mine field; 2 – Contour lines of coal-rock massif structural coefficient, m·m (a) and contour lines of accumulated methane technogenic reserves density, m³/m² (b); 3 – Borders of worked-out area in terms of coal seam; 4 – Expendable wells.

Similar results confirming abovementioned association are obtained in V.M. Bazhanov mine. According to calculations, minimums of structural coefficients are for well No. 1811 (73.2 m·m), well

No. Щ-42 (111.6 m·m) depending upon availability of the only grists in a zone of "slow" gas. Maximum being 339.4 m³/m² is determined in a well No. Щ-817. Such a value is a result of the five coal grists availability in a zone of "slow" gas: m_4^1 (thickness (m) is 0.45 m; distance from coal seam m_3 (M) is 88.6 m); m_5'' ($m = 0.62$ m; $M = 160.4$ m); m_5^6 ($m = 0.2$ m, $M = 170.2$ m); m_5^1 ($m = 0.7$ m, $M = 178.0$ m); and m_5^3 ($m = 0.2$ m, $M = 207.2$ m). The calculations results are involved to prepare a map in Figure 6(a) demonstrating contour lines of accumulated methane technogenic reserves density in coal grists. Minimums of accumulated methane tech-

nogenic reserves density in coal grists are in the neighbourhood of wells No. 1811 and Щ-62 being 11.3 and 14.6 m³/m² respectively. Maximums are in the neighbourhood of such wells as Щ-817 (48.8 m³/m²) and No. 3873 (49.5 m³/m²). Hence, comparison of the two maps distinctly shows coincidence of zones with heightened and lowered values in terms of both factors.

(a)   (b)

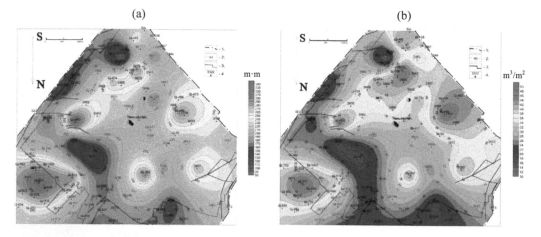

Figure 6. A Map of Coal-Rock Massif Structural Coefficient for Coal Grists (a) and Contour Lines of Accumulated Methane Technogenic Reserves Density in Coal Grists (b) above m_3 seam in V.M. Bazhanov Mine Field: 1 – Boundaries of a mine field; 2 – Contour lines of coal-rock massif structure, m·m (a) and contour lines of accumulated methane technogenic reserves density, m³/m² (b); 3 – Boundaries of worked-out area in terms of coal seam; 4 – Expendable wells; 5 – Tectonic fault.

Comparison of coal-rock massif structural coefficient for coal grists with accumulated methane technogenic reserves density for coal grists in terms of three mines in Donets'k-Makiivka geological and industrial area explains that increase in accumulated methane technogenic reserves density in coal grists depends on increase in their number and thickness; it is connected with coal formation thickness increase.

logical and industrial area the tendency of increasing north-east density of accumulated methane technogenic reserves is available. As for sandstones, such dependence can be explained by the fact of coal strata deepening in which methane technogenic accumulations form; as for coal seams and interlayers, it increases in Donets'k-Makiivka area coal formations south-west to north-east.

4 CONCLUSIONS

Hence, obtained dependences of methane redistribution in sandstones and coal grists as a result of their undermining help to draw following conclusions. Density of accumulated methane technogenic reserves in undermined coal-rock mass of Donets'k-Makiivka area experiences northeast extension. That depends on regional changes in geology of the coal field; namely, increase in thickness and stratification depth of coal formations; increase in total thickness of collectors (grists and sandstones) and gas pressure in them. It is determined that in Donets'k-Makiivka geo-

REFERENCES

Lukinov, V.V., Klets, A.P., Bobryshev V.V. et al. 2002. *Filtration Parameters of Collector – Coal-Rock Massif Undermined by Mine Workings*. Geotechnical mechanics: Interdepartmental collection of scientific papers / IGTM NAS of Ukraine. Publication, 37.

Lukinov, V.V., Shkuro, L.L. & Bezruchko. K.A. 2010. *Effect of Technogenic Factor on Physics of Sandstones*. Scientific Messenger of the NMU.

Lukinov, V.V. & Bezruchko. K.A 2010. *Formation of Porosity under the Effect of Technogenic Fac*tor. Coal of Ukraine, 6.

Airuni, A.T., Galazov, R.A., Sergeiev, I.V. et al. 1990. Gas *Abundance of Bituminous Coal Mines in the USSR, Integrated Development of Gas-Bearing Coal Deposits.* Moscow: Nauka.

Bulat, A.F. 2000. *Opening Remarks by a Director of Institute of Rock Mechanics A.F Bulat, Corresponding member of NAS of Ukraine, Doctor of Engineering Science.* Geotechnical mechanics: Interdepartmental collection of scientific papers / IGTM NAS of Ukraine. Publication, 17.

Lidin, G.D., Airuni, A.T., Bessonov, Y.N. et al. 1969. *Analysis of Degassing Rules for Coal Seams in the Process of their Mining and Undermining.* Publishing House of Institute of Geophysics of AS of the USSR.

Zabigailo, V.E., Shirokov, A.Z., Belyi, I.S. et al. 1974. *Geological factors of Donbass Rocks Outburst.* Kyiv: Naukova Dumka.

Khanin, A. A. 1969. *Oil and Gas Collecting Rocks and their Analysis.* Moscow: Nedra.

Alekeseiev, V.D. 1963. *Donetsk-Makeievka Coal Area // Geology of Coal and Slate Coal Deposits in the USSR.* Gosgeolotekhizdat. Vol. 1.

Lukinov, V.V. & Zhikaliak, N.V. 2005. *Prognosis Evaluation of Sandstones Gas Content Maximum Depth.* Geotechnical mechanics: Interdepartmental collection of scientific papers / IGTM NAS of Ukraine. Publication, 53.

Progressive Technologies of Coal, Coalbed Methane, and Ores Mining – Bondarenko, Kovalevs'ka & Ganushevych (eds)
© 2014 Taylor & Francis Group, London, ISBN: 978-1-138-02699-5

Equalization of traffic flows of benches in the working area with excavator-truck complexes

S. Moldabayev, Zh. Sultanbekova & Ye. Aben
Kazakh National Technical University after K.I. Satbayev, Almaty, Kazakhstan

I. Gumenik
National Mining University, Dnipropetrovs'k, Ukraine

ABSTRACT: The technology of pit construction and exploitation with the efficient use of powerful excavator-truck complex (ETC) in great depths with a minimum spacing of pit walls is developed. The equal length of working front on all benches and steep angles of pit walls are ensured by choosing the direction of the working front perpendicular to the highwall developing direction. Equal traffic flows and varying levels of working area will pit completion date even if cap rock has big height and ensure achieving and maintaining maximum productivity during the operation until the end of ore and coal deposits. High performance of ETC and optimal schedule of stripping operations using excavators of equal capacity increases mine productivity and greatly expand the boundaries of the effective application of opencast method.

1 INTRODUCTION

Nowadays reserves of many solid minerals occurring at shallow depths from the surface are almost exhausted. About 20 new mines are planned to build in Kazakhstan. So mining works will be developed in difficult geological and technical conditions. For example, cap rocks height of Lomonosov iron deposit is more than 120 m. Pit completion date minimization and increasing efficiency of opencast mining directly depends on the adoption of economical decisions of the complex mechanization structure and mining technology. It should also take into account the fact that the reduction of extraction front is making difficult to maintain high productivity during pit sinking.

The efficiency of opencast mining of steep deposits to great depths achieved through the use of powerful excavator-truck complexes. The best technical and economic performance is achieved when the time of reaching of the ultimate pit depth at the first stage with transition to the internal dumping in extended deposits.

G.G. Sakantsev (UB RAS) developed a technology for mining such deposits with the transition to the internal dumping (Sakantsev 2002), which reduces the number of transport sites on the pit wall. The trains are allocated along the pit length. The first train is mined to design depth with external dumping. The next train is mined with internal dumping. However, safety of mining works is reduced and haulage is hampered when embankment is constructed along the road side. K.N. Trubestkoi, A.G. Shapar, V.G. Pshenichnyi have written about efficiency of internal dumping technology.

The technology of mining the steep deposits (Drizhenko 1987) is known, in which mining efficiency is achieved by shortening working out stages, improving the mining equipment using and reducing the volume of mining and construction works. The technology is based on mining the benches by horizontal layers. Another one technology of mining steep deposits when stripping costs are reduced by mining the overburden by steeply inclined layers. Working area is formed of steeply inclined layers arranged across to ore deposit with a horizontal width equal to the width of the working area. Working off of benches in each layer is developed by panels along the longitudinal pit wall. After the panel mined out mining works moved to the underlying bench in the same layer. Due to this technology the maximum volumes of overburden shifted to the final period of operation. The conditions for internal dumping appear in end of working level when pit gets to ultimate depth. Working front moves to the opposite direction.

However, reducing of working front in these technologies with mining works sinking make difficult to use same type of equipment on each bench and limits the pit productivity. Otherwise, same excavator operates at different levels, which increases the amount of preparatory work and leads to unproductive use.

2 MINING TECHNOLOGY USING HUGE EXCAVATOR-TRUCK COMPLEXES

The double subbench technology of mining the inclined and steep deposits with varying levels of working area was developed in KazNTU after K.I. Satbayev (Rakishev 2012). Feature of this technology is that firstly the upper subbenches are mined simultaneously, then the lower subbencehs are mined. Such sequence of mining by high benches using cross panels and temporary ramps construction in pit ends allows to leave only the safety berms between benches and subbenches. Temporary ramps are constructed by excavator-truck complexes in pit ends when the upper subbenches are mined from transport berms. After they are mined out excavator-truck complexes work in working area of the lower subbenches. However working front of each bench decreases when mining works sink. It doesn`t allow to evenly distribute the volumes of mining operations and leads to excavators downtime with decreasing length of the panels.

Ensuring equal length of work front in benches and subbenches working areas and effective use of powerful excavator-truck complexes are encouraged to implement as follows (Figures 1 and 2). Initially, the final depth H_P is reached in pit of first train (Rakishev 2013), technological layers 3 are isolated in pit of the second train. They have width equal to the length of the panels L_{bl}.

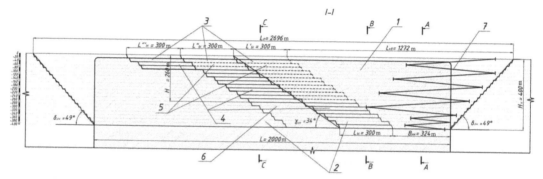

Figure 1. Final pit: 1 – the first pit train; 2 – the second pit train; 3 – technological layers; 4 – panels, which are being mined; 5 – mined panels; 6 –panels envisaged for mining; 7 – opening trace in the pit of the first train.

Transverse two-wall mining system will increase productivity of the second train pit. The next bench is stripped by stripping trench (initial cut) 8 (Figure 2). Equal length of the working front of benches and subbenches in two working areas 9 and 10 is achieved by the direction of the working front advance perpendicular to the direction of working side advance 11 in the second train pit.

Construction of only two-lane transport berms 12 and only safety berms 13 between subbenches provide steep angle of pit wall.

Transport communication between working areas 9 and 10 through external and internal transport incline is conducted by transport pit wall 15. Traffic flows generated from the upper subbenches of working area 9, which are adjacent to the longitudinal transport wall. Traffic flows move to transport berms through transport of technological layers. Also they are moving through opening 17 when the upper subbenches are being mined in working area 10, which is far from transport pit wall. Openings are partially located on the sides of the initial cuts of this technological layer

Openings will gradually be oriented in the same direction to the place of mining and transport berms will be used for overburden handling into internal dump. Dump is formed in the first train pit, which minimizes transportation distance.

Pit productivity rises by increasing the bench height. Also it allows to minimizing of overburden current volumes and reducing the area of land for external waste piles. So open pit mining costs will be reduced.

The working area width between benches shall be equal to the sum of the panel width and safety berm width by base option 1 (Rakishev 2012). Width of the mining stages must be multiple to several panels width.

The proposed mining technology (option 2) engages selection of mining stages contours by simultaneous mining of benches in several technological layers (maximum – three). Since mining operations are conducted by two-wall system, each mining stage corresponds to a certain amount of panels.

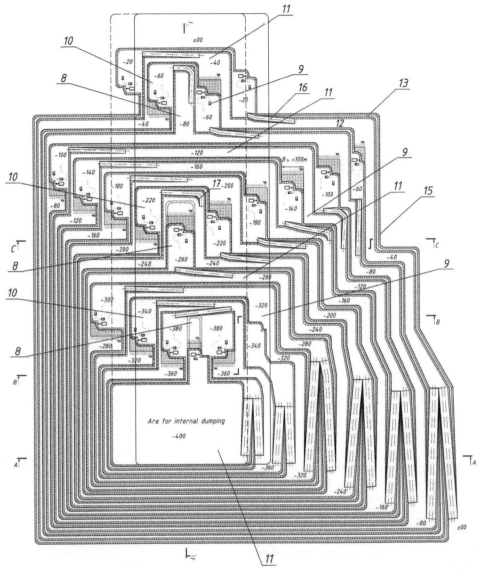

Figure 2. Mining technology with ensuring equal length of working front on each horizon (the position of mining only the upper subbenches): 8 – cutting trenches; 9 and 10 – working areas of technological layers adjacent to the longitudinal transport pit wall and placed through the cutting trenches; 11 – highwall of the second train pit; 12 and 13 – transport and safety berms; 14 – mined-put space; 15 – longitudinal transport pit wall; 16 and 17 – openings in working areas 9 and 10.

Mining operations are ensured by simultaneous development of several technological layers 3 throughout the pit depth of the second train. Height of working areas which mined by panels 4 can be kept constant at H_P. Then the next bench is mined in adjacent technological layer. Therefore, panels envisaged for mining 6 are pre-allocated and prepared (Rakishev 2012).

Communication with each horizon is provided through transport berms of highwall (and longitudi-nal transport pit wall) and opening traces 7 in the first train pit for ore haulage and overburden stock-piling in dumps (Figure 2).

Implementation of double subbench technology (mining the lower after mining the upper subbench) of steep deposits by cross panels with varying level of working areas provides the best results when huge excavator-truck complexes of equal capacity are applied on each bench.

3 ADVANTAGES OF NEW TECHNOLOGY

Productivity of proposed option 2 is more than productivity of base option. So a graphical comparison of stripping ratio (Figure 3), overburden volumes and productivity (Figure 4) of two options is satisfied only by the second train.

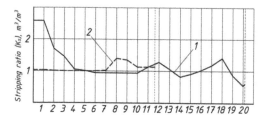

Figure 3. Stripping ratio of the second train pit: 1 – option 1; 2 – option 2.

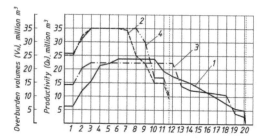

Figure 4. Mining mode of the second train pit: 1, 2 – ore volumes, 3, 4 – volume of overburden by options 1 and 2.

The stripping ratio changes from a smaller value to a larger – from 1.03 to 1.4 m³/m³, by option 1 – from 2.25 to 0.94 m³/m³. Analysis of the graph (Figure 4) shows that option 2 is more preferred in the second train pit. The designed productivity is reached faster and there are not any abrupt changes of stripping volumes. Also maximum productivity is provided until the end of the development.

4 APPROBATION OF TECHNOLOGY IN DESIGN STUDIES

Preliminary study of the feasibility of the technology using powerful excavator-truck complexes in northwestern part of the Lomonosov iron deposit with integrated mining-geological systems based on software LLP "ArdProekt-Kazyna" shown in Figures 5 and 6.

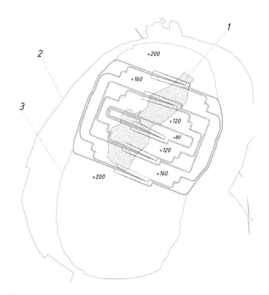

Figure 5. Commission of the first train pit in northwestern part of the Lomonosov iron deposit: 1 – deposits boundaries; 2, 3 – Designed and recommended contours of the first train pit.

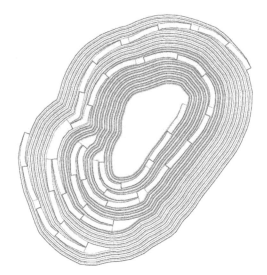

Figure 6. Final pit of the the Lomonosov iron deposit.

Comparison of final contours of the first train pit shows that the proposed mining double subbench technology in Lomonosov iron deposit by cross panels allows shortening of pit commission due to the smaller spacing of pit walls more intensive mining by transverse two-wall system, as well as reduction of the average stripping ratio (19.4%).

5 CONCLUSIONS

Efficient development of steep deposits using powerful excavator-truck systems is achieved by implementation of the proposed mining technology, which differs from the known technologies mainly for the following signs:

1) The technological layers are highlighted from the first train pit boundaries along the long axis of pit. They have same width, which equal to the panels length. The maximum productivity of the second train pit is reached when double subbench technology is used and mining works are conducted in 2–3 technological layers at the same time. The next bench is stripped by stripping trench (initial cut), which is situated in the middle of deposit.

2) Equal length of the working front of benches and subbenches is achieved by the direction of the working front advance perpendicular to the direction of working side advance 11 in the second train pit.

Implementation of double subbench technology (mining the lower after mining the upper subbench) of steep deposits by cross panels with varying level of working areas provides the best results when huge excavator-truck complexes of equal capacity are applied on each bench.

Improving the efficiency of double subbench technology of mining the steep deposits using powerful excavator-truck complexes with varying levels of working area and transverse two-wall system will increase the second train pit productivity and reduce the separation of pit walls.

Achieving the equal traffic flows allow to use excavators with same capacity. Steep angle of highwall allows reducing current volume of stripping works. Productivity can rise to 47%.

REFERENCES

Drizhenko, A.Y. & Bogdanov, V.M. 1987. *Method of opencast mining the steep deposits. Copyright certificates on inventions of the USSR.*

Drizhenko, A.Y. 2011. *Staged development of overburden of iron ore pits by steeply inclined excavation layers.* Mining magazine, 2: 25–28.

Shapar', A.G., Yakubenko, L.V., Piven', V.A. & Romarenko, A.V. 2009. *New technological decisions of mining the steep deposits with internal dumping for pits developing steeply sloping deposits.* Miners forum: 78–86.

Pshenichnyi, V.G. 2012. *Determination of rational mining mode and internal dumping for pits developing steeply sloping deposits.* National University of Kryvyi Rih. Bulletin, 31: 3–6.

Rakishev, B.R. & Moldabyaev, S.K. 2012. *Method of opencast mining the steep deposits. Innovation patent.*

Rakishev, B.R., Moldabayev, S.K. & Aben, Ye. 2013. *Minimizing the Time of reaching the Final Depth in the Pit of the First Train When Extracting the Elongated Steep-Dipping Deposits.* Mine Planning and Equipment Selection. Vol. 1: 209–216.

Sakantsev, G.G. 2002. *Opencast mining of steep deposits with internal dumping.* RF patent.

Trubetskoi, K.N., Peshkov, A.A. & Matsko, N.A. 1994. *Defining the application of development methods of steep deposits using preformed mined-out space of pit.* Moscow: Mining magazine, 1: 51–59.

Progressive Technologies of Coal, Coalbed Methane, and Ores Mining – Bondarenko, Kovalevs'ka & Ganushevych (eds)
© 2014 Taylor & Francis Group, London, ISBN: 978-1-138-02699-5

Defining the parameters of the atmospheric air for iron ore mines

I. Mironova & O. Borysovs'ka
National Mining University, Dnipropetrovs'k, Ukraine

ABSTRACT: The regularities of change of surface air concentration of ecologically hazardous substances around the ventilation shafts from the annual specific consumption of explosives are established. Regression dependences between changes of bio-indicators state and the concentration of atmospheric pollutants as well as the distance to the emission source are obtained. Method for determining the parameters of the atmospheric air for iron ore mines is developed.

1 INTRODUCTION

The mining industry is the main source of raw materials for metallurgical enterprises. Unfortunately, it poses an ecological threat to environmental objects.

The underground mining of iron ore is carried out by drilling and blasting with the use of TNT containing explosives, in which mine air, contaminating by the explosion products and iron ore dust is emitted into the air without any treatment and it constitutes a menace for the environmental components in areas of enterprises disposal. Long and large-scale production of iron ore led to higher levels of pollution of air, water bodies, land as well as to the accumulation of significant amounts of industrial waste.

The level of environmental safety of underground mining of iron ore remains low due to insufficient study of the mechanism of mine and atmospheric air pollution by harmful emissions of blasting and the lack of effective means of influence on these emissions. To date, TNT containing analogues are used as the main explosive in the iron ore mines in Ukraine. Analysis of technical indicators of production activity of Kryvyi Rih basin and Belozersky iron ore district revealed that mines of JSC "Zaporizhzhya iron ore combine" (JSC "ZIOC") use in an average of 2.9 million kg of TNT containing explosives in a year for the extraction of iron ore, that is 5–7 times larger than each iron ore mine separately (Khomenko 2011). Therefore production site of JSC "ZIOC" and adjacent territory are relevant ground for research to establish the level of air pollution.

The aim of this paper is to develop a methodology for determining the parameters of the atmospheric air state at a location nearby to the iron ore mine. To achieve this goal the following tasks were set and solved:

– determining the concentration of environmentally hazardous substances emanating from the ventilation shafts of the mine;
– calculation of hazardous materials dispersion around the ventilation shafts of the mine;
– assessment of toxic and mutagenic state of atmospheric air and changes of biological indicators of a winter wheat in the first generation in areas adjacent to mine;
– development of a methodology for calculating the parameters of the atmospheric air in this area.

2 INVESTIGATION OF THE ATMOSPERIC AIR STATE IN THE TERRITORY OF IRON ORE MINE

Measurement of the concentration of environmentally hazardous substances emanating from the ventilation shafts of the mine was done by the rapid method of physical and chemical analysis. Sampling of analyzed air was carried out in the mine after blasting in the break between shifts. Measuring devices were used to measure the concentration of environmentally hazardous substances in air samples – gas analyzer "Palladium-3M" and gas detector GC-M. Measurements of concentration of carbon monoxide, nitrogen oxides and sulfur dioxide were carried out with these instruments for the three vent shafts of mine. The obtained results of hazardous materials concentrations in the mine air samples collected during 2006–2010 are presented in Table 1 and 2.

Data on annual productivity, general and specific consumption of explosives were collected to determine the correlation between the concentration of harmful substances in air samples and indicators of mine for the period 2006–2010, obtained data are presented in Table 3.

Table 1. Results of concentration measuring of environmentally hazardous substances in the outgoing air stream of channels of main fans.

Gas	Concentration of environmentally hazardous substances, mg/m³				
	2006	2007	2008	2009	2010
Northern ventilation shaft (NVS)					
CO	36	40	32	39	38
NOₓ	2.3	2.4	2.1	2.4	2.3
SO₂	3.2	3.3	3	3.3	3.3
Drain ventilation shaft (DVS)					
CO	39	43	35	43	41
NOₓ	1.2	1.4	1.1	1.4	1.3
SO₂	3.2	3.3	3.1	3.3	3.3
Southern ventilation shaft (SVS)					
CO	34	38	31	37	36
NOₓ	2.1	2.3	2	2.2	2.2
SO₂	3.2	3.4	3.1	3.4	3.4

Table 2. Results of calculation of emission intensity.

Year	Fan capacity, m³/s	Speed of air movement, m/s	Emission intensity					
			CO		NO_x		SO_2	
			g/s	kg/h	g/s	kg/h	g/s	kg/h
Northern ventilation shaft (NVS)								
2006	217	14.47	7.812	28.12	0.499	1.797	0.694	2.498
2007			8.680	31.25	0.521	1.876	0.716	2.578
2008			6.944	25.00	0.456	1.642	0.651	2.344
2009			8.463	30.47	0.521	1.876	0.716	2.578
2010			8.246	29.69	0.499	1.797	0.716	2.578
Drain ventilation shaft (DVS)								
2006	232	14.5	9.048	32.57	0.278	1.001	0.742	2.671
2007			9.976	35.91	0.325	1.170	0.766	2.758
2008			8.120	29.23	0.255	0.918	0.719	2.588
2009			9.976	35.91	0.325	1.170	0.766	2.758
2010			9.512	34.24	0.302	1.087	0.766	2.758
Southern ventilation shaft (SVS)								
2006	257	14.95	8.738	31.46	0.540	1.944	0.822	2.959
2007			9.766	35.16	0.591	2.128	0.874	3.146
2008			7.967	28.68	0.514	1.850	0.797	2.869
2009			9.509	34.23	0.565	2.034	0.874	3.146
2010			9.252	33.31	0.565	2.034	0.874	3.146

Table 3. Annual productivity, general and specific consumption of explosives of JSC "ZIOC" for the period 2006–2010.

Год	Annual productivity, million tons/year	General consumption of explosives, kg	Specific consumption of explosives, kg/t
2006	4.31	2794943	0.648
2007	4.40	3003133	0.683
2008	4.50	2768779	0.615
2009	4.30	2909648	0.677
2010	4.50	3000078	0.667

Measurement of the concentration of environmentally hazardous substances in the outgoing air stream of fans channels allowed establishing the correlation between concentration of hazardous materials and the annual specific consumption of explosive for each ventilation shaft of mine (Gorova 2011).

For NVS these dependencies have the form:
- for carbon monoxide

$$C = 115.1 \cdot q - 38.71, \qquad (1)$$

where q – annual specific consumption of explosive, equal ≥ 0.4 kg/m³.

- for nitrogen oxides

$$C = 4.31 \cdot q - 0.54 \, ; \qquad (2)$$

- for sulfur dioxide

$$C = 4.61 \cdot q + 0.187 \, . \qquad (3)$$

For DVS these dependencies have the form:
- for carbon monoxide

$$C = 121.4 \cdot q - 39.65 \, ; \qquad (4)$$

- for nitrogen oxides

$$C = 4.64 \cdot q - 1.75 \, ; \qquad (5)$$

- for sulfur dioxide

$$C = 3.18 \cdot q + 1.146 \, . \qquad (6)$$

For SVS these dependencies have the form:
- for carbon monoxide

$$C = 100.8 \cdot q - 31 \, ; \qquad (7)$$

- for nitrogen oxides

$$C = 4 \cdot q - 0.46 \, ; \qquad (8)$$

- for sulfur dioxide

$$C = 4.94 \cdot q + 0.05 \, . \qquad (9)$$

Determination of dispersion of environmentally hazardous substances in the atmosphere around the ventilation shafts were performed using an automated system for calculating air pollution "EOL 2000 [h]". Consider the results of the formation of ground level concentration isolines of the total impact of environmentally hazardous substances around each emission source (Figure 1) (Mironova 2013).

The performed analysis of surface concentrations values of environmentally hazardous substances with cumulative impact revealed that with increasing distance up to 1500–2000 m from the emission source the concentration values decrease in 3–5 times. The main factor affecting the original data will be the average specific consumption of explosives, and the distance from the source.

Further studies have strengthened the empirical formulas that determine cumulative impacts of surface concentration of hazardous materials taking into account the specific consumption of explosives and distances from the emission source which have the form:
− for NVS:

$$C_{cum.imp.} = 1.39 \cdot q^{1.65} \cdot e^{-0.001 \cdot L} \, , \qquad (10)$$

where L is the distance from the emission source.
− for SVS:

$$C_{cum.imp} = 1.07 \cdot q^{1.24} \cdot e^{-0.0009 \cdot L} \, ; \qquad (11)$$

− for DVS:

$$C_{cum.imp.} = 0.72 \cdot q^{2.34} \cdot e^{-0.0008 \cdot L} \, . \qquad (12)$$

(a)

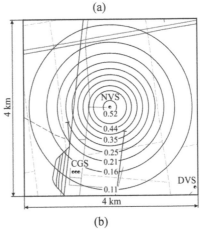

(b)

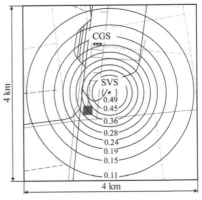

(c)

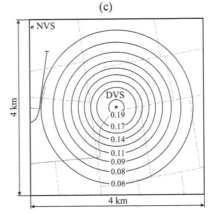

Figure 1. Character the concentration distribution of environmentally hazardous substances with cumulative impacts around NVS (a), SVS (b) and DVS (c) of mine.

(a)

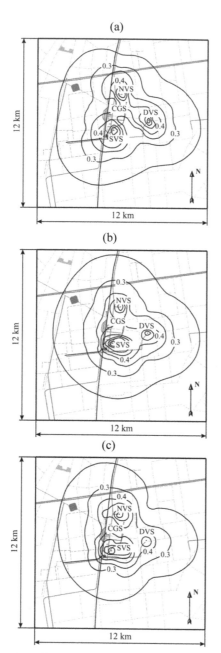

(b)

(c)

Figure 2. Isolines of conditional indicators of damageability of bioindicators at the production site and in the territory adjacent to the JSC "ZIOC" in 2009 (a) 2010 (b) and
2011 (c).

To assess the toxic and mutagenic activity of atmospheric air test polygons were determined, which
were located in the four cardinal directions from
three ventilation shafts at a distance of 50, 100, 300,
500, 1000 and 2000 m. These distances were taken

such a way to investigate the most dangerous man-
made areas near the emission sources. Plant pollen
sampling was performed during the spring-summer
season (April-September) during 2009–2011. Assessment of potential toxicity and mutagenicity of
atmospheric air was performed by the test "Plant
Pollen Sterility" (Gorova 1996 & Mironova 2013).

Calculation of conditional indicators of damageability (CID) of the environment allowed visualizing toxic and mutagenic activity of atmospheric air
around the emission sources (Figure 2).

Further studies allowed receiving regression dependence between CID, specific consumption of
explosives and the distance from the emission
source:

$$CID = 0.41 \cdot q^{-0.53} \cdot e^{-0.0003 \cdot L} . \qquad (13)$$

Further research allowed obtaining the correlation
dependence of the CID changes and parameters of
ground level concentration of active substances
(Figure 3).

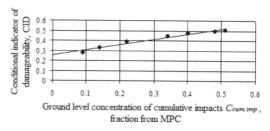

Figure 3. Graph of CID changes depending on the value of
ground level concentration of cumulative impact.

After spending approximation of maximum values an empirical equation of dependence between
the value of CID and ground level concentration of
cumulative impacts was obtained. It has the form:

$$CID = 0.53 \cdot C_{cum.imp.} + 0.25 , \qquad (14)$$

where $C_{cum.imp.}$ – value of ground level concentration of cumulative impact.

To estimate the anthropogenic influence on the
processes of ontogenesis of a winter wheat the selection of wheat sheaves was performed from test
sites with in the area of 1 m², located along the
north-east direction at distances of 50, 100, 300,
500, 1000 and 10000 m from the emission source
(NVS). Hereinafter selected test sheaves of wheat
were transported to the laboratory where the linear
dimensions of wheat were measured.

Further investigations of biological features of a
winter wheat allowed establishing the correlation

dependence of the biological productivity and the value of ground level concentration of cumulative impacts shown in Figure 4.

After spending approximation of maximum values an empirical equation of dependence between the biological productivity and ground level concentration of cumulative impacts was obtained. It has the form:

$$B_{biol.} = 82.21 \cdot e^{-0.986 \cdot C_{cum.imp.}} \ . \qquad (15)$$

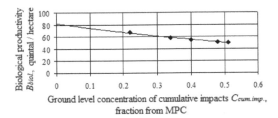

Figure 4. Dependence graph of changes of the biological productivity and the value of ground level concentration of cumulative impact.

To assess the impact of technogenesis on winter wheat in the first generation grain of this plant were grown. For this purpose, 100 prepared grains were placed on a filter paper in a laboratory dish. Sprouting grains of a winter wheat was carried out for 72 hours, with maintaining a constant ambient temperature equal to 25°C, and after every 12 hours a number of germinated seeds was determined to assess their germination. After 3 days the average length, wet and dry mass of roots of plantlets was determined; plantlets were selected on test plots at a distance of 50, 100, 300, 500 and 1000 m. Than obtained data were compared with indicators of grains of control area (10,000 m) to find statistically significant differences.

The executed analysis of values of biological characteristics of sprouted wheat grains revealed that environmentally hazardous substances emanating from the ventilation shafts have a significant impact on winter wheat in the first generation, and they contribute to the increase of technogenesis when approaching the emission source.

Further investigations of technogenic influence on agrophytocenosis cultures in the first generation of winter wheat allowed to establish correlation changes between phytotoxic effect and the value of ground level concentration of cumulative impact that are shown in Figure 5.

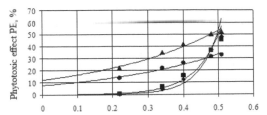

- on length of tops;   ■ on length of roots;
▲ on wet weight;   ● on dry weight.

Figure 5. Graph of depending changes of phytotoxic effect and the value of ground level concentration of cumulative impact.

After spending approximation of maximum values empirical equations of dependence between the phytotoxic effect and ground level concentration of cumulative impacts were obtained. They have the form:

- on length of tops

$$PE_{tops} = 0.008 \cdot e^{17.53 \cdot C_{cum.imp.}} \ ; \qquad (16)$$

- on length of roots

$$PE_{roots} = 0.046 \cdot e^{13.96 \cdot C_{cum.imp.}} \ ; \qquad (17)$$

- on wet mass

$$PE_{wet.mass} = 12.16 \cdot e^{2.95 \cdot C_{cum.imp.}} \ ; \qquad (18)$$

- on dry mass

$$PE_{dry.mass} = 7.64 \cdot e^{2.94 \cdot C_{cum.imp.}} \ . \qquad (19)$$

3 METHOD FOR DETERMINING THE PARAMETERS OF THE ATMOSPHERIC AIR FOR IRON ORE MINES

By results of research of physical and chemical analysis and biological assessment of the atmospheric air in the areas adjacent to the iron ore mine, it was found that the mine air coming from the ventilation shafts and saturated with environmentally hazardous substances after blasting has a significant impact on the environment.

The obtained results of physical and chemical analysis made possible to establish that the concentration of environmentally hazardous substances coming into the atmosphere, varied depending on the specific annual consumption of explosives in blasting work in the mine, as well as with increasing distance from the emission source. Analysis of the results of the biological assessment, biological characteristics changes and biotesting revealed that the total ground level concentration of environmentally hazardous substances significantly affects the value of CID, biological productivity of winter wheat and phytotoxic effect.

The foregoing analysis of the results of ecological condition of air allowed proposing a methodology for calculating its parameters in the areas adjacent to the mine, by average dependencies of three ventilation shafts in the following sequence:

1. Concentration of environmentally hazardous substances emanating from the ventilation shaft of the mine:

- carbon monoxide

$$C_{CO} = 112.43 \cdot q - 36.45 ; \qquad (20)$$

- nitrogen oxides

$$C_{NO_x} = 4.32 \cdot q - 0.92 ; \qquad (21)$$

- sulfur dioxide

$$C_{SO_2} = 4.24 \cdot q + 0.46 . \qquad (22)$$

2. Ground level concentration of cumulative impact of environmentally hazardous substances

$$C_{cum.imp.} = 1.06 \cdot q^{1.74} \cdot e^{-0.0009 \cdot L} . \qquad (23)$$

3. CID of indicators

$$CID = \frac{0.205}{q^{0.53} \cdot e^{0.0003 \cdot L}} + 0.265 \cdot C_{cum.imp.} + 0.125 \cdot \qquad (24)$$

4. Winter wheat productivity

$$B_{biol.} = 24.9 \cdot e^{0.0003 \cdot L} + \frac{41.105}{e^{0.986 \cdot C_{cum.imp.}}} . \qquad (25)$$

5. Phytotoxic effect
- on length of tops

$$PE_{tops} = 0.004 \cdot e^{17.53 \cdot C_{cum.imp.}} + \frac{31.38}{e^{0.0054 \cdot L}} ; \qquad (26)$$

- on length of roots

$$PE_{roots} = 0.023 \cdot e^{13.96 \cdot C_{cum.imp.}} + \frac{28.26}{e^{0.0043 \cdot L}} ; \qquad (27)$$

- on wet mass

$$PE_{wet.mass} = 6.08 \cdot e^{2.95 \cdot C_{cum.imp.}} + \frac{27.32}{e^{0.0009 \cdot L}} ; \qquad (28)$$

- on dry mass

$$PE_{dry.mass} = 3.82 \cdot e^{2.94 \cdot C_{cum.imp.}} + \frac{17.08}{e^{0.0009 \cdot L}} . \qquad (29)$$

Results of multiyear studies of ecological condition of production site air, and areas adjacent to the mine, made possible to compile the table on which the environmental assessment of atmospheric air around the source is determined. The necessary data to determine the environmental assessment of air around the ventilation shaft of the mine are presented in Table 4.

As seen from Table 4, the level of damage of bioindicators and the condition of air within the production site and the territory adjacent to the emission source (ventilation shaft) of mine are determined by the value of ground level concentration of cumulative impact, conditional indicator of damageability as well as winter wheat productivity obtained by the formulas (23)–(29).

Table 4. Environmental assessment of atmospheric air around the emission source.

Ground level concentration of cumulative impact, $C_{cum.imp.}$, fraction from MPC*	Conditional indicator of damageability CID, share units	Biological productivity $B_{biol.}$, q/ha	Level of damageability of indicators	State of atmospheric air
≤ 0.095	0…0.150	≥ 74.9	low	favourable
	0.151…0.300		below the average	disturbing
0.096…0.378	0.301…0.450	74.8…56.7	average	conflictual
0.379…0.661	0.451…0.600	56.6…42.9	above-average	threatening
0.662…0.944	0.601…0.750	42.8…32.4	high	critical
0.945…1.415	0.751…1.000	≤ 32.3	maximum	dangerous

Note: MPC – maximum permissible concentration

338

5 CONCLUSIONS

Dependences of the concentration of environmentally hazardous substances from the annual specific consumption of explosives for each mine ventilation shaft, that allow to determine the expected concentrations of pollutants emitted into the atmosphere, are set.

The empirical formulas determining ground level concentration of cumulative impact of environmentally hazardous substances, taking into account the specific annual consumption of explosives and the distance from the source are obtained. Conditional indicator of damageability of bioindicators is calculated. It allowed visualizing toxic and mutagenic activity of atmospheric air around the emission sources of mine. Comparison of the obtained calculated values allowed establishing the dependence of the CID changes from the ground level concentration of cumulative impact.

Complex parameters of winter wheat condition are defined. Comparison of its productivity with the ground level concentration of cumulative impact allowed establishing the relationship between changes in the biological productivity and the surface concentration of cumulative impacts.

The dependence of phytotoxic effect for winter wheat sprouted in a polluted atmosphere is established and relations on the value of surface concentration of cumulative impacts are obtained. The method for assessing the ecological condition of air within the mine territory adjacent to the sources of its emissions is proposed.

REFERENCES

Khomenko, O.E., Kononenko, M.M., Vladyko, O.B. & Maltsev, D.V. 2011. *Ore mining in Ukraine Internet.* Dnipropetrovs'k: National Mining University.

Gorova, A.I. & Mironova, I.G. 2011. *Determination of the concentration of harmful substances in outgoing jet of the mine air.* Dnipropetrovs'k: Collection of scientific works of NMU, 36. Vol. 2: 192–200.

Mironova, I.G. & Pavlychenko, A.V. 2013. *Analysis of air pollution levels in underground mining of iron ores.* The annual scientific and technical collection "Development of deposits": 261–266.

Gorova, A.I., Bobyr, P.F., Skvortsova, T.V., Digurko, V.M. & Klimkina, I.I. 1996. *Methodological aspects of assessing mutagenic background and genetic risk to humans and biota from the mutagenic action of environmental factors.* Cytology and Genetics, 6. Vol. 30: 78–86.

Mironova, I.G. 2013. *Ecological status of of atmospheric air in areas of placing of enterprises of underground mining of iron ores.* Dnipropetrovs'k: Collection of scientific works of NMU, 40: 204–209.

Progressive Technologies of Coal, Coalbed Methane, and Ores Mining – Bondarenko, Kovalevs'ka & Ganushevych (eds)
© 2014 Taylor & Francis Group, London, ISBN: 978-1-138-02699-5

Government control of mines adaptation to infrastructure of coal regions

I. Lozynskyi & O. Ashcheulova
National Mining University, Dnipropetrovs'k, Ukraine

G. Bondarenko
National Commission of State Regulation of Public Service, Kyiv, Ukraine

ABSTRACT: The paper proposes technical approaches to reach a compromise between expediency of unprofitable enterprise subsidizing on the basis of production diversification potential and low-capacity mines construction within new sites.

1 UNTRODUCTION

Central task of government control as for unprofitable mine capacity support is, first of all, selection of corresponding coal region, a level of its depressiveness, and mine facilities condition from the viewpoint of reserve supportability. Basically some authors (Vagonova 2005; Raikhel & Pavlenko 2006) reject such an idea as they suppose that it is impossible to develop optimum criterion for such a problem. That very time the criterion is the vector field orienting rather various mines or regions. Hence a conclusion can be drawn that government control should be based on certain optimum criterion.

A problem definition concerning analysis of specific coal-mining region as well as its connections with other regions creates background to make the examination from the viewpoint of foreign trade theory. However, in this context the theory application is not expedient as excess of product import over product export is advantage. It means investment for the region as which territory the products manufactured in other parts of the country are applied. Thus, uniform progress of each region of Ukraine is basic principle of government control. In other words living standards of those inhabiting this or that coal-mining region is the most important development factor.

For regions which population majority activities depend on coal mine operation, living standards will be characterized by full employment. That means that first of all it is required to select a region where there is a necessity to restore coal enterprises being decayed through chronic deficit of government dotations. It is the government that has to decide on the priority and finance targeting for specific coal mines.

– following factors are widely used as the basic ones while comparing (Economical and mathematical... 1982):

– gross product per capita;
– real income per capita (especially it concerns a share of minimum consumer basket in terms of real income per capita);
– each budget income at the territory per capita;
– living standards criteria involving: living conditions, system of education and health, social infrastructure, life expectancy, increase or decrease of population.

According to these criteria monofunctional towns and industrial centers in which enterprise of military-industrial complex, coal-mining, chemical, and machine building enterprises are concentrated belong to depressed territories. Suspension of one or two basic enterprises results in total degradation of such a town and its social infrastructure, unemployment, and lack of sources of local budgetary recharge. Regional depressed situation is characterized by impossibility to have simple restoration let alone expanded one as well as economic, demographic, and other processes. Depressed regions turn to be centers of political and social stresses in the country. Depressed territory is small separate monofunctional town within which production dynamics, living standards, and employment level of population are quite lower to compare with average figures in the state (Amosha 2002). Mining town should have:

– coal-mining enterprises;
– prevailing specific share of coal industry products in total output of the town;
– employment of the majority of employable population at coal-mining enterprises;
– considerable specific share of industrial production facilities of coal-mining enterprises in the town industry.

2 STUDY OF LATELY RESEARCH AND PUBLICATIONS

A number of home scientists consider the problem of stable development of coal-mining enterprises (Amosha 2002, 2005 & Pivnyak 2004). Their study shows that scientifically based plans and programs of any enterprise development are required for stable market activities of unprofitable mines. They should involve not only the efficient use of material and finantional resources but alsoways of production diversification to improve environmental situation in mining regions (Koval 2003 & Salli 1994). Many papers consider decrease of governmental dependence that is free search and selection of consumers, participation in corporate mining etc. However, current scientific plan has no clear concept of expediency regulation and ways to preserve unprofitable enterprises as for the situation in a certain coal-mining region. The paper proposes a number of concepts to control resource potential of unprofitable enterprises in the context of the region infrastructure transformation.

2.1 Formulation of the paper objective

The objective of the paper is to generalize and develop scientific and methodological background, to develop tools and practical recommendations as for regulation of streams of governmental support for unprofitable enterprises; it aims at improving potential of specific coal-mining region.

2.2 Statement of basic material

Lack of real plan for crisis overcoming rather than economic problems ruins today industrial region which used to powerful. Unfortunately, there are many people who like the situation as for beneficial expenditures of budgetary dotations, exploitation of slave labour in "dug mines". Nowadays the problem of Donbas is not to ruin old unprofitable enterprises which used to work for defense industry of the USSR but to develop real plan to overcome the crisis.

For post-socialist states of Eastern Europe market reforms resulted in the fact that new small private capital-based companies using new economic technologies became important in the industry rather than old giant integrated works. They raise the economy of these countries making them unachievable. Today Ukraine has not such new productions in the sites of closed mines and plants. Currently there are 450 towns in Ukraine 334 of which belong to small ones. Their population is less than 50 thousand inhabitants. On the whole almost 6.5 mln people live in them to be 13.5% of the total population.

Only 7% of small towns feel more or less comfortable. All other belongs to so-called depressed as they are monofunctional settlements which life was supported by one or two industrial enterprises. Some Donbas towns are typical depressed regions today. Specifically, low incomes (50.3% of respondents) and unemployment are crucial problems for citizens of Shakhtars'k, Dzerzhyns'k, Ukrains'k, Torez, Snizhnoie, and many others.

Like Donbas, Ruhr also had its industrial boom as high-grade coal was mined there and millions tons of pig iron and steel were melted. But now metallurgical centers of Germany situated mainly in seaports use imported raw materials. As for Ruhr it experienced a series of large-scale restructurizations which started long before than so-called transformations in coal-mining industry of Ukraine after collapse of the USSR. Such word combination "depressed mining towns" is familiar both for Ukraine and Germany. But if depressed mining towns and settlements of Donbas are dying, German government turned to be more experienced. Starting from 1950-s, economy of Ruhr experienced a number of serious changes. First, its government decided to stop coal-mining operations after world coal (1958) and steel (1975) crises. Germany spent 130 mlrd Euros for gradual closing of mines in Ruhr and preparing the territory for living without coal mining. They had to retrain about 500 thousand people. Still government donates remains of coal mining having only strategic ideas in mind: net cost of German fuel is three to four times as expensive as in neighboring countries. Nowadays there are only 4 mines and 3 coke-producing plants in Ruhr; government donates them. Almost 500 thousand jobs died away in mining industry from 1980 to 2002. That very time more than 300 thousand of new jobs appeared in service industry. New productive (car- and machine-building, electrical technology, high-precision mechanics) and non-productive (banking, IT) industries started. It is worth noting that only 4.5 mln people live in Donets'k region at the territory of more than 26.5 thousand square kilometers when population of Ruhr is 5.3 mln people and its territory is 4435 square kilometers only. Thus, unemployment rate at the territory being almost 6 times less is by 18% higher.

To compare with budgets of Lugans'k and Donetsk being 2680–3378 UAH per capita, possibilities of many depressed Donbas towns an districts trail by 2.5 to 3.4 times. As a result, governmental subvensions and dotations are 50 or even 60–80%. Moreover, in 2010–2013 the situation got worse. In addition, people are be concerned about poor medicine, efforts of community services and situation with housing resources; 71.9% of town dwellers can not see future for their towns considering them as dying. The majority of town dwellers are not concerned with the environmental problems, and problems of worthful education. By the way, it should be noted that annual budget of Essen being Ruhr's centre is 3 billion Euros (that is 50000 UAH per capita).

As before, jobs stay to be one of the crucial unsolved problems in Donbas. Its solution could help to avoid a number of other issues. Thus, survey results show that the majory of respondents (48.6%) feel the lack of finance for social support, and 22% are in want of their town economy. In Ukraine a law "on stimulation and development of territories" operates. Among other things, settlements at which territories coal-mining and coal-oil-gas refinery either have been closed or have been in the process since 1996 are considered as depressed territories. Kirovs'k where five mines operated in the earliest days of our Independence can be a bright example of such a territory in Lugans'k region. None of the functions today and unemployment rate is 75%.

It should be noted that Russian Donbas is depressed and totally subsidized region too. However, if Ukrainian Donbas stays to be considered as one the country supports, Russia has not such a political idea; that is why Eastern Donbass is Russia's adverse and unwelcome for life territory. Mining towns are regarded as poor ones even for Rostov region, and their inhabitants try to move to Novocherkas'k, Taganrog or Rostov if it is possible. Restructurization of coal industry in Russia took place long before than far in Ukraine; hence, practically no one coal enterprise is in the region. Their liquidation resulted in job displacement. As a site of Ministry of Industry and Energy of Rostov region informs, up to the mid of 1990-s enterprises of coal industry were located in nine municipal entities of the region and their population was more than 1.3 million inhabitants. Coal industry dominated within the region territories; as for such towns as Gukovo, Zverevo and Novoshakhtins'k it was practically the only one. Execution of mining territory budgets, situation with their social and public infrastructure, and living standards of population depended on the coal-mining enterprises operation. The majority of labour force of the territories was employed in mines and at auxiliary production.

Rostov region is among regions where wide-range closure of unprofitable enterprises was performed. The majority of mines in the region operated for 40–50 and more years. Basically, favorable reserves had been mined. Besides, the mines had multistage transport chain and complicated mine ventilation systems. From the beginning of restructurization process, the number of productive mines in Eastern Donbas decreased from 64 in 1995 to 13

in 2012. All mining enterprised were closed in such towns as Novoshakhtins'k and Shakhty. Similarly, during the same period the number of labour force employed in coal companies decreased from 117.3 thousand to 9.4 thousand people.

In many Donbas regions involved in a wave of mass closure of mines (sometimes the process was unreasonable), essential production decline resulted in adequate decrease in living standards. Following figures demonstrate unemployment scale: only in Torez-Snezhnoie region 5 productive mines and 2 preparation plants left to operate of 20 mines and 5 preparation plants; more than 35000 people are unemployed of 48000. Sneznoie plant of chemical machine-building ("Snezhnianskhimmash") and Sneznoie machine-building plant ("Motor-Sich" OJSC) are not on full power.

Thousand of people left destitute have to find way out. Most of all, it results in sacking and further soling material values which belong both to closed enterprises and operating ones. The situation is responsible for arising places where coal is mined illegally; those, who left unemployed after their mines were closed, find work there. Such employment means aberration from safe professional behavior resulting in higher level of accident rate and in fatal cases. Almost 226 places of illegal coal mining are located at the territories of Torez and Snezhnoie; they mine about 4000 tons of coal a day to be comparable with mining capacity of officially operating coal-mining enterprises of the region. Miner's labour stopped to be prestigious and well-paid; now it belongs to problem professions with relatively low salary level (18th place in Ukraine), and pay pauses reaching months and sometimes years.

However, positive practice of attractive coexistence of stable operating mine and settlement inhabited by the majority of the enterprise workforce. Take for example Kirovs'ke, a small town in Donetsk region, which territory is 7.2 km^2. 31.1 thousand people lived there for the 1st of January 2012. However, it is among donor towns. That is to say, at the expense of local fiscal revenue the town meets costs delegated by government; besides, it allots money to maintain other towns. There are 31 donor towns and cities in Ukraine. Kirovske's feature is its economy diversification: coal industry is 99.2% of its overall production (three mines operate, and "Komsomolets Donbasu", being the largest of them, is actually a budget-forming enterprise). Since 2009 the town has started implementing a project which objective is to optimize performance and expenses for social support municipal system functioning within current legal base and budget system. It should be done at the expense of improving the targeting and transparency of budgetary funds use for social needs making it more qualitative and accessible for the town.

We believe that Torez-Snezhnoie region also can stay among Ukrainian producers of anthracite playing vital role for power industry, ferrous and nonferrous metallurgy, chemical industry and electrical industry. Taking into account the importance of anthracite and loss ratio of the majority of mines extracting it, one may say that it is required to work out scientifically grounded mechanism to develop enterprises of the region. Hence, our government should reach a compromise between the expediency of anthracite mines subsidizing, making decision on their liquidation necessity, and developing new sites with favorable bed positions. Such a mechanism developing should involve objective assessment of each certain anthracite mine as well as new low-capacity mines taking into consideration macro- and microeconomic factors in terms of their construction financing. These same things have determined the paper main objective, specifically development of a model to improve economic potential of enterprises within specific coal-mining region.

The necessity to build and put into operation of two mines (their conventional names are "Vira" and "Nadia") within "Miusky 1–2" depends on unique edge anthracite bed position with outcrop under quaternary deposition at a depth of 20 m below surface. In due time, "Richkova" mine was rapidly built. It had operated efficiently during 20 years. In the early 21st century it was flooded for reasons unknown. Sufficient seam thickness, lack of gasdynamic manifestations at moderate depths provided insignificant for edge seams face output within 350-400 tons per day. For comparison, day face output in terms of edge seams of Donbas Central district is hardly 100 tons. Capacity of the two mines proposed to be built is initially 300000 tons with possible following increase up to 450.000 t/a. On the one hand it depends on forecast rates concerning slowdown of mining within the region, and on the other hand – on the necessity to complete the construction process as quick as possible. The mines are rated for four levels mining during 26 years to extract balance reserves to the extent of 24.9 million tons. The most important reason for "Vira" and "Nadia" mines construction is social situation – the requirement to employ workers dismissed from other mines in the district.

Operation schedule provides two incline shafts and incline conveyor cross-drift (Salli 1994). Incline shafts are located along h_7 seam with 56° slope angle; conveyor cross-drift is located across to the rock strike with 14° angle to the level; its collar is located in the neighborhood of h_{10} outcropping.

Such location depends on calculating minimum distance from other industrial site of incline shafts and inclines cross-drift to the first haulage level within a suite of workable seam. Each level is provided with surface cross-drifts; required facilities are built. Total capacity for all operations concerning seams is 47.3 thousand m^3. Basic construction period is 20.3 months.

Application of method of mine development modeling according to concepts of dynamic progress helped to analyze changes in economical strength of enterprises on terms of extra financing the processes of production diversification at the expense of mine waste processing, and thermal power of mine water and methane using (Table 1, Figure 1). The analysis shows that if current losses against EVA in this regions are more than 180 mln UAH; if diversified productions are developed (it especially concerns mine waste processing) with 50 mln UAH financing, then negative values of EVA will cut back to 36 mln UAH supposing that unprofitable mines operate.

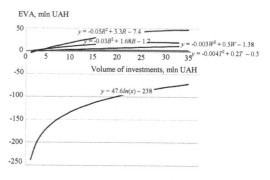

Figure 1. Changes in the region economic potential rate in terms of production diversification financing.

Improvement of the region condition is forecasted to be more striking if "Vira" and "Nadia" mines within "Miusky–2" site are put into operation. Owing to their intensive construction (24 months), annual output of the region will experience 600000 tons increase. If we take into account the fact that cost value of mining in such mines will not be more than 550–600 UAH / t, then the fact will excerpt immediate positive influence on level of wages in this region (Figure 2) and budget revenue in terms of extra taxes and recently built enterprise profits tax (Figure 3, Table 2).

Today less and less specific formulae are proposed to improve the situation in depressed Donbas regions. More attention is paid to getting the better of coal-mining industry without their problems solving. Dying of coal industry state sector allows allocating investment support for unprofitable mines as if investing heavily in mine liquidation. During last 20 years not really transparent system for the money distribution and redistribution was developed. That very time, in 2013 Donbas demonstrated a depth of living standards quality in Ukraine. Donets'k and Lugans'k Regions are among outsiders of Ukrainian regions (*Coal of Ukraine: future.* Available at www.experts.in.ua).

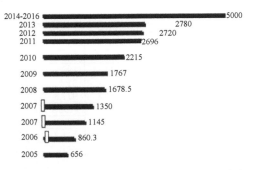

Figure 2. Wages growth dynamic in term of coal mining increase.

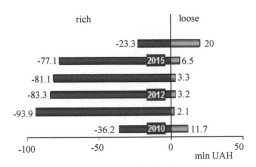

Figure 3. Changes in budget structure for Torez and Sneznoie in terms of profitable coal mining.

Stagnation of coal industry is crucial reason. Coal was a basis for Donbas economy as well as for infrastructure of numerous settlements and even districts: housing, medical care, roads etc. Projects on coal enterprises closure provided solving social and economic, and environmental problems. However, neither governmental programs nor local ones can get through the situation.

Table 1. Output data for funds disposition dynamic modeling.

Annual financing volume, mln UAH	Profits (EVA) by diversified productions, mln UAH / year				
	Coal mining	Coal quantity in the process of mine waste processing	Ceramics manufacture	Use of thermal pumps for mine drainage	Restoration and production of road materials
0	-180	0	0	0	0
2	-180	0	0	0	0.10
4	-179	5.0	1.0	0.3	0.30
6	-178	6.0	3.0	0.5	0.60
8	-177	8.0	8.0	0.9	0.90
14	-160	11.0	10.2	1.8	1.10
16	-154	12.0	11.3	2.2	1.30
38	-90	38.5	17.8	7.5	2.71
40	-80	38.9	18.0	7.9	2.80
42	-70	40.2	18.5	8.5	2.85
48	-42	42.1	19.6	9.3	2.99
50	-40	42.6	20.8	9.6	3.05
52	-36	43.0	21.0	9.8	3.20
0	-180	0	0	0	0
2	-180	0	0	0	0.10
4	-179	5.0	1.0	0.3	0.30
6	-178	6.0	3.0	0.5	0.60
8	-177	8.0	8.0	0.9	0.90
14	-160	11.0	10.2	1.8	1.10
16	-154	12.0	11.3	2.2	1.30
38	-90	38.5	17.8	7.5	2.71
40	-80	38.9	18.0	7.9	2.80
42	-70	40.2	18.5	8.5	2.85
48	-42	42.1	19.6	9.3	2.99
50	-40	42.6	20.8	9.6	3.05
52	-36	43.0	21.0	9.8	3.20

The situation has apparent way out: it is required to apply urgent measures to save everything that can improve economic potential of the industry enterprises. Objective is to bring back to life those enterprises which can give guaranteed return with minimum government investments. Many times it was confirmed that mines having quality reserves for 15–20 years of operation could be profitable. World demand for coal will grow at least up to 2020. Price trend demonstrates its increase for five years already.

As for internal market, it gives a promise of consumption expansion for the reason that unreasonable gas prices make it profitable to use coal for power generation sector.

Table 2. Parameters of Torez and Snezhnoie budget sphere transformation when "Vira" and "Nadia" mines are built within "Miusky 1-2" site.

Parameters of the mines development	2013	2014	2015	2016	2017
Roadheading, thousand m³	–	–	20	40	40
Expenditures connected with the mine construction, mln UAH	–	–	110	320	170
Mining, thousand tons	1500	1600	1600	1900	2200
Labour productivity, UAH / t	21.1	21.3	21.4	31.1	41.4
Production cost of mining, UAH / t	1007.4	1150.1	1154.8	977.8	844.5
Annual return, mln UAH	-187.8	-190.7	-196.2	-136.4	-106.3
Monthly wages, UAH	2800	2900	3500	4500	5000
Local budget contribution, mln UAH	93.6	94.9	108.8	121.8	137.9
Budget revenue per capita (Torez and Snezhnoie)	2312	2380	2400	2490	2525
Creation of new jobs, units	–	–	100	1000	700
Export of coal products per an inhabitant, USD	937	1000	1000	1187	1312

3 CONCLUSIONS

1. To a great extent, Donbas loses its chance. For continuing (not to speak of developing) coal industry, extraordinary measures should be applied to support facilities mines having relatively favorable mining and geological conditions. Integrally, general policy of mine structure improvement should be based on the necessity to mine effective reserves still limited in Ukraine. Those pseudooptimistic forecasts that coal reserves will be sufficient for hundreds years are wide open to criticism.

2. East of Donetsk region has unreasonably dormant known "Miusky 1–2" and "Grabivsky rudnyk" sites. There upper boundary of coal seams is at insignificant depth (30–80 m). Practice of construction and operation of "Richkova" and "Yablunivka" mines within "Miusky 1–2" site shows that small enterprises have right to exist; they are adaptable for market conjuncture owing to low production cost of coal.

3. Putting into operation two mines within "Miusky 1–2" site and expansion of production diversification as for mine waste processing will help to cut annual loss of Torez-Snezhnoie region by 100 mln UAH; average monthly wages will increase up to 5000 UAH; payments to local budgets will practically double.

REFERENCES

Vagonova, A.G. 2005. *Economic problems of supporting capacity and investment of coal mines in Ukraine.* Dnipropetrovs'k: National Mining University: 287.

Raikhel, B.L. & Pavlenko, I.I. 2006. *Investments and potential sources of their financing in coal industry.* Development problems of external economic relations and attraction of foreign investments: regional aspect. Donetsk: DonNU, Part 3: 1175-1179.

Economical and mathematical modeling of manufacturing locations. 1982. [Edited by Y.A. Shatalin]. Moscow: Nauka: 144.

Amosha, A.I., Iliashov, M.A. & Salli, V.I. 2002. *System analysis of a mine as investment.* Donets'k: IEP of NAS of Ukraine: 68.

Amosha, A.I., Kabanov, A.I. & Starichenko, L.L. 2005. *Prospects of native industry developing and reforming at the background of the world tendencies.* Scientific report. IEP of NAS of Ukraine. Donetsk: 32.

Pivnyak, G.G., Amosha O.I., Yashchenko Y.P. et al. 2004. *Restoration of mines and investment processes in coal industry of Ukraine.* Kyiv: Naukova dumka: 331.

Koval, V.M. & Koval, D.V. 2003. *Depressed region development programming.* Old-industrial regions of Western and Eastern Europe in terms of integration. Collection of scientific papers of Donets'k National University. Donets'k: DonNU: 161-171.

Salli, V.I., Malov, V.I. & Bychkov, V.I. 1994. *Support of coal mine capacity in terms of limited possibilities of new construction.* Moscow: Nedra: 272.

Coal of Ukraine: future. Available at www.experts.in.ua

Progressive Technologies of Coal, Coalbed Methane, and Ores Mining – Bondarenko, Kovalevs'ka & Ganushevych (eds)
© 2014 Taylor & Francis Group, London, ISBN: 978-1-138-02699-5

Modern technologies of bolting in weakly metamorphosed rocks: experience and perspectives

V. Lapko, V. Fomychov & L. Fomychova
National Mining University, Dnipropetrovs'k, Ukraine

ABSTRACT: The article is dedicated to the experience of two-level scheme usage of anchor bolting using resin-grouted anchor support and deep-level rope anchor support to complete different tasks of coal mine and colliery underground working support. Results of testing, computation and combined anchor bolting usage under the conditions of the West Donets'k Basin are presented.

1 INTRODUCTION

Currently the bulk of coal workings, coal mines and collieries is driven using anchor support. Meanwhile technological opportunities of underground mineral extraction are being increased. Both new anchor bolting tasks appear.

Russian and Ukrainian mines as like as Kazakhstanian workings have gathered great experience of two-level anchor support usage to complete the following tasks (Lapko & Fomychov 2013):

– maintenance of workings and face-ends up to 12 m width;

– previously driven and formed disassembling chambers;

– reinforcement of drift support for reusing and for extracting the coal without pillars;

– reinforcement of drift support in order the working face to operate without face-end powered bolts;

– reinforcement of working support to maintain them in order them to be used for gas control, drainage and as an escape ways;

– reinforcement of drift support in the area of front abutment pressure;

– assurance of shall depth border zone mine working rocks stability, unstable rocks, rocks in disturbance areas;

– suspended monorail installation;

– foundationless installation of belt conveyor stations etc.

2 FORMULATING THE PROBLEM

Stress redistribution in rock massif during driving and maintaining of boards and their face-ends is followed by great horizontal and vertical rock displacements both along the border of the working and along the bearing massif. Deformation zone appearance during massif scrubbing promotes increasing of factual supported span of the working width (from B to $B+\Delta l$) and delamination height to $(0.5–0.7)B$, which makes working maintenance impossible without implementation of additional support reinforcement (Fomychov 2012).

Prop, composite and frame supports are traditionally used in boards for support reinforcement. Anchor support has following advantages over them:

– low costs and steel intensity;

– low labour intensity during transportation and installation;

– doesn't prohibit people and equipment transportation;

– doesn't require additional installation of the support during equipment installation in the working.

That is why the most reasonable choice for stable state provision of workings and their face-ends during the whole exploitation period is to use two-level scheme of anchor bolting with not only 3 m. anchor bolts (I level) but also deep-level anchor bolts (II level). The second level anchor bolts are fastened in stable roof rocks outside the arch of natural cavity with account of weakness and deformation of working sides (Figure 1). Unstable roof rocks are bound by the first level anchor bolts and "hooked" on the anchor bolts of the second level to the stable roof rocks outside the arch of natural cavity.

3 STATEMENT OF BASIC RESEARCH MATERIAL

That is why it becomes necessary to analyze in detail setting angle influence on rope anchor bolting efficiency in mining conditions. So far the change of the setting angle of the anchor bolting includes redistribution of the influence on its behavior by different stress components, the most preferably is to use stress intensity diagram σ (Figure 1) during computation result analyzing.

(a)

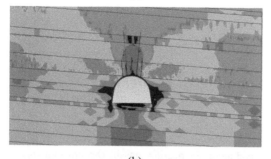

(b)

(c)

Figure 1. Stress intensity at depth of $H = 800$ m with anchor bolting angle setting of 85° (a), 80° (b) and 75° (c).

With near-vertical angles of rope anchor bolting setting (from 85 to 88°) intensity distribution diagram has number of peculiarities caused mainly by factors of geological structure of rock massif. Figure 1(a) manifests that rope anchor bolts don't spread the load on 1 m sandstone layer. In this way the roof suspension scheme of the direct working roof to the main roof is not implemented. As the result – increased difference of main stresses in the lying wall (Figure 1a).This can cause intensive rock heaving.

However, there are relatively small areas of high intensity stresses in working sides σ. This creates illusion of rope anchor bolting efficiency. Yet, such stress redistribution leads to great value of maximal difference of main stresses in frame-anchor support. With such mining conditions high angles of rope anchor bolting setting cause quick growth of working support rebuff reaction.

Starting from the angle of 83° of rope anchor bolting setting, the intensity distribution diagram of rock massif leaps. Firstly, it's clear (Figure 1b and c) that the rope anchor bolts fulfill their main function of direct working roof support. Stress intensity rises sharply in the part of anchor bolts, where the sandstone layer is. Secondly, the stress intensity in the lying wall is minimal and so the possibility of rock heaving mechanism starting is also minimal. Thirdly, there are high stress intensity zones formed in sides of the working, which means appearance of great shear forces. This is caused by the reduction of the pressure on borders of lithological varieties and it promotes side pressure increase on frame props and decreases the meaning of vertical element of rock pressure.

Slight and almost linear increase of stress intensity in working sides appears with rope anchor bolting setting angles from 72 to 83°. This growth goes with discharge zone increase in direct working roof. Please, compare corresponding areas of stress diagrams on Figures 1 (c) and (b). This demonstrates growth of efficiency of rope anchor bolts in the context of increasing setting angle. However growth of this indicator comes along with efficiency reduction of resin-grouted anchor supports. That means that components of bearing construction are being changed in "frame-anchor support – rope anchor support" system and this causes growth of frame stresses with rope anchor bolting load reduction. Skew happens in internal forces distribution of support construction, which can cause premature failure of one of its elements.

Thus, the most efficient angle of rope anchor bolting setting should be considered in terms of two statements: first – influence of the setting angle on stress distribution in border massif; second – interaction of construction elements of the support with different angles of rope anchor bolting setting. Firs statement analysis results can be presented as diagram set which shows influence level of rope anchor boltings on the maximal values of stress intensity (Figure 2).

The second statement requires independent analysis of interaction in two systems: "rope anchor boltings-resin-grouted anchor boltings" and "rope anchor boltings-frame support". First of all let us consider different anchor bolting type interaction (Figure 3).

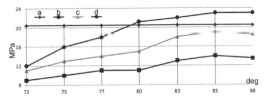

Figure 2. Stress intensity maximum change in rock massif for test computation $H = 600$ m (a) and with different anchor bolting setting angles on depth: $H = 400$ m (b); $H = 600$ m (c) and $H = 600$ m (d).

(a)

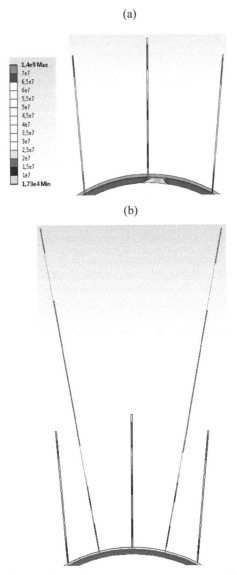

(b)

Figure 3. Stress intensity σ in frame-anchor support in test computation (a) and during usage of rope anchor boltings (b).

The base for the analysis is stress intensity diagram of the test computation (Figure 3a) and main calculation with rope anchor boltings, set at the angle of 80° (Figure 3b) on the depth of working location $H = 600$ m.

Given diagrams (Figure 3) demonstrate that the influence of rope anchor bolts on resin-grouted anchor bolts causes stress intensity reduction in the last ones. In the illustrated example discharge of side resin-grouted anchor bolts makes up about 14%, which in magnitude equals to about 21 megapascal. The main zone of stress intensity reduction is situated directly along the rock border of the working. This allows to conclude stress gradient reduction in working arch, which means decrease of active splintering possibility in this area of the computed model. On the other hand, stress-strain state changes in the central resin-grouted anchor bolt in demonstrated computation result made 4% against the test computation. This equals to computational error.

However, starting from rope anchor bolting setting angle of 83°, the diagram of interaction with resin-grouted anchor boltings begins to change. The central resin-grouted anchor bolting is discharged and with the angle of 88° of anchor bolting setting reaches maximum of 12%. It comes along with 9% load growth on side anchor bolts against rope anchor bolting setting angle of 83°. Figure 4 demonstrates connection between resin-grouted anchor bolting state change, working depth and rope anchor bolting slope angle.

(a)

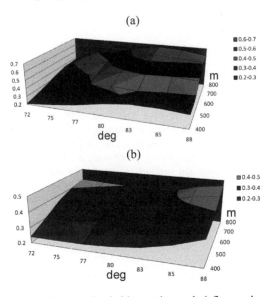

(b)

Figure 4. Rope anchor bolting setting angle influence degree of stress distribution in side (a) and central (b) resin-grouted anchor boltings considering working depth.

Obtained dependences (Figure 4) together with stress diagram (Figure 3) allow to make clear conclusion about the non-linear qualitative change of bearing capacity of anchor bolting system (I level) under rope anchor bolting (II level) location change. Resin-grouted anchor bolting design features (length and slope angle) in a less degree affect internal forces distribution in deep-level rope anchor bolts than in reverse. Rope anchor bolting slope angle influence degree increases proportionally to maintained working depth and is almost linear.

4 CONCLUSIONS AND PERSPECTIVES OF FOLLOWING RESEARCHES

Anchor support use as one kind of support is not allowed in workings, located in disturbance areas, watered grounds, areas with thin layers of coal, calcite or carbon mudstone within anchor bolting fixing depth.

1. Combined support (anchor support with other types of envelope or supporting supports) is used in workings, located in disturbance areas, watered grounds, areas with thin layers of coal, calcite or carbon mudstone within anchor bolting fixing depth, permanent mine workings of high responsibility with long exploitation term and development workings under intensive rock pressure.

2. Combined support is used in following combinations:
– bolt and shotcreting supports;
– bolt support with metal holdings, clouts or base plates with weld metal grating tie and shotcreting support;
– bolt and reinforced support.

3. Combined support options and setting are accepted and computed depending on working stability (Table 1).

Table 1. Rock stability categories.

Rock steadiness category	Rock steadiness evaluation	Working roof displacement U, mm
I	Stable	Up to 50
II	With medium stability	50–200
III	Unstable	200–500
IV	Very unstable	Over 500

In terms of non-seamy rocks of the I stability category, when roof displacements don't exceed 50 mm, working is fixed with one anchor support with base plates and grating tie or only with reinforced or shotcreting support which wide is 40–50 mm.

In terms of seamy roof rocks of the I stability category it is necessary to set anchor support with base plates metal mesh combined with reinforced or shotcreting support which wide is 40–50 mm.

In terms of workings with rocks of the I stability category, when roof displacements reach 100–200 mm, it is necessary to use combined support, containing metal holdings, grating tie and reinforced or shotcreting support which wide is 80–100 mm.

4. Combined support is installed in the following way: The first support to install is anchor one, then reinforced or shotcreting one. In III or IV category of rock stability the first support to install is reinforced one. As temporary solution, shotcreting layer is put on roof border and working sides. Metal mesh or weld grating are set over. Then the second layer comes.

REFERENCES

Lapko, V.V. & Fomychov, V.V. 2013. *Peculiarities of rope bolts usage during maintenance of stopes at mines of Western Donbass.* Scientific herald of National Mining Ukraine, 5: 31–36.

Fomychov, V.V. 2012. *Bases of calculation models plotting of bolt-frame support considering non-linear characteristics of physical environment behavior.* Scientific herald of National Mining Ukraine, 4: 54–58.

Progressive Technologies of Coal, Coalbed Methane, and Ores Mining – Bondarenko, Kovalevs'ka & Ganushevych (eds)
© 2014 Taylor & Francis Group, London, ISBN: 978-1-138-02699-5

Creating the effective implementation of double subbench mining technology

S. Moldabayev, Zh. Sultanbekova & Ye. Aben
Kazakh National Technical University after K.I. Satbayev, Almaty, Kazakhstan

B. Rysbaiuly
International Information Technologies University, Almaty, Kazakhstan

ABSTRACT: An analytical interpretation of the highwall construction method of double subbench mining stages is proposed. The technology provides optimization of mining mode through the solution of a nonlinear programming. The current stripping ratio of mining stages should be changed from the lowest to the highest value with a uniform distribution of accessed reserves. The method allows to build the stages of mining the steep deposits of complex configuration with providing the best mining mode and uniform distribution of accessed reserves between the two higwalls.

1 INTRODUCTION

The methods of mathematic modeling of deposits and mining system parameters are described in studies of K.E. Vinnitsky, N.B. Tabakman, E.I. Reyentovich, D.G. Bukeikhanov and etc. In this studies modeling methods classification and model creation methods by dynamic programming is proposed. Stripping works schedule studies allowed to develop the theory of the gradual pit development (V.V. Rzhevsky, V.S. Hohryakov, A.N. Shilin, B.P. Yumatov and etc.). Also huge excavator-truck complexes technologies and stripping method impacts on final pit are studied (Kuleshov 1980, Yakovlev 1989; Stolyarov 2009, Drizhenko 2011). However existing methods of determining the pit boundaries doesn`t consider such an important factor as stripping scheme constructed in last stages of pit development, which substantially corrects overburden volumes towards their increase in final stages (Shpansky 1998).

Objective of this study is providing a stepwise increasing graph of mining mode in the implementation of double subbench technology with varying levels of working areas (Rakishev 2012). Value of the current stripping ratio by mining stages should vary from the lowest value to the highest value in ascending order with a uniform distribution accessed reserves. Also the accessed reserves volume must be equally distributed to both sides until the end of mining the ore deposits. Variable is the width of the block-panels B_b, its value will be constant after its justification in both highwalls and mining stages. The block-panel width should provide predetermined optimization criteria (Rakishev 2012 & Moldabayev 2013). Width depends on used excavator type. The more powerful the excavator, the greater width . The transport berm width depends on the type of dump trucks, which are chosen for the respective types of excavators. Therefore, the transport berm width increases with increasing the blocks panels width. Safety berm width shall be equal to at least 1/3 of the subbench height (designed to ensure slope stability).

2 OBJECT INTERPRETATION

Pit cross-section of steep deposits represented as trapezoids *ADFE* (Figure 1). Left hanging pit wall is designated through *ABE*, and right lying pit wall is designated through *CDF*. Mineral deposits are designated through *BCFE*, excluding cap rocks with width *M*. Slope angles of pit wall are δ_r and δ_l. Mining deposit *BEF* angle is β, and mining stages are $t = 1, 2, 3, \ldots, 2t_0$.

Stages number equal to the number of horizons, each horizon comprises subhorizons – upper and lower.

Sequence number of subhorizons is j starting from the upper horizon. So $j = 1, 2, 3, \ldots 2t_0$. The upper horizon is $2t - 1$ and the lower horizon is $2t$ for stafe t. Pit depth is H, height of the each subbench is h_b.

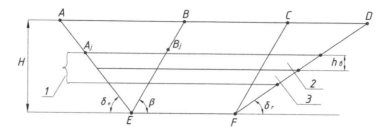

Figure 1. Pit cross section with steep deposit contours without cap rocks: 1 – horizon t; 2 – the first subhorizon $j = 2t - 1$; 3 – the second subhorizon $j = 2t$.

The hanging pit wall is considered separately to identifyof the geometric characteristics of the cross-section of deposit (Figure 2). A trigonometric relations are made from triangles $A_j KE$ and $B_j EL$

$$|KE| = (H - jh_b)ctg\delta_l , \quad |EL| = (H - jh_b)ctg\beta , \quad (1)$$

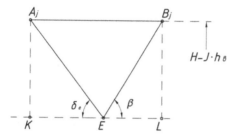

Figure 2. Pit by hanging pit wall.

The length of side $A_j B_j$ of triangle $A_j B_j E$ is

$$|A_j B_j| = (H - jh_b)(ctg\delta_l + ctg\beta) . \quad (2)$$

Similarly, the length of side $C_j D_j$ of triangle $C_j D_j E$ is defined by the following expression

$$|C_j D_j| = (H - jh_b)ctg\delta_r . \quad (3)$$

Note that the length of the $B_j C_j$ is initially constant and equal to M.

3 SIMULATION OF STUDY OBJECT

Proceed to describe the distribution of block-panels and transport berms widths by horizons. The length of $\bar{B}B_{2t-1}$ side is determined from triangle $\bar{B}B_{2t-1}B_{2t}$

$$\left| \bar{B}B_{2t-1} \right| = h_b(ctg\delta_l + ctg\beta) . \quad (4)$$

We introduce the notation: $\alpha = \delta_l$ – slope angle of subbench, degree. For subhorizon j amount of the parallelogram Equal to

$$\left[\frac{(H - jh_b)(ctg\delta_l + ctg\beta)}{B_b} \right] . \quad (5)$$

Square of triangle $\bar{B}B_{2t-1}B_{2t}$ is

$$S_\Delta = h_b^2(ctg\delta_l + ctg)/2 . \quad (6)$$

And the area of the parallelogram determined by the formula

$$S_b = B_b(k)h_b . \quad (7)$$

Figures obtained after partition numbered ascending from left to right. The total number of figures in each subhorizon is

$$n_j = \left[\frac{(H - jh_b)(ctg\delta_l + ctg\beta)}{B_b} \right] + 1 . \quad (8)$$

The obtained formulas may be written by the system as follows

$$S_b(1, j) = S_\Delta ; \quad S_b(i, j) = S_b ;$$

$$i = 2,3,...,n_j , \quad j = 2,3,...,2t_0 . \quad (9)$$

Proceed to describe the partitioning of ore deposits $BCFE$ into small pieces. Oblique lines of the pit wall from the hanging side of deposit continue until mid-field of $BCFE$. The right half of the field $BCFE$ partitioned into parallelograms by parallel lines. The even-numbered horizons contain whole parallelograms and horizons with odd numbers also split into parallelograms, but shifted to right from parallelograms of odd horizon. Shift distance is $B_b / 2$. Par-

352

titions are made so that the right side of the last parallelogram of the second subhorizon of odd horizon passes through the *CF*. This breakdown continues to *CDF* (pit wall by lying side of deposit). The result is a grid shown in Figure 3.

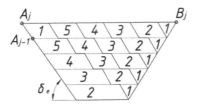

Figure 3. Distribution of block-panels and transport berms widths by horizons.

So we have the following pattern: parallelograms, parallel *AE* form a chain of continuous polylines (left) and parallel *BE* form the chain of discontinuous polylines (right). Note that height of 0 is between left and right parts. Quantity of left lines are denoted by *L*, and right oblique lines are denoted by *R*. A cross-section of initial cut has a trapezoid shape. Its left side rests against the left side of the left continuous broken line, and right side – the right discontinuous broken line. All figures located to the left and right of the trapezoids will be numbered in ascending order. The left half of the trapezoids refers to left figures group and the right half refers to right figures group of *BCFE*. The results obtained are written in the form:

$S_l(i,j;t)$ – area of the shape of the pit wall from the hanging side of deposits;

$S_r(i,j;t)$ – area of the shape of the pit wall from the lying side of deposits;

$S_{r,l}(i,j;t)$ – area of the left half of *BCFE*;

$S_{r,r}(i,j;t)$ – area of the right half of *BCFE*.

where i is sequence number of subhorizon j figures; t is sequence number of deposit development stage.

It is necessary to ensure the following equality to provide equal volumes of mining works in each block-panel of *BCFE*

$$\sum_i \sum_k S_{r,l}(i,j;k) = \sum_i \sum_k S_{r,r}(i,j;k),$$

$$j = 2,3,...,2t_0. \tag{10}$$

Furthermore, we know that the distance between the oblique lines is the width of the block-panel $B_b(k)$, k is the serial number of oblique line. It should be noted that transport berm width can be different in each subhorizon, that is $B_r(j)$.

The current stripping ratio by stages defined by the formula

$$k(t) = \frac{\sum_{i,j} S_l(i,j;t) + \sum_{i,j} S_r(i,j;t)}{\sum_{i,j} S_{r,l}(i,j;t) + \sum_{i,j} S_{r,r}(i,j;t)}. \tag{11}$$

4 SOLUTION OF THE OBJECT BY NONLINEAR PROGRAMMING METHOD

Ensuring stepwise increasing graphics of stripping works mode with a uniform distribution of volumes of accessed reserves in both pit walls at each mining stage is achieved with the following criteria of evaluating the efficiency of double subbench mining technology:

$$k(t) \le k(t+1), \quad t = 1,2,...,t_0, \tag{12}$$

$$\sum_i \sum_k S_{r,l}(i,j,k) = \sum_i \sum_k S_{r,r}(i,j,k), \tag{13}$$

$$B_{0,b} \le B_b(k) \le \overline{B}_{0,b} \text{ and } B_{0,r} \le B_r(k) \le \overline{B}_{0,r}, \tag{14}$$

where left pit wall in the left half $k = 1,2,...,L$; $j = 1,2,...,2t_0$, and right pit wall in the right half $k = 1,2,...,R(j)$.

Even

$$\sum_{i,j} S_l(i,j,k) + \sum_{i,j} S_r(i,j;t) = A(t;B_b,B_r),$$

$$\sum_{i,j} S_{r,l}(i,j;t) + \sum_{i,j} S_{r,r}(i,j;t) = B(t;B_b,B_r). \tag{15}$$

where B_b and B_r – vectors determined from the relations (12)–(13). In the future we seek value of B_b and B_r.

To reduce the relations (1)–(3) to the standard form, we transform (12) into

$$\frac{A(t;B_b,B_r)}{B(t;B_b,B_r)} \le \frac{A(t+1;B_b,B_r)}{B(t+1;B_b,B_r)}; \tag{16}$$

$$A(t;B_b,B_r)B(t+1;B_b,B_r) \le$$

$$\le A(t+1;B_b,B_r)B(t;B_b,B_r). \tag{17}$$

Introduce the notation

$$A(t+1;B_b,B_r)B(t;B_b,B_r)-$$

$$-A(t;B_b,B_r)B(t+1;B_b,B_r) = V(t,t+1). \tag{18}$$

Introduce the notation

$$J(B_b, B_r) = \sum_{i=1}^{t_0-1} V^2(t, t+1). \qquad (19)$$

Unknown variables B_b and B_r include in the formula of area of S; this values include linearly in formulas of A and B. A and B are multiplied together in the formula $V(t, t+1)$, so $V(t, t+1)$ comprises a second-order nonlinearity. Finally, variable $V(t, t+1)$ squared in functional $J(B_b, B_r)$, it means that the functional $J(B_b, B_r)$ contains a fourth-order non-linearity. As a result, we obtain nonlinear programming problem

$$J(B_b, B_r) = \sum_{i=1}^{t_0-1} V^2(t, t+1)^{B_b, B_r} \to min \qquad (20)$$

with constraints

$$\sum_i \sum_k S_{r,l}(i, j; k) = \sum_i \sum_k S_{r,r}(i, j; k), \qquad (21)$$

$$B_{0,b} \le B_b(k) \le \bar{B}_{0,b} \text{ and } B_{0,r} \le B_r(j) \le \bar{B}_{0,r}, \qquad (22)$$

$V(t, t+1) \ge 0$, $t+1,2,...,t_0-1$ in the left pit wall $k = 1,2,...,L$; $j = 1,2,...,2t_0$ in the right pit wall $k = 1,2,...,R(j)$,

where

$$A(t; B_b, B_r) = \sum_{i,j} S_l(i, j; t) + \sum_{i,j} S_r(i, j; t), \qquad (23)$$

$$B(t; B_b, B_r) = \sum_{i,j} S_{r,l}(i, j; t) + \sum_{i,j} S_{r,r}(i, j; t), \qquad (24)$$

$$V(t, t+1) = A(t+1; B_b, B_r)B(t; B_b, B_r) - $$
$$- A(t; B_b, B_r)B(t+1; B_b, B_r). \qquad (25)$$

Algorithm for solving the problem: the initial solution of the problem is shown in Figure 3. The concavities are provided at each stage of the deposit development. This is a basic requirement for the development of steep deposits by double subbench technology. The second requirement is providing the ratio (12). This is achieved by solving a nonlinear programming problem (15)–(20). That is determined by the width of such panels and the transport blocks berms to provide correlation (12) and (13).

Placement (which pit wall) of the initial cut in new horizon depends on amount of accessed reserves in each subsequent stage and providing simultaneous mining of block-panels on both pit walls. The best schedule of mining mode is obtained by trenches (initial cuts) forming along the strike of the mineral deposits at each horizon. If trenches are formed across the strike of the mineral deposits ac-

cessed reserves reduction takes place until progress in depth of mining works. It occurs due to the displacement of trenches and mining the block-panels to side lying mineral deposits.

Block-panels of same horizon will be mined in both pit walls or pit bottom maximizes in certain horizon when pit gets to final contours. Such techniques allow a wide and narrow initial cut, which will pass closer to the mineral deposits. It doesn't significantly affect to the volume of overburden volume during this period.

Thus, it is advisable to build stages of double subbench mining technology by cross panels with varying levels of work area with reference of mining stages design to mineral deposits and final pit contours from its hanging and lying sides.

5 CONCLUSIONS

Solution of the double subbench technology parameters optimization with varying levels of working are by nonlinear programming method allows to build the mining of stages deposits of complex configuration with providing the best mining mode and uniform distribution of accessed reserves between the two pit walls. All the formulas are assigned and attached to the selected coordinate system and will help solve the problem, which needs to determine some parameters. The specified function does not exceed the fixed values and dedicated objective function reaches a global minimum. Test results show that the width of the panels should be determined for mining stages and by the hanging and lying sides of deposits when the double subbench technology with varying levels of working area is designed. It provides obtaining the optimal mining mode with uniform distribution of accessed reserves on both pit walls.

REFERENCES

Kuleshov, A.A. 1980. *Huge excavator-truck complexes of pits*. Moscow: Nedra.

Yakovlev, V.L. 1989. *The theory and practice of choosing transport in deep pits: monograph*. Novosibirsk: Siberian Division of the Russian Academy of Sciences.

Stolyarov, V.F. 2009. *The theory of mining systems development: monograph*. Ekaterinburg: Ural Division of the Russian Academy of Sciences.

Drizhenko, A.Y. 2011. *Pit technological mining and transport systems: monograph*. Dnipropetrovs'k: National Mining University.

Shpansky, O.V. & Ligostsky, D.N. 1998. *Nomogram of determining the final pit parameters*. Moscow: Mining magazine, 12: 78–86.

Rakishev, B.R. & Moldabyaev, S.K. 2012. *Method of opencast mining the steep deposits. Innovation patent.*

Rakishev, B.R. & Moldabyaev, S.K. 2012. *Resource saving technologies in coal mines: monograph.* Almaty: Kazakh National Technical University.

Moldabaev, S.R., Rysbaiuly, B. & Sultanbekova, Zh.Zh. 2013. *Justification of Countours Belonging to the Stages of Mining Steeply Dipping Deposits Using the Solution of the Problem of Nonlinear Programming.* Proceeding of the 22nd MPES Conference. Dresden, Germany. Springer, Volume 1: 125–132.

Moldabaev, S.R., Rysbaiuly, B. & Sultanbekova, Zh.Zh. 2013. *Justification of mining stages of steep deposits with solution by nonlinear programming.* International scientific and technical collection ed. Bondarenko. Dnipropetrovs'k: LizunovPress: 241–246.

Moldabaev, S.R., Rysbaiuly, B., Sultanbekova, Zh.Zh. & Aben, Ye. 2013. *Stages optimization of steep deposits mined by double subbench technology using cross panels with minimum pit wall spacing.* Dnipropetrovs'k: Miners forum, 1: 134–139.

Progressive Technologies of Coal, Coalbed Methane, and Ores Mining – Bondarenko, Kovalevs'ka & Ganushevych (eds)
© *2014 Taylor & Francis Group, London, ISBN: 978-1-138-02699-5*

Legal regime of energy resources of European Union: ecological aspect

S. Gryschak & R. Kirin
National Mining University, Dnipropetrovs'k, Ukraine

ABSTRACT: Legal aspects of "ecologization" of energetic resources extraction sphere in European Union are considered. Situation with appliance of alternative fuels and also legal ideas and principles of steady development, loss prevention, split of responsibilities, precaution is analyzed. It is proposed to use in Ukrainian legislation mechanisms that can be serve for realization of EU energy policy, especially ecological certification, agreements, management, audit and design.

1 INTRODUCTION

From the mid-to-late century rational use of natural resources including energy ones and environmental conservation have become one of the crucial political, social, economic problems of Ukraine.

Interconnection between energy provisions, subsurface resources management and ecology is multifaceted. Weighty ecological aspect is an attribute of contemporary mining engineering of the EU and its legal coverage. Key task of new energy program of the EU, formulated in the document of EU Commission, is to "limit climate change to 2 °C – the choice of EU and world policy till 2020 and in future" (Commission of the European... 2007).

2 THE "ECOLOGIZATION" OF THE SPHERE OF ENERGY RESOURCES

It goes without saying that the source of "ecologization" of the sphere of energy resources winning has economic character: increasing of dependence of Europe on the import of raw hydrocarbons. With following the path "business as usual", energy dependence of Europe on import from 50% of total final energy consumption as of today will increase to 65% in 2030. Dependence on import of gas, according to experts, will increase from 57 to 84 %, on oil – from 82 to 93%. Such a tendency carries serious political and economic risks. Besides, hasty growth of oil and gas industry, requirement of these resources is not an optimum alternative of development of energy from the point of view of social and ecologic interests of society, the value of which has briskly increased in recent decades. Usage of renewable energy sources, development of advanced technologies and energy efficiency enhancement is the way out within the frame of energy policy of the EU. Resolution of these tasks has important social and ecological "component" – decreasing of load upon natural resources, enhancement of environment's quality (Vylegzhanina 2011).

Special document of European Commission – Renewable Energy Roadmap (Communication from the Commission... 2006) is devoted to renewable energy sources. Imperative task assigned in it is to enhance the level of energy from renewable sources from 7 to 20% in 2020. The third of electric power consumed in the EU could be generated by renewable sources by 2020. Nowadays wind farms produce a profit of 20% of electric power in Denmark, 8% in Spain, 6% in Germany. New technologies of heating and cooling are developed. For instance, more than 185 000 of units for geothermal energy are used in Sweden. In German and Austria requirements of heating are partially covered by solar power. According to experts, if other countries-members of the EU follow this path, consumption of energy from renewable energy sources in this sector will increase to 50%. Biofuel is actively used in Sweden and Germany. By 2020 it may run at 14% of transport fuel.

The Law of Ukraine "Of alternative fuels" that determines legal, social, economic, ecologic and organizational basis of production and usage of alternative fuels, dictates stimulation of increasing of their usage to 20% from total consumption of fuel in Ukraine to 2020.

Legal tools, providing ecologic "component" in the sphere of energy resources extraction in the EU countries, form within the framework of the main principles and priority tendencies of development of integrative ecological regulation.

In spite of that ecologic law of the EU is relatively new and complex legal unit of European integration, it is unique by its content.

Its development is characterized by the search of innovative and optimal approaches to resolution of problems of nature conservation: head for ecologization of all the EU policy directions including sphere of energy, stimulation of involving entrepreneurial communities into tasks' fulfillment.

The Fifth ecologic program of the EU is a unique boundary in the development of ecologic regulation. Its novelty is that it is focused on five fields of activity that have a great impact on the environment. Mining engineering and energy hold key positions.

Conceptual basis of the Fifth program, developed in existing the Sixth one "Environment 2012: our future, our choice" is an idea that ecological problems are to be studied as the result of incorrect management. Real problems, conditioning degradation of environment and natural resources loss, cause models of consumption and behavior existing in human society. On this basis, legal ideas of sustainable development, prevention from damaging, shared responsibility, principle of precaution are of the priority. Sustainable development understands harmonization of social, ecologic, economic parameters of development of society in the perspective of "enhancement of general quality of life of European Union's citizens".

Special emphasis lays on the necessity of integration of ecologic "dimension" in the policy within economic fields, in energy, in the first instance. Climate change on global level, conservation of nature and biodiversity, environmental protection and people's health support, sustainable usage of natural resources usage and recycling management were rated as supreme. For their resolution is suggested:

– integration of ecologic aspects into other spheres of policy;

– work with entrepreneurs with the aim of enhancement of ecological requirements fulfillment;

– provision of community with multifaceted and scientifically justified information.

It worth mentioning that indicated approaches in law regulation in the context of the sphere of extraction of energy resources correspond to those approaches and principles, that have been formed nowadays in the international law and directed to harmonization of ecologic, social and economic interests and long-term perspective, provision of ecologic security of mining and energy cycles. They received reflection in the Protocol of Energy Charter (Energy Charter Treaty... 2002).

Analysis of separate specific mechanisms, applied to ecologic regulation of the EU and to which an important part is assigned in realization of energy policy, aimed at decreasing of dependence on import of fuel and gas and environment quality enhancement by renewable energy resources are topical for Ukraine. Steps aimed at the development of new technologies (best available technologies), internalization, the so-called voluntary agreements, state support is among such mechanisms, prescribed in the documents describing new energy policy of EU (Commission of the European... 2007, Communication from the Commission... 2006; 1998; Directive 2005/32/EC... 2005).

As already mentioned, the key part in the increasing of energy efficiency and decreasing of dependence on the import of oil and gas belongs to new technologies' development. This direction has exact legal regulation. The question is the Guideline 96/61 on the complex prevention from environmental pollution from October, 10, 1996 (Directive 96/61 C... 1996), characteristic of which is the legal framework of key concepts, usage of which have economic and legal consequences. In this particular case the question is the concept of the best available technologies (best available technologies). Definition of the concept "best available technologies" is formulated very correctly: it is the most efficient and "advanced" types of activity and methods of production; their presence speaks for the existence of concrete technologies, on the basis of which values of wastes may be defined or in the cases when it is not rationally.

We use the term "technologies" both technologies themselves and the way of construction, building and operation and are demounted economic entities.

"Available" (or "present", that is more exact translation of word in best available this context) – those which are developed to the level, letting their application in corresponding sphere of production, in economically and technically acceptable irrespective of whether these technologies are used in certain state-member of European Union, until they are available for operator.

"Best" – are technologies that efficiently provide high level of environment protection. The Guideline understands duties of states-members to provide corresponding bodies with information on such technologies' development.

Maximum permissible emission, questions of responsibility, insurance, licensure and technical steps are determined by reference to the best available technologies, taking into consideration technical characteristics of concrete object, its position and local natural conditions.

In realization of ecologic aspect of program of extraction of energy resources the principle of "shared responsibility" received development. The mining of this principle is that everyone is responsible for environment conservation and all the layers of society are to be involved in the resolution of the problems. Final objective of this process is creation of balance between shot-terms interest in benefits of business community and long-terms interests of society.

Shared responsibility principle understands integration of ecological factor into the development and realization of economic and social policy, state bodies' decisions, productive processes and individual behavior of every person.

In the field of energy this principle understands gradual elimination of existing legislative barriers for extension of possibilities of companies to invest into energy efficiency enhancement.

In legal terms, this direction is well developed in ecologic regulation of the EU. In 1994 the EU Commission approved the document named "General instructions of the Community about state aids for environment conservation". The most frequently such aids were directed to the sector of energy and obtained the forms of taxes decreasing and tax exemption.

With due regard to such tendencies and with the aim of alignment of criteria of aids providing Commission elaborated new document in 2001 (2001/C 37/03).

The document emphasizes that in correspondence with article 6 of Agreement about the EU, the aims of ecologic policy are to be integrated into the policy of the Commission of the EU in the field of control of the aid in ecologic sector that must serve the achievement of sustainable development. In light of long-term tasks of environment conservation not every type of state aid is recognized legal. The Guidelines offers parameters of legacy of provision of aids. This is the result of the impact of state aids usage on the achievement of sustainable development and on the application of the principle "polluter pays".

Indicated parameter corresponds to such aid provision when the highest level of environment protection is achieved and the principle of "internalization" of prices isn't defied. Other forms of aid may contradict "polluter pays" principle and prevent from sustainable development.

Commission set the task regarding to experience of application of previous "General instruction" to formulate criteria of qualification of state aid as necessary for environment protection and sustainable development without negative impact on the economic growth.

Environmental protection is determined as any action aimed at elimination or prevention from the damage, caused by material object to environment or natural resources or development of effective usage of these resources. From the point of view of the Commission energy efficient steps and usage of renewable energy resources are measures aimed at environment protection.

State aid is regulated by "General instructions" of Community. Concrete forms of aid depend on the type of renewable energy resource and the series of other factors. For instance, in order to facilitate access to the market of renewable energy resources, the state may compensate to the manufacturer the margin between manufacture cost and market price. Another form of aid is usage of market mechanisms in the form of green certificates or tenders. These systems understand the receiving by all the manufacturers of such energy demand on their energy at higher prices than market ones. The price of green certificates isn't fixed beforehand but depends on the submission of energy and demand on it.

Another legal mechanism used in realization of energy policy is "voluntary agreements" – voluntaryagreements between state bodies and business for resolution including ecologically significant tasks (more active usage of such agreements with the aim of extension of production of economy cars). It is comparatively new ecological and legal tool applied in the EU. National experience of resolution of environmentally friendly problems through conclusion of the agreements between state bodies and entrepreneurial circles was generalized with the aim of further promotion on the level of the EU. December 9, 1996 Commission approved recommendation on Community guidelines' execution by means of conclusion of ecologic agreements (96/733/EC: Commission Recommendation... 1996).

Today such agreements mean the agreements between state bodies and industrial companies in order to contribute achievement of ecologic aims supporting and encouragement of corresponding activity of the companies. Conclusion of such agreements is one of the ways of realization of Guidelines of Community. States-members are obliged to guarantee achievements of the results that is why ecologic agreements are to have obligatory character and answer requirements of "transparency, authenticity and reliability". Guidelines must contain the points that may be realized by means of such agreements with obligatory requirements. Each ecologic agreement must correspond to the Agreement on European Community, requirements of internal market, rules of competitiveness and Guidelines 83/189 EEC from March 28, 1983 (with amendments of Decision of Commission 93/139/E) that understands the procedure of provision of information in the field of technical standards and rules.

Commission formulated the series of recommendations on the practice of conclusion ecologic agreements. In all the cases they are:

– to be concluded in the form of contracts, fulfillment of which is provided on the basis of civic or public law (enforceable under civil or under public law);

– contain the aims and intermediate steps with indication of time parameters;

– published in the national official news paper;

– understand monitoring of achieved results for regulatory report before state bodies and provision corresponding information to people;

– be open for all possible partners wishing become parties of the agreement.

Agreements may contain the following:

– agreements on the collection, evaluation and examination of results;

– requirements to the companies-participants to provide with any third party with access to the information on agreement execution on the same conditions as state bodies according to Guidelines 90/313/EEC from June, 7, 1990 on free access to the information on environment;

– applying sanctions in forms of fines and permission annulation.

Agreement must contain the point on the right of competent state bodies to evaluate achieved results and taking additional steps if necessary.

3 CONCLUSIONS

It goes without saying that the variety of means and mechanisms applied with the aim of resolution of ecologic tasks of mining engineering policy of the EU, are not completed with mentioned above. Procedure of evaluation of impact, ecology management and audit, institute of responsibility is of important status.

New and interesting direction in legal coverage of energy sphere is regulation of Eco design of energy consuming production, modern house building (passive houses) etc. Ecological component has an increasing importance in mining engineering and energy sphere of the EU, that reflects general process of integration in economic sphere of interests of conservation of environment, objectively leads to decreasing of anthropomorphic impact on natural resources, coal, oil and gas and biosphere in general.

REFERENCES

Commission of the European Communities. 2007. *Communication from the Commission to the European Council and the European Parliament An Energy Policy for Europe.* Brussels.

Vylegzhanina, E. 2011. *Ecological angle of reserves management: generalization of international-law experience*: educational book. Moscow: Prospect: 128.

Communication from the Commission to the European Parliament and the Council. 2006: Renewable Energy Roadmap of renewable Energies in the 21st century; building a sustainable future – COM.

Energy Charter Treaty. 2002: The way to investments and trade for East and West / edited from Russian by Konoplyanik A.M.

Communication from the Commission to the European Parliament and the Council. 2006: Renewable Energy Roadmap of renewable Energies in the 21st century; building a sustainable future – COM. Communication from the Commission Action Plan for Energy Efficiency. 2006: Realizing the Potential COM 545 final, Brussels, 19.10.2006; Communication from the Commission of 14 October 1998 Strengthening environmental integration within Community energy policy, COM(98)571; Directive 2005/32/EC of the European Parliament and of the Council of 6 July 2005 establishing a framework for the setting of ecodesign requirements for energy using products and amending Council Directive 92/42/EEC and Directives 96/57/EC and 2000/55/EC of the European Parliament and of the Council, etc.

EC Directive 96/61 C. 1996. Concerning Integrated Prevention of Environmental Pollution, OJ Ns L 257.

96/733/EC: Commission Recommendation of 9 December 1996 Environmental Agreements implementing Community directives. Official Jou 21/12/1996/P. 0059-0061.

Progressive Technologies of Coal, Coalbed Methane, and Ores Mining – Bondarenko, Kovalevs'ka & Ganushevych (eds)
© 2014 Taylor & Francis Group, London, ISBN: 978-1-138-02699-5

Substantiation of ventilation parameters and ways of degassing under bleeding of methane

O. Mukha & I. Pugach

National Mining University, Dnipropetrovs'k, Ukraine

ABSTRACT: In the article the question of increase of safety of mining operations on extraction sites due to development of actions for the prevention of emergence of local methane congestions and their elimination is considered.

1 INTRODUCTION

Coal mining on coalmines in many respects depends on a gas situation in productive workings and development heading. Thus, safety of mining operations directly depends on a type of methane layering and intensity of intake of explosive gas in the mine air.

At this stage, questions of formation of local methane layering in the conditions of coal mines with high gas content are insufficiently studied and demand carrying out additional researches for definition of their influence on load of a clearing face by gas factor.

2 FORMULATING THE PROBLEM

The purpose of this work is increase of mining operations safety on extraction sites due to development of measures for the prevention of emergence of local methane layering and their elimination.

For achievement of a goal, it is necessary to solve the following problems:

– to analyze places of emergence of local methane layering in underground working of working area;

– to establish influence of intensity of arrival of methane in the mine air on cutter-loader production in a breakage face depending by gas factor;

– to define limits of change of air supply in developments of an working area depending on methane emission;

– to develop measures for the prevention and elimination of local methane layering in mine working;

– to give a social and economic assessment of proposed solutions.

3 RESEARCH

Object of research – local methane layering in mine working of working area of A.F. Zasyadko mine.

Subject for study – influence of intensity of methane emission on working area on the maximal permissible stope output by gas factor.

The aim of the given work is increase of mining safety due to air distribution regulation in mine working of working area and application of measures for the prevention of local methane layer when providing the set of stope output.

According to the direction and character of objectives when performing researches the following methods of researches are used:

– method of the scientific analysis and generalization – during the studying and the analysis of literary data about ways of fight against methane layering on coal mines;

– experimental method – at measurement of air expenses, methane concentration in mine working;

– methods of mathematical statistics – for processing of experimental data and obtaining mathematical dependences;

– methods of mathematical modeling – at determination of parameters of coal production taking into account norms (Safety rules ... 2010) of methane concentration.

The analysis of the mine air of A.F. Zasyadko mine showed that the main reasons for formation of local and layering methane are:

– low speed of air movement in mine working (less than 0.5–1 m/s);

– existence in the footwall and top wall of mine working gas-bearing a seam or the carboniferous breeds located at distance to 10 m;

– existence of bleeding in mine working;

– existence of geological dislocations.

The greatest dangers from the point of view of

excess of maximum-permissible methane concentration represent bleeding.

Preceding from the above in work the special attention is paid by questions of research of influence bleeding on coal production when ensuring admissible concentration of methane in mine working.

The key task of studies is to determination of ventilation parameters of a working area when providing specified load on a stope. The calculated scheme of the working area is submitted in Figure 1. On the scheme the length of a stope and an extraction pillar are specified. Numbers of knots and branches of network are put.

Mining deepening as well as increase in gas concentration of coal beds and rocks mined and contiguous results in the problem that it is impossible to keep non-hazardous methane concentrations in a mine air with the help of ventilation only. Under such conditions, degasification of gassy coal beds and mined-out space is important procedure favouring increase in mining safety, and load increment on a stope.

In conditions A.F. Zasyadko mine there is a need to degassing of underworked seams and host rocks for providing the planned load for a stope

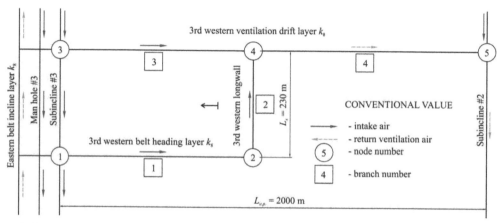

Figure 1. The calculated scheme of working area by layer k_8.

Consider influence gas blower on coal production in the stope by the example of third west longwall of layer k_8.

Ventilation scheme of a working area is 3-V-N-v-pt (Guidance... 1994) (the isolated dilution of noxiousness with delivery of a return ventilation air on the worked-out space with independent the airing, ascending movement of a stream in a stope and direct-flow movement of air current on an working area).

Air supply of working area is carried out at the expense of a mine ventilating pressure drop.

The analysis of materials on gas blower at extraction seam of k_8 showed that emergence it is possible in any mining of a working area. In this regard for further researches, the working area (Figure 1) breaks into fourth characteristic zones:

The 1st zone – intake air, moving on the 3rd western belt heading layer k_8 (a branch 1 with hubs 1–2);

The 2nd zone – intake air, moving on the 3rd western longwall layer k_8 (a branch 2 with hubs 2–4);

The 3rd zone – additional air, moving on the 3rd western ventilation drift layer k_8 (a branch 3 with

hubs 3–4);

The 4th zone – return ventilation air, moving on the 3rd western ventilation drift layer k_8 (a branch 4 with hubs 4–5).

A measure of fight against gas blower it is possible in two main directions:

1) By means of ventilation (diluting of methane by means of giving of additional amount of air in mine working);

2) By degassing of methane-release sources.

Consider possible ways of fight against methane at various places of emergence bleeding.

At emergence of methane, bleeding in the first branch all emission gas will arrive into a stope. In these conditions, the situation when to stope air with the concentration of methane exceeding according to (Safety rules... 2010) maximum-permissible norm (0.5% on volume) will arrive is possible.

Possibility of its dilution by additional amount of fresh air to demanded concentration (Safety rules... 2010) for the 3rd western belt heading layer k_8 is absent as the increase in a consumption of air will lead to increase in speed of movement in a stope.

According to Table 1 it will occur at once since load on a face was limited on a gas factor (speed in the stope according to measurements makes 3.96 m per second).

Table 1. Observation results.

Branch	Hubs			Q, m³/s	S, m²	V, m/s	C, %	I, m³/s
1	1	–	2	30.62	15.4	1.99	0.07	0.0221
2	2	–	4	18.83	4.75	3.96	0.19	0.0366
3	3	–	4	6.18	15.4	0.40	0.05	0.0031
4	4	–	5	36.79	15.4	2.39	0.37	0.1353

For fight against gas blower in work application of a tapping cap, which is connected, to the existing degasification pipeline is offered. Emergence of gas blower in the second branch will lead to operation of methane automatic control sensors in a return ventilation air from stope. In this case, it is necessary to stop coal-face works until gas emission from cracks will stop and actions for a degassing of mine working according to (Instruction book... 2003) won't be made.

Occurrence of gas blower in the third branch will not influence conducting clearing works in the stope. However, emergence of an additional methane-release sources in an additional air on the 3rd western ventilation drift layer k_8 can lead to that in a return ventilation air from a working area concentration of methane will exceed maximum-permissible value (1.3% on volume).

For prevention of gassed ventilating drift, it is necessary to give additional amount of air on a branch 3. There will be thus a restriction: speed of air movement in the third and fourth branches is limited to value of 6.0 m/s according to (Safety rules... 2010).

Amount of air in an additional air of the third western ventilation drift if concentration of methane won't exceed 0.5% (Figure 1) must define.

The increase in an air consumption in an additional air is carried out by adjustment of ventilating window section in a ventilating construction (a ventilating door with an adjustable window), established in a hub 3.

Intensity of gas blower is accepted in the range from 1 to 20 m³/min.

The dependence on Figure 2 is described by the following equation

$$Q_{f.a.3-4} = 3.33 I_{g.b},$$

where $I_{g.b}$ – intensity of gas blower, m³/min; $Q_{f.a.3-4}$ – amount of in addition air, m³/min.

The value of the approximation accuracy $R^2 = 1$.

If intensity of gas blower does not exceed 20 m³/min at methane dilution to demanded concentration the magnitude of air velocity won't exceed the set limit in 6 m/s (the maximum of air velocity makes 4.7 m/s) (Figure 3).

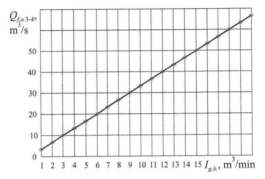

Figure 2. Dependence of air consumption necessary for additional supply from intensity of gas blower.

Change of air velocity in a branch 3 depending on intensity of gas blower it is presented in Figure 3. This dependence is described by the equation

$$V_{m.w.3-4} = 0.2165 I_{g.b} + 0.401.$$

The value of the approximation accuracy $R^2 = 1$.

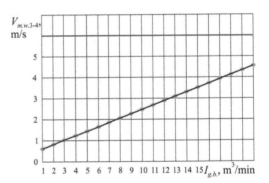

Figure 3. Dependence of air velocity in an additional air stream in ventilation drift layer k_8 from intensity of gas blower.

If intensity of gas blower does not exceed 20 m³/min at methane dilution to demanded concentration the magnitude of air velocity won't exceed the set limit in 6 m/s (the maximum of air velocity makes 4.7 m/s) (Figure 3).

Dependence of air velocity in an additional air stream of ventilation drift layer k_8 from intensity of gas blower is presented on Figure 4

$$V_{m.w.4-5} = 0.2165 I_{g.b} + 2.3892.$$

The value of the approximation accuracy $R^2 = 1$.

According to dependence on Figure 4 possibility of dilution of the methane arriving from gas blower, in a return ventilation air is limited. At intensity of gas blower in 16.7 m³/min. velocity air in the fourth branch reaches maximum-permissible value (6 m/s). Therefore, at great values of intensity gas blower application of degassing of methane sources is necessary.

Emergence of gas blower in the fourth branch will have impact on methane concentration in a stream proceeding from a stope of working area. Thus, influence of gas blower on consumption of additional air it will be similar to the dependence presented on Figure 4.

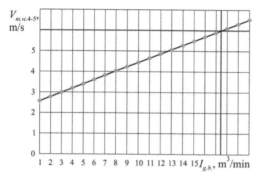

Figure 4. Dependence of air velocity in return ventilation air of ventilation drift layer k_8 from intensity of gas blower.

Influence of gas blower in the third western belt heading, on the maximal permissible load on a face is shown on Figure 5.

Dependence can be described by a polynomial of the 6th degree

$$A_{day} = 0.0007 I_{g.b}^6 - 0.0493 I_{g.b}^5 + 1.3889 I_{g.b}^4 -$$

$$- 20.127 I_{g.b}^3 + 161.92 I_{g.b}^2 - 744.36 I_{g.b} + 19235.3.$$

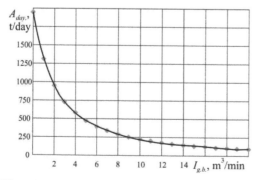

Figure 5. Dependence of a maximum permissible load on a face by gas factor from intensity of gas blower.

The value of the approximation accuracy $R^2 = 0.9997$.

The presented dependence allows to predict coal mining on working area depending on intensity of gas blower.

Proceeding from it, the following dependences are received:

1) Dependence of velocity shearer from intensity of gas blower (Figure 6)

$$V_s = 3 \cdot 10^{-6} I_{g.b}^6 - 0.0002 I_{g.b}^5 + 0.0058 I_{g.b}^4 -$$

$$- 0.079 I_{g.b}^3 + 0.575 I_{g.b}^2 - 2.2 I_{g.b} + 4.2.$$

The value of the approximation accuracy $R^2 = 0.9997$.

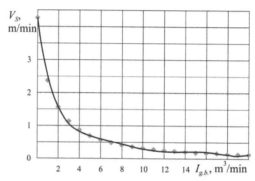

Figure 6. Dependence of shearer velocity from intensity of gas blower.

2) Dependence of a maximal permissible stope output from the shearer velocity (Figure 7)

$$A_{day} = -0.016 V_s^6 - 0.27 V_s^5 + 2.44 V_s^4 - 17.2 V_s^3 +$$

$$+ 114.1 V_s^2 - 750.4 V_s + 0.0012.$$

The value of the approximation accuracy $R^2 = 1$.

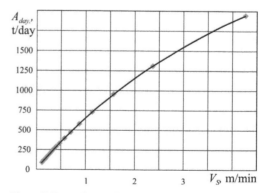

Figure 7. Dependence of a maximal permissible stope output from the shearer velocity.

The received equations allow determining the shearer velocity depending on intensity gas blower and loading on a face.

Thus, the conducted researches allowed to establish influence of gas blower on a gas situation on a working area and to receive dependence of load on a face from intensity of methane emission.

One of negative consequences of gas blower is emergence of local (layered) methane congestions. For fight against them, used of the turbulent pipeline with the booster fan of VM-4 is offered (Figure 8).

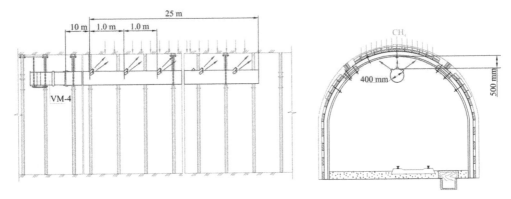

Figure 8. The scheme of placement of the turbulent pipeline in the mine working.

We define costs of elimination of a gas blower in the 3rd western ventilation drift layer k_8

Maintenance costs:
– Expense of the electric power consumed by the booster fan of VM-4.

Initial expenses:
– Cost of linings; cost of VM-4; cost of the rigid steel pipeline; production of shaped parts for the turbulent pipeline; delivery and assembling-dismantling;

We define the necessary volume of materials.

Length of the rigid steel pipeline L_p= 35 m.

Linings:
– Steel chains for suspension bracket a pipeline to a roof: L_{ch} = 25 m;
– Metalware: n_h = 20 pcs.

$$C_m = C_p L_p + C_{ch} L_{ch} + C_h n_h ,$$

where C_p – cost of one long metre of the rigid steel pipeline, UAH/m; C_p = 500 UAH/m; C_{ch} – cost of one long metre of steel chains for suspension bracket a pipeline to a roof, UAH/m; C_{ch} = 280 UAH/m; C_h – cost of one set of the metalware, UAH/pc; C_h = 200 UAH/pc.

$$C_m = 500 \cdot 35 + 280 \cdot 25 + 200 \cdot 20 = 28500 \text{ UAH.}$$

The costs of the booster fan: C_{BF} = 75000 UAH.

Production of shaped parts for the turbulent pipeline ($C_{sal}^{p.s.p.}$). Shaped parts for the turbulent pipeline are made by of two workers within five working days

$$C_{sal}^{p.s.p.} = n_{man.} n_{w.d.} C_{sal} ,$$

where C_{sal} – tariff rate for one worker, UAH/day; C_{sal} = 175 UAH/day.

$$C_{sal}^{p.s.p.} = 5 \cdot 2 \cdot 175 = 1750 \text{ UAH.}$$

A team of workers from four people will carry out delivery and assembling-dismantling works during two shifts

$$C_{sal}^{d.,a-d} = n_{man.} n_{w.d.} C_{sal} ,$$

$$C_{sal}^{d.,a-d} = 4 \cdot 2 \cdot 175 = 1400 \text{ UAH.}$$

We determine an expense of the electric power consumed by the booster fan of VM-4 for elimination of a gas blower by a formula

$$E_{BF} = \frac{Q_p h}{102\eta} \cdot 24 \cdot 10, \text{ kW}$$

where Q_p – consumption of air on an air pipeline, m^3/s; h – pressure loss in the pipeline, kgf/m^2; 24 – coefficient considering a number of hours of continuous the fan operation; 10 – coefficient considering number of days on liquidations of a layered congestion; 102 – coefficient of the translation of miscellaneous units of measure in SI system; η – efficiency of ventilator installation.

The expense of the electric power will make:

$$E_{BF} = \frac{2 \cdot 123}{102 \cdot 0.73} \cdot 24 \cdot 10 = 792.9 \text{ kW}.$$

Expected costs of the electric power consumed by the booster fan of VM-4 at elimination of a layered congestion

$$C_e = E_{BF} c_e, \text{ UAH},$$

where c_e – cost 1 kW·h of the consumed electric power, UAH/kW; for the enterprises of the mining industry $c_e = 0.973$ UAH/kW;

$$C_e = E_{BF} c_e, \text{ UAH},$$

$$C_e = 792.9 \cdot 0.973 = 771.5 \text{ UAH}.$$

The general costs of implementation of the design decision for elimination of a layered congestion in air working will make:

$$\Sigma C = C_m + C_{BF} + C_{sal}^{p.s.p.} + C_{sal}^{d.,a-d.} + C_e, \text{ UAH}.$$

$$\Sigma C = 107422 \text{ UAH}.$$

4 CONCLUSIONS

As a result of performance of researches dependences of a consumption of air on a working area, the maximal permissible load on a face by gas factor and shearer velocity from intensity of gas blower have been determined.

Novelty of the received results:

1. The technique of definition of the maximal permissible stope output by gas factor taking into account intensity of gas blower has been stated.

2. Influence of the gas blower intensity on a velocity shearer in a stope was defined.

The practical value of the received results consists in:

1. Establishment of nature of change of load on a face by a gas factor from intensity of gas blower in conditions A.F. Zasyadko mine.

2. Decrease in probability of a gas-pollution of working area at application of actions for the prevention of local methane congestions.

3. Increase of work safety for account of application air distribution in mine working.

Scope of the researches results – design of system ventilation of a gas mines.

REFERENCES

Safety rules in coal mines. NPAOP 10.0-1.01-10. 2010. Kyiv. Derzhgirpromnaglyad: 432.
Guidance for designing of coal mines ventilation. 1994. Kyiv: Osnova: 311.
Instruction book for Safety rules in coal mines. Vol. 1. 2003. Kyiv: Osnova: 480.

Impact of face advance rate unevenness on methane release dynamics

V. Okalelov, L. Podlipenskaya & Yu. Bubunets
Donbass State Technical University, Alchevs'k, Ukraine

ABSTRACT: Methane releases at excavation sites of mines in dynamically unstable depending on number of elements like advance rate of face. United analysis of dynamic row main components of methane release and advance rate of face has shown correlation between components of low rhythm and 1–2 days delay in methane release change comparing to advance rate change of a face. Random fluctuations of that factor determined by Hirst's index allow setting correlation between the randomness level of face movement and a level of absolute methane release. Herewith the more stable is face excavation, the less is the level of methane release at production site.

1 INTRODUCTION

Advance rate of stoping face refers to number of factors affecting the geomechanical processes in rock body. The latter, in its turn, impact greatly the level of methane release at excavation sites. Except this the advance rate of face determines the coal production capacity which influences the number of methane released out of broken coal mass (Briuzgina 2004, Polevshchikov & Kozyreva 2002).

Operational experience of Donbass coal mines shows that daily advance rate of a producing longwall is changeable. So it is absolutely natural to be supposed that fluctuations of this factor lead to unevenness of methane release degree in the extraction blocks.

Till now this hypothesis has not been checked properly. Statistic data have been obtained though determining the links between average advance rates of a face and average methane release values, which allow evaluate the dynamics of the process while cutting extraction blocks and develop activities on methane release control at extraction blocks.

2 CORRELATION BETWEEN UNEVENNESS OF FACE ADVANCE RATE AND METHANE RELEASE DYNAMICS

To check the above hypothesis the analysis of methane release dynamics (I) has been made, as well as acoustic emission (AE) and average daily advance rate (V) in 25th Orlovskiy face of Molodogvardeiskaya mine at PJSC "Krasnodonugol" for the whole period of its operation. Figure 1 shows diagrams of changes of methane release, acoustic emission intensity and advance rate.

Dissociation of the above rows into components using SSA method (Elsner 1996) and analysis of the components obtained has showed the following:

1. There is rather close correlation between trend components of methane release and acoustic emission. Herewith, if intensity of AE increases one can observe a significant I growth, that proves the existence of correlation between methane release and geomechanical processes in rock massif (Figure 2).

2. Correlation between trend-cyclic components I and AE, according to which one can see delay in global maximum of methane release to global maximum AE as 25 days (Figure 3). To advance rate trend and trend-cyclic components have not been found.

3. Matching of cyclic components with 60–70 days period has shown as in a previous case, there occur delay of methane release index related to AE index and advance rate index of a face in average about 25 days (Figure 4). Herewith it is established the accordance in the changeability format of fluctuations for advance rate and acoustic emission of the rocks.

4. Cyclic components of a short period (6–7 days) are marked in the rows of methane release and advance rate of a face, and any significant fluctuation component for acoustic emission with the above period has not been found (Figure 5). This fact testifies a low sensibility of fluctuations of methane release with a short period to changes in acoustic emission. Analysis of I and V components has shown insignificant delays of 1–2 days for methane release component related to advance rate component. These testify increase or reduce of methane release at extraction block 1–2 days after the due change of its advance rate.

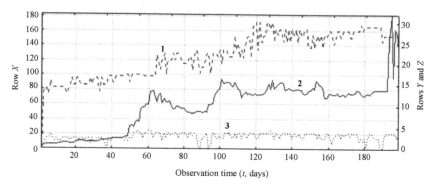

Figure 1. Rows Y25 – methane release (curve 1), X25 – acoustic emission (curve 2), Z – advance rate (curve 3).

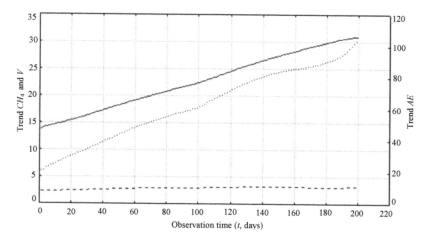

Figure 2. Diagram of trend components of methane release, acoustic emission and advance rate of stoping face: —— trend CH_4; - - - - trend V; ·········· trend AE.

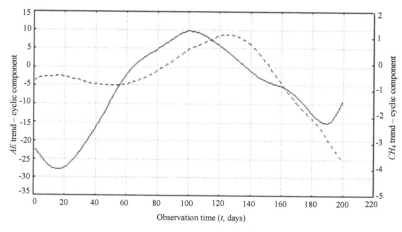

Figure 3. Diagram of trend-cyclic components of AE and methane release: —— AE trend – cyclic component; - - - - CH_4 trend – cyclic component.

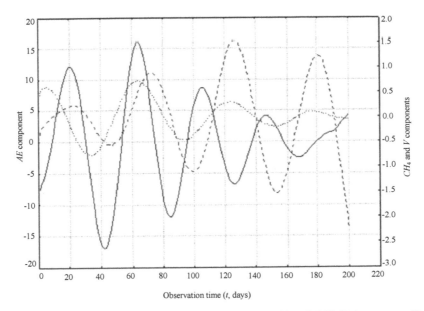

Figure 4. Diagram of fluctuations of *AE*, methane release and advance rate with period 60–70 days: ——— *AE* component; - - - - *CH₄* component; ·········· *V* component.

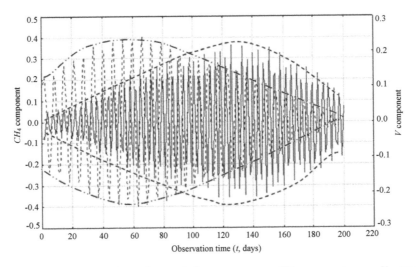

Figure 5. Diagram of components of methane release and advance rate: — ·· — *CH₄* component; - - - - *V* component.

Generalizing obtained results one should claim that advance rate fluctuations of a stoping face form components with a short cyclic period while trend, trend-cyclic and cyclic ones with a long period indicate the effect of *AE* intensity as indicator of geomechanical processes in rock massif.

As long as fluctuations of face advance rate while breaking is difficult to predict using the known computing methods the task arose to substantiate an index which in general could assess the unevenness

impact of advance face rate onto methane release fluctuations.

Hurst index has been offered to be used as such an index (Kalush & Logikov 2002). The index has been like indicator for chaotic process and shows ration of trend force (determined component of a process) to noise level (casual component of a process) for the rows of methane release and face advance rate. Hence unlike autocorrelation a Hurst index (*H*) is able to analyze dynamic rows containing both determined and casual component of fluctuations of values of the investigated process.

Calculations of Hurst index have been made on the data from zonal absolute methane redundancy and advance rate in #25 production face at Molodogvardeiskaya mine and 1st Eastern production face at Samsonovskaya-West mine of PJSC "Krasnodonugol". Herewith it has been used a method of consequent R/S–analysis considering the first residuals of advance rate instead of their initial values (Zinenko 2012).

Calculated results for H and the relevant values of zonal absolute methane redundancy are presented on Figures 6 and 7.

Visual analysis of constructed correlation fields testifies the ration between I and H.

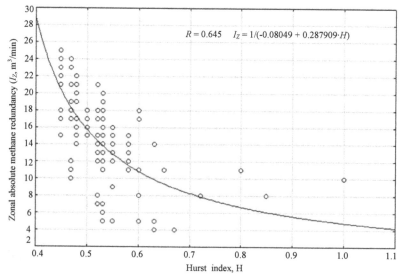

Figure 6. Dependence I_z on Hurst index – H for time row of the first residuals of advance rate for the 1st eastern face of Samsonovskaya-West mine of PJSC "Krasnodonugol".

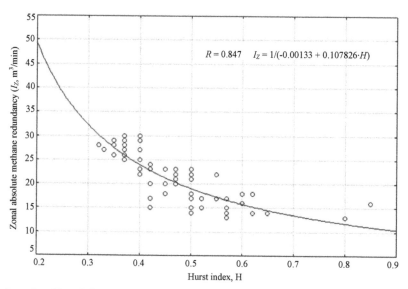

Figure 7. Dependence I_z on Hurst index – H for time row of the first residuals of face advance rate for the 25th Orlovskaya Molodogvardeiskaya mine of PJSC Krasnodonugol.

More accurate this ratio is described by equation $y = \dfrac{1}{a+bx}$. Values a and b for data shown in Figure 6 are 0.08 and 0.288 accordingly, and for figure 7 they are 0.0013 and 0.108 accordingly. Correlation ration are statistically valuable and equal for the first case 0.645 and for the second one 0.847.

According to impact assessment to Hurst index of the first residuals of advance rate of a face it has been studied a direct impact of absolute values V onto I. As a result they have found that connection between these variables is direct and is described as linear dependence with correlation indexes 0.2 and 0.3. These indexes in spite their small values at the same time are statistically valuable at selection with 285 observations as threshold of criteria $t = 1.96$ (when $\alpha = 0.05$), and computed 3.52 and 5.56.

3 CONCLUSIONS

Analysis of obtained dependences indicates that when H is increased methane redundancy is reduced, but when it reduces it in contrary grows. It testifies that if chaos of advance face rate fluctuations grow ($H < 0.5$) one can observe an increase tendency for absolute release of methane at stoping face. At the same time, at fixation of advance rate values of indicating the strengthening of determined component of investigated relation ($H > 0.5 \div 0.6$) leads to I reducing.

Such impact character onto I linear unevenness residuals ΔV follows from impact character of the index onto methane release from broken off coal, coal massif and produced space according to which in some cases increase of V leads to increase of volume of broken off coal in a unit of time and consequently to the growth of methane release from coal. Simultaneously one can observe decreasing the area of bearing pressure which leads to degradation of rock fractioning, reducing the gas penetrating ability of a coal layer and thus to reduce in gas release share.

It is this ambiguous character of advance face rate impact onto I that can explain the determined dependences between this index and Hurst one.

Given results allow to summarize that absolute values V in great extent form a trend (determined) component I, and fluctuations of the index determined according to H index allow assessing its changeability impact onto fluctuations I. So when predicting I it is reasonable to consider simultaneously absolute values V and H index of the first residuals of face advance rate.

REFERENCES

Briuzgina, O.V. 2004. *Impact assessment for distribution of gas capacity of coal-methane deposit onto parameters of gas kinetic pattern of a massif.* Conference proceedings "Natural resources of Kuzbass region": 26–27.

Polevshchikov, G.Y. & Kozyreva, Y.N. 2002. *Gas-dynamic pattern of excavated rock massif.* Mining information-analytical bulletin, 11: 117–120.

Elsner, J.B. & Tsonik, A.A. 1996. *Singular Spectrum Analysis.* A New Tool in Time Series Analysis. New York. London: Plenum Press: 164.

Kalush, Y.A. & Logikov, V.M. 2002. *Hurst index and its hidden properties.* Siberian journal of Industrial Mathematics. 402/10-12: 29–37.

Zinenko, A.V. 2012. *R/S analysis at stock exchange.* Business information, 3(2): 24–29.

Progressive Technologies of Coal, Coalbed Methane, and Ores Mining – Bondarenko, Kovalevs'ka & Ganushevych (eds)
© 2014 Taylor & Francis Group, London, ISBN: 978-1-138-02699-5

Substantiation of parameters of filtration flows in mining collapsed areas

I. Oshmyansky & L. Yevstratenko
Kryvyi Rih National University, Kryvyi Rih, Ukraine

ABSTRACT: The authors present their studies of the dynamics of filtration flows through a porous medium of aerodynamically active collapse areas under different physical conditions. The possibility of a mathematical description of filtration processes occurring in porous media of different types is also presented. According to the theory of one-dimensional filtering and on the basis of the solution of differential equations of motion of gas through a porous medium the authors obtained theoretical dependencies for defining basic parameters of filtration flows in terms of one-dimensional rectilinear parallel and plane-radial flows of collapse areas.

1 INTRODUCTION

The use of high-performance systems of mining development that includes caving of ore and overlying rock during extraction of mineral resources also distorts the equilibrium that exists in them, causing the development of displacements and deformations. Depending on the dimensions of the areas under development, shifting process can reach terrestrial surface. As we already know from our experience, changes in the states of stress and shifting of the rock mass and terrestrial surface occur at considerable distances that sometimes exceed the size of developed plots by several times. The main feature of the development of shifting process is the inevitable occurrence of collapse area (Figure 1) (Oshmyansky & Lapshin 1974) conditioned by the parameters of the ore bodies, the applied development technology and the properties of the host rocks.

Formation of collapse areas is connected with the occurrence of an active aerodynamic connection between underground works and surface, which in case of the suction method of ventilation is the cause of penetration of harmful and toxic contaminants along with the filtration flows of air into the underground development areas. Furthermore, the formation of breathable collapse areas can lead to partial or total disruption of the required ventilation in underground mining areas, units, chambers and certain places.

When developing measures destined to improve the efficiency of ventilation systems in conditions of occurrence of collapse areas it is necessary to determine reliable parameters of filtration flows in these areas.

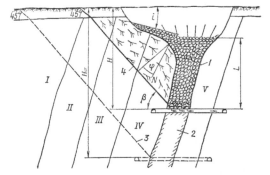

Figure 1. Scheme of shifting of the terrestrial surface and formation of a collapse area in the progress of extraction of ore deposits: 1 – collapse rocks; 2 – deposit; 3 – the shifting line of rocks of the latest collapse; 4 – the shifting line of rocks of penultimate collapse; 5 – cracks that occurred due to collapse; I – rocks of the upper formation; II – shales; III – martite hornfels; IV – hematite-martite hornfels; V – shales.

To justify the basic parameters of filtration flows it is necessary to clarify the scheme of their movement in the collapse areas, which will help develop methods of calculation that include all the basic phenomena occurring in the process of filtering the air through a porous medium.

2 MAIN SECTION

As a result of the performed research (Oshmyansky et al. 1975) it was found that the movement of filtration flows through collapse areas takes place according to plane-parallel and plane-radial flow patterns (Figure 2).

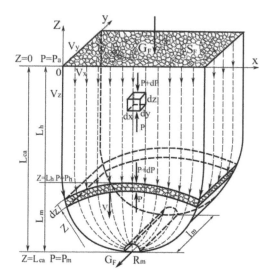

Z=0 P=P$_a$

Z=L$_h$ P=P$_h$

Z=L$_{ca}$ P=P$_m$ G$_F$ R$_m$

Figure 2. Model of filtration flows through a porous medium of collapse area.

The simplest rectilinear parallel circuit of air flow through collapse area is located in L_h. In case of this scheme, cross-sectional area of moving air flow remains constant, all air particles trajectories are parallel, and the filtration rates at all points in each transverse (perpendicular to the flow lines) section are equal to each other. The laws of motion along all paths in such filtration flow are the same, and therefore it is sufficient to study the motion along one of the paths that can be taken as a coordinate axis – the axis Z.

In an area with plane-radial flow the filtration rate varies with height and depends on the radius r of the hemisphere of filtration area relatively to the axis of developments adjacent to the collapse space in its bottom location (Oshmyansky et al. 1979). Air particles will move in vertical paths, radially converging to an adjacent mine place. The figure for flow lines in any horizontal plane will be the same, and for full flow characteristics it is sufficient to study the movement of the fluid in one horizontal plane. In one-dimensional plane-radial flow the pressure and filtration rate at any point depend only on the distance r of such taken point from the mining axis.

The described schemes of one-dimensional filtration flows allow to create simple models of real flows arising from the formation of the collapse areas during field development and to solve practical problems. The task of investigating the set filtration flow is to determine the following characteristics: rate of flow (or loss), pressure, velocity of filtration at any point, as well as the establishment of the law of motion of the particles of liquid or gas along their trajectories and determining a weighted average

volume of pore pressure space in the collapse rock.

To calculate the above-listed characteristics of one-dimensional filtration flow of fluid we apply an approach based on the use of differential equations and their solutions separately for rectilinear parallel and plane-radial flows of liquid and gas.

Filtration properties of the porous medium of collapse area depend on the granulometric composition, rate of rock pressure, breed fraction size, shape, roughness of the pore channels and are characterized, as is known, by the porosity coefficient K_p, and the permeability coefficient K. Actual shape and size of rock particles of real collapse areas, blown up rock in chambers, as well as the size and directions of the pore channels are sporadic, that is why it is common to use average values of velocity, pressure and flow during studies of the filtration properties of the porous medium in collapsed areas. Under this assumption, the actual flow of air is replaced by a fictitious filtration flow having the same aerodynamic parameters as the real one. There is a dependence between filtration velocity υ_F and average velocity υ_a in the pore channels

$$\upsilon_F = K_p \upsilon_a ,$$

where K_p – porosity coefficient.

When the question is posed in such way, in case of study of air motion through collapse areas it becomes possible to apply differential equations of air flow motion through porous media used in fluid dynamics.

The main task of aerodynamics is to find the fields of velocity, pressure and density in the flow of air moving under the influence of given external forces, i.e. finding the following five functions of coordinates and time

$$\upsilon'_x = f_1(x,y,z,t) ; \quad \upsilon'_y = f_2(x,y,z,t) ;$$

$$\upsilon'_t = f_3(x,y,z,t) ; \quad P = f_4(x,y,z,t) ; \quad P = f_5(x,y,z,t) .$$

The Equations of motion and continuity are sufficient to solve the main problem of fluid dynamics of any liquid in which density ρ and both viscosity coefficients depend only on the pressure P. In particular, this is true for an ideal incompressible fluid, ideal barotropic fluid, and isothermal movement of viscous liquid. In all other cases, to solve the basic problem of fluid dynamics it is necessary to use the extended system of equations consisting of the equations of motion, continuity, energy, and the state of fluid as well as equations expressing the dependences of the dynamic and kinematic viscosity on the fluid state parameters.

In order to solve the basic problems of filtration of fluid flow through porous media, let's use basic equations of motion in vector form in projections on coordinate axes for an ideal incompressible fluid (Aravin & Numerov 1953).

$$
\left.
\begin{array}{l}
\dfrac{\partial \upsilon_x}{\partial t}+\upsilon_x\dfrac{\partial \upsilon_x}{\partial x}+\upsilon_y\dfrac{\partial \upsilon_x}{\partial y}+\upsilon_z\dfrac{\partial \upsilon_x}{\partial z}=K_p\!\left(F_x+R_x\right)-\dfrac{1}{\rho}\dfrac{dP}{dx} \\[2mm]
\dfrac{\partial \upsilon_y}{\partial t}+\upsilon_x\dfrac{\partial \upsilon_y}{\partial x}+\upsilon_y\dfrac{\partial \upsilon_y}{\partial y}+\upsilon_z\dfrac{\partial \upsilon_y}{\partial z}=K_p\!\left(F_y+R_y\right)-\dfrac{1}{\rho}\dfrac{dP}{dy} \\[2mm]
\dfrac{\partial \upsilon_z}{\partial t}+\upsilon_x\dfrac{\partial \upsilon_z}{\partial x}+\upsilon_y\dfrac{\partial \upsilon_z}{\partial y}+\upsilon_z\dfrac{\partial \upsilon_z}{\partial z}=K_p\!\left(F_z+R_z\right)-\dfrac{1}{\rho}\dfrac{dP}{dz}
\end{array}
\right\}, \qquad (1)
$$

where F_x, F_y, F_z – vector F Cartesian coordinate system projections of the resultant field strength of mass forces having effect upon the fluid, N/kg; R_x, R_y, R_z – projections of resistance force of air movement through a porous medium on vector R axis, per mass unit of fluid, N/kg; P – pressure of the fluid in Pa; υ_x, υ_y, υ_z – projections of the fluid filtration velocity υ_F on the coordinate axis, m/s.

If we assume that there is a set motion that takes place during fluid filtration, i.e.

$$
\frac{d\upsilon_F}{dt}=0 \text{ and } \frac{d\upsilon_x}{dt}=\frac{d\upsilon_y}{dt}=\frac{d\upsilon_z}{dt}=0,
$$

then the Equations (1) shall have such form

$$
\left.
\begin{array}{l}
K_p\!\left(F_x+R_x\right)-\dfrac{dP}{\rho}dx=0 \\[2mm]
K_p\!\left(F_y+R_y\right)-\dfrac{dP}{\rho}dy=0 \\[2mm]
K_p\!\left(F_z+R_z\right)-\dfrac{dP}{\rho}dz=0
\end{array}
\right\} \qquad (2)
$$

For the conditions of the problem, the solution of which is shown by the scheme in Figure 2, 0Z axis is pointing upwards, so $F_z = F = -g$, $F_x = F_y = 0$ and the Equations (2) in vector form shall have the following look

$$
K_p\!\left(R-g\right)-\frac{dP}{\rho}dZ=0. \qquad (3)
$$

The resistance force R of air flow through the porous medium per mass unit of liquid during filtering may be expressed by binomial formula, wherein the force vector is directed in the opposite direction to the filtration rate

$$
R=-a\upsilon_F-b\upsilon_F^2, \qquad (4)
$$

where a and b – coefficients, determined experimentally at fluid filtration through a porous medium.

Then after substitution of (4) into (3) we shall obtain

$$
\frac{dP}{dz}=K_p\rho\!\left(-a\upsilon_F-b\upsilon_F^2-g\right).
$$

The results of numerous experimental studies performed on units, simulating a porous medium, as well as industrial experiments allowed to establish that the most appropriate way to describe filtering in violation of the law of Darcy is to apply binomial formula with the following values of coefficients (Oshmyansky et al. 1975, Charny 1956, Minsky 1951)

$$
-\frac{dP}{dz}=\mu\frac{\upsilon_F}{K}+u\rho_a\frac{\upsilon_F^2}{K}+\rho_a gK_p, \qquad (5)
$$

where K is permeability of the porous medium, m; u – channel shape parameter, depending on the geometrical characteristics of fractions of rocks in collapses, determining a violation of Darcy's law, m (u value is assumed to be constant when the pressure P is changeable and is determined experimentally); ρ_a – average density of air in the collapse space, kg/m³.

In case of the suction method of ventilation of mines through aerodynamically active collapse area under the influence of pressure generated by the fans, the air flow moves from top to bottom in the plane X 0Z in the direction of the axis 0Z (Figure 2), that is why $\upsilon_x = \upsilon_y = 0$, and $\upsilon_z = -\upsilon_F$.

In the area of rectilinear-parallel filtration flow with the length L_h from the surface we assume that

$\dfrac{dP}{dZ}=\dfrac{dP}{dL}=const$, and the filtration rate υF in this area in the vertical direction is determined by the expression

$$v_F = \frac{G_F}{S_F \rho_a}, \tag{6}$$

where G_F – air mass flow rate, kg/s; S_F – filtration area on a plot of rectilinear-parallel flow, m.

In view of (6) the basic equation of filtration motion (5) for rectilinear-parallel flow has the following form

$$dP = \left(\frac{\mu G_F}{S_F K \rho_a} + \frac{u G_F^2}{S_F^2 K \rho_a} - K_p \rho_a g \right) dl. \tag{7}$$

For the conditions of the considered problem in the area of rectilinear-parallel flow in case of a stationary motion of air with constant density ($v_F = const$ and $\rho_a = const$) the initial and boundary conditions will be as follows at $t = 0$, $Z = 0$, $P = P_a$; at $Z = L_h$, $P = P_h$, where P_a is the air pressure on the free collapse surface equal to atmospheric pressure, Pa; P_h – air pressure at the level of the border between rectilinear parallel and plane-radial flows of air in the collapse space, Pa.

At the same time we assume that the permeability of collapsed rock in this area is constant ($K = const$).

By integrating (7) for the given boundary conditions, we obtain

$$P_h - P_a = \frac{\mu L_h G_F}{S_F K \rho_a} + \frac{u L_h G_F^2}{S_F^2 K \rho_a} - K_p \rho_a g L_h. \tag{8}$$

Let's denote $K_p \rho_a g L_h = \Delta P_a$ and

$$P_h - P_a = h'ca,$$

where ΔP_a – increment of aerostatic pressure at the area of rectilinear-parallel filtration flow, Pa; $h'ca$ is the depression of collapse area in case of rectilinear-parallel filtration flow, Pa.

Given the adopted notation expression (8) has the following form

$$h'_{c.a.} = \frac{\mu L_h G_F}{S_F K \rho_a} + \frac{u L_h G_F^2 sign(G_F)}{S_F^2 K \rho_a}, \tag{9}$$

where sign (G_F) is the sign character of filtration flow rate; at sign (G_F) = -1 we have air incoming from the surface of the mine in case of the suction method of ventilation; at sign (G_F) = $+1$ we have outgoing air from the mine to the surface at the discharge or combined methods of ventilation.

On the basis of (9) the value of mass flow of filtration leakage through collapse area on the plot

with a rectilinear-parallel flow is determined from the expression

$$G_F = \frac{\mu S_F}{2u} \left(\sqrt{1 + \frac{4u \rho_a K h'_{c.a.}}{\mu L_h}} - 1 \right).$$

On a plot with plane-radial flow the filtration rate varies with height and depends on the radius r of the hemisphere of filtration area relatively to the axis of developments adjacent to the collapse space in its bottom location (Figure 2)

$$v_F = \frac{G_F}{\pi r l_m \rho_a}, \tag{10}$$

where l_m – length of pit, m.

In view of (10) filtration law equation for this section of collapsed space in differential form will have the following form

$$dP = \left(\frac{\mu G_F}{\pi r l_m K \rho_a} + \frac{u G_F^2}{\pi^2 r^2 l_m^2 K \rho_a} - K_p \rho_a g \right) dr. \tag{11}$$

By integrating the Equation (11) with initial conditions $P(r = L_m) = P_h$, $R = R_m$ = P_m, we obtain

$$P_m - P_h = \frac{\mu G_F}{\pi l_m K \rho_a} ln \frac{L_m}{R_m} + \frac{u G_F^2}{\pi^2 l_m^2 K \rho_a} \times$$

$$\times \left(\frac{1}{R_m} - \frac{1}{L_m} \right) - K_p \rho_a g (L_m - R_m). \tag{12}$$

Denoting in the formula (12)

$$K_p \rho_a g (L_m - R_m) = \Delta P_a'',$$

$$P_m - P_h + \Delta P_a'' = h''_{c.a.},$$

and neglecting the value of $1/L_m$ compared to $1/R_m$ we obtain

$$h''_{c.a.} = \frac{\mu G_F}{\pi l_m K \rho_a} ln \frac{L_m}{R_m} + \frac{u G_F^2 sign(G_F)}{\pi^2 l_m^2 R_m K \rho_a}. \tag{13}$$

In the area of plane-radial flow from $R = L_m$, to $R = R_m$ the pressure gradient h/L and the filtration rate v_F are not permanent, as the filtration area $S_F' = \pi r l$ decreases as the air flow approaches the adjacent workings.

Then on the basis of formulas (9) and (13) general equation for the law of filtration through the collapses of mines shall be

376

$$h_{c.a.} = h'_{c.a.} + h''_{c.a.} = \frac{\mu G_F}{K\rho_a}\left(\frac{L_h}{S_F} + \frac{ln\frac{L_m}{R_m}}{\pi d_m}\right) +$$

$$+ \frac{uG_F^2 sign(G_F)}{K\rho_a}\left(\frac{L_h}{S_F^2} + \frac{1}{\pi^2 l_m^2 R_m}\right). \quad (14)$$

If in formula (14) we denote

$$R_l = \frac{\mu}{K\rho_a}\left(\frac{L_h}{S_F} + \frac{ln\frac{L_m}{R_m}}{\pi d_m}\right),$$

$$R_K = \frac{u}{K\rho_a}\left(\frac{L_h}{S_F^2} + \frac{1}{\pi^2 l_m^2 R_m}\right),$$

then the formula for calculating depression of collapsed mining space in a simplified form shall be

$$h_{c.a.} = R_l G_F + R_K G_F^2. \quad (15)$$

The solution of Equation (15) relatively to mass flow of filtration air leakage through the collapses in the combined motion allows to obtain the following expression

$$G_F = \frac{\mu\left(\frac{L_h}{S_F} + \frac{ln\frac{L_m}{R_m}}{\pi d_m}\right)}{2u\left(\frac{L_h}{S_F^2} + \frac{1}{\pi^2 l_m^2 R_m}\right)}\left[\sqrt{1 + \frac{4K\rho_a u h_{c.a.}\left(\frac{L_h}{S_F^2} + \frac{1}{\pi^2 l_m^2 R_m}\right)}{\mu^2\left(\frac{L_h}{S_F} + \frac{ln\frac{L_m}{R_m}}{\pi d_m}\right)^2}} - 1\right]. \quad (16)$$

Equations (14) and (16) are obtained from the condition that permeability of the porous medium K shall not alter at mine collapse areas between $Z = 0$ and $Z = L_r$, since it is impossible to detect the borders of the individual layers of differing permeability within collapses. Adoption of $K = const$ allows to solve a system of Equations, although the assumptions affect the accuracy of mathematical calculations. Expressions (9) and (13) were obtained for one single collapse and at the presence of one adjacent mine pit. If there are several aerodynamically unrelated collapses at one mine, then the parameters L_h, S_F, K, l_m, u, μ, ρ_a, R_m of each one of them should be taken separately. In those cases when at each ventilation area of a pit there are extensive aerodynamically active collapses with a large number of former worksites of haulage, ventilation and transportation horizons, then the calculations by formulas (14) and (16) should include average values of aerodynamic parameters (L_h, K, u, μ, ρ_a), the total value S_F, the total length of all pits l_m at the average value of R_m. In this case, at some average value h_{ca} for the entire vent area, by the formula (16) we may calculate the total filtering air flow through the collapses.

Reliability of the values G_F, h_{ca}, v_F calculated by the proposed formulas will be determined by reli-

ability of such initial aerodynamic parameters as permeability coefficient K and channel shape coefficient u, which should be determined in experimental way.

3 CONCLUSIONS

As a result of the performed research, the basic regularities of the filtration movement of air through the aerodynamically active collapse areas at mines of Kryvbass with rectilinear parallel and plane-radial flow, taking into account the properties of the porous medium as well as gas motion mode existing therein.

Offered are the dependencies required to justify the basic parameters of filtration flows in areas of mine collapses and allowing to predict accurately enough the filtration processes which may occur in conditions of mining operations.

REFERENCES

Oshmyansky, I.B. & Lapshin, A.E. 1974. *Calculation of filtration air leakage through collapse areas in conditions of mines of Kryvbass*. Higher Education News. Mining Journal, 1974, 4: 69–73.

Oshmyansky, I.B., Lapshin, A.E. & Los, O.V. 1975. *Resistance law the air for moving through collapse areas at the mines of Kryvbass*. Higher Education News. Mining Journal, 6: 62–67.

Oshmyansky, I.B., Lapshin, A.E. & Ovchinnik, I.T. 1979. *Aerodynamic parameters of the filtration movement of air in mine collapses of Kryvbass*. Higher Education News. Mining Journal, 4: 73–76.

Aravin, V.I. & Numerov, S.N. 1953. *Theory of movement of liquids and gases in non-deformable porous environment*. Moscow: Gosgortekhizdat.

Charny, I.A. 1956. *Fundamentals of underground hydraulics*. Gostoptekhizdat.

Minsky, E.M. 1951. *On turbulent filtration in porous media*. Reports of the Academy of Science of the USSR, Vol. 78, #3.

Statistical analysis and numerical modelling of dust pollution. Case study the city Dniprodzerzhyns'k

D. Rudakov & A. Lyakhovko
National Mining University, Dnipropetrovs'k, Ukraine

ABSTRACT: We analysed monitoring data on air pollution in the Dniprodzerzhyns'k city during 2008–2012 in terms of changes of on-ground dust concentration depending on meteorological conditions and emissions into the atmosphere. The most of particulate matter in the city (over 95%) is released by Dniprovs'kyi metallurgical plant named after Dzerzhyns'kyi. The statistical analysis carried out with five-year records collected at four monitoring points has revealed no correlation between mean monthly values of dust concentration and wind directions. Thus, there is the need for applying a mathematical model to describe long-time dust dispersion with account for time scale of air quality monitoring and wind distribution over a long period of time. Such a model developed earlier has been applied to describe the patterns of on-ground dust distribution in urban areas; the model has been identified using the time series of sampled and averaged monthly dust concentrations at the monitoring points. Numerical modelling allowed predicting the environmental effect of upgrading gas cleaning facilities at the plant as a result of reducing particulate matter emissions in the city.

1 INTRODUCTION

Most of the particulate matter released into the atmosphere in Ukraine falls within a few industrialized regions, with the Dnipropetrovs'k region having a significant share of emissions (Ministry of Ecology... 2012). As a rule, dust is emitted by numerous industrial enterprises in the region using outdated production technologies; there are a large number of process and fugitive emissions. In fact, available gas cleaning facilities have insufficient capacity, they are often obsolete, and now are unable to operate at high cleaning rates. This causes enormous damage to the environment, both near their location and in surrounding areas and, sometimes, even in neighbouring countries due to transboundary transport of contaminants in the atmosphere (United Nations 1979).

To control air quality and reduce air pollution efficiently environmental engineers need to have the reliable data on contaminant emissions, their compositions and concentrations as well as meteorological data. These data are commonly required in predictive models to estimate contaminant dispersion in air over large densely populated areas for long periods of time. Thus, multivariant predictive simulations of dust dispersion will enable assessing the feasibility and environmental impact of engineering solutions reducing the air pollution level in industrial cities like upgrade or replace of dust cleaning facilities.

The up-to-day models of pollutant dispersion in the atmosphere applied in the practice (Babkov & Tkachenko 2011) allow quantifying, primarily, the transport of contaminant in the atmosphere under stable meteorological conditions, which meets the conditions of short-time accidents and dispersion during some hours or days. However, reliable assessment of long-term effects as a result of facility modernization on industrial enterprises has to be based on applying the models using wind parameters statistically averaged over a relatively long period, commonly a year or a month. This will allow comparing the average dust concentrations before and after taking technological measures and making well-founded conclusions, f.e. whether the planned investments to upgrade dust cleaning facilities will be environmentally efficient and acceptable.

A case in point is the ambient air pollution in the city of Dniprodzerzhyns'k that is characterized by frequently occurring high concentrations of contaminants. Particularly, dust concentration in the city permanently exceeds 1.1 of a maximum allowed concentration (State department of environmental protection... 2012). Among more than 50 industrial enterprises of the city Private Limited Company Dniprovs'ky metallurgical plant named after Dzerzhinsky (DMP) remains the main source of air pollution with dust because it releases about 96% of local dust emissions. Currently the gas emissions at the plant are cleaned with facilities and scrubbers at 66.5% of the total amounts with the efficiency ranging from 85 to 98% (State environmental protection ... 2012). Meanwhile, the small particulates of the size less than 10 microns highly dangerous to human health (Zhidetsky et al. 2000) are currently trapped by available cleaning facilities insufficiently and mostly emitted into the atmosphere.

Thus, there is the urgent need for upgrading dust cleaning facilities of DMP, which is proposed to perform by upgrading the unit" Electrical cyclone filter" (Patent 77681 Ukraine... 2012). This unit has the higher cleaning rate (95 to 99% depending on dust fraction partitioning) than available facilities at the plant; the unit can be especially effective to settle harmful fine particulates. Yet, the large-scale introduction of this unit has to be assessed in terms of the expected environmental effect.

2 FORMULATING THE PROBLEM

The study aims to identify recent trends and patterns of dust pollution in the atmosphere of Dniprodzerzhyns'k and assess the reduction of on-ground concentration of dust in the city in case of upgrading gas cleaning facilities of Dniprovs'kyi metallurgical plant with a unit "Electrical cyclone filter".

Achievement of this aim requires, first of all, statistical analysis of air quality monitoring data, particularly, dust concentration measured at the monitoring network points in different locations of the city. To solve this task we used the data of environmental monitoring collected by the state Laboratory for monitoring atmospheric pollution (LMAP) in Dniprodzerzhyns'k. Then we identified the mathe-

matical model of long-time dispersion of particulate matter in the atmosphere above the heterogeneous land surface and distributed parameters of wind (Rudakov 2011). This allowed predicting the reduction of particulate matter concentration in air in case of upgrading the available gas cleaning facilities with the unit "Electrical cyclone filter".

3 ANALYSIS OF THE ATMOSPHERE MONITORING DATA IN DNIPRODZERZHYNS'K

Four stationary points of LMAP are positioned in different districts of Dniprodzerzhyns'k (Figure 1) at the distances of 1.6 to 7.7 km away from the main source of dust emission that consists of two closely located pipes, each 100 m high, of the agglomeration process in the metallurgical plant on the bank of the river Dnipro. Positioning of the observation points in different areas of the city – two points in the downtown, one in the northwest residential area on the left bank of the river and one in the industrially charged southeast area –allows getting a valuable data on air pollution in the city as well as evaluating the contributions of other local dust sources. The laboratory LMAP is sampling air for analysing concentrations of harmful species on the daily basis. Then the measured concentrations are processed and summarized to calculate monthly concentrations, besides, the mean maximum concentrations for the same period are derived. This allows clearer understanding of long-time air pollution trends over a studied period of time.

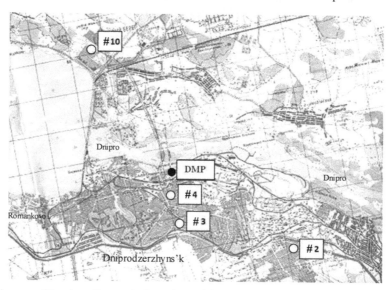

Figure. 1. Location map of the air quality monitoring points and the emission pipes of Dniprovs'kyi metallurgical plant.

The annual mean dust concentrations determined during 2008–2012 at the monitoring network points of LMAP have been compared with the annual average emissions of dust released by the metallurgical plant (State management of protection... 2013) (Figure 2). Despite some decrease in particulate matter amounts emitted by the plant during this period there is the tendency of growing dust concentrations in air last three years. The likely reasons for this trend may be the multifactor influence of meteorological conditions and non-estimated contributions of other particulate emissions (e.g., generated by cars) appeared or increased last time both within and outside the city.

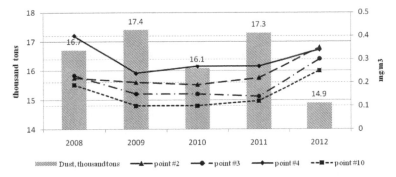

Figure. 2. Dynamics of dust emissions of DMP and dust concentration in air in Dniprodzerzhyns'k during 2008–2012.

4 CORRELATION ANALYSIS OF MONITORING RESULTS

The visual analysis of variations in monthly mean dust concentration (Figure 3) reveals no evident or apparent interrelations between the wind direction from the emission source to an observation point and concentration measured at the same point. Under normal conditions we could assume that the concentration at the point reaches a maximum when the wind blows from the emission source to the observation point.

As can be seen from Figure 3 the monthly maximal concentrations at all points exceed mean values by 2–4 times with clearly visible correlation in time. However, the increases in concentration at the monitoring points seem to be independent on dominating wind direction during the same period. Moreover, dust concentration in the ambient air has been growing last years despite the reduction in dust releases by the plant.

The cause of this situation may be the exposure to other sources of dust emissions into the atmosphere, likely, located outside the city. However, according to the recent data on dust emissions in the city (State department of environmental protection... 2012) 96.4% of all particulate matter in Dniprodzerzhyns'k has been released by DMP; therefore the hypothesis on external impacts on dust concentration within the city is apparently unlikely. The nearest similar industrial sites releasing big amounts of particulates comparable to that emitted by DMP are located in Dnipropetrovs'k on the distance of 30-50 km from the monitoring points.

The cumulative impact of wind parameters on dust concentration during a long period of time depends on many constituents that are difficult to estimate. Thus, the statistical analysis has been conducted in order to try correlating monthly mean concentrations of dust at all observation points and the angle between the wind direction and the ray from the emission pipes to the measurement point. The statistical hypothesis checked is that the dust concentration reaches a maximum when this angle tends to zero. This trend could appear for all wind velocities, likely at different correlation rates. The results of calculations are demonstrated in Table 1.

The analysis of results reveals the absence of any statistically significant correlation between mean monthly values of dust concentration and wind direction at the confidence limit of 95%. The absence of correlation between these parameters averaged within a month demonstrates the necessity to apply the dust dispersion models with statistically averaged wind parameters on the time scale conforming with the time scale air quality monitoring.

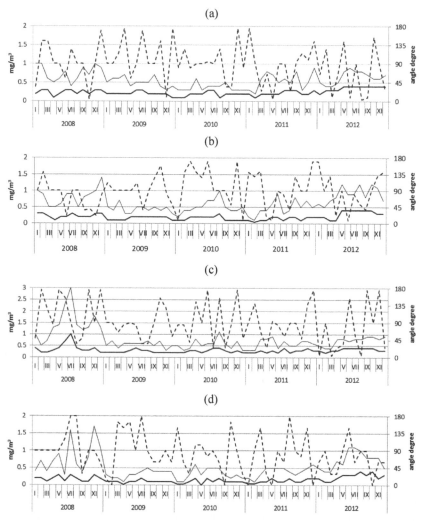

Figure 3. Variations of dust concentration and wind directions versus time at the monitoring network points during 2008–2012. Mean concentration (solid curve), maximum concentration (thick line) at the network points: (a) point #1, (b) point #2, (c) point #3, (d) point #4. The values of dust concentrations are in mg/m³ and attributed to the left axis. The angle between the direction of the wind and the direction from the emission source to the observation point is shown with a dashed line; its values are in angle degrees and attributed to the right axis.

Table 1. The results of correlation analysis of the air quality monitoring.

Point #	Distance between the monitoring point and emission pipes, km	Direction from the emission pipes to the monitoring point, degree (bearings)	Correlation coefficient
2	6.7	145 (SSE)	−0.125
3	3.5	189 (S)	−0.101
4	1.6	185 (S)	−0.002
10	7.7	330 (NNW)	−0.172

0° corresponds to the direction to North; SSE is the direction to South-South-East; SSW is the direction to South; NNW is the direction to North.

5 MODELLING OF LONG-TIME DUST DISPERSION FROM THE EMISSION PIPES

The model of long-time dispersion of particulates in the atmosphere (Rudakov 2011) has been applied for calculating the mean annual dust concentration on the territory of Dniprodzerzhyns'k and surrounding areas. The model is based on the advection-dispersion transport equation and uses the annual average distribution of local directions and frequencies of wind; also it takes into account the elevated emission height due to effluent gas initial velocity and higher temperature of gas in comparison to ambient air.

Two pipes of the agglomeration factory of DMP are assumed to be the single source of dust emitted by the plant because of their close proximity to each other (about 100 m) on the scale of the city. Actually, we estimated the contributions of DMP into the increase in dust concentration; therefore, the calculated concentrations have to be close to those derived from the monitoring data because emissions of DPM are still overriding in the city.

The particulate fractions of medium size (less than 100 microns) dominate in the emitted dust and constitute about 78% of its total amount. The mass content of particulates smaller than 40 microns is estimated at 14–15%. In case of modernization and/or replacing the available cleaning facilities of only agglomeration production with the developed "Electrical cyclone filter" (Rudakov & Lyakhovko 2013) one could expect the reduction of dust releases by 5913 thousand tons or about 40% of the emitted amount in 2012; the mass of finest particles less than 10 microns can be reduced by 1508.7 thousand tons.

To identify the dust dispersion model we used the data on mean monthly dust concentration presented in Section 4 (Figure 3). We calculated mean annual concentrations of dust for each point within the specified time interval using the model. To assess the reliability of the calculated concentrations the standard deviations and variances have been calculated for the experimental time series. The results are brought together in Table 2.

Table 2. Estimates of mean annual dust concentration at the monitoring points in Dniprodzerzhyns'k for 2008–2012.

Point #	Mean concentration derived from monitoring data, mg/m^3	Standard deviation of measured concentration, mg/m^3	Variance of mean monthly concentration, %	Calculated concentration by the model for current dust releases, mg/m^3	Predicted concentration after applying "Electrical cyclone filter", mg/m^3	Expected reduction of dust concentration, %
2	0.24	0.09	37.5	0.15	0.10	34
3	0.19	0.09	47.4	0.22	0.14	37
4	0.30	0.30	100	0.36	0.11	59
10	0.15	0.09	60	0.12	0.10	17

Then, we calculated dust concentration on urban areas using the data on particulate fraction distribution and averaged wind parameters. The mean annual distribution of on-ground concentration of the dust emitted by DMP is shown in Figure 4(a). After that we assessed how on-ground concentration would change in case of hypothetical replacing the available cleaning facilities of DMP with the unit "Electrical cyclone filter", which would reduce dust emissions and change the distribution of particle size. The map of predicted on-ground dust concentration in case of applying this unit at the plant is shown in Figure 4(b).

Comparative analysis of the modelling results and the experimental data allows drawing the following conclusions.

• the area where mean dust concentration in the air exceeds the sanitary limit of 0.15 mg/m^3 covers urban areas completely, which is in a good agreement with the monitoring data (State department of

environmental protection... 2012). The calculated dust concentrations at the monitoring points are within the range of concentrations derived from the air quality monitoring data taking into account mean values and variances; at the time being this concentration ranges from 1.6 to more than 3 sanitary limits within the city borders;

• the area of maximum dust concentrations exceeding 0.3 mg/m^3 is situated in the downtown on the distance of 1.5–2 km from the emission pipes to the South-South-West direction, which can be explained by dominating wind direction from North-East;

• the variance of mean monthly dust concentrations at the monitoring points ranges from 37.5% to 100%, the variance falls with increasing distance from the emission pipes;

• the model predicts the maximum concentration at the point #4 in downtown (0.36 mg/m^3) and the minimum concentration at the points #2 (0.15

mg/m³).and #10 (0.12 mg/m³) that are located in suburbs on the left and right banks of Dnipro. These values are slightly less than those measured, which can be explained by additional contributions of other minor sources of particulate matter in the city except DMP (cars, other industrial enterprises);

• the modelling results demonstrate the significant positive environmental impact after upgrade of gas cleaning facilities at the agglomeration factory of DMP; the estimated reduction of dust concentrations at the monitoring points would range from 17% in suburbs to about 60% in downtown.

(a)

(b)

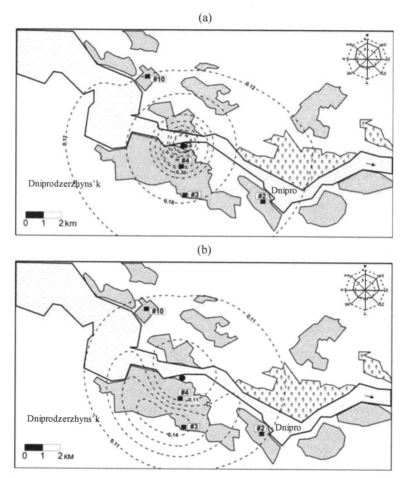

Figure 4. Calculated on-ground dust concentration in the city: (a) estimation for current dust emissions, and (b) after hypothetic application of Electrical cyclone filter: P – emission pipes of the plant; ⧄ residential areas; green zones; rivers; – 0.11 – . isolines of dust concentration, mg/m³.

6 DISCUSSION AND CONCLUSIONS

1. The overall study aim was to identify the previously developed model of long-time dust dispersion in the atmosphere on the example of the city of Dniprodzerzhyns'k based on the available data of air quality monitoring. The model is based on the advection-dispersion equation and uses mean annual distributions of local directions and frequencies of wind velocity; besides, it takes into proper account the emission parameters and particulate size distribution.

2. We have analyse the time series of mean monthly dust concentrations measured at the points operated by the laboratory LMAP and the data on particulate matter amounts released by DMP during the period 2008–2012. The obtained results have evidenced inaccessible monthly and annual mean dust concentrations at the monitoring points exceeding 0.3 mg/m³ in downtown (2 times more than the

sanitary limit for dust) despite some reduction in the amounts of particulates released by DMP from 18.6 to 15.7 thousand tons. The statistical analysis has established no significant correlation between monthly mean values of dust concentration and wind direction at the monitoring points; the concentration variance diminishes three times with the increasing distance from the emission pipes.

3. The model of long-time dust dispersion in air was used to clarify the patterns of on-ground concentration of dust in the city and predict the reduction of this concentration in case of hypothetical application of the unit "Electrical cyclone filter" developed by the authors. The model allowed calculating mean concentrations of particulates in air for long periods of time using the statistically averaged wind parameters adhered to the time scale of air quality monitoring. The identified model yielded the concentration values close to those derived from monitoring data with account for minor contributions of other particulate sources.

4. As a result of modelling the pattern of dust dispersion in the city and suburbs from the emission pipes has been estimated. The area of maximum concentration exceeding 0.3 mg/m^3 is located in the downtown on the distance of 1.5–2 km to South-South-West from the emission point. Mean annual concentration decreases to the city borders and does not exceed the sanitary limit on the distances far from 7 km of the emission pipes.

5. The identified model allowed estimating the effect of reducing the dust concentration in Dniprodzerzhyns'k and surrounding areas in case of modernizing the available gas cleaning facilities of the agglomeration factory of DMP. The reduced dust concentration is expected to range from 0.1 to 0.14 mg/m^3 over urban areas, which is slightly less than the sanitary limit.

6. The model is proposed to use for more detailed estimations of the environmental effect on air quality in Dniprodzerzhyns'k and surrounding areas in case of possible upgrade of gas cleaning facilities at DMP and other industrial enterprises with similar technological processes.

REFERENCES

Ministry of Ecology and Natural Resources of Ukraine. 2012. National Report on the State of Environment in Ukraine in 2011. Kyiv.

United Nations. 1979. Convention on Long-range Transboundary Air Pollution. [Online] Available from: http://www.un.org/ru/documents/decl_conv/conventions/transboundary.shtml [Accessed from 13th November 1979].

Babkov, V. & Tkachenko, T. 2011. *Analysis of mathematical models of the spread of species from point sources.* Scientific works of Donetsk National Technical University, 13(185): 147–155.

State department of environmental protection in the Dnepropetrovsk region. 2012. Environment conditions. Information and analytical review. Dnipropetrovs'k.

State environmental protection public administration in Dnepropetrovsk region. 2012. Ecological passport of Dnepropetrovsk region. Dnipropetrovs'k.

Zhidetsky, V., Dzhigirey, V. & Melnikov, A. 2000. *Basics of labour protection.* Lviv: Afisha.

Patent 77681 Ukraine, МПК B03 C 3/15 Electrical cyclone filter / Declared by Lyakhovko, A.D. & Rudakov, D.V. Owner National Mining University. №201209233; declared on 27.07.2012; published 25.02.2013, Bulletin, 4.

Rudakov, D.V. 2011. *Estimation of Mitigation Measure Effects on Air Pollution in Industrial Cities.* Proc. NATO ARW "Stimulus for human and social dynamics in the prevention of catastrophes", Yerevan, 5-8 October 2010. 257–264.

State Management of protection of the environment in Dnipropetrovs'k region. 2013. Regional report on the state of the environment in Dnipropetrovs'k region in 2012. Dnipropetrovs'k.

Rudakov, D.V. & Lyakhovko, A.D. 2013. *The prospects of application of Electrical dust cleaner to reduce the amounts of dust emitted by metallurgical enterprises. Case study Dniprodzerzhyns'k.* Sci. Proc. of National Mining University, 41: 134–139.

Progressive Technologies of Coal, Coalbed Methane, and Ores Mining – Bondarenko, Kovalevs'ka & Ganushevych (eds)
© 2014 Taylor & Francis Group, London, ISBN: 978-1-138-02699-5

New approaches to the problem of sudden outburst of gas and coal dust

V. Portnov, R. Kamarov & I. Shmidt
Karaganda State Technical University, Karaganda, Kazakhstan

V. Yurov
Karaganda State University, Karaganda, Kazakhstan

ABSTRACT: The questions of detection zones with increased coal seam gas recovery, related to the problem of sudden release of gas and coal are considered. Theoretical models based on the theory of catastrophes, statistical physics, thermodynamics are offered. The relationship between gas content of coal seams and experimentally identified by a geophysical method studies are received. Based on this, the substantiation of the use of geophysical methods for detecting zones with increased coal seam gas recovery is presented. The obtained results possess an interest to specialists in the field of commercial methane production.

1 INTRODUCTION

Zones with high gas recovery, in most cases, emerge with the sudden emission of methane in mines at the opening of the coal seams. The most likely cause of the sites with high gas recovery is the local development in coal seams of fluid conversion processes of coal in conjunction with stress-state state of coal and rocks (Trufanov et al. 2004).

According to the author of the work (Gritsko 2007), the study of this problem is now almost completely absent from modern geological, geophysical and geochemical aspects, with all their most powerful techniques and accumulated materials. The formation kinetics processes, selection and release of methane from the coal bed are explored. Remain unexplored and geomechanical processes leading to geomechanical education or the release of methane. It is unclear why some tectonic emissions take place, and an overwhelming majority of violations – do not.

In this work, we considered the use of geophysical methods for detecting zones with high gas recovery. Also we have considered some model presentation on the sudden release of gas and coal.

2 HIGHLY EXPLOSIVE ZONES AND CATASTROPHE THEORY

Sources of catastrophe theory are a theory of smooth mappings of the Whitney and the theory of Bifurcations of dynamical systems the Poincare and Andronova (Arnold 1990).

Disasters are called intermittent changes in the form of a sudden response to smooth changes in the external environment.

Catastrophe theory can be used in the technical tasks for the mathematical modelling of processes, which can be a disaster. Sudden release of gas and coal are also a disaster.

For the mathematical modelling of the possible consideration of three objects:

– purpose of functioning
– one or two coordinates
– one or more control options

In our consideration the purpose of the operation is destruction of hazardous area (zones with higher gas recovery) and coordinate the process of destruction speed of the dissipation of free energy F of this zone, due to the formation of (N) elementary lesions destruction, proportional number of cracks, pores, i.e. proportional to the fracture and porosity of zones with high gas recovery.

Here you must make the following observation. Sorption ability of coal chemical zone with high gas recovery σ here must be understood within the model of coal fluidization and coal containing rocks.

For single coordinates $\dot{F}$ and two governors of the parameters in the theory of catastrophes, there is only one standard, canonical correlation for writing based on the function

$$\dot{F}(N) = 0.25N^4 - 0.5\rho N^2 - \sigma N , \qquad (1)$$

where $\dot{F}(N)$ – potential function, which is the energy education N pockets of devastation, the Disaster. the potential function is called the disaster a "build". the Assembly has in the undercritical field of a steady state equilibrium (one hole a potential function) and in the aftercritical area, two stable and one unstable equilibrium State (i.e. two poles separated by a hill).

Undercritical and aftercritical fields are set to the following set of options

$$\rho < 0, \quad -\infty > \sigma > \infty$$

$$\rho > 0, \quad |\sigma| > \frac{2\rho}{3}\sqrt{\frac{\rho}{3}} \quad \rho > 0, \quad |\sigma| < \frac{2\rho}{3}\sqrt{\frac{\rho}{3}}. \quad (2)$$

undercritial field   aftercritical field

Equation (1) sets the static model of fracture zones with high gas recovery. For a dynamic model, we assume a zone with high gas recovery gradient system. This means that the potential function $\dot{F}(N)$ seeking for an extremum. The gradient of the function $\dot{F}(N)$ is equal to

$$grad\dot{F}(N) = \frac{\partial\dot{F}(N)}{\partial N} = N^3 - \rho N + \sigma. \quad (3)$$

Further, to a first approximation, we assume that the rate of increase in free energy is proportional to the rate of growth of pockets of destruction. Then you can see that

$$\beta\frac{dF}{dt} = \frac{\partial\dot{F}(N)}{\partial N} = N^3 - \rho N + \sigma, \quad (4)$$

where β – factor of proportionality; $(d)^2$ $(F)/dt^2$ – second derivative of the free energy.

In the mechanics of such an equation of motion for the viscous friction. In this case, the process of transition from one equilibrium State to another is smooth, like a S-shaped logistic curve.

Figure 1 shows the results of solving differential equations (4) graphics for the respective cases are marked with numbers 1 and 2 setting for both cases, $\rho = 3 = const$. Choice of parameter values is optional in this case and our aim is to illustrate the general approach. The coefficient $\beta = 10$, it defines a time scale and agrees with the dimension variables. Parameter $\sigma = 3 = const$.

In the first case, as can be seen from the graphs, speed of the dissipation free energy is growing around the S-curve, and reaches its maximum, approximately equal to 9 units.

In the second case, the modifier is uniformly reduced by the law $\sigma = c - \gamma t$.

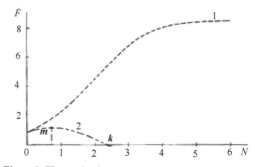

Figure 1. The result of solving differential Equations (4).

As can be seen from the graph, to the point m dissipation rate of free energy still growing, though not as fast as in the first case. Then the dissipation rate of free energy starts to decrease, and at k becomes zero, i.e. area with increased gas recovery is a sudden release. Obviously, with such a relatively simple model of fracture zones with enhanced gas recovery the possibility of transition from one S-curve to another can be qualitatively assessed.

3 STATISTICAL CLUSTERING MODEL OF ELEMENTARY LESIONS DESTRUCTION IN THE AREA WITH HIGH GAS RECOVERY

Zones with high gas recovery, as already was mentioned above, are a highly defective carbon structure with pores, cracks, fluids, etc. increase the sorption capacity of zones with high gas recovery contributes to the clustering structure defects (Consolidation) into larger aggregates. We will look at this issue from the point of view of statistical physics.

Consider a zone with high gas recovery number of m defect. Let the distance between the defects is equal and equal to R. Describe around each defect 0 sphere of radius r. Let the number of defects density in this area is n_0, then the probability of $W_0(r)$ the closest the defect is located from defect 0, it is not difficult to get from classical statistical physics, and it is equal to

$$W_0(r) = 4\pi n_0 r^3 \exp\left[-4\pi n_0 r^3 / 3\right]. \quad (5)$$

The probability of finding the N_0 defects in the 0 zone radius r is

$$W_{N_0}(r) = \prod_{k=1}^{N_0} W_k(r) =$$

$$= (4\pi n_0)^{N_0 r^3 N_0} \exp\left[-4\pi N_0 n_0 r^3 / 3\right]. \quad (6)$$

388

The probability (6) on the other hand, is defined define as the ratio of the number of defects N_0 defect zone 0 to the total number of defects in the leased area – $Q_0 = 4/3\pi(n)_0 R^3$

$$p_0 = \frac{N_0}{Q_0} = (4\pi n_0)^{N_0 r^3 N_0} \exp\left[-4\pi n_0 r^3 /3\right]. \quad (7)$$

For a system of m defects are

$$p_0 = (4\pi n_0)^{N_0 r^3 N_0} \exp\left[-4\pi N_0 n_0 r^3 /3\right] = \frac{N_0}{Q_0},$$

$$p_1 = (4\pi n_1)^{N_1 r^3 N_1} \exp\left[-4\pi N_1 n_1 r^3 /3\right] = \frac{N_1}{Q_1}, \quad (8)$$

- -

$$p_m = (4\pi n_m)^{N_m r^3 N_m} \exp\left[-4\pi N_m n_m r^3 /3\right] = \frac{N_m}{Q_m}.$$

For the entire zone with high gas recovery to the number of defects 0, 1, 2, ...(m) we have

$$P = \prod_{i=0}^{m} p_i = \prod_{i=0}^{m} (4\pi n_i)^{N_i r^3 N_i} \times$$

$$\times \exp\left[-4\pi N_i n_i r^3 /3\right] \frac{\prod\limits_{i=0}^{m} N_i}{\prod\limits_{i=0}^{m} Q_i} \quad (9)$$

system of Equations (8) and (9) is a system of transcendental equations that can only approximate numerical methods.

Therefore, you can do a numerical rating, based on the real situation and Equation 1 (8)

$$ln N_0 - ln Q_0 = N_0 ln(4\pi n_0) +$$

$$+ 2N_0 ln r - \frac{4\pi}{3} N_0 n_0 r^3. \quad (10)$$

The assessment provides that the first term on the left side of the Equation (10) and the first two members of the right side is negligible. The result will be

$$N_0 = \frac{3 ln Q_0}{4\pi n_0 r^3}. \quad (11)$$

Since $4\pi r^3 /3 = (V)$ and $ln n_0 > ln R$ from (10) we have

$$V_0 = \frac{ln n_0}{n_0 N_0}. \quad (12)$$

Taking the response as a function of the volume of the cluster

$$V_0 = c \frac{kT}{G_0} \cdot N_0. \quad (13)$$

Comparing (12) and (13), we get

$$N_0 = \left(\frac{1}{c} \cdot \frac{ln n_0}{n_0} \cdot \frac{G_0}{kT}\right)^{1/2}. \quad (14)$$

Formula (14) is the equilibrium value of the number of defects in the cluster.

Thus, the defectiveness of the zone with high gas recovery (porosity, fracture, etc.) depends primarily on its thermodynamic State

$$N_0 = const(G^0)^{1/2}. \quad (15)$$

Gibbs Energy $G^0 = U - TS + PV$, where U – internal energy zone with high gas recovery; T is temperature; S – entropy, P – pressure; V – the volume.

From here, you can see that defining parameters are the internal energy zones with high gas recovery and external (mountain press). Internal energy zones with high gas recovery is defined as the Intermolecular forces of the coal, and energy of the elastic deformation mode.

Laboratory measurement G^0 cannot produce the true meaning of the coal seams. Therefore, other parameters-gas permeability, Young's modules, etc. remain unknown.

Still, the situation is not as hopeless as shown in (Portnov 2003), many geophysical methods of testing mineral deposits are dealing with real response functions of the ore body, useful component, etc. in the form of specific measured values (electrical, magnetic, etc.), which contain information about the G^0.

Use of integrated geophysical research as well, and with a daily surface, can provide the most reliable information on the State of zones with high gas recovery. Below we will consider this question in detail.

4 ENERGY MODEL OF SUDDEN RELEASE OF GAS AND COAL

Energy theory of sudden emissions was developed in Hodotom as the first attempt to reveal the mechanism of sudden coal and gas burst based on the modern ideas of mechanics of the coal bed, his strength and reservoir properties, stress state and gas-dynamic regime of bottom hole formation zone of (Hodot 1961).

Mapping of extensive statistical material of mine survey with the results of the study the properties and State of the gas coal has led to the following conclusions, which is a base of energy theory.

The origin and development of gas dynamic phenomenon of sudden release of coal and gas, enough potential energy of coal and methane in it and the speed of the energy above critical conditions are necessary, but not sufficient; the latest development release and sufficient condition is grinding in the destruction of coal for gas output in quantities and at a speed sufficient to mine coal from the front of the destroyed garbage.

If to examine the methane trapped in coal seam with infinite parallel walls and in which there is a release of energy. In the present case, the allocation of energy in coal seam due to stress relaxation of energy entrainment, and on a wall of the reservoir due to thermal conductivity. Our task is to find the conditions under which the energy is so high that the stationary solution of energy balance Equations. The most convenient approach in this case is to find the stationary solution and determine conditions when it ceases.

Imagine the energy balance equation in coal seam in the form

$$\chi \frac{d^2 T}{dx^2} + f(T) = 0,$$

(16)

where χ – coefficient of energy transfer (in the particular case is the coefficient of thermal conductivity) of the coal bed; x – direction that is perpendicular to the seam, for the origin, we choose the middle of the reservoir.

Here t is the total energy of coal and the particular case we mean by it the temperature of the reservoir, which is proportional to the kinetic energy of the molecules of methane.

Value $(f)(T) = N * NK_{rel} \acute{C}\omega$ is the energy per unit volume per unit of time ($N*$ is the density of vibrationally excited molecules of methane, N – density of molecules, which results in vibrational relaxation; $\acute{C}\omega$ – energy vibrational quantum). as the oscillatory temperature is compared with the progressive, we do not take into account the reverse process leading to the excitation of molecules. significantly, the rate constant of the vibrational relaxation. The $K_{rel}(t)$ depends on the temperature. Enter the amount of $\beta = d \ln k_{cut}(T)/dT$ so in a narrow temperature region

$$f(t) = f(T_0) exp[-\beta(T_0 - T)],$$

(17)

where T_0 – temperature at the Centre of the sheet.

Solve the Equation of energy balance (16), by introducing a new variable $y = \beta(T_0 - T)$, here is the equation

$$\frac{d^2 y}{dx^2} - Ae^{-y} = 0,$$

(18)

where $A = \beta f(T_0)/\chi$ and $y(0) = 0$ the temperature distribution (energy) in seam is even on x, i.e. $t(x) = t(-x)$.

It follows that they $y(x) = y(-x)$ and $\left.\frac{dy}{dz}\right|_{x=0} = 0$. With this in mind, we will consider the Equation (18) only for $x > 0$.

The solution to the Equation (18) with boundary conditions $y(0) = \left.\frac{dy}{dz}\right|_{x=0} = 0$ has the form

$$y = 2 \ln chx\sqrt{A/2}.$$

(19)

In the beginning there are two curves described by the left and right portions of the ratio (19). The tangent point is equal to the value, and their derivatives. This gives

$$\Delta T = \frac{2}{\beta} \ln chZ, ZthZ = 1,$$

(20)

where $Z = l\sqrt{\beta f(T_0)/2\chi}$.

Solution of the second Equation (20) $Z = 1.2$, where the threshold for the collapse of the coal we find

$$l^2 \beta f(T_0)/\chi = 2.88.$$

(21)

In addition, the ratios for the temperature difference are

$$\Delta T = 1.19/\beta.$$

(22)

Rewrite condition of sudden blast through heat settings near the walls.

$$f(T_{cm}) = f(T_0) exp(-\beta \Delta T) = 0.30 f(T_0).$$

(23)

Based on this, the condition of sudden ejection form

$$\ell^2 \beta f(T_{cm})/\chi = 0.44.$$

(24)

The parameter definition β condition (24) may be represented as follows

$$\frac{\ell^2}{\chi} \left| \frac{df(T = T_{cm})}{dT} \right| = 0.44.$$

(25)

Equation (25) has a simple physical meaning. It includes attitude of the energy flow, made to a volume of the reservoir through the internal process of energy release, to the flow of energy that escapes from the volume at the expense of the thermal conductivity and mass transfer. As soon as this ratio exceeds some critical value of units, there is a sudden ejection (Figure 2).

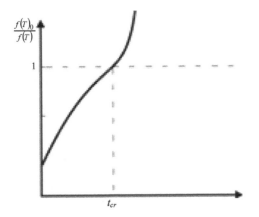

Figure 2. Illustration of the sudden release.

5 GAS PRESSURE IN COAL SEAM

If we as a response function f from work (Portnov 2003) take the relative out gassing and linearize this expression, we will obtain the formula

$$c = \frac{kT}{C_1} \cdot \frac{A}{G^0} \cdot c_0^2, \qquad (26)$$

where G^0 is the initial concentration of methane in coal (coal seam), i.e. methane content; and-work (energy) of the external forces (field); with the C_1 constant.

Formula (26) gives a quadratic dependence of methane emission from methane content reservoir (Figure 3).

The amount of natural gas pressure in microtime and natural microcracks in coal seams is one of the key indicators that define the amount of methane in absorbed and free states in the seam (Ajruni 2002). With increasing the mining depth, methane in loose condition increases and can reach 10–12% (Ajruni 2002).

Since $G^0 = U - TS + PV$ then from (26), with c_0 we have

$$c_0 = C_2 G^0 = C_2 (U - TS) + C_2 PV .$$

Thus, the pressure in the reservoir of methane is

$$P = [c_0 - C_2(U - TS)]/(C_2 V). \qquad (27)$$

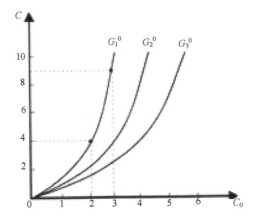

Figure 3. Schematic of the dependence of the gassing of methane content of coal seam.

Equation (27) shows a complicated dependence of gas pressure in the reservoir from its concentration and properties of the coal. However, all the parameters in (27) experimental even identified what is important from a practical point of view.

6 RATIONALE FOR THE USE OF GEOPHYSICAL METHODS FOR DETECTING ZONES WITH HIGH GAS RECOVERY

Consider the question of how to use the results of electrical logging for measuring gas content of coal scams. To do this we will use the theory of conductivity of heterogeneous media, as set further in the work (Yurov et al. 2010).

Resistivity of rocks and minerals varies primarily due to changes in the conductivity of the solution, saturating in the pores, and the pore parameter.

Let us look first at the homogeneous isotropic environment that contains conductivity electrons and is characterized by thermodynamic potential of Gibbs G°. The emergence of current density (j) the environment is a response system of non-integrate of electrons in the outer margin and has the form

$$j = \frac{kT}{C_1} \cdot \frac{eE}{G^0} \cdot \overline{N}. \qquad (28)$$

When $\overline{N} = const$, from (28) have Ohm's law in differential form

$$j = \sigma E , \qquad (29)$$

where $\sigma = \frac{kT}{C_1} \cdot \frac{e\overline{N}}{G^0}. \qquad (30)$

Conductivity s linked to specific resistance r value

$$\rho = 1/\sigma = C \cdot G^0 / e\overline{N} , \quad C = \frac{C_1}{kT}. \tag{31}$$

Constant characterizes the process of transition from the excited state electron system in basic and is similar to many substances. The exception can be only those substances with specific mechanisms of dispersion. Thus, the heterogeneity of the environment will affect its conductivity through the Gibbs energy.

Now comparing formula (26) and (30), we find the relationship between gas-emitting characteristics and specific resistance in the linear approximation

$$c_0 = const / \rho. \tag{32}$$

Thus, the results of electrical logging can be used to measure the gas content of coal seams.

Use of gamma testing methods in rocks and ores is based on linear gamma-ray attenuation coefficients and coefficients of primary radiation conversion in the secondary on the density of rocks and ores and mineral content of the component.

In the diffusive approximation of dependence of the intensity of gamma radiation scattering substance is expressed explicitly (Ochkur 1976)

$$J / J_0 = const \frac{\rho}{R} e^{-\xi} , \tag{33}$$

where ρ is the density of the substance, $\xi = \overline{\mu} \rho R$, $\overline{\mu}$ – attenuation of gamma-radiation; R – probe length (distance between the source and the detector of gamma radiation).

In an extreme case ($R \to 0$) asymptotic expression has the form

$$J / J_0 = const \frac{1}{\overline{\mu}_{ef}} , \tag{34}$$

where $\overline{\mu}_{ef} = \mu / \rho$.

Taking as a response function f the relative intensity of the diffuse gamma radiation with an energy of $(E)_\gamma$, we get

$$1 - I / I_0 = -B \frac{C_{Fe}}{G^0 E_\gamma} , \tag{35}$$

where $B = (kT)^2 / C$, $C = 2\Delta S / k$ is constant for a given element and source of gamma-radiation; ΔS – change of entropy in a quantum transition from the excited state in equal $\Delta S = \overline{N} E_\gamma^2 / 2kT^2$,

where $\overline{N}$ is the average number of atoms of the element in the rock or ore; $(G)^0$ is Gibbs energy of rock or ore.

Using (26) and (35) of the coal seam gas content are

$$c_0 = const(1 - I / I_0). \tag{36}$$

Thus, results of the gamma-ray logging can be used to measure the gas content of coal seams.

In many areas of technology and physics there are phenomena which are direct analogues of the processes in potential fields. Table 1 presents the analogy between electric and acoustic variables and parameters (Olson 1947).

Table 1. Analogy between electric and acoustic variables and parameters.

Electrical system	Speaker system
Voltage U	Pressure P
Current I	Particle velocity v
Charge (e)	The Offset u
Inductance (L)	Medium density ρ
Container With	Acoustic tank with $1/\tau$
Resistance R	Impedance $(R)_{(A)}$

Refer to the table 1 and Ohm's law, it is not difficult to get the speed of propagation of acoustic waves associated with the gas content of the seam by the Equation

$$c_0 = const / v . \tag{37}$$

Thus, the results of the acoustic method can be used to measure the gas content of coal seams.

In a similar way, you can get the connection between the gas content of coal seams and any option of a geophysical method studies. The most promising methods for detecting zones with high gas recovery, i.e. ejection threaten zones are ground-based geophysical research methods.

The above relation between the gas content of the seam and the measured value in any method of geophysical studies are ongoing to determine empirically using correlation analysis.

The authenticity of the received results will increase with the proximity of various geophysical methods.

7 CONCLUSIONS

Above we have touched only some aspects of the complex problem of sudden coal and gas emissions. This problem is related to the detection of zones with high gas recovery. The nature of the technology of extraction of methane from coal seams of such deposits is given, and the important task of identifying local zones of high natural gases of coal are set.

In such deposits the abnormally high methane concentrations are located, and also in the so-called fluid-active integrated zones (zones of Fluidization) caused by fluid-metasomatic transformation of coal seams in areas of tectonic disturbances. Forecasting and detection of such methane areas and nonstructural gas reservoirs: one of the most difficult and important tasks in solving the problem of coal bed methane.

REFERENCES

Trufanov, V., Gamov, M. & Rilov, V. 2004. *Fluidization hydrocarbon deposits the Eastern Donbass coals Rostov-on-Don*. Rostov State University: 270.

Gritsku, G. 2007. *Sudden emissions of methane in mines in Siberia*, 32–33: 2617-2618.

Arnold, V. 1990. *Catastrophe theory*. Moscow: Nauka: 128.

Portnow, V. 2003. *Thermodynamic approach to the challenges of the geophysical testing of iron ore deposits:* 178.

Hodot, V. 1961. *Sudden coal and gas emissions*. Moscow: Gosgortehizdat: 363.

Ajruni, A., Baimukhametov, S., Prezent, G. etc. 2002. *Problems of developing methane coal seams of industrial extraction and use of coal mine methane in the Karaganda coal basin*. Moscow: Publishing House of the Academy of mining Sciences Russian Federation: 318.

Yurov, V., Halenov, O. & Zakamolkin, V. 2010. *Model of electrical conductivity of solid solutions*. Herald of the development of science and education, 3: 7–10.

Ochkur, A. 1976. *Gamma methods in mining geology*. Moscow: Nedra: 320.

Olson. 1947. *Dynamic analogies*. Moscow: Silt: 276.

Influence of the dynamic loading on stress-strain state of subsiding soils under buildings and structures

O. Solodyankin & N. Ruban
National Mining University, Dnipropetrovs'k, Ukraine

ABSTRACT: Feature of construction on loess soils is shown; it is dealt with problem long-term dynamic influence on subfoundation; capabilities of two computation model of soil, such as Mohr-Coulomb and Hardening soil small in Plaxis are discussed; results of computation analyzes are described in short; in conclusions recommendations on the choice of the computational model of soil with the ability to take into account the dynamic loadings are given.

1 INTRODUCTION

With the development of cities and growing population there is an urgent need for new areas for the constructions of public and industrial facilities in dense urban areas. Some buildings and facilities require reconstruction, reprofiling and reequipment. It often leads to increased payloads on buildings and structures. Due to the intense development of the cities, there are difficulties with their functional zoning, whereupon boundaries between residential areas, industrial and transport zones are erased.

In such circumstances, it is increasingly difficult comply to regulatory loadings on existing buildings. There is an urgent need for new methods of construction, reconstruction and strengthening of buildings on urban areas, as well, as the research to account for all possible static and dynamic effects on new and existing buildings.

Vibration and vibro-impact loads on buildings and structures have a significant influence on their strength and durability. At the same time, effects of long-term loads associated with the construction works, road traffic, particularly rail, can lead to equally serious consequences for the bearing capacity of structures, than single powerful vibro-impact loading, such as seismic. Details of the dynamic loads from the traffic sources are shown in Table 1.

Many writings devoted to the study of dynamic effects on the ground. Among them are works by the following authors: O.A. Savinov, A.D. Krasnikov, D.D. Barkan, S.S. Grigorian, G.M. Lyahov, H.A. Rahmatulin, E.A. Voznesenskiy. However, existing methods for calculating the influence of dynamic loads on the foundation practically ignores real properties and behavior of soil bases under the influence of the vibration load, namely nonlinear deformation, creep, vibrocreep and absorption energy by soil (Chirkov 2010).

Table 1. Characteristics of transport sources dynamic loadings (Voznesenskiy 1997).

Sources	Dominant frequency	Vibration velocity of soil particles		Vibration acceleration of soil particles		Influence zone
	Hz	10^{-3} m/s	dB	10 m/s^2	dB	m
railway	10–70	16–50	110–120	1–22	70–97	150–300
tramline	20–45	1.6–160	90–130	0.5–45.2	56–103	150–300
subway	30–60	0.3–300	75–135	10–1800	90–135	6–120
motorway	10–20	0.005–0.07	40–65	0.0003–0.011	to 31	40–100

More than 30% of urban buildings and structures erected on the subsidence loess soils in Dnipropetrovs'k. According to the classification (Osipov 2010), loess is a weak soil and does not have a structural strength. It belongs to the most dangerous group number IV, which is very sensitive to dynamic influence. Therefore, for the urbanized, developing city Dnipropetrovs'k is very important problem of rational use of local land resources, taking into account the influence of dynamic loads.

According to Ukrainian national construction regulation (B.1.1-5-2000), structure on the subsiding soils, should be calculated based on joint work with the subfoundation, design features, type of load and destination of building. Strength, stability and durability of buildings and structures erected on loess soils, largely depend on the complete exception of the possibility of soaking the subfoundation while it is used. Ingress of water into the soil reduces the strength and deformation soil characteristics. By soaking, loess soil, even without increasing the external load on the foundation, goes into the category of weak soils and uneven subsidence occurs, which leads to the destruction building and structures. Feature compression curve loess soil is shown in Figure 1. This feature of loess soils requires consideration the development of additional deformations by the influence of water, depend on time.

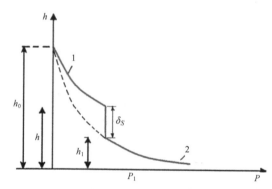

Figure 1. Compression curve of loess soil sample: h_0 is the sample initial height, h is the sample natural humidity, was compressed pressure by P_1 without the possibility lateral extension, h_1 is the height of the same sample when water was passing through it, δ_S is the additional subsidence of sample when water was passing through it.

Currently, when the dynamic impacts are considered, the following models are the most widely used: perfectly elastic model (linear and nonlinear), dilatational model, elastoplastic and viscoplastic models.

The parameters of elastic and absorbing soil properties under dynamic impacts of small intensity (not exceeding the limit of elasticity): Young's modulus E, Poisson's ratio ν, damping ratio a and other equivalent characteristics, for example, propagation velocity and absorption coefficient of the elastic waves.

Soils characteristics under dynamic loads of significant intensity (exceeding the elastic limit), necessary for designing objects: curves "stress-strain", modulus of deformation under loading and unload-

ding, dynamic characteristics of the resistibility to distortion (shearing) and limit state (strength) of the soil, and the stiffness of their structure during liquefaction (Tsytovych 1983).

2 COMPUTER SIMULATION THE STRESS-STRAIN STATE OF SOIL

Solution of various engineering problems with the large number of influencing factors in the analytical formulation is very difficult, and sometimes impossible. At the present stage, complex calculation justification of construction projects, exploitation and reconstruction facilities in difficult geological condition became possible using modern computer programs. However, with the development of information technology and a large amount of computing programs, there is the problem of choosing an adequate computational model for more complicated tasks. For example, in calculating the stress-strain state of foundation soil constructions erected on subsiding soils under the influence of dynamic loads and consideration of subsidence developing over time. Certainly, the choice of the computational model depends on the physical and mechanical characteristics, the most comprehensive reflecting the soil properties. However, designers often are not enough standard set of parameters obtained in geotechnical investigations for exploring more complicated models.

Consider the example modeling the stress-strain state of loess subfoundation, which exposed vibration from the moving train. The calculation scheme is shown on Figure 2.

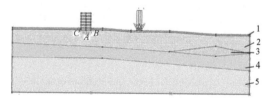

Figure 2. Calculation scheme. 1 – loamy loess, 2 – sandy loam loess, 3 – clay loams, 4 – plagiogranit, 5 – plagiogranit creviced.

Referring to building regulations of Ukraine, principle of the engineering by admissible deformations can be fully realized only using elastoplastic soil models. These models allow describe the stress-strain state on entire range of variation loads, up to the maximum (destroying) values. Therefore, modeling and calculation were done using a software PLAXIS. It allows solve problems in the elastoplastic formulation and has successfully established itself among both domestic and foreign specialists. Comparative analysis was performed with use of Mohr-Coulomb model and Hardening Soil Small model.

Covered problems:
– the possibility to consider dynamic influence;
– distribution of soil motion under dynamic effects;
– uniformity of subsidence under building;
– influence of load application sequence.

The settlement scheme includes several layers of loess soil. Loads, which were taken into account: own weight of soil; own weight of building; static load B from weight of train and dynamic load A (harmonic) from train traffic.

Mohr-Coulomb model. This model uses the strength condition Mohr-Coulomb: It includes five basic input parameters: Young's Modulus E; Poisson's ratio v; Friction angle φ; Cohesion and Dilatancy angle ψ. Input parameter are shown in Table 2. Basic defining Equations: Hooke's law $\sigma = E\varepsilon$; summability deformations $\varepsilon = \varepsilon^e + \varepsilon^p$ and $\gamma = \gamma^e + \gamma^p$; associated plastic flow rule; plastic hardening rule (Plaxis 2D 2012, Tutorial Manual).

Table 2. Estimated parameters for the Mohr-Coulomb model.

Symbol and unit of measure	Parameter	Loamy loess	Sandy loam loess	Clay loams
E, kN/m^2	Young's Modulus	$15 \cdot 10^3$	$30 \cdot 10^3$	$25 \cdot 10^3$
$v, -$	Poisson's ratio	0.3	0.3	0.35
C, kN/m^2	cohesion	6	6	5
φ, degree	friction angle	33	26	24
γ_{unsat}, kN/m^3	soil unit weight above phreatic level	17	16.7	17.5
γ_{sat}, kN/m^3	soil unit weight below phreatic level	20	19.1	18
$e_{init}, -$	initial void ratio	0.7	0.4	0.4

Mohr-Coulomb model uses the most verified strength criterion, which has been repeatedly confirmed by tests of soils in tension and compression. However, this model is quite conservative in the definition of the intermediate stress-strain state not reached limit value. This is a significant factor influencing the choice of soil model, since, according to Ukrainian national construction regulation (B.2.1-10-2009), the foundation must be calculated not only on I group limit states (failure limit state), but also on group II (limit state of deflection). Particular attention should be paid to input value of elasticity module. Throughout calculation it remains unchanged. For example, for closely-packed clays and other types of rock showing linear elastic properties in wide load range, adequate results can be obtained using initial slope of stress-strain curve E_0. However, for sands and normally consolidated clays more often used the secant modulus at 50% strength designated as E_{50}.

Hardening Soil Small model. Hardening soil model with small strains stiffness is advanced elastoplastic model. In addition to strength parameters c and φ, it requires introduction additional soil stiffness parameters, such as: tangent modulus of elasticity E_{50}; modulus of elasticity under action unloading – repeated loading E_{ur}; odometric mo-

dule of elasticity E_{oed}; the power for stress-level dependency of stiffness m. In order to follow history of the stress-strain state depending on the time, two additional parameters are introduced: the initial modulus C_0 and the shear strain level $\gamma_{0.7}$, at which the secant shear modulus C_s is reduced to about 70% of C_0. Listed input characteristics of stiffness correspond to certain reference value, which was performed in triaxial and odometer soil tests. The real stiffness characteristics are calculated by program PLAXIS taking into account the development of the stress-strain state of the soil in a hyperbolic relationship

$$E_{50} = E_{50}^{ref} \left(\frac{c \cos \varphi - \sigma_3' \sin \varphi}{c \cos \varphi + p^{ref} \sin \varphi} \right)^m,$$

$$E_{ur} = E_{ur}^{ref} \left(\frac{c \cos \varphi - \sigma_3' \sin \varphi}{c \cos \varphi + p^{ref} \sin \varphi} \right)^m,$$

$$E_{oed} = E_{oed}^{ref} \left(\frac{c \cos \varphi - \sigma_1' \sin \varphi}{c \cos \varphi + p^{ref} \sin \varphi} \right)^m,$$

$$C_0 = C_0^{ref} \left(\frac{\sigma_3 + c \cot \varphi}{p^{ref} + c \cot \varphi} \right)^m,$$

where m is the determined by the compression tests (Ohde 1939).

Hardening Soil Small model describes nonlinear behavior of the material during unloading, and inelastic behavior of soil during reloading.

Main equations: hyperbolic relation, summability of strains, not associated flow rule for plastic shear strain and associated flow rule for plastic volumetric strain.

Using the model HS Small, should pay attention to the possibility of accounting initial stress state of the soil. Preconsolidation soil in geologic time, in addition to the initial porosity and initial water saturation plays an important role in the assessment of the current stress-strain state of the soil mass. Not taking into account this factor, can negatively affect on predicted data of displacements and deformations. Additional parameters for the Hardening Soil Small models are shown in Table 3.

Table 3. Additional parameters for the Hardening Soil Small models.

Symbol and unit of measure	Parameter	Loamy loess	Sandy loam loess	Clay loams
E_{50}^{ref} kN/m²	reference stiffness for triaxial compression	$15 \cdot 10^3$	$30 \cdot 10^3$	$25 \cdot 10^3$
E_{oed}^{ref} kN/m²	reference stiffness for primary oedometer loading	$15 \cdot 10^3$	$30 \cdot 10^3$	$25 \cdot 10^3$
E_{ur}^{ref} kN/m²	reference stiffness for triaxial unloading	$30 \cdot 10^3$	$60 \cdot 10^3$	$60 \cdot 10^3$
v_{ur}	Poisson's ratio for unloading – repeated loading, by default $v_{ur} = 0.2$	0.2	0.2	0.2
m (power)	the power for stress-level dependency of stiffness	0.5	0.5	0.5
C_0 kN/m²	initial shear modules	$60 \cdot 10^3$	$120 \cdot 10^3$	$120 \cdot 10^3$
$\gamma_{0.7}$	shear strain	0.001	0.001	0.001
K_0	coefficient of earth pressure	0.429	0.561	0.538
p^{ref} kN/m²	reference stress for stiffness, by default $p^{ref} = 100$	100	100	100

Determination of input parameters. Modulus of elasticity, cohesion, friction angle, void ratio, weight in water-saturated and natural state are accepted according to engineering and geological research executed Research Institute. The purpose of modeling was to determine the impact of railway on existing buildings. Poisson's ratio was accepted according to Ukrainian national construction regulation (B.2.1-10-2009), lateral earth pressure coefficient in the natural state according to formula $K_0^{nc} = v / (1 - v)$ (Jaky 1944), m – was accepted according to sources (Ohde 1939 & Strokova 2008). Shear modulus $C_0 = 2E_{ur}$ and shear strain $\gamma_{0.7} = 0.001$ was appointed in accordance with (Strokova 2008), because their definition is not included in the standard definition of parameters mechanical properties of soil and requires complex and expensive equipment.

According to the engineering and geological surveys, geological cross-section consists of loess soils, which underlies a thick layer of granite. Behavior of such rocks cannot be described by the equations for elastoplastic models. Therefore, *Linear elastic* model was chosen for modeling underlying layers of

rock. The input parameters for the model: the modulus of elasticity and Poisson's ratio.

Appointment of loads: 1) Own weight of building; 2) Own weight of soil; 3) Dynamic load $A = 1/10 \cdot B = $ kN/m²; 4) Oscillation frequency $f = 50$ Hz; 5) Static load from train is approximately equal to $B = 30$ kN/m²; static load from laden train was accepted as $2 \cdot B$.

3 RESULTS OF THE INITIAL CALCULATION

Consider three points of subfoundation A, B and C marked on Figure 2. Within the Figure 3 is contained the amplitudes of displacement of point A, calculated on the Mohr-Coulomb model and on Hardening soil small model. Deformation amplitude obtained by the Mohr-Coulomb model is more than HSSmall model. This can be explained by the fact, that the model Mohr-Coulomb uses the constant modulus of elasticity, not taking into account the changes of stiffness at the stages loading, unloading and re-loading.

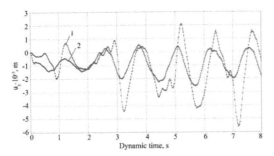

Figure 3. Total displacement point A: 1 – Mohr-Coulomb model, 2 – HSSmall model.

Within the Figure 4 is represented the curves of displacement of the points A, B and C for the three phases "loading-unloading-reloading". According to the results it is seen that the displacement of point B, located closer to dynamic loads, has a maximum value. This indicates the appearance uneven subsidence takes place.

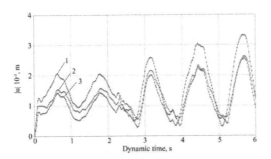

Figure 4. Total displacement of points A, B, C. Point A is designated #2, point B is designated #1, point C is designated #3.

The Figure 4 also depicts the effect of loading history. Three stage of loading was performed. The first stage includes forced oscillations, second stage includes free vibration of building after unloading, and the third stage includes repeated forced oscillation. All stages lasted for 2 seconds. When forced oscillations pass into free oscillations, amplitude deformation increases. Next step shows amplitude increase of ground motion after applying additional forced oscillations.

Figures 5–10 show results of a calculation using the model Hardening Soil Small.

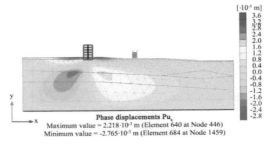

Figure 5. Diagram of displacement along the x axis under load by own weight of building and train.

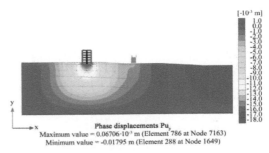

Figure 6. Diagram of displacement along the y axis under load by own weight of building and train.

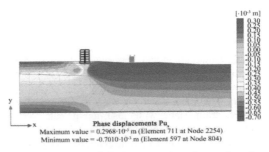

Figure 7. Diagram of displacement along the x axis under dynamic loading with amplitude $A = 3$ kN/m^2, $f = 50$ Hz.

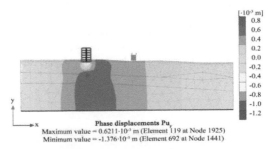

Figure 8. Diagram of displacement along the y axis under dynamic loading with amplitude $A = 3$ kN/m^2, $f = 50$ Hz.

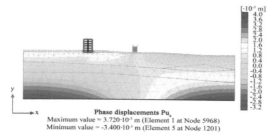

Phase displacements Pu$_x$
Maximum value = 3.720·10^{-3} m (Element 1 at Node 5968)
Minimum value = -3.400·10^{-3} m (Element 5 at Node 1201)

Figure 9. Diagram of displacement along the x axis under dynamic loading with amplitude A = 3 kN/m^2, f = 50 Hz.

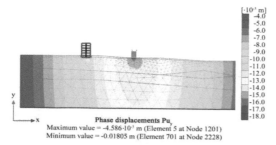

Phase displacements Pu$_y$
Maximum value = -4.586·10^{-3} m (Element 5 at Node 1201)
Minimum value = -0.01805 m (Element 701 at Node 2228)

Figure 10. Diagram of displacement along the y axis under dynamic loading with amplitude A = 3 kN/m^2, f = 50 Hz, B = 60 kN/m^2.

4 CONCLUSIONS

Mohr-Coulomb model performs well small deformations, characterizing the elastic behavior of the soil or strains much greater than the elastic limit state. The model can be used to pre-analysis of stress-strain state with insufficient data on the physical and mechanical properties of soils.

When the Mohr-Coulomb model is using for dynamic analysis, it is necessary to pay special attention to selecting the stiffness parameter. This value for all calculations is accepted constant and should correctly display velocity of the waves in soils. Plastic deformation will occur only when failure criterion of Mohr-Coulomb will be reached. Not reaching to this criterion, will be generated only elastic deformation without damping, without generating pore pressure and soil liquefaction.

HSSmall model can show prelimit stress-strain state. Due to additional stiffness parameters, allows us to estimate unacceptable deformation. There is possible to take into account the initial stress state. In the calculations on the dynamic effects HHSmall model shows the wave damping, which depends on the amplitude of the applied load and strain amplitude.

Mandatory step in creating a computational model should be simulated laboratory and field tests, which is possible in the PC Plaxis. An important point of this step is validation of selected equations and determining the correct destination input design parameters. Validation criterion of selected model is coincidence of calculated results and experimental data.

Effect of water on the stress-strain state of the soil mass, composed of loess soils, in further, can be considered by researching properties of subsiding soil under dynamic loads.

REFERENCES

Voznesenskiy, E.A. 1997. *Behavior of soils under dynamic loads.* Textbook. Vol. 288.
Chirkov, A.L. 2010. *State of question the assessment of vibration influence on buildings.* Building and technotropic security, 33–34. Vol. 15: 117–131.
Osipov, O.F. 2010. *Classification of buildings and foundations for resistance to dynamic effects and changes in the stress-strain state.* Construction technology in urban conditions, 30.Vol. 14: 137–150.
Tsytovych, M.A. 1983. *Soil mechanics (short course).* Textbook for engineering universities. Vol. 288.
Plaxis 2D 2012. Tutorial Manual.
Ohde, J. 1939. *Zur Theorie der Druckverteilung im Baugrund.* Der Bauingenieur, 20. Vol. 9: 451–459.
Strokova, L.A. 2008. *Determination of parameters for numerical simulation of soil behavior.* Journal of Tomsk Polytechnic University, 1.Vol. 6: 69–74.
Jaky, J. 1944. *The Coefficient of Earth Pressure at Rest.* Journal for Society of Hungarian Architects and Engineers. Vol. 4: 355–358.

Progressive Technologies of Coal, Coalbed Methane, and Ores Mining – Bondarenko, Kovalevs'ka & Ganushevych (eds)

Features of using numerical experiment to analyze the stability of development workings

V. Sotskov
National Mining University, Dnipropetrovs'k, Ukraine

O. Gusev
PrJSC "DTEK Pavlogradugol", Pavlograd, Ukraine

ABSTRACT: The scrutiny degree of the determining problem of rational parameters of the development drift support construction in overworking conditions were analyzed. The expediency of conducting extensive scientific researches using the computer simulation by finite element method. A series of calculations on the approaching the working face to the development drift in the elastic and elastoplastic formulation, taking into account the physical and mechanical properties of rocks and layered rock mass were conducted. Analyzed the results of the experiment, identifies areas of stress concentration and discharge in the massif, as well as the degree of rock pressure influence on the frame and anchors. Detected the characteristic features of the results obtained in the solution of linear and bilinear tasks, allowing an objective assessment the stress- strain state of the rock massif and fastening. The results can be used to develop a methodology for determining the rational parameters of fastening of development drift in overworking conditions.

1 INTRODUCTION

In today's International Economics fuel and energy sector occupies an extremely important place. Cumulative net energy production share in the structure of the world's GDP is currently estimated at an average of 10–12%, or about 1.8 thousand dollars a year per capita.

The main sources of energy today are geological energy resources: oil, coal, gas, oil shale, peat, uranium, etc. They accounted for 93% of the world's energy. Global industrial coal reserves amount to over 1 trillion tons, significantly outperforming the reserves and resources of all other forms of energy. This fact makes the coal major energy source that can meet the ever increasing demand for energy, not covered by other available sources.

Ukraine's leading coal-mining enterprise is PJSC "DTEK Pavlogradugol", which includes 10 mines, total production which in 2013 amounted to 38.6 million tons Moreover, according to the company's long-term strategy to 2030 coal production should reach 50 million tons. In connection with this core work is aimed at carrying out a series of measures to ensure a stable production growth. The main ways of improving performance indicators is the development and introduction of new technological schemes and the use of modern high-performance equipment . Most common in recent years gained a tendency to increase the lengths of lava and extrac-

tion panel that the intensification of the extraction works with the help of modern technology makes it possible to significantly increase the production of coal.

However, the resulting significant increase of production volumes contains a number of difficulties. In complex geological conditions of the Western Donbass, where coal fortress is often more than a fortress of enclosing rocks, with increasing length of the excavation panel simultaneously increases and the length of development workings, which leads to serious fastening problems. Given the production trend for the reuse of development workings, need to initially use the most rational scheme of support to avoid having to spend refastening workings, thereby avoiding additional production costs.

One of the most illustrative examples are the difficulties associated extraction coal seam C_5 "Samars'ka" mine is PJSC "DTEK Pavlogradugol" where to eliminate the effects of severe water influx into the lava was held airway drainage drift at a depth of 8–9 m from the coal seam. As a result of production due to a number of technological factors negatively affecting its stability. First, water which is supplied by the drain wells to the drift from overlying horizon promotes soil and rocks soaking and provoking swelling, secondly, a part of excavation panels which extracted the coal seam passes directly over the drift, and results the drift entering in the bearing pressure zone around the lava.

The result is a situation where the choice of rational parameters of drainage drift fastening systems can not be done on the SOU recommendations to support development workings. Need to develop new options that fully meet the specifics of developing drifts in complex geological conditions, taking into account the overworking influence and availability of water in the roof.

2 FORMULATION OF THE PROBLEM.

The lack of degree of the problem scrutiny and the lack for similar problems solutions for specific geological conditions, becomes necessary to continue researches on the definition of rational parameters of development drifts fastening for the geological conditions of the Western Donbass using modern methods of stress-strain state calculations of rock massif in overworking conditions.

At the current stage of computer systems development, conducting numerical experiments of great complexity becomes quite affordable way to solving the various applications. Numerical experiment is a method for studying complex problems based on the construction and analysis mathematical models of object being studied using computer technologies. Unlike analytical methods for solving numerical methods allow the use of specially designed software solution for a large range of applications (Samarskiy 1989). The basis is the using for computer simulation the finite element method as the most perfect and suitable for solving the problems of geomechanics. This numerical mesh method, which is based on the representation of the area, which stress- strain state is necessary to determine, as a set of flat or spatial elements such as core or frame structures. Feature is a clear physical interpretation of solving tasks. Ability to determine the physical and mechanical properties of each element takes into account the inhomogeneity of the deformed region that blends perfectly with the need to simulate the thin-layered massif. Technology of spatial simulation allow opportunity for you to create elements of varying complexity to maximize the model fit to the real conditions. Extensive possibilities for changing boundary conditions much closer to the model of the necessary conditions.

Using modern computer-aided engineering analysis (Computer Aided Engineering – CAE) today is one of the most effective ways to solve such tasks. One of the most common of these systems today is the program ANSYS, using finite element method. Multipurpose orientation of the program, independent of hardware (from PCs to workstations and su-

percomputers), geometric modeling tools based on B splines (technology NURBS), full compatibility with CAD / CAM / CAE systems leading manufacturers and "friendly" interface allowed ANSYS to become a leader in the specialized engineering software. Currently ANSYS adapted and used for scientific researches in geomechanics (Jhidkov 2006).

The basic approach to the choice of the initial parameters of the model was the maximum correspondence to real geological conditions of a particular area, as well as fastening passport of drainage airway drift. Estimated the sizes of the model along the vertical and horizontal coordinates Y and X, which are quite sufficient to describe the presentation of the stress distribution picture around the working face and the underlying development opening. As a result, a model consists the 25 layers of rock with real physical-mechanical characteristics of rocks, and its dimensions were: 55 m in width and 48 m in height. The incidence angle of the coal seam 3°, Poisson's ratio $\mu = 0.3$. For coal seam development is modeled the working face with mechanized fastening that is built to facilitate the calculation of a rectangular block with sizes and physico-mechanical properties the corresponding real. At a depth of 11 m from the coal seam held drainage drift whose section is modeled under the frame KSHPU. In opening inscribed the frame support with profile SVP–22, as well as roof bolting, consisting the 7 anchors in the roof length of 2.4 m and 2 anchors in the sides working in length 1.5 m.

3 THE RESULTS OF THE EXPERIMENT

Numerical experiment in elastoplastic permutation consisted of two stages: the location of the working face at a distance of 14 m from the drift and directly over the roadway. For the results analysis of the calculation used the diagrams of the intensity σ, of vertical σ_y and horizontal σ_x stress. In an article for the possibility of visual comparative analysis, the diagrams of the intensity σ of rock massif in the vicinity of the drainage drift and roadway's support constructure presented in pairs in the elastic and elastic-plastic formulation with variable distance approximation stope by overlying the coal seam.

Results of numerical experiments obtained in the elastic formulation with the description of zones of increased stresses, deformations, unloading areas, as well as the characteristic singularities of the stress distribution are presented in (Sotskov & Saleev 2013). In contrast to the linear task, for solving nonlinear rask the stresses are reduced in the massif, but deformation continue to increase, which corresponds to the concept of the plasticity law. This confirms the objectivity of the results presented in Figure 1, as well as in the linear calculation (Figure 1a) the bearing pressure zone is formed around the working face, which extends to a considerable distance in the roof and the sill of coal seam and reaches in height 20 m and in width 7 m. The intensity of the stresses in the roof rocks reach 50MPa, which exceeds the tensile strength of rock uniaxial compression and thus reveals significant deformation of

the layers up to the formation of cracks and the integrity of the massif. A characteristic feature of the calculation is the close proximity to the overworking development working to the extracted coal seam. In this regard, the concentration of stresses derived from excavation works and around drainage drift connected while on the breed layers lying between them, is the impact of two zones of bearing pressure. This area covers 13-16 m in width and up to 10 m in height, the stresses in the range of 20–30 MPa, which is sufficient for the formation of cracks in weak rocks of the Western Donbass.

(a) (b)

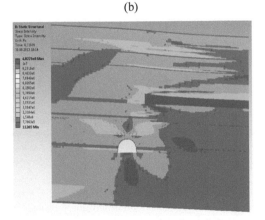

Figure 1. The diagram of the rock massif stress intensity σ at the location the working face over drainage drift: (a) elastic production, (b) elastic-plastic formulation.

Compared with the elastic formulation, the results of solving the bilinear task (Figure 1b) give a completely different picture of the stress distribution in the rock massif. Stresses around the working face and development working are decreased significantly and only 10–15% higher than the equilibrium state of the massif. This preserves the qualitative distribution of bearing pressure zones in the sides and unloading zones at the roof and sill of drift. In the working face roof and over the mined-out space occurs rock layers inflection, due to the gradual subsidence of the overlying layers on the previously tumbled consoles of the main roof.

To obtain a complete picture of the overworking impact on the development drift is necessary to analyze the stress-strain state of the frame and anchor fastening. On the Figure 2(a) shows the stress intensity diagram σ obtained during linear numerical experiment. When analyzing diagrams necessary to allocate a symmetrical distribution of stresses over 270 MPa in both frame legs that characterizes the excess steel yield strength due to intense exposure of rock pressure in the working sides. At the same

time frame relative unloaded roof bar, the maximum stresses does not exceed 170 MPa. Anchors more loaded by the working face approach, which is quite natural. Wherein the anchors are not uniformly loaded along its length, there is an obvious pinching anchors in the rock layers, which leads to a significant increase in the stresses on the individual sections of rebar, up to the yield point is exceeded, indicating that the deformation of resin-grouted roof bolt. In more detail the influence of rock pressure heterogeneity on the stress-strain state of the support constructure is described in (Sotskov 2013).

As a result of the bilinear calculation (Figure 2b) distribution of reduced stress in the support constructure has changed significantly. Side legs of the frame subject to high stresses not the entire length, but mainly in the central part, while there was a reduction of stress concentrations at the bottom and top of the legs. At the same time stresses in the roof bar of the frame increased until the contrary quantities, let's talk about trough roof bar inside working due to significant pressure from the misuse of the roof rocks. Analysis of anchor support stress distribution shows a steady downward stresses trend. Stresses close to the yield strength of steel present dot or a much smaller part of the anchors. For the rest, there was a significant stresses decrease in the average to 135 MPa. The only exception is the side anchor from the working face approaching, 80% of which are subject to stresses in excess of 270 MPa.

(a)   (b)

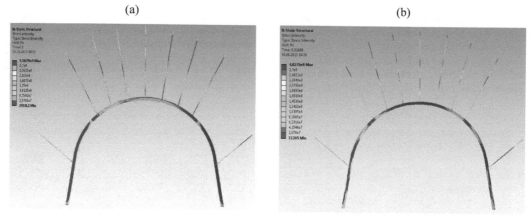

Figure 2. Stress intensity diagrams σ of the frame and roof bolting at the location the working face over the drainage drift: (a) elastic production, (b) elastic-plastic formulation.

4 CONCLUSIONS

The analysis results of the fastening system behavior problems of development workings considering overworking revealed that there are parameters of support do not provide workings operational state. In the computational experiment were obtained diagrams of the stresses distribution and deformations of the coal-bearing rock massif and support system of the drainage drift for two cases of the calculation. Zones of critical reduced in the massif are derived from extraction works and around the drainage drift, which are connected, while on the rock layers lying between them, is the impact of two zones bearing pressure. The resulting area takes 13–16 m in width and up to 10 m in height, the stresses in the range of 20–30 MPa. The zones of plastic flow in the sides and roof bar frame support reaching 270 MPa, as well as in some parts of the anchors. Comparative analysis of the solution of problems in the elastic and elastoplastic formulation showed that the optimal condition for obtaining objective findings is their share in terms of continuity solutions.

REFERENCES

Samarskiy, A.A. 1989. *Numerical methods.* Moscow: Nauka: 429.
Jhidkov, A.V. 2006. *Using ANSYS system for solving geomechanical and finite element tasks of modeling* Nighniy Novgorod: 115.
Sotskov, V. & Saleev, I. 2013. *Investigation of the rock massif stress strain state in conditions of the drainage drift overworking.* Mining of mineral deposits. Netherlands: CRC Press / Balkema: 197–201.
Sotskov, V. 2013. *Investigation of influence of the drainage drift overworking on the stress strain state of support system.* Dnipropetrovs'k: Mining of Mineral Deposits: 317–322.

Progressive Technologies of Coal, Coalbed Methane, and Ores Mining – Bondarenko, Kovalevs'ka & Ganushevych (eds)
© 2014 Taylor & Francis Group, London, ISBN: 978-1-138-02699-5

Preparation of filler-stabilizer for composite materials

O. Svetkina
National Mining University, Dnipropetrovs'k, Ukraine

ABSTRACT: The filler-stabilizer on the basis of limestone is received by the method of the vibration loading. It is shown that in the machining of the material change in energy characteristics of the filler takes peace which leads to structural changes in the surface layer of the filler. The energy barrier layers prevent chemical interaction of matrix and filler-stabilizer at high temperatures, as a result thermo stable characteristics of composites are increased.

1 INTRODUCTION

A new epoch in development of materials was begun with development and application of composites, having such combination of physical-mechanical properties, which is unattainable in the traditional structural materials. According to the determination of material composition is the heterogeneous system consisting numbers of phases with different physical-chemical nature, which is characterized by the presence of the developed internal boundary surfaces, gradients concentrations and internal stresses.

Any composition material can be presented as a combination of matrix and filler. The success of creation of composition materials depends on ensuring of the controlled physical-chemical interaction of matrix and filler.

Application of composition materials based on polymers and copolymers of vinyl chloride is restrained due to their low thermal stability. In addition, the introduction of different target additives in PVC-compositions reduces the resistance of the polymer.

Traditional filler, which also acts as a flame retardant-stabilizer is an antimony oxide, used in a number of 1–7%. New extender-retardant is "Duosonic" – a complex hydrated salt of carbonate sodium-aluminum.

Although the bicarbonate of sodium and potassium and ammonium carbonates are part of the driest powder extinguishing materials, they can't be used in most polymeric materials due to the low temperature of decomposition.

2 FORMULATING THE PROBLEM

The aim of this work is to obtain effective stabilizers by activation of limestone ($CaCO_3$) in the verti-cal vibrating mill (MVV), engineered in National Mining University.

3 RESULTS

Calcium carbonate widely used as a cheap inert filler for many polymeric materials is quite effective fire retardant mean mainly due to the effect of dilution.

The effectiveness of $CaCO_3$, as a stabilizer-flame retardant is negligible at low concentrations and temperatures below 1000 °C. It is because at a temperature below the 1000 °C calcium carbonate (limestone, swept) used in composite materials, starts to decompose, however, the rate of emission of carbon dioxide at this is so great that it does not have time to prevent burning. In addition, decomposition of filler leads to the destruction of a composite material in general and can contribute to the combustion of polymeric matrixes.

An important factor, influencing on effectiveness of the grinding process in MVV, is a gap (Δ) between the upper layer of balls and grinding chamber cover installed at the preparation of tests depending on the degree of filling camera with balls. It depends on the energy intensity of technological mass, participating in the process of grinding.

Mechanochemical activation of materials was performed in vertical vibrating mill, the main advantage of which is the reduction of time grinding powders – on one hand, and on the other –shocking impact on destructible material allows to influence on destroyed material with the penetration of deformation zone on the whole volume of the particle. This will ensure, in its turn, not only surface activation of the material, but also the violation of the internal structure of the particles and, therefore, the penetration of active zone in great depth.

For determining constant of dispersion we applied method of granulometric composition evaluation of powders by a specific surface, which is a convenient feature of dispersion.

The specific surface was determined by resistance of filtering rarefied gas. The method is based on developed B.N. Deryagin theory of flow of rarefied gas through a system of hard balls, when free path of gas molecules is much larger than the distance between the balls (Knudsen flow regime). Method bias is 1%.

The average size of the crushed particles was determined by the data obtained on "SK MICRON LAZER".

Figure 1 presents the kinetic curve of grinding $CaCO_3$.

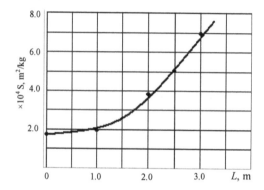

Figure 1. Kinetic curve dispersion limestone.

The formation of the activated state is closely connected with the energy characteristics of material. In this regard we used universal method of study of energetic characteristics of materials by potentiometric measurements and suspensions with indifferent electrode. Method of the potentiometer measuring allows to expect potential curve and total adsorption potential.

Figure 2 shows a general view of the potential curve of calcium carbonate.

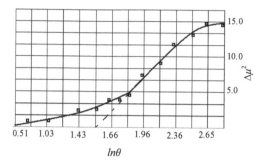

Figure 2. Potentiometer curve limestone.

Potential curve is plotted in the coordinates $\Delta\mu^2 - ln\theta$, where $\Delta\mu$ is the conditional adsorption capacity, which is calculated from experimental data; θ the degree of surface coverage for ions of iron, also designed from the experimental data. Curve consists of several linear sections, each of which corresponds to adsorption on the specific adsorption-active centers. The curve allows to calculate the followings parameters:

1. Real ground, per adsorbed ion Fe (III), nm^2;

2. θ_{01} – the value θ in the point intersection of the first section of the potential curve with the x-axis (see Figure 2), and also θ_{0i} – the degree of fillings of surface by this type of adsorption-active centers, corresponding to the intersection of this area with the x-axis;

3. θ_{ai} – value of adsorption potential in a maximum of distributing of this type of centers, kJ/mol;

4. $\theta_{\Sigma i}$ – total adsorption potential of every type of centers, kJ/mol;

5. θ_{Σ} – total adsorption potential of surface, kJ/mol.

The results of calculations of parameters are presented in a Table 1.

Table 1. Information of calculation of parameters of active centers of surface.

Time of grinding down	$\delta\theta_i$	$\Delta\mu_{ai}$, kJ/mol	Type of adsorption-active centers
$CaCO_3$, initial	0.24	0.46 0.75	Active oxygen adsorbed water
$L = 1$ m	0.09 0.14 1.97	2.09 2.75 1.49	OH^- is a group Co-ordinating-linked water hydrated
$L = 2$ m	0.60 1.2 2.3	2.3 2.07 −1,68	Co-ordinating-linked water OH^- is a group CaO – waterless
$L = 3$ m	0.75 2.26	2.07 −1.68	OH^- is a group CaO – waterless

On the basis of information of Table 1 we can conclude, that at grinding down of sample $CaCO_3$ on the sample surface, there are 2 types of adsorption-active center:

$- CaO$ – waterless ($\Delta\mu_a = -1.68$);

$- OH^-$ – group ($\Delta\mu_a = 2.07$).

Data shows how result of machining compounds with endothermic effects is formed. On the surface of the filler calcium carbonate under the action of fibrinogen a special barrier layers are formed. That prevents chemical interaction with the matrix at high temperatures.

Everybody knows that during thermal decomposition occur both monomolecular and radical chain reactions. During mechanochemical activation on surface of absorbent material takes place unmarshaling surface layers with significant excess of free energy. Localization of surface atoms in unusual provisions leads to changing of their electrons distribution. On a number of marks, the structure of surface of such materials is similar to that achieved by processing of solids and by irradiation in reactors, i.e. there is a configuration change of surface atoms status.

Study of freely-radical properties of upper layer of particles in combination with study of its adsorption capacity was held by Butyaginym, and also Nikitinoy and Kisilevym. Studies of these scientists showed that at a room or at more high temperature on a surface, sustainable long-lived free radicals are not formed. However, these studies do not exclude the possibility of formation of short-lived free radicals in the grinding process and ambient temperature. In this regard, we investigated the inhibition of the process of dehydrochlorination (DHC) and burning with help of activated $CaCO_3$.

We investigated the influence of activated calcium carbonate on the process DHC and burning of composite material on the basis of emulsion PVC paste, brand E 66P and plasticizers. As a plasticizer, we used dialkylftalat – DAF-clamps 7–8–9; triethyleneglycoldimethacrylat – TGM-3.

Thermostability was determined by decomposition of temperature (T_p K) at temperature uplift to 8 °C/min, the induction period (τ, s) before the allocation of hydrogen chloride at 448 K and number of emission hydrogen chloride (mg/g PVC) after the end of induction period at 448 K.

Widely known stabilizers in chemical industry are salts of various metals, but in works (Svetkin 1979) it is shown that more stabilizing effect can be reached in the case if these salts used in combination with heterocyclic compounds.

Therefore, to consolidate effect of activation and improvement of physico-mechanical properties of materials were used complex stabilizers, prepared by the method described work.

Figure 3 presents the kinetic curves of the process DHC, which indicate that the process DHC slowed down considerably.

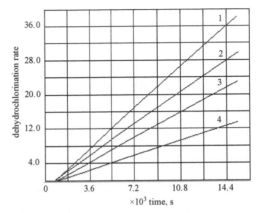

Figure 3. The kinetic curves of process of dehydrochlorination: 1 – without a stabilizer; 2 – industrial stabilizer; 3 – activated limestone; 4 – an integrated connection.

Shredding or activation of materials at MVV is connected with the heat in the vibro-impact loading and partial increase in the temperature at the point of impact leads to dehydroxylation of surface and is accompanied by formation of coordinatively unsaturated cations. This effect was first described in the work and confirmed by IR-spectroscopy and NMR. The formation of such coordinatively unsaturated cations leads to increasing of activity of compounds, thus, the data coordinatively unsaturated ions with half-filled electron shells can be considered as free valence on the surface.

Thus, since the destruction of PVC goes in two directions, the use of supplements with mixed functions can achieve a higher thermal stability based on PVC.

This phenomenon can be explained by the fact that in the process of polymerization spatially cross-linked structure formed, thanks to the existence free of double bonds, on the one hand, and with another – owing to the formation of the active surface, containing hydroxyl group (as shown by potentiometric study).

It is known that obtaining thermostable composite materials based on calcium carbonate is to replace part of the filler on the stabilizer aluminum hydroxide. As it was found out, activated $CaCO_3$ is not only as a filler, but also as stabilizer, since it does not require the introduction of additional stabilizer.

Table 2 shows the effect of mechanical activation calcium carbonate on the burning of thermoplastics.

Table 2. Properties of thermoplastics.

Composition of filler	Index of closeness of smoke, %	Rate of burning, m/mines
Unfilled polyethylene	4	0.79
40% $CaCO_3$	4	0.95
40% $Al(OH)_3$	2	extinguishing
10% activated $CaCO_3$	2	0.32

According to the results of kinetic studies we calculated constants of the reaction DGH. Table 3 shows the results of the research of activated $CaCO_3$ and its complexes with 3(5)-methylpyrazole on the processes of destruction and burning of composites.

As it is seen from Table 3, the introduction of complex stabilizers increase thermal stability, compared with just activated stabilizer and significantly improve these indicators in the case of the application of industrial stabilizers on the basis of carbonates.

Being notincresing filler, $CaCO_3$ as an aluminum hydroxide reduces durability. In this regard, it is interesting to use calcium carbonate to obtain complex stabilizers. Generally, industrial stabilizers impair physical-mechanical properties of composite materials. However, in the case of complex stabilizers usage that based on activated limestone observes increasing of these indicators. The data is given in Table 4.

Table 3. Results of tests of composition materials.

Stabilizator	Decomposition temperature (T_p K)	Induction period, $\tau \cdot 10^3$ s	Constant of DHC, $k \cdot 10^{-6}$ s^{-1}	Fire-resistance on the method of "fire pipe"	
				time of the free burning, s	a loss is in weight, %
Without a stabilizator	427	0.6	4.6	100	10.2
Industrial stabilizator	440	1.2	3.5	80	8.0
Activated limestone	460	1.9	2.0	10	4.0
Complex salt on the basis of the activated limestone	480	3.4	1.44	0	0.8

Table 4. Physical-mechanical characteristics of composite materials.

Properties of composition materials	Without a stabilizer	Industrial stabilizers	Activated stabilizers	Complex stabilizer
Tensile strength, MPa	19.6–20.2	19.6–21.6	20.0–23.0	28.5–35.6
Tensile elongation, %	1.0–1.2	1.2–1.5	1.3–1.5	2.1–2.9
Brinell hardness (B.h.), MPa	49.5–55.5	50.0–57.0	51.0–60.0	58.5–98.0

Assessment of stabilizing effect was performed with help of UV-spectroscopy. One of the main indicators of efficiency of stabilizers is their ability to prevent formation of double bonds. Formation of polyene sequences is a consequence of hydrogen chloride removal from macrochain polymer.

It is impossible to completely to stop this process. So, the activity of stabilizers is judged according to what length of polyene sequence are formed in polymer destructions. The more stabilizing efficiency of input stabilizer, the more contribution of short polyene sequences in the chain pair of double bonds.

Presence of polyene sequences and their lengths was determined according to the Table 5.

Table 5. The ratio of the wavelength and lengths mates.

Wavelength, 10^{-9} m	308	320	340	365	388	412	434	454	470
Lengths mates, n	4	5	6	7	8	9	10	11	12

Figure 4 represents electronic absorption spectra of composite materials.

Figure 4 shows that comprehensive stabilizer reduces the length of conjugated chain to 4–5 (range 1), $CaCO_3$, activated by the method of vibroactivation to 6–7 (range 2), industrial stabilizer to 7–9 (range 3), and the range of composite material without the stabilizer chain with forms of conjugated length of 10–12 links.

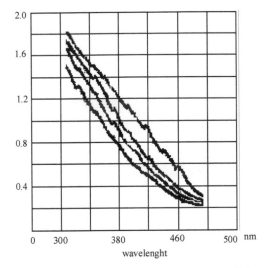

Figure 4. UV spectra: 1 – without a stabilizer; 2 – industrial stabilizer; 3 – activated limestone; 4 – an integrated connection.

4 CONCLUSIONS

Mechanochemical activation of calcium carbonate $CaCO_3$ leads to change of energy characteristics of the source material. The change of physical-chemical structure $CaCO_3$ leads to formation in near-the-surface layer active particles which are inhibitors of DHC process. Thus, by vibroactivation in MVV of calcium carbonate we can obtain filler-stabilizer for composite materials on the basis of PVC, which is more effective than widely applied in industry regulator on the basis of carbonates.

REFERENCES

Svetkin, Yu.V. & Rud, L.G. 1979. *Stabilization of PVC by complex compounds*. Donets'k: Ukrplastmash: 13.

Progressive Technologies of Coal, Coalbed Methane, and Ores Mining – Bondarenko, Kovalevs'ka & Ganushevych (eds)
© 2014 Taylor & Francis Group, London, ISBN: 978-1-138-02699-5

Control of stability of mine workings equipped with roof bolting

R. Tereschuk, O. Grigoriev, L. Tokar & V. Tikhonenko
National Mining University, Dnipropetrovs'k, Ukraine

ABSTRACT: A problem of elastoplastic stability of hydrostatically pressurized structurally heterogeneous rock mass broken by circular working is considered. Effect of internal and external dimensions area of inelastic deformations and values of relative volume on radial displacements within the working contour is analyzed. Effect of roof bolting in a mine working on its stability is studied.

1 INTRODUCTION

R. Fenner and A. Labass originated research of rock mass stress-and-strain state in the neighbourhood of underground mines based on elastoplastic medium model. They considered long mine working with circular cross-section driven in elastic isotropic rock mass where horizontal stress is equal to a unit.

Research of R. Fenner and A. Labass was followed by K.V. Ruppeneit. He detailed a problem concerning equal state of rock medium involving separate mine working. The problem has been solved applying a method of small parameter for conditions of non-hydrostatically pressurized rock mass in terms of rectilinear envelope of boundary circles of basic stresses.

Use of linear criteria of Coulomb-Mohr or Tresca-Saint-Venant type as a breaking factor with substantial simplification of mathematical procedures during the problem solving results in linear dependence of radius of inelastic deformation area on a mine working depth that cannot be confirmed by full-scale measurements. To great extent the fact has resulted in a reason for setting elastoplastic problems in geomechanics; in this problems breaking factor was assumed as a curved envelope of basic stress boundary circles.

Physically a problem of determining field of stresses and displacements in the neighbourhood of long horizontal mine workings is the study of stress-and-strain state of an unponderable plate weakened by proper number of specified-diameter holes under known boundary conditions. Such a formulation makes it possible to apply well-developed mechanical approaches for deformable solid body where success was made owing to application of complex variable theory.

D.D. Ivlev was the first who solved approximately the problems in displacements using an approach of small parameter. Then N.I. Ostrosablin gave correct formula for displacements within elastic area basing on a condition of material noncontractibility (Haar-Karman theory). A.S. Kosmodamiansky and V. Mirsalimov considered elastoplastic problems for a medium containing infinite series of similar circular holes. B.D. Annin and N.I. Ostrosablin found approximate solution for problems concerning elastoplastic distribution of stresses around a finite number of circular holes. L.M. Kurshin and I.D. Suzdalnitsky solved elastoplastic problems for a plane weakened by biperiodic system of circular holes.

All the problems assume that plastic area covers the whole contour of a mine working. P.I. Perlin applying his own numerical method solved a number of elastoplastic problems in terms of partial coverage of a hole with plastic area.

Applying method by P.I. Perlin, V.I. Sazhin researched elastoplastic distribution of stresses in the neighbourhood of non-circular mine workings located in hydrostatically pressurized rock mass.

Recently physical model of rock medium has been generalized for structurally heterogeneous solid body. A.I. Kuznetsov has solved one of the first elastoplastic problems for specific heterogeneity. Then research of elastoplastic distribution of stresses and deformations in the neighbourhood of mine workings in weaken rock medium was carried out in (Glushko 1980 & Baklashov 1986) papers and by a number of other authors.

Complication of physical models being the basis for solving a problem of determination of parameters of rock mass elastoplastic state is required to reach maximum adequacy of analytical results for full-scale measurements.

The paper objective is to research a degree of roof bolt effect in a mine working on its stability.

2 MAIN PART

Generally geomechanical research starts from the development and validation of the three models: rock medium, loads applied to the boundaries of area under study, and a mine working contour. Development of adequate mechanical model of rock medium formed in the process of loading is the most complicated problem which solution always depends on proper laboratory and full-scale tests.

When mining depth is sufficient, destruction of rocks in the neighbourhood of the given mine working, takes place as a result of elastic potential energy of gravitationally pressed rock mass effect. Rock mass loading towards formed cavity is rather slow and very inelastic process. Experiments and analytical research by G.M. Mankovskiy, N.S. Bulychiov, Y.A. Veksler, K.V. Ruppeneit, I. Winter and P. Stassen, R. Parashkevov, B. Schwartz and others confirm that. All of them demonstrate that resistance of enclosing and standing supports effects insignificantly on a value of mine working contour displacement and dimensions of inelastic deformation area.

Thus, we assume that structural model should reflect peculiarities of inelastic deformations of rock mass having some fundamental mechanical features.

Consider a problem of elastoplastic stability of hydrostatically pressurized structurally heterogeneous rock mass broken by a circular mine working.

For one-dimensional problem in a polar coordinate system, initial ratios will be as follows
– balance equations

$$\frac{d\sigma_r}{dr} - \frac{\sigma_\theta - \sigma_r}{r} = 0, \tag{1}$$

– displacement compatibility

$$\frac{d^2\varepsilon_\theta}{dr^2} + \frac{2}{r} \cdot \frac{d\varepsilon_\theta}{dr} - \frac{1}{r} \cdot \frac{d\varepsilon_r}{dr} = 0, \tag{2}$$

– ratios of Hooke

$$\varepsilon_r = \frac{1}{2G}\left[(1-\mu)\sigma_r - \mu\sigma_\theta\right], \tag{3}$$

$$\varepsilon_\theta = \frac{1}{2G}\left[(1-\mu)\sigma_\theta - \mu\sigma_r\right]. \tag{4}$$

And ratios of Cauchy

$$\varepsilon_r = \frac{dU}{dr}; \qquad \varepsilon_\theta = \frac{U}{r}, \tag{5}$$

where σ_r, σ_θ and ε_r, ε_θ – are radial and tangential components of stress and deformation respectively; U – radial displacement, G – shear modulus, and μ is Poisson ratio.

Boundary conditions and conjunction conditions are

$$\sigma_r = \sigma_\theta = \gamma H \qquad \text{if} \qquad r \to \infty, \tag{6}$$

$$\sigma_r = P_0 \qquad \text{if} \qquad r = 1, \tag{7}$$

$$\sigma_r = \sigma_r^{(1)}, \; U_r = U_r^{(1)} \text{ if} \qquad r = r_L. \tag{8}$$

All the components of stresses and displacements in elastic area will not have any index and those in plastic one will have index 1.

Introduce stress function into elastic area as follows:

$$\sigma_r = \frac{1}{r} \cdot \frac{dF}{dr}; \; \sigma_\theta = \frac{d^2F}{dr^2}.$$

Then expression (1) is satisfied identically and from (2) we obtain Euler equation using ratio of Hooke

$$\frac{d^4F}{dr^4} + \frac{2}{r} \cdot \frac{d^3F}{dr^3} - \frac{1}{r^2} + \frac{d^2F}{dr^2} + \frac{1}{r^3} \cdot \frac{dF}{dr}. \tag{9}$$

Having developed its solution and satisfying outer boundary conditions (6) we obtain formula to determine stress components in elastic area

$$\sigma_r = \gamma H - \frac{C}{r^2}, \qquad \sigma_\theta = \gamma H + \frac{C}{r^2}, \tag{10}$$

where C – unknown integration constant determined from the conditions of radial stress conjunction within L contour (8).

Find extra radial displacements in elastic area from equation of Hooke law (4) using ratios of Cauchy

$$U = \frac{C}{2Gr}. \tag{11}$$

Physical equation $\sigma_\theta - \sigma_r = 2k\left(\frac{A}{r^2} - B\right)$ is true in elastic deformation area. Solving it together with balance Equation (1) we obtain expressions for stress components in plastic area taking into consideration boundary conditions (7)

$$\sigma_r^{(1)} = -2k\left[0.5A\left(r^{-2} - 1\right) + B\ln r\right] + P_0, \tag{12}$$

$$\sigma_\theta^{(1)} = -2k\left[0.5A\left(r^{-2} + 1\right) + B\ln r\right] + P_0. \tag{13}$$

412

If $r = r_L$, we obtain a value of unknown integration constant $C = kr_L^2$ taking into account radial stresses equality determined by formulas (10) and (12).

Hence, stress components are determined in elastic and plastic areas. Then, using (8) and (12) obtain transcendence expression to determine radius of inelastic deformation area

$$0.5A\left(r_L^{-2} - 1\right) + B\ln r_L = \frac{\gamma H - P_0}{2k} - \frac{1}{2}, \qquad (14)$$

where A and B – coefficients are determined by

$$A = \frac{r_L^2}{1 - r_L^2}\left(1 - k_{res}\right); \quad B = \frac{r_L^2 - k_{res}}{1 - r_L^2},$$

where k_{res} – residual strength.

Obtained formula (14) follows from the ratio

$$\varepsilon_\theta^{(1)} + \varepsilon_r^{(1)} = \varepsilon_v(r). \qquad (15)$$

Using Cauchy ratio (5) and $f'(r) = 1 + B - Ar^{-2}$ expression for weakening function we obtain the initial heterogeneous differential equation

$$\frac{dU}{dr} + \frac{U}{r} = \varepsilon_v^*\left(1 + B - \frac{A}{r^2}\right). \qquad (16)$$

Solution for an adequate homogeneous equation is

$$U^{(1)} = Cr^{-1}. \qquad (17)$$

Varying the constant and taking into consideration the equality of radial displacements within L contour we obtain the expression for displacements in plastic area

$$U = \frac{\varepsilon_v^*}{2r}\left[(B+1)\left(r^2 - r_L^2\right) - 2A\ln\frac{r}{r_L}\right] - \frac{kr_L^2}{2G_r}. \qquad (18)$$

Research shows that elastic radial displacements within L contour can be neglected for a value of a mine working contour displacements; hence if $r = 1$ then $\frac{kr_L^2}{2G_r}$ value in a right side of (18) equation can be neglected as well. In this case taking into consideration $A = \frac{r_L^2}{1 - r_L^2}\left(1 - k_{res}\right)$, $B = \frac{r_L^2 - k_{res}}{1 - r_L^2}$ and $k = 0.5\sqrt{R_c^2\psi + (1 - \psi)R_c(\sigma_1 + \sigma_3)}$ we obtain fol-

lowing expression for displacements within a mine working contour

$$U_0 = \varepsilon_v^*\left[0.5 - \left(\left(\frac{R_c k_c}{\gamma H}\right)\psi + 2(1 - \psi)\frac{R_c k_c}{\gamma H}\right)^{-0.5}\right], \qquad (19)$$

where k_c – coefficient of rock mass structural weakening.

Expressions (14) and (18) as well as their conclusions are true if $\frac{R_c k_c}{\gamma H} < 1$.

As $\psi = 0$ for brittle rocks in the abovementioned dependences one may conclude that expressions (18) and (19) will be

$$U_0 = \varepsilon_v^*\left(0.5 - \sqrt{\frac{\gamma H}{R_c k_c}}\right). \qquad (20)$$

Thus, analysis of basic formulas has shown that initial physical model rather adequately explains complex mechanical processes taking place in a border zone.

Abovementioned basic dependences for elastoplastic state of rock mass parameters in the neighbourhood of separate mine working can help to determine certain (point) values of probabilistic values: radius of inelastic deformation area r_L and on mine working contour U_0. Their true values in each mine working section will change randomly due to the fact that such determining parameters as uniaxial compression strength, rock density, and coefficient of volumetric dislodgement are random as well.

Mining results in unbalancing of virgin mass state and developing mechanical processes when rocks are displaced into cavities formed.

At certain depth a level of stress concentration in the neighbourhood of a mine working results in enclosing rock destruction and closed area of broken rocks is formed around the mine working. The area is called a zone of inelastic deformations. Process of the zone formation is practically instant or to be exact at the velocity of sound for the environment.

Outer dimensions of inelastic deformation area within a mine working neighbourhood cannot depend on a type of its support as while using standard technique it is impossible to fix a support when border mass is outcropped; instead they are determined by a factor of mining conditions (strength characteristics of enclosing rocks and mine working depth

$$r_L = exp\left(\sqrt{\frac{\gamma H}{2R_c k_c}} - 0.5\right)).$$ Figure 1 (Shashenko 1988) demonstrates dependence of inelastic defor-

mation area value on mining factor for various mining and geological conditions.

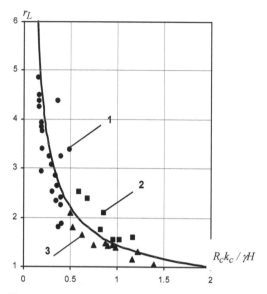

Figure 1. Dependence of inelastic deformation area radius on mining factor value: 1 – data by N.E. Kostomarov (Kostomarov 1975); 2 – data by Y.Z. Zaslavskiy (Zaslavskiy1966) and 3 – data by E.V. Streltsov (Streltsov 1972).

Mine working contour displacements depend on a value of relative increase in rock volume in terms of plastic dislodgement ε_v^*, mining factor, and a degree of enclosing rocks brittleness ψ (Figure 2)

$$U_0 = \varepsilon_v^* \left[0.5 - \left(\left(\frac{R_c k_c}{\gamma H} \right) \psi + 2(1-\psi) \frac{R_c k_c}{\gamma H} \right)^{-0.5} \right].$$

Inner dimensions of inelastic deformation area depend greatly on a type of a support fixed in the mine working as in a plastic area rock mass weakens and dislodges gaining certain residual strength and adequate value of relative capacity increase ε_v^* providing mine working contour displacement. Control of border mass state is to reduce ε_v^* value either by rock strengthening within the area with the help of shotcreting methods or with the help of timely roof bolting being an artificial structural component.

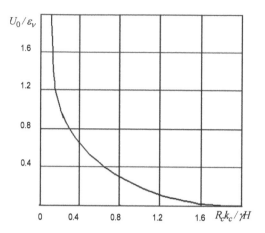

Figure 2. Dependence of mine working contour displacement on mining factor (if $\psi = 0.1$).

Timely and correct support fixing in a mine working changes structure of border rock mass resulting in decrease of relative capacity value ε_v^*. Depending on support type, a value of relative capacity will decrease for bearing support less and for rock mass strengthening supports (roof bolting) more (Figure 3).

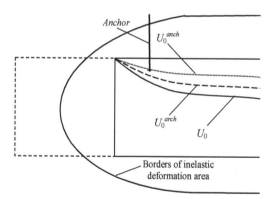

Figure 3. Description for various support types effect on a value of border mass displacement value.

When ε_v^* decreases expressions (19) and (20) help to obtain decrease in radial displacements within a mine working contour by a value

$$\Delta U_0 = {}_\Delta \varepsilon_v^* \left[0.5 - \left(\left(\frac{R_c k_c}{\lambda H} \right) \psi + 2(1-\psi) \frac{R_c k_c}{\lambda H} \right)^{-0.5} \right].$$

In turn decrease in contour displacements will result in decrease of mine working stability by $\Delta \omega$ value Figure 4 (Solodiankin 2009).

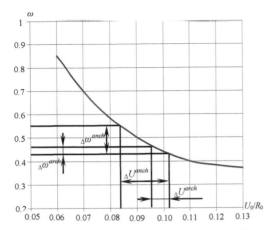

Figure 4. Dependence of a mine working stability factor $\omega = S^*/S$ (S^* is the total length of maintenance-free sites, S is the total mine working length) on relative displacements of mine working contour U_0/R_0 (U_0 is the radial contour displacements and R_0 is the mine working radius).

3 CONCLUSIONS

Thus, availability of roof bolting will result in displacements minimization within mine working contour which in turn effect mine working stability. To identify efficiency of roof bolting systems with various parameters on mine working stability it is expedient to perform further laboratory and mathematical simulation as only complex approach allows studying and evaluating roof bolting effectiveness.

REFERENCES

Glushko, V.T. & Vinogradov, V.V. 1980. *Rock failure and mine pressure manifestation predicting.* Moscow: Nedra: 214.

Baklashov I.V. & Kartozia, B.A. 1986. *Mechanical processes in rock masses.* Moscow: Nedra: 272.

Shashenko, A.N. 1988. *Underground mine working stability in heterogeneous rock mass.* Dr. Sc. En. 05.15.04; 05.15.11. Dnipropetrovs'k: 507.

Kostomarov, N.E. & Pushkariov, V.I. 1975. *On the determination of rock mass strength.* Kolyma, 12: 9–12.

Zaslavskiy, Y.Z. 1966. *Analysis of mine pressure manifestation in permanent workings of deep mines in Donetsk coal field.* Moscow: Nedra: 180.

Streltsov, E.V. 1972. *On the supports of permanents workings in mines of Donbass Central Region.* Mine construction, 9: 17–21.

Solodiankin, A.V. 2009. *Geomechanical models in the system of deep coal mines geomonitoring and methods to provide stability of extended mine workings.* Dr. Sc. En. 05.15.04; 05.15.09. Dnipropetrovs'k: 426.

Progressive Technologies of Coal, Coalbed Methane, and Ores Mining – Bondarenko, Kovalevs'ka & Ganushevych (eds)
© 2014 Taylor & Francis Group, London, ISBN: 978-1-138-02699-5

Petrographic characteristic of middle carboniferous coal of Bashkirian formation in Lozovskoi coal area

V. Savchuk, V. Prykhodchenko, D. Prykhodchenko & V. Tikhonenko
National Mining University, Dnipropetrovs'k, Ukraine

ABSTRACT: The article gives detailed petrographic characteristics of Middle Carboniferous coal of Bashkirian formation of Lozovskoi coal-bearing Carboniferous period of Western Donbass. Classification according to petrographic types of continuous and economically efficient minable formations of the area is developed. Peculiarities of petrographic composition changes in stratigraphic sequence of coal-bearing series of seams are determined.

1 INTRODUCTION

Historically Donbass basin area is subdivided into five groups according to the coal industry location: Old, Western, Northern, Eastern, and Southern Donbass. Old Donbass covers basic groups of geological and industrial coal areas. Intense coal mining within this area was accompanied by a number of operations to have thorough study of coal quality and composition as well as to substantiate their efficient industrial use. Northern Donbass and partially Western Donbass are the least studied areas. Mississippian carboniferous coal of Western Donbass being mined intensely today and has been studied in detail. Productive deposits of Middle carboniferous coal have been found northward from Mississippian carboniferous deposits. The following geological prospecting operations marked new coal area that called Losovskoi. Considerable reserves of mineral coal (7.8 bln tons) of low carbonization phase are concentrated here at comparatively low depth (140 – 775 m).

2 FORMULATING THE PROBLEM

Depletion of industrial coal reserves in Old Donbass requires extended use of coal of other regions including the one of Losovskoi coal area. Problem of their efficient industrial use is of high importance nowadays. The technology of advanced complex processing that have replaced coal combustion in furnaces requires more detailed coal petrographic study of coals with finding out their genetic peculiarities that are taken into account while developing raw material base for chemical coal conversion. Such studies have not only practical but also theoretical importance (Resources of hard fuel of Ukraine 01.01.2001).

3 MATERIALS UNDER ANALYSIS

Industrial coal-bearing characteristic within the area is limited to middle carboniferous deposits of series of seams $C_1^2 - C_2^7$ included natural northern-western extension of Old Donbass productive thicknesses.

Moscovian formation is defined by the highest coal-bearing characteristics with maximum minable coalfield being limited to series of seams C_2^6 and C_2^7. Maximum of general and minable coalfield within Bashkirian formation is limited to series of seams C_2^2. Industrial Moscovian coal seams are found mostly in northern area while seams of Bashkirian formation are found in southern part of Lozovskoi coal area being limited to series of seams C_2^1, C_2^2, C_2^3, and C_2^4.

Series of seams C_2^1 is represented by sandy and clay thickness with thick sandstone benches and quite frequent limestone and coal seams. Series of seams thickness grows towards the north-west within 190–250 m. The series of seams contains 8–10 limestone seams and 14 coal seams and interlayers. Seams f_0^7 and f_1^3 with the thickness varying within 0.60–1.30 m are of industrial importance. Seam f_0^7 is characterized by the highest areal extent.

Seam f_0^7 is made up by semilustrous coal, finely and medium banded striated-based coal; sometimes its semi-dull finely striated variations can be found. There can be minor leather coats and fusain lenticles along deposition plane; thin mudstone bands and lenticles can be detected less frequently. Coal is fissured. Fissure surfaces are caused by calcite and pyrite coating.

Microscopically the coal is humic, mostly glance, sometimes duroclarain one.

Gelified substance is mostly reddish-brown, sometimes brown-orange and red in transmitted light. Prevailing structural elements are as follows: structural vitrain, xylovitrain, and lumpy ground mass. Rare vitrain lenticles are inconsistent in thickness. Vitrain bands thickness is often within 500–1500 mcm. They are usually yellowish preserving the cellular structure. There can be gellified fragments which cell interior is filled with yellow substance. Cell walls are flattened and elongated along the bedding. Sometimes there can be oval gellified bodies of unclarified nature inside the vitrain lenticles. There can be relatively rare large vitrain lenticles with poor cellular structure. They are non-uniform and mottled. Such vitrains are often delineated by more fusainized substance with cellular structure. As a rule, these are semixylovitrain-fusain transforming into fusain. Apart from these fusain variations, there can be its smaller fragments with well-marked parenchymal structure. Such vitrains are often delineated by cuticles. The less vitrain fragment is, the better cuticle preservation is. Xylovitrain is quite rare. Ground transparent mass is represented by lumpy substance of leaf parenchyma gelification. There is also vitroattritus along with parenchoattritus.

Semivitrinite group is more often represented by relatively large fragments with well-preserved structural remains. There are sometimes structureless semivitrinite fragments in the form of small fragments. Minor amounts of fusainized components can be found in the form of fusain-vitrain lenticles as well as in the form of structural fusain. Both their fine- and large-cellular variations are available often being with deformed cellular structure. Their pores are made up either by thin-disperse clay substance or by calcite that is less frequent occurrence. Vitrain-fusains are usually limited to heterogeneous bands of duroclarain or sometimes clarain. There can be also fusainized tissues with cellular structure being delineated by cuticles. Small amounts of scleretinite can also be available.

Minor amounts of microspores, cuticles, and macrospores can be singled out among cutinized coal elements. There can be single micro- and macrosporangia mostly of poor preservation. Large macrospores are rare. Microspores are of linear-type form being up to 100 mc long, sometimes with structural ornaments. Macrospores are more often thin-walled with medium thickness exine sometimes containing perispore remains. Cuticle is mainly thin-walled being sometimes medium-walled. They often occur angle-wise as for the bedding. There are small amounts of elongated straw-coloured resin bodies. Cutinized elements are quite granulated, partially up to attritus. They are mostly orange-yellow. Substance composition of seam f_0^7 is generally clarain with the content of vitrinite (87.0%), inertinite

(6.0%), and liptinite (5.5%) group components (Table 1). According to classification of A.P. Karpinsky Russian Geological Research Institute (Petrographic types of USSR coal 1975) as for petrographic composition the coal belongs to class of gelitolites, subclass of gelites being represented by two coal types – fusinite-gelites (60%) and lipoid-fusinite-gelites (40.0%).

Series of seams C_2^2 is located in central and southern part of Lozovskoi coal area. Series of seams thickness is 160 m. It is characterized by wide development of thick sandstone masses limited mostly to the middle part of the series of seams. Upper and lower series of seams parts are characterized by the development of siltstones and mudstones. Series of seams section shows 6 limestone levels without ubiquitous extent. There are up to 11 coal seams in the series of seams though only seams g_1^3, g_1^2, and g_1 have minable thickness of 0.60–2.10. As for morphology persistence the seams belong to the group of irregular ones; only seam g_1^3 is characterized as relatively continuous one. All the coal seams are subjected to intraformational washout. Levels of washed-out seams are replaced by sandstone.

Seam g_1^3 is made up by semudull finely striated coal with the interlayers of striated and banded coal. Vitrain bands thickness reaches up to 2 mm. Fusain lenticles can be observed along the bedding. Coal is fissured. Fissure surfaces are covered by calcite coating. Coal is black with pitch glance. Fracture is uneven.

Microscopically coal is represented by clarain of mixed composition. The coal is vitrinite-attrite and fragmented, homogeneous and lumpy. There are often vitrain fragments with partially preserved structure of plant tissue. As a rule, their thickness is within 300–1000 mcm. Parenchymal tissues are marginated by cuticles. There is gradual transition from homogeneous to lumpy areas. Vitrinite colour is brownish-red. Moreover, vitrinite fragments show some mottling.

Attrite prevails among fusainized components. It is most often represented by xylovitrain-fusain; it is less often represented by semixylainvitrain-fusain. There are separate straticules which composition shows increased content of xylovitrain-fusain lenticles with the thickness of 100–500 mcm and length of 500–3000 mcm. Lipoids do not have wide extent being represented by lipoid-attritus, rarely by microspores of linear form of 40–60 mcm long and microspores with thickened exine of 100–120 mcm long. Megaspores with the exine of 8–10 mcm thick and 1000–1500 mcm long are rare. There are minor amounts of macrosporangia with thin cutinized vessels and resinoid bodies. All the cutinized elements except macrospores are lemon-yellow. Macrospores are usually reddish.

Among mineral impurities clay matter straticules of hydromicaceous composition interlaying with vitrinite are most often to be found. Separate coal straticules contain large amount of pyrite and calcite. Quartz grains are most often to be seen in cutinized elements; less often, they are found in fusain pores. In terms of total petrographic composition, coal of seam g_1^3 has no practical differences from seams g_1 and g_1^2.

Generally, material composition of the series of seams coal is clarain with the content of some components of vitrinite (88.3%), inertinite (5.8%), and liptinite (4.3%) groups. According to classification of A.P. Karpinsky Russian Geological Research Institute as for petrographic composition the coal belongs to class of gelitolites, subclass of gelites (96.2) 5 being represented by three coal types – fusinite-gelite (41.8%) and lipoid-fusinite-gelite (57.0%), and lipoid-gelite (0.6%). Coal of gelite subclass is available in small amounts (3.2%) being represented by lipoid-fusinite-gelite type.

Table 1. Petrographic composition of Bashkirian formation coal.

Series of seams	Petrographic composition, %				Substance and petrographic classification according to A.P. Karpinsky Russian Geological Research Institute		
	$Vt + Sv$	I	L	ΣOK	Class	Subclass	Type
C_2^4	74.8	15.5	9.7	16.2	Gelitolites	Gelites 31%	fusinite-gelites 27% lipoid-gelites 0% fusinite-lipoid-gelites 4%
						Gelitites 69%	fusinite-gelites 57% lipoid-gelitites 0% fusinite-lipoid-gelitites 12%
C_2^3	87.7	7.1	5.2	8.6		Gelites 96.2%	fusinite-gelites 71% липоидо-гелиты 0% fusinite-lipoid-gelites 25.2%
						Gelitites 3.8%	fusinite-gelites 0% липоидо-гелититы 0% fusinite-lipoid-gelitites 3.8%
C_2^2	89.9	5.8	4.3	7.1		Gelites 99.4%	fusinite-gelites 41.8% lipoid-gelitites 0.6 % fusinite-lipoid-gelites 57 %
						Gelitites 0.6%	fusinite-gelites 0% lipoid-gelites 0% fusinite-lipoid-gelitites 0.6%
C_2^1	88.5	6.0	5.5	7.0		Gelites 100%	fusinite-gelites 60% lipoid-gelites 0% fusinite-lipoid-gelites 40%
Over the area	85.3	8.5	6.2	9.7		Gelites 81.6%	fusinite-gelites 49.9% lipoid-gelites 0.2% fusinite-lipoid-gelites 31.5%
						Gelitites 18.4%	fusinite-gelites 14.3% lipoid-gelitites 0% fusinite-lipoid-gelitites 4.1%

Series of seams C_2^3 is characterized by development of thick (up to 30–50 m) sandstone benches limited to lower and upper series of seams parts. Thickness between sandstones is made up by siltstones and mudstones with coal and limestone bands. The series of seams contains up to 3–5 limestone bands and up to 8–10 coal seams and bands. Seams h_2, h_5, and h_{10} are the main minable ones with the thickness of 0.60–1.40 m. If we take total area, according to morphology the seams belong to the group of irregular ones.

Seam h_{10} is made up mostly by semilustrous homogeneous and banded coal with the areas of lustrous and semidull striated one. There can be often found thin fusain lenticles, pyrite coating, mudstone layers, and calcite lenticles along the deposition plane.

Microscopically coal is clarain with duroclarain areas. Vitrain bands and fragments are usually reddish-brown with the remains of cellular structure. Darker gellified substance sometimes encrusts the cavities. Vitrinite is often non-uniform and mottled. There can be oval gellified bodies of unclarified nature. Vitrinite of heterogeneous layers is lumpy being often crushed to attrite. Vitrinite areas being remains of parenchymal tissue are quite often as well.

Fusainized tissues are scattered within vitrinite non-uniformly in the form of small pieces and attrite; less often there can be large lenticles of structural fusain and xylofusain. Duroclarain layers show increased content of nontransparent ground mass. Parenchyma of such layers is represented as pieces of dark brown colour. Pedicellate tissues prevail among gellified components. Large vitrain lenticles in duroclarain are more often structureless with soft edges.

Liptinite group is represented by cuticle, mega- and macrospores, mega- and microsporangia, and lipoid attrite. Cuticles are most often thin and smooth in the form of pieces of various sizes. It most often edges vitrain fragments. Pieces of thicker and dentate cuticles are less often to be found. Large-size megaspores up to 1200–15000 mcm long are characterized by granulation tissue. Microspores are thin and fine (40-60 mcm). They are quite often preserved badly. There can be micro- and megasporangia and their pieces. There are a lot of oval resin bodies of different length. They sometimes underlay angle-wise to the bedding. They are lemon-yellow. Macrospores are mostly orange-yellow while micro- and macrospores are yellow. Coal pyrite is globular being dissipated within vitrinite non-uniformly. Its amount is minor. Generally, material composition is clarain with the content of some components of vitrinite (85.5%), inertinite (7.1%), and liptinite (5.2%) groups. According to classification of A.P. Karpinsky Russian Geological Research Institute as for petrographic composition the coal belongs to class of gelitolites, subclass of gelites (96.2)5 being represented by three coal types – fusinite-gelite (71.0%) and lipoid-fusinite-gelites (25.2%). Coal of gelite subclass are available in small amounts (3.8%) being represented by lipoid-fusinite-gelite type.

It should be noted that coal seams of this series of seams are characterized by the highest variability of petrographic composition. Thus, average content of inertinite group along series of seams section varies from 4.0 (seam h_4) up to 11.5% (seam h_1) and content of vitrinite group varies from 81.1 (seam h_1) up to 87.0% (seam h_{10}). Content of liptinite group on this series of seams varies within the highest degrees – from 4.5 (seam h_4) up to 7.0% (seam h_2).

Series of seams C_2^4 is located in central and northern parts of the area. Its deposits are represented mostly by mudstones, aleurites with coal and limestone bands. Average series of seams thickness is 170 m. Series of seams thickness decreases gradually from the north-east to the south-west.

Contrary to series of seams C_2^3, it is characterized by drastically less sandstone thicknesses. Series of seams deposits contain 10 coal seams and bands among which minable thickness of two seams (i15 and i3) is 0.60–1.25 m being main industrially important ones. Minable thickness of coal seams on separate boreholes is 0.60–0.70 m.

Seam i_1^5 is made up mostly of semilustrous thin and non-uniform banded coal with small fusain lenticles. Sometimes there can be transitions to semidull coal.

Semilustrous coal is clarain and duroclarain both spore and mixed spore-fusain one. Generally, material composition is duroclarain with the content of some components of vitrinite (71%), inertinite (14%), and liptinite (12%) groups. Mineral inclusions are represented by clay minerals and iron sulphides.

When the microscope is used it is seen that semidull coal is the interlaying of gellified and lipoid substance with clay one.

Seam i_1^5 is the alternation of semilustrous finely banded coal layers and bands of semidull coal and rare bands of lustrous one with the thickness up to 3 mm. Fusain lenticles occur along the bedding. Coal is dull because of microscopic lenticules of clay matter or fusain lenticles and microspores.

Coal microstructure is clarain and duroclarain, sometimes clarodurain.

Microscopically coal is clarain with ultraclarain bands. Coal is made up by finely banded vitrain and heterogeneous bands. Vitrinite is homogeneous, sometimes lumpy. It is reddish-brown. Gellified substance is mostly semistructured. Components of inertinite group are represented by small and large fusain lenticles being distributed non-uniformly. In the upper part their cavities are often filled with calcite while in the middle and lower seam parts they are filled mostly with finely dispersed clay matter. Bands of semidull coal are represented by the interlaying of vitrinite and fine lenticles of finely dispersed clay. Liptinite components are shown by linear microspores, small amount of macrospores, high content of resin bodies especially in the upper seam part, fine and cellular cuticle. Small fragments of cutinized elements and their attritus are most often to be found. They are orange-yellow. Generally material composition of the seam is duroclarain: vitrinite – 71.5%, inertinite – 17.0%, and liptinite – 10.5%. The seam is characterized by high content of resin bodies, available cellular cuticle, large macrospores with the exine up to 40 mcm, and high values of durain coefficient.

Overall, material composition of the series of seams is duroclarain with the content of some components of vitrinite (73.8%), inertinite (15.5%), and liptinite (9.7%) groups. According to classification of A.P. Karpinsky Russian Geological Research Institute as for petrographic composition the coal belongs to class of gelitolites, subclass of gelitites (56.0%) and gelites (44.0%) being represented by two coal types – fusinite-gelite (57.0 and 27% correspondingly) and lipoid-fusinite-gelites (12.0 and 4.0% correspondingly).

4 CONCLUSIONS

1. As for basic material coal of Bashkirian formation belongs to humus coal with rare inclusions of low-thickness sapropelite bands.

2. Macroscopically coal is semilustrous thin-banded and striate-based with the inclusions of fusainized tissues in the form of striae and lenticles of low thickness.

3. Macroscopically coal is fragmented and attrital with different degree of gellified substance preservation.

4. According to petrographic composition coal belongs to the class of gelitolites, subclass of gelites and gelitites with prevailing fusinite-gelite and fusinite-gelitite types.

5. In stratigraphic section, petrographic coal composition shows the decrease of the amount of vitrinite group components and the increase of the amount of inertinite and liptinite group components from the series of seams C_2^1 to series seams C_2^4.

6. Coal of Bashkirian lower cycle ($F_1 - H_4$) is characterized by more stable petrographic composition with the prevailing thin-walled cuticle, thin-walled macrospores, fine fusain lenticles, and low values of durain coefficient.

7. Coal of Bashkirskii upper cycle ($H_4 - K_1$) are characterized by higher variability of petrographic composition, increased content of thick-walled megaspores, thick-walled cuticle, large fusain lenticles, numerous yellow and brownish-red oval elongated bodies, and increased values of durain coefficient.

REFERENCES

Resources of hard fuel of Ukraine (01.01.2001). Kyiv.
Petrographic types of USSR coal. 1975. Moscow: "Nedra": 248 (With figures).
Drozdnik, I. & Shulga, I. 2009. *On professional use of slightly metamorphized coal. Mineral dressing*. Issue 36 (77)– 7(78).
Yeriomin, I. & Bronovets, T. 1994. *Grade coal composition and its reasonable use*. Moscow: Nedra.
State Standards of Ukraine 3472-96. 1997. Lignite, hard coal and anthracite. Classification. Public Standard of Ukraine.
AUSS 25543-82. 1983. Lignite, hard coal and anthracite: Classification by genetic and technological parameters. USSR National Committee on standards. Moscow.
International codification system of coals of high and middle ranks. 1988. New-York. United Nations.

Progressive Technologies of Coal, Coalbed Methane, and Ores Mining – Bondarenko, Kovalevs'ka & Ganushevych (eds)
© 2014 Taylor & Francis Group, London, ISBN: 978-1-138-02699-5

Substantiation of the rational longwall length methodology

S. Vlasov
National Mining University, Dnipropetrovs'k, Ukraine
Vyatka State University, Kirov, Russia

ABSTRACT: Rational longwall length determination methodology is substantiated based on the convergence factor on the landing line of the mechanized support props. Economic effect determination methodology is proposed which is dependent on the longwall length increase due to decrease of cost per unit share for drivage of development mine workings in the general structure of coal extraction prime cost.

1 INTRODUCTION

Coal extraction efficiency increase from thin coal seams on mines of Western Donbass in Ukraine under modern conditions of economic crisis is an important macroeconomical task. Effective use of modern high-production mechanized complexes in a stope is impossible without taking into account rock massif geomechanics. All this predefines the necessity to substantiate longwall rational length.

2 FORMULATING THE PROBLEM

Longwall length is one of the main parameters of the mining method that influences technical-economic indices of a stope and production unit in total (Bondarenko et al. 2003). There is an optimal longwall length for concrete mining-geological conditions at which expenditures for 1 ton of coal will be minimal. Solution of such tasks, as a rule, is in plotting economic-mathematical model of the unit as a function of a longwall length with its following realization or by the way of the function differentiation or by the method of variants selection (Sapytskyy et al. 1999). However, except economic feasibility, the longwall length is also significantly influenced by mining-geological factors (geological faults distribution within the extraction field, quality and thickness of the mineral and so on) and technical possibilities of the used equipment.

In a general case, longwall length increase leads to increase of machinery parts of the stope and decrease of face end mechanization in the general duration of the extraction cycle that, in its turn, promotes load increase on the longwall and extracted coal prime cost decrease.

Thus, when there are no restrictions by mining-geological factors, as a rule, longwall length is accepted based on technical possibilities of a face equipment (generally, face conveyor length is the restricting factor). At the modern level of mining complex development, Ukrainian and Russian equipment producers for stoping can realize delivery of mechanized complexes for coal extraction with length of 250–300 m, import analogues (complexes DBT, OSTROJ, JOY and other) have maximal length of 400–500 m.

At the modern stage of a mining complex development, Ukrainian and Russian producers of stoping equipment can implement delivering mechanized complex for coal extraction of 250–300 m, import analogues (complexes DBT, OSTROJ, JOY and others) have maximal length of 400–500 m.

3 THE PRESENTATION MATERIALS AND RESULTS

The results of computer modeling and natural studies show that rocks convergence in a stope is described with a curve that has local extremum (maximum) in the point $x = 0$, that is in the middle of the longwall. It means that at a definite longwall length along the extraction pillar there is a possibility when rocks convergence at the props landing line of mechanized support in the middle part of the longwall will become greater than allowed yielding property of hydroficated props. In this case the support goes on a rigid base that makes it impossible to move sections of the mechanized support and is an emergency (Figure 1).

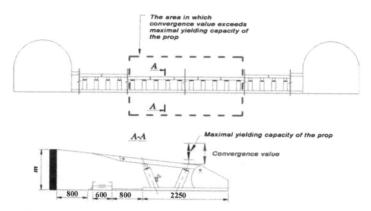

Figure 1. The most possible area of the support transformation of a rigid base.

From the plates and covers bending theory it is known that sagging of plates is directly proportional to the length to the fourth power and inversely proportional to the elasticity module multiplication and the plate thickness to the third power (Feodos'yev 1979). However, analytical solution gain for determination of the ultimate longwall length in the given mining-geological conditions for a given type of mechanized support is quite complicated. This is explained by the necessity of the task solving in volumetric setting considering complex influence of force parameters of the support of stoping and development galleries, change of physical and mechanical properties of rocks in the bedding plane and its perpendicular plane, geometrical parameters of the limit state zone in the worked-out seam roof, formation mechanism of the caving area starting from the extracted seam up to the height of a gradual sagging of rocks without discontinuities and others.

The solution of such a multi-factor task is possible only using numerical methods. Determination of the longwall rational length in the considered mining-geological conditions for the accepted face equipment is necessary to carry out by computer modeling of an extraction area with following solution of a series of direct tasks up to the moment of a strict periodicity definition during limit state zones formation of rocks around a mine gallery. Further, within the limits of an established period the maximal rocks convergence is determined at the landing line of mechanized support props (most possible place of a section landing on a rigid base). Comparing gained results with yielding ability which is allowed by the technical characteristics of used supports, the summary is made about the possibility of a modeled longwall length as a project one. Considering miscalculation of the used numeric method (LEM-limited elements method) and physical-mechanical properties of rocks within the extraction area, the modeling results should be considered within the interval of not less than 15%.

In the case when possible convergence of rocks in the longwall exceeds maximally possible constructive yielding capacity of hydroficated props, recommendations development on the correct technological decisions making is needed. As possible variants of the present situation solution the following recommendations can be used:

1. Longwall length decrease down to a sage value by the factor of rocks convergence at the line of the landing row of mechanized support props. The main disadvantage of the given variant is the use of the stoping equipment not in full that, as a result, negatively influences the extracted coal prime cost. As a variant, change of some joints and aggregates the face complex can be considered that are designed for longwall normal work ensuring with already corrected (shorter) length.

2. Use of the sections with increased bearing capacity in the zone of possible support landing on a rigid base. Bearing capacity increase of some supports (KD-80, KD-90, KD-99 and other.) is possible by the way of adjusting props valves EPV (electro pneumatic valves). Also there is a possible way of various type supports use, mutual use of general sections and sections of a "T" type (heavy type).Disadvantage of this variant is that valves EPV (electro pneumatic valves) adjusting can increase bearing capacity for not more than 20%, further increase of nominal pressure inside the hydrocylinder leads to an increased deformation of seals and, as a result, to fast failure of props. When using sections of various bearing capacity at the place of transferring from one type to another the tangent stresses in the roof are observed that can lead to an increased fracturing and local rocks fall from roof in the zone of transformation.

3. Use of mechanized support with the value of an allowed yielding capacity of hydraulic props exceeding maxima value of rocks convergence in the longwall. This solution can be realized by the way of selection of the corresponding standard dimension type of mechanized support or an order to the plant-producer of the support with necessary constructive yielding capacity.

4. Application of the roof control methods presuming compulsory caving of rocks in the worked-out area of the longwall (various types of torpedoing). The disadvantage of this method is that it is necessary to use additional expensive equipment in case of roof hydrofracture and increase of safety rules in case of drill and blast works conduction and also general complexity and labor intensity of the conducted operations (Safety rules… 2005).

5. Increase of the extracted thickness due to undercut of wall rocks in the zone of possible landing on a rigid base both along the longwall length and extraction pillar. Disadvantage of this proposal is impossibility of use when using plow complex and extracted coal quality decrease due to ash content rise with all following consequences.

Possibility of each of the mentioned variants use or their combinations depends on concrete mining-technical conditions and must be followed by feasibility study.

After making a decision it is necessary to conduct checking modeling taking into account corrected conditions and only after receive of positive result the final (rational) longwall length can be put into the project together with the face equipment type and other technological parameters.

According to the proposed methodology, the spatial model of an extraction area was plotted for the mining conditions of the 157th longwall (C_6 coal seam) at the "Stepnaya" mine. The initial data for the modeling are presented in the Table 1.

Table 1. Initial data for the model plotting.

Technological parameters	Values
Longwall length, m	303
Mechanized support	OSTROJ 70/125T
Coal seam dip angle, degree	2–4
Extracted thickness, m	1.00
Roof control	Full caving
Coal seam title	C_6
Stoping depth, m	300–490
Mining method	Pillar mining

During spatial computer modeling the rocks convergence distribution laws were received at the line of hydroficated support props – OSTROJ 70/125T

depending on the location of the longwall face along the extraction pillar.

The analysis of received dependences has allowed to establish that for the point of the distance from the face entry at 85–95 meters, the distribution of rocks convergence along the longwall length on the hydraulic props installation line is described by the following equation

$$K = 6 \cdot 10^{-10} z - 2 \cdot 10^{-5} z^2 + 0.251,$$

and starting from 115–125 m retreat and more – the following equation

$$K = 7 \cdot 10^{-10} z^4 - 3 \cdot 10^{-5} z^2 + 0.43,$$

where K – convergence, m; z – longwall length, m, $z \in [-151.5; +151.5]$.

Graphically these dependences are shown on the Figure 2.

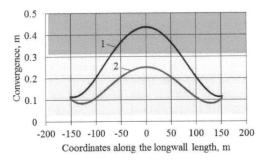

Figure 2. Convergence distribution along the longwall length: 1 – convergence distribution along the longwall length at the retreat being more than 75 m; 2 – convergence distribution along the longwall length at the retreat being 55-65 m; ▨ – area of the support normal functioning; – area in which the support transformation on rigid base is possible; ▨ – area in which the support goes on a rigid base.

From Figure 2 it is seen that in modeled mining-geological conditions and mining-technical conditions when the longwall retreats from the face entry for 55–65 m and goes on a rigid base.

The modeling results are confirmed by the experience of the 157th longwall development at the "Stepnaya" mine on which in May 2009 when moving from the face entry for 56 m there was an accident connected with the mechanized support transformation on the rigid base, although, according to the methodology that was used when making technical pattern of the extraction area, the yield factor coefficient of OSTROJ 70/125T support made up $K_{sf} = 7.35$ based on the working resistance of the support.

To provide failure-safe work of the extraction unit under the considered conditions with accepted type of the equipment (OSTROJ 70/125T mechanized support) it is necessary:

– to decrease longwall length down to the safe value (200–210 m);

– to increase the extracted thickness up to 1.2–1.25 m.

To provide failure-safe work of the extracted unit under the considered conditions with accepted longwall length (303 m) it is necessary:

– to adopt mechanized support with yielding capacity of hydraulic props not less than 0.45 m (with section resistance being 3500–4500 Kn);

– to increase the section resistance up to the value that will provide set yielding capacity of the props.

As it has been already mentioned, for any adopted technical and technological decision there must be conducted an additional volumetric computer modeling for concrete extraction unit and specific laws of convergence distribution along the longwall length depending on the face located.

Economic effect from the longwall length increase basically is in the decrease of cost per unit for drivage works in the general structure of coal extraction prime cost. Based on this, the dependence function of cost per unit for drivage on longwall length can be expresses as

$$f(z) = \frac{nLQ}{Z},$$

where Q – cost of drivage of 1 meter of a development working, UAH/m; L – length of development working, m; n – necessary number of development mine workings; Z – longwall length, m.

Economic effect from the longwall length increase in this case can found by the next formula

$$E = \int_{0}^{Z_2} \int_{\frac{nLQ}{Z_2}}^{\frac{nLQ}{Z_1}} dz d(f(z)) = \int_{0}^{Z_2} \left[\frac{nLQ}{Z_1} - \frac{nLQ}{Z_2} \right] dz =$$

$$nLQ \left[\frac{1}{Z_1} - \frac{1}{Z_2} \right] Z_2 = \frac{nLQ(Z_2 - Z_1)}{Z_1}, \text{ [UAH]},$$

where Z_1 – longwall length by the base variant (standard length), m; Z_2 – longwall length by the factor of rocks convergence in the longwall, m.

For example, taking the following averaged values for the conditions of Western Donbass: $Q = 10^4$ UAH/m; $L = 1500$ m; $n = 2$; $Z_1 = 200$ m, economic

effect from the longwall length increase for 50 m, i.e. up to $Z_2 = 250$ m, will make

$$E_{200}^{250} = \frac{2 \cdot 1500 \cdot 10^4 (250 - 200)}{200} = 7.5 \cdot 10^6 \text{ [UAH]}$$

Graphically received decision is presented on Figure 3.

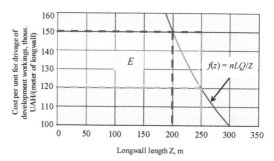

Figure 3. Determination of the economic effect from the longwall length increase.

4 CONCLUSIONS

1. Rational longwall length determination methodology is substantiated based on the convergence factor on the landing line of the mechanized support props.

2. The modeling results are confirmed by the experience of the 157th longwall development at the "Stepnaya" mine of the PJSC "Pavlogradugol".

3. Economic effect determination methodology is proposed which is dependent on the longwall length increase due to decrease of cost per unit share for drivage of development mine workings in the general structure of coal extraction prime cost.

REFERENCES

Bondarenko, V.I., Kuz'menko, A.M. & Gryadushiy, Y.B. 2003. *Technology of layered deposits underground mining*. Dnipropetrovs'k: Poligraphist: 707.

Sapytskyy, K.F., Prokof'yev, V.P. & Yarembash, I.F. 1999. *Book of problems on underground mining of coal deposits*. Donets'k: Donets'k State University: 194.

Feodos'yev, V.I. 1972. *Materials resistance*. Moscow: Science: 544.

Safety rules in coal mines. NPAOP10.0-1.01-05. 2005. Kyiv: Vidlunnia: 400.

Modeling slope stability of internal overburden dumps and quarry faces on equivalent materials

K. Seituly & B. Rakishev
Kazakh National Technical University named after K.I. Satpayev, Almaty, Kazakhstan

O. Kovrov
National Mining University, Dnipropetrovs'k, Ukraine

ABSTRACT: The paper deals with the modeling stability of internal dumps and quarry faces at the open pit "Maikubenskiy" (Kazakhstan) by using method of equivalent materials. The physical model of an internal overburden dump and quarry face with different angles of slope inclination are simulated in specially designed flat stand with using equivalent mix of sand and spindle oil. Types of deformations and the shape of sliding surface inside the rock massif are investigated. Critical values of inclination angles for the slopes of equivalent model are determined.

1 INTRODUCTION

Stability of internal overburden dumps poses one of the most essential issues while surface mining technology. It depends on the range of interdependent factors: open-pit technological operations, physical and mechanical properties of the rock mass, climatic factors, groundwater levels, shape and geometry of the foundation, and external loads. Complex influence of these factors leads to the emergence and outspreading geomechanical deformations in the dump core with the formation of rockslides, which complicate the mining operations and result in an increase of the specific mining capital expenditures. Therefore, precise estimation and effective management of open-cast mining technology and internal overburden dumping is an important engineering issue.

Physical modeling as an instrument of engineering research remains reliable technique for resolving numerous geomechanical issues including stability of slopes and internal dumps while surface mining. It is based on the theory of similarity of an artificially prepared equivalent material to the natural geotechnical object in-situ. Basic principles of this method consider replacement of natural rocks or soils by artificial materials with specific physical and mechanical properties in accordance with the scale of modeling and correlation with the properties of in-situ object. In spite of some inaccuracies inherent to physical models still widely used to calibrate numerical models and better understand the processes in rocks and soils.

2 THE SLOPE STABILITY ISSUES WHILE SURFACE MINING

Open-cast "Maikubenskiy" is a coal mine company specializing in mining brown coal of Shoptykolskoye coalfield with an annual project capacity of 25 million tons of coal per year. At present, the actual output is about 4 million tons per year and it is planned to increase an annual coal output up to 8 million tons. Brown coal seams are mined by the advanced open-cast technology with internal dumping.

Shoptykolskoye coalfield is characterized by flat and slightly inclined coal seams in the range of 4–10°. The overburden is deposited both inside and outside of the open-cast area by using traffic (Eastern tract) and non-traffic mining systems (Central tract). Considerable volumes of overburden reduce the technological and economic performance of the enterprise and negatively influence to the territory adjacent to the national park "Jasybay".

Flat bedding of brown coal seams (4-12°) allows optimize mining operations by using internal space of the open-cast for disposal of overburden mass. Internal dumping reduces outlay on transportation of overburden into external dumps and minimize negative environmental load on this territory.

The objective of this paper is estimating slope stability of internal overburden dumps and quarry faces in application to geological and mining conditions of open-cast "Maikubenskiy" (Kazakhstan) via physical modeling with equivalent materials.

3 PHYSICAL MODELING ON EQUIVALENT MATERIALS

Modeling on equivalent materials as a time-proved research technique allows investigate in detail the mechanism that being occurred inside the rock or soil massif particularly the processes of deformations with breaking continuity that is normally impossible by using other methods of physical modeling. Therefore, the method of equivalent materials is the most effective, so it is widely used to solve various geomechanical issues concerning rock mechanics.

The method of modeling on equivalent materials proposed by G.N. Kuznetsov (Kuznetsov 1959) is widely used for investigation of geotechnical issues for rock or soil behavior under impact of external and internal forces. Its advantage is the possibility of determining stress-strain state in the areas of failure which cannot be observed in natural conditions. The principle idea of this method is based on the replacement of the natural object by artificially prepared materials with certain physical and mechanical properties which correspond to the scale and mechanical characteristics of the natural model in accordance with the theory of similarity.

In the elementary case the model reproduces the studied phenomenon saving physical nature and geometrical similarity, and it differs from original object only by size and intensity of processes investigated.

The following equalities should be taken into consideration for meeting requirements of similarity conditions (indexes m and n are used for model and natural object respectively)

$$C_m = \frac{l_m}{l_n} \cdot \frac{\gamma_m}{\gamma_n} \cdot C_n, \qquad (1)$$

$$\tan \varphi_m = \tan \varphi_n, \qquad (2)$$

$$E_m = \frac{l_m}{l_n} \cdot \frac{\gamma_m}{\gamma_n} \cdot E_n, \qquad (3)$$

$$\mu_m = \mu_n, \qquad (4)$$

$$\gamma_m = 0.7\gamma_n, \qquad (5)$$

where l_m, l_n – linear sizes; γ_m, γ_n – volume weights; C_m, C_n – cohesive strengths; φ_m, φ_n – angles of internal friction; E_m, E_n – values of deformation modulus; μ_m, μ_n – Poisson's ratio.

Mechanical processes in a rock mass are considered similar in all the range of stresses if the following condition is fulfilled

$$\frac{\varepsilon_{p.m}}{\left(\varepsilon_e + \varepsilon_p\right)_m} = \frac{\varepsilon_{p.n}}{\left(\varepsilon_e + \varepsilon_p\right)_n}, \qquad (6)$$

where ε_e, ε_p – elastic and plastic strains.

Experiments with using equivalent materials are carried out in Geotechnical laboratory of the National Mining University. The flat testing stand has been specially designed for these purposes (Figure 1). It allows modeling the geo-mechanical processes which take place in slopes. The stand represents the hollow tray with working space formed by two glass partitions of 6 mm in thickness. Partitions are rigidly pressed by screw clamps to U-shaped restrictive frames of 60 mm in thickness. Thus, the volume of equivalent material is limited by the stand frame dimensions and surfaces of glass partitions.

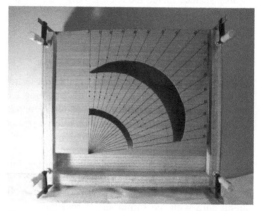

Figure 1. Flat stand for modeling with equivalent materials.

The mix of the washed river sand sifted through a sieve with cells of 0.5 mm was applied as an equivalent material. Using recommendations concerning modeling mechanical processes in soils the spindle oil was added as a binding component to initial mix (3% by volume). Main advantage of liquid oil application in equivalent materials is maintenance of working mixture permanent humidity and desired mechanical characteristics. The typical relation of volume weights was accepted as $\gamma_m / \gamma_n = 0.7$.

Considering only the internal space of the flat stand with external dimensions 630×545 mm and the width of limiting frames of 40 mm the overall weight of the sandy material model can be calculated according to the formula

$$P = (l - 2n) \cdot (b - n) \cdot m \cdot \gamma_m \cdot k, \qquad (7)$$

where P – overall weight of the model material, g;

l – model length, cm; b – model width, cm; n – width of limiting frames; m – model thickness, cm; γ_m – volume weight of the material, g/cm³; k – material loss factor, traditionally accepted in the range 1 10–1.15, in this case it equals 1.10.

The equivalent mixture ratio is presented in the Table 1.

Table 1. The equivalent mixture ratio.

Components	Content		
	% by volume	% by weight	by weight, g
Sand	97.0	98.89	19911.1
Spindle oil	3.0	1.11	406.3
Total:	100	100	20317.4

Taking into consideration the and the model scale 1:150 calculation of strength properties for the material being equivalent to original rocks was made according to the formulas

$$C_m = \frac{l_m}{l_n} \cdot \frac{\gamma_m}{\gamma_n} \cdot C_n = \frac{1}{50} \cdot 0.7 \cdot 32 \text{ kPa} = 0.15 \text{ kPa};$$

$$\tan \varphi_m = \tan \varphi_n = \tan 21° = 0.384;$$

$$E_m = \frac{l_m}{l_n} \cdot \frac{\gamma_m}{\gamma_n} \cdot E_n = \frac{1}{150} \cdot 0.7 \cdot 20000 \text{ kPa} = 93.3 \text{ kPa};$$

$$\mu_m = \mu_n = 0.35;$$

$$\gamma_m = 0.7\gamma_n = 0.7 \cdot 1.75 \text{ t/m}^3 = 1.225 \text{ t/m}^3.$$

The procedure of the slope failure test consists of the following operations. Initially the flat stand is replaced in horizontal position. In the stand workspace is filled with the mix of sandy equivalent material up to the brim of limiting frames without compaction. After leveling the surface by putty knife, the slope profile is cut off according to full-scale dimensions of the modeled slope at the scale 1:150. The angles of slope inclination are set by a protractor with the step 1°.

To document mechanical changes in the slope the coordinate grid was applied on the leveled surface of the equivalent material. For these purposes, a rigid cardboard with longitudinal slots with the size of 3 mm in width and 20 mm distance between slots was used as a stencil. This stencil was applied on horizontal surface of the formed model, and the coordinate grid was sequentially dispersed by paint spray. Within 10 minutes, the paint dried and the stand was covered by glass front wall then pressed by clamps to the limiting frame. Then the model was lifted slowly in the vertical position. This was accompanied by the deformation process of the sandy material. The sliding surface inside the massif is visually observed via transparent front wall and

recorded with the camera. Deformations occurring inside the slope were analyzed according to displacements of lines in coordinate grid.

It is well known that flat models with equivalent materials have inaccuracies arising due to frictional forces between sandy mixture and lateral faces resulting in an additional resistance to forces tending to shift the slope. To reduce the friction forces on the boundaries between equivalent material and glass walls, their working surfaces were powdered with talcum before filling with sandy material.

To simulate the sliding process occurred in the slope an equivalent material with different angles of inclination in the range from 40 to 70° with the step of 5° inside the stand has been formed. For the purposes of documenting mechanical changes in the slope, the coordinate grid is applied on the frontal side of the model. Displacements of the equivalent material in the plane model are fixed by camera.

Figure 2 shows the key stages of modeling slope stability on the equivalent material.

Modeling slope failure with equivalent materials revealed that under inclination angle up to 50 ° the slope structure has no significant discontinuities. Nevertheless slight deformations in the form of curved lines are visually observed along the lines of coordinate grid on the lateral surface. It suggests starting the process of the massif deformation but overall slope stability is still maintained. Substantial displacements emerge at the slope angle 55°. Integrity of the massif is failed and some pieces of the material detached from the slope surface and slide down to the toe. At the same time, there is a little part of the slope surface is covered with vertical cracks. By stepwise increasing the slope angle up to 60-70 ° these cracks spreading far into massif causing considerable deformations and gradual displacement of the failure prism on circular cylindrical surface. It should be noted that increasing the angle of slope inclination causes massif deformations that affect the deeper layers of the model. The sliding surface itself has arched-convex shape

(Shashenko 2010) that requires more detailed investigation of this phenomenon in-situ and via contemporary techniques of numerical simulation.

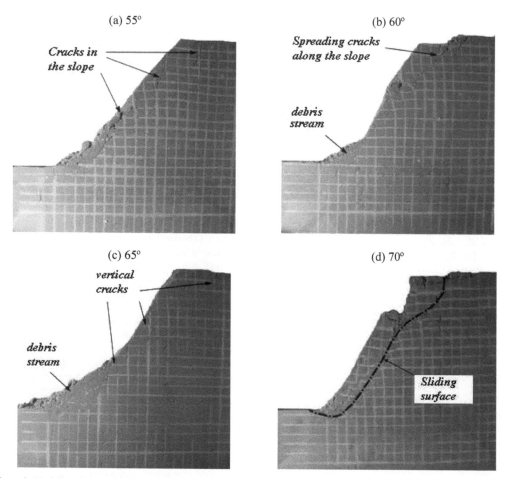

Figure 2. Modeling slope stability on equivalent materials.

4 STUDY OF CONTROLLED SLOPE FAILURE ON EQUIVALENT MATERIALS

For a more in-depth study of geomechanical processes occurred in the slope and the shape of the sliding surface, modeling quarry faces using elements
of proved in surface mining practice the technology
of controlled failure is fulfilled. The feature of this
technology is undermining the toe of the slope and
creating cutting slit in the crest area (Shashenko
2010).

As a case study the real slope failure at the quarry
was considered (Kovrov 2009). For this case the following geometry parameters are known: height of the
slope $H = 43.0$ m, the slope angle $\alpha = 42°$, failure
step at the toe $a_1 = 1.54$ m , the distance from the
crest to the cutting slit $a_2 = 10.39$ m, depth of cutting
slit $h = 9.54$ m; physic-mechanical properties of the
rock: bulk density $\gamma = 1900$ kg/m^3, friction angle
$\rho = 22°$, the cohesion $C = 60$ kPa, deformation modulus $E = 20000$ kPa, Poisson's ratio $\nu = 0.35$.

The Figure 3 presents results of simulating slope
controlled failure by the method of equivalent material. At the set slope angle $\alpha = 42°$ and without any
external forges and loads the slope surface is in a stable condition. To reduce cohesion forces at the crest
the vertical slit was cut off from the top side of the
model on the distance of 4 cm from the crest area.
Taking into consideration the model scale, the vertical height of the cutting slit is taken as $h = 9.54$ m
$/150 = 0.064$ m $= 6.5$ cm.

To initiate slope displacement, vertical slices were sequentially cut off in the toe area with the step of 1 cm, which corresponds to the distance of 1.5 m for natural object. The sliding surface was fixed and compared with the results of numerical simulation and traces of possible sliding surfaces described by Y.I. Solovyov (Shashenko 2004).

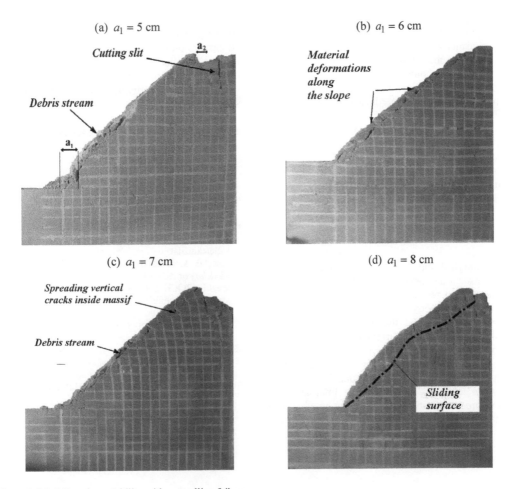

(a) a_1 = 5 cm

Cutting slit

a_2

Debris stream

a_1

(b) a_1 = 6 cm

Material
deformations
along
the slope

(c) a_1 = 7 cm

Spreading vertical
cracks inside massif

Debris stream

(d) a_1 = 8 cm

Sliding
surface

Figure 3. Modeling slope stability with controlling failure.

By vertical cutting the slope at the crest area on the fixed distance a_2 and vertical cutting at the toe area on a changing distance from the bottom edge a_1, the sliding surface with complex profile was visually observed.

Visual deformations begin with the vertical cutting the slope at the distance a_1 = 5 cm from the toe. The process is accompanied by vertical cracks along the slope. When the width of the extracted vertical slice at the toe increases up to 6–7 cm, the sliding surface occurs. It originates from the vertical cutting slit in the crest and has the shape far from traditional circular cylindrical profile (Figure 3d). It begins as

circular line with the convex in the middle part and concave in the bottom part.

5 RESULTS

Modeling slope failure with equivalent materials revealed that under inclination angle up to 50° the slope structure has no significant discontinuities. Nevertheless slight deformations in the form of curved lines are visually observed along the lines of coordinate grid on the lateral surface. It suggests starting the process of the massif deformation but overall slope stability is still maintained. Substantial displacements emerge at the slope angle 55°. Integrity of the massif is failed and some pieces of the material detached from the slope surface and slide down to the toe. At the same time, there is a little part of the slope surface is covered with vertical cracks. By stepwise increasing the slope angle up to 60-70° these cracks spreading far into massif causing considerable deformations and gradual displacement of the failure prism on circular cylindrical surface. It should be noted that increasing the angle of slope inclination causes massif deformations that affect the deeper layers of the model. The sliding surface itself has arched-convex shape.

Modeling slope stability with controlling failure provided some valuable information concerning mechanism and the shape of the sliding surface occurred in the massif. It originates from the vertical cutting slit in the crest and has the shape far from traditional circular cylindrical profile.

6 CONCLUSIONS

For geological and technological conditions of the open-cast "Maikubenskiy" slope stability issues in application to technology of internal dumping and optimizing geometry of pit edges is an actual engineering decision. But the most problematic issue while surface mining operations is inclined bedding of brown coal deposits. As a result, all the mined out area inside the open-cast has tilting surface. Disposal of the overburden on such foundation can cause to instable condition of the internal dumps. Therefore justification of rational geometry for internal dumps and estimation their stable conditions under assigned physical and mechanical properties is an important stage in engineering design.

Using the method of modeling on equivalent materials allowed analyze geomechanical processes occurring inside the rock or soil massif of the slope, and establish general laws of its failure. With well-chosen parameters of equivalent mixtures the method ensures quite satisfactory results of slope failure mechanism that correlate with those obtained via numerical simulations and observed in situ.

Nevertheless the sliding surfaces obtained in considered technique have arched-convex shape that contradicts common theories of slope failure mechanism. So this phenomenon requires more detailed investigation both in-situ and via contemporary techniques of numerical simulation.

REFERENCES

Kuznetsov, G.N., Budko, M.N. & Filippova, A.A. 1959. *Study of rock pressure occurencies in models.* Moscow: Ugletekhizdat.

Shashenko, A.N., Sdvizhkova, E.A. & Kovrov, O.S. 2010. *Modeling of the rock slope stability at the controlled failure.* "EUROCK 2010: Rock Mechanics in Civil and Environmental Engineering": Proceedings and monographs in engineering, water and earth sciences, Lausanne, Switzerland. – London: CRC Press / Balkema: 581–584.

Kovrov, O.S. 2009. *Modeling failure phenomenon of the rock benches on equivalent materials.* Scientific Bulletin of the National Mining University, 9: 27–30.

Shashenko, A.N. & Pustovoitenko, V.P. 2004. *Rock Mechanics.* Kyiv: Novyi druk.

Progressive Technologies of Coal, Coalbed Methane, and Ores Mining – Bondarenko, Kovalevs'ka & Ganushevych (eds)
© 2014 Taylor & Francis Group, London, ISBN: 978-1-138-02699-5

Changes of water inflow to underground gasifier depending on hydroheomechanical state of coal seam

V. Tishkov, V. Tymoshchuk & Ye. Sherstuk
National Mining University, Dnipropetrovs'k, Ukraine

ABSTRACT: The formation of zones with different dilatation is found by results of calculations based on numerical modeling of geomechanical state of massif containing degasified coal seam. An adjusting of filtration parameters in the roof and sole of gasifier channel was carried out according to the identified zones. Quantitative changes of inflow to the gasifier, which directly depend on the typical lithological composition of rocks and residual hydraulic pressures were detected by filtration modeling. Recommendations for improving conditions of stability of the process of underground coal gasification in hydrological conditions specific to deposits of Dniprovskiy basin were proposed.

1 INTRODUCTION

The UCG process has the potential to cause significant hydrologic and geomechanical changes in the area surrounding the coal seam. Estimation of the environmental threat posed to groundwater resources as a result of UCG involves consideration of several elements, including (Burton): Enhanced vertical hydraulic conductivity of the rock matrix above the burn chamber as a result of collapse and fracturing; Thermally-driven upward flow of groundwater resulting from in situ burning of coal; Generation of the contaminants within the burn chamber; Potential for bioattenuation of contaminant compounds that migrate into potable water aquifers. To directly address the issue of environmental risk posed to groundwater, the parameter spaces associated with these relevant processes need to be explored to identify, in a quantitative context, those scenarios are most favorable for UCG and which are least favorable. Output from such an effort will facilitate comparison of risk scenarios.

Spatial filtration parameters of the rock mass in their quantitative context when gasifying primarily depend on hydrogeomechanical changes occurring in this mass. In this paper, the need to differentiate the most influential factors among the major active geological factors on the stability of gasification process in conditions of Dniprovskiy basin was defined:

– depth of the coal seam, which will be subject to exploitation by underground coal gasification;

– tectonic and paleogeological conditions of occurrence of the producing formation;

– the presence of sand-and-clay or clay layer with a thickness of about 2 meters in the roof of the coal seam;

– reduction of flooded rock thickness in the roof strata to 2–10 m depending on the gas-hydrodynamic mode of gasifier operation;

– if there are flooded rocks in the bottom of coal strata so – to ensure the removal of hydraulic pressures to 6 m;

– the thickness of coal layer is at least 1, and preferably 1.5–2 m or more;

– ash content and moisture of coal in natural occurrence should not exceed 25–40%, and preferably be within 30%.

There is no doubt that within the above positions for underground gasifier site just some of natural elements are subject to significant influence during the technological process – the thickness of flooded rock layers in the roof and the bottom of coal seam and the moisture of the coal seam.

If take into account the fact that the change of moisture of coal seam is less sharply defined within a particular deposit it is necessary to say that the change in hydrogeomechanical condition of rocks containing coal seam, is a factor that has a major influence on the quality and calorific value component of extracted gas.

International experience (Burton et al., Kreinin 1982, 2004) shows that there is an optimum specific discharge of vapor for a normal operation of process of underground coal gasification. It is caused by the difference of physical and chemical properties of coal and technology of management of inflow into the gasifier. In this case the rate of steady gasification process with minimal vapor flow (in conversion on water) is 0.35–0.40 m^3 with a maximum its value 0.50–0.55 m^3 per 1 ton of actually gasified fuel in the gasifier channel.

2 STUDYING OF HYDRODYNAMIC REGIME IN LAYERED MASSIF USING NUMERICAL SIMULATION

Geomechanical numerical model of rocks was created for studying seepage fields and regularities of their changes in conditions affected the gasification process. The model describes the rock mass typical of brown coal deposits of Ukraine and represented by sandy-clayey sediments containing coal seam.

The basis of the numerical finite element model of soil mass is an elastic-plastic model, which is a generalization of elastic and rigid-plastic medium with internal friction and combine two principal theories of modern rock mechanics: theory of elasticity and the theory of limit state (Fadeev 1987). Software implementation of the algorithm allows obtaining the elastic-plastic solution in terms of plane strain and elastic solutions in terms of plane stress in a homogeneous and inhomogeneous medium.

The result of numerical solution is to define displacements of model nodes based on specified nodal forces due to external load and own weight of model elements (contour and volume forces). Axial and main deformation and corresponding stresses calculated by determined values of displacement.

Consideration of limit state of simulated rock mass and implementation of plastic flow in the field of behind limit strain in numerical model are based on the method of initial stresses. The numerical solution achieved with initial specified elastic properties of medium and constant stiffness matrix of the system.

In conditions of plane deformation when tensions in the environment do not exceed a specified limit level, the relation of stress and strain is described by Hooke's law

$$
\left.\begin{aligned}
\sigma_1 &= E_n(\varepsilon_1 + \nu_n \varepsilon_3)/(1 - \nu_n^2) \\
\sigma_3 &= E_n(\varepsilon_3 + \nu_n \varepsilon_1)/(1 - \nu_n^2)
\end{aligned}\right\}, \tag{1}
$$

where σ_1 and σ_3 – maximum and minimum principal stresses in accordance; ε_1 and ε_2 – largest and smallest principal deformations in accordance; $E_n = E/(1 - \nu^2)$, $\nu_n = \nu(1 - \nu)$ – "plane" analogs of deformation modulus E and Poisson's ratio ν.

In the area of a stretching limit strain are limited by tensile strength T ($T < 0$)

$$
\sigma_3 = T, \tag{2}
$$

and in the field of compression – by Mohr-Coulomb criterion

$$
\sigma_1 = S + \sigma_3 Ctg\varphi, \tag{3}
$$

where $S = 2CCtg(45 - \varphi/2)$,

$Ctg\varphi = (1 + Sin\varphi)/(1 - Sin\varphi)$, C and φ – cohesion strength and friction angle in accordance.

When strained state of environment element goes to the zone of elasticity limit the tension in the element is controlled by the equation of state, which for behind limited strain are defined relatively to the principal stress σ^T_1 and σ^T_3

$$
\sigma_3^T = [E_n(\varepsilon_1 + \varepsilon_3) + S(\nu_n - 1)]/
$$

$$
(1 - \nu_n Ctg\beta + Ctg\beta - \nu_n), \tag{4}
$$

$$
\sigma_1^T = S + Ctg\psi \sigma_3^T. \tag{5}
$$

Angle β in Equation (4) determines the law of plastic flow: at value of β, set equal to ψ for the used model, flow is associated that allows to consider the dilatation of rocks in the field of extraordinary strain.

Principal stress values σ^T_1 and σ^T_3 are theoretical stresses corresponding to achieved level strain, numerical solution is striving at the i-th step of iterative process.

Reducing the strength characteristics of soil in behind limit deformations is taken into account in numerical model on the basis of consideration of strain strength criterion presented in the main stresses

$$
\sigma_{comp} = \sigma_1 - (2\lambda + 1)\sigma_3, \tag{6}
$$

$$
\lambda = sin\varphi/(1 - Sin\varphi), \tag{7}
$$

where σ_{comp} – ultimate uniaxial compression strength.

In the area of elastic deformation and behind limit deformation the stress-strain state of the model elements is clearly determined from the analysis of main deformations by equations:

– in the area of elastic deformations ($\varepsilon_1 \le \varepsilon_1^y$)

$$
\sigma_1^T = (2\lambda + 1)\sigma_3 + \sigma_{comp}; \tag{8}
$$

– when strained state of the element goes beyond tensile strength, subject to conditions for ε_1:

$$
\varepsilon_1^y < \varepsilon_1 < \varepsilon_1^y + (\sigma_{comp} - \sigma_{res})/E_c:
$$

$$
\sigma_1^T = (2\lambda + 1)\sigma_3 + \sigma_{comp} - E_c(\varepsilon_1 - \varepsilon_1^y); \tag{9}
$$

– when the strength properties of the soil are reduced to the value of residual strength σ_{res}

$$
(\varepsilon_1 \ge \varepsilon_1^y + (\sigma_{comp} - \sigma_{res})/E_c):
$$

$$
\sigma_1^T = (2\lambda + 1)\sigma_3 + \sigma_{res}. \tag{10}
$$

In the above equations: ε_1 – the longitudinal deformation; ε_1^y – limit elastic deformation; E_c – modulus of decrease.

In the numerical algorithm reducing the strength of rocks in the area of behind limit deformations is taken into account when calculating the theoretical main stress by putting Equation (9) or (10) instead of (5) according to the achieved deformation level. The results of numerical modeling of the stress-strain state of rock mass obtained by solving the problem in elastic-plastic formulation. Estimated parameters of geomechanical model were set based on the materials of previously performed research and study the mechanical properties on undisturbed and disturbed samples of rocks in conditions of volume stress state (Tishkov 2012 & Tymoshchuk et al. 2012).

Previously obtained by the simulation results analytical dependences (Kreinin 1982, 2004) proves the defining the amount of zones where the most active geomechanical changes occur in the massif.

The study of rock mass shear when underground coal gasification in different geological conditions with in-situ data defined, that parameters of shear during gasification of coal seam differ little from those when underground mining of coal seam, shift process at the coal gasification occurs more smoothly, and deflection zone of rocks with a break of continuity (cracking zone) is much smaller.

It was proposed to determine the height of rupture cracks during gasification of flat-lying layers based on field research by the following dependence (Tishkov 2012)

$$H_T = n \cdot m \text{ when } H_T < H, \qquad (11)$$

where H_T – height of crack development, m; H – depth of the coal seam, m; m – thickness of degasificated coal seam, m; n – empirical coefficient considering the development of cracks in roof rocks at the process of gasification.

For deposits where is an experience of underground coal gasification coefficient n is known and ranges from 6 to 15 (Kreinin 1982, 2004). For new deposits coefficient n can be adjusted on the basis of our testing of rocks and mathematical modeling.

Considering the determined in laboratory conditions dependence of technogenic permeability of sand-and-clay deposits from the achieved level of strain (Tishkov 2012 & Tymoshchuk et al. 2012) and transforming it into a diagram obtained by the simulation results it becomes possible to get a thorough understanding of the nature and zoning of seepage fields around the gasifier area.

3 CALCULATION RESULTS

By the results of calculations based on numerical modeling of geomechanical state of massif and forming of zones of different filtration permeability in it revealed the following:

a) if there is no strata of waterproof rocks in roofing volumetric deformations of deconsolidation are formed within the range 5–10 m directly above the central part of roofing of gasification channel (Figure 1). The size of these areas varies from 5.5% down to zero depending on changes in lithology of rocks.

b) additional compression areas, located on the edges of the gasification channel form zones of additional consolidation of rocks in the range of 2 to 8%.

c) the main inflow formes within the inner area contour of the gasifier, and reaches a value of about 0.43 m^3 per one meter of gasification channel depending on the strain. The distribution of the field of hydraulic pressures is illustrated in Figure 2.

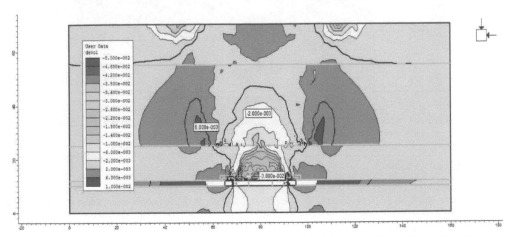

Figure 1. Distribution of relative volumetric strain in layered massif, which violated with the gasifier, unit fraction.

As can be seen from the figure, in the presence of permeable rocks in the roof of coal seam with the permeability of 0.8–1.65 m/day a deep depression which is accompanied by significant flooding of degasification area is formed in gasifier contours.

The specific value of the inflow that is calculated due to field velocity and hydraulic gradient reaches 0.26–0.78 m²/day and distributed quite unevenly. Average groundwater inflow is about 0.37–0.48 m³ per 1 ton of gasified coal, excluding natural moisture in the working layer. Changing the inflow into channel of gasification will directly depend on the residual pressure in the immediate roof of the coal seam.

In the above conditions for sustainable gasification process it is appropriate to use a drainage with single-contour scheme – for draining water-bearing strata mainly off-site the gasifier.

If there are waterproof clay rocks in the roof and sole of coal seam, even with coaly component, and with their characteristic permeability – up to 0.01 m/day (Figure 3), the value of specific inflow is within 0.16–0.25 m²/day in conditions of shifts of strata when gasifier working.

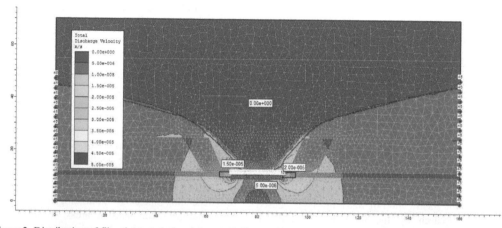

Figure 2. Distribution of filtration rate in layered massif, disrupted by gasifier, in the absence of impermeable layer in the roof of the coal seam, m/s.

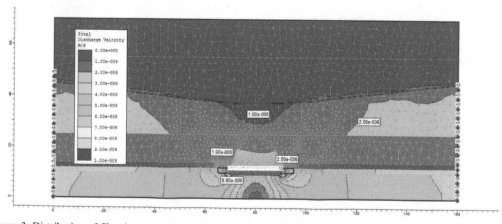

Figure 3. Distribution of filtration rate in layered massif, disrupted by gasifier, in the presence of impermeable layers in the roof of the coal seam, m/s.

Under these conditions, if necessary, it is appropriate to use the double-contour (relative to gasifier boundaries) drainage which at one level intercepts inflow on the gasifier area and at the second level – intercepts the contour inflow formed in coal seam.

Executed prognosis estimation of changes of hydrodynamic regime of rock strata that contains gasifier showed that formation of filtration fields depends on changes in rock lithology and residual hydraulic pressures.

4 CONCLUSIONS

1. Research of changes in stress-strain state of rock mass at underground coal gasification in different geological conditions found that the shift parameters during of coal seam gasification are similar when underground mining of coal seams, but the shift process in coal gasification occurs more slowly, and deflection zone of rocks with a break of continuity is much smaller.

2. Found that formation of the zone of high hydraulic permeability is typical for central area of gasifier, and the most intensive increase of permeability in conditions of shift the rock strata, containing coal seam, corresponds 2–5 meter interval within the immediate roof and contact of lithology changes. Representations of the nature and zonation of seepage fields are obtained based on the analysis of horizontal, vertical and volumetric deformations of rocks around the gasifier area.

3. Estimation of inflows that are formed into the gasifier in conditions of sustainable filtering mode, indicating that the hydrodynamic factor has a significant impact on the quality of generating gas, especially in the absence of impermeable rocks in the roof of the coal seam. In the studied conditions, the magnitude of inflows to the gasifier area reaches 0.37–0.48 m^3 per 1 ton of gasified coal, excluding natural moisture in the working layer, in the specific terms it is about 0.26–0.78 m^2/day.

4. To ensure stable gasification process in complicated geological conditions it is appropriate to use the drainage of water level, which, depending on the occurrence of coal seam and gasification schemes can be single-contour or double-contour in relation to gas generator area. Availability of drainage provides increase the quality of gas generator by reducing the vapor flow to 0.20–0.30 m^3 per 1 ton of actually gasified fuel in the gasification zone.

REFERENCES

Burton, E., Friedmann, J. & Upadhye, R. *Practices in Best Practices in Underground Coal Gasification.* Lawrence Livermore National Laboratory.

Kreinin, E. 1982. *Underground coal gasification.* Moscow: Nedra: 394.

Kreinin, E. 2004. *Nontraditional thermal mining technologies of difficult to extract fuels: coal, hydrocarbon feedstock.* Moscow: IRC Gazprom: 301.

Fadeev, A. 1987. *The finite element method in geomechanics.* Moscow: Nedra: 221.

Tishkov, V. 2012. *Features of formation of technogenic permeability at the top of the coal seam in underground coal gasification.* Dnipropetrovs'k: Scientific Bulletin of National Mining University, 1: 35–39.

Tymoshchuk, V., Tishkov, V., Inkin, O. & Sherstiuk, E. 2012. *Influence of layers gasification on bearing. Geomechanical processes during underground mining.* Dnipropetrovs'k: Proc. of the School of underground mining / CRC Press, Taylor and Francis Group, London: 109–113.

Progressive Technologies of Coal, Coalbed Methane, and Ores Mining – Bondarenko, Kovalevs'ka & Ganushevych (eds)
© 2014 Taylor & Francis Group, London, ISBN: 978-1-138-02699-5

Development of recovery technology for broken shotcrete lining with use of fiber reinforced concrete

V. Kovalenko
National Mining University, Dnipropetrovs'k, Ukraine

ABSTRACT: A large number of mine workings, fastened with sprayed concrete lining requires repairs and restoration of lining. To avoid critical situations involving violations of sprayed concrete lining it requires immediate implementation of measures for its restoration. Most appropriate to restore the broken sprayed concrete lining is using fiber-reinforced concrete (FRC). In the article it is cosidered standard a three point flexural test for detirmine perfomance and ultimate bending loads of recovered with macro synthetic FRC the damaged specimens of sprayed concrete. During experiments it was studied the ability of the joint work of the ground and recovering coatings. To determine the stresses arising in FRC layer, formula for determine bending stresses was adjusted to the real conditions of operation of support, taking into account the availability of the damaged base layer of shotcrete. Recovery of shotcrete can provide restore the original strength characteristics of the specimens by increasing the cross-section with fiber reinforced concrete. As a result, deformation curve changes in the diagram "load-deflection". Comparing the ultimate loads may be noted that after the restoration of the specimen was able to resist a 1.56–2.69 times greater loads than the check sample. The positive effect of restoring the disturbed pattern of sprayed concrete coating with fiber-reinforced concrete was observed when as a reinforcing element was used the macro synthetic rigid fibers. An increase of bearing load at 63% was due to use of FRC and the redistribution of stresses.

1 INTRODUCTION

When fastening the capital and, in some cases, the main development workings there was widespread use of sprayed concrete lining. The high degree of mechanization and other benefits of spray technology provided this type of roof support with opportunities for use in coal mine construction. One of the main drawbacks of this lining is its limited field of work – in the mode of steady pressure and the limit displacement, defined by different researchers in a wide range from 20 to 160 mm. This shortcoming limits the actual life of linings.

At mines operates a large number of mine workings, fastened with sprayed concrete lining. Some of these mines require repairs, restoration of lining.

In the case of cracks appearance sprayed concrete loses integrity, and in a short time lining loses its stability. Although the formation of cracks in the sprayed concrete does not mean complete exhaustion of its bearing capacity, in the process of cracking may occur threatening the collapse of rock in some sites of mine workings. To avoid critical situations involving violations of sprayed concrete lining it requires immediate implementation of measures for its restoration. This will protect operating personnel and ensure the continued stability for mines.

Most appropriate to restore the broken sprayed concrete lining is using fiber-reinforced concrete (FRC) (Kovalenko 2011).

Terms of use of this material is determined by the ability to perceive the displacement of rocks without breaking, i.e. its great deformative ability .FRC is able to work under considerable strain from the host rock and provide safe conditions of work and sustainable state of the mine working.

As a result of restoring the broken shotcrete lining by application on top of it the layer of FRC and after FRC hardening in the new restored lining occurs redistribution of significant bending stresses. Impact of bending forces takes itself not sprayed concrete, but the outer layer from FRC. After the restoration of damaged sprayed concrete lining, the tensile strength in bending should increase in whole for the entire combined lining (Kovalenko 2010).

2 THE AIM OF THIS PAPER

To determine the character of the deformation and the bending strength of restored with the use of macro synthetic FRC the broken shotcrete specimen.

3 DESCRIPTION OF THE EXPERIMENT

For test operation there was used a standard a three point flexural test. The main difference was that in the pre-weakened by the crack pattern was applied a coating of FRC. Depending on the thickness of the coating was determined the ability of FRC to resist bending forces in conjunction with a concrete specimen. Testing scheme is shown in Figure 1. Previously, samples prepared from a cement-sand mortar, were tested in bending. After their destruction by the formation of the cleavage plane there was carried out reconstruction of the specimen.

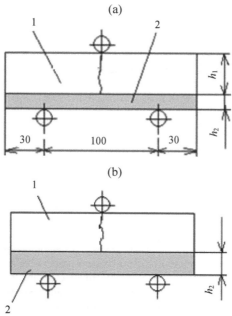

(a)

(b)

Figure 1. Scheme of testing the specimens in dependence on thickness of the layer of macro synthetic FRC (h_2): (a) $h_2 = 20$ mm, (b) $h_2 = 30$ mm: 1 – damaged shotcrete layer, 2 – restoring layer of FRC.

Both the cleavage surface covered with a thin layer of cement mortar. Then on the horizontal surface of the specimen was deposited a layer of FRC. After compaction of FRC with use of vibrator, the specimens were kept for 28 days.

For the preparation of FRC were spread two types of macro synthetic fibers – a rigid and tortuous. Type fibers to a large extent influence the strength and efficiency of fiber-reinforced concrete. To determine the most effective type of fiber, which provides higher strength and deformation parameters of FRC were carried out preliminary experiments. They showed significantly better parameters for samples reinforced with rigid fibers.

On the images (Figure 2 and 3) can be seen joint work of sprayed concrete and FRC. First layer on the top is the basic coverage of sprayed concrete, on which is applied below the restoring cover of FRC. During experiments it was studied the ability of the joint work of the ground and recovering coatings. Laboratory experiments simulate the work of sprayed concrete lining in the actual operating conditions, when the lining and restoration can be performed through the same mechanisms, what used to apply sprayed concrete on rock surface – sprayed concrete gun. With the difference, that sprayed concrete is reinforced with macro synthetic fibers. Therefore, simulated in laboratory experiments a restorative coating made up of a sprayed concrete, reinforced with tough fiber.

Figure 2. Test of bilayer specimen with 20 mm height of macro synthetic FRC: disrupted shotcrete (top), restored coating of FRC with a rigid macro synthetic fibers (bottom).

Figure 3. Test of bilayer specimen with 30 mm height of macro synthetic FRC: disrupted shotcrete (top), restored coating of FRC with a rigid macro synthetic fibers (bottom).

In most cases, a new strain crack was held out of the area prior violation (crack). Cleavage of FRC consisted of a few strain cracks. In all cases, laboratory tests completed prior to time, when model exhausted its deformative ability, what confirms the feasibility of restoring the sprayed concrete coverings and their high operating capabilities to ensure the bearing capacity of support in the stage of cracking, as well as conditions for maintaining the stability of mine working.

4 ANALISYS OF RESULTS

The presented photos of laboratory tests of restored and check specimen show that the deformation process of the specimen, which is restored with a 28 mm layer of fiber-reinforced concrete, on the diagram (Figure 4) is characterized by the formation of the peak load limit of 6.65 kN and a further decline in two phases. On the very specimen there was observed the formation of a large network of small cracks with low intensity of their disclosure.

Diagrams of loading of the samples are presented in the load- deflection curves in Figures 4–6.

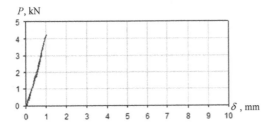

Figure 4. Load-deflection curve for shotcrete specimen.

When studying the test results of the check specimen made from sprayed concrete, it can be noted the linear character of deformation of the sample – elastic deforming and destruction of the specimen after reaching the limit of the bending strength of 4.26 kN. Unlike to check samples, damaged prisms of sprayed concrete recovered with FRC show greater deflections and may resist bigger flexural loads.

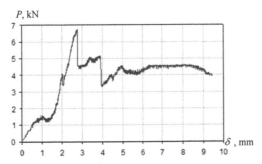

Figure 5. Load deflection curve for bilayer specimen with 20 mm height of macro synthetic FRC coating.

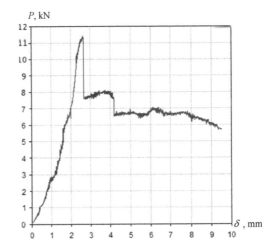

Figure 6. Load deflection curve for bilayer specimen with 30 mm height of macro synthetic FRC coating.

In order to determine the maximum stresses that occur in cross-section of specimens we use standard expression for determine flexural stress in a rectangular cross section (1)

$$\sigma_f = \frac{3PL}{2bh^2}, \qquad (1)$$

where σ_f – stress in outer fibers at midpoint, (MPa); P – ultimate load, N; L – support span, mm; b – width of test beam, mm; h – height of tested beam, mm.

To determine the stresses arising in FRC layer, we adjust the formula (1), taking into account the availability of the damaged base layer.

$$\sigma_f = \frac{3PL}{2bh_2(h_1 + h_2)}, \qquad (2)$$

where h_1 – height of the base layer, mm; h_2 – height of the recover layer, mm.

Based on the limit loads operating on specimens, as well as using formula (2) all obtained data about flexural stresses were dropped into the Table 1.

Samples 1–3 were made of shotcrete without recover layer of FRC. They are used as check samples. Specimens 4–6 presented themselves the damaged shotcrete covered with 20mm layer of macro synthetic FRC. Accordingly, specimens 7–9 were covered with 30 mm of macro synthetic FRC.

Table 1. Main testing parameters of specimens.

Sample	P, kN	L, mm	B, mm	h, mm		σ_f, MPa	Change of bearing ability, %	Ratio of increase in ultimate load $\dfrac{P_{recover}}{P_{check}}$
				Base layer, i.e. damaged shotcrete, mm	Recover layer, i.e. FRC, mm			
1-3	4.26	100	40	40	–	1.0	0	1
4-6	6.65	100	40	41	20	2.04	204	1.56
7-9	11.50	100	40	40	30	2.05	205	2.69

As can be seen from Table 1, the tensile strength of the samples on a bend in the event of recovery is greatly increased and may reach from 1.56 (specimens with 20 mm height of FRC) to 2.69 (specimens with 30 mm height of FRC) times increase in ultimate load in comparison with check samples (specimens 1–3) (Figure 7).

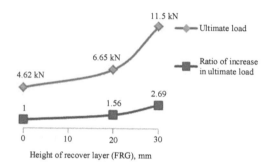

Figure 7. Dependence of ultimate load and increase in ultimate load on height of recover layer – macro synthetic FRC.

It should be noted that in the recovery of the specimens achieved an increase in strength properties. When you restore a damaged sprayed concrete beams with use of fiber-reinforced concrete it was carried out increasing the thickness of the specimen, meanwhile the cross-section of fiber-reinforced concrete has been shifted into the region of tensile stress, where the fibrous concrete works best on a bend. As a result, the sample was able to resist the destruction of a greater strength in bending.

Proceeding from the presented data, it follows that recovery of shotcrete can provide restore the original strength characteristics of the specimens by increasing the cross-section with fiber reinforced

concrete. As a result, deformation curve changes in the diagram "load-deflection". Comparing the ultimate loads may be noted that after the restoration of the specimen was able to resist a 1.56–2.69 times greater loads than the check sample.

5 CONCLUSIONS

Restoration of sprayed concrete lining can be successfully realized with the use of fiber-reinforced concrete. Using of fiber-reinforced sprayed concrete provides increase in efficiency for specimen, recovered with the use of rigid macro synthetic fibers. The positive effect of restoring the disturbed pattern of sprayed concrete coating with fiber-reinforced concrete was observed when as a reinforcing element was used the macro synthetic rigid fibers. Moreover, an increase of bearing load (tensile strength in bending) at 63% was due to use of FRC and the redistribution of stresses.

This technology extends the use of sprayed concrete, and can be used at greater depths in comparison with the depth at which the currently used shotcrete technique.

REFERENCES

Kovalenko, V.V. 2011. *Study of efficiency of recovered fiber concrete at various reinforcement fibers*. Dnipropetrovs'k: Scientific bulletin of National Mining University. Vol. 9-10: 26–30.

Kovalenko, V.V. 2010. *Laboratory study of influence of reinforcement character on efficiency of recovered fiber reinforced concrete*. Dnipropetrovs'k: Collection of research papers National Mining University, 34. Vol 2: 120-125.

Progressive Technologies of Coal, Coalbed Methane, and Ores Mining – Bondarenko, Kovalevs'ka & Ganushevych (eds)
© *2014 Taylor & Francis Group, London, ISBN: 978-1-138-02699-5*

Geotechnical schemes to the multi-purpose use of geothermal energy and resources of abandoned mines

I. Sadovenko, D. Rudakov & O. Inkin
National Mining University, Dnipropetrovs'k, Ukraine

ABSTRACT: The analysis of energy consumption in Ukraine and the growing needed for wider use of alternative sources including those accumulated in abandoned mines is shown. Regarding to the rapid growth of various applications during last decades geothermal energy is considered by worldwide as one of promising energy sources, with the significant share and prospects of heat pumps. The paper describes concepts of combined use of ground source heat pumps, the resources of abandoned mines, and various geotechnologies including underground combustion of residual coal seams, mine drainage, water flow regulation in mining and post-mining areas, underground hydropower plant, mine and ground water treatment and water supply. The four geotechnical designs are economically feasible because of combining low-grade heat recovery or/and power generation with addressing economical and environmental challenges, primarily, in mining areas.

1 INTRODUCTION

Last decades have been demonstrating the growthing geothermal energy share in the world energy resource balance (Lund et al. 2010). This process is also typical for Ukraine with some domestic features depending on economical, climatic, and geological factors. The most common ways of using geothermal resources in Ukraine are deep boreholes and ground source heat pumps; besides, mine and municipal sewage waters can be leveraged as a technically achievable resource. Some of these sources have much bigger potential especially in the areas changed as a result of industrial activities and intensive building.

Rock temperature within the Ukraine's territory on the depth of 1000 m varies from 20 to 70 °C, and on the depth of 3000 m it ranges from 40 to 135 °C. Heat flux density ranges from 25–30 to 100–110 mW/m^2 (National Atlas of Ukraine 2007). The maximums of temperature and heat flux were measured in mountainous areas in Crimea peninsula and Carpathians. Potential geothermal resources amount to 27.3 million m^3/day of thermal waters. Taking into account cogeneration capacity and thermal water specifics these resources can be estimated at 84 million GCalories per year (National report... 2010). The big thermal water resources are concentrated in West side of Carpathians (490 MW), Black Sea coastal area (4900 MW), and Crimea (37600 MW).

About 9.3 million houses on homesteads in Ukraine have the total heated area of more than 510 million m^2. They need roughly 160 million MWh for heating and hot water supply annually. In principle, this demand can be met by ground source heat pumps. The total country's potential that can be used by ground source heat pumps amounts 157530 MWh/year, with technically achievable resource being estimated at 71.4%, but only 6.7% of that is profitable to use nowadays. Major limitations on putting heat pumps into wider practice are high installation costs and long payback time. F. e., the costs for a heat pump device of 4–5 kW power range currently from 3000 to 7000 €; an increase in power to 10–15 kW raises the costs up to 5000–10000 €. Passive cooling mode would be more profitable, which would ensure the savings 90–95% of costs; however this it is possible in summer time only during 100–150 days and is applicable to southern regions of Ukraine. Maximal estimated savings owing to reduction of fuel consumption amount 700–1000 € annually, which is not profitable under the conditions of high interest rates.

Besides, Ukraine has the vast resource of low-potential heat in mining lands. Annually more than 500 million m^3 of mine water are pumped just in Donetsk coal basin and discharged into ponds and rivers. The temperature of this water ranges from 16 to 22 °C depending on the season; the mine waters temperature deeper than 800 m may reach 30–33 °C. The annual low-potential heat loss is estimated at 5 million GCalories (Unlimited resource... 2007, Nechitailo 2011). Use of this resource is restricted by many technical problems including high salt content (up to 60 g/l) and the requirement to isolate mine water regarding in order to protect surface and ground water.

There are now probably only a few examples of mine water heat recovery in Ukraine. A heat pump system implemented at the mine "Blagodatnaya" of "DTEK Pavlogradugol" (Central Ukraine) has heat output of 800 kW at mine water flow rate 200 m^3/h (Samusia et al. 2012). Mine water temperature of 16 °C is increased by heat pumps up to 42–45 °C. In case of additional using low-potential heat of waste water from baths at 30 °C the rate of heat transformation growths to 7.0–8.0.

The total annual volume of municipal waste waters is estimated at about 3.74 million m^3 (National report... 2010). The waste water temperature ranges from 12 to 20 °C depending on the season. The annual technically achievable quantity of this kind of heat energy equals to 18 million tons of s. f. However, there are technical difficulties to recover sewage water heat regarding to obsolete underground infrastructures and water supply facilities.

Nowadays the total potential resource of geothermal energy is not used in Ukraine properly (Table 1) and is still one of the lowest among the industrialized countries of the world.

Table 1. Use of geothermal resources in 2010 (Lund et al. 2010).

Country	Capacity, MW	Annual Use, GWh/yr	Annual Use per capita, KWh/(yr person)*	Capacity Factor
Germany	2 485.4	3 546.0	43.35	0.16
Russia	308.2	1 706.7	11.9	0.63
Ukraine	10.9	33.0	0.72	0.35
U.S.A.	12 611.4	15 710.16	49.87	0.14
World	50 583	121 696.0	17.38	0.27

*Calculated using (Lund et al. 2010).

2 FORMULATING THE PROBLEM

The analysis of technical and geological conditions reveals the fact that use of geothermal energy would often become profitable in case of combination with other geotechnologies or environmental protection measures, particularly in man-changed environments, primarily in mining lands. The multifaceted and multi-purpose engineering solutions could significantly increase the effectiveness of geothermal applications and facilitate putting into practice these technologies in Ukraine; these would be of interest also for the countries that have similar natural conditions and extensive mining.

Multi-purpose geo-technological schemes considered in this paper include:

1. Coupled regulation of water flows and low-grade heat extraction at an abandoned mine.

2. Using heat energy of residual coal reserves in flooded mines.

3. Combining an underground hydropower plant (UHPP) with heat pumps.

4. Combining heat recovery and water supply.

This paper describes the principal features of these promising engineering solutions in terms of geotechnological design.

3 DESIGNS FOR USE OF LOW-GRADE HEAT RESOURCES

3.1 Coupled regulation of water flows and low-grade heat extraction at an abandoned mine

Mined out rocks are commonly quite permeable due to underground mining and generated mine workings, man-made fissures and other disturbances (Yakovlev 2007). This intensifies ground water flow and results in rising ground water head and flooding the land surface. At the same time, natural river beds are typically draining the shallow aquifer that accumulates low-potential heat fluxes and dissolved toxic compounds applied in mining. In this regard, it is reasonable to combine heat recovery with water treatment and purification technologies e.g. membrane distillation that is economically efficient in the presence of low-grade heat.

The idea of this scheme consists in simultaneous water-regulating drainage, water withdrawal for treatment and heat pumps by the same gravity-driven drainage system (Sadovenko et al. 2002). The major elements of technological scheme are shown in Figure 1. Here Q_1 is the discharge of mine water withdrawn by the drain and delivered to pre-treatment facilities; T_1 is the temperature of mine water withdrawn by the drain; Q_2 is the discharge of mine water withdrawn by the drain and delivered to the consumer(s) for heat recovery and cleaning; T_2 is the temperature of water heated by heat pumps; $Q_{2,c}$ is the discharge of water to the natural drain after use; $T_{2,c}$ is the temperature of cooled water; K_1, K_2, K_3, and K_4 are hydraulic conductivities of rocks disturbed by mining, naturally permeable rocks, fluvial deposits, and aquitard respectively (K_1> K_2> K_3> K_4).

Mine water at the temperature T_1 of about 15–20 °C is transported to heat pumps at a rate Q_2; then it is heated up to the operational temperature T_2 of 40 to 55 °C depending on the season and used for heating residential and industrial buildings. Some portion of heated mine water can be used for producing potable water in membrane water treatment units effective in the presence of low-grade waste heat. After cooling to temperature $T_{2,c}$ of 10 to 12 °C water is discharged to a natural drain at a rate $Q_{2,c}$. In this case, the specific flux of heat recovered from the drainage system amounts 33.5 MJ/day.

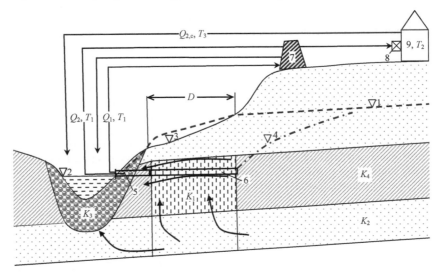

Figure 1. Water flows and heat fluxes in a mined out area: 1 – ground water level in the unconfined aquifer, 2 – water level in the surface water body, 3 – ground surface elevation, 4 – ground water level in the confined aquifer, 5 – the underground drainage gallery, 6 – main haulage road, 7 – water treatment facilities, 8 – heat pumps, 9 – heat consumer(s).

The drained area "D" affected by underground mining requires systematic surface drainage; it can be enhanced by vertical boreholes having self-drainage effect. These boreholes are most effective when combined with longwall mine workings of a closed mine. This way of heat recovery excludes re-circulation of water as heat transfer agent transported from flooded mine workings to the consumers. Therefore, this design allows preventing from ground water rising; it provides also cleaning of mine drainage water that may contain toxic compounds after seepage through mined out rocks. Such combination increases the efficiency of heat pump performance.

3.2 Using heat energy of residual coal reserves in flooded mines

The available residual coal reserves in flooded mines of Ukraine are estimated at 2 billion tons; most of them are suitable for underground combustion (Bottomless supplies... 2013). Besides, there are natural and man-made underground reservoirs (storage capacities) in mining lands where mine water as heat transfer agent can be temporally stored. Due to high thermal capacity mine water can be used for heating and hot water supply. Although the initial temperature of pumped water from deep mine workings (more than 800 m below the ground) attains 30–33 °C, the use of this water for hot water supply requires additional heating to the temperature at least of 45 to 55 °C. This can be conducted either by heat pumps, which is linked with significant investments or by means of underground coal combustion.

The second option is preferable in the presence of low-grade coal seams or residual coal reserves that are unprofitable to use in one of the conventional ways. The abundance of residual coal in flooded and closed mines makes in-situ combustion of coal an attractive way for heat and hot water supply in post-mining lands. The successful experience of underground coal gasification with extraction of over-heated steam at the site in Rocky Mountains, USA.

(Gash et al. 1989, Lindblom & Smith 1993) evidenced the prospects of this geotechnology.

The main feature of this scheme consists in using low-grade heat of mine waters of flooded mine with periodical heating by combustion of low-quality coal seams regarding to seasonal variations of energy consumption. This approach allows heightening the water temperature, eliminating boiler-houses as heat generating facilities for local needs, and, possibly, creating compact modules for specific purposes, f. e. to heat greenhouses.

The hydrothermal module concept and feasibility studies (Gholudev 2003, Sadovenko & Inkin 2002, 2013, Sadovenko et al. 2012) imply using both residual resources of coal and geothermal energy (Figure 2). Here Q_1 is the gas flow rate from the gasification channel; q_1 is the heat flux of the gas extracted from the gasification channel; q_2 is the heat flux from the gasification channel to surrounding rocks; T_1 is the temperature of the gas in the gasification channel; Q_2 is the pumping rate of heated ground water withdrawn from the aquifer over the bed confining the gasification channel above; T_2 is the temperature of ground water above the heated aquitard overlying the coal seams burned; T_3 is the temperature of withdrawn heated ground water delivered to the consumer.

When coal is burned at a temperature T_1 of 900 to 1000 °C produced gas is diverted to the surface and then delivered to the consumer(s). The combustible components of the diverted gas bring the heat flux q_1 generated in the gasification channel. The rest part of emerging heat disperses from the channel to surrounding rocks; mostly it is transporting with gas leaks through a low-permeable confining bed to the overlying aquifer and heats ground water. Ground water temperature T_2 may reach 100 °C under certain conditions depending on the combustion intensity and ground water pumping rate. The hot water delivered to the consumer is cooling and discharged into unused aquifers within the underground gasification site. This technology has to be applied in winter; thus, coal combustion has to be started just before the heating season. The specific heat flux received by the consumer from heated ground water per 1 m^3 of coal burned daily is estimated at 146 MJ/day.

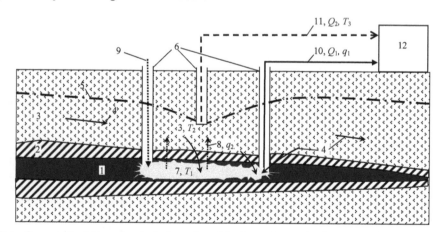

Figure 2. Water flows and heat fluxes in a hydrothermal module: 1 – coal seam, 2 – low-permeable rocks, 3 – aquifer, 4 and 5 – ground water head and flow direction, 6 – wells, 7 – gasification channel, 8 – heat flux to the aquifer, 9 – air blowing, 10 – gas cleaning and transportation, 11 – hot water transportation, 12 – heat consumer.

3.3 Combining an underground hydropower plant with heat pumps

Electrical power supply networks in mining lands are known to be susceptible to daily overloads; this requires additional power generating facilities working periodically. The energy of water flowing down from upper to lower mine workings in a drained mine could be utilised as an additional resource for this purpose. This became feasible as a result of draining and flooding of adjacent mines and significant changes of ground water heads in comparison to the pre-mining hydrogeological regime. This resource is reasonable to use in combination with heat pumps and underground mine workings served as heat storage reservoirs in adjacent flooded mines. Performance of a single UHPP in a partially flooded mine has been discussed in details in (Sadovenko & Inkin 2002).

The proposed combination of the UHPP and heat pumps is demonstrated in Figure 3. Here Q_1 is the rate of pumping the water withdrawn from deep mine workings and discharged to the retention pond;

T_1 is the temperature of mine water pumped; Q_2 is the flow rate of the water delivered to heat pumps in winter; T_2 is the temperature of the water delivered to heat consumers; $Q_{2,c}$ is the discharge of cooled water to mine mine workings; $T_{2,c}$ is the temperature of cooled water to be discharged to mine mine workings; Q_3 is the emergency discharge of water from the retention pond to the turbines of the UHPP in case of peak consumption of electricity; Q_4 is the discharge of the water from the retention pond to the surface water bodies or watercourses; Q_5 is the discharge of the water pumped from the mine in case of flooding of an adjacent active mine.

During the periods of the minimum electricity consumption the UHPP operates in the pumping mode, delivering water to the surface at a rate Q_1 and the temperature T_1 of 30 to 33 °C to the retention pond. The maximum load on the power supply system is attained when pumps are operated in the hydraulic turbine mode and generate electricity. The turbine pump is rotating under the pressure of drainage water with a discharge Q_2, taking water from the sediment pond on the ground surface. Implementation of this proposal requires increasing the capacity of water tanks and the pumping rate, as well as installation of reversible hydraulic turbine engines with an electric machine capable of running either as an engine or a generator.

During the heating season from November to March a portion of mine water at the temperature T_1 is transported directly to heat pumps with the dis-

charge Q_2; then it is heated to the operational temperature T_2 (45–55 °C) sufficient for heating and hot water supply. Cooling of the water circulating in buildings lowers its temperature to $T_{2,c} = 10$–15 °C; then water is discharged to the mine. Specific heat flux at a flow rate of 1 m³ per day recovered from mine water for heating buildings averages 83.6 MJ/day. Under the risk of flooding mine drainage operates in the emergency mode at a higher pumping rate Q_5, which allows controlling the safe elevation of the ground water level.

The main feature of this geotechnical design is that the storage and heat resources of disturbed rocks in mines to be flooded are utilised by coupling the UHPP in the shaft with the equipment using mine water heat. Water is pumped from flooded mine workings in a daily cycle with alteration of intake and discharge. The module is positioned depending on the volumes of discharge into surface water bodies and pipeline characteristics. The hydropower plant generates electricity during the periods with maximum daily loads in electrical networks.

The UHPP as the tool for smoothing peak loads in electrical networks doubles the effectiveness of heat generating equipment recovering mine water heat. The economical effect increases by the difference in prices for electricity in day and night. Environmental effect is reached by keeping the safe elevation of the ground water level, which prevents from flooding the land surface without additional efforts.

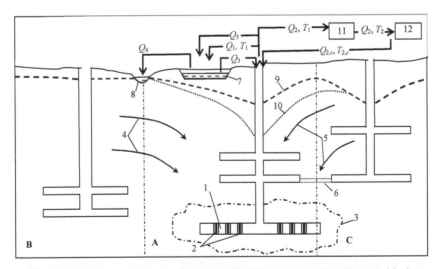

Figure 3. The principal scheme of using mine water heat and generating power in the underground hydropower plant on the area of three mines that are (A) mine with the UHPP, (B) flooded mine, and (C) active mine: 1 – underground mine workings, 2 – water permeable dams, 3 – zone of hydrodynamic impact of UHPP on water outflow, 4, 5 – ground water flow to the mine with the UHPP from a flooded and active mine, 6 – connecting mine workings, 7 – retention pond, 8 – surface water body, 9, 10 – ground water level under normal and emergency operation of the UHPP, 11 – heat pumps, 12 – heat and power consumer(s).

3.4 *Combining heat recovery and water supply*

Ground water reserves in southern regions of Ukraine are poor and do not meet the local needs for supplying potable water without shortages. This is due to commonly brackish ground water in the shallow aquifer that does not meet drinking quality standards. Regarding to population density in these regions the centralized water intake and water treatment f. e. desalinization of this water will be not profitable.

The major feature of this design consists in coupling of water intake and re-circulating facilities in one module that provides utilisation of low-potential heat and partial extraction of water for desalinization that should meet drinking quality standards after treatment. This design implies modular arrangement of heat and water supply facilities for building as well as the option to use ground water of low quality for producing potable water, which makes such a technology economically more attractive. The design implies separated subsurface storage of the water heated in summer through a "warm" well and the water cooled in winter through a "cold" well. The notations in Figure 4 are following: $Q_{1,s}$ is the pumping rate of ground water withdrawn for cooling the building in summer from the "cold" well; $T_{1,s}$ is the temperature of ground water withdrawn from the "cold" well; $Q_{1,w}$ is the pumping rate of ground water withdrawn for heating the building in winter from the "warm" well; $T_{1,w}$ is the temperature of ground water withdrawn from the "warm" well; $Q_{2,s}$ is the discharge of water after conditioning the building in summer to the "warm" well; $T_{2,s}$ is the temperature of water after conditioning the building in summer; $Q_{2,w}$ is the discharge of water cooled after heating the building in winter to the "cold" well; $T_{2,w}$ is the temperature of water cooled after heating the building in winter.

In summer ground water is withdrawn from the "cold" well in the aquifer at the initial temperature $T_{1,s}$ of 10 to 12 °C. After conditioning the building the water temperature raises to $T_{2,s} = 30$–32 °C. Then, some portion of this water is stored in the aquifer through the "warm" well; the rest can be additionally heated, e. g. by available heat pumps or solar panels and then used for producing potable water. Water desalinization and treatment is recommended to carry out in membrane distillation units especially effective in the presence of cheap low-grade waste heat.

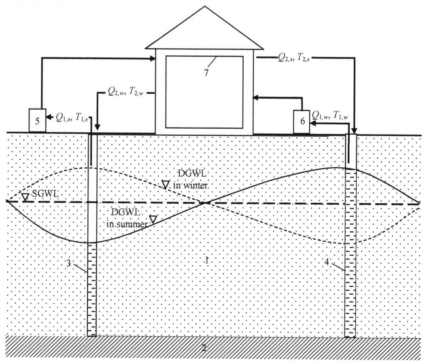

Figure 4. Design of combined low-grade heat recovery and water treatment and supply: SGWL and DGWL – static and dynamic ground water levels, respectively, 1 – shallow aquifer, 2 – aquitard; 3 – "cold" well, 4 – "warm" well, 5 – unit for preparing cold ground water to condition building, 6 – heat pump, 7 – heat transfer loop in the building.

In winter ground water is withdrawn from the "warm" well at the initial temperature $T_{1,w}$ of 20 to 23 °C and then heated by heat pumps to the operational temperature $T_{2,h}$ of 45 to 55 °C and used for heating and hot water supply. A portion of heated water can also be used for water treatment in membrane units. Cooled water at a temperature $T_{2,w}$ is stored in the aquifer through the "cold" well.

Estimated specific heat flux gained as a result of applying this scheme amounts in summer 82 MJ/day and in winter 54.3 MJ/day.

4 CONCLUSIONS

Analysis of power generation and energy consumption as well as actual environmental conditions in coal-mining lands of Ukraine demonstrates the growing need for larger exploitation of alternative energy sources, particularly, geothermal energy and available resources of residual coal and man-made heat and water storages in abandoned mines. Besides, the problems linked with ground water management in post-mining lands, particularly, controlling the ground water level in abandoned and adjacent active mines, have to be solved simultaneously. In addition, shrinking amounts of fossil fuel that can be profitably extracted as a result of mine closure necessitate using alternative sources of thermal energy. This demand can be met by leveraging the vast amounts of natural and man-made energy resources in Donetsk coal basin formed as a result of long-term underground mining, primarily, warm mine waters and residual low-grade and thin coal seams.

Three developed engineering designs aim to use mine water waste heat of residual coal reserves in flooded mines in combination with environmental geotechnologies. These design include (1) coupled regulation of water flows and low-grade heat recovery at an abandoned mine, (2) using residual coal reserves of a flooded mine, and (3) combining an underground hydropower plant and heat pumps. The fourth geotechnical design combines heat recovery and water treatment for domestic supply by leveraging heat resource accumulated in summer for heating and cold resource accumulated in winter for conditioning the building in summer. This geotechnology is recommended to application in southern areas of Ukraine that characterized by poor availability of drinking water and some excess of onground heat or cold that can be temporarily stored underground.

The main feature of all proposed designs is combination of power generation or heat extraction with addressing environmental challenges in the man-made environment, mostly in coal-mining lands. Economical feasibility of all developed designs is enhanced due to their multi-purpose nature.

REFERENCES

Lund, J.W., Freeston, D.H. & Boyd, T.L. 2010. *Direct Utilization of Geothermal Energy 2010 Worldwide. Review.* Proc. of World Geothermal Congress 2010. Bali, Indonesia (25-29 April 2010).

National Atlas of Ukraine. 2007. Kyiv: Institute of Geography: *Kartographia.*

National Report about Implementation of the Energy Efficiency State Policy. 2010. Kyiv: National Agency of Ukraine on Ensuring of Efficient Use of Energy Resources Management: 246

Unlimited resource: Ukrainian potential of renewable energy resources. 2007. Fuel and power complex, 8: 40–46.

Nechitailo, O.N. 2011. *Utilization of mine water heat using heat pumps.* KiEM, 4.
http://ukrrosmetall.com.ua/content/1125

Samusia, V.I., Oksen, Yu.I., Komissarov, Yu.O. & Radiuk, M.V. 2012. *Utilization of the mine water exhaust heat by means of the heat pump.* Hot-water supply plant. *Sci. Bull. of National Mining University,* 4: 143-147.

Yakovlev, E.A. 2007. *Regional man-made changes of geological environment of Donbas under the influence of mining works.* Kyiv: Dumka Publ.

Sadovenko, I.A., Timoshyk, V.I. & Zagritsenko A.N. 2002. *Solution of hydrogeological problems in restructuring the coal industry.* Bull. of Zhytomyr state technological university, 3 (22): 164–167.

Bottomless supplies. Coal as basis of source of raw materials of Ukraine. 2013. Available at: http://www.geonews.com.ua/news/detail scribd.com/doc/1034528/ (accessed 15 February 2013).

Gash, B.W., Hill, V.L., Barone, S.P. & Martin, J.W. 1989. *Rocky Mountain 1 Underground Coal Gasification Test: In-Situ Coal Conversion Using Directionally Drilled Wells. Society of Petroleum Engineers.* Annual Technical Conference and Exhibition (8-11 October 1989, San Antonio, Texas).

Lindblom, S.R. & Smith, V.E. 1993. *Rocky Mountain 1 Underground Coal Gasification Test Hanna, Wyoming Groundwater Evaluation.* Final Report June 10, 1988 – June 30, 1993.

Gholudev, S.V. 2003. *Calculation of the thermal mode of gas-generator during underground coal gasification.* Announcedr of the Dnipropetrovs'k university. Series are geology, geography, 7: 11–20.

Sadovenko, I.A. & Inkin, A.V. 2002. *Underground hydroelectric power station as ecological and power regulator.* Coal of Ukraine, 5: 32–34.

Sadovenko, I.A., Rudakov, D.V. & Inkin, A.V. 2012. *Numerical study of thermal field features around an underground gas generator.* Bull. of National Mining University, 39: 11-20.

Sadovenko, I.A. & Inkin, A.V. 2013. *Estimation of efficiency of the thermal module on the basis of resource potential of the flooded mine.* Bulletin KNU, Vol. 80, 3: 123–127.

Wieber, G. A. *Source of Geothermal Energy – Examples from the Rhenish Massif.* 2008. Mine Water: In: Technical University of Ostrava Faculty of Mining and Geology, In: Proceedings of the 10th IMWA Congress – 2008 in Karlovy Vary, Check Republic: 113–116.

Malolepszy, Z. 2003. *Low temperature, man-made geothermal reservoirs in abandoned workings of underground mines.* Proc. 28 Workshop on Geothermal Reservoir Engineering Stanford University, Stanford, California (January 27-29): 259–265.

Progressive Technologies of Coal, Coalbed Methane, and Ores Mining – Bondarenko, Kovalevs'ka & Ganushevych (eds)
© 2014 Taylor & Francis Group, London, ISBN: 978-1-138-02699-5

Concerning CAE systems development of hydraulic hoists within ship mining complexes

V. Kyrychenko, E. Kyrychenko, V. Samusya & A. Antonenko
National Mining University, Dnipropetrovs'k, Ukraine

ABSTRACT: The CAE development features of hydro-transport systems within marine mining enterprise are considered. The block-hierarchical structure for CAE of air-lift and pump hydraulic hoists is offered, that is realized by means of research software package, being in the stage of active development within the mathematical and programmatic framework of CAE.

1 INTRODUCTION

Projecting tasks of equipment development for industrial exploration of deep-water ocean mineral deposits are pioneer in fact. In view of operating condition complexity of the deep-water hydraulic hoists, comparable with space branch, it is necessary to refuse the established stereotypes of design for traditional mining equipment. Of course, to speak about building the computer-aided engineering (CAE) systems for marine mining enterprise (MME), is prematurely. This process needs huge expenses of intellectual work and time. On the other hand, some projects just today can serve as a basis for mathematical foundations and software development of the CAE, especially for research programs package.

2 BLOCK-HIERARCHICAL STRUCTURE DEVELOPMENT FOR MARINE MINING ENTERPRISE

The goal of the work is development the CAE block-hierarchical structure for airlift and pump units within the Marine Mining Enterprise, and the respective mathematical foundation and software as well. The problems of mathematical foundations (MF) and software development have some features, specified by peculiarity of marine mining complex that is classified as an object of "high complexity", according to CAE Classification Document.

It is obvious that CAE MF must describe physical processes adequately. At the same time, realization of already developed MF elements, generally speaking, is possible by various ways (various numerical methods at Micro, Macro and Meta levels) that predetermines structure and functionality of the software substantially. Therefore, for a new area of the

mining equipment exploitation, it is expedient to build MF with taking into account possibilities and restrictions of modern computer hardware in respect of correct program realization (rationality, reliability, speed) of all software components. For today, such approach, applicable to CAE task solutions of the unique machine-building constructions, is represented as the most effective.

Functionally, technologically and kinematically a hydraulic hoist is a link between ground and surface MME blocks that is a fundamental factor of CAE structure (Kyrychenko 2001). Thereby, hydraulic hoist parameters a priori depend on the chosen mining operation technology, abilities of the ground block mining equipment, and, first of all, productivity of seabed harvester (SH). On the other hand, the geometry of the pipeline and its equilibrium form, which is directly determining construction strength and transportation process stability by means of aerodynamic coefficients, considerably influences on the ship movement speed (especially at maneuvers) and on the power of cruise engines. Besides, the pipeline form "is responsible" kinematically for direct contact of SH work tool with exploited ledge that is especially actual for the towed SH. And finally, the hydraulic hoist productivity depends on power and type of beneficiation equipment, as well as concentrate storage capacity and other factors defining basic configuration of machines and units for ship mining complexes (Figure 1) (Kyrychenko 2009).

The variety and complexity of the interconnected tasks define the general structure of hydraulic hoist design as multilevel hierarchical object that found reflection in morphological structure of hydraulic hoist design as a part of MME (Figure 2). Realization of the general structure was reached by the level decomposition of the design process due to the modular principle of the elasticity theory tasks, as well as theory of hydroaeroelasticity and hydrodynamics of multiphase flows in the adjacent formulation.

It is necessary to notice, that the subject of author's inventions is the hydraulic hoist only (the blocks at the diagram that are marked with regular font). The blocks, marked with cursive, are given for the goal of identification of their places and links in the general design structure of whole MME.

Designing of hydraulic hoist is connected with a list of tasks (Kyrychenko 2001):
• definition of static and dynamic characteristics for transport pipeline (TP) deflected mode;
• calculation of primary constructive, consumed and energetic system parameters;
• space form definition of TP during its towing through water mass.
• development of local regulation systems for control of hydraulic hoist flow-rate parameters etc.

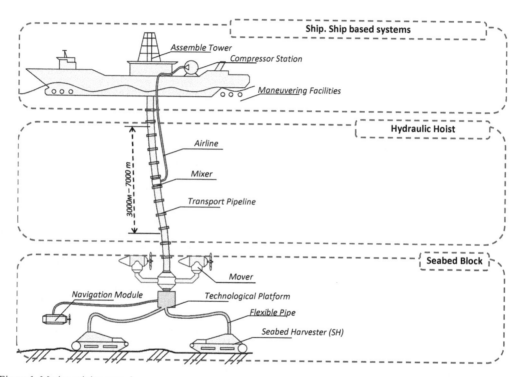

Figure 1. Marine mining complex.

Definition of estimated loading – is the most responsible stage of design process, stimulated by improvement of computer equipment and mathematic methods. Mathematical models for hydroelastic interaction of pipeline elements with the surrounding marine environment and a flowing pulp are based on modeling of the pipeline as the closed set of shafts with constant length of mechanical and geometrical characteristics and consider the following design features (Kyrychenko 2009):
• the piecewise-constant through the TP length the law of change of geometrical, mechanical and physical parameters, and also the existence of concentrated elements;
• distributed and concentrated static and dynamic loadings that form the spatial scheme of forces;
• asymmetry of cross section of the lifting pipe system concerning a central TP axis;
• vertical and cross (following and lateral) pipeline fluctuations near a balance position;
• the stepwise deformed form of the floating transport pipeline during the calculation of hydraulic parameters for pulp streaming.

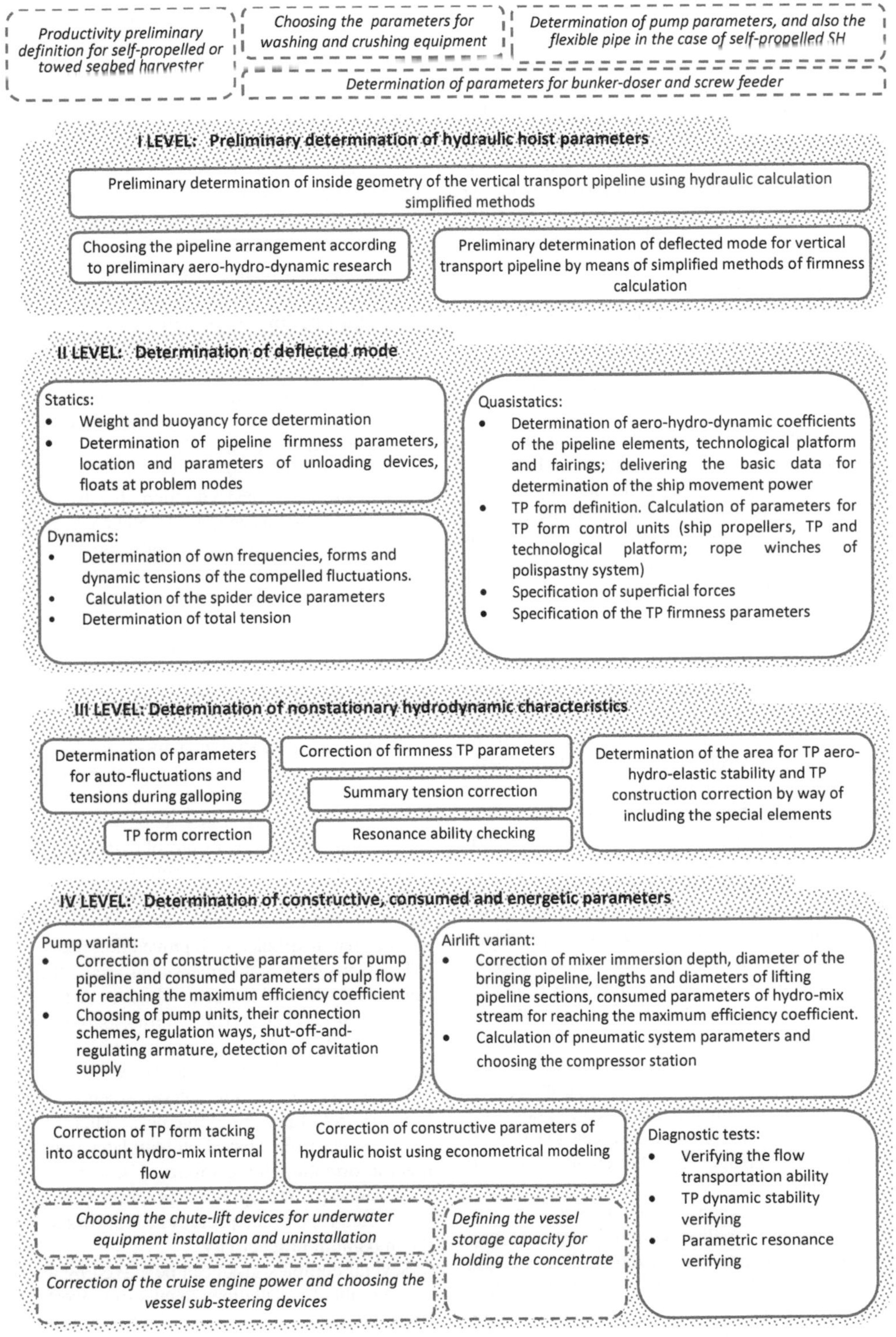
Productivity preliminary definition for self-propelled or towed seabed harvester
Choosing the parameters for washing and crushing equipment
Determination of pump parameters, and also the flexible pipe in the case of self-propelled SH
Determination of parameters for bunker-doser and screw feeder
I LEVEL: Preliminary determination of hydraulic hoist parameters
Preliminary determination of inside geometry of the vertical transport pipeline using hydraulic calculation simplified methods
Choosing the pipeline arrangement according to preliminary aero-hydro-dynamic research
Preliminary determination of deflected mode for vertical transport pipeline by means of simplified methods of firmness calculation
II LEVEL: Determination of deflected mode
Statics:
• Weight and buoyancy force determination
• Determination of pipeline firmness parameters, location and parameters of unloading devices, floats at problem nodes
Dynamics:
• Determination of own frequencies, forms and dynamic tensions of the compelled fluctuations.
• Calculation of the spider device parameters
• Determination of total tension
Quasistatics:
• Determination of aero-hydro-dynamic coefficients of the pipeline elements, technological platform and fairings; delivering the basic data for determination of the ship movement power
• TP form definition. Calculation of parameters for TP form control units (ship propellers, TP and technological platform; rope winches of polispastny system)
• Specification of superficial forces
• Specification of the TP firmness parameters
III LEVEL: Determination of nonstationary hydrodynamic characteristics
Determination of parameters for auto-fluctuations and tensions during galloping
Correction of firmness TP parameters
Summary tension correction
Determination of the area for TP aero-hydro-elastic stability and TP construction correction by way of including the special elements
TP form correction
Resonance ability checking
IV LEVEL: Determination of constructive, consumed and energetic parameters
Pump variant:
• Correction of constructive parameters for pump pipeline and consumed parameters of pulp flow for reaching the maximum efficiency coefficient
• Choosing of pump units, their connection schemes, regulation ways, shut-off-and-regulating armature, detection of cavitation supply
Airlift variant:
• Correction of mixer immersion depth, diameter of the bringing pipeline, lengths and diameters of lifting pipeline sections, consumed parameters of hydro-mix stream for reaching the maximum efficiency coefficient.
• Calculation of pneumatic system parameters and choosing the compressor station
Correction of TP form tacking into account hydro-mix internal flow
Correction of constructive parameters of hydraulic hoist using econometrical modeling
Diagnostic tests:
• Verifying the flow transportation ability
• TP dynamic stability verifying
• Parametric resonance verifying
Choosing the chute-lift devices for underwater equipment installation and uninstallation
Defining the vessel storage capacity for holding the concentrate
Correction of the cruise engine power and choosing the vessel sub-steering devices

Figure 2. Block-hierarchical structure of the hydraulic hoist CAE inside MME.

The structure of hydraulic hoist CAE contains four levels. The first level provides determination of preliminary geometrical and expense parameters of deep-water hydraulic hoist (HH), providing construction integrity and transportation stability of hydraulic mix, in the simplified formulation without including several factors. These parameters are corrected at following three iteration levels already including the specific of marine hydraulic hoist exploitation. At the second level the spatial-deformed form of pipeline is determined and the tasks of providing of its construction strength inclusive of vibrations on marine heaving are examined. The solution of these tasks, however, does not guarantee the transportation of pulp in the area of parameters optimal on energy intensity. Information on a frequency spectrum is needed for rebuilding from the possible resonance modes by changing of mass and elastic characteristics of constructions. The third level determines form and dynamic stability of pipeline including aero-hydro-elastic auto-oscillations, because of co- interaction with a marine environment. After that total tensions are specified. The fourth level determines construction and expense parameters of hydraulic hoist for the achievement of maximal efficiency coefficient of airlift.

The offered structure organizes the consistent task solution of all hieratic levels with passing the results on the next levels as the proper limitations or terms of these levels (Kyrychenko 2008). The analysis of the demonstrated structure proves that the basic transfer function of hierarchy different levels is the form of pipeline, which is determined by interaction of elastic, hydrodynamic (external flowline, flow of pulp) and power parameters.

In addition, in the process of development of such elaborate constructions with the uncertainty of inputs (parameters of the wind-wave mode, undercurrents etc.) it is necessary already on the stage of planning to foresee methods and facilities of the local operating systems of exploitation parameters of airlift, that is a new scientific-applied result. Therefore for realization of basic principles of hydraulic hoist design as part of MME and the technological process controls within the united conception in a sectional-hieratic structure are foreseen the following facilities of management, that could be divided on four groups:

1. Adjusting of expense characteristics of pulp flow in a transport pipeline with the purpose of specific energy demands minimization. Density and speed mix flow will be the ruled parameters. These parameters are regulated by the rotation speed changing of screw feeder, which gives hard material from a bunker-metering device, and by expense of salt water.

2. Adjusting of the spatial pipeline form, approaching it to vertical, can be provided by the following methods:

• changing of speed and trajectory of ship motion in the case of the flexible pipeline connection with the seabed harvester;

• placing of special screw propellers in the certain places of pipeline;

• placing of powerful sub-steering device on the technological platform;

• the rope winches of the reeving system, connecting vessel with different areas of pipeline.

3. Adjusting of auto-oscillations processes parameters with the purpose of decline of vibrations intensity is taken to development of the special measures, directed on violation of eddy formation periodicity, reduction of unstationary hydrodynamic influence and increase of hydrodynamic amortization (Kyrychenko 2009).

4. During the forced vibrations the frequency rebuilding is most acceptable. In other words, changing of own frequency of the pipeline vibrations to frequency of excitant force. To this end it is necessary to do the pipeline construction the ruled by model. For example, to handle the resilient joint leaning with variable inflexibility in the spider device, to provide a technological platform with the ruled fairwaters for the changing of its inertia descriptions and etc.

At the heart of block-hierarchical structure iterative approach is put. Transition to the following iteration happens by return to LEVEL II including the sequential passing of all its points. This structure assumes further improvement by means of expanding theoretical inventions and accumulation of experimental results. For an embodiment of difficult mathematics into a program code it was decided to make use of several free-distributed well proved libraries.

For work with multidimensional arrays the Armadillo library is well suitable, which is distributed by GNU LGPL license (http://arma.sourceforge.net). Thanks to Armadillo's integration with MKL package the high efficiency of calculations is reached.

Blitz++ is applied for specific scientific calculations (http://blitz.sourceforge.net). This library is open source project based on templates that provide meta-information for compiler. Using it, the compiler generates the optimum program code specific to selected type of the processor. It allows achieving the maximum productivity.

Several complex tasks can be solved by a method of symbolical calculations thanks to powerful C ++ library GiNaC (http://www.ginac.de), that is intended for creation of integrated systems where symbolical manipulations are combined with numerical methods. The library is free (license GNU GPL) and well known by its powerful API (Application Layer Interface). The main advantage of this library is use of the CLN (Common Lisp Numbers) technology for solving the differential equations.

Taking into account that all chosen libraries are written in C ++ programming language, it looks expedient to development the package of research programs using the same language. It is known that a weak place of C ++ language with its libraries is poor or inconvenient tools for graphic interface creation. There are several ways for overcoming this trouble.

The first way consists of refusal of Native-UI (the "native" graphics and toolkits of an operating system) in favor of the Web interface. Today it is popular approach in software market, making easy to port the final software product on the majority of widespread platforms, including mobile OS. The sense of approach is that the software product shares on two separate modules: server part (application logic) and client part (the user interface). While porting on each new platform the only client module needs to be changed by way of simply replacing of WEB-engine, which is the most suitable for a target platform. However this way is rather effective for software that is focused on the mass consumer, and, perhaps, is hardly applicable for unique machine-building complexes.

The second way is that C ++ code always can be wrapped into packages and modules of the Microsoft.Net Framework platform (or its Linux analog MONO (http://www.mono-project.com), but of course it's not a simplest way.

The third way consists in creation of the user interface by means of the free version of C++ cross-platform framework Qt 4.x. (http://qt.digia.com). The choice was made in favor of the last option though it has also a number of shortcomings concerning the second.

The given reasons are used during developing a package of the research programs as a part of CAE for deep-water hydraulic hoists.

3 CONCLUSIONS

The block-hierarchical structure of the CAE for deep-water hydraulic hoists is developed as a part of the Marine Mining Enterprise taking into account the corresponding communications and restrictions from linked systems. In the presented structure the principles of the hydraulic hoists design and technological process control are realized within the united concept. Original approach for solving the problems of automated design of the pump and air-lift systems, based on the development of mathematics and software in the mated definition is offered. The software modules for calculation the pipeline deflected mode and hydrodynamic parameters of heterogeneous flows are developed.

REFERENCES

Kyrychenko, E.A. 2001. *Scientiffic grounding of the pipe system parameters for hydraulic lifting of the minerals.* Author's abstract on doctor's dissertation. Dnipropetrovs'k: Thesis: 38.

Kyrychenko, E.A. 2009. *Mechanics of deepwater hydrotansport systems in the marine mining deal.* Monograph: 334. Dnipropetrovs'k.

Kyrychenko, V.E. 2008. *According to a question of method development for automated control of the transients within deepwater airlifts.* Scientific bulletin of National Mining University, 11: 71–75.

http://arma.sourceforge.net. Armadillo. C++ mathematic library.

http://blitz.sourceforge.net. Blitz++. C++ library for fast mathematic calculations.

http://www.ginac.de. GiNaC. Powerful mathematic library with rich possibilities.

http://www.mono-project.com. MONO. Cross-platform open source framework.

http://qt.digia.com QT. Cross-platform C++ open source framework.

Progressive Technologies of Coal, Coalbed Methane, and Ores Mining – Bondarenko, Kovalevs'ka & Ganushevych (eds)
© 2014 Taylor & Francis Group, London, ISBN: 978-1-138-02699-5

Corporate governance of vertically integrated coal and methane mining holdings

A. Bardas & O. Horpynych
National Mining University, Dnipropetrovs'k, Ukraine

ABSTRACT: The rules of consolidation and corporatization of Ukrainian enterprises which belong to national fuel and energy complex are described in the paper. It has been substantiated some principles and directions to develop organizational and economical mechanism for vertical integrated manufacturing and distribution corporations which mine and utilize methane from coal seams. The modified model of corporate governance for Ukrainian corporation DTEK is suggested there.

1 INTRODUCTION

One of the most important factors of Ukrainian economy and industries sustainability is appropriate functioning of the country fuel and energy complex the development of which is mostly provided by coal mining. At the same time the world practice demonstrates the great interest to the alternative sources of energy that is caused by depletion and cut of the traditional energy sources. During last years the special attention was paid to coal mining methane (CMM) known as coal methane. The International Symposium on the Alternative Sources of Gas Raw Materials (1992) recognized methane gas, i.e. hydrocarboneous gases, as the most promising source of energy to be used in the nearest future. It is estimated that methane resources to be extracted from coal beds will take the third place among the world reserves of all fuel resources after coal and natural gas, if they are calculated as conditional fuel.

The fact that such kind of energy resources will be the most demanded in the near future is proved by the ongoing methane extraction and utilization carried out abroad that is rather successful. Thus, during last decades USA, Japan, France, Germany and China successfully mined and used carboneous gases in different branches of their economies. During last 15–18 years industrial mining of carboneous gases in the USA reached 35-40 bln. m³/year that is equal to the half of natural gas consumption in Ukraine (Ancyferov et al. 2004, Horpynych 2011).

Ukraine has significant methane resources in its coal deposits, which are actually has not been developed. Overall resources of coal in Ukraine are 117.5 bln. tons with only 46.7 bln. tons have been explored. Each ton of hard coal contains from 5 to 25 m³ of methane, the content of methane in each ton of anthracites and semi-anthracites is 35–40 m³. Calculations demonstrate that the reserves of methane in the explored conditional coal beds up to 1.800 m deep are within the range of 450–550 bln. m³. That is why Ukraine takes the fourth place in the world CMM deposits. 95% of Ukrainian mines from 219 mines, on whole, are gaseous, 70% of them are dangerous by coal dust explosions, 45% – dangerous for their gas dynamic phenomena, 30% - dangerous by spontaneous combustion of coal (Lytvynenko 2010).

Industrial mining and use of coal bed methane positively influence the national economy, environment and coal mining operations. The State Complex Programme on Coal Bed Methane Development implementation will allow to:

• use the alternative energy resources that will lead to Ukrainian independence from energy resources imported, first of all, from natural gas imported from Russia and other CIS countries;

• decrease the amounts of methane emissions, that is the considered the most harmful gas among 'greenhouse gases', into atmosphere by coal enterprises;

• increase the production volumes of the primary (commodity coal) and additional (gas, heat, electricity) products;

• increase mine safety.

Methane utilization and its further industrial extraction and exploitation are closely connected with coal mining in Ukrainian mines where coal is a commodity product. This brings Ukrainian mines to production diversification and mining enterprise expansion in different sectors of the single energy market.

The major problem on the way to improvement the mechanism of Ukrainian fuel and energy complex functioning is caused by establishing and development of organisational structures and economic conditions which could contribute to coal mine enterprises activities. In other words, the future of the Ukrainian fuel and energy industry is aimed to establish strong consolidated production and service units (PSU) which function on the market principles and to provide their competitiveness on the home and international markets of fuel and energy resources. Such subjects of business entities shall provide realisation of the following principal strategic objectives:

• to raise competitiveness;
• to use synergy effect;
• to minimise social costs on restructuring fuel and energy complexes;
• to provide country energy safety;
• to raise added value (Lulchak et al. 2012).

The dramatic changes in Ukrainian economy structure and its amounts of capital have predetermined the need of developing international standards generally accepted that will help to identify investment attraction of enterprises, to build a system of their effective management and control. All the mentioned above lead to establishment and development of corporate governance, where the corporate governance is seen as an autonomous economic and legal institute and/or system with the help of which joint-stock companies are directed and controlled. Although various forms and approaches to some problems of corporate sector functioning have been developed by researchers and practitioners so far, there has been no unified model of effective corporate management yet. This can be explained by the nature of relationships which form the basis for corporate governance and objectively applied to the state legal system selected by the country.

Functioning practice of the majority of home joint-stock companies points out a wide range of problems in company management: lack of and in some cases absence of scientifically grounded approaches to planning, organising and controlling company activities; low effectiveness of stimulating enterprise managers and staff; existence of interest conflicts among company founders, stakeholders and CEO; absence of any specific missions and goals. Formation of the original regulations and provisions for implementation and realisation corporate governance, i.e. development of corporate governance principles, strict compliance with them and their improvement, is a prerequisite of levelling the mentioned problems. Principles of corporate governance are evolutionary by their nature. Therefore, they must be subjected to substantial revision in connection with changes in the environment of functioning of one or another corporate structure.

2 MATERIALS UNDER ANALYSIS

There are many large corporate structures functioning in different countries of the world. They are based on the principle of integrating ownership, resources and areas of their activities.

The most widely spread ones are vertically integrated associations known as business concerns. They are different, and differ from one another by autonomy of the enterprises included in them. For example, if US concerns are established on the basis of complete ownership of their branches, Western European and Japanese ones encompass formally autonomous enterprises managed by the head company which is the owner of their controlling shares (holding). Relationships inside corporations are regulated by administrative rules and CEO decisions, even when degree of decentralisation is high. The degree of control of the parent company over its branches depends on the share of its participation in the capital and the legislation adopted in the country.

Strengths of such vertical integration is stability of economic relationships, 'linked' diversification, guarantee of supply, control over resources, accelerated capital turnover and recoupment of costs, access to innovative technologies (Sapytska 2011).

The other form of associations appears when there is unlinked diversification within one juridical person. These are diversified corporations or conglomerated companies. In the process of merging level of autonomy of such business units can differ. They can be given the status of a juridical person, i.e. a branch company, that will need to establish financial holding company for their effective management.

From the perspective of foreign and national experts, corporate method of interest consolidation is considered to be the most effective. Moreover, practice demonstrates a great variety of company types able to meet different interests and needs in the form of corporation (Thompson & Streakland 2007).

Creating corporate structures is appropriate to mining industry, first of all, due to the advantages of large structures of this industry, among which:

• overcome of specialised product market limitations and rapid entry into new market segments;
• reduction of business risks because of expansion of income sources distribution;
• compensation of losses during the period of structural changes, reorganising and short-term fluc-

tuation;

- possibility of flexible manoeuvring of investment and subsidized resources donation resource receipt.

National experience and world practice prove that the most effective corporative structures are not those integrated on the territory basis, but those which are integrated on the basis of technological links where the processes of industrial and financial integration are performed on the mutual benefit principles. Such processes are used when establishing organisation and economic corporations which integrate the enterprises both of mining industry and its adjacent branches. Such kind of complexes have all technological links on the way starting from mining of coal and finishing with the end products in the form of electric and heat energy, coke, metals, synthetic motor fuel etc. As a result, they will have a significant economic effect.

Thus, 'integrated corporate structure' is referred to in the modern economic theory as a group of juridically or economically autonomous enterprises (organisations) which perform common business on the basis of their assets consolidation or contract relationships aimed at achievement common goals. Vertical integration is an association of commercial organisations functioning in the branches which represent different stages of a single production process. Vertically integrated company (VIC) is a complicated organisational and production structure of holding type with a single managing Centre. The VIC company integrates the enterprises which consistently take part in production, sales and consumption of a commodity product that makes them interdependent by commodity and cash flows (Gorpynych & Nykytiuk, 2012, Bardas et al. 2012).

Holding company management should be based on the up-to-date and fundamental principles of governance.

Generally accepted international principles of corporate management are considered to be the following:

- corporate governance structure should provide protection of shareholder's rights. It should be the only principal method of the preventive settlement of emerging conflicts of interests;
- corporate governance regime should provide equal attitudes to all groups of stakeholders, including foreign stakeholders and stakeholders with small share in the corporation. The rights of all shareholders should be provided and protected effectively in case of their violence;
- corporate governance should provide compliance of stakeholders' rights with the established law and encourage cooperation of all the subjects of

corporate governance in the corporation development;

- corporate governance should provide openness of corporation information, in-time full disclosure of information about principal issues of corporation's financial and economic activities;
- corporate governance structure should provide effective performance of CEO and accountability of corporation management bodies to their stakeholders (Corporations and integrated structures... 2007).

World experience has accumulated a significant number of standards (Principles, Recommendations, Codes) in the area of corporate management and governance at the international, national, local levels as well as at the corporation level. The Principles of Corporate Governance of the Organisation of Economic Cooperation and Progress (OECP), Principles of Corporate Governance recommended by Reconstruction and Development Bank (RDB), Principles developed by Confederation of European Association of Stakeholders "European Stakeholders" (Brussels) are the most known world standards.

General principles of OECP corporate governance regulate national standards of corporate governance and management. They raise the role of investors (stakeholders) in managing companies and corporations which they invest their money in.

Principles of corporate governance developed by RDB are aimed at facilitating understanding between corporations and creditors and/or investors when adopting decisions on crediting or investing through introducing rational business regulations in corporate practice. The special attention in the RDB principles is drawn to the relationships between stakeholders (beneficiaries) and creating a balance of their interests in the activity of joint-stock company.

The principles developed by Confederation of European Association of Stakeholders "European Stakeholders" (Brussels) as well as those developed by OECP and RDB are aimed at improving legal, institutional and regulatory basis of corporate governance. The first ones are more specific and detailed than the latter two ones. Principal recommendations developed by "European Stakeholders" group cover the following: corporation missions and goals; distribution of profits/incomes; influence of stakeholders on key issues of company/corporation life (corporation reorganisation by mergers and takeovers); realisation of the stakeholder's right to vote, to get the proper information; the role of management body etc. (Korpan 2011).

3 RESULTS

It is expedient to describe mechanism of company corporate governance as a system of interrelated principles, methods, management algorithms which are aimed at achievement holding company's strategic goals focused on maximising market capitalisation and increasing the profitability of shares.

The Ukrainian holdings' practice demonstrates that from the perspective of potential investors, increase of corporate governance efficiency is being considered an important resource for raising holdings' competitiveness and their investment attraction. Firstly, introducing market principles of corporate governance creates a positive image of a company/corporation for its potential investors. Secondly, good corporate governance decreases risks of investors, therefore, encourages funding.

Two circumstances should be considered when developing a system of corporate governance in a holding.

First of all, change in a holding structure is a necessary condition for creation an effective system of corporate governance.

The most wide-spread version of restructuring aimed at forming transparent corporate structures is establishing the Corporate Centre, i.e. a head company of a holding specially established. The role of the Corporate Centre can be performed by:

• company with limited liability (CLL), closed joint-stock companies. (In case holding owners are aimed at avoiding any possibilities of ownership structure dilution, minimising costs on information disclosure, easifying the process of making management decisions);

• open joint-stock companies (This alternative is used for realisation strategies with significant funding needed or when the company enters the international market of borrowing capital, sells its business on market conditions).

Secondly, when restructuring, holdings face great resistance to unfriendly mergers, but the process of restructuring possesses certain risks.

There is a necessity to use systematic approach to the research of corporate governance mechanism. This need is rooted in the fact that a holding company, from the one hand, represents a company, business portfolio of which consists of controlling shares of a group of companies (financial approach), and from the other, it is a system including both tangible and intangible assets (accounting approach). In other words, there is a set of elements that constitute the holding company with these elements being in structural interrelations and forming certain integrity of this system.

It should be mentioned that when treating the corporation as a complex system with a great number of elements, such form of enterprise integration requires effective system of management that is implemented through the mechanism of corporate governance of the holding company.

Structural features essential for the proposed principles of corporate governance of a vertically integrated company (VIC) involved in extraction and use of CMM are given in Table 1 (see below).

Special attention in the proposed principles of corporate governance should be drawn to qualitative planning, organising, stimulating, controlling and regulating corporation activities, i.e. performing classic functions of management; existence of the developed system of economic, technological, socio-psychological and administrative methods of corporate governance. Existence of efficient mechanisms of making and taking managerial decisions and their optimization provide complex and effective corporate governance.

Let us have a look at the process of corporatization, management and governance of methane (CMM) extraction and utilization in the conditions of Donets'k Fuel and Energy Company (DTEK) which is the largest vertically-integrated energy company in Ukraine and which is a part of "System Capital Management" company (SCM that correspond to its Ukrainian name CKM). Assets of the Group within the frameworks of separate branches were gathered in specialised holdings. This new business structure, on the one hand, contributes to the realization of potential synergies within the frameworks of each industry, and on the other hand, makes business of SCM Group more understandable and transparent to all stakeholders-shareholders, investors, employees, partners and clients of the Group, representatives of local communities and the state. SCM as a majority shareholder, i.e. the major investor, manage industry holdings by delegating representatives in their Supervisory Boards.

SCM approved single continental system of corporate governance which provides clear separation of functions of the Executive Directors (Executives) from control functions (Non-executives) between two separate bodies: Board of Directors and Supervisory Board. For some business areas, where there are no branch holdings, system of corporate governance functions via the Supervisory Boards of directly operating companies. The Corporate Centre is responsible for the financial and production results. Uniform management processes have been introduced in all the enterprises of DTEK that allowed to arrange effective work between all the structural subdivisions of the company enterprises.

Table 1. Structure and essence of corporate governance principles.

Corporate governance principle	Feature
1. Systematization	Complex coverage of all the structural elements of corporate governance and interrelationships between them, their nature. Forming corporate governance system structure, selection of its sub-systems, elements and components in organisations should be performed consistently with the goals identified, taking into account priority directions, strength and weaknesses, impacts of the functioning environment, evident limiting factors etc. Ignoring the principle of systematization may result in the loss of strategic vision of corporate governance system functioning.
2. Information transparency	Corporation should provide its own activities with the information transparency by putting all the operations in accounting data (not limited to the national standards and procedures, international ones are used too) to avoid shadow operations. All the information disclosed to public at the consumers' and suppliers' markets, for the state power bodies or competitors, should correspond to the following parameters: reliability, objectivity, realism and completeness. In other words, transparency of financial and economic practice is indispensable condition for a corporate structure functioning.
3. Creating a balance of interests of all the subjects of corporate governance and preventing corporate conflicts.	Subjects of corporate governance encompass: founders of corporation, owners of corporate rights (stakeholders), CEO, managers, members of Supervisory Board, Audit Committee members, employees and workers of the system under governance. When implementing corporate governance, the balance of economic interests of all the subjects of corporate governance should be provided in order to prevent conflict situations and protect corporation from their negative consequences.
4. Qualitative realisation of planning, organising, motivating, controlling and regulating corporation activities.	Qualitative corporate governance is based on the effective implementation of management technology in joint-stock company, first of all. In other words, any management process should be taken in particular logical consequence, namely: planning corporate structure activity, organising, staff motivating, controlling and regulating. Each corporation should have certified document about sharing powers, responsibilities and rights of all the subjects of corporate relationships, in particular of corporate governance and management bodies; stimulus for each of the subjects of corporate governance should be created in accordance with the aim of effective fulfilment of (delegated) responsibilities and avoiding unscrupulous behaviour by employees of the joint-stock company/corporation.
5. Existence of the developed system of economic, technological, social and psychological, administrative methods of corporate governance.	Implementation of management functions should finish with the management methods have been developed. As a rule, methods exist in the form of documentation, where administrative methods formalise all the others.
6. Existence of effective mechanisms, preparing and making managerial decisions and their optimization.	This is about effective mechanisms of taking managerial decisions by all the joint-stock company/corporation management bodies (i.e. general meeting of stakeholders, Board of Directors, Audit Committee, Supervisory Board, managers of lower levels).
7. Clear separation of ownership rights and managerial functions.	This principle provides objectivity and commitment when implementing corporate governance and delegates the authority to the Board of Directors to provide on behalf of corporate structure founders and owners of corporate rights (shareholders) adequate market management aimed at strengthening company's/corporation's competitive position and market capitalization.
8. Providing financial efficiency and solvency of a holding.	A corporation should function with financial efficiency, being solvent, using limited resources and creating new value.
9. Taking into account risks when managing corporation.	Appropriate and adequate methods of dealing with potential risks: diversification, insurance, avoidance, reservation etc, are involved.

Vertical integration is one of the major competitive advantages of the company on the market. DTEK development strategy preserves a sustainable balanced chain 'coal – electric energy – distribution'. The main areas of company's activities is mining of mineral resources (coal), its processing and enrichment, sales, electricity generation and its sales and supply to electricity consumers. Production capacity of DTEK coal mining segment is represented by three largest enterprises in the branch: JSC "DTEK Pavlogradvugillia" (10 mines), LLC "DTEK Dobropilvugillia" (5 mines) and JSC "DTEK "Donbass Komsomolets Mine".

DTEK vertical integration in addition to favourable geographical location of its assets allows the Company guarantee quality and reliability of the whole production chain and manage its efficiency and control the whole value chain. Decrease of business risks

of branch blocks of the Company is one of the advantages of vertical business integration.

Synergy of coal mining, energy generating and distribution enterprises, implementation of the advanced technologies, professional management provide DTEK to sustain its leading positions on the fuel and energy market of Ukraine.

Taking into account the mentioned above, organisational and economic mechanism of corporate governance of holding companies involved in CMM extraction and utilization is presented in the form of model given in Figure 1.

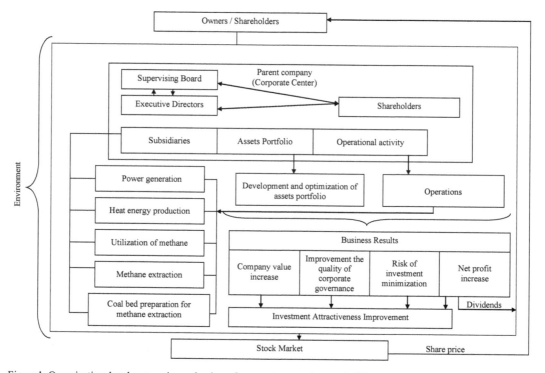

Figure 1. Organizational and economic mechanism of corporate control over a holding company.

The presented mechanism allows to provide complex management of effective holding structure activities, using the up-to-date mechanism of corporate governance.

The developed mechanism is specified by the possibility to take into account combinatory (synergic) advantages of the Company holding enterprises.

Distribution and use of the additionally acquired advantages are performed by the parent company (the Corporate Centre) of the holding.

Parent companies within the holding governance system are delegated the following functions:

• the association-wide industrial-technological, investment and trade policy;

• capital centralisation and coordination of financial, material and information flows;

• development of cooperation at various levels with the aim to use internal and external environment more effective;

• coordination and communication relations between members of the holding company.

Financial and economic functions of the Corporate Centre within the organisational and economic mechanism of corporate governance are the following:

• strategic management of the corporate enterprises' portfolio on the basis of calculation of the integral criterion of the corporate enterprise portfolio attractiveness;

• monitoring financial activity results of the corporate enterprises and the entire holding company on whole;

• identifying a single investment, marketing policy, etc.;

• use of the new investment and credit sources.

The formation of economically-analytical software for implementation the model of operational and financial management of the holding company will develop and implement a unified production and economic, scientific and technical, technological, financial, and investment policy

So, the proposed organisational and economic mechanism will provide comprehensive management of the diversified vertically integrated holding activities, using the up-to-date mechanisms of strategic, operations and finance management.

Holding enterprise activity management are proposed to carry out in two directions: across subdivisions and via technological sub-chains (see Figure 2).

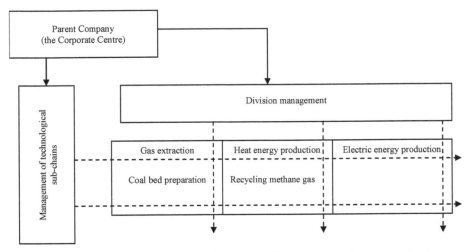

Figure 2. Types of corporate governance of a vertically integrated holding on the extraction and use of methane.

The main task of management across subdivisions is to budget and control activities of the enterprises of each of the divisions. The main task of management of technology sub-chains is production planning and coordination of product flows of each of the sub-chain. These tasks are solved by the managers of the company.

So called divisionalization involves allocating the following duties to the Corporate Centre:

– identifying autonomous business operations within the Group and separating them in the form of divisions;

– strategic planning of the Group activities, including the distribution of general financial resources;

– developing the control system for implementation and execution of the regulations and directives;

– appointing and firing managers of the holding enterprises;

– controlling the division activities;

– coordinating technological sub-chain operations;

– creating centralized services (such as marketing, public relations, corporate relations etc.).

4 CONCLUSIONS

Ukraine has significant, not actually developed resources of CMM, but this strategic resource can't be extracted for some reasons of legal, economic and organisational character. There is no governance in CMM preparation, extraction, production, processing and distribution operations.

The national experience and world practice prove that the most effective in production and economic activity corporate structures are those which combine enterprises by the level of their technological links and use the process of industrial and financial integration on mutually beneficial basis.

Nowadays corporations which combine enterprises of both coal mining and its adjacent industries are being established in Ukrainian fuel and energy complex. Such complexes cover all technological chains from coal mining to the end product manufacturing. The process of organising and managing methane extraction and utilization to be carried out on the corporate principles by such organisational structures as holdings, which also use the principles of synergy, linked and non-linked diversification, is proposed.

Analysis of market positioning and the state of the corporate governance of vertically integrated holding company "Donets'k Fuel and Energy Company" indicates the availability of appropriate infrastructure and production factors for the industrial coal mine methane gas extraction and utilization. At the same time there is no organisational support of realisation this process within the vertical integration of the company.

The methodology of development organisational and economic mechanism for managing vertically integrated holdings is based on the generally accepted international principles. The principles of corporate governance which take into account legal specificity of fuel and energy complex holding activity are proposed. The substantiated principles have contributed to formation the areas of improvement of vertically integrated holding company corporate management, where the holding encompass businesses that carry out preparation, extraction, processing and sale of methane products obtained from this raw material.

The organisational and economic mechanism of holding corporate governance is proposed in the form of model which forms direct, reverse and inter-relations between the holding enterprises and the holding corporate governance members. The holding enterprise activity management is carried out in two directions: across divisions and within the technological sub-chains.

Possibility of taking into account synergetic advantages of holding enterprises and companies distinguishes the developed and disclosed method. Sharing, division and use of the additional benefits is provided by the holding parent company (i.e. the Corporate Centre).

REFERENCES

Ancyferov, A.V., Golubiev, A.A. & Ancyferov, V.A. 2004. *Prospects of development the Donbas as a complex coal and gas basin*. Coal of Ukraine, 8: 4–5.

Horpynych, A.V. 2011. *Elaboration of coal methane extraction projects*. School of Underground Mining. V[th] International Conference: 340–348.

Lytvynenko, A. 2010. *Ukraine can be free from Russian gas for 50 years at least*. Web: // www.metinvestholding.com/ru/csr/news/2010/2/689.

Lulchak, Z.S., Karpiy, O.P. & Ilchuk, G.I. 2009. *Principles of enterprise consolidation in energy and power sector of Ukraine*. Bulletin of the National University "Lviv Polytechnic", 649: 109–120.

Sapytska, I. 2011. *Corporate structures as the basis of mining enterprises transformation*. Economic, 4: 64–67.

Thompson, A. & Streakland, A. 2007. *Strategic Management*. Moscow: Banks and Stock Exchanges.

Bulat, A.F. & Chemeris, I.F. 2008. *Perspectives of an energy complex development based on a coal-mining enterprise*. Coal of Ukraine, 2: 3–6.

Bulat, A.F., Perepelytsia, V.P. & Chemeris, I.F. 2001. *Diversification of coal mines business operations in the context of coal mining sector restructuration*. Coal of Ukraine, 1: 5–7.

Horpynych, O.V. & Nykytiuk, O.O. 2012. *Corporate governance over vertical integrated methane-mining and utilization company*. Modern management: Problems of theory and practice. Scientific and practical conference for higher schools in Kryvyi Rih, Vol. 6: 48–49.

Bardas, A.V., Horpynych, O.V. & Nykytiuk, O.O. 2012. *Corporate control over diversified vertical integrated company*. Problems and perspectives of innovative development of Ukrainian economy: Theses of International Scientific and Practical Conference in Dnipropetrovs'k. Vol. 2: 32–33.

Corporations and integrated structures: Scientific and practical problems. 2007. Kharkiv: VD Izhek.

Korpan, O.S. 2011. *Principles of corporate management: national and global practice*. Bulletin of Khmelnitsky National University, 6. Vol. 3: 33–38.

Progressive Technologies of Coal, Coalbed Methane, and Ores Mining – Bondarenko, Kovalevs'ka & Ganushevych (eds)
© 2014 Taylor & Francis Group, London, ISBN: 978-1-138-02699-5

Application of fine-grained binding materials in technology of hardening backfill construction

O. Kuzmenko & M. Petlyovanyy
National Mining University, Dnipropetrovs'k, Ukraine

A. Heylo
LLC "Krasnolimans'ke", Rodins'ke, Ukraine

ABSTRACT: The expediency of differential formation of high filling massif chambers based on a developed technology with usage of fine-grained binding materials. Two-stage grinding circuit domain of granulated slag and limestone to produce fine particles, allowing to improve the structural and mechanical properties of filling massif, increase the proportion of rocks in mixture and reduce material consumption by 25–75% are considered.

1 INTRODUCTION

Annually underground mining of ore deposits penetrates deeper into the Earth's interior, leading mining in complex geological conditions, applying a hardening stowing chambers. This ensures completeness recess ore with minimum losses and mining safety, reduced environmental pressures on industrial region. Flow diagrams for filling mixtures and their components, as well as equipment for grinding raw material, have not been significant transformation since introduction of this method of rock pressure control. And it's despite the fact that there is a constant decrease in the depth of mining and increases the economic component for the purchase of filling materials. To reduce the cost of filling operations, in hardening backfill add crushed rock from underground development.

To increase the strength of artificial local massif of stress concentration arising at the individual elements of the chambers in the filling mixture-comprising increasing proportion of cement. Under the conditions of intensive mining operations and manifestations of rock pressure, and the use of massif explosions necessary mechanical properties achieved by increasing the flow of cement in filling mixture is not always possible. Partial collapse of the filling massif to the chamber in the second stage of mining floor leads to the contamination of the ore massif. Increases of filling operations costs and risks associated with downtime due to hardening of mixture in pipes, their mounting/dismounting works.

One of the ways to improve the curable compositions of backfill is mechanical or chemical activation of components of backfill (Voloschenko 1985). It was established that dispersion state in the curable binder system affected on density and porosity of the stone monolithic: it increases with increased density, reduced porosity, increase in strength occurs. Mechanical activation was seen as setting a two-stage grinding circuit binder.

At present time in underground mines CIS technology of filling operations provides single-stage crushing circuit binder to give the final product particle size of 50–60% of the particles -0.074 mm, which corresponds to a specific particle surface 2000 cm^2/g with an average diameter of 35–40 microns.

Application in stowing mixtures of materials possessing binding properties or inert solve a number of issues, but have limited opportunities because of insufficient knowledge of interaction of chemical elements mixed with water according to disclosure of relationships and particle surface. Here is a powerful energy potential, which could give a considerable addition of strength artificial massif generated by the internal connections of different mode of action upon the application of external loads.

Technological direction on strength increasing of created artificial massif is in renewal stage of, but has a great scientific and practical importance, because at large volumes of filling operations decreasing of prime cost and improving of filling massif quality becomes actual issue.

2 MAIN MATERIAL OF RESEARCH

The criterion for technological schemes development of fine binder specific surface 2800, 4300, 5500 cm²/g serves ultimate size of the product. Particles studied compositions of filling mixtures of size 15 microns have a maximum specific surface area 5500 cm²/g In terms of technological schemes of filling complexes mines CIS achieve specified product size is only possible by improving the grinding circuit. The advantage of hardening compositions developed by the authors based on backfill fine particles along with high strength is the development of forms of internal structures in the artificial massif. Its hardness increases, reduces consumption of blast furnace slag by 25–75% and increases share-part mixed breed. Under laboratory conditions, Zaporizhzhya iron plant (ZZHRK) studies have been conducted on the impact of highly blast-furnace slag and limestone flux (S_{sp} = 2000–6600 cm²/g) on the strength and rheological properties of filling mixtures. Considered structural changes occurring at different filling massif particles dispersed state.

It has been established that experimental values of the specific surface of particles are slightly different from theoretical. This is due to a variety of physical and mechanical properties of crushed materials. Analysis of the distribution of particle size fractions of blast-furnace slag as a main binder per unit is realized on device Multisizer-3 and the results are shown on Figure 1. Rounding actual specific surface of blast furnace slag, resulting in changing steps – 2000, 2800, 4300, 6600 cm²/g. The maximum increase in the specific surface area compared to the traditional (ZZHRK) was 3.3 times.

Increase of the specific surface of the particles results in a reduction of their diameter, which is necessary for the hardening of filling massif. When the specific surface area of the slag 2000 cm²/g up to 50% of the particles are in the inert state, without forming new compounds. Starting with a specific surface of 4300 cm²/g all particles completely react with water molecules involved in the formation of the structure and form strong bonds in the backfill massif. Aggregates of ultrafine grinding lend out scatter fractions, whereas there are ball milling fractions of from 4 to 100 microns. To maximize the effectiveness of finely ground blast furnace slag is necessary to cut down shredder scatter particles.

Distribution histograms of class fractions of slag and limestone found between average particle diameter and specific surface, which is shown on Figure 2. Using representation of dependencies it is possible to predict the average particle diameter at a different specific surface area of the slag.

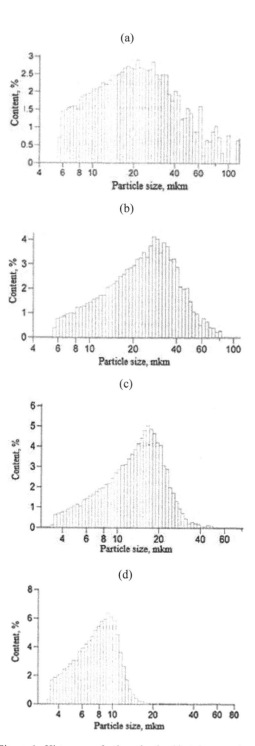

Figure 1. Histograms factions in the blast-furnace slag with (a) S_{sp} = 1999 (2000) cm²/g; D = 35 microns; (b) S_{sp} = 2831 (2800) cm²/g, D = 26 microns; (c) S_{sp} = 4258.8 (4300) cm²/g, D = 14.5 microns; (d) S_{sp} = 6592.8 (6600) cm²/g, D = 8.2 microns.

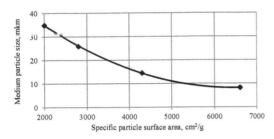

Figure 2. Interconnection average diameter and specific surface area of particles of blast furnace slag.

The results showed the effect of specific surface area increasing of the slag and limestone composition of filling mixture to improve the quantitative and qualitative characteristics of artificial massif. Zaporizhzhya iron ore plant in hard mine-geological conditions develops rich in hematite-martite ores of South Belozersky deposit (Fe content > 60%) Num-

ber of floors-camera system development with sub-level ore breaking deep wells and then filling goaf hardening mixture. Difficult conditions are defined at the top of occurrence of ore-crystalline massif aquifers unstable rocks hanging wall of the fortress with a coefficient scale prof. M.M. Protodyakonov f = 4–8, depth development (330–940 m), fracture massif. These factors lead to increased demands on the strength of filling massif. Actual dumped filling massif to chamber was observed in the middle and sides of the camera connection areas with its roof. Composition of filling mixture as follows: granulated blast furnace slag – 18.1%, flux production waste – 47.5%, the breed – 16.3%, water – 18.1%.

For full disclosure of binding properties of granulated blast furnace slag and flux of the waste production as well as improve the structural strength properties of filling massif proposes the use of two-stage crushing circuit (Figure 3).

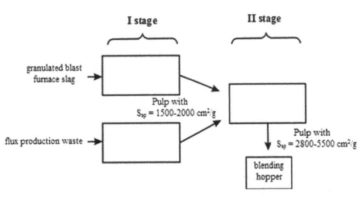

Figure 3. Two phase slag and limestone grinding for preparing curable backfills.

Chamber filling technology involves usage of filling mixture with varying amounts of binders to ensure durability artificial massif in a weakened field (bottom, location sublevel workings). Actually, it provides differentiation distribution strength in technological features. Instead expensive binder (Portland) technology is proposed differential chambers filling mixture for the production of fine filling operations, but with varying degrees of grinding. Depending on the size of the camera, the amount of sublevel openings in the chamber, as well as the number of daily flow chambers filled grinding slag and limestone in an amount will vary. For conditions ZZHRK need to fine mixture will be approximately 1150 tons/day. Ball mill MSHTS-36-55 number one source produces pre-grinding (up to 10 mm) fractions of granulated blast furnace slag, a ball mill number 2 – grinds fluxing limestone (50%

of the flow of slag). In its usual form used size up to 5 mm as an inert filler to activate binding properties. Components tabs 55–60% are brought to a particle size class of –0.074 mm, which corresponds to a specific surface of 2000 cm^2/g further crushed slag and limestone slurry received in the second stage mill. The second stage provides regrinding slag and limestone to the desired fineness. In our case, the specific surface area is required variation of slag and limestone within three modes 2800, 4300, 5500 cm^2/g for the formation of the necessary structural formations in filling massif.

For reliable quality control of pulp consisting of ultrafine fractions are encouraged to use the grain analyzer Multisizer-3. It allows for a short period of time (up to 5 min) obtaining the distribution diagram class fractions of crushed material with high reliability. With the construction of filling massif in the first stage of the chambers mining ore reserves recommended 65% of the total mixture to fill a specific surface area of binder 2800 cm²/g (the space between the sublevel) 30% (sublevels, reinforcing layers) – 4300 cm²/g , 5% (bottom) – 5500 cm²/g. Application of two-step milling scheme reduces fuel cost blast-furnace slag. So for the chamber with a volume of 110 thousand m³ level 840–940 m ZZHRK amount crushed binder decreases from 61.600 to 39.600 tonnes, or 1.55 times. Differentiated filling chamber with backfill by different specific surface makes ultrafine grinding process effective, since the volume of the chamber gob finely slag amount (4300–5500 cm²/g) is negligible.

Completed studies of dispersion effect on the properties of binder filling mixture showed that the promising direction of development is usage of filling works of fine grinding. This is achieved by using a new technological level mills type Vertimill, SMD (USA) and IsaMill (Australia) in the circuit grinding apparatus filling mixture in the mining industry over the past decade. They refer to the ball and pebble mills, but have fundamentally different physical and kinematic basis achieve the desired class crushing.

In the process of preparation of the complex chain of stowing hardening backfill based on fine particle binding recommended IsaMill M3000, with sufficient capacity (up to 100 t/h), the possibility of achieving the final product size to 5 microns (6000 cm²/g).

In order to provide an adequate economic assessment should be guided grinding energy performance depending on the size of the output product (Stirred milling technology....). In this paper, noted that with increasing fineness varies the percentage of passing particles and the specific energy consumption. The feed material to the mill in the experiment has IsaMill fineness grade of 60% -0.08 mm particles, which in general corresponds fineness ball mill under the conditions of the packing ZZHRK complex. Based on the indicative timetable obtain specific energy costs three modes in the second stage grinding unit cost of grinding to S_{sp} = 2800 cm²/g – 19 kW/t , S_{sp} = 4300 cm²/g – 33 kWh/t , S_{sp} = 5500 cm²/g – 45 kWh/t.

Costs for slag grinding in ball mills stowing complex "ZZHRK" is 15 kW/t. As a result of lower fuel chamber slag ball mill load the first stage is reduced, which would entail a reduction of energy consumption. These control fineness stowing complex show a wide variation in producing particles of size -0.074 mm from 46 to 57%. To receive uniform in size with ball milled slag grinding is problematic. This factor has a direct influence on the formation of filling massif with uniform strength, which is possible under the same terms of dispersion of slag

and constant chemical composition.

Despite the fact that for two-stage binding circuit increases the specific power consumption by reducing the flow of blast furnace slag at the camera at the final cost of one-and two-stage grinding scheme will differs slightly.

To calculate the economic impact of proposed recommendations we compared two versions of cooking hardening backfill with a single-stage (base case) and two-step (recommended) grinding binder. The main criterion of economic efficiency cost of components adopted hardening backfill as the cover story in the production costs and the cost of filling works on energy grinding binder.

In the basic version the spent an average of 110 camera thousand m³ is filled to a height of 130 m with a flow of blast furnace slag 560 kg/m³ (including moisture) when the specific surface area of up to 2000 cm²/g, fluxing limestone – 950 kg/m³, rock – 550 kg/m³.

In preferred variant, the total volume of chamber is filled with a hardening backfill specific surface area of the blast furnace slag and limestone (50% of the flow of slag) 2800 cm²/g – 70 thousand m3, 4300 cm²/g – 34 thousand m3, 5500 cm²/g – 6 thousand m³ of filling mixture composition: blast furnace slag – 240 kg/m³, ground limestone – 110 kg/m³, limestone – 1050 kg/m³, the breed – 650 kg/m³.

Backfill materials and energy costs at base and preferred variant will be 92.3 UAH/m³ and 88.9 UAH/m³. During usage of the technological recommendations, costs of filling works of cleaning chamber on the first stage on materials and energy could be reduced to 3.9%. With increasing strength of filling massif up to 10 MPa (traditional 7.5 MPa) at the age of 180 days increases its resistance to outcropping and ore dilution rates are dropping from 4.5 to 1.8% (Figure 4).

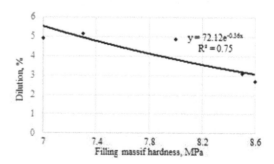

Figure 4. Dependence of dilution during ore chambers mining of second stage from backfill hardness.

Loss from falling 1% backfill material is accompanied by decrease in iron content of mined ore is equal to 28 UAH/t. Additional economic benefit from reduced dilution is 2.7·7.2 = 19.44 UAH/t.

3 CONCLUSIONS

1. Technology for differential filling chambers of the first stage of filling mixture with a specific surface binder 2800 cm^2/g 65% of its volume (the space between the sublevel) 30% (sublevels, reinforcing layers) – 4300 cm^2/g , 5% (bottom) – 5500 cm^2/g. This will relieve the outcrop filling massif height and form a filling massif – needled fibrous structure in the zones of tensile stresses on the contour extraction chambers, as well as layered lamellar structure under compressive stresses.

2. Two-stage circuit grinding granulated blast slag and limestone waste, allowing reducing their consumption by 25–75%, improving the structural and mechanical properties of filling massif and increasing the proportion of mixed breed. The first stage is represented by ball mills, the second stage of the ultrafine grinding mill series IsaMill (Australia).

3. The economic estimation of the proposed recommendations filling high chambers and hardening filling massif, which will be about 3.4 UAH/m^3 of mixture and 19.44 UAH/t of ore mined from the reduction of its dilution is given.

REFERENCES

Voloschenko, V.P. 1985. *Geomechanical basis for development of iron ore deposits powerful systems with a backfill*: Abstract of dis. Doctor. Tech. Scien., Dnipropetrovs'k: 314.
Stirred milling technology: http://myxps.ca/EN/ Presentations/Documents/xps_speech_0905112_mcgill_Seminar. pdf

Progressive Technologies of Coal, Coalbed Methane, and Ores Mining – Bondarenko, Kovalevs'ka & Ganushevych (eds)
© 2014 Taylor & Francis Group, London, ISBN: 978-1-138-02699-5

Mechanism of additional noxious fumes formation when conducting blasting operations in rock mass

V. Soboliev & N. Bilan
National Mining University, Dnipropetrovs'k, Ukraine

O. Kirichenko
State Enterprise Research-Industrial Complex "Pavlograd Chemical Plant", Pavlograd, Ukraine

ABSTRACT: The physical and mathematical model of the destruction mechanism of stable molecules as components of explosion products is suggested. The kind of particles and their possible role as catalysts for chemical reactions on the fracture face in rock under the explosion of a blasting charge, and the whole problem of the stability of the chemical bond of molecule are discussed. In particular, it is assumed that due to the interaction of explosion products with solid surfaces of the forming and developing fractures most probable particles are ions, where in their fields the chemical bonds of either molecules are broken and new movable chemical compounds are formed. Using quantum-chemical regularities, the authors showed the possibility of such dynamic chemical act occurrence.

1 INTRODUCTION

Detonation of explosives is accompanied by gas formation, in which composition the nitrogen oxides and carbon dioxide are toxic components. Especially they are highly dangerous in the mine atmosphere of underground openings. At the moment, having knowledge of the elementary composition of either explosive, you can calculate the concentration of the components in the resulting explosion products.

The results of theoretical thermochemical calculation to cartridge emulsion explosives "ERA-P3" with water-ratio 6.92–6.95% are given as an example (Table 1).

Table 1. Composition of the explosion products (in % wt).

Defined indices	Parameters	
	Composition #1	Composition #2
Carbon monoxide (CO)	2.624	1.139
Carbon dioxide (CO_2)	10.973	12.575
Water (H_2O)	48.252	48.057
Nitrogen (N_2)	25.153	28.213
Aluminium oxides (Al_2O_3)	3.703	3.703
Sodium carbonate (Na_2CO_3)	6.285	6.304
Explosive characteristics		
Explosion heat, kcal/kg	875.17	887
Explosive temperature, K	2946	2975.4
Gas volume, l/kg	901.9	896.3
Oxygen balance, %	-1.26	-0.42

Charge density, the value of oxygen balance, the availability of sensitizing or phlegmatizing agents and flame-arresters, the dispersability of explosive particles, the incompleteness of chemical transformation, the quenching rate of explosion products (the rate of heat exchange with the environment), the conditions of dispersion of explosion products and peculiarities of their interrelation, chemical composition, the concentration of additives, the properties of rocks and other factors have a great influence on the composition of explosion products.

It is known that the components of explosive transformation are the most thermodynamically stable compounds. The composition of explosion products is always different from the theoretic one in the process of borehole or blast-hole charges blasting in rock. Experiments proved that while blasting of various explosives in one rock the total amount of noxious fumes as compared to those in the case of blasting in the air increases to 220%; while blasting of one type explosive in various rocks – to 1000%. Carrying out of blasting in the sulfur-containing ores leads to the formation of sulfurous gases and hydrogen sulphide; carbon dioxide passes into the oxide while blast-hole charges blasting in coal; in fragmented rock total number of noxious fumes is scarcely different from the analogous index in the case of blasting in the air, but at the same explosive charges blasting in the crushed coal the quantity of noxious fumes is doubled.

Ratio CO/CO_2 increases with increasing rock-hardness ratio. There are different points of view on the physical-chemical way for increasing the yield of noxious fumes upon the rock hardness. Taken as a whole, according to Melvin A. Cook, CO/CO_2 ratio increase is due to the shift from state of shot firing time in the direction of a more rapid decrease in the fugacity coefficient than K-value.

NO_2 concentration in the blasted rock mass increases with depth, at that the CO2 concentration rapidly reduces with increasing depth. The true values of noxious fumes concentrations depends mainly on the accuracy, conditions and location of measurement (mine atmosphere, blasted rock mass, rock and coal dust, various surfaces of openings, etc.).

Experiments carrying out in National Mining Research Center "A.A. Skochinsky Institute of Mining" and in the State Safety in Mines Research Institute (MakNII, Makeyevka) suggest that the influence of rock surrounding the explosive charge is much stronger than the effect of explosives composition on the total amount of noxious fumes in explosion products.

Thus, the analysis of experimental results comes to clear conclusion about that explosion products depend not only on chemical composition of explosives, blasting conditions, but also on the rocks whose physicochemical properties have a catalytic effect on chemical processes of formation of new gas molecules in detonation products.

Despite extensive experimental data there are few scientific publications, wherein the mechanism and physicochemical features of interaction, e.g., of new-formed surfaces in breakable rock mass with explosion products were discussed.

Ideally, the detonation products of the explosive charge are the most thermodynamically stable compounds, composition and quantity of which can be estimated with reasonable accuracy. However, to make any evaluation of explosion products after interaction with rocks is impossible. The reason is that mechanism of additional noxious fumes formation is still unknown. But as practice testifies to the appearance of both an additional amount of toxic and new-formed gas in explosion products, it is quite obvious the role of secondary chemical reactions, whose reasons and mechanism of activation still do not have a reasonable model.

Thus, the physicochemical process of the first stage of explosion gas formation is quite predictable, follows the basic laws of chemical thermodynamics, and is confirmed by experimental data and reliably modelled. The second stage of chemical transformations is mainly debatable, but is well-founded. The catalytic activity of rocks depends not only on thermodynamic parameters, but also on the functional cover of solid rock surface, its structural features, metallogenic trend, mineral composition, energy potential, the density distribution of active surface states and other parameters.

2 FORMULATING THE PROBLEM

In this paper we propose the physical-mathematical model of the mechanism of destruction of stable molecules as components of explosion products and the possible role of the catalytic action of the new-formed surface on the stability of molecule's chemical bond. It is assumed that as a result of interaction of detonation products with new-formed surfaces ions may be most likely particles, which have the potential to actively participate in the formation of mobile chemical compounds.

Interfacial surfaces in rock and boundaries between every rock in a wide temperature range are characterized by a certain difference of electric potentials, due to the difference in electron densities in each of the phases. Destruction of rock initiates electric field between surfaces of formed fractures, electric field intensity reaches a maximum about 108 V/cm at the forefront of fracture. The occurrence of high temperatures and electric fields leads to a highly nonequilibrium state of surfaces, to high surface conductivity, to a new functional state and chemical activity of surfaces. Fractures widening is most likely to interfacial and inter-mineral boundaries. Electric charges are chemical active centres, catalysts of chemical reactions on the surfaces of fractures (Soboliev 2010).

According to various scientific publications on the example of blasting operations the speed of fractures widening in coal is about 360–900 m/s, the speed of explosion products in fractures is about 300–400 m/s at gas pressure $(25...50) \cdot 10^5$ Pa (with the same gas pressure the speed of fractures widening in the rock reaches 2.000 m/s). Penetrating components of the gas stream at such parameters actively interact with charges on solid surfaces, i.e. there is a high probability of their convergence on the distance at which will be breaking of bonds in molecules formed as a result of detonation.

We assume that the research of stability of chemical bonds between the particles in the solid of explosives and between the atoms in the molecules of detonation products considers a single charge or the local group of point charges as the main reason initiating the first elementary act of chemical reactions. If we use a Coulomb center, it is possible through the physical and mathematical modelling to assess the characteristics of behavior of the chemical bond in dynamics, i.e. to assess the regularities in change of the binding energy as a function of distance and the value of a charge.

We also assume that the development of physico-chemical processes in rocks is mainly due to the influence of kinetic parameters that determine the catalytic properties of rock, which include surface and the features of its functional state. It is also possible that the activity of local areas on the surface is due to the presence of electric charges. In this case the initiation of chemical reaction does not require any additional stimulation by external influences.

3 RESULTS AND DISCUSSION

Let us consider one of the possible case of chemical bonds opening on the example of a system consisting of small molecules (CO, N_2, CO_2, HCl, H_2, HF, etc.), which are located in the space bounded by solid surfaces. Put the system under investigation as a statistical model and assume that due to the approach of a molecule and a free point charge or a surface having electric charges (incomplete and distorted bonds, dislocation, adsorbed ions, etc.), the unit of molecules enters the field charge effect. Electric field intensity is about $\sim 1.5 \cdot 10^7$ V/cm at a distance of 10^{-9} m from the monovalent ion at room temperature. If the energy of the molecule is high enough so that molecule could move in on some critical distance to the ion, the activation of the molecule occurs, but at some distance – bonds opening and this process can have several ways.

The minimum energy that energizes, for example, the CO opening, conforms to the reaction $CO \rightarrow C + O$. Hence we consider only this reaction. We also assume that the surface charge (ions) density is constant.

For the performance of the task quantum-mechanical calculation method is used. Its main advantage is that the chemical bond energy terms easily calculated in both ground and excited state. In this case it is possible to observe the development process (in dynamics), ranging from conditions characterized by a stable chemical bond and its gradual "opening" up to the moment of rupture.

In general, in the Schrödinger steady equation the complete nonrelativistic Hamiltonian of the molecule can be represented as the sum of three operators

$$H = T^{Nucl}(R) + T^{el}(r) + U(r, R),$$

where $T^{Nucl}(R)$ and $T^{el}(r)$ – kinetic energy of nuclei and electrons, $U(r, R)$ – potential energy of the molecule, including the attractive energy of electrons to nuclei, as well as the internuclear and interelectron repulsion energy, R and r – nuclear and electron coordinates.

Properly, the movement of electrons and nuclei in a molecule are linked. However, given the fact that the masses of nuclei in 10^3–10^5 times as the mass of the electron, and hence the speed of nuclei is much less than the speed of electrons further calculations will be carried out in the adiabatic approximation by dividing these intramolecular motions, that is, assuming their independent nature. Then the subsystem of electrons and nuclei are characterized by their own Schrödinger equation and own wave function.

It should be noted that the accurate solution of the Schrödinger equation can be obtained only for the system consisting of two particles: an electron and a proton (hydrogen atom). Many approaches are developed at present to calculate the molecular terms; the most popular of these is the technique for calculating of a linear combination of atomic orbitals (LCAO). Calculations for a number of small molecules point to a reasonable agreement with the experimental data. However, using this technique, as well as other ones, requires considerable numerical calculations, the analysis of which is hampered by the lack of analytical solutions. Therefore, we used the model to calculate the molecular terms based on the solution to the state of the electron in the field of two Coulomb centers – the so-called three-center task. Two-central wave functions in ellipsoidal coordinates are set as the basis; all functionals are calculated analytically, and constructed expressions in closed form for the Green functions allow to solve the quantum-mechanical task with perturbations. LCAO technique makes it impossible to carry out such calculations because wave function is defined by the expression $\psi I = \sum C_{ij} \cdot \varphi_j(i)$, where C_{ij} – coefficients calculated at each point in space. Difficulties also rise when calculating the energy of electron-electron interaction.

Assume a diatomic molecule as two positively charged Coulomb centers located in the shell of electrons that are given by each atom to chemical bonding. The wave function of the valence electron describes its condition in the field produced by bound electrons and atomic nuclei charges forming the molecular core. Let us introduce an effective potential, which means the nuclear potential, the screened electron core of bound electrons. When calculating the electron-electron interaction we use the screening constant that takes into account the fact that the charges of bound electrons are not point ones and "dispersed" in the space near the nucleus. In the case of large cores of bound electrons they will be polarized by interaction with each other at short internuclear distances. This polarization must also be taken into account in the calculations.

To assess the supposed case use the task in which some perturbation $W(\tau)$ influences to certain chemical bond. We will consider the system, consisting of three particles with the charges Z_1, Z_2, Z_3 and the masses M_1, M_2, M_3, interactive by Coulomb's law. Here charges Z_1, $Z_2 > 0$, and $Z_3 < 0$. Hamiltonian for an electron in the field of two Coulomb's centers, at introduction of coordinates Jacobi, written down as follows

$$H = -\frac{\hbar^2}{2M_0}\Delta_R - \frac{\hbar^2}{2M}\Delta_R - \frac{\hbar^2}{2m}\Delta_r +$$

$$+\frac{Z_1 \cdot Z_2}{R} - \frac{Z_1 \cdot Z_3}{r_a} - \frac{Z_3 \cdot Z_2}{r_b},$$

where M_0 – mass of three interacting particles; $M = M_1 M_2/(M_1 + M_2)$; M_1 and M_2 – the masses of molecule kernels; Z_1 and Z_2 – charges, $Z_2 \geq Z_1$; Z_3 – electron charge; $M_0 = M_1 + M_2 + M_3$;

$$\frac{1}{M} = \frac{1}{M_1} + \frac{1}{M_2} \; ; \; \frac{1}{m} = \frac{1}{M_3} + \frac{1}{M_1 + M_2} \; ;$$

$$\left. \begin{array}{l} r_a = |R_3 - R_1| \\ r_b = |R_3 - R_2| \end{array} \right\} \text{ – distances from the ions a and b to}$$

the examined point of space of valence electron; R – interkernel distance.

The first member of hamiltonian describes motion of centre-of-mass of the three particles system, the second one – relative motion of particles M_1 and M_2.

We will consider that $M_3 \ll (M_1; M_2)$, i.e. a particle with M_3 mass and Z_3 charge can be an electron. Thereafter the task is examined with participation an electron.

The Shroedinger equation in the task about motion of electron in the field of two immobile charges (kernels), remote on distance R, in the atomic system of units $\hbar = Z_3 = M_3 = 1$ in the ellipsoid system of coordinates looks like

$$\left\{ \frac{4}{R^2 \cdot (\lambda^2 - \mu^2)} \cdot \left[\frac{\partial}{\partial\lambda}(\lambda^2 - 1)\frac{\partial}{\partial\lambda} + \frac{\partial}{\partial\mu}(1 - \mu^2)\frac{\partial}{\partial\mu} + \frac{4}{R^2 \cdot (\lambda^2 - 1) \cdot (1 - \mu^2)} \cdot \frac{\partial^2}{\partial\varphi^2} \right] \right\}\psi + 2 \cdot [E + U(\lambda,\mu)]\psi = 0 \text{ , (1)}$$

where $\lambda = \dfrac{r_a + r_b}{R}$, $1 \leq \lambda \leq \infty$; $\mu = \dfrac{r_a - r_b}{R}$, $-1 \leq \lambda \leq 1$; r_a and r_b – distances from the kernels Z_1 and Z_2 to the electron;

$$U(\lambda,\mu) = \frac{2}{R} \cdot \left(\frac{Z_1}{\lambda + \mu} + \frac{Z_2}{\lambda - \mu} \right) \text{ – potential energy of}$$

electron in the field of charges Z_a and Z_b.

The Equation (1) can be represented as truncated (model) equations and some component. The solution of the truncated equation is the first approximation to the solution (1). Solution has the form:

$$\psi = F = [t(t+2)]^{-\frac{1}{2}} \cdot y^{\frac{\Lambda+1}{2}} \cdot e^{\frac{y}{2}} \times$$

$$\times \Phi\left(\frac{\Lambda}{2}k + \frac{1}{2}, \Lambda = 1, y \right) P_n^\Lambda(\mu) e^{i\Lambda\varphi} \text{ , (2)}$$

where Φ – confluent hypergeometric function; P_n^Λ – associated Legendre functions; $|\Lambda| = 0$, 1, 2,..., $n = 0, 1, 2, \ldots$ – quantum numbers; $t = \lambda - 1$; $y = 2\sqrt{-\varepsilon} \cdot t$; $\varepsilon = ER^2/2$.

On the basis of solutions of model task get energies of the states, corresponding to the quantum numbers of $k = 5/2$, $3/2$ and $1/2$, $\Lambda = 0$, $n = 0$. The expression for the electron-electron interaction for the case when two electrons are in the same quantum state.

$$E(ee) = \frac{4}{R} \cdot \left[\left(\frac{3}{40b^2} + \frac{1}{20b} \right)(c + ln\,2b) + e^{8b}E_i^2(-8b) \cdot \left(\frac{3}{40b^2} - \frac{11}{20b} + \frac{7}{5} - \frac{8b}{15} \right) + \right.$$

$$\left. + e^{8b} \cdot E_i^2(-4b) \cdot \frac{4b^2}{15} + e^{4b} \cdot E_i(-4b) \cdot \left(-\frac{3}{20b^2} + \frac{1}{2b} - \frac{1}{5} \right) + \frac{1}{8b} - \frac{1}{10} \right] \cdot \left[\frac{1}{2b} - \frac{4}{3} \cdot b \cdot e^{4b} \cdot E_i(-4b) \right]^{-2} \text{ (3)}$$

where c – Euler constant; $b_i = -2k \pm \sqrt{4k_i + 2L}$ ("+" sign is chosen for bound state $b_i > 0$ in the for-

mula); $L = \dfrac{\Lambda^2 - 1}{4} + n(n+1) + \dfrac{Z^+}{2}$.

Figure 1 shows the results of calculation according to (3).

474

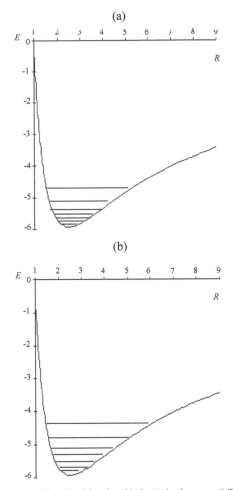

(a)

(b)

Figure 1. Vibrational levels of *NO*: (a) in the state (1/2, 0, 0) and (b) (3/2, 0, 0).

When calculating the nucleus-nucleus interaction the methodology used in our work, obtained in (Soboliev 2010), which takes into account the screening of nuclear potentials by electrons, which is not involved in the formation of chemical bond. This was done due to the fact that they were obtained solutions of Poisson's equation for the potentials produced by bound electrons.

At the first approximation, the corrections to the electronic term of molecules due to the action of Coulomb center, were determined from the expression

$$\Delta E(R) = \sum_{j=1}^{n} \sum_{i=1}^{N} \frac{\left\langle \psi_j \left| \frac{Z_i}{r_{ij}} \right| \psi_j \right\rangle}{\left\langle \psi_j | \psi_j \right\rangle}, \tag{4}$$

where Z_i / r_{ij} – Coulomb potential produced by the i-th charge Z_i, acting on the j-th electron of molecule; n – the number of valence electrons on the bond.

Coulomb potential in ellipsoidal coordinates can be represented in the form of Neumann expansion

$$W(\lambda, \mu, \varphi) = \frac{2 Z_3}{R} \sum_{k=0}^{\infty} \sum_{m=-k}^{k} (-1)^m \times$$

$$\times (2k+1) \cdot \left[\frac{(k-|m|)!}{(k-|m|)!} \right]^2 \cdot P_k^{|m|}(\lambda_<) \cdot Q_k^{|m|}(\lambda_>) \times$$

$$\times P_k^{|m|}(\mu_i) \cdot P_k^{|m|}(\mu) \cdot e^{im(\varphi-\varphi_i)}, \tag{5}$$

where $\lambda_<$, $\lambda_>$ – smallest and the highest coordinates; λ and λ_i; $P_k^{|m|}$ and $Q_k^{|m|}$ – Legendre function of the first and the second kind; $\mu_i = \dfrac{R_{1i} - R_{2i}}{R}$,

$$\lambda_i = \frac{R_{1i} + R_{2i}}{R}.$$

If the wave functions do not depend on coordinates, for example, for the first three main states ($k = 5/2$, 3/2 and 1/2, $\Lambda = 0$, $n = 0$.), the matrix element $< \psi_i | W | \psi_j >$ can be simplified

$$\left\langle \psi_i | W | \psi_i \right\rangle = \frac{\pi Z_i R^2}{2} \left[Q_0(\lambda_i) I_1 + I_2 - \frac{10}{3} \cdot P_2(\mu_i)(Q_2(\lambda_i) I_3 + P_2(\lambda_i) I_4) \right] \tag{6}$$

where

$$I_1 = \int_{1}^{\lambda_i} |\psi_i|^2 \left(\lambda^2 - \frac{1}{3} \right) d\lambda; \quad I_2 = \int_{\lambda_i}^{\infty} |\psi_i|^2 \left(\lambda^2 - \frac{1}{3} \right) \cdot \frac{1}{2} \ln \frac{\lambda+1}{\lambda-1} d\lambda;$$

$$I_3 = \int_{1}^{\lambda_i} |\psi_i|^2 \frac{(3\lambda^2 - 1)}{2} d\lambda; \quad I_4 = \int_{\lambda_i}^{\infty} |\psi_i|^2 \left[\frac{(3\lambda^2 - 1)}{2} \ln \frac{\lambda+1}{\lambda-1} - \frac{3}{2} x \right] d\lambda. \left. \right\} \tag{7}$$

Solving Equation (7), we obtain expressions for the corrections to the energy of the electron term of molecule that use the action of "third" centers on the molecule.

Figure 2 shows the results of calculations of the electronic terms of molecule N_2, which is in the Coulomb field, placed at different distances H along the normal, which runs through the center axis of the molecule. Electric field intensity of the monovalent ion is $\sim 1.5 \cdot 10^7$ V/cm at a distance of 10^{-9} m! As the distance from the ion decreases the intensity increases. The bond of any molecule is breaking in the field of Coulomb centre even at 0 K (in the case of convergence for a critical distance).

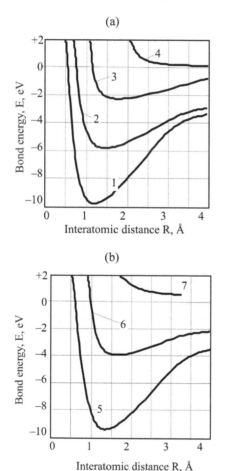

Figure 2. Change of potential energy $E(R)$ of molecule N_2 in the field of negative charge Z: (a) from Z (distance H from a charge to chemical bond $2.5 \cdot 10^{-10}$ m); (b) from distance H at $Z = (-4)$; 1 – a nonexcited molecule; 2, 3 and 4 –molecule in the field of charge Z (-2), (-4) and (-8) accordingly; 5, 6 and 7 – molecule in the distance H (m) 10^{-9}, $5 \cdot 10^{-10}$ and $1.5 \cdot 10^{-10}$ from a charge.

Action of the charge field on the molecule leads to the "loosening" of the chemical bond. Influence of the field of the external charge (Coulomb center) decreases with increasing distance from the molecule. Regularities of change in the bond energy of molecules O_2, CO_2, and CO depending on the sign of value of a point charge and the distance from the charge have the qualitative analogy with the behavior of the molecule N_2 (Figure 2). The nature of change in the state of chemical bond is weakly dependent on the sign of a point charge (Figure 3).

Verification of the mathematical model was carried out by calculating the ground and excited electronic terms of various diatomic molecules. The obtained values of the energy minima of the terms and interatomic distances are in good agreement with the experimental results. This is used as a basis for solving the main task – the fundamental researches of rupture of molecule bond under the action of external charge – the third center.

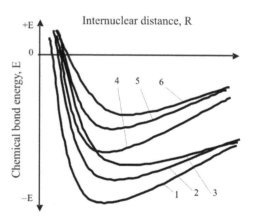

Figure 3. Indicative dynamics of the change of potential energy $E(R)$ of chemical bond of any molecule as it gets close to the point charge: 1, 4 – excited molecule; 2, 3 – negative charge (-1) and (-2); 5, 6 – positive charge (+1) and (+2).

Probability of bond rupture all other things being equal increases either with decreasing distance from the molecule to the ion, or increasing in the valence of the ion. Stability of the molecule is determined by the function $E(Z^{\pm}; R)$ in case if it affects the charge field. Strong activation factor that increases the probability of bond rupture is temperature. Rupture of CO bond in the field of divalent ion occurs at room temperature. Thus, the probability of rupture of molecule bond with increasing temperature, all other things being equal, increases significantly. A similar influence has the increase of pressure.

As CO molecule (or NO_x) gets close to a point charge it is appeared a gradual "loosening" of the chemical bond, and bond breaks at some distance from the charge. Since the actual temperature of the high-energy system (explosive products) generally range from about 1000 to 3000 K, it is not difficult to understand that functions $E(Z^{\pm}; R)$ will tend to have a very strong flattening. And if the presence of an ion with a charge (–3) can significantly reduce the binding energy of the CO molecule, converting it into the unstable state, then at the initial temperature, corresponding to the value of the specified range, it should be expected the rupture of the chemical bond of the molecule. The increase in temperature will lead to the fact that the CO molecule bond rupture at a distance from a point charge, which is several times higher than the typical interatomic distances in solids.

Bond rupture energy (D) of the CO molecule is characterized by the highest value ~1074 kJ/mole; this value is ~810 and ~540 kJ/mole for CO^+ and CO_2 respectively. For nitrogen oxides such as NO, NO^+ and NO^- bond rupture energy will be ~628, ~1045 and ~500 kJ/mole respectively. These values of bond rupture energy for toxic molecules formed in explosion product suggest that the probability of molecules disrupting increases at a decrease in D, and the transition from stable state to disrupting is not changed naturally, completely corresponding to the CO bond behavior in the field of a point charge.

4 CONCLUSIONS

Established regularities among other things indicate on the activity of a point electric charge, and this activity is exerted in ability of its field to cause "loosening" and disrupting of chemical bonds of molecules. Bond is opening due to the approach of molecules to the charge on certain critical distance, i.e. molecules are destroyed on excited atoms. For all other things being equal (distance to a Coulomb center, the pressure) dissociation probability increases with increasing temperature.

In the technological cycle of manufacture of emulsion explosives related to the chemical processes of dissolution, crystallization, creating composites from mutually insoluble phases of different aggregate state, anyway, chemical processes are realized due to the interaction of chemical bonds of particles of different components with charges. The destruction of a large number of neighbor bonds leads to the formation of local area, provoking a degradation of the microstructure of emulsion explosives. Supersaturation of microstructure by defects comes at some point in this process and thus overcome the threshold beyond which a chemical process developed tragically fast.

The pattern of radial fractures is formed when explosive destruction of rocks, where pressurized explosion products are penetrated. The active centers are located on the surfaces newly formed fractures and primarily react with the gases of explosion. The processes of formation of new chemical compounds occur in a time of 10^{-14} s.

REFERENCES

Soboliev, V. 2010. *The evolution of molecules chemical bonds stability in the field of surface charges*. Scientific Reports on Resourse Issues. Volume 1. Freiberg (Germany): TU Bergakademie Freiberg.

Soboliev, V. 2010. *Regularities of a change in the chemical bond energy in the field of a point charge*. Reports of the National Academy of Sciences of Ukraine.

Theoretical aspects of the potential technological schemes evaluation and their susceptibility to innovations

S. Salli, V. Pochepov & O. Mamaykin
National Mining University, Dnipropetrovs'k, Ukraine

ABSTRACT: The direction of the parameter estimates of technological schemes of coal mines to innovation susceptibility are described.

1 INTRODUCTION

Over the past century in Ukraine mined 9.4 billion tons of coal that is almost one third of the available reserves. It is quite natural that were worked out formations with respect to the most favorable conditions. Ukraine today inherits not only the explored reserves of the coal, but mine fund with enterprises of pre-revolutionary, pre-war and post-war construction. Large mines are the new buildings are calculated by ones, as evidenced by the level of average production capacity of the Donbass mines – 500 thousand t/year. Donbass is faced with the alternative closure of most unprofitable mines, which has depleted reserves. At the same time, fully manifested harmful consequences age underground coal mining with storage rock on the surface, unsystemic dumping of highly mineralized mine water and other aspects of activity without regard for the environment. As a natural final of the mass closure of mines, increasingly began to sound the category of "problem regions" and "unstable areas". In relation to the region with a fairly worn mine fund and the presence of significant amounts of industrial wastes considers two alternatives. One of them provides for the deployment of large-scale work on the processing of waste mines and processing plants, and the second is the continuation of unprofitable mines work (Amosha et al. 2005, Bondarenko et al. 2009). In each of these alternatives is necessary to invest approximately equal to the capital. Then the economic effect per unit of the final product will be different depending on susceptibility natural quality of the remaining reserves and technogenic sites. This is fundamental question.

2 STUDY OF LATELY RESEARCH AND PUBLICATIONS

In relation to coal mines almost no system of quantitative assessment of technological schemes and existing fragmented characteristics schemes inherent fundamental errors, namely advantage of the extensive restored to the detriment of innovation preferences. Internal potential of the technological networks is one of the most important parameters of the assessment of coal mines. Its formation is the result of a complex factors influence determining the efficiency of underground mining, and above all, the property of a coal mine - development in area. This property has an objective character, the pace of development is determined by human activities and depends on many factors, in particular, the level of scientific and technical progress, but the necessity of development set by nature and cannot be removed or replaced, even if it is changed, the technology of the production process.

The desire of the technological scheme of any mines in the conditions of reserves completion in mine fields in the state of bifurcation stability (Naman & Morgenstern 1970) due to attempts to mine of a mine field left earlier areas or move to mine of the non-commercial reserves. I.e., research of technological network should occur within the framework of the structural-dynamic theory from the viewpoint of sustainable development and change. So, on the one hand, for the specific problem decision of planning development of mining operations may correspond to several elements of the system, on the other hand – one element of the technological scheme can provide the solve of several tasks. On this basis, the efficiency of implement tasks is determined not only on the efficient functioning of the mine subsystems, but also by relations in the system between tasks to be solved.

Thus, formation of potential in technological networks mines is a symbiosis of the interaction factors of concentration mining operations, stability ventilation and influence on the formation of the company productive flows. Each of the factors is characterized by one of the indicators of "correlation of bandwidth technological links", "limited power of the factor ventilation" and "density productive flows", respectively. Optimization of target indicators of "techno-economic stability of the technological network of mine", that characterizes the potential of the mine to innovation is the result of interaction these factors and factors of the second order, which determine the formation of several important indicators. It can be argued that the stability of a technological network is ability to preserve its integrity and mission supplier of coal products to function in the desired regime in conditions of uncertainty of internal and external factors, adjusting its throughput in simple regime or extended reproduction. Therefore, it can be argued that the technical potential of technological network mine is a result of multiple concurrent and interrelated factors of the first and subsequent orders.

2.1. Formulation of the paper objective

To establish their influence on the formation of this parameter, it is advisable to use the method of statistical analysis and to establish the relationship between the indicator technical potential and a number of independent parameters of production and economic activity of the mine. Obtain analytical dependences describing the measure "technical potential", based on the investigation of dependencies between this indicator as effective geological and technological parameters, and also factor. The investigation of dependencies and obtain regression equations of the indicator "technical potential", as integral evaluation of innovative technological network of mines based on the use of optimal programming and decision making under uncertainty.

2.2 Statement of basic material

As the source of factors to study the formation of the parameter "technical potential" were adopted following indicators from different groups that characterize production activity, geological and technological conditions of Donbass mines: the level of concentration at mining operations L, productivity P, longwall advance V, the per-ton cost of coal S.

First of all, you need to define the concept of "development of mine", i.e., remove from the central shafts along strike and to the across strike seams and removal (reduction) vertically from the first horizon of origin exploitation of the mine. Mine is developed in three dimensions, but not equally. If you take a fairly long period of time (one or a few decades), there will be mine development in all three dimensions, over a short time period (one to two years), as a rule, mine is developing along strike, and two other dimensions within such period of time the situation may remain stable.

To determine the internal potential of technological networks were considered the main factors that can determine the formation of technical potential in topological network mine, as an integrated assessment of the potential mine in innovation part. As for productivity worker extraction, monthly longwall advance, and the per-ton cost of coal, that their values were adopted on actual data on the activities of the anthracite mines, members of the "Sverdlovanthracite", "Rovenkianthracite" and "Donbasanthracite" for 2010–2012.

To build the multiple regression equation, describing the indicator "technical potential", used step-by-step method for incorporating variables. Thus, determination of the maximum achievable value of the economic value created by the mine is one of the varieties of multicriteria problem with the four criteria that must be combined to a one-criterion problem with an objective function (1)

$$k_k = -\alpha L + \beta P - \mu S \to max, \qquad (1)$$

where k_k – summary indicator of the internal potential of the network workings; L – parameter characterizing the length of workings and the length of longwalls; P – worker productivity extraction, ton/month; V – annual longwall advance, m; S – the per-ton cost of coal.

As formation the technical potential of the technological network in mine described by Equation (1), it is obvious that reaching its maximum value depends primarily on the ratio of the values of L, V, P, and S. You should also consider the fact that the maximization of the indicator "technical potential" is achieved in conditions of limited capacity factor ventilation and density productive flows. So, the task of maximization parameter k_k, which is the main measure of the potential of the technological scheme of the mine, is to find a compromise between the values of the four main factors (1).

Economically marginal status is a function of time, then the task should be formulated in a dynamic setting. With this graphical interpretation of the economic-mathematical model is written thus: to determine the optimal objective function values according to the following criteria (Salli et al. 2009)

$$
\left.\begin{array}{l}
L(X,Y) \to min \\
Pi(X,Y,Z) \to max \\
Vi(X,Y,Z) \to max \\
Si(X,Y,Z) \to min
\end{array}\right] \qquad (2)
$$

The most appropriate way of solving this problem is multicriterion method of Pareto used in problems of this type, which are distinguished by originality: for them it is impossible to build the concept of extremum, but you can enter no improve situation ("Pareto optimum") (Boychenko 2002).

That prioritization impact on the strategic parameters of production and economic activities of the mine should compare their actual values to the optimum values defined solution of the equations system that are included in the model (1)–(2). Optimum parameters values of the technological scheme implies the full realization of the mine economic potential, that is very achievable (reference) levels since the achievement of this level makes the technological scheme of mine susceptibilities to innovation.

Now for the coal mines evaluation are using the assessment system, including up to twenty-five in-

dicators. Such variety of indicators doesn't provide opportunities strict quantitative assessment of those or other alternatives, because it violates one of the basic requirements of mathematical modeling is the unity of criteria. Most generally, any production is considered so (Pivnyak et al. 2004). A number of persons M through N machines during the time T do the work for the extraction of coal in the volume A with a certain quality of α. For the production are necessary resources in the capacity K. For the maintenance of people and machines, as well as on mining spent some funds S. Both for people and machines necessary favorable conditions of work (quantitatively it can be expressed by the ratio operability people ΣR and machines Σq, respectively).

Any of these values are dependent on the conditions and the purpose of solving the problem can serve as a characteristic of the effectiveness of this technological scheme of mine, i.e., may be taken as a criterion of optimality. The latter characteristics or are permanent, or allowed them fluctuations within certain limits (Table 1).

Table 1. Effectiveness characteristics of coal mining.

Indicators	Form of criterion		Form of restrictions
The number of workers, people	$M \to min$	(3)	$0 \leq M \leq M_{max}$
The number of mining equipment	$N \to min$	(4)	$1 \leq N \leq N_{max}$
The volume of material resources, UAH.	$A \to min$	(5)	$A_{min} \leq A \leq A_{max}$
Operation time, hours	$T \to min$	(6)	$0 \leq T \leq T_{max}$
	$T/T_{pl} \to 1$	(7)	
The volume of the extracted coal, tons	$Q_t \to min$	(8)	$Q_{min} \leq Q_T \leq Q_{max}$
	$Q_t/Q_{pl} \to 1$	(9)	
The average ash content of coal, %	$\alpha \to min$	(10)	$\alpha_{min} \leq M \leq \alpha_{max}$
	$\alpha_s \to min$	(11)	
Expenses for 1 ton, UAH.	$S \to min$	(12)	$S_{min} \leq S \leq S_{max}$
Safety and usability of people	$\Sigma R_t / \Sigma R_0 \to 1$	(13)	$M_T/M_0 = 1$
			$\Sigma R_t/\Sigma R_0 \geq k_{min}$
Technical reliability of the mining equipment	$\Sigma q_t/\Sigma q_0 \to 1$	(14)	$\Sigma q_t/\Sigma q_0 \geq k_{min}$

However, for specific organizational objectives are sufficiently stringent conditions of the use of certain criteria:

1. Four of the value of the named (M, N, A, T) describing the basic components of coal extraction: production resources, technology and labor costs (M and T). Obviously, as a criterion of optimality they should be minimized (expression (3)–(6) in the Table 1), but with the obligatory requirements specified volume of coal: $Q_t \geq Q_{pl}$. Otherwise, the task is meaningless. Such variant of the task expedient if physically possible saving of resources spent on extraction. For coal mine this is true mainly at the stage of designing, reconstruction and long-term

planning. In addition, this approach makes sense for individual sections of the mine in the case of transfer of reserves to another mine or association of mines by mining operations.

2. Indicators ΣR and Σq could theoretically serve as the criteria and even be applied in a separate short-term situations. However, essentially the expression (13) and (14) have contradictions sense of active work, so that suppresses most cases, these indicators are used as an explicit or implicit restrictions when solving for organizational and other problems.

3. Indicators Q_t, α and S characterize the results of production, but they are different in their essence, so let's consider each of them separately.

Maximizing volumes of extraction Q_t greatest extent correspond with the aim of extraction under condition of observance of all necessary limitations with resources, working conditions and quality of raw coal. This criterion can't be used only when power limit of related links. Then it is better to use the expression (9), where Q_{pl} – volume of extraction.

In some cases, system capability (q) can be accepted constant for discrete systems appropriate level. Then $Q_t = qT$, and in these cases, to facilitate the formulation and solving the problem instead of variable Q_t as the criteria it can apply coefficient of use equipment $k_u = T/T_{pl}$. For mine it is expedient in the short-term, operational organization tasks, where without a large error can be taken $q = const$. The consistency of expressions (7) and (9) are obvious.

Quality indicators of rock massif are determined by the topology of the network workings and the compliance mining techniques conditions of bedding layers. For coal production are often ash, moisture, harmful components, (then the criterion $\alpha \rightarrow min$). Quality indicators are possible the other. However, in practice, in most cases through technological features and general objectives of the mining industry, the main indicator is the amount of saleable coal in concentrate after processing. Quality indicators may also a role restrictions on the tasks for which cannot be accepted $\alpha = const$. These include primarily the task of planning the mining hard of sites of a mine field (substandard in quality stocks).

Expenses S productions of a certain volume of extraction are derived indicator

$$S = \varphi(N, M, T, A, Q_m).$$

As generalizing economic indicator, this value can serve as a criterion of optimality provided a fixed production $Q = const$. A fairly accurate cost calculation can be implemented only with a relatively long period of time (ten days, a month), so the solution of operative tasks this criterion apply impractical because of its low sensitivity. Moreover, in practice only cost is difficult to estimate the actual efficiency of production, because at different organizational alternatives $Q_t \neq const$. Therefore more likely to use economic criteria derived from S.

In particular, in the tasks of the current (monthly) planning and management as a criterion are widely used unit costs.

$$c = S / Q_m. \tag{15}$$

In the tasks that cover a large number of subsystems coal mine and longer periods, some widely criterion of profit (for private mines), which in the most general form is

$$D = C \cdot Q_m - S, \tag{16}$$

where C – the selling price of production unit.

One of the most important criteria of the efficiency of coal production is labor productivity P. As considered above criteria, this indicator should be used for long periods of time when there is a real possibility of changes number of workers M. With the aim of interrelation operational and long-term tasks, and to compare the effectiveness solutions of different systems this indicator is useful and in the strategic and current organization tasks.

The most common and critical conditions of selection criteria for a specific task are character of the variables and their relationship with the criterion. In addition to known lack criterions is possible mutual compensation values are part of them, for organizational tasks, these criteria are ineffective because of the complexity of calculation and reducing sensitivity to the desired criterions.

By this time at coal mines main and practically the only optimality criterion of the technological parameters was the implementation of coal mining in current month. Indicators of quality fuel and economic indicators implemented mostly, the role of limitations. In the operational tasks the main criterion is the fulfillment of shift jobs or better useful equipment. According to the aims of the work are necessary to develop a mechanism of ratio quality of the technological scheme of mine comparative to innovations. Therefore it is very important to the intensity of productive flows, which are formed during coal extracting. However, these flows are often variable intensity, whether the change in demand for coal, or the decline of rock volumes is transported to the surface. The efficiency analysis of indicators technological schemes of coal mining and consideration in fact system of situation indicators help to develop recommendations for the optimality criterions use in the organizational tasks of the assessment of the potential technological scheme of mine. To interaction criterions operating on different time levels, use the appropriate restrictions.

3 CONCLUSIONS

1. The main property of a coal mine, which defines all elements of its activity – development in area. This property has the objective character, as she caused a fundamental category of coal – it isn't re-

peatable. The temp of development of mining works identify by human activities and depends from a lot of factors, including the level of scientific and technical progress, but the necessity of development set by nature and cannot be removed or replaced, even if it is changed, the technology of the production process.

2. The existing criterions don't allow to take into a number of provisions in the strategic management perspective of extracting reserves, such as the decrease of investment costs for each selected innovative projects and don't take into the most likely risks, the appearance of which is explained by the fact of attracting new technologies in a circumstances of changing structure of bedding layers in the mine field.

3. The potential formation technological networks of mines is a symbiosis of the interaction of concentration factors of the mining operations, resistance ventilation and influence on the formation of productive flows in the company. Optimization of aim indicator "technical and economic potential of the technological network of mine" that characterizes the mine susceptibility to innovation is the result of interaction between these factors and factors of the second order, which determine the formation of the several main indicators.

REFERENCES

Amosha, A.I., Kabanov, A.I. & Starichenko, L.L. 2005. *Prospects of development and reforming of the domestic coal industry against the background of global trends.* Scientific report. INRM of NAS of Ukraine: 32. Donets'k.

Naman, J. & Morgenstern, O. 1970. *Theory of games and economic behavior* / Trans. from English. Moscow: Nauka: 708.

Salli, S.V., Bondarenko, Y.P. & Tereshchenko, M.K. 2009. *Management of the technical-economic parameters of coal mines* (National Mining University). Dnipropetrovs'k: Gerda: 150.

Boychenko, N.V. 2002. *To the issue of economic sustainability of the coal mines in the depressed areas of Donbass.* Economy: issues of theory and the practice. Dnipropetrovs'k: National Mining University. Issue 159: 157–164.

Pivnyak, G.G., Amosha, A.I., Yaschenko, Y.P. etc. 2004. *Reproduction of mine fund and investment processes in the coal industry of Ukraine.* Kyiv: Naukova Dumka: 331.

Bondarenko, V.I., Ilyashov, M.A., Rudenko, N.K. & Salli, S.V. 2009. *Organization and planning of stoping and preparatory works: Textbook for Universities.* Dnipropetrovs'k: National Mining University: 327.

Progressive Technologies of Coal, Coalbed Methane, and Ores Mining – Bondarenko, Kovalevs'ka & Ganushevych (eds)
© 2014 Taylor & Francis Group, London, ISBN: 978-1-138-02699-5

Evaluation of the prospects of comprehensive development of mineral resources in Belozersky iron-ore region

M. Ruzina, N. Bilan, O. Tereshkova & N. Vavrysh
National Mining University, Dnipropetrovs'k, Ukraine

ABSTRACT: The results of a comprehensive study of accompanying types of mineral raw materials in Belozersky iron-ore district are given in the paper. In particular, research was carried out for the presence of chromite, apatite, chrysotile-asbestos and copper-pyrite mineralization. Deposits of low-grade free-milling iron ore are still undeveloped part of the main iron horizon of productive Zaporozhye iron ore suite. Ore-formation type and rather stable occurence of indicators of ore-bearing zone along the strike are productive and encourage to take them into consideration when by-passing strip-mining in the upper shale horizon of zaporozhye suite or exploration of magnetite quartzite. It is probable the discovery of small deposits of chalcopyrite ores and medium-scale talc-magnesite deposit within the South Belozersky mass.

1 INTRODUCTION

Relevance of research is determined by the necessity of comprehensive development of the mineral resources base of Ukraine. An integrated approach is particularly important in the areas of mining enterprises with developed economic infrastructure, which include Zaporizhzhya iron ore factory located in the center of Belozerskaya greenstone structure in Middle Pridneprovie megablock of the Ukrainian shield. This essentially throw open to integrated development of the Belozersky iron-ore district through exploration and exploitation of additional to iron ore, but little-known types of minerals without the associated environmental implications and major capital spending.

2 MATERIALS AND RESULTS

Chromite ore manifestation within the Belozersky iron-ore district are identified and specified in the drill-hole core throughout South Belozersky serpentinite mass. According to the research ore manifestation is presented as stripe-like, lenticular and rare disseminated chromite accumulations in apodunite, rarely in apoharzburgite serpentinites. Their width does not exceed 1–5 m; the cubic contents of chromite are up to 20–30%, contacts of lentils with the enclosing rocks are gradual, unsharp. Size of chromespinelide crystals does not exceed hundredths and tenths of a millimeter. The oval shape of the grains and growths corresponds to nodular microtexture, which allows to assume the segregation character of ore accumulations. Primary silicates (olivine, orthopyroxene) are entirely replaced

by chrysotile, antigorite and bastite. A widespread and usually intensive replacement of chromite by magnetite (Figure 1) or its complete dissolution in serpentinization is extremely unfavorable on the practical side.

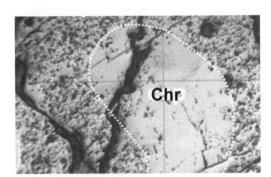

Figure 1. Replacement of chromite (Chr) grains by magnetite. Reflected light. Magnification 100x.

The limbate, pseudomorphic and boxy structure (Figure 2) or the structure of residues from the replacement arises as a result of these phenomena, as well as regeneration of pseudomorphic magnetite occurs with the formation of inherent octahedral forms. According to V.D. Ladieva, the continued presence of magnetite margin, from which it is impossible to remove during the fractionation of chromespinelide monofractions, makes it difficult to assess the quality of their concentrates.

Comparison of these features of chromite ore manifestations with the characteristics of chromium ore formations allows to refer them to chromite ore formation in dunite-peridotite complex of ultramafic rocks with mild signs of commercial kempirsay formation type.

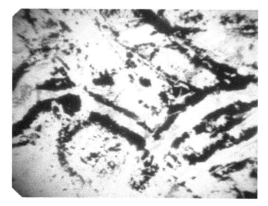

Figure 2. Boxy structure of replacement of olivine by hysterogenous magnetite. Transmitted light, parallel nicols. Magnification 100x.

Apatite in available forms of syenite mineralization was found out on the contact of konkskaya and belozerskaya series in the southwestern part of the southern block of Belozerskaya greenstone structure. Ore manifestation is enclosed in small hypabyssal intrusive bodies of metamorphosed syenite. Apatite with the content of 3–5% and in the form of prismatic crystals of 0.5–1 mm in length is uniformly distributed in the groundmass. Its presence is revealed in thin sections. Given data allows to establish the magmatic genesis of apatite manifestations and presumably to refer them to two probable ore formations. Their matching to formation of apatite ores in alkaline rocks is more authoritative, though approximate. Apatite manifestations can not be referred to formation of apatite ores in carbonatites due to the complete absence of additional signs.

G.F. Guzenko & Z.V. Turobova mentioned about apatite mineralization and single apatite ore manifestation within terrigenous rocks of mikhailovskaya suite at first, but V.I. Ganotskiy investigated it in detail. Manifestations are represented by gray quartz-sericite schists with increased to 5–10% apatite content. They form four interlayers with thickness of 0.5–2.6 m within strata of metaterrigenous rock with total thickness about 120 m. These strata take a clear stratigraphic position in the upper part of mikhailovskaya suite. Interlayers of apatite schists are paragenetically associated with all metasedimentary components of mikhailovskaya suite. There are acidic metaeffusive rocks in the upper interlayer. Interlayers of intraformational metabreccia (thickness of 0.2–2 m) with fragments of quartz sideritolites make great originality of composition of discussed strata. The essential minerals of apatite schists are quartz (35–40%), sericite (30–35%), carbonates such as sideroplesite and ankerite (5–35%), chlorite (7–10%) and apatite (5–10%).

Apatite in the form of prismatic, often zoned crystals up to 0.5–1 mm in length with a hexagonal cross-sections are relatively uniformly dispersed in the form of single crystals and fan-shaped growths. The great majority of its grains split into 2–3 pieces. Fragments are divided by accumulations of carbonates, chlorite, and rarely pyrite. Structure of schists is fine-grained, microporphyroblastic with a grain size of groundmass 0.015–0.03 mm and segregations of apatite on average 0.5 mm. Heteroblastic structure is developed under intensive carbonation. Texture of schists is predominantly lenticular-eutaxic due to constrictions of bands of carbonate aggregates.

Composition of secondary minerals, streaky texture and signs of cataclastic structures are similar to composition and structure of listvenite-beresites of mikhailovskaya suite. This indicates that the band of apatite schists is in one of the stripe-like zones of dislocation metamorphism, which had a transformative effect on apatite mineralization. The chemical composition of apatite schists is characterized by analyzes of chip samples representing the maximum concentration of P_2O_5 about 1.31–2.05. Its enrichment opportunities have not been studied, but it is rated presumably as prospecting resources by analogy with the process of apatite-type silicate ores containing 3–4 P_2O_5 due to the relatively large apatite segregations of fine-grained schist mass. Considerable difficulties caused by the choice of the ore-formational appurtenances of mineralization. Most probably it can be identified among well-defined homogeneous group of phosphate volcano-sedimentary metamorphosed siliceous-schist formations of Precambrian strata of the East Urals and Mugodzhar. N.S. Shatskiy highlighted them at first. According to these data the numerous manifestations of apatite in siliceous-schist strata are metamorphosed under greenschist to amphibolite facies and contain from 0.5 to 8% of P_2O_5 (often up to 4–6%). Of course, in this case there is not a complete analogy between ore and enclosed geological formations. The closest match is in the similarity of lithofacies of discussed formation with distal volcanism of siliceous-schist type formations that paragenetically associated with felsic volcanics. Analogy is confirmed by predominant role of terrigenous rocks in the composition of metapsammites and metapelites among which there are also black carbonaceous shale. Microquartzites and quartz sideritolites are silica (jasper) components, the number of which is small in remote formations perform.

It is believed about the genesis of phosphorite that close paragenesis of apatite schist with terrigenous rocks, planparallel tabular or lenticular shape of their occurrence and the great length of the enclosing band along the strike (more than 7 km) indicate primary sedimentary metamorphosed origin.

Deposits of low-grade free-milling iron ore (magnetite quartzite) are still undeveloped part of the main iron horizon of productive zaporozhye iron ore suite, which consists of five embayed blocks (unoxidized outliers), separated deep oxidation zone. All five blocks represent the sovereign deposit of magnetite quartzite and are explored by wells at the exploratory stage of assessment or preliminary exploration, and the block between South Belozerskoye and Pereverzevskoy rich ore deposits are explored in detail.

Pyrite (sulfide) ore manifestations are represented by two formations. Among the common features that allow to refer them to the genetic group of pyrite deposits within the meaning of V.I. Smirnov it should be noted the high content of sulfides, mainly pyrrhotite-pyrite ore composition, distinctly stratiform – regular bedding of ore lenses and mineralized zones with enclosed rocks, matching of isotopic composition of sulfur from sulfides with its extreme values in pyritic formations (from -0.9 to +4.2%).

Pyritaceous manifestations and occurrences of pyrrhotite-pyrite mineralization were encountered in the key section of mikhailovskaya suite in belozerskaya series and were available for the study of thin sections and polished sections. They are placed in the upper part of the suite, where are associated with interlayers of carbonaceous shale.

Manifestations have sericite- siliceous-pyrite-pyrrhotite composition and occur as rare regular interlayers with thickness of 5–20 cm. They have a massive, shaly-banded or brecciated texture and contain at least 50% pyrrhotite and pyrite. Sulfides of iron are recrystallized in varying degrees (grain size up to 0.1–0.4 mm), but many individuals preserved within their relics fine-grained (less than 10–50 μm) spongeous mass. Signs of cataclastic microbreccia textures are clearly marked in some interlayers.

Manifestations of lean pyrite mineralization in black shales with sulphide content of 5–10% (mainly pyrrhotite) occur frequently. Pyrite and chalcopyrite are major among ore minerals, the total number reaches 5–10%, and greater values in massive interlayers. Pyrrhotite has subordinate value. Each of the minerals is represented by two generations. While pyrrhotite and a part of chalcopyrite preserved in the form of poikiloblast in pyrite;

newly formed magnetite in the form of margin overgrown pyrite, or in association with chalcopyrite crosses its grain. Such interrelation is quite clearly fixed sequence of segregation of ore minerals. Predominant size of ore grains lies in the range 0.1–0.4 mm.

Given data allow to determine matching of characterized manifestations to formation of chalcopyrite (copper-sulphide) ores. In this distinctly stratiform nature and paragenesis with acidic metavolcanics admit volcanogenic hydrothermal-sedimentary way of deposit of sulfide components. There are not distinct analogues of this formation. It matches to deposits of ore-Altay type by association with acidic metavolcanic rocks and chalcopyrite ore type, but the closest type is pyrite formation of shale horizons in Kostomukshskoe magnetite quartzite deposits. It should be noted that ore-formation type and rather stable occurrence of indicators of ore-bearing zone along the strike are productive and encourage to take them into consideration when by-passing strip-mining in the upper shale horizon of zaporozhye suite or exploration of magnetite quartzite. This does not exclude the possibility of discovery of small deposits of chalcopyrite ores.

Comprehensive geochemical anomalies of noble metals have presumably syngenetic origin, are revealed by their monometallic or joint manifestations among unaltered by metasomatic processes black and phyllitic shales, metasandstones, metagravelstone of mikhailovskaya suite. All of them are characterized by fine outward look and low metal content, which indicates the low degree of accumulation in the sedimentary process. This type of mineralization can not have independent commercial value in Belozerskaya greenstone structure, but it is important in genetic terms, it indicates one of the possible sources of metal complex.

Greater commercial importance is put to the prospects for the development of complex epigenetic manifestations of mineralization of precious metals identified within the section of mikhailovskaya suite in belozerskaya series spatially and genetically associated with metasomatic zones of altered rocks.

V.P. Zhulid discovered and tested the manifestations of cross-fibrose chrysotile-asbestos in serpentinite. Serpentinites are fractured; fractures are filled up chrysotile-asbestos, less carbonate. Thickness of veinlets is 1–2 mm up to 2–3 cm. Cross-fibrose chrysotile-asbestos is mainly dominated, but there is also slipfibre one. Veinlets volume reaches 1–8%, on average about 5%, suggesting the mineral ore occurrences. Ore-bearing interval is confined to tectonic zone of crustal weakness. It is revealed in thin sections that the enclosed serpentinites refer to apodunit-harzburgite formation, the main ore manifestations are confined to apoharzburgite serpentinites in contact with apopyroxenite ones; serpentinization of ultrabasites is almost complete, relics of the original rocks are rare and insignificant. Serpentine is represented by several generations, among which dominates early antigorite, partially replaced by lizardite.

Hydrothermal-metamorphic origin of chrysotile-asbestos can be considered as generally recognized with a variety of options. Thus, as R.L. Beyts and I.F. Romanovich independently of each other, stress leading role of structurally metamorphic differentiation of serpentinite material under removal of stress load that promotes penetration of the hydrothermal solutions, which deposit chrysotile-asbestos. It is these conditions are typical for the intersection of faults that contains all Belozerskaya greenstone structure and it is acceptable to specify the genetic class of ore manifestation as dislocation-metamorphic one. Despite the limited information, formation type and analogues of chrysotile-asbestos manifestations in South Belozyorsk mass is quite reliably established using classification and prospect evaluation criteria. In accordance with it manifestations are certainly formational type of deposits of cross-fibrose chrysotile-asbestos in dunite-harzburgite formations; they are most likely labinskiy or bredynskiy subtypes. At the same time, the main evaluation criteria of asbestos content in ultrabasites allow to evaluate positively the background of asbestos presence in South Belozyorskiy serpentinite mass. This is indicated by its position in the jointing of fault structures, the presence of harzburgite, greenschist facies metamorphism, and pre-ore character of massive antigoritisation.

The result is acceptable to assume that the discovery of several small deposits of cross-fibrose chrysotile-asbestos of non-textile applications within the main body of serpentinite in Belozerskaya greenstone structure is probable. Expected size of deposits was identified by their belonging to labinskiy or bredynskiy morphogenetic type. Their thickness can reach 2–10 m, and the length along the strike of several hundred meters. To find out the potential ore bodies it is necessary to specify the morphogenetic type and petrochemical subtype of revealed mineralization, to perform the failed tests of washability and to proceed to detailed search, guided mainly formational and structural factors of geological control.

Many researchers have drawn the attention to numerous manifestations of talc and magnesite in serpentinites South Belozersky mass. They are found out in drill-hole cores throughout its stretch. Considered manifestations are located in the zones of intense schistosity and hydrothermal-metasomatic al-

teration of serpentinite, determining the shape, size and modes of occurrence of talc-magnesite bodies with thickness 5–30 m. Dip at high angles up to 70–90° and predominantly diagonal orientation to the strike of the mass with the azimuth of about 150° (SE) is characteristic throughout the mass. Contacts with the enclosed serpentinite bodies are gradual, due to a decrease in the degree of rocks alteration, especially serpentinite carbonatization. Usually, contact zones are complicated by the sliding surfaces. A strict dependence talc-magnesite zone on the initial composition of serpentinite is not installed. In thin sections revealed that the main minerals are talc, magnesite, breunnerite (Figure 3), serpentine and chlorite (prohlorit).

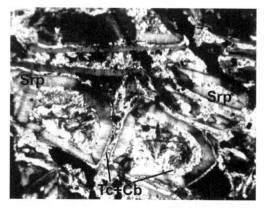

Figure 3. Replacement of serpentine (Srp) by talc-carbonate aggregates (Tc+Cb). Transmitted light, crossed nicols. Magnification 100x.

Amount of talc and magnesite varies within wide limits and is 40–60% in each mineral. Together they make about 40–95% of the rock volume. Residual chromite, magnetite and later pyrite connected with talc-magnesite gold-bearing zones are minor impurity. Talc is presented by ferrous phase – minnessotait, and magnesite often replaced by breunnerite. Texture of rocks is schistose, rarely banded. Structure is lepidogranoblastic and porphyroblastic with carbonate having a grain size up to 0.2 mm, and talc flakes – 0.05 mm in length.

Essentially described hydrothermal manifestations are presented by schistose hydrothermal metasomatites of listvenite formation that formed with the active addition of silica and carbonic acid in high-Mg medium of deformed serpentinite. In this respect, the process of their formation is the same, not only in fact, but at the time of origin with zones of more thick listvenite-beresites within terrigenous rocks of mikhailovskaya suite. With no signs of spatial and genetic relation with granitoids – common to most talc deposits, and the leading role of fault-fracture structures, this gives reason to connect processes of listvenitization in general cases with dislocation metamorphism. Discussed manifestations correspond to talc-magnesite stones with the content of talc 35–75% and magnesite 33–42% by type of mineral raw materials. They represent the ore formation of talc-carbonate rocks in serpentinites of dunite-harzburgite and other apoultrabasite formations and belong to either shabrovskiy or medvedevskiy morphogenetic subtype. Such deposits can be small, medium, and large-sized objects that are placed in completely serpentinized ultrabasites having no connection with granitoids and therefore referred to the metamorphogenic formations. Talc-magnesite bodies have the form of steeply dipping lenses, veins, tabular deposits with thickness from a few meters up to 100–300 m length and a length of 400–650 m and even 2–3 km. Dimensions of accumulations depend on the size of mass of parent rocks. On this criterion the South Belozersky mass recedes to Pravdinsk mass and therefore we can expect the presence of small and medium-sized accumulations.

3 CONCLUSIONS

Talc-magnesite deposits are types of mineral raw materials in very short supply for Ukraine. Demand for it is provided only due to import. So, detailed searches talc-magnesite in Belozerskaya greenstone structure is very topical task. Leading criteria in exploratory work should be regarded as a structural and mineralogical, which are expressed in close connection of talc-magnesite listvenites with schistosity zones in serpentinites. As a result, within the South Belozersky mass it is probable the discovery of medium-scale talc-magnesite deposit, which is readily accessible in conditions of Zaporozhye Iron Ore Factory.

V.M. Kravchenko et al. described in detail the small bodies of rich quartz-magnetite and sideroplesite-magnetite ore with iron content up to 52–55‰, occurring in ferruginous quartzites of the same composition in Pereverzevskoe deposit of ferruginous magnetite quartzite and later they were met in the North Belozerskoye deposit. Such ores form small low-angle cross or concordant bodies and located on the flanks of large ore bodies of the main industrial type – dispersive hematite-martite ores. In fact, they are remnants of earlier large deposits, where on their place the ore-shoots of oxidized richer ores formed due to hypergene oxidation and enrichment. Relatively low iron content in the non-oxidized rich ores (52–54%) is explained by its iron

content in ore bands of ferruginous quartzite, which approach due to dissolution of quartz bands during metamorphogenic shrinkage.

Epigenetic series of minerals is also characterized by two different origins, but on the same material composition deposits of rich dispersive hematite-martite ores that formed in place of ferruginous quartzites. The differences lie in the different concentrations of iron in the original magnetite ores and its increase in varying degrees under the influence of hypergene processes. Considered ore deposits are two ore formations – belgorodskiy and saksaganskiy types. Comparison of their geology and genesis are described in detail in the monograph of Ya.N. Belevtsev et al.

Small subsurface deposits of hypergene dispersive hematite-martite ores of belgorodskiy type of formation occurs on the tops of horizon of ferruginous quartzite outcropping on the crystalline basement. This type of deposits is most clearly expressed in the North Belozerskoye deposit. These deposits were formed in the now buried weathering crust directly from ferruginous quartzite. They are related to genetic class of residual hypergene ore. Rich ore of formation play a minor role when small amounts of deposits.

Large deposits of dispersive hematite-martite ores of saksaganskiy type – South Belozerskoye and Pereverzevskoe deposits constitute the main wealth of Belozerskaya greenstone structure. They not only have been thoroughly explored, but also operated by Zaporozhye Iron Ore Factory. In accordance with the leading role, geology of iron ores studied in detail and highlighted in numerous articles, scientific and industrial reports. According to them the rich ore deposits are classic ore-shoots flattened shape that fit entirely within the horizon of ferruginous quartzite of zaporozhye suites and with them go down to a depth of 1500 m. Thickness of accumulations is about tens of meters, sometimes they hold all the thickness of iron ore horizon. Such bodies can be traced intermittently throughout the stretch of the west wing of the South Belozerskaya folds and everywhere surrounded by a halo of oxidized ferruginous quartzite – deep oxidation zones. As in the previous formation of hypergene ore rich ores have dispersive hematite-martite composition, have high porosity and semi-earthy composition. Due to its large scale and detailed exploration of deposits of this formation, they do not need any more practical evaluation.

Under any circumstances, the main direction of the region development over the next decade is the development of deposits of rich and low-grade iron ore. However, the economic indicators of the current mine of Zaporozhye Iron Ore Factory can grow much due to the involvement of additional mineral raw materials – precious metals, talc, magnesite, asbestos, and others. To do this, along with the continuation of the search and audit work it is necessary to carry out detailed searches and search evaluation of discovered new types of minerals at separate places.

Attractive technological and economic aspects of Ukrainian coal production development

O. Vagonova, V. Prokopenko & A. Kyrychenko
National Mining University, Dnipropetrovs'k, Ukraine

ABSTRACT: Having in mind Ukrainian experience in innovation activity with dramatically insufficient output of science-intensive products, and basing on examples from developed countries of the world the authors have analyzed favorable possibilities to realize innovative development of Ukrainian economy with the help of "strategy of the borrowed knowledge". The article shows economic effectiveness and promising aspects of technological upgrading of engineering environment in Ukrainian mines by own efforts of the mining operators through implementation of available innovative models of the roadway supports.

1 INTRODUCTION

Today about 90% of Ukrainian products are manufactured with no state-of-the-art scientific basis. Research departments in the factories has been shortened by quarter in number. A great majority of industrial enterprises are impetuously reducing their innovative activity. Totally 6.5% of all sold products can be classified as innovative and only by some characteristics, though in European Union this figure exceeds 60% (Karakay 2006). Growth of Ukrainian GDP at the expense of introduced up-to-date practices is 0.5–0.7% if compare with 60–90% in developed countries of the world.

Among the most critical problems, manufacturers consider lack of own funds (80% of respondents) and great expenses needed for innovations (57%) as well as insufficient financial support of government (54%) and high level of risks (41%). However, the author believe that the above mentioned prime problem – lack of own funds – is actually a consequence of the fact that national industries produce out-of-date products that consumers are not eager to buy.

Law of Ukraine (Law of Ukraine) "About First-Priority Fields for Innovative Activity" lists 41 fields of national economy starting from space-rocket hardware and aeronautic engineering and ending with technologies to preserve agricultural products to which Ukrainian government will set the greatest store and, consequently, will allocate budgetary funds. To compare, this year Russians have compiled a list of their scientific priorities that accounts only eight positions, and in Germany such list includes totally five items. From all world-known structures involved into innovative activity only technological parks are available in Ukraine to

this or that extent. These parks have produced science-intensive products in amount of USD 1.5 billion and created 3 000 new working places, and this is tens times less than certain technological parks in China or India.

Today among the key factors threatened to innovative safety of Ukrainian coal-producing industry we should mention a poorly developed infrastructure for sharing advanced technologies, growing dependence on imported science-intensive goods and unsatisfactory informational support. As for the coal production technology and economy, all above mentioned problems could be successfully solved today with no essential time and money consumption if basing on available national scientific and technological support. This article focuses attention exactly on these target tasks: to generalize scientifically existing thoughts and opinions concerning strategy of the coal production development; to choose a priority model for the industry economic growth; and to establish proper approaches to improve informational safety of the industry.

2 FORMULATION OF THE TASK

Technical innovations are one of the most important instruments that can raise economic effectiveness and improve competitive strength of the coal mines through realization of the following innovation basic components(Savytska 2004):

– to provide high level of qualification and purposeful training of personnel in order to gain universal knowledge of how a mine operates;

– to choose optimal scheme for secondary workings with taking into account concrete geological conditions and technical characteristics of equipment;

– to introduce advanced and economically highly effective practices for preparatory operations in the mines.

Analysis of strategic measures that have been taken in the coal industries of Great Britain, Germany and Poland in order to exit from the crisis shows that principle of effective operation of up-to-date machinery and technologies was in the heart of the processes of their production concentration and underground production intensification. In these countries, technological practices were continuously oriented to utilization of full potential of state-of-the art machineries, and, at the same time, the machineries were continuously improved to be maximally adapted to specific mining, geological and technical conditions of concrete mines.

Condition of the Ukrainian coal industry is still unsatisfactory in spite of the fact that the country has managed to stabilize annual coal output at the level of 80 million tons. The critical state of the industry is explained by two key factors: difficult mining and geological conditions for developing the coal reserves and obsolete mine facilities caused by funds chronically not allocated in a volume needed for capital reconstruction and modernization of production. Privatization is in process, and majority of the mines are still public.

Following the above mentioned, the authors have considered concrete approaches worked out by researchers and manufacturers of the industry that concern opportunities of Ukraine to realize innovative development of national economy basing on a "strategy of the borrowed knowledge". A many-sided estimation of this strategy was made by I. Egorov, O. Popovych and V. Soloviov in their work (Egorov et al. 2003) and supported by the Governmental Fund for Fundamental Researches of Ukraine. The authors analyze different models of economic growth and role of science and technology in these processes in various countries. This analysis is an evidence of the fact that a key factor of success is harmonization of all aspects of industrial, social, scientific and technological policies.

Orientation to the "strategy of the borrowed knowledge", i.e. focusing of national production mainly on applying of up-to-date practices worked out in other countries of the world is much more difficult process as it could be thought on the face of it. Thus, in this article we are setting more store on the issue: what exactly and to what extent it is necessary to borrow knowledge today.

3 RESULTS

Advocates of strategy of maximal orientation to the foreign experience often mention Japan and new industrial countries as an example and, as a rule, direct our attention to the aspects that are evidence in favour of policy of borrowing. However, addressing to Japan and South Korea the strategy of borrowings advocates neglect that fact that both countries had to buy licenses for foreign technological products due to the lack of own technologies and, at the same time, they were and are intensively increasing their scientific and technical potential (Kiyasko et al. 2006). In 70s of the last century, when this strategy was mostly presented in policy of Japan this country each year invested additionally 6.4% of funds into science, and after less than 10 years the allocated funds had reached 2.2% of national GDP. Reformation of technological structure of economy was impossible in the country without own research and technological potential. So, beginning from 80s, government of Japan always keeps the emphasis on necessity to increase financial support of fundamental researches and to actively prepare specialists. During 90s, total number of research institutions in the Science and Technology Agency of Japan was doubled.

During 1971–1988, South Korea was increasing funds spent on science 6 times quicker than gross domestic product was growing (Egorov et al. 2003). It should be noted that expenditures of private sector of Korean economy on scientific researches and projects during this period of time was increased 16 times quicker than the national GDP.

During the last decades such not great Asian countries as Taiwan, Singapore and Hong Kong demonstrate a phenomenal progress of their economies. These countries, in spite of their different initial specialization in the world market, step-by-step moved from manufacture of technically simple products towards hi-tech goods with much greater additional value. Priorities of researches and developments (RED) were declared at the highest governmental levels and were realized by government programs of economic, scientific and technical development.

Licenses for the best practices were intensively bought; but at the same time part of the GDP that was invested by Taiwan, Singapore and Hong Kong (as well as South Korea) to develop own science was 3–5 times increased during the period 1980–2000. Such policy on development of own researching sector could not but give positive results – within this period of time number of patents that were registered in these countries was 3–6 times more than during the previous 20 years, while number of patents received by residents of other countries shows stable tendency to reduction. Increased number of authors from South Korea and Taiwan published their works in the USA during the last twenty years is really impressive: 100 times – for authors from South Korea and about 70 times – for authors from Taiwan.

The most important feature of all mentioned countries is high and trending to further growing level of population education especially in the field of technical and natural sciences. In South Korea it has been grown by almost 4 times.

Thus, we can to conclude that all new industrial countries intensively realize a policy that focuses on creation of own developed scientific and technical potential able not only to receive and adapt some borrowed foreign advanced practices and innovations but also generate own innovative technologies with very high economic results after their implementation. It means that "strategy of borrowings" was just an episode in the long-term innovation policy in these countries, a step that is justified by initial weakness of their national scientific and technological potential.

To meet competition at the world market, Ireland adopted a strategy of European integration. The most important elements of this strategy are participation in EU projects and programs on regional development and focusing on scientific and technological factors of the modernization. But above all these, the key component of the governmental innovation policy the government considers essential strengthening of the Irish national science and research system as only with the help of this instrument it is possible for the country to provide adaptation of technological innovations in its economy.

Thus, none of the above mentioned countries that successfully used foreign scientific and technological developments to innovate their own economies never did it at the account of destruction of their national scientific potential (Egorov et al. 2003). All of them proceeded from the understanding that no real breakthrough is possible without development of national science. Studies (Pavitt 1993) show that countries that have no own developed scientific basis face essentially more difficulties with assimilation of revolutionary technologies and innovative achievements of the world contemporary science.

In the present conditions of scientific innovative development of industrial infrastructure of Ukrainian economy a problem of renovating and strengthening of technological component of production process in the coal industry is the most important as this branch features high intellectual potential of researchers and engineers and highly qualified labour resources who have been prepared for work in hi-tech productions. Besides, today our financial-industrial holdings orient to innovative development, and some mines have already their own scientific, researching and technological developments based on strong engineering school.

Continuous improvement (safety and reliability in the first place) of machinery and coal-mining practices is necessitated for successful development of the coal mines. In judgment of O. Ruban (Ruban 2005), a contrast between obsolete technological level of the coal production and high innovative potential of researching sector is especially striking in Donbass. Today output of the longway faces is 800 t/day in average in Ukraine. The most weak link that should provide proper reliability and resource parameters for highly technological organization of secondary workings is designing and introduction of coaling plants of new technical generation. First of all, it is necessary to improve mechanized facilities for mining operations and for keeping the roadways in an operating state. Great depth of the coal bedding plus complicated mining conditions for the roadway driving lead to lost stability even of preparatory roadways, gateroads, slops, passways, etc. in the Donbass mines requiring repairs as earlier as at the drivage stage, i.e. far long before the mining.

Achieving of high working resistance of the mine supports is possible only with the help of new types of supporting frames, new materials, additional strengthening of arch frame provided by new-type supporting, and a set of technological measures. For example, preparation and development of a working horizon with the help of arch supports КМП – А5С – 18.6 essentially improved condition of the gateroad and made possible to refuse from re-setting of support until the gateroad extinction. Rejected operation on arranging of machine stables has increased daily output of the logway up to 2300–2800 t. Totally, no-stable mining practice used instead of standard technology gives a raise of monthly output and reduces expenditures by 2–3 times (see Table 1).

"Geomehanika" Zapadno-Donbassky Research-and-Production Center (Pavlograd, Dnipropetrovs'k region, Ukraine) designs various models of the steel supports of new technical level that advantageously differ from those that are widely used today. Design of these new models is based on scientific achievements with legislative validation that orient to further development and governmental support of national science.

When estimating innovation effectiveness the following aspects should be considered. Chosen supports and their setting could be not expensive; however, their repair during the operation could require great expenditures. Or, on the contrary, support setting as one of operations of preparatory development could cost big money but in future the installed supports would be rarely damaged and, thus, would not require essential funds for their repair. Thus, when making decision on choosing supports of the new technical level (the NTL supports) it is necessary to take into consideration not only primary expenditures on strengthening the mine tunnels when the supports are set but future expenditures on maintaining the installed supports in a stable state as well.

Table 1. Expenditures on driving and repair operations with different types of supports, UAH/r.m.

Pos.	Expenditures	A.G. Stakhanov Mine				A.F. Zasyadko Mine	
		Gateroad		Boundary passway		Gateroad	Gateroad
		AP3-15.5	KMP A3 P2	AP3-15.5	KMP A3 P2	AP3-18.3	KMP A5C-18.7
I.	**Operations**						
1.	Support setting	2210	1440	2210	2030	1944	2542
2.	Cost of supporting	168	126	217	176	65	86
II.	**Support maintenance**						
1.	Resetting of supports	5211	564	2320	0	4360	282
2.	Undermining of the floor						
	up to 1200 mm					510	0
	up to 1000 mm	836	836	886	886	370	400
	up to 500 mm	214	214	0	0	148	148
3.	Cost of new supports for re-setting	2210	241	740	0	1302	0
III.	**Totally, incl:**	**10849**	**3421**	**6373**	**3092**	**8699**	**3458**
1.	Setting of supports: I(1+2)+II(3)	4588	1807	3167	2206	3311	2628
2.	Maintenance and repair: II (1+2)	6261	1614	3206	886	5388	830

If a new support improves stability of the mine tunnel then intensity of failures r of the support and time period t_p needed to repair this support are reduced. Volume of the coal production O_b grows in direct proportion to reduction of the failure intensity and time period needed for the repair. In this case the following Equation

$$\frac{O_n}{O_m} = \frac{\left(T_b - r_b \cdot t_{p.n}\right)}{\left(T_b - r_n \cdot t_{p.m}\right)},\qquad(1)$$

operates where O_n, O_m – coal output, t/m, with the NTL supports and traditional supports, correspondingly; T_b – fund of working time for the coal mining and transporting, h/month; r_b, r_n – failure intensity for new and traditional supports, correspondingly, in operation, month^{-1}; $t_{p.n}$, $t_{p.m}$ – time periods to repair, hours per frame, with the NTL supports and traditional supports, correspondingly.

Possible economic effect from innovative supporting due to the less costs and faster driving of the mining workings would be equal to:

$$E_{n.b} = B_{n.m} - B_{n.n} , \text{[UAH/month]},$$

where $B_{n.m}$, $B_{n.n}$ – costs of haulage roadway (air roadway) arrangement with traditional supports and the NTL supports, correspondingly, UAH/month.

To calculate value of $B_{n.m}$, costs should be divided along the whole length of the mine tunnel strengthened by the NTL steel frames, and the length should be adjusted to productivity of the mining district taking into account possible speed of the roadway driving, capacity of the mine transports and throughput capacity of the mine tunnels.

As soon as the coal is mined costs calculated by the Equation (1) for the haulage and air roadways at panel mining should be transferred to the coal cost price by the item "Extinction of preparatory development". Thus, economic effect $\Delta B_{n.b}$ can be received simultaneously with the effects ΔB_0, ΔB_m and ΔB_b. These effects mean value of possible reduction of the current inputs of the mine into its production process when the NTL supports are used. Total economic effect from the innovative facilities used for the mining operations is equal to

$$E_z = \Delta B_0 + \Delta B_m + \Delta B_b + \Delta B_{n.b} , \text{[UAH/month]} \quad (2)$$

With the pillar mining, haulage and air roadways from the very beginning are driven till the boundary of the mine field, and coal is extracted by retreat mining. In this case economic effect from the innovating model of supports is created at the account of (1) the same factors shown in the Equation (1) the shorter time period for the coal-pillar preparation. At the pillar length L_c and speed V_κ of the mine tunnel drivage time period T_n for the pillar preparation is

$$T_n = \frac{L_c}{V_\kappa} = \frac{L_c}{\mu_\kappa \cdot A_\kappa} , \text{[month]} \quad (3)$$

The Equation (3) assumes that if traditional supports are used speed of the mining is V_m, and coal pillar is cut during the time period T_m; when innovating supporting is used speed of the mining is V_n, and time period for the pillar preparation is T_n. If monthly expenditures on the mining operations are equal to $B_{n.m}$ and $B_{n.n}$ accordingly to the supporting methods then expenditures needed during the tunnel driving till the end of the mining operations on preparation of the coal pillar are equal to

$$B_{ct.t} = \sum_{t=1}^{T_m} B_{n.m} \cdot t \cdot (1 + R_a)^{t-1} , \text{[UAH]} \quad (4)$$

$$B_{ct.n} = \sum_{t=1}^{T_n} B_{n.n} \cdot t \cdot (1 + R_a)^{t-1} , \text{[UAH]} \quad (5)$$

In result of the tunnel driving with the help of NTL supports it becomes possible to receive economic effect equal to

$$\Delta B_{ct} = B_{ct.t} - B_{ct.n} , \text{[UAH]} \quad (6)$$

The faster preparation of a coal pillar means earlier beginning of the coal extraction. When duration of the preparation is shortened by $(T_m - T_n)$ it is possible to produce coal ahead of schedule in amount

$$O_{d.n} = (T_m - T_n) \cdot I_d \cdot A_d \cdot L_l \cdot m_b \cdot \gamma_b , \text{[t]} \quad (7)$$

where I_d, A_d are, accordingly, intensity of the coal seam mining by the set of winning machines, cycle/month, and drivage of the gateroad per one cycle, m.

After selling the coal the mine can receive gross profit in amount of

$$P_{sd} = (P_b - C_{bn}) \cdot O_{dn} \cdot (1 - C_{n.n}) , \text{[UAH]}. \quad (8)$$

Money received ahead the schedule could give the mine additional profit P_{sd} thanks to their turn-over in other spheres. This profit can be planned by formula

$$P_{sb} = P_{ds_t} \sum_{T_{n+1}}^{T_m} \left((1 + R_a)^t - T_m + T_n\right) , \text{[UAH]} \quad (9)$$

where P_{dst} – profit of the mine after selling the coal produced, thanks to the faster preparation, ahead of the schedule in comparison with the schedule of fulfilling the secondary workings by traditional way in t-th month, UAH/month.

This analytical models (1)–(9) help to estimate how innovating facilities and, in particular, supports of a new technical level (NTL supports) used for secondary workings in the mining district can influence the coal mine's economic figures. Summarized effect from the NTL support implementation is estimated by the scale of their implementation (quantity of the mine tunnels and their length). Below are peculiarities of a mechanism for creating a profit for the enterprise when a significant volume of traditional supports are replaced by the NTL supports.

Let's analyze impact of the NTL supports on the profit on the example of O.G. Stakhanov Mine. The Mine annually produces 1 mln. t of coal from the seam with thickness 1.2 m bedded at the depth 836 m. To be prepared to mine this volume of coal the Mine should develop in advance about 7 km of the tunnel. Cost of 1 km is UAH 4.5 mln. in average. We assume that the NTL supports are used everywhere along the preparatory roadway. These supports helps to arrange greater space between the arch frames, and, thank to this, metal consumption in the tunnel supports is reduced. And, consequently, costs of support driving, supporting, maintenance and repair are also reduced (Table 2).

Through implementation of the innovating facilities to strengthen the mine tunnels it is possible to increase a gross profit of the Mine by the value of economic effect E_{in} that is calculated by formula

$$E_{in} = \Delta B_m + \Delta B_p + \Delta B_b - B_y , \text{[UAH/year]}, \quad (10)$$

where ΔB_m, ΔB_p, ΔB_b, – reduction of operational expenses thanks to the reduced metal consumption of supporting system, less costs of repairing and resetting operations and cost of the working space expansion, UAH/year; B_y – purchasing cost of special equipment and machines to build tunnels, UAH/year.

$$\Delta B_m = P_\kappa \cdot \Delta M_z \cdot L_b , \text{[UAH/year]}, \quad (11)$$

where P_κ – cost of 1 kg of special metal profiles, UAH; ΔM_z is a reduced metal consumption in the tunnel supporting system, kg/r.m; L_b – length of the tunnel strengthened by metal arches, r/m/year.

Table 2. Costs by factors that constitute effectiveness of the NTL supports in the O.G. Stakhanov Mine.

Pos.	Expenditures	Expenditures, UAH/r.m	
		traditional supports AP 3–15.5	NTL supports KMP A3P2 (16.1)
I	*Setting*		
1	Support setting	2210.0	1440.0
2	Cost of supporting	168.0	126.0
II	*Maintenance*		
1	Re-setting	5211.0	564.0
2	Undermining of the floor up to 1200 mm up to 1000 mm up to 500 mm	– 836.0 214.0	– 836.0 214.0
3	Cost of new supports for re-setting	2210.0	231.0
III	*Totally*		
1	Supporting	4588.0	1797.0
2	Maintenance and repair	6261.0	1614.0

After the components ΔB_p and ΔB_b have been calculated we can define dependence of economic effect on above mentioned factors of implementation of the NTL supports (with index i) instead of traditional supports (mp)

$$E_n = L_b \left(P_\kappa \times \Delta M_z + (1 - P_{bt}) \cdot \left(C_{p.mp} - C_{p.n} \right) + \left(C_{b.mp} - C_{b.n} \right) \right) - B_y \text{, [UAH/year],} \tag{12}$$

where $(1 - P_{bt})$ – part of supporting sets that are damaged, of their total volume (this factor reflects length of districts where the supports are repaired and re-set and working space is expanded); $C_{p.mp}$, $C_{p.n}$, $C_{b.mp}$, $C_{b.n}$ are, accordingly, cost of support repair and maintaining in a stable state and cost of the working space expansion with traditional supports and NTL supports, UAH/r.m.

With the aim to calculate economic effect, we assume that cost of 1 t of metal profile is UAH 2 500 for traditional supports and supports is UAH 2 800 for the NTL.

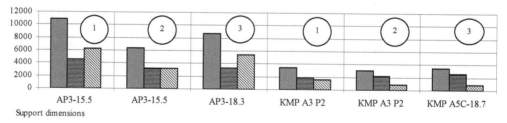

(*totally; supporting; maintenance and repair).

Figure 1. Expenditures on the mine-tunnel supporting, maintaining and repairing using traditional supports and supports of the SPC Geomechanics design: 1, 2, 3 are for favourable, restrictedly favourable and complicated conditions.

Possible mining and geological conditions for the working driving in the Mine are divided into three groups: favourable (weight of 1 r.m of the frame with special profile is 22kg), restrictedly favourable (the same but 27 kg) and complicated (the same but 33 kg) (Kyrychenko 2009). According to these groups, spaces between the installed metal frames are assumed as 0.8; 0.5; 0.33 m for traditional supports and 1.0; 0.8; 0.5m for the NTL supports. Portion of damaged supports is assumed as 0.5 for traditional supports and 0.4 for the NTL supports. Costs of the active opencast expansion are assumed at the level 20% of the original costs of the tunnel supporting system. For these items, expenditures on preparatory roadway maintenance and repair are shown in Figure 1.

In case the Mine replaces traditional supports by the NTL supports profit increases in direct proportion to the economic effect which changes in accordance with mining conditions in the following way:

Operational conditions	Traditional supports	NTL supports	Economic effect
Favourable	22890	10647	12243
Restrictedly favourable	25480	16968	8512
Complicated	25480	9800	20632

We have analyzed reduction of summarized expenditures spent on purchasing and repairing metal profiles and established the following regularity: the more complicated are conditions of the mining labour the greater is economic effect from implementation of innovative supports in the mine tunnels. With traditional supports the mine's profit is UAH 5 mln. at output 1 mln. t per year. All-round replacement of these supports by innovative ones gives a raise to the mentioned profit to UAH 8.5–20.6 mln. depending on the labour conditions.

Our practical experience in O.G. Stakhanov Mine confirms high economic effect of the NRL supports used for the tunnel strengthening and maintaining in a stable state and for the working space expansion. Basing on the above method, operation of the NTL supports were also studied in mines of Joint-Stock Company "Pavlogradvygillya". The mines continuously increases their output showing growth of the coal production from 10.22 to 12.06 mln. t, i.e. by 15%, during 2000–2006. The growth was possible thanks to increased production load of the longway face (from 838 up to 912 t/day) and advanced preparatory development. Without major modification of the mining technologies further complication of the mining conditions due to entering the greater depth would have to be resulted in significantly grown expenditures.

All above data are an evidence of good opportunities to improve economy of the coal mines through innovative supports in the mine tunnels. However, process of replacement of traditional supports by the innovative ones is associated with essential expenses of labour hours and facilities, and duration of the transition period depends upon content of a strategy chosen for the company development in whole.

4 CONCLUSIONS

1. Ukrainian scientific, technological and innovation policy should differ from strategy of developed countries. Role of Ukrainian government in transforming of scientific and innovative sphere should be more weighty.

2. Ukrainian coal-producing industry features high intellectual potential of researchers, engineers and technicians who can ensure scientific-and-technological development of infrastructure in industrial sector of economy; Ukrainian existing financial-industrial holdings orient themselves towards innovating development, and mines have organized their own scientific centers with own technological projects and strong engineering school.

3. Promising direction of the coal mines transition to technological innovations is introduction of different models of the steel supports designed and manufactured by "Geomechanics" Research-and-Production Center. By technical characteristics and operation experience, these models are not second to any world models.

REFERENCES

Karakay, Yu. 2006. Future Is Impossible Without Innovative Development. Thoughts about results of hearings "Innovative Activity in Ukraine: Problems and Ways of Their Solving" in Education and Science Committee of Verkhovna Rada of Ukraine. [Electronic resource] // Golos Ukrainy, 245. – Access to the article: http://patent.km.ua/ukr/articles/group11/i1001.

Savytska, I.K. 2004. *Management Innovations in Donbass Coal Mines.* Economic Herald of National Mining University, 1: 96–101.

Kiyasko, Yu.I., Kosarev, V.V. & Kyrychenko, A.V. 2006. *About Effectiveness Of Putting Into Operation Additional Development Headings And Longwalls.* Ugol Ukrainy, 11: 15–18.

Egorov, I., Popovych, O. & Soloviov, V. 2003. *Strategy of the borrowed knowledge and science development.* Herald of NAS of Ukraine, 5: 3–14.

Pavitt, K. 1993. *What Do Firms Learn from Basic Research?* // Foray D., Freeman C. (eds.) Technology and Wealth of Nations. London: Macmillan: 115–143.

Ruban, O. Machines, Money And Brain. 2005. [Electronic resource]. Access to the article: http://patent.km.ua/rus/articles/group11/i731.

Law of Ukraine "About First-Priority Fields for Innovative Activity".

Kyrychenko, A.V. 2009. *Improved Economic Methods for Managing Innovative Activity of a Coal-Mining Company.* Diss....Cand. Econ. National Mining University. Dnipropetrovs'k: 184.

Progressive Technologies of Coal, Coalbed Methane, and Ores Mining – Bondarenko, Kovalevs'ka & Ganushevych (eds)
© 2014 Taylor & Francis Group, London, ISBN: 978-1-138-02699-5

Analytical determination of stress-strain state of rope caused by the transmission of the drive drum traction

D. Kolosov & O. Dolgov
National Mining University, Dnipropetrovs'k, Ukraine

A. Kolosov
Moscow State University of Technology and Management, Moscow, Russia

ABSTRACT: The stress-strain state of the rubber-rope having one cable under the action of distributed shear force is determined. This case reflects the stress-strain state which takes place in extreme cable in the rope on the drive drum lifting machine. It is established that the distribution of stresses and displacements of the rubber cover of the rope in almost all sections does not essentially change. This corresponds to the deformation conditions of the rope with a cable that is not deformed and indicates the reliability of the results. The obtained dependence of the of the stress concentration coefficient on the parameters of the rope allows to take into account the distribution of stresses within rubber in designing of lifting-and-shifting machines with the traction member. This makes it possible to increase the safety of machine using.

1 INTRODUCTION

Expanding the scope rubber-rope cables in industry requires the solution of a set of knowledge-intensive tasks and development of scientific foundations and principles for the new, more durable and reliable lift installations. In this regard, the development of a new generation of high-tech lifting equipment with rubber-rope cables is an urgent problem of applied research.

The most critical components of a lift machine are steel cables. In rubber-rope cable the latter is vulcanized into rubber shell that protects it from their external environment and interaction with machine elements. The rubber cover destruction eliminates its protective properties and roping cable protection required. Accordingly, the definition of the shell destructive conditions allows to take into account the operating conditions to ensure the safety of the ropes and their use as a traction rope elevators, and in the future, as the main ropes in the mine hoisting machines.

Rubber-rope cables allow the use of small diameter wire ropes. As a result lifting machines may have drums with overall dimensions. The reduced dimensions of the drive drums allow more efficient use of elevator drives.

2 PROBLEM STATEMENT

Interaction of the flat tractive member with the drum is considered in (Blokhin et al. 2009; Kolosov 2007, Belmas et al. 2009, Bilous & Kolosov 2010, Kolosov et al. 2013). In these studies the impact of the construction of the rope on stress distribution in its rubber cover at the site of interaction with the drum was not investigated. Determination of the impact of rubber-rope cable design on its stress concentration factor, when interacting with the drum of lifting machine, is the current scientific and technical problem. Its solution will allow reasonably choose rubber-rope cable for elevators of a new generation. Solving this problem is directly related to the experimental determination of the strength of the rubber and cable connection.

In general, there are forces of friction and pressure between rope and drum. The pressure along the length of the rope can be considered as uniformly distributed. Along the width of the rope (drum) the pressure forces are distributed cyclically with a pitch equal to the pitch location of cables. The forces of tension in the rope cables, due to their mutual connection with the rubber shell, are distributed almost evenly (Kolosov 1987). This feature allows us to consider not all the rope, but only its part.

However, it is known that at the local loading of the samples of large widths, such as rubber-rope cables, the character of deformation in the middle of the sample will be different from the nature of extreme deformation. It is difficult to investigate the nature of deformation of rope with a given width (with a given number of cables). This task can be simplified by considering the rope with one cable and the rope with infinite number of cables. The first case will determine the stress-strain state (SSS), which has extreme rope. The second one – cable, situated at a considerable distance from the edge. Consider these cases separately. In this work the SSS of the rope, which has one cable, under the action of distributed shear force is determined.

3 PROBLEM SOLUTIONS

Consider the diagram of the experimental test of a single cable loading by distributed shear forces shown in Figure 1.

Figure 1. Diagram of experimental loading of one cable by distributed shear forces.

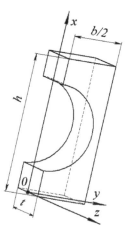

Figure 2. Symmetric part of the regularly recurring element of the rubber shell.

This rope is symmetric with respect to the diametric plane of cable parallel to its edges. Whereas last, consider symmetrical rubber cover segment of the

rope that has one cable (Figure 2).

Consider the stress state caused by the shift. Neglect the bend of the rope. We assume that along the plane $x = h$ the rope is fixed. Relate a sample to cut with orthogonal coordinate system, as shown in Figure 2. Denote the width of the rope by the letter b, thickness – h, length of the cut section – t, the diameter of the rope – d.

Find a solution for such a complex element shapes. Simplify the problem neglecting the presence of cables. Consider a prismatic sample of the sizes $h \times b / 2 \times t$. The body will be considered as one made of a rope shell material – rubber. We assume that the surfaces of said sample points are located in the planes $z = \pm t / 2$ corresponding to the ropes cross-sectional shape and move synchronously in the direction of the x axis.

In order to study the effect of different factors, consider separately two types of force interaction of a drum and rope. Assume that the rope acts with the drum along surface $x = h$. On this surface the displacements of the rope relative to the drum should be zero

$$u_y = u_x = u_z = 0. \tag{1}$$

On the surface $x = 0$ there are no external forces, so

$$X_x = 0, \; X_z = X_y = 0. \tag{2}$$

Boundary conditions for surface $y = 0$ is

$$u_y = 0, \; X_y = Z_y = 0. \tag{3}$$

On the surface $y = b / 2$

$$Y_y = 0, \; Z_y = 0, \; X_y = 0.$$

On the surfaces $z = \pm t / 2$ should not act shear stress, so the conditions are

$$X_z = Z_y = 0. \tag{4}$$

In addition, there must be provided equal displacements of all points corresponding to the location of the cable and the absence of pressure on rubber placed between cables

$$\begin{cases} y^2 + \left(x - \dfrac{h}{2}\right)^2 \le \left(\dfrac{d}{2}\right)^2, u_x = 1; \\ y^2 + \left(x - \dfrac{h}{2}\right)^2 > \left(\dfrac{d}{2}\right)^2, Z_z = 0. \end{cases} \tag{5}$$

In the above boundary conditions displacements are indicated by the letter u, the direction of movement – by the index, which corresponds to the coordinate axis. Shear stresses are marked by Z_x and Z_y. Normal stresses are denoted by two identical letters that coincides with the appropriate direction, for example Z_z.

To solve the problem apply a biharmonic function φ. In this case, the displacements and stresses are defined by dependencies

$$2Gu_x = -\frac{\partial^2 \varphi}{\partial x \partial z}, \; 2Gu_y = -\frac{\partial^2 \varphi}{\partial y \partial z}, \; 2Gu_z = \left[2(1-\mu)\Delta^2 - \frac{\partial^2}{\partial z^2}\right]\varphi, \; X_x = \frac{\partial}{\partial z}\left[\mu\Delta^2 - \frac{\partial^2}{\partial x^2}\right]\varphi,$$

$$X_y = -\frac{\partial^3 \varphi}{\partial x \partial y \partial z}, \; Y_y = \frac{\partial}{\partial z}\left[\mu\Delta^2 - \frac{\partial^2}{\partial y^2}\right]\varphi, \; Y_z = \frac{\partial}{\partial y}\left[(1-\mu)\Delta^2 - \frac{\partial^2}{\partial z^2}\right]\varphi, \; Z_z = \frac{\partial}{\partial z}\left[(2-\mu)\Delta^2 - \frac{\partial^2}{\partial z^2}\right]\varphi,$$

$$Z_x = \frac{\partial}{\partial x}\left[(1-\mu)\Delta^2 - \frac{\partial^2}{\partial z^2}\right]\varphi, \tag{6}$$

where $\Delta^2(...) = \frac{\partial^2(...)}{\partial x^2} + \frac{\partial^2(...)}{\partial y^2} + \frac{\partial^2(...)}{\partial z^2}$.

Assume that biharmonic function has a following form

$$\varphi = A\left[ch(Cz) + Dz\,sh(Cz)\right]\cos(Nx)\cos(My), \tag{7}$$

where A, D, N, M – arbitrary constants;
$C^2 = N^2 + M^2$.

The value of a constant D determine from the condition of absence of shear stresses on the sample boundaries $z = \pm\frac{t}{2}$. Specified condition is valid when

$$D_{n,m} = -\frac{1}{\dfrac{2\mu}{C_{n,m}} + zth(C_{n,m}z)}. \tag{8}$$

In order to satisfy all the boundary conditions in an isolated volume of material, values of the constants A, D, N, M will be regarded as arrays. Define the constants N and M by the following expressions

$$N_n = \frac{(n+1/2)\pi}{h};$$

$$M_m = \frac{2(m+1/2)\pi}{b}.$$

Prescribe the constant C by array from condition

$$C_{n,m}^2 = N_n^2 + M_m^2.$$

Taking into account the boundary conditions, assume biharmonic function (Erie function) for the considered scheme of the rope loading in the following form

$$\varphi = \sum_{n=1}^{K}\sum_{m=1}^{K} A_{n,m}\left[ch(C_{n,m}z) + D_{n,m}z\,sh(C_{n,m}z)\right]\times$$
$$\times \cos(N_n x)\cos(M_m y). \tag{9}$$

In this equation summation is limited by some value K. In the rope SSS determining we take several components. Displacements along the cables, z-axis, for the chosen function, according to (6), have the form

$$u_z = \sum_{n=0}^{K}\sum_{m=0}^{K} A_{n,m}\left[\frac{ch(C_{n,m}z)}{2} + D_{n,m}\left(\frac{ch(C_{n,m}z)}{C_{n,m}}(2\mu-1) + \frac{zsh(C_{n,m}z)}{2}\right)\right]C_{n,m}^2\cos(N_n x)\cos(M_m y). \tag{10}$$

Normal pressure along the z axis is

$$Z_z = \sum_{n=0}^{K}\sum_{m=0}^{K} A_{n,m}\left[sh(C_{n,m}z)C_{n,m} + D_{n,m}\left(sh(C_{n,m}z)(2\mu-1) + zch(C_{n,m}z)\right)C_{n,m}\right]C_{n,m}^2\cos(N_n x)\cos(M_m y). \tag{11}$$

Required shear stresses in planes parallel to the conventional plane of friction are

$$Z_x = \sum_{n=0}^{K} \sum_{m=0}^{K} A_{n,m} \left[ch(C_{n,m}z) + D_{n,m}\left(2\frac{ch(C_{n,m}z)}{C_{n,m}}\mu + zsh(C_{n,m}z) \right) \right] C_{n,m}^2 \, sin(N_n x) cos(M_m y) n. \qquad (12)$$

Boundary conditions (5) are mixed. So instead, we assume the following conditions for $z = t/2$

$$Z_{z=t/2} = \begin{cases} 1, & \text{when} \quad y^2 + \left(x - \dfrac{h}{2}\right)^2 \le \left(\dfrac{d}{2}\right)^2; \\ \\ 0, & \text{when} \quad y^2 + \left(x - \dfrac{h}{2}\right)^2 > \left(\dfrac{d}{2}\right)^2. \end{cases} \qquad (13)$$

To do this, expand the relation (13) in a double Fourier series

$$f(n,m) = \frac{8\int_0^{h}\int_0^{b/2} Z_{z=t/2}\, cos(xN_n)cos(yM_m)}{bh}. \qquad (14)$$

Equation (12) will be satisfied when, according to (13), are valid following expression

$$A_{n,m}\left[sh\left(C_{n,m}\frac{t}{2}\right)C_{n,m} + D_{n,m}\left(sh\left(C_{n,m}\frac{t}{2}\right)(2\mu - 1) + \frac{t}{2}ch\left(C_{n,m}\frac{t}{2}\right) \right) \right]C_{n,m} = f(n,m).$$

Said allows finding the unknown constants, which we denote by additional subscript 1

$$A_{n,m} = \frac{f(n,m)}{\left[sh\left(C_{n,m}\dfrac{t}{2}\right)C_{n,m} + D_{n,m}\left(sh\left(C_{n,m}\dfrac{t}{2}\right)(2\mu - 1) + \dfrac{t}{2}ch\left(C_{n,m}\dfrac{t}{2}\right) \right) \right]C_{n,m}}.$$

Computed constants provide the specified character of model deformation u_z on the surface $z = \pm t/2$. Displacement distribution is shown in Figure 3. This figure shows also the change of normal pressure forces.

Curve 1 in Figure 3(b) corresponds to the displacements; curve 2 shows the contours of conventional cable. Parameters h_0 and b_0 are determined by dependencies

$$h_0 = 10\left(\frac{h - 1.1d}{d}\right) \text{ and } b_0 = 10\left(\frac{b - 1.1d}{d}\right).$$

A form of solution and given values of arrays components N_n and M_m do not ensure full compliance with conditions (1), (2) and (3).

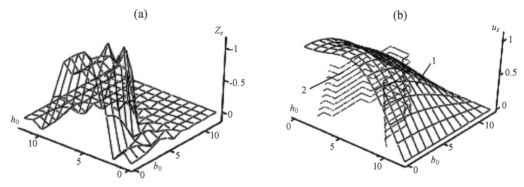

(a)   (b)

Figure 3. Diagram of distribution of normal pressure forces Z_z and displacements u_z on the surface z = ± t/2 of the sample in the plane.

Condition $u_x = 0$ is not fulfilled on the surface $x = h$, condition $X'_x = 0$ – on the surface $x = 0$ and condition $X_y = 0$ – on the surface $y = b/2$. These conditions are satisfied only when $z = 0$. Taking this into account, SSS of the sample is considered in a plane $z = 0$. At the same time, it should be noted that the displacements in a plane $z = 0$ coincides with displacements in a plane for samples of the length less than cable diameter.

Using the relationship (12) the shear stress in the plane tangent to the surface of the cable by the joining of real rubber sample, in cross-section $x = h - \dfrac{h-d}{2}$, are determined. Distribution diagram of shear stresses for unit displacement of a cable of the length $10d$ is shown in Figure 4.

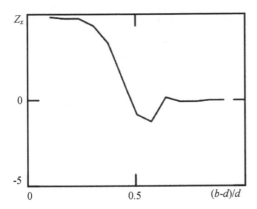

Figure 4. Diagram of distribution of shear stresses along the y-axis in relative coordinates.

The graph shows that the shear stresses vary across the width of the sample and different from zero only in front of the cable. Maximum values they take at the minimum distance from the tether to the surface of the rubber shell to which the shear stress is applied. However, the distribution of stress depends on the geometrical parameters of the rope. In calculating the strength of machine parts designers use stress concentration factor, which is the ratio of maximum stress to the medium. Figure 5 shows the dependence of the stress concentration factor on the parameters of the rope with a cable of unit diameter.

The dependence of stress concentration factor k on the parameters of the rope is almost linear. The only exception is the case when $0 < b_0 < 1$. Approach of parameter b_0 to zero corresponds to the reduction of the cover thickness to zero. This reduction is associated with the unlimited growth of shear

stresses between the cable and the surface that interacts with a driving drum. This shows the compliance obtained solutions to analyzed physical phenomenon. The latter can be considered as a validity confirmation of the results.

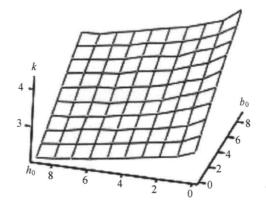

Figure 5. Dependence of stress concentration factor on the parameters of the rope (h_0 and b_0)

The case $b_0 = 1$ corresponds to the excess of step making cables over their diameter by 10% (rope parameter, the ratio of the thickness of the rope to cable diameter, proposed by (Kolosov 1987), is 1.1). Ropes, tapes of this parameter in practice do not apply. The latter is caused by mining conditions. These conditions include that tapes and ropes are used to transport mining masses, which are characterized by the presence of lumps of considerable size, formed by random destruction during extraction.

In using of rubber-rope cables for lifting and rope manufacturing with small diameter of cables (1-10 mm), the rope parameter equal to 1.1 must lie within 0.05–0.5 mm. Sheets production of this thickness of raw rubber is actually impossible.

Accordingly, with sufficient accuracy for practical use, we can take a linear dependency of stress concentration factor of geometrical parameters of rubber-rope cable, when the thickness of the rope is greater than drum diameter at least by 20%.

Using the obtained values of the concentration factor produce empirical dependency of the form

$$k(h,b,d) = A\frac{h}{d} + B\frac{b}{d} + C\frac{b}{d}\cdot\frac{h}{d} + D.$$

Having solved the system of algebraic equations of the fourth order for the values of the rope parameters, it was determined $A = 0.324$; $B = -0.945$; $C = -0.003$; $D = 1.467$.

The calculated values of stress concentration factor k or the resulting graph allows finding it as a function of rope parameters h_0 and b_0, and summarizing the experimentally established values of the samples strength.

4 CONCLUSIONS

These results suggest that the nature of the distribution of stresses and displacements of the rubber cover of the rope on the sample surface and in the plane $z = 0$ are close among themselves, indicating that almost at all sections of the sample SSS varies slightly. This corresponds to the conditions of deformation of the sample with a rigid cable. Therefore, the results can be considered as reliable. The stress state of a cable in a rope with single cable is more dangerous than the cables of infinitely wide rope. In the ropes of limited width the extreme cables work at conditions which are closer to the load conditions of the rope with one cable. Therefore, experiments should include testing of such ropes.

The dependence of the stress concentration factor on the parameters of the rope allows taking into account the distribution of stresses in the rubber in the design of lifting-and-shifting machines with a traction rubber-rope and increasing the safety of using of these machines.

REFERENCES

Blokhin, S.E., Kolosov, D.L. & Kolosov, A.L. 2009. *The stress strain state of the flat traction rubber-rope on the drum.* Journal of Dnipropetrovs'k National University of Railway Transport named after academician V. Lazaryan, 30: 88–91.

Kolosov, D.L. 2007. *Finite element model of the flat rubber-rope cable on the driving belt pulley.* Scientific Bulletin of National Mining University, 2: 58–60.

Belmas, I.V., Kolosov, D.L. & Bobyleva, I.T. 2009. *Taking into account the shear stress in the automatic design of stepped rope.* Steel ropes. Collection of Scientific Papers, 7: 147–152.

Bilous, O.I. & Kolosov, D.L. 2010. *Stress state of a rubber shell of the tape on the drum of the conveyor feeding to the blast furnace charge.* Metallurgical and Mining Industry, 4 (262): 113–114.

Kolosov, D.L., Dolgov, O.M. & Kolosov, A.L. 2013. *The stress-strain state of the belt on a drum under compression by flat plates.* Annual Scientific-Technical Collection. Mining of Mineral Deposits. Leiden, Netherlands: CRC Press/Balkema: 351–357.

Kolosov, L.V. 1987. *Fundamentals of research and application of rubber-rope cables of winders in deep mines.* Dnipropetrovs'k: PhD thesis: 426.

Progressive Technologies of Coal, Coalbed Methane, and Ores Mining – Bondarenko, Kovalevs'ka & Ganushevych (eds)
© 2014 Taylor & Francis Group, London, ISBN: 978-1-138-02699-5

Creation of gas hydrates from mine methane

K. Ganushevych, K. Sai & A. Korotkova
National Mining University, Dnipropetrovs'k, Ukraine

ABSTRACT: Mining of coal deposits nowadays in Ukraine has to include not only coal extraction but following coalbed methane, taking into account that Ukraine are amongst ten leading world countries by reserves of coalbed methane. Releasing from degassing wells gas either goes into atmosphere or burnt having negative influence on the environment. The present paper deals with the creation of gas hydrates from mine gas that releases from degassing wells. It is proposed to use surface active substance (SAS) of various concentrations to intensify formation process. Also the possibility of application of various mineralogical content mine water for gas hydrates creation is studied. The results of laboratory experimental research are presented in the form of graphs, formulas and tables. The conclusions are drawn at the end of the paper with perspective trends for future studies.

1 INTRODUCTION

The analysis of coal deposits mining in modern conditions shows the necessity of new technological solutions development for complex exploitation of mineral resources and environment protection. Mine methane utilization is amongst such problems. Huge amounts of gas releases to the surface by various means of degassing or with ventilation current of the mine.

Present paper presents the results of studies of gaseous mixture conversion into solid gas hydrate (GH) state and methods of intensification of this conversion by means of surface active matters application.

During conduction of studies the analysis and generalization of modern research of gas hydrate formation process were used together with conduction of laboratory experiments, graphical and analytical methods of data processing.

More than 20 artificial samples of gas hydrates were created to establish laws of formation and influence on it various technological and thermobaric parameters (Ovchynnikov et al. 2013).

2 INTENSIFICATION OF THE PROCESS

The main problem that occurs when creating gas hydrates is the time factor. Thus, there are many studies directed to intensification of the gas hydrate formation process from gaseous mixture of various content using materials accelerating the formation.

The preliminary gaseous mixture received from a degassing well of mine management "Pokrovs'ke",

Ukraine, was used as a sample gas for studies. The content of the mixture is presented on Table 1.

Table 1. Content of used gaseous mixture (molar parts).

Methane CH_4	Ethane C_2H_5	Propane C_3H_8	Butane C_4H_{10}	Carbon dioxide CO_2	Nitrogen N_2
0.952	00425	0.0161	0.005	0.0053	0.0043

As the surface active substance (SAS) presumably increasing the kinetics of gas hydrates formation the DB (dibutylphenol treated with ethylene oxide) was used.

Ten experiments with adding SAS of various concentrations into the water were conducted to check the influence degree on the GH formation kinetics with temperature inside the reactor $T = +2...+3\,°C$ and pressure $P = 5$ MPa. The results of the experiments are presented in the Table 2, Figure 1, 2 and 3.

Table. 2. Amount of SAS and its influence on GH formation time.

Amount of SAS, ml	GH formation time, min
0.3	220
0.5	200
1	100
2	50
3	20
4	35
5	50
6	90
7	120
8	200

From Figure 1 it is seen that maximal speed of formation – 20 minutes occurs when adding 0.3% of SAM into water. SAS concentrations less than this value have a lower influence on the formation speed: at SAS concentration of 0.1% the GH forms in approximately 100 minutes, at 0.2% – in 45 minutes. At that, increase of time needed for GH formation takes place by logarithmic dependence

$$y = -1.128 \cdot ln(x) + 6.3533$$

with the approximation $R^2 = 0.9905$, where y – amount of SAS, %; x – time of formation, min.

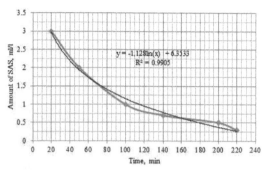

Figure. 1. Dependence of GH formation time on added amount of SAS in the range from 0.3 to 3 ml/l of water.

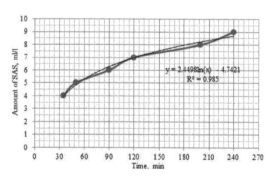

Figure 2. Dependence of GH formation time on added amount of SAS in the range from 4 to 9 ml/l of water.

From Figure 2 it is obvious that with SAS concentrations from 0.4% and more, kinetics of the process decreases and with concentrations more than 0.9% the creation of GH becomes not reasonable in terms of industrial application of given intensification method, as GH formation time makes more than 3 hours. The connection between the amount of SAS and the process duration in this case is also described by logarithmic dependence

$$y = 2.4498 \cdot ln(x) - 4.7421$$

with approximation equal to $R^2 = 0.985$, where y –

amount of SAS, %; x – time of formation, min.

Figure 3 presents general dependence of GH formation kinetics on added amount of SAS.

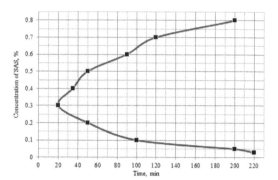

Figure 3. General dependence of GH formation time on added amount of SAS in the range from 0.01 to 0.9% of water at temperature $T = +2...+3°C$ and pressure $P = 5$ MPa.

Dependence presented on Figure 3 is described by the following system of equations

$$\begin{cases} 0.3 \leq SAS \leq 3 \\ y = -1.128\,ln(x) + 6.3533 \\ 4 \leq SAS \leq 9 \\ y = 2.4498\,ln(x) - 4.7421 \end{cases}$$

The peak of GH formation speed is 20 minutes when adding 3 ml of SAS per 1 liter of water (0.3%). At this, under the same thermobaric conditions but without adding any SAS the GH formation time makes 420 minutes (7 hours). That is, the results of conducted experiments are decrease of GH formation time from 7 hours (without SAS) down to 20 minutes that is 21 times faster than the initial result!

After reaching the maximal speed of formation, i.e. shortest time needed, the speed begins decreasing and, respectively, formation time becomes greater according to logarithmic dependence.

Also, as the result of laboratory research, the following dependence was established: GH formation speed is almost equal when adding 1 ml and 6 ml of SAS. Similar situation is observed at 2 and 5 ml; 7 and 0.8 ml; 8 and 0.5 ml (Figure 3).

Thus, having analyzed received results, it is obvious that there exist the optimal value of added SAS per a definite quantity of water at which GH formation speed is maximal. When adding quantity of SAS exceeding optimal value, the GH formation speed starts decreasing and when adding more than 8 ml of SAS the formation process becomes impossible because excessive quantity of SAS solved in water prevents formation of crystallization centers and, as the consequence, serves as an inhibitor decreasing GH formation speed making crystal lattices impossible to form.

Based on the received results, it is possible to calculate the revenue from using gas hydrate technology on the example of mine management "Pokrovs'ke", Ukraine. The average debit of one degassing well is 20000 m³ of gas a day with concentration of methane equal to about 95–98%. It is possible to create about 125 m³ of gas hydrate a day from one degassing well using 93.75 m³ of water. The amount of SAS needed to form 1 m³ of GH is 0.00225 m³ (at 3% of SAS). There are 10 wells used at the enterprise. Taking into account all necessary factors, including price of the equipment, salary for the personnel, amortization etc., the revenue of coalbed methane utilization using GH technology for this enterprise will make 252000 $/year (considering price of 1000 m³ of natural gas for Ukraine). And this amount considers only one coal enterprise! Total amount of coalbed methane released in Ukraine makes about 1.2 billion m³ a year.

3 USE OF MINE WATER TO CREATE GAS HYDRATES

As it is known, to form gas hydrates it is necessary to have two major components – fresh water and gas. It is also known that natural gas hydrates form in marine environment from salty water. Thus, to implement the proposed method of coal methane utilization with help of gas hydrate technology, it is proposed to use mine water of this enterprise with various mineralogical content with purpose to examine the influence of mine water content on parameters of GH formation. The main difficulty occurring when using mine water is its mineralogical composition.

When creating GH using mineralized water it is necessary to consider mineralization degree of water that significantly changes thermobaric conditions of gas hydrates formation. The formation mechanism from mineralized water is as follows: when mixing water and salt, the formation of gas hydrate takes place but this hydrate does not include salt as it stays in the solution since its molecules size is too large to fit the cavities formed by water molecules.

To research mineralogical water influence degree on the GH formation process the water of various composition was used (Table 3).

Table 3. Chemical composition of used mineralized water for laboratory studies.

Number of mixture	Mineralization, ml/l	$K + Na$, ml/l	Ca, ml/l	Mg, ml/l	Cl, ml/l	SO_4, ml/l
1	5	0.3	0.5	2	2	0.2
2	10	0.5	1	3	5	0.5
3	15	0.5	3	3	3	6.5
4	20	0.5	2	2	10	5.5
5	25	1	3	5	6	10
6	30	0.5	1	3	15	10
7	35	1	3	4	12	15
8	40	1	3	3	20	13

Gas with methane concentration of 95% and water with total salts content from 5 to 40 ml/l were used for conduction of the experiments. 8 water samples were used. The main results are shown on Figure 4.

From above-shown figures it is obvious that presence of salts in water moves equilibrium conditions of GH formation towards higher pressure. The most significant increase of pressure occurs when adding from 20 to 40 ml of salts. At temperature $T = -2\,°C$ and addition of 20 ml of salts, the needed pressure is 3.3 MPa. At the conditions but adding 30 ml of salts, the pressure rises to 7.4 MPa, and when adding 40 ml – up to 22 MPa. Received results are described by exponential dependence

$$y = 28.4 \cdot e^{0.1144 \cdot x}$$

with the approximation accuracy of $R^2 = 0.989$,

where y – pressure, MPa; x – temperature, °C.

Starting from the temperature of 0 °C and addition of 40 ml of salts it was impossible to create GH on the used equipment, since maximally possible pressure created in the reactor – 25 MPa was not enough to form gas hydrates at this concentration of salts and temperature. Hence, considering exponential dependence according to which the pressure increases, the task of its value determination for temperatures in the range from 0 to +9 °C and with addition of 4% SAS has been conducted by calculation method. Thus, taking into account received data: 28.4 MPa (at 0 °C) and up to 79.5 MPa (at +9 °C), it is obvious that the creation of such great pressures is a technically unrealizable task considering economic feasibility of the proposed technology.

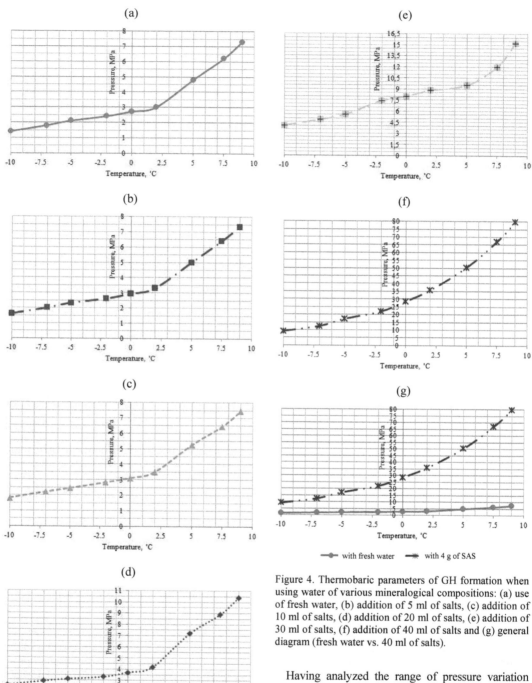

(a)

(e)

(b)

(f)

(c)

(g)

with fresh water   with 4 g of SAS

(d)

Figure 4. Thermobaric parameters of GH formation when using water of various mineralogical compositions: (a) use of fresh water, (b) addition of 5 ml of salts, (c) addition of 10 ml of salts, (d) addition of 20 ml of salts, (e) addition of 30 ml of salts, (f) addition of 40 ml of salts and (g) general diagram (fresh water vs. 40 ml of salts).

Having analyzed the range of pressure variation when adding various quantity of salts into water, the following conclusion can be made: difference in pressure when adding 5 and 10 ml of salts is in average – 0.23 MPa; between 10 and 20 ml – 0.7–2.5 MPa; between 20 and 30 ml – 1.5–4.5 MPa; between 30 and 40 ml – 5–20 MPa.

508

The phenomenon of pressure increase when creating gas hydrate with large quantity of salt can be explained by the fact that when adding salts into water they get dissolved in it "worsening" quality of water, thus making it impossible to form gas hydrates. That is, with increase of salts in water the chances gas hydrates will form are becoming smaller.

Hence, it is clear that SAS application for increasing GH formation kinetics is a very effective method of intensification allowing to speed up the process by 21 times. Application of mineralized water for GH creation negatively influences the kinetics and, particularly, leads to pressure increase according to exponential dependence.

An important fact while conducting experiments is the so-called memory of water, i.e. its property to memorize its structure after the first hydration (Bondarenko et al. 2011). Thus, it is expedient to repeatedly use water in GH formation process cyclically. This fact is subjected to further detailed scientific and laboratory analysis.

4 CONCLUSIONS

1. Use of surface active substances (SAS) for gas hydrate creation process intensification is a very effective method that allows to speed up the process kinetics by 21 times when using optimal quantity of SAS – 3 ml/l.

2. Application of mineralized water for gas hydrates creation slows the formation process and leads to pressure increase by exponential dependence: the more salts we add, the slower the process goes on. At 40 ml/l too high pressure is needed to form GH that is inexpedient in industrial conditions.

3. With average debit of one degassing well equal to 20000 m^3/day, it is possible to make 125 m^3 of gas hydrates a day and to gain a profit of about 252000 $ a year from one average Ukrainian coal mine.

REFERENCES

Ovchynnikov, M., Ganushevych, K. & Sai, K. 2013. *Methodology of gas hydrates formation from gaseous mixtures of various compositions.* Balkema, CRC Press: Vol. 4: 203–205.
Bondarenko, V., Ganushevych, K. & Sai, K. 2011. *Concerning borehole underground recovery of gas hydrates.* Scientific bulletin of National Mining University, 1 (121): 60–66.

Progressive Technologies of Coal, Coalbed Methane, and Ores Mining – Bondarenko, Kovalevs'ka & Ganushevych (eds)
© 2014 Taylor & Francis Group, London, ISBN: 978-1-138-02699-5

Optimization of room-and-pillar method parameters under conditions of limestone rocks

N. Ulanova, O. Sdvyzhkova & V. Prikhodko
National Mining University, Dnipropetrovs'k, Ukraine

ABSTRACT: The optimization criterion is developed to optimize the parameters of room-and-pillar method in terms of Inkerman limestone mine providing the long-term stability of mine workings. Boundary elements method is used to determine the rock stress strain state at a stage of direct problem. Multiextremum algorithm of nonlinear programming is applied at inverse problem (optimization).

The method takes into account spread in physical and mechanical properties of rocks.

1 INTRODUCTION

The rational use of natural resources and landscape preservation at increasing building material production is a great challenge. Rational land use and environmental protection is topical task for the Crimean peninsula. About 5% of prospected limestone rocks used for building in Ukraine is concentrated there. The combined underground extraction method is developed for limestone mining in Crimea. This method has to ensure minimum of building stone losses and long-time stability of excavations. The mined-out space is supposed to be used for industrial objects later.

The given work is devoted to determination of optimum chamber shape and size in terms of Inkerman geological conditions. The parameters are determined by solving multiextremum problem of nonlinear programming. The target function considering the shape and geometrical sizes of chambers and pillars is accepted to be an optimization criterion. Minimization of criterion is carried out under technological demands and restrictions on rock strength. The optimum design algorithm considers a spread in initial data concerning the physical and mechanical properties of rocks.

To expose the post-peak straining of rock mass the model of two mediums interacting is used. According to this model the medium within regions of inelastic stress is considered to be elastic one, but its Young's modulus corresponds to the rock post-peak behavior (post-peak straining modulus).

Many publications are devoted to studying the rock pressure manifestations while using the room-and-pillar mining system. Due to complexity and variety of geological and mining conditions the different design schemes are used to determine the bearing elements. For example, G.S. Esterhuizen (Esterhuizen & Dollinar 2008) studied parameters of room-and-pillar extracting method, insuring roof stability in US underground limestone mines. Some authors assumed pillars to be separate from a roof and floor and a roof is considered as a beam or a plate and a pillar shape and depth of mining are taken into account by means of empirical coefficients. However, owing to equal rigidity of pillar and surrounding rocks, it is necessary to determine the stress-strain state of a system "pillar – rock mass". Therefore in this study numerical simulation is used for modeling the stress-strain state around excavations with complex geometry. Besides, according to up-to-date ideas, post-failure straining of rock mass is of importance. Thus, numerical model has to consider the post-peak behavior of rocks.

In given paper boundary elements method is used at a stage of direct problem. As to an inverse problem (optimization), the known solutions are related only to an economic-mathematical problem and they are solved or by classical analysis methods, or by methods of linear and nonlinear programming. For example, the method of alternative analysis is often used in non-linear programming problems. The same method is applied for geo-mechanical problems to determine support rational parameters, but the works of such kind are not numerous. The effect of spread in physical-mechanical characteristics upon the results is ignored mostly. The given algorithm considers the probability nature of input data.

The combined method of mineral extraction expects building the industrial platform near a mouth of underground development openings. The productive seam is dissected by system of parallel galleries. A floor of a gallery is a roof of sawn layer, and chambers are formed by layers in a direction from top to down. The beginning of "sawn" layer makes a bank in the opening quarry bordering upon an underground excavation. The galleries are driven in two pulls, therefore they has an arch outline. According to equipment requirements the berms are provided (Figure 1). Governing parameters are size and shape of chambers and pillars between chambers. In given paper their optimum size is determined by solving a multiextremum problem of nonlinear programming. Restrictions are dictated by technological demands and mined-out space stability.

2 SIMULATION OF ROCK STRESS-STRAIN STATE

Simulation was carried out by a boundary element method in the form of "fictitious loads". The method was realized in additional stresses. The solution given by S. Crauch and A. Starfild (Crauch & Starfild 1987) is used as a basic one.

To detect the most effective governing parameters the design scheme including three rectangular excavations and berms are considered (Figure 1). Here $2b_1$ is a chamber breadth, h is a chamber altitude, h_1 is a crest altitude, l is a berm breadth, h_0 is a distance from berm to roof, c is a pillar breadth, R is a roof radius, h_b is a maximal chamber altitude.

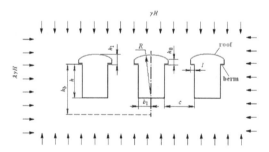

Figure 1. Design scheme.

The shape of excavations, their number and a sequence of mining are varied. The input data for simulations are: compressive strength $\sigma_c = 10$ MPa, Young's modulus $E = 2.14 \cdot 10^3$ MPa, Poisson's ratio $\nu = 0.27$, volume gravity $\gamma = 21$ kN/m^3, depth of mining $H = 100$ m. Lateral thrust coefficient λ was changed within 0.3–0.5. Geometrical sizes b_1/h varied from 0.3 to 0.5 and h_1/h varied from 0 to 0.23.

Simulation results show, that only two next chambers placed on the left and on the right of the central one and a number of extracted limestone layers influence essentially on rock stress state near the excavation. That is why the design scheme containing three parallel excavations was accepted. The size of chamber rectangular part and pillar width are chosen as driving parameters.

The simulation shows that the least level of stresses is attained at roof radius $R^* = b_1 + l$, where b_1 is a chamber half-breadth, l is a berm breadth.

3 SIMULATION OF ROCK STRESS-STRAIN STATE CONSIDERING A PLASTIC ZONE

A two-section diagram of limestone compression is used to determine failure zones in rocks. The solution is achieved iteratively. On each step of iterations a boundary line of inelastic strain area is determined according to criterion offered by L. Parchevskiy and A. Shashenko (Shashenko 2008) with using a long-term strength coefficient k_D. In terms of East Inkerman can be assumed $k_D = 0.73$. Within the yielded area the rock strains are characterized by post-peak modulus (for limestone $M = 1.28 \cdot 10^3$ MPa). Boundary condition for each boundary element is null of stresses and the simultaneity of strains is assumed on the elastic-plastic zone border. The maximum extent of failure area R_1 depending on coefficient $\dfrac{k_D \sigma_c}{\gamma H}$ and geometrical size of excavation h/b_1 looks like

$$R_1 = 1.37 (k_D \sigma_c / \gamma H)^{-0,46} (1.59 - 0.02 h / b_1). \quad (1)$$

It is valid for $k_D \sigma_c / \gamma H = \{2\}, \{2,5\}, \{3\}, \{3,5\}, \{4\}$ and $h/b_1 = \{1\}, \{2\}, \{3\}, \{4\}, \{5\}$.

The magnitude R_1 is used to determine an anchor length for excavation stability improvement. Simulation under Inkermansky mine geological conditions have demonstrated, that the anchors installed apart $h_2 = 0.3$ m downwards from a berm at the distance of $l_0 = 1.65$ m and with step $l_3 = 0.5$ m along excavation reduce maximum stresses in berm in 1.3 times.

4 THE OPTIMIZATION CONSIDERING INPUT DATA SPREAD

The target function depends on multi-dimensional vectors of casual and determined arguments

$$K_1 = K_1 (\overline{X}, \overline{B}), \quad (2)$$

where $\overline{X}$ – vector of the determined variables, $\overline{B}$ –

casual vector of uncontrollable variables. The magnitude $K_1 = S_{o.}/S_i$ characterizes the mineral losses ($S_{i.}$ – mined-out space in a cross-sectional area, $S_{o.}$ – cross-sectional area of unfinished reserves within the panel).

Components of controlled vector $\overline{X}$ are parameters $b_1{}^i$, h^i, R^i, c^i, P^i, $l_0{}^i$, and components of a casual vector $\overline{B}$ are parameters σ_c, λ, γ, ν.

The problem is put as follows. It is required to determine components of vector $\overline{X}$, ensuring function (2) minimum value under limitations on maximum stresses, occurring on excavation contour (the long-term strength requirement)

$$max\,\sigma_e \leq k_D \sigma_c, \qquad (3)$$

where σ_e – equivalent stress according strength theory (Shashenko 2008).

Mining technology demands (limitations on varied parameters) looks like

$$\overline{X}^n \leq \overline{X} \leq \overline{X}^b. \qquad (4)$$

Ranges for $\overline{X}^n$ and $\overline{X}^b$ are assigned from technical possibilities of used extraction devices.

To define the error, induced by initial data spread, the numerical experiment is provided. The $K_1{}^j$ ($j = \overline{1,n}$) sample which refers to defined component of a vector $\overline{B}$ is considered.

Unknown parameter $a(K_1)$ is a mean of the coefficient which characterizes mineral losses. It is estimated with confidence interval and serves as measure of precision. This confidence interval for supposed normal distribution law looks like

$$a^*(K_1) - t_{q,n-1}\,s/\sqrt{n} < a(K_1) < a^*(K_1) +$$

$$+ t_{q,n-1}\,s/\sqrt{n}, \qquad (5)$$

where s – variance, $t_{q,n-1}$ – chosen under the table of Student's density function for corresponding levels of confidence $q = 1 - \alpha$ and n – sample size. The distribution can be converted to normal by one of formulas

$$y = K_2 = 1/K_1;\ y = lg\,((\,K_2 \pm d\,)\cdot 10^b\,);$$

$$y = 1/\sqrt{K_1}, \qquad (6)$$

where K_2 – characterizes extracted volume of lime-

stone; d, b – constants.

Optimum designs for which the mineral loss factor is located within the confidence interval, are considered as the equivalent ones.

Thus, to search the optimum design matching to certain geological conditions, the series of optimization problems should be solved varying vector $\overline{B}$ components. The solution is carried out by random searching method with self-training. The self-training consists of a purposeful action on a random changing of a working step direction, which stops to be an equal probability and gains advantage in the direction of the best step according to the formula

$$\overline{\xi}' = \overline{X} + \frac{\overline{\xi} - \overline{X}}{1 + e\left(\overline{\xi} - \overline{X}\right)^2},$$

where $\overline{\xi}$ – vector of uniformly distributed random quantities, $\overline{X}$ – best value, computed on the previous step of searching, e – self-training parameter (in the beginning $e = 0$). In the course of searching e varies by formula

$$e' = e + \frac{t}{\left(\overline{X}^b - \overline{X}^n\right)^2},$$

where t – input parameter defining motion within a domain. As initial approximation value $\overline{X} = 0.5\left(\overline{X}^b + \overline{X}^n\right)$ is accepted.

Proposed approach was used for a software development implementing the problems of optimum design concerning the rock mass structures and bearing elements. The numerical experiments executed using these codes for different restrictions and mining conditions testify the satisfactory convergence of the algorithm.

The target function (2) related to optimum parameters of sawn limestone mining under Inkerman conditions looks like

$$K_1 = \frac{S_{rem.}}{S_{extr.}} = \frac{hc + (2b_1 + c)(h_b - h)}{2b_1 h}.$$

Limitations on stresses has the form (3), and constructive restrictions (4) are represented by inequalities 24 m $\leq h \leq 34$ m (on altitude of the rectangular part of chamber) and 5 m $\leq b_1 \leq 7$ m (on chamber breadth). Besides, proceeding from geometrical parameters of existing mining methods, limitation on pillar width is put: 7.5 m $\leq c \leq 25$ m. Initial values of varied parameters are accepted as follows:

$b_1 = 5.1$ m, $h = 33$ m, $c = 24$ m. Values $h_0 = 1.2$ m and $l = 1$ m are not vary during calculations. The radius of roof is put as $R = b_1 + l$. Results of calculations for various components of a vector $\overline{B}$ are given in the Table 1.

Table 1. Results of calculations for various components of a vector $\overline{B}$.

$k_D \sigma_c$, MPa	λ	b_1, m	c, m	h, m	K_1	K_2	$lg(100K_2)$
	0.3	5.0	17.7	32.9	1.863	0.537	1.729
6.20	0.4	5.0	12.3	33.8	0.858	1.166	2.066
	0.5	5.0	10.9	33.7	0.864	1.158	2.063
	0.3	5.0	15.8	34.0	1.331	0.751	1.875
6.57	0.4	5.0	10.0	34.0	0.839	1.191	2.075
	0.5	5.0	8.5	33.4	0.880	1.136	2.055
	0.3	5.0	14.8	33.8	1.160	0.862	1.934
6.94	0.4	5.0	9.6	33.9	0.845	1.183	2.072
	0.5	5.0	7.5	34.0	0.846	1.181	2.072
	0.3	5.0	13.8	33.7	1.010	0.990	1.990
7.30	0.4	5.0	9.1	33.3	0.893	1.120	2.049
	0.5	5.2	7.5	34.0	0.766	1.306	2.115
	0.3	5.0	12.6	33.9	0.849	1.178	2.071
7.67	0.4	5.0	8.2	34.0	0.847	1.181	2.072
	0.5	5.1	7.5	34.0	0.811	1.233	2.091
	0.3	5.0	11.8	32.7	0.921	1.086	2.035
8.03	0.4	5.0	8.2	33.7	0.863	1.158	2.063
	0.5	5.6	7.5	34.0	0.640	1.562	2.193

The parameter related to $y = lg(100K_2)$ and obtained from (6) at $d = 0$, $b = 2$ and sample characteristics according to Table 1 are: a mean $a^*(y) = 2.052$, a variance $s = 0.072$, a coefficient of variation $V = 3.5\%$. The hypothesis about the K_2 logarithmic-normal distribution can be accepted on 5% level. According to (5) at $n = 17$, $q = 0.95$, $t_{q,n-1} = 2.12$ the mean $a(y)$ is in a confidence interval $2.15 \le a(y) \le 2.089$. The density function looks like

$$g(y) = \frac{1}{s\sqrt{2\pi}} exp\left(-\frac{(y - a(y))^2}{2s^2} \right).$$

Transferring to random value K_2, we gain a confidence interval $1.035 \le a(K_2) \le 1.227$ and a density function

$$g(K_2) = \frac{1}{ln10 \cdot K_2 \cdot s\sqrt{2\pi}} \times$$

$$\times exp\left(-\frac{(lg(100K_2) - a(K_2))^2}{2s^2} \right).$$

The density function profiles for random values $y = K_2$ and $y = lg(100K_2)$ are shown on Figure 2.

The obtained results testify that the error of the optimal design determination makes $\approx 8.5\%$. Thus for equivalent designs (in the table they are underlined) the width of chambers b_1 fluctuates within 5.0–5.02 m and altitude h varies from 33.3 to 34.0 m and interchamber pillars width changes from 7.5 to 12.3 m.

According to the used criterion the maximum stresses are located in berms in the optimal design. Their values are determined by expression

$$max\, \sigma_{e.} = \frac{1}{2\psi}\left(\sqrt{(1-\psi)^2(\sigma_{xx} + \sigma_{yy})^2 + 4\psi((\sigma_{xx} - \sigma_{yy}) + 4\tau_{xy}{}^2)} - (1-\psi)(\sigma_{xx} - \sigma_{yy}) \right),$$

which do not exceed the value $k_D\sigma_c$. Here $\psi = \sigma_p/\sigma_c$ and σ_p, σ_c – tension and compressive strengths respectively.

The experimental studying was carried out to verify the simulation results. Photo elasticity method was used. Object of modeling was stress state around the rectangular excavations with berms. The excavations were mined at the depth of $H = 100$ m in limestone rock with volume weight $\gamma = 21$ kN/m^3 and Young's modulus $E = 2.14 \cdot 10^3$ MPa and lateral thrust coefficient $\lambda = 0.33$. Geometric parameters of model corresponded to optimal project sizes on the scale 1:400. Isochromatic lines obtained experimentally are shown on Figure 3. Comparison of theoretical and modeling results showed that maximum difference in σ_{xx}, σ_{yy} and τ_{xy} values is 10%.

(a)

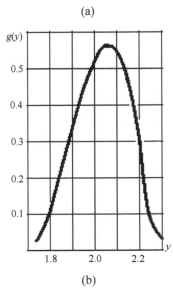

(b)

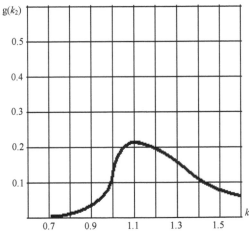

Figure 2. The theoretical density functions profiles for random values: (a) $y = lg(100K_2)$ and (b) $y = K_2$.

Figure 3. Isochromatic lines of stresses around the excavations obtained with photo elasticity method.

5 CONCLUSIONS

The algorithm of multiextremum non-linear programming is featured. This algorithm optimizes both geometrical plans of various mining methods and support constructions.

The optimization criterion K_1 considering the shape and sizes of chambers and interchamber pillars is offered. Its minimization at the requirements of excavation long-time stability and technological demands allows obtaining the optimal design considering the spread in rock properties.

The distribution law characterizing limestone extraction volume in the case of optimal design is obtained and dependence between input data probability and optimal parameters is established. In terms of Inkerman geological conditions (at spread in compressive strength of limestone 8% and in lateral thrust coefficient 25%) the optimal sizes of chambers and pillars with reliability of 95% are in the following limits: interchamber pillars width is 7.5–12.3 m and chamber altitude is 33.3–34.0 m, and chamber width practically does not change and makes 10 m.

The analytical function (1) is determined. It links the greatest expansion of failure zone in rocks with physical and mechanical properties and depth of mining. This function allows predicting the rock pressure manifestations around excavations and justifying the parameters of supports ensuring excavation long-time stability.

REFERENCES

Esterhuizen, S & Dollinar, D. 2008. *Roof Stability Issues in Underground Limestone Mines in the United States.* NIOSH Annual Report.

Crauch, S.& Starfild, A. 1987. *Boundary element methods in solid mechanics*. Moskow: Myr: 328.

Shashenko, A., Sdvyzhkova, O. & Gapiev, S. 2008. *Rock mass deformability and strength*. Dnipropetrovs'k: National Mining University: 224.

Progressive Technologies of Coal, Coalbed Methane, and Ores Mining – Bondarenko, Kovalevs'ka & Ganushevych (eds)
© 2014 Taylor & Francis Group, London, ISBN: 978-1-138-02699-5

The frictional work in pair wheel-rail in case of different structural scheme of mining rolling stock

K. Ziborov & S. Fedoriachenko
National Mining University, Dnipropetrovs'k, Ukraine

ABSTRACT: An impact on exploitation characteristics of mining vehicle by means of variation of design scheme is studied in the article. A value of energy consumption is estimated, which depends on transmission of motion by friction. Values of friction force components for electric locomotives in coal mine conditions are defined. Graphical relations, which were obtained mathematically, represent the results of the study.

1 INTRODUCTION

The development of Ukrainian mining-metallurgical industrial strategy aims to supply economical demands in mineral deposits. Clearly, its quality must be of international standard and its low-cost production is an essence of industry competitiveness.

Thus, along with development of mining, sinking, lifting machines and equipment for gatehead and intercharge points the demands of their requirements have grown as well. Therefore, it is necessary to use a mining rolling stock, which was developed taking into account modern and prospective technologies and control systems, to solve strategy issues.

The problem of insufficient usage of friction properties of rolling stock output members and rails causes unreasonable energy loss, reduction of its exploitation characteristics. Exploitation indexes of railway transport show that to overcome friction spends up to 30 % of all consuming energy, and loss of material of friction pair amounts to 15% of producing metal (Isaev & Luznov 1985, Harris 2001).

To reduce wear intensity to acceptable values developed certain technical and technological-organizational arrangements (Shur et al. 1995) (lubrication, improving the profile geometry of the rolling profile of wheelsets and rails, improving its metal quality, etc.) Unfortunately, none of these actions has solved the issue fully. Existing for the moment tendencies of development of mining railway transport is an evidence of development and active implementation of a new technical solutions, and alternative configurations of a several units (Figures 1 and 2). However, use of a new technical solution without scientific substantiation reduces to difficulties along exploitation of mine railway vehicles (insufficient negotiation in rail track curvatures, increased energy consumption, poor safety indexes while motion along sign-variable rail track etc.).

To refuse these shortcomings it is necessary to define the degree of influence of structural, design parameters on dynamical characteristics of mining vehicle, on wear of frictional pair and lifetime.

The purpose of the work is to detect exploitation characteristics and energy consumption changings subject to structural and design parameters of mining vehicle, in order to find out rational parameters of transmitting motion by friction with minimal energy loss.

Possible edges of exploitation indexes of transportation vehicles, which detect competitiveness end economic efficiency of mining transport (lading weight, velocity of locomotive haulage), all other thing being equal define by tractive ability of locomotive and dynamical characteristics of mining wagons.

2 MATHEMATIC MODELING OF THERMOCHEMICAL PREPARATION PROCESS

In real conditions, operation of locomotive is described by big cyclicity, which is specified by a significant part of transient processes. During them both values of normal force are changing (especially while jerking and hunting) and the friction coefficient too. In addition, operating of mining transport in transient regime is characterized by skid. Presence of denoted occurrences provides raised wear of interaction surfaces of frictional pair, especially in transition curve and power consumption.

After each wheel spin, on a rail pressure pad (Figure 1) arise plastic and elastic deformations, caused by external forces and internal material energy. Therefore, frictional elements start to act on a finite contact area Φ.

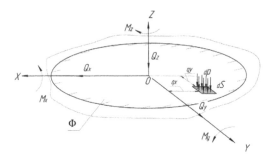

Figure 1. External forces which acts on contact area.

The resulting force that transmits from rail surface to wheel expands as normal reaction Q_z, which acts along general normal; tangential Q_{xy}, which is under influence of frictional forces reaction.

The value of frictional force Q_{xy} must be less or equal of interfacial friction, i.e.

$$Q_{xy} \leq \mu \cdot Q_z , \qquad (1)$$

where μ – coefficient of interfacial friction.

Under the term "tangential reaction" in the theory of frictional interaction we will mean the resultant of longitudinal Q_x and lateral Q_y reactions of contact surface xOy. It is functionally connected with wheel slip speed δV by the rail surface (Figure 2). Evidently, that vector of line velocity of wheel set V_1 consists of wheel's rolling velocity V_2 and its slip δV. The last one is characterized by compound influence of deformations and slip in contact points.

The wheel while moving on the rail can be in several regimes: free ($Q_{xy} = 0$), braking ($Q_{xy} < 0$) or tractive ($Q_{xy} > 0$). Thus, the characteristic of relative slip S necessary to define taking into account different velocities of wheel set V_1 and angular speed V_2

$$S = \begin{cases} |\delta V| / |V_2| & \text{at } Q_{xy} > 0; \\ |\delta V| / |V_1| & \text{at } Q_{xy} < 0. \end{cases} \qquad (2)$$

The tangential component of reaction Q_{xy} we will mean a dimensionless relation

$$\xi = \frac{Q_{xy}}{Q_z} \leq \mu , \qquad (3)$$

where Q_z – normal reaction in pressure pad; μ – coefficient of interfacial friction.

While slipping of a locked wheel $S = 1$ frictional forces at each i contact point are parallel to δV and are equal to product of frictional coefficient μ with normal load Q_{zi} at this point. In this case, $Q_{xy} = \mu \sum Q_{zi} = \mu \cdot Q_z$, and $\xi = \mu$.

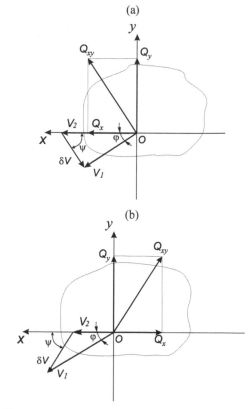

Figure 2. Calculation scheme of forces and velocities: (a) wheels are in tractive regime ($Q_{xy} > 0$) and (b) wheels are in braking regime ($Q_{xy} < 0$).

While wheel skidding $S = 1$ also. However, owing to lateral slip of contact points of rotating wheel, which tread has conicity, slip velocities δV_i and tangential forces Q_{xyi} in contact points are not parallel. Thus, for current regime $\xi < \mu$.

During free wheel motion $S = 0$ and $Q_{xy} = 0$. Therefore, defined regime arises only in a case of flat motion.

The presence of defined events causes increased wear of a frictional pair, especially in transient curvatures (Figure 3), and additional consumption of a tractive power.

Wearing mechanism, based on several scientists (Isaev & Luznov 1985, Kragelskiy et al. 1977) works, consists of metal collapse, which follows by plastic deformation of a surface layer of a wheel during its rolling on a rail and wheel scuffing during skid.

It is considered that the scuff during locomotive wheel skid dominates the collapse during rolling and is defined by longitudinal force component of elastic slip (creep) and friction on the roll surface, which are variable because of changeable interaction surfaces for differently worn wheels and rail.

Figure 3. Wear of inner surface of the rail on a bend of a switch.

The properties of polydisperse polluters on frictional surfaces of wheel and rails can significantly vary on different segments of mining course. Exploitation experience shows, that frictional characteristics of interacting pairs strongly depend on composition and rheological properties of surface layer of polluters (Rengevich 1961, Ziborov 2014 & Ziborov et al. 2008) and depend on tractive effort of mining locomotive.

Table 1. Relation of ψ depending on rail conditions.

Rail condition	Value ψ
Covered with liquid coal and rock mud	0.07–0.08
Damp, clean	0.09
Wet, clean	0.12–0.13
Dry, almost clean	0.17
Sprinkled with sand	0.18–0.24
Sprinkled with sand, crushed after previous haulage	0.14–0.18

Meanwhile surface pollution depending on humidity level gets properties of colloid solution like paste or liquid properties so can be assumed about hydroabrasive nature of interacting pair wear. However, in conditions of frequent grip breaking and following boundary film breaking hydroabrasive wear can modify into abrasive, which zone will be detected by area of short-time contact taking into account deformation. Abrasive and hydroabrasive wear significantly supplements with fatigue wear of surface layers, which is a result of high contact tensions. These factors do not allow obtain the exact analytical relation for wheel-rail wear estimation.

In the paper (Derugin 2000) one of the authors of the article used experimental method to study the wear. It comes to definition of a real roll profile of the wheelsets of several locomotives: K10 with

frame design, and module Э10. Experiment was provided by plaster casting and following comparison of worn profiles with original, standard (Figure 4). Further, the data about layer thickness and volume of worn material were processed with use of appropriate mathematical methods for locomotives with different structures.

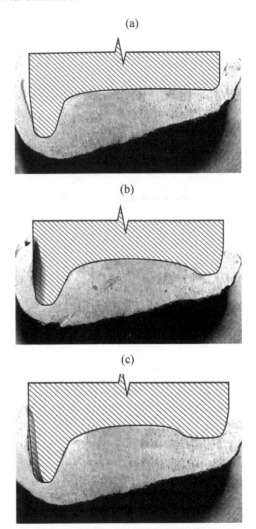

Figure 4. Plaster castings of roll surfaces of wheelsets: (a) original profile, (b) worn profile of Э10 locomotive and (c) worn profile of K10 profile.

Friction work of frictional pair wheel-rail subject to geometric and physical imperfections of work surfaces defined by formula

$$A_{mp} = \int_0^T Q_z(t)\psi(S)SV(t)dt ,\qquad (4)$$

where Q_z – normal wheel load; S – relative slip; $V(t)$ – velocity.

From Equation (4) it follows that friction work defines by physical parameters, which depend on duration of each component of working cycle. The duration of speed diagram at the average equals: acceleration – 10–15%; quasiuniform motion – 65–75%; deceleration – 15–20%.

Contact of the friction pair does not spread on the hole surface of wheel and rail. The profile of wheel and rail as it is prevents the full contact. The contact is limited by areas, depicted on Figure 5, and suposses changing of its profile form both wheel and rail during exploitation.

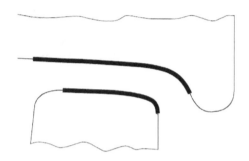

Figure 5. The most possible areas of wheel-rail contact.

Wear of frictional pair runs, as usual, on the track segments where the power interaction between rubbing bodies occurs and defines the real contact surface.

For driving wheels of mining locomotives standard tread profile keeps its geometry and is defined by roll path on it. Further, it runs until the profile obtains rails parameters, which are called "milling" (Figure 4b, c). For output members of mining wagon contact surface (Figure 5) are distributed on wheel profile more irregularly, owing to high dependence on loading parameters (loaded or empty) and is defined by smaller wheel size.

When direction of wheelset in rail track cannot be provided by creep force, flange contact rises, and lateral forces on a flange emerge that prevent derailment. With forces on the flange connected friction work elements. Derailment rises, especially in curvatures with small radius. Owing to this fact, large forces in a contact of flange and rail arise, which causes wear of correspondent wheel and rail regions (Figure 6).

Denoted forces, in its turn, depend on structural scheme of wheelset and parameters of elastic-dissipative linkages between members (Harris 2001, Derugin 2000).

During wheel and rail wear wagon motion characteristics change – motion resistance rises. Additional dynamical component, which is connected with track irregularities, causes increase of lateral force on a leading wheel, flange tightness to the rail in curved track and stability loss that increases the risk of derailment.

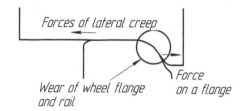

Figure 6. Wear of wheel and rail while lateral creeping.

One of the ways to reduce resistance of motion is the lubrication of working zones by dispersion of lubricant on side edge of rail. However, for mining conditions it is not appropriate owing to the fact it will be like a fixant of coal-rock abrasive dust and worsen interacting conditions and exploitation characteristics of friction pair.

Therefore, to increase the motion stability against flange mounting on a rail, and consequently, exploitation safety can be provided by the use of self-stabilizing design of wheelset with additional kinematical moveability of leading wheel. Thus, additional angular movability of cylindrical hinge of wheels around vertical axle in 2 degrees can increase safety factor till 20%, which is real for curvilinear rail track sections and high rail joints. Also, it provides reduction of wear up to 15%.

To estimate efficiency of design changes the above mentioned approach can be used Total work of friction forces can be represented as two components: one of them, most of the work, spends (efficiently) for generation a tangential grip force (for locomotive) or for a move (for wagon); another, smaller, spends (insufficiently) for wheel slip along rail surface, that results in wheelset's tread and rail wear.

To calculate numerous values of each component it is necessary simultaneously solve Equation (1) and equation of motion for each design scheme subject to special features of wheelset both locomotive's and wagon's. For this task can be used set of modern applied software. The results of solution depicted by graphical relations of friction force work and relative slip in time (Figure 7). Analysis shows that area, which lies between curves 1 and 2, represents that second part of friction force work that spends on skid (pure wear).

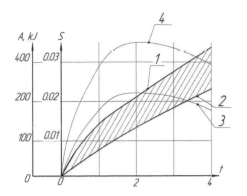

Figure 7. Relation of friction force work to relative slip in time: 1, 2 – work of friction force of wheelset for locomotives with frame and module structure correspondingly; 3, 4 – relative slip of locomotive's wheelset of frame and module structure correspondingly.

4 CONCLUSIONS

Wheel slip change during transient motion (feature of work process of mining rail transport) is characterized by significant growth of friction forces work between wheel and rail. During stable motion, when wheel slip takes up minimal constant value, friction force work changes insignificantly. The change of wheel slip during acceleration till skidding for locomotive (from null up to constant value in transient period) or while driving in curvatures of small radius for mining wagon, is characterized by significant growth of frictional work between wheel and rail, which causes wear of frictional pair. Structural scheme (frame or module) and design of wheelset influence both on wear characteristics, and on dynamical quality of mining vehicles.

REFERENCES

Isaev, I.P. & Luznov, Ju.M. 1985. *Problems of locomotive's adhesion.* Moscow: Machinery building: 238.

Harris, J. 2001. *Guidelines to best practices for heavy haul railway operations: wheel and rail interface issues.* Virginia: Virginia Beach: 408.

Shur, E.A., Bichkova, N.Ya, Markov, D.P. & Kuzmin, N.N. 1995. *Durability of rail and wheel steels.* Friction and wear, 1, Vol. 16: 80–91.

Kragelskiy, I.V., Dobichin, M.N. & Kombalov, V.S. 1977. *Bases of friction and wear calculations.* Moscow: Machinery building: 526.

Rengevich, A.A. 1961. *Coefficient of friction of mining electric locos.* Issues of mining transport. Moscow: Gortechizdat: Vol. 5: 227–247.

Ziborov, K.A. 2014. *Characteristics of frictional pair wheel-rail of mining locomotive while kinematical and force imperfections.* Dnipropetrovs'k: Mining equipment and electromechanics, 3 (100): 26–32.

Ziborov, K.A., Derugin, O.V. & Matsiuk, I.N. 2008. *On wheelset tread wear while different structural scheme of drive unit.* Mining of ore deposits, 2, Vol. 92: 137–140. Kryvyi Rih.

Derugin, O.V. 2000. *Substantiation of rational parameters of elastic-dissipative bonding of mining locomotives*: Dis. of candidate of sciences (tech.) sp. 05.05.06 "Mining Machines": 173. Dnipropetrovs'k.